Basic College Mathematics
Sixth Edition

Margaret L. Lial
American River College

Stanley A. Salzman
American River College

Diana L. Hestwood
*Minneapolis Community and
Technical College*

Addison
Wesley

Boston   San Francisco   New York
London   Toronto   Sydney   Tokyo   Singapore   Madrid
Mexico City   Munich   Paris   Cape Town   Hong Kong   Montreal

Publisher	Jason A. Jordan
Executive Editor	Maureen O'Connor
Editorial Project Management	Ruth Berry and Suzanne Alley
Editorial Assistant	Melissa Wright
Managing Editor/Production Supervisor	Ron Hampton
Text and Cover Design	Dennis Schaefer
Supplements Production	Sheila C. Spinney
Production Services	Elm Street Publishing Services, Inc.
Media Producer	Lorie Reilly
Associate Producer	Rebecca Martin
Marketing Manager	Dona Kenly
Marketing Coordinator	Heather Rosefsky
Prepress Services Buyer	Caroline Fell
Technical Art Supervisor	Joseph K. Vetere
First Print Buyer	Hugh Crawford
Art Creation & Composition Services	Pre-Press Company, Inc.
Cover Photo Credits	Photos and Photos/Index Stock Imagery; Walter Bibikow/Index Stock Imagery; Tina Buckman/Index Stock Imagery
Photo Credits	All photos from PhotoDisk except the following:

Bill Aron/PhotoEdit, p. 386 right; Bettmann/CORBIS, p. 264 bottom; Bongarts/A. Hassenstein/SportsChrome, pp. 465, 503; Robert Brenner/PhotoEdit, p. 407; Christopher Briscoe/Photo Researchers, Inc., p. 464; John Cancalosi/Stock Boston, pp. 310 middle, 310 right; Cindy Charles/PhotoEdit, p. 28; CORBIS, p. 497 right; Rob Crandall/The Image Works, p. 238; Bob Daemmrich/The Image Works, p. 302; © 2000 DaimlerChrysler. All rights reserved., pp. 91, 144, 456 top; Mary Kate Denny/PhotoEdit, p. 8 right; Paul Dols/Press Publications, pp. 521, 567 left; Duomo/CORBIS, p. 287; Eastcott-Momtiuk/The Image Works, p. 623; Kathy Ferguson/PhotoEdit, p. 107; Tony Freeman/PhotoEdit, p. 543 right; Judy Gelles/Stock Boston, p. 443; Robert Ginn/PhotoEdit, p. 247; Francois Gohier/Photo Researchers, Inc., p. 216; Spencer Grant/Stock Boston, p. 256; John Griffin/The Image Works, p. 149; William Johnson/Stock Boston, p. 144; Carol Kaelson/CORBIS, p. 491 top; Layne Kennedy/CORBIS, p. 504; Earl & Nazima Kowall/CORBIS, p. 2; Craig Lovell/CORBIS, p. 373; Chris Marona/Photo Researchers, Inc., p. 288; Maslowski/Photo Researchers, Inc., p. 362; Will & Deni McIntyre/Photo Researchers, Inc., pp. 617, 624; NASA, p. 295; Bill Nation/CORBIS-Sygma, p. 264 top; Michael Newman/PhotoEdit, pp. 1, 365, 395 left, 476 bottom, 512 left, 566; Boyd Norton/The Image Works, p. 696; Omni Photo Communications/Index Stock Imagery, p. 723; David Parker/Science Photo Library/ Photo Researchers, Inc., p. 703; PhotoEdit, p. 493; Pictor/Uniphoto, p. 567 right; Procter & Gamble Products, p. 395 right; Neal Preston/ CORBIS, p.111 top; Pascal Quittmelle/Stock Boston, p. 434; Roger Ressmeyer/CORBIS, p. 578; Gary Retherford/Photo Researchers, Inc., p. 476 top; Mark Richards/PhotoEdit, p. 491 bottom; N. Richmond/The Image Works, p. 417 right; Phil Schermeister/CORBIS, p. 474; Rhoda Sidney/PhotoEdit, p. 61; WD-40, p. 417 left; Ben Woods/CORBIS, p. 532; David Young-Wolff/PhotoEdit, pp. 386, 484, 456 bottom, 512 right.

Library of Congress Cataloging-in-Publication Data

Lial, Margaret L.
    Basic college mathematics.—6th ed./Margaret L. Lial, Stanley A. Salzman, Diana L. Hestwood.
        p.   cm.
    Rev. ed. of: Basic college mathematics/Margaret L. Lial . . . [et.al.]. 5th ed. ©1998.
    Includes index.
    ISBN 0-321-06457-7 (Student Edition)
    ISBN 0-321-06241-8 (Annotated Instructor's Edition)
    ISBN 0-321-09730-0 (Hardback)
1. Mathematics.   I. Salzman, Stanley A.   II. Hestwood, Diana L.   III. Basic college mathematics.   IV. Title.
QA37.2 .M55 2002
513'.1—dc21   00-068917

1 2 3 4 5 6 7 8 9 10   WC   04 03 02 01

Contents

List of Applications

List of Focus on Real-Data Applications

Preface

The sixth edition of *Basic College Mathematics* continues our ongoing commitment to provide the best possible text and supplements package that will help instructors teach and students succeed. To that end, we have addressed the diverse needs of today's students through an attractive design, updated applications and graphs, helpful features, careful explanation of concepts, and an expanded package of supplements and study aids. We have also responded to the suggestions of users and reviewers and have added many new examples and exercises based on their feedback.

The text is designed to help students achieve success in a developmental mathematics program. It provides the necessary review and coverage of whole numbers, fractions, decimals, ratio and proportion, percent, and measurement, as well as an introduction to algebra and geometry, and a preview of statistics. This text is part of a series that also includes the following books:

- *Essential Mathematics* by Lial and Salzman

- *Prealgebra,* Second Edition, by Lial and Hestwood

- *Introductory Algebra,* Seventh Edition, by Lial, Hornsby, and McGinnis

- *Intermediate Algebra with Early Functions and Graphing,* Seventh Edition, by Lial, Hornsby, and McGinnis

- *Introductory and Intermediate Algebra,* Second Edition, by Lial, Hornsby, and McGinnis.

WHAT'S NEW IN THIS EDITION?

We believe students and instructors will welcome the following new features.

▶ *New Real-Life Applications* We are always on the lookout for interesting data to use in real-life applications. As a result, we have included many new or updated examples and exercises throughout the text that focus on real-life applications of mathematics. Students are often asked to find data in a table, chart, graph, or advertisement. (See pp. 79, 143, and 269.) These applied problems provide an up-to-date flavor that will appeal to and motivate students. A comprehensive List of Applications appears at the beginning of the text.

▶ *New Figures and Photos* Today's students are more visually oriented than ever. Thus, we have made a concerted effort to add mathematical figures, diagrams, tables, and graphs whenever possible. (See pp. 82, 208, and 319.) Many of the graphs use a style similar to that seen by students in today's print and electronic media. Photos have been incorporated to enhance applications in examples and exercises. (See pp. 104, 287, and 302.)

▶ *Increased Emphasis on Problem Solving* Introduced at the end of Chapter 1, our six-step process for solving application problems has been refined and integrated throughout the text. The six steps, *Read, Plan, Estimate, Solve, State the Answer,* and *Check,* are emphasized in boldface type and repeated in specific problem-solving examples in Chapters 2, 3, 5, 6, 7, and 9. (See pp. 85, 145, and 167.)

○ *Study Skills Component* Poor study skills are a major reason why students do not succeed in math. A few generic tips sprinkled here and there are not enough to help students change their behavior. So, in this text, a desk-light icon at key points in the text directs students to a separate *Study Skills Workbook* containing carefully designed activities that correlate directly to the text. (See p. 241.) This unique workbook explains *how* the brain actually learns and remembers so students understand *why* the study skills activities will help them succeed in the course. Students are introduced to the workbook in a new To the Student section at the beginning of the text.

○ *Focus on Real-Data Applications* Each one-page activity presents a relevant and in-depth look at how mathematics is used in the real world. Designed to help instructors answer the often-asked question, "When will I ever use this stuff?," these activities ask students to read and interpret data from newspaper articles, the Internet, and other familiar, real-world sources. (See pp. 62, 168, and 354.) The activities are well-suited to collaborative work or they can be completed by individuals or used for open-ended class discussions. Instructor teaching notes and activity extensions are provided in the *Printed Test Bank and Instructor's Resource Guide*.

○ *Diagnostic Pretest* A diagnostic pretest is now included on p. xxiii of the text and covers all the material in the book, much like a sample final exam. This pretest can be used to facilitate student placement in the correct chapter according to skill level. The pretest also exposes students to the scope of the course content.

○ *Chapter Openers* New chapter openers feature real-world applications of mathematics that are relevant to students and tied to specific material within the chapters. Examples of topics include finding the best buy on cell phone service, home improvements, recipes, medical tests, fishing, and work/career applications. (See pp. 107, 185, and 311—Chapters 2, 3, and 5.)

○ *Calculator Tips* These optional tips, marked with a calculator icon, offer basic information and instruction for students using calculators in the course. (See pp. 252, 268, and 323.) In addition, an Introduction to Calculators is included as an appendix.

○ *Test Your Word Power* This new feature, incorporated into each chapter summary, helps students understand and master mathematical vocabulary. Key terms from the chapter are presented along with four possible definitions in a multiple-choice format. Answers and examples illustrating each term are provided. (See pp. 94, 231, and 349.)

WHAT FAMILIAR FEATURES HAVE BEEN RETAINED?

We have retained the popular features of previous editions of the text, including the following:

○ *Learning Objectives* Each section begins with clearly stated, numbered objectives, and the material within sections is keyed to these objectives so that students know exactly what concepts are covered. (See pp. 129, 163, and 321.)

○ *Cautions and Notes* These color-coded and boxed comments, one of the most popular features of previous editions, warn students about common errors and emphasize important ideas throughout the exposition. (See pp. 66, 153, and 155.) There are more of these in the sixth edition than in the fifth, and the new text design makes them easier to spot; Cautions are highlighted in bright yellow and Notes are highlighted in green.

○ *Margin Problems* Margin problems, with answers immediately available on the bottom of the page, are found in every section of the text. (See pp. 79, 186, and 321.) This key feature allows students to immediately practice the material covered in the examples in preparation for the exercise sets. Based on reviewer feedback, we have added more margin exercises to the sixth edition.

◗ *Ample and Varied Exercise Sets* The text contains a wealth of exercises to provide students with opportunities to practice, apply, connect, and extend the skills they are learning. Numerous illustrations, tables, graphs, and photos have been added to the exercise sets to help students visualize the problems they are solving. Problem types include skill building, writing, estimation, and calculator exercises as well as applications and correct-the-error problems. In the *Annotated Instructor's Edition* of the text, the writing and estimation exercises are marked with icons for writing 🖋 and for estimation ≈ so that instructors may assign these problems at their discretion. Exercises suitable for calculator work are marked in both the student and instructor editions with a calculator icon ▦. (See pp. 26, 70, and 253–256.)

◗ *Relating Concepts Exercises* These sets of exercises help students tie concepts together and develop higher level problem-solving skills as they compare and contrast ideas, identify and describe patterns, and extend concepts to new situations. (See pp. 72, 162, and 256.) These exercises make great collaborative activities for pairs or small groups of students.

◗ *Summary Exercises* There are four sets of in-chapter summary exercises: fractions, percent, geometry concepts, and operations with signed numbers. These exercises provide students with the all-important *mixed* practice they need at these critical points in their skill development. (See pp. 229, 435, 569, and 651.)

◗ *Ample Opportunity for Review* Each chapter ends with a Chapter Summary featuring: Key Terms with definitions and helpful graphics, New Formulas, Test Your Word Power, and a Quick Review of each section's content with additional examples. Also included is a comprehensive set of Chapter Review Exercises keyed to individual sections, a set of Mixed Review Exercises, and a Chapter Test. Beginning with Chapter 2, each chapter concludes with a set of Cumulative Review Exercises. (See pp. 231–246, 297–310, and 445–464.)

WHAT CONTENT CHANGES HAVE BEEN MADE?

We have worked hard to fine-tune and polish presentations of topics throughout the text based on user and reviewer feedback. Some of the content changes include the following:

• In Chapter 1, Whole Numbers, reading and understanding tables is now introduced in the first section, and a new section (1.9, Reading Pictographs, Bar Graphs, and Line Graphs) contributes additional application skills early in the course.

• In Chapter 4, Decimals, the sections on adding and subtracting decimals have been combined to provide more interesting options for application exercises.

• In Chapter 6, Percent, shortcuts for finding 200%, 300%, 10%, and 1% have been added to the shortcuts for finding 100% and 50%. The topics of percent proportion and identifying the parts in a percent problem are now treated in one section (6.3) rather than two sections so that students have an easier time relating these concepts.

• In Chapter 7, Measurement, Section 7.1 on English measurement now focuses on solving application problems as well as converting among units.

WHAT SUPPLEMENTS ARE AVAILABLE?

Our extensive supplements package includes an *Annotated Instructor's Edition,* testing materials, solutions manuals, tutorial software, videotapes, and a state-of-the-art Web site. For more information about any of the following supplement descriptions, please contact your Addison-Wesley sales consultant.

FOR THE STUDENT

○ **Student's Solutions Manual (ISBN 0-321-09061-6)** The *Student's Solutions Manual* provides detailed solutions to the odd-numbered section exercises and to all margin, Relating Concepts, Summary, Chapter Review, Chapter Test, and Cumulative Review exercises.

○ **Study Skills Workbook (ISBN 0-321-09185-X)** A desk-light icon at key points in the text directs students to correlated activities in this unique workbook by Diana Hestwood and Linda Russell. The activities in the workbook teach students how to use the textbook effectively, plan their homework, take notes, make mind maps and study cards, manage study time, review a chapter, prepare for and take tests, evaluate test results, and prepare for a final exam. Students find out *how* their brains actually learn and remember, and what research tells us about ways to study effectively. A new To the Student section at the beginning of the text introduces students to the *Study Skills Workbook*.

○ **Addison-Wesley Math Tutor Center** The Addison-Wesley Math Tutor Center is staffed by qualified college mathematics instructors who tutor students on examples and exercises from their textbook. Tutoring is provided via toll-free telephone, toll-free fax, e-mail, and the Internet. White Board technology allows tutors and students to actually see problems being worked while they "talk" in real time over the Internet during tutoring sessions. The Math Tutor Center is accessed through a registration number that may be bundled free with a new textbook or purchased separately.

 ○ **Web Site: www.MyMathLab.com** Ideal for lecture-based, lab-based, and on-line courses, MyMathLab.com provides students with a centralized point of access to the wide variety of on-line resources available with this text. The pages of the actual book are loaded into MyMathLab.com, and as students work through a section of the on-line text, they can link directly from the pages to supplementary resources (such as tutorial software, interactive animations, and audio and video clips) that provide instruction, exploration, and practice beyond what is offered in the printed book. MyMathLab.com generates personalized study plans for students and allows instructors to track all student work on tutorials, quizzes, and tests.

○ **InterAct Math® Tutorial Software   (ISBN 0-321-09056-X)** This interactive tutorial software provides algorithmically generated practice exercises that are correlated at the objective level to the content of the text. Every exercise in the program is accompanied by an example and a guided solution designed to involve students in the solution process. For Windows users, selected problems also include a video clip to help students visualize concepts. The software tracks student activity and scores and can generate printed summaries of students' progress. Instructors can use the InterAct Math® Plus course-management software to create, administer, and track tests and monitor student performance during practice sessions. (See For the Instructor.)

○ **InterAct MathXL: www.mathxl.com** InterAct MathXL is a Web-based tutorial system that enables students to take practice tests and receive personalized study plans based on their results. Practice tests are correlated directly to the section objectives in the text, and once a student has taken a practice test, the software scores the test and generates a study plan that identifies strengths, pinpoints topics where more review is needed, and links directly to InterAct Math® tutorial software for additional practice and review. A course-management feature allows instructors to create and administer tests and view students' test results, study plans, and practice work. Students gain access to the InterAct MathXL Web site through a password-protected subscription, which can either be bundled free with a new copy of the text or purchased separately with a used book.

○ **Real-to-Reel Videotape Series (ISBN 0-321-09088-8)** This series of videotapes, created specifically for *Basic College Mathematics*, Sixth Edition, features an engaging team of math instructors who provide comprehensive lectures on every objective in the text. The videos include a stop-the-tape feature that encourages students to pause the video, work the presented example on their own, and then resume play to watch the video instructor go over the solution.

◆ ○ *Digital Video Tutor* **(ISBN 0-321-09089-6)** This supplement provides the entire set of Real-to-Reel videotapes for the text in digital format on CD-ROM, making it easy and convenient for students to watch video segments from a computer, either at home or on campus. Available for purchase with the text at minimal cost, the Digital Video Tutor is ideal for distance learning and supplemental instruction.

FOR THE INSTRUCTOR

○ *Annotated Instructor's Edition* **(ISBN 0-321-06241-8)** The *Annotated Instructor's Edition* provides immediate access to the answers for all text exercises by printing them in color next to the corresponding problems. To assist instructors in assigning home-work problems, icons identify writing 📝, estimation ≈ , and calculator 🖩 exercises.

○ *Instructor's Solutions Manual* **(ISBN 0-321-09060-8)** The *Instructor's Solutions Manual* provides complete solutions to all even-numbered section exercises.

○ *Answer Book* **(ISBN 0-321-09058-6)** The *Answer Book* provides answers to all the exercises in the text.

○ *Printed Test Bank and Instructor's Resource Guide* **(ISBN 0-321-09059-4)** The *Printed Test Bank* portion of this manual contains two diagnostic pretests, six free-response and two multiple-choice test forms per chapter, and two final exams. The *Instructor's Resource Guide* portion of the manual contains teaching suggestions for each chapter, additional practice exercises for every objective of every section, a correlation guide from the fifth to the sixth edition, phonetic spellings for all key terms in the text, and teaching notes and extensions for the Focus on Real-Data Applications pages in the text.

○ *TestGen-EQ with QuizMaster EQ* **(ISBN 0-321-09057-8)** This fully networkable software enables instructors to create, edit, and administer tests using a computerized test bank of questions organized according to the chapter content of the text. Six question formats are available, and a built-in question editor allows the user to create graphs, import graphics, and insert mathematical symbols and templates, variable numbers, or text. An "Export to HTML" feature allows practice tests to be posted to the Internet, and instructors can use QuizMaster-EQ to post quizzes to a local computer network so that students can take them on-line and have their results tracked automatically.

○ *InterAct Math® Plus* **(ISBN 0-201-72140-6)** This networkable software provides course-management capabilities and network-based test administration for Addison-Wesley's InterAct Math® tutorial software. (See For the Student.) InterAct Math® Plus enables instructors to create and administer tests, summarize students' results, and monitor students' progress in the tutorial software, providing an invaluable teaching and tracking resource.

○ *MathPass* MathPass helps students succeed in their developmental mathematics courses by creating customized study plans based on diagnostic test results from ACT, Inc.'s Computer-Adaptive Placement Assessment and Support System (COMPASS®). MathPass pinpoints topics where the student needs in-depth study or targeted review and correlates these topics with the student's textbook and related supplements. The study plan can be saved as an HTML file that, when viewed on the Internet, links directly to text-specific, on-line resources. Instructors can add their own custom Web links to the HTML study plan. The MathPass learning system provides diagnostic assessment, focused instruction, and exit placement all in one package.

○ *Web Site:* www.MyMathLab.com In addition to providing a wealth of resources for lecture-based courses, MyMathLab.com gives instructors a quick and easy way to create a complete on-line course based on *Basic College Mathematics*, Sixth Edition. MyMath-Lab.com is hosted nationally at no cost to instructors, students, or schools, and it provides access to an interactive learning environment where all content is keyed directly to the text. Using a customized version of Blackboard™ as the course-management platform, MyMath-Lab.com lets instructors administer preexisting tests and quizzes or create their own, and it provides detailed tracking of all student work as well as a wide array of communication tools for course participants. Within MyMathLab.com, students link directly from on-line pages of their text to supplementary resources such as tutorial software, interactive animations, and audio and video clips.

ACKNOWLEDGMENTS

The comments, criticisms, and suggestions of users, nonusers, instructors, and students have positively shaped this textbook over the years, and we are most grateful for the many responses we have received. The feedback gathered for this revision of the text was particularly helpful, and we especially wish to thank the following individuals who provided invaluable suggestions:

George Alexander, *University of Wisconsin Colleges*
Sonya Armstrong, *West Virginia State College*
Solveig R. Bender, *William Rainey Harper College*
Ernie Chavez, *Gateway Community College*
Terry Joe Collins, *Hinds Community College*
Martha Daniels, *Central Oregon Community College*
Donna Foster, *Piedmont Technical College*
Joe Howe, *St. Charles County Community College*
Rose Kaniper, *Burlington County College*
Douglas Lewis, *Yakima Valley Community College*
Wayne Miller, *Lee College*
Thea Philliou, *College of Santa Fe*
Richard D. Rupp, *Del Mar College*
Ellen Sawyer, *College of DuPage*
Lois Schuppig, *College of Mount St. Joseph*
Mary Lee Seitz, *Erie Community College—City Campus*
Kathryn Taylor, *Santa Ana College*
Sven Trenholm, *North Country Community College*
Bettie A. Truitt, *Black Hawk College*
Jackie Wing, *Angelina College*

Our sincere thanks go to these dedicated individuals at Addison-Wesley who worked long and hard to make this revision a success: Maureen O'Connor, Ruth Berry, Ron Hampton, Dennis Schaefer, Dona Kenly, Suzanne Alley, and Melissa Wright. Steven Pusztai of Elm Street Publishing Services provided his customary excellent production work. We are most grateful to Peg Crider for researching and writing the Focus on Real-Data Applications feature; Paul Van Erden for his accurate and useful index; Becky Troutman for preparing the comprehensive List of Applications; Abby Tanenbaum for writing the new Diagnostic Pretest; and Janis Cimperman, Luanne Galt, Mary Lee Seitz, Sam Tinsley, Thea Philliou, and Ellen Sawyer for accuracy checking the manuscript.

The ultimate measure of this textbook's success is whether it helps students master basic skills, develop problem-solving techniques, and increase their confidence in learning and using mathematics. In order for us, as authors, to know what to keep and what to improve for the next edition, we need to hear from you, the instructor, and you, the student. Please tell us what you like and where you need additional help by sending an e-mail to math@awl.com. We appreciate your feedback.

In appreciation of your lasting support and never-ending enthusiasm: family, colleagues, and more than a generation of motivated students.

Stan Salzman

This book is dedicated to my dad, who always told me when I was young that girls could learn math, and to my students at Minneapolis Community and Technical College, who keep me in touch with the real world.

Diana L. Hestwood

Feature Walk-Through

Ratio and Proportion 5

5.1 Ratios
5.2 Rates
5.3 Proportions
5.4 Solving Proportions
5.5 Solving Application Problems with Proportions

New! Chapter Openers New chapter openers feature real-world applications of mathematics that are relevant to students and tied to specific material within the chapters.

Nearly ⅓ of the people in the United States now own a cell phone. (*Source: Scientific American.*) Everyone likes to talk, but no one likes to pay the bills! Now you can get the best possible deal on cellular phone service by finding unit rates. (See Section 5.2, Exercises 47–50.)

MyMathLab.com
You're Connected

311

---

80 Chapter 1 Whole Numbers

❷ Use the bar graph to find the approximate number of fans who picked each sport as their favorite.

(a) Pro football

(b) Pro baseball

(c) Pro basketball

(d) College basketball

(e) Golf

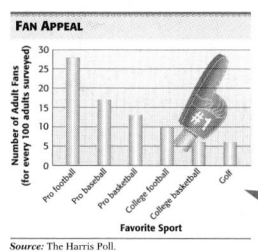

FAN APPEAL

Source: The Harris Poll.

Example 2 Using a Bar Graph

Use the bar graph to find the number of fans who picked college football as their favorite sport.

Use a ruler or straight edge to line up the top of the bar labeled "College Football," with the numbers on the left edge of the graph, labeled "Number of Adults Fans." We see that 10 out of 100 adult fans picked college football as their favorite sport.

Work Problem ❷ at the Side.

❸ Use the line graph to find the predicted population of the United States for each year.

(a) 2050

(b) 2075

3 Read and understand a line graph. A **line graph** is often used for showing a trend. The following line graph shows the U.S. Bureau of the Census predictions for U.S. population growth to the year 2100.

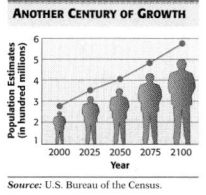

ANOTHER CENTURY OF GROWTH

Source: U.S. Bureau of the Census.

Figures and Photos Today's students are more visually oriented than ever. Thus, a concerted effort has been made to add mathematical figures, diagrams, tables, and graphs whenever possible. Many of the graphs use a style similar to that seen by students in today's print and electronic media. Photos have been incorporated to enhance applications in examples and exercises.

New! Relating Concepts These sets of exercises help students tie together topics and develop problem-solving skills as they compare and contrast ideas, identify and describe patterns, and extend concepts to new situations. These exercises make great collaborative activities for pairs or small groups of students.

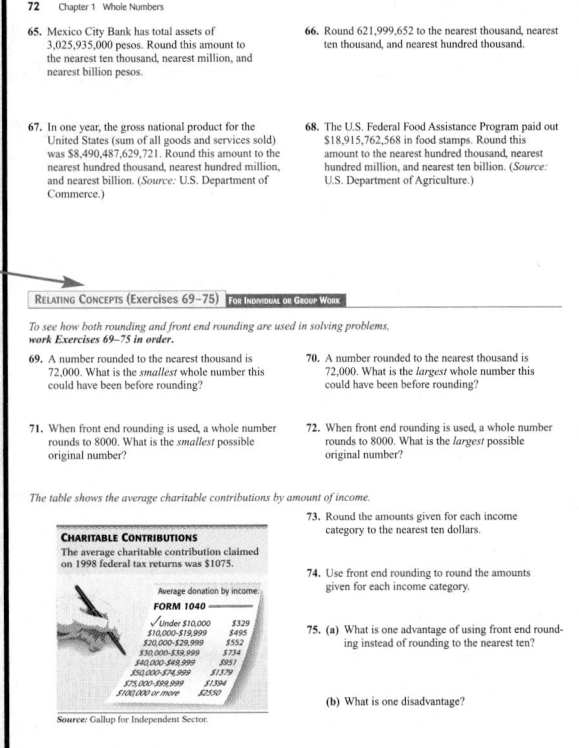

72   Chapter 1   Whole Numbers

65. Mexico City Bank has total assets of 3,025,935,000 pesos. Round this amount to the nearest ten thousand, nearest million, and nearest billion pesos.

66. Round 621,999,652 to the nearest thousand, nearest ten thousand, and nearest hundred thousand.

67. In one year, the gross national product for the United States (sum of all goods and services sold) was $8,490,487,629,721. Round this amount to the nearest hundred thousand, nearest hundred million, and nearest billion. (*Source:* U.S. Department of Commerce.)

68. The U.S. Federal Food Assistance Program paid out $18,915,762,568 in food stamps. Round this amount to the nearest hundred thousand, nearest hundred million, and nearest ten billion. (*Source:* U.S. Department of Agriculture.)

RELATING CONCEPTS (Exercises 69–75) FOR INDIVIDUAL OR GROUP WORK

To see how both rounding and front end rounding are used in solving problems, work Exercises 69–75 in order.

69. A number rounded to the nearest thousand is 72,000. What is the *smallest* whole number this could have been before rounding?

70. A number rounded to the nearest thousand is 72,000. What is the *largest* whole number this could have been before rounding?

71. When front end rounding is used, a whole number rounds to 8000. What is the *smallest* possible original number?

72. When front end rounding is used, a whole number rounds to 8000. What is the *largest* possible original number?

The table shows the average charitable contributions by amount of income.

CHARITABLE CONTRIBUTIONS
The average charitable contribution claimed on 1998 federal tax returns was $1075.

Average donation by income:

FORM 1040

✓ Under $10,000	$329
$10,000–$19,999	$495
$20,000–$29,999	$552
$30,000–$39,999	$734
$40,000–$49,999	$951
$50,000–$74,999	$1379
$75,000–$99,999	$1394
$100,000 or more	$2550

Source: Gallup for Independent Sector.

73. Round the amounts given for each income category to the nearest ten dollars.

74. Use front end rounding to round the amounts given for each income category.

75. (a) What is one advantage of using front end rounding instead of rounding to the nearest ten?

(b) What is one disadvantage?

Focus on **Real-Data Applications**

Currency Exchange

When you travel between countries, you will exchange U.S. dollars for the local currency. The exchange rate between currencies changes daily, and you can easily find the updated rates using the Internet or any major newspaper. The table shown below has been extracted from the Bloomberg Currency Calculator Web page.

NORTH AMERICA/CARIBBEAN CURRENCY RATES

| | | | Currency per 1 unit of USD | |
Currency	Symbol	Value	Net Chg	Pct Chg
Canadian Dollar	CAD	1.4967	+0.0024	+0.1606
Cayman Islands	KYD	0.8282	—	—
Jamaica Dollar	JMD	45.1	+0.1	+0.2222
Mexican Peso	MXN	9.761	−0.014	−0.1432
United States Dollar	USD	1.00		

On February 4, 2001, the currency exchange rate from U.S. dollars to Mexican pesos was given as follows:

$1.00 U.S. was equivalent to 9.761 Mexican pesos

You can set up a proportion to convert dollars to pesos. For example, suppose you want to determine the number of pesos that is equivalent to $50.00.

$$\frac{\$1}{9.761 \text{ pesos}} = \frac{\$50}{x \text{ pesos}} \quad \text{or} \quad \frac{1}{9.761} = \frac{50}{x}$$

$$1 \cdot x = (9.761) \cdot 50$$
$$x = 488.05 \text{ pesos}$$

So $50 buys 488 pesos and 5 centavos.

1. Based on the currency exchange rates for February 4, 2001, find the amount of each local currency that is equivalent to $50 U.S. and find the number of U.S. dollars that is equivalent to 200 units of each local currency. Round your answers to the nearest hundredth.

(a) $50 = _____ Canadian dollars, and 200 Canadian dollars = _____ U.S. dollars.

(b) $50 = _____ Cayman Island dollars, and 200 Cayman Island dollars = _____ U.S. dollars.

(c) $50 = _____ Jamaican dollars, and 200 Jamaican dollars = _____ U.S. dollars.

2. Set up a proportion to find the number of U.S. dollars that was equivalent to 1 Mexican Peso. 1 Mexican peso was equivalent to $ _____ (U.S.).

3. From Problem 2, you should recognize the conversion rate based on 1 Mexican peso as the expression $\frac{1}{9.761}$. What is the mathematical word that describes the relationship between the conversion rates 9.761 and $\frac{1}{9.761}$?

354

New! Focus on Real-Data Applications These one-page activities, found throughout the text, present even more relevant and in-depth looks at how mathematics is used in the real world. Designed to help instructors answer the often-asked question, "When will I ever use this stuff?," these activities ask students to read and interpret data from newspaper articles, the Internet, and other familiar, real sources. The activities are well suited to collaborative work and can also be completed by individuals or used for open-ended class discussions.

Calculator Tip When using a calculator to find unit prices, remember that division is *not* commutative. In Example 3 you wanted to find cost per ounce. Let the *order* of the *words* help you enter the numbers in the correct order.

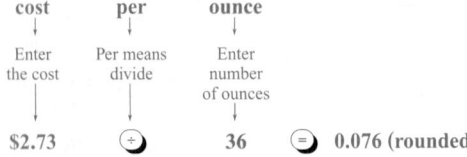

cost	per	ounce
↓	↓	↓
Enter the cost	Per means divide	Enter number of ounces
↓	↓	↓
$2.73	÷	36	= 0.076 (rounded)

If you entered 36 ÷ 2.73 =, you'd get the number of *ounces* per *dollar*. How could you use that information to find the best buy? (*Answer:* The best buy would be to get the greatest number of ounces per dollar.)

Calculator Tips These optional tips, marked with calculator icons, offer basic information and instruction for students using calculators in the course.

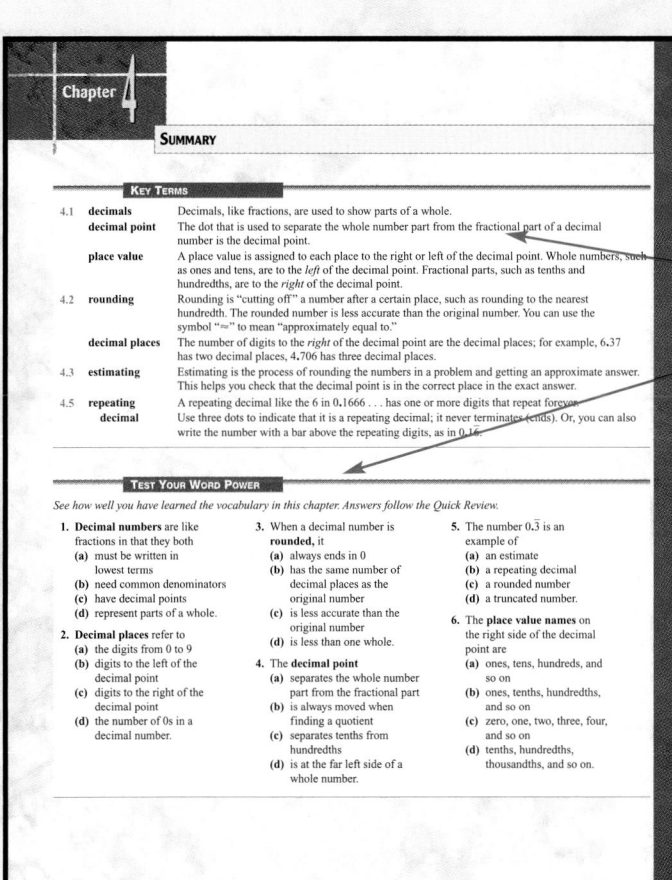

End-of-Chapter Material   One of the most admired features of the Lial textbooks is the extensive and well-thought-out end of chapter material. At the end of each chapter, students will find:

Key Terms are listed, defined, and referenced back to the appropriate section number.

New! Test Your Word Power   To help students understand and master mathematical vocabulary, Test Your Word Power has been incorporated in each Chapter Summary. Students are quizzed on Key Terms from the chapter in a multiple-choice format. Answers and examples illustrating each term are provided.

A Chapter Test helps students practice for the real thing.

New! Study Skills Component   A desk-light icon at key points in the text directs students to a separate *Study Skills Workbook* containing activities correlated directly to the text. This unique workbook explains how the brain actually learns, so students understand *why* the study tips presented will help them succeed in the course.

Quick Review sections give students not only the main concepts from the chapter (referenced back to the appropriate section), but also an adjacent example of each concept.

Review Exercises are keyed to the appropriate sections so that students can refer to examples of that type of problem if they need help.

Chapter 3   REVIEW EXERCISES

[3.1] *Add or subtract. Write answers in lowest terms.*

1. $\dfrac{5}{9} + \dfrac{2}{9}$ 2. $\dfrac{2}{7} + \dfrac{3}{7}$ 3. $\dfrac{1}{8} + \dfrac{3}{8} + \dfrac{2}{8}$ 4. $\dfrac{5}{16} - \dfrac{3}{16}$

5. $\dfrac{3}{10} - \dfrac{1}{10}$ 6. $\dfrac{5}{12} - \dfrac{3}{12}$ 7. $\dfrac{36}{62} - \dfrac{10}{62}$ 8. $\dfrac{68}{75} - \dfrac{43}{75}$

Solve each application problem. Write answers in lowest terms.

9. Nurse Suzie Brasher screened $\dfrac{7}{16}$ of her patients in her first hour on duty and $\dfrac{5}{16}$ of her patients in the second hour. What fraction of her patients did she screen in the two hours?

10. The Koats for Kids committee members completed $\dfrac{5}{8}$ of their Web-page design in the morning and $\dfrac{3}{8}$ in the afternoon. How much less did they complete in the afternoon than in the morning?

[3.2] *Find the least common multiple of each set of numbers.*

11. 4, 3   12. 5, 4   13. 10, 12, 20

Chapter 3   Review Exercises   **239**

Simplify by using the order of operations.

63. $6 \cdot \left(\dfrac{1}{3}\right)^2$ 64. $\left(\dfrac{2}{3}\right)^2 \cdot 15$ 65. $\left(\dfrac{3}{4}\right)^2 \cdot \left(\dfrac{8}{9}\right)^2$

66. $\dfrac{7}{8} \div \left(\dfrac{1}{8} + \dfrac{3}{4}\right)$ 67. $\left(\dfrac{1}{2}\right)^2 \cdot \left(\dfrac{1}{4} + \dfrac{1}{2}\right)$ 68. $\left(\dfrac{1}{4}\right)^3 + \left(\dfrac{5}{8} + \dfrac{3}{4}\right)$

MIXED REVIEW EXERCISES

Simplify by using the order of operations as necessary. Write answers in lowest terms and as whole or as mixed numbers when possible.

69. $\dfrac{5}{6} - \dfrac{1}{6}$ 70. $\dfrac{7}{8} - \dfrac{3}{4}$ 71. $\dfrac{29}{32} - \dfrac{5}{16}$ 72. $\dfrac{1}{4} + \dfrac{1}{8} + \dfrac{5}{16}$

73. $6\dfrac{2}{3}$
$-\ 4\dfrac{1}{2}$

74. $9\dfrac{1}{2}$
$+\ 16\dfrac{3}{4}$

75. 7
$-\ 1\dfrac{5}{8}$

76. $2\dfrac{3}{5}$
$8\dfrac{5}{8}$
$+\ \dfrac{5}{16}$

77. $32\dfrac{5}{12}$
$-\ 17$

78. $\dfrac{7}{22} + \dfrac{3}{22} + \dfrac{3}{11}$

79. $\left(\dfrac{1}{4}\right)^2 \cdot \left(\dfrac{2}{5}\right)^3$

80. $\dfrac{3}{8} \div \left(\dfrac{1}{2} + \dfrac{1}{4}\right)$ 81. $\left(\dfrac{2}{3}\right)^2 \cdot \left(\dfrac{1}{3} + \dfrac{1}{6}\right)$ 82. $\left(\dfrac{2}{3}\right)^3 + \left(\dfrac{2}{3} - \dfrac{5}{9}\right)$

Mixed Review Exercises require students to solve problems without the help of section reference.

Cumulative Review Exercises gather various types of exercises from preceding chapters to help students remember and retain what they are learning throughout the course.

Cumulative Review Exercises   CHAPTERS 1–5

Name the digit that has the given place value.

1. 216,475,038
   thousands
   tens
   millions
   hundred thousands

2. 340.6915
   hundredths
   ones
   ten-thousandths
   hundreds

Round each number as indicated.

3. 9903 to the nearest hundred

4. 617.0519 to the nearest tenth

5. $99.81 to the nearest dollar

6. $3.0555 to the nearest cent

First use front end rounding to round each number and estimate the answer. Then find the exact answer.

7. *Estimate:*   *Exact:*
   28
   5206
   $+\ 351$

8. *Estimate:*   *Exact:*
   63.1
   $-\ 5.692$

9. *Estimate:*   *Exact:*
   4716
   $\times\ 804$

10. *Estimate:*   *Exact:*
   0.982
   $\times\ 17.8$

11. *Estimate:*   *Exact:*
   $53\overline{)48{,}071}$

12. *Estimate:*   *Exact:*
   $4.5\overline{)1638}$

13. *Estimate:*   *Exact:*
   $\_\_ \cdot \_\_ = \_\_$ $1\dfrac{5}{6} \cdot 3\dfrac{3}{5}$

14. *Estimate:*   *Exact:*
   $\_\_ \div \_\_ = \_\_$ $5\dfrac{1}{4} \div \dfrac{7}{8}$

15. *Estimate:*   *Exact:*
   $\_\_ - \_\_ = \_\_$ $2\dfrac{4}{5} - 1\dfrac{5}{6}$

16. *Estimate:*   *Exact:*
   $\_\_ + \_\_ = \_\_$ $2\dfrac{9}{10} + 10\dfrac{1}{2}$

To the Student: Success in Mathematics

There are two main reasons why students have difficulty with mathematics:

- Students start in a course for which they do not have the necessary background knowledge.

- Students don't know how to study mathematics effectively.

Your instructor can help you decide whether this is the right course for you. We can give you some study tips.

Studying mathematics *is* different from studying subjects like English and history. The key to success is regular practice. This should not be surprising. After all, can you learn to play the piano or ski well without a lot of regular practice? The same is true for learning mathematics. Working problems nearly every day is the key to becoming successful. Here is a list of things that will help you succeed in studying mathematics.

1. *Attend class regularly.* Pay attention to what your instructor says and does in class, and take careful notes. In particular, note the problems the instructor works on the board and copy the complete solutions. Keep these notes separate from your homework to avoid confusion when you review them later.

2. Don't hesitate to *ask questions in class.* It is not a sign of weakness but of strength. There are always other students with the same question who are too shy to ask.

3. *Read your text carefully.* Many students read only enough to get by, usually only the examples. Reading the complete section will help you solve the homework problems. Most exercises are keyed to specific examples or objectives that will explain the procedures for working them.

4. Before you start on your homework assignment, *rework the problems the teacher worked in class.* This will reinforce what you have learned. Many students say, "I understand it perfectly when you do it, but I get stuck when I try to work the problem myself."

5. Do your homework assignment only *after reading the text* and reviewing your notes from class. Check your work against the answers in the back of the book. If you get a problem wrong and are unable to understand why, mark that problem and ask your instructor about it. Then practice working additional problems of the same type to reinforce what you have learned.

6. *Work as neatly as you can.* Write your symbols clearly, and make sure the problems are clearly separated from each other. Working neatly will help you to think clearly and also make it easier to review the homework before a test.

7. After you complete a homework assignment, *look over the text again.* Try to identify the main ideas that are in the lesson. Often they are clearly highlighted or boxed in the text.

8. *Use the chapter test at the end of each chapter as a practice test.* Work through the problems under test conditions, without referring to the text or the answers until you are finished. You may want to time yourself to see how long it takes you. When you finish, check your answers against those in the back of the book, and study the problems you missed.

9. *Keep all quizzes and tests that are returned to you,* and use them when you study for future tests and the final exam. These quizzes and tests indicate what concepts your instructor considers to be most important. Be sure to correct any problems on these tests that you missed, so you will have the corrected work to study.

10. *Don't worry if you do not understand a new topic right away.* As you read more about it and work through the problems, you will gain understanding. Each time you review a topic you will understand it a little better. Few people understand each topic completely right from the start.

Reading a list of study tips is a good start, but you may need some help actually *applying* the tips to your work in this math course.

Watch for this icon as you work in this textbook, particularly in the first few chapters. It will direct you to one of 12 activities in the *Study Skills Workbook* that comes with this text. Each activity helps you to actually *use* a study skills technique. These techniques will greatly improve your chances for success in this course.

• Find out *how your brain learns new material.* Then use that information to set up effective ways to learn math.

• Find out *why short-term memory is so short* and what you can do to help your brain remember new material weeks and months later.

• Find out *what happens when you "blank out" on a test* and simple ways to prevent it from happening.

All the activities in the *Study Skills Workbook* are brain-friendly ways to enjoy and succeed at math. Whether you need help with note taking, managing homework, taking tests, or preparing for a final exam, you'll find specific, clearly explained ideas that really work because they're based on research about how the brain learns and remembers.

Diagnostic Pretest

[Chapter 1]

1. Use digits to write "eighty-nine million, twenty-three thousand, five hundred seven."

   1. _____

2. Subtract. $\begin{array}{r} 7009 \\ -\ 2678 \end{array}$

   2. _____

3. Divide. $20{,}213 \div 29$

   3. _____

4. Round 88,658 to the nearest thousand.

   4. _____

[Chapter 2]

5. Write $\dfrac{235}{8}$ as a mixed number.

   5. _____

6. Write the prime factorization of 392 using exponents.

   6. _____

7. A cake recipe calls for $2\dfrac{1}{4}$ cups of flour. How much flour is needed to make 5 cakes?

   7. _____

8. First estimate the answer. Then multiply to find the exact answer. Write the exact answer as a mixed number.

   $$5\frac{3}{4} \cdot 2\frac{1}{8}$$

   8. *Estimate:* _____

   Exact: _____

[Chapter 3]

9. Find the least common multiple of 5, 8, 12, and 30.

   9. _____

10. Subtract. Write your answer in lowest terms.

    $$\frac{9}{10} - \frac{4}{15}$$

    10. _____

11. First estimate the answer. Then add to find the exact answer. Write the exact answer as a mixed number.

    $$8\frac{2}{9} + 12\frac{5}{6}$$

    11. *Estimate:* _____

    Exact: _____

12. Use the order of operations to simplify.

    $$\left(\frac{3}{4}\right)^2 + \left(\frac{5}{6} - \frac{2}{3}\right)$$

    12. _____

13. _____

14. _____

15. _____

16. _____

17. _____

18. _____

19. _____

20. _____

21. _____

22. _____

23. _____

24. _____

[Chapter 4]

13. Round $1.3852 to the nearest cent.

14. Write $6\dfrac{5}{9}$ as a decimal. Round to the nearest thousandth if necessary.

15. Use the order of operations to simplify $4.5^2 - 3.2 + 0.6 \cdot 12$.

16. Find the cost (to the nearest cent) of 5.3 pounds of chicken at $1.59 per pound.

[Chapter 5]

17. Write the ratio "50 minutes to 4 hours" in lowest terms. Change to the same units if necessary.

18. Determine whether the proportion is true or false.

$$\frac{16.2}{23.6} = \frac{5.4}{8}$$

19. Find the unknown number in the proportion. Write your answer as a mixed number.

$$\frac{2\frac{1}{2}}{x} = \frac{\frac{3}{4}}{8}$$

20. Martha earns $59.50 in 7 hours. How much will she earn in 40 hours?

[Chapter 6]
Write each number as a percent.

21. 0.582

22. $8\dfrac{3}{4}$

23. The price of a cell phone is $128 plus 6% sales tax. Find the total cost of the phone including sales tax.

24. Roberto's annual salary was $24,500. After his first year on the job, he received a raise to $26,215. Find the percent of increase.

[Chapter 7]
Convert each measurement.

25. (a) 36 quarts to gallons

    (b) $4\frac{3}{4}$ pounds to ounces

26. (a) 0.675 meters to centimeters

    (b) 4528 grams to kilograms

27. Jonathan weighed 3 kg 170 g at birth. When he was 5 months old, he weighed 6 kg 90 g. How much weight had he gained in kilograms?

28. Write the most reasonable metric unit in each blank. Choose from km, m, cm, mm, L, mL, kg, g, and mg.

    (a) Leanne took a 325 _____ pill for her headache.

    (b) The cover of this textbook is about 21 _____ wide.

25. (a) _____

    (b) _____

26. (a) _____

    (b) _____

27. _____

28. (a) _____

    (b) _____

[Chapter 8]

29. Find the perimeter.

29. _____

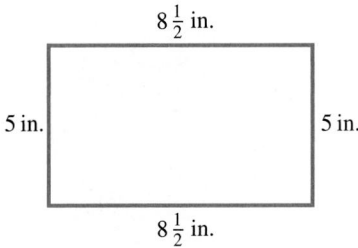

30. Find the area.

30. _____

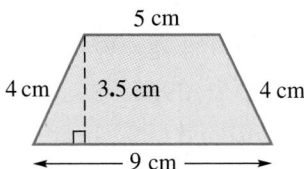

31. Find the volume. Use 3.14 as the approximate value for π.

31. _____

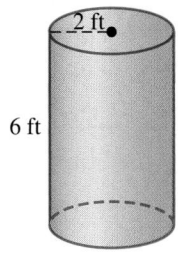

32. _____

32. Find the unknown length.

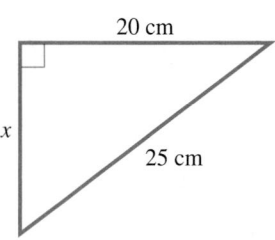

[Chapter 9]

33. _____

33. Subtract. $12 - (-13)$

34. _____

34. Multiply. $(-0.4)(11.5)$

35. _____

35. Find the value of $7p - 8q$ if $p = -3$ and $q = -4$.

36. _____

36. Solve. $8r + 9 = 2r - 3$

[Chapter 10]

37. _____

37. The circle graph shows the budget for the Song family. The total budget for one month is $3200. Find the amount budgeted for food in one month.

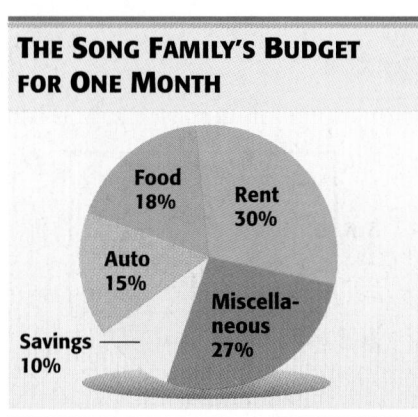

38. _____

38. Find the mean for the following ages of members of an extended family:

15, 52, 16, 51, 29, 55, 26, 85, 29, 55

39. _____

39. Find the weighted mean. Round to the nearest tenth if necessary.

Quiz Score	Frequency
10	3
9	8
8	10
7	9
6	4
5	1

40. _____

40. Find the median for the following hourly wages:

$7.50, $6.25, $9.80, $8.10, $13.75, $11.00

Whole Numbers

1

A typical Starbucks location has between 20 and 25 employees. Each of these employees is involved with work schedules, keeping track of their work hours, computing sales and sales taxes, and ordering inventory. An understanding of whole numbers is essential because each of these employees will be using mathematics as part of their jobs. (See Example 2 and Exercise 2 in Section 1.10.)

You're Connected

1.1 READING AND WRITING WHOLE NUMBERS

OBJECTIVES

1 Identify whole numbers.

2 Give the place value of a digit.

3 Write a number in words or digits.

4 Read a table.

Study Skills Workbook
Study Skills **Activity 2**

1 Identify the place value of the 4 in each whole number.

(a) 341

(b) 714

(c) 479

Knowing how to read and write numbers is an important step in learning mathematics.

1 **Identify whole numbers.**   The **decimal system** of writing numbers uses the ten digits

$$0, 1, 2, 3, 4, 5, 6, 7, 8, 9$$

to write any number. For example, these digits can be used to write the **whole numbers:**

$$0, 1, 2, 3, 4, 5, 6, 7, 8, 9, 10, 11, 12, 13 \ldots$$

The three dots indicate that the list goes on forever.

2 **Give the place value of a digit.**   Each digit in a whole number has a **place value,** depending on its position in the whole number. The following place value chart shows the names of the different places used most often and has the whole number 29,022 entered.

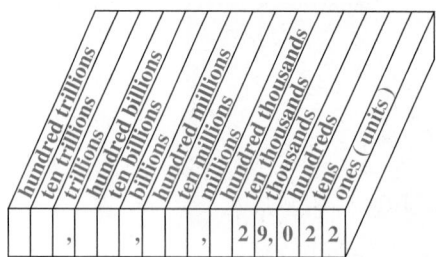

At 29,022 ft, Mt. Everest is the world's tallest mountain.

The tallest mountain in the world is Mt. Everest at 29,022 ft. Each of the 2s in 29,022 represents a different amount because of its position, or *place,* within the number. The *place value* of the 2 on the left is 2 ten thousands (20,000). The place value of the middle 2 is 2 tens (20), and the place value of the 2 on the right is 2 ones (2).

Example 1 **Identifying Place Values**

Identify the place value of 8 in each whole number.

(a) 28   **(b)** 85   **(c)** 869
        └ 8 ones   └ 8 tens   └ 8 hundreds

Notice that the value of 8 in each number is different, depending on its location (place) in the number.

Work Problem 1 at the Side.

Example 2 **Identifying Place Values**

Identify the place value of each digit in the number 725,283.

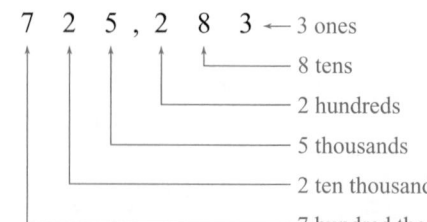

7 2 5 , 2 8 3 ← 3 ones
                └ 8 tens
                  2 hundreds
                  5 thousands
                  2 ten thousands
                  7 hundred thousands

Notice the comma between the hundreds and thousands position in the number 725,283 in Example 2.

Work Problem ❷ at the Side.

Using Commas

Commas are used to separate each group of three digits, starting from the right. This makes numbers easier to read. (An exception: commas are frequently omitted in four-digit numbers such as 9748 or 1329.) Each three-digit group is called a **period.** Some instructors prefer to just call them **groups.**

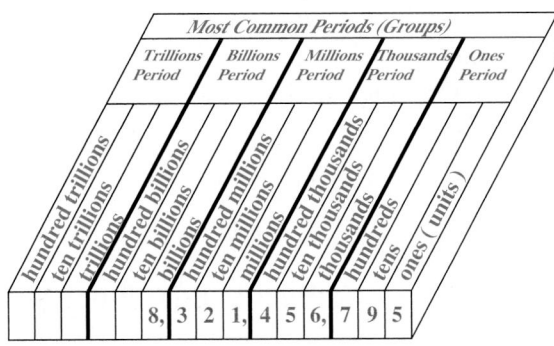

> **Example 3** **Knowing the Period or Group Names**

Write the digits in each period of 8,321,456,795.

$$8,321,456,795$$

8 billions ◄
321 millions ◄
456 thousands ◄
795 ones ◄

Work Problem ❸ at the Side.

Use the following rule to read a number with more than three digits.

Writing Numbers in Words

Start at the left when writing a number in words or saying it aloud. Write or say the digit names in each period (group), followed by the name of the period, except for the period name "ones," which is *not* used.

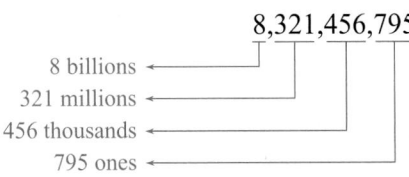 **Write a number in words or digits.** The following examples show how to write names for whole numbers.

> **Example 4** **Writing Numbers in Words**

Write each number in words.

(a) 57
    This number means 5 tens and 7 ones, or 50 ones and 7 ones. Write the number as

                        fifty-seven.

Continued on Next Page

---

❷ Identify the place value of each digit.

(a) 24,386

(b) 371,942

❸ In the number 3,251,609,328 identify the digits in each period (group).

(a) billions period

(b) millions period

(c) thousands period

(d) ones period

④ Write each number in words.

(a) 15

(b) 68

(c) 293

(d) 902

⑤ Write each number in words.

(a) 7309

(b) 95,372

(c) 100,075,002

(d) 11,022,040,000

⑥ Rewrite each number using digits.

(a) one thousand, four hundred thirty-seven

(b) nine hundred seventy-one thousand, six

(c) eighty-two million, three hundred twenty-five

(b) 94

ninety-four

(c) 874

eight hundred seventy-four

(d) 601

six hundred one

CAUTION

The word *and* should never be used when writing whole numbers. You will often hear someone say "five hundred *and* twenty-two," but the use of "and" is not correct since "522" is a whole number. When you work with decimal numbers, the word *and* is used to show the position of the decimal point. For example, 98.6 is read as "ninety-eight *and* six tenths." Practice with decimal numbers is the topic of **Section 4.1**.

Work Problem ④ at the Side.

Example 5 ▶ **Writing Numbers in Words by Using Period Names**

Write each number in words.

(a) 725,283

seven hundred twenty-five **thousand,** two hundred eighty-three

Number in period Name of period Number in period (not necessary to write "ones")

(b) 7835

seven **thousand,** eight hundred thirty-five

Name of period No period name needed

(c) 111,356,075

one hundred eleven **million,** three hundred fifty-six **thousand,** seventy-five

(d) 17,000,017,000

seventeen **billion,** seventeen **thousand**

Work Problem ⑤ at the Side.

Example 6 ▶ **Writing Numbers in Digits**

Rewrite each number using digits.

(a) seven **thousand,** eighty-five

7085

(b) two hundred fifty-six **thousand,** six hundred twelve

256,612

(c) nine **million,** five hundred fifty-nine

9,000,559

Zeros indicate there are no thousands.

Work Problem ⑥ at the Side.

Calculator Tip   Does your calculator show a comma between each group of three digits? Probably not, but try entering a long number such as 34,629,075. Notice that there is no key with a comma on it, so you do not enter commas. A few calculators may show the position of the commas *above* the digits, like this ────────

┌─────────────────┐
│   **34′629′075** │
└─────────────────┘

Most of the time you will have to write in the commas.

4   **Read a table.**   A common way of showing number values is by using a **table.** Tables organize and display facts so that they are more easily understood. The following table shows some past facts and future predictions for the United States.

NUMBERS FOR THE NEW CENTURY
These estimated numbers give us a glimpse of what we can expect in the new century.

Year	1990s	2001	2020
U.S. population	261 million	281 million	338 million
Births	16 million	14 million	13 million
Household income	$42,936	$45,069	$53,375
Average salary	$21,129	$23,411	$28,050

Source: Family Circle magazine.

If you read from left to right along the row labeled "U.S. population," you find that the population in the 1990s was 261 million, then the population in 2001 was 281 million, and the estimated population for 2020 is 338 million.

Example 7   **Reading a Table**

Use the table to find each number, and write the number in words.

(a) The predicted household income in the year 2020
Read from left to right along the row labeled "Household income" until you reach the 2020 column and find $53,375.

Fifty-three thousand, three hundred seventy-five dollars

(b) The average salary in the 1990s
Read from left to right along the row labeled "Average salary." In the 1990s column you find $21,129.

Twenty-one thousand, one hundred twenty-nine dollars.

═══════════════ **Work Problem 7 at the Side.**

7 Use the table to find each number, and write the number in digits, or write the number in words.

(a) The number of births in 2000

(b) The predicted population in 2020

(c) Household income in the 1990s.

(d) The predicted average salary in 2020

ANSWERS
7. (a) 14,000,000
   (b) 338,000,000
   (c) forty-two thousand, nine hundred thirty-six dollars
   (d) twenty-eight thousand, fifty dollars

'Til Debt Do You Part!

Bride's magazine recently released statistics comparing average wedding costs in the United States from 1990 and 1997. In 1990, the cost of the average wedding was $15,208. The average number of wedding guests had grown to 200 in 1997 compared to 170 in 1990.

Category	Average Cost in 1997
Misc. expenses (stationery, clergy, gifts, limousine)	$1260
Bouquets and other flowers	$ 756
Photography and videography	$1311
Music	$ 830
Engagement and wedding rings (bride and groom)	$4060
Rehearsal dinner	$ 698
Bride's wedding dress and headpiece	$ 989
Bridal attendants' dresses (average of 5 bridesmaids)	$ 790
Mother of the bride's apparel	$ 231
Men's formalwear (ushers, best man, groom)	$ 544
Wedding reception	$7635
Grand Total	

Source: Bride's 1997 Millennium Report: Wedding Love & Money.

1. What is the grand total of expenses shown in the chart?

2. How much more expensive was a wedding in 1997 compared to 1990?

3. The groom pays for the bouquets and flowers, the rehearsal dinner, the bride's engagement and wedding rings ($3500), the clergy ($232), and the groom's formalwear ($95). What is the total amount spent by the groom?

4. If you budgeted $38 per person for the wedding reception and you invited 200 guests to a wedding in 2001, how much money would you have spent compared to the 1997 wedding reception costs?

5. If you budget $5000 for the reception and the caterer charges $38 per person, how many guests can you invite? How much of your budget is left over?

6. If you budget $8500 for the reception and the caterer charges $38 per person, how many guests can you invite? How much of your budget is left over?

7. What type of arithmetic problem did you work to get the answers to Problems 6 and 7? What is the mathematical term for the "left over" budget?

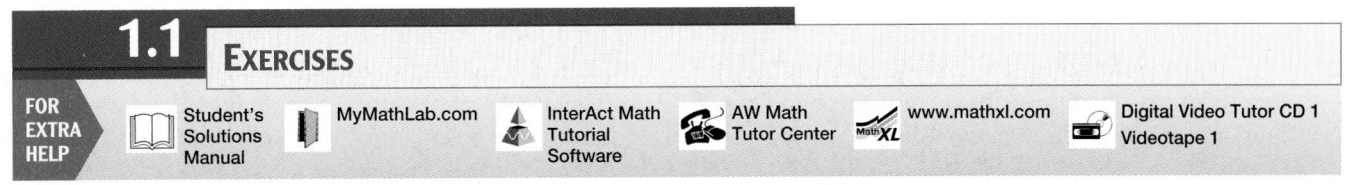

FOR EXTRA HELP

Student's Solutions Manual   MyMathLab.com   InterAct Math Tutorial Software   AW Math Tutor Center   www.mathxl.com   Digital Video Tutor CD 1 Videotape 1

Study Skills Workbook
Study Skills **Activity 3**

Write the digit for the given **place value** *in each whole number. See Examples 1 and 2.*

1. 2078

thousands

tens

2. 5139

thousands

ones

3. 18,015

ten thousands

hundreds

4. 75,229

ten thousands

ones

5. 7,628,592,183

millions

thousands

6. 1,700,225,016

billions

millions

Write the digits for the given **period** *(group) in each whole number. See Example 3.*

7. 2,768,543

millions

thousands

ones

8. 28,785,203

millions

thousands

ones

9. 60,000,502,109

billions

millions

thousands

ones

10. 100,258,100,006

billions

millions

thousands

ones

11. Do you think the fact that humans have four fingers and a thumb on each hand explains why we use a number system based on ten digits? Explain.

12. The decimal system uses ten digits. Fingers and toes are often referred to as digits. In your opinion, is there a relationship here? Explain.

Write each number in words. See Examples 4 and 5.

13. 23,115

14. 37,886

15. 346,009

16. 218,033

17. 25,756,665

18. 999,993,000

Write each number using digits. See Example 6.

19. thirty-two thousand, five hundred twenty-six

20. ninety-five thousand, one hundred eleven

21. ten million, two hundred twenty-three

22. one hundred million, two hundred

Write the numbers from each sentence using digits. See Example 6.

23. The Sunday newspaper says that it contains one hundred ninety-eight dollars in "money-saving" coupons. (*Source: Sacramento Bee.*)

24. The United States Postal Service set a record of two hundred eighty million, four hundred eighty-nine thousand postmarked pieces of mail on a single day. (*Source:* U.S. Postal Service.)

25. The Binney & Smith Company in Pennsylvania makes about two billion Crayola crayons each year. (*Source:* Binney & Smith Company.)

26. There will be four hundred seventy-three thousand new jobs for registered nurses in the year 2005. (*Source:* U.S. Department of Labor.)

27. In the year 2005, there will be five hundred thirty-two thousand new jobs for retail salespeople. (*Source:* U.S. Department of Labor.)

28. The total Temporary Assistance for Needy Families (TANF) in the United States in 1998 was twenty billion, five hundred twenty-eight million, four hundred ninety-one thousand dollars. (*Source:* U.S. Department of Health and Human Services.)

29. Rewrite eight hundred trillion, six hundred twenty-one million, twenty thousand, two hundred fifteen by using digits.

30. Rewrite 70,306,735,002,102 in words.

The table at the right shows various ways people get to work. Use the table to answer Exercises 31–34. See Example 7.

31. Which method of transportation is least used? Write the number in words.

32. Which method of transportation is most used? Write the number in words.

33. Find the number of people who walk to work or work at home, and write it in words.

34. Find the number of people who carpool, and write it in words.

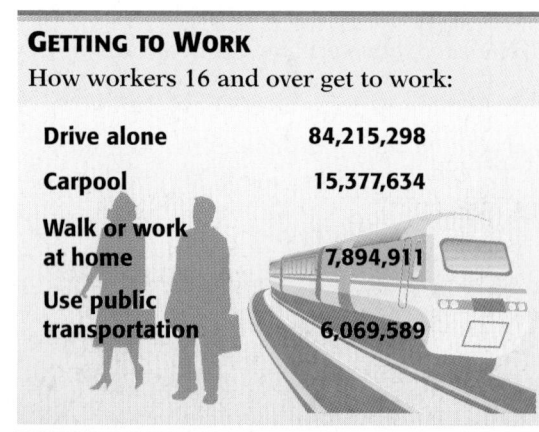

GETTING TO WORK

How workers 16 and over get to work:

Drive alone	84,215,298
Carpool	15,377,634
Walk or work at home	7,894,911
Use public transportation	6,069,589

Source: U.S. Bureau of the Census.

1.2 ADDING WHOLE NUMBERS

There are four dollar bills at the left and two at the right. In all, there are six dollar bills.

The process of finding the total is called **addition.** Here 4 and 2 were added to get 6. Addition is written with a + sign, so that

$$4 + 2 = 6.$$

1 Add two single-digit numbers. In addition, the numbers being added are called **addends,** and the resulting answer is called the **sum** or **total.**

$$
\begin{array}{r}
4 \leftarrow \text{Addend} \\
+\ 2 \leftarrow \text{Addend} \\
\hline
6 \leftarrow \text{Sum (total)}
\end{array}
$$

Addition problems can also be written horizontally, as follows.

$$
\underset{\text{Addend}}{4} \;\; + \;\; \underset{\text{Addend}}{2} \;\; = \;\; \underset{\text{Sum}}{6}
$$

Commutative Property of Addition

By the **commutative property of addition,** changing the order of the addends in an addition problem does not change the sum.

For example, the sum of $4 + 2$ is the same as the sum of $2 + 4$. This allows the addition of the same numbers in a different order.

Example 1 Adding Two Single-Digit Numbers

Add, and then change the order of numbers to write another addition problem.

(a) $6 + 2 = 8$ and $2 + 6 = 8$

(b) $5 + 9 = 14$ and $9 + 5 = 14$

(c) $8 + 3 = 11$ and $3 + 8 = 11$

(d) $8 + 8 = 16$

Work Problem **1** at the Side.

Associative Property of Addition

By the **associative property of addition,** changing the grouping of the addends in an addition problem does not change the sum.

For example, the sum of $3 + 5 + 6$ may be found as follows.

$$(3 + 5) + 6 = 8 + 6 = 14 \quad \text{Parentheses tell what to do first.}$$

Another way to add the same numbers is

$$3 + (5 + 6) = 3 + 11 = 14.$$

Either grouping gives the answer 14.

---

1 Add, and then change the order of numbers to write another addition problem.

(a) $6 + 3$

(b) $7 + 9$

(c) $7 + 8$

(d) $6 + 9$

ANSWERS
1. (a) $9; 3 + 6 = 9$ **(b)** $16; 9 + 7 = 16$
(c) $15; 8 + 7 = 15$ **(d)** $15; 9 + 6 = 15$

❷ Add each column of numbers.

(a)
```
    6
    5
    3
    2
 +  9
```

(b)
```
    4
    6
    5
    3
 +  2
```

(c)
```
    8
    7
    9
    2
 +  1
```

(d)
```
    3
    8
    6
    4
 +  8
```

2 ▭ **Add more than two numbers.**   To add several numbers, first write them in a column. Add the first number to the second. Add this sum to the third digit; continue until all the digits are used.

Example 2 **Adding More Than Two Numbers**

Add 2, 5, 6, 1, and 4.

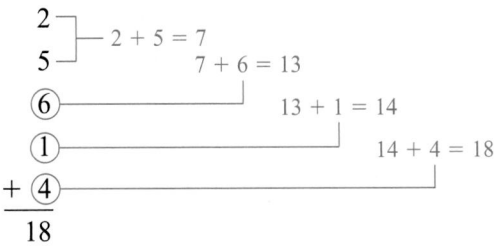

NOTE

By the commutative and associative properties of addition, numbers may also be added starting at the bottom of a column. Adding from the top or adding from the bottom will give the same answer.

Work Problem ❷ at the Side.

3 ▭ **Add when carrying is not required.**   If numbers have two or more digits, you must arrange the numbers in columns so that the ones digits are in the same column, tens are in the same column, hundreds are in the same column, and so on. Next, you add column by column starting at the right.

Example 3 **Adding without Carrying**

Add 511 + 23 + 154 + 10.

First line up the numbers in columns, with the ones column at the right.

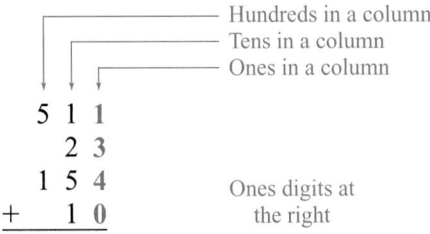

Now start at the right and add the ones digits. Add the tens digits next, and finally, the hundreds digits.

```
    5 1 1
      2 3
    1 5 4
 +    1 0
    6 9 8
```
— Sum of ones
— Sum of tens
— Sum of hundreds

The sum of the four numbers is 698.

Work Problem ❸ at the Side.

4▭   **Add with carrying.**   If the sum of the digits in any column is more than 9, use **carrying.**

┌
| **Example 4**   **Adding with Carrying**
|
| Add 47 and 29.
|
| Add ones.
|
| $$\begin{array}{r} 47 \\ +\ 29 \end{array}$$
|
| Sum of ones is 16.
|
| Because 16 is 1 ten plus 6 ones, place 6 in the ones column and carry 1 to the tens column.
|
| $$\begin{array}{r} 1 \\ 47 \\ +\ 29 \\ \hline 6 \end{array}$$ $7 + 9 = 16$
|
| Add the tens column, including the carried 1.
|
| $$\begin{array}{r} 1 \\ 47 \\ +\ 29 \\ \hline 76 \end{array}$$
|
| Sum of digits in tens column

Work Problem ❹ at the Side.

┌
| **Example 5**   **Adding with Carrying**
|
| Add $324 + 7855 + 23 + 7 + 86$.
|
| Add the digits in the ones column.
|
| $$\begin{array}{r} 2 \\ 324 \\ 7855 \\ 23 \\ 7 \\ +\quad 86 \\ \hline 5 \end{array}$$
|
| Sum of the ones column is 25.
| Carry 2 to the tens column.
| Write 5 in the ones column.
|
| In 25, the 5 represents 5 ones and is written in the ones column, while 2 represents 2 tens and is carried to the tens column.
| Now add the digits in the tens column, including the carried 2.
|
| $$\begin{array}{r} 12 \\ 324 \\ 7855 \\ 23 \\ 7 \\ +\quad 86 \\ \hline 95 \end{array}$$
|
| Sum of the tens column is 19.
| Carry 1 to the hundreds column.
| Write 9 in the tens column.
|
| **Continued on Next Page**

❸ Add.

(a) $\begin{array}{r} 34 \\ +\ 62 \end{array}$

(b) $\begin{array}{r} 478 \\ +\ 221 \end{array}$

(c) $\begin{array}{r} 42{,}305 \\ +\ 11{,}563 \end{array}$

❹ Add by carrying.

(a) $\begin{array}{r} 76 \\ +\ 19 \end{array}$

(b) $\begin{array}{r} 76 \\ +\ 18 \end{array}$

(c) $\begin{array}{r} 56 \\ +\ 37 \end{array}$

(d) $\begin{array}{r} 34 \\ +\ 49 \end{array}$

Answers
3. (a) 96   **(b)** 699   **(c)** 53,868
4. (a) 95   **(b)** 94   **(c)** 93   **(d)** 83

5 Add by carrying as necessary.

(a)
$$\begin{array}{r} 396 \\ 87 \\ 42 \\ + 651 \end{array}$$

(b)
$$\begin{array}{r} 4271 \\ 372 \\ 8976 \\ + 162 \end{array}$$

(c)
$$\begin{array}{r} 57 \\ 4 \\ 392 \\ 804 \\ 51 \\ + 27 \end{array}$$

(d)
$$\begin{array}{r} 7821 \\ 435 \\ 72 \\ 305 \\ + 1693 \end{array}$$

6 Add by carrying mentally.

(a)
$$\begin{array}{r} 816 \\ 363 \\ 17 \\ 2 \\ 5 \\ + 7654 \end{array}$$

(b)
$$\begin{array}{r} 3305 \\ 650 \\ 708 \\ 29 \\ 40 \\ 6 \\ + 3 \end{array}$$

(c)
$$\begin{array}{r} 15,829 \\ 765 \\ 78 \\ 15 \\ 9 \\ 7 \\ + 13,179 \end{array}$$

Add the hundreds column, including the carried 1.

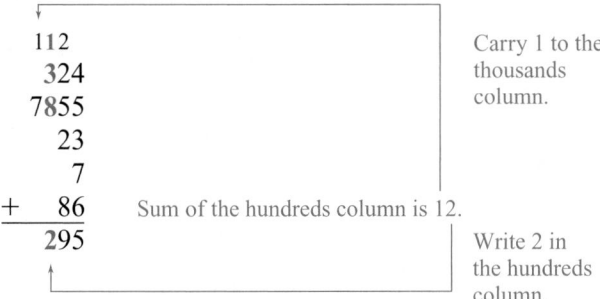

Carry 1 to the thousands column.

Sum of the hundreds column is 12.

Write 2 in the hundreds column.

Add the thousands column, including the carried 1.

$$\begin{array}{r} 112 \\ 324 \\ 7855 \\ 23 \\ 7 \\ + 86 \\ \hline 8295 \end{array}$$

Sum of the thousands column is 8.

Finally, $324 + 7855 + 23 + 7 + 86 = 8295$.

Work Problem 5 at the Side.

NOTE

For additional speed, try to carry mentally. Do not write the carried number, but just remember it as you move to the top of the next column. Try this method. If it works for you, use it.

Work Problem 6 at the Side.

5 **Solve application problems with carrying.** In Section 1.10 we will describe how to solve application problems in more detail. The next two examples are application problems that require adding.

Example 6 **Applying Addition Skills**

On this map, the distance in miles from one location to another is written alongside the road. Find the shortest distance from Altamonte Springs to Clear Lake.

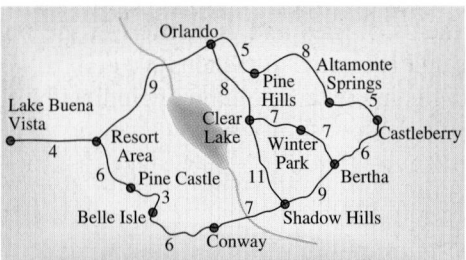

Continued on Next Page

ANSWERS
5. (a) 1176   (b) 13,781   (c) 1335
   (d) 10,326
6. (a) 8857   (b) 4741   (c) 29,882

1.3 SUBTRACTING WHOLE NUMBERS

Suppose you have $8, and you spend $5 for gasoline. You then have $3 left. There are two different ways of looking at these numbers.

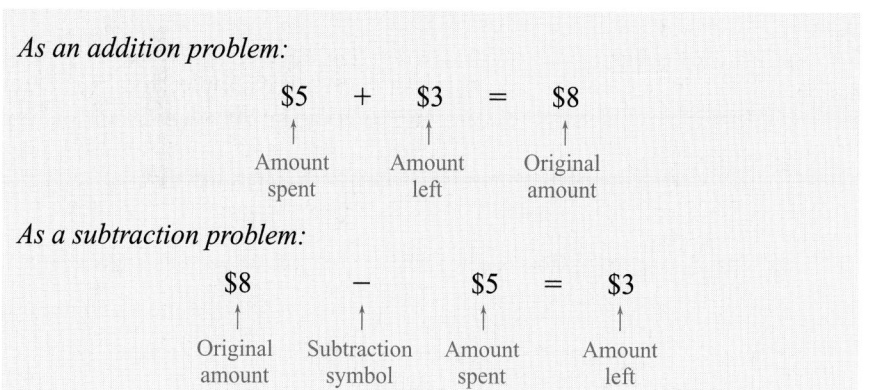

As an addition problem:

$$\$5 \quad + \quad \$3 \quad = \quad \$8$$

Amount spent   Amount left   Original amount

As a subtraction problem:

$$\$8 \quad - \quad \$5 \quad = \quad \$3$$

Original amount   Subtraction symbol   Amount spent   Amount left

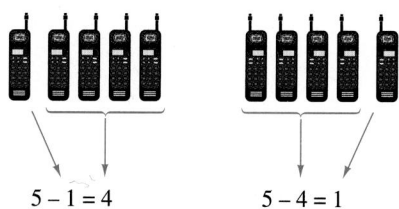

1 **Change addition problems to subtraction and subtraction problems to addition.** As shown in the preceding box, an addition problem can be changed to a subtraction problem and a subtraction problem can be changed to an addition problem.

Example 1 **Changing Addition Problems to Subtraction**

Change each addition problem to a subtraction problem.

(a) $4 + 1 = 5$

Two subtraction problems are possible:

$$5 - 1 = 4 \quad \text{or} \quad 5 - 4 = 1$$

These figures show each subtraction problem.

$$5 - 1 = 4 \qquad\qquad 5 - 4 = 1$$

(b) $8 + 7 = 15$

$$15 - 7 = 8 \quad \text{or} \quad 15 - 8 = 7$$

================ **Work Problem ❶ at the Side.**

Example 2 **Changing Subtraction Problems to Addition**

Change each subtraction problem to an addition problem.

(a) $8 - 3 = 5$

$$8 = 3 + 5$$

It is also correct to write $8 = 5 + 3$.

========= **Continued on Next Page**

❶ Write two subtraction problems for each addition problem.

(a) $6 + 1 = 7$

(b) $7 + 4 = 11$

(c) $15 + 22 = 37$

(d) $23 + 55 = 78$

❷ Write an addition problem for each subtraction problem.

(a) $6 - 4 = 2$

(b) $9 - 4 = 5$

(c) $21 - 15 = 6$

(d) $58 - 42 = 16$

❸ Subtract.

(a) $\begin{array}{r} 63 \\ -22 \\ \hline \end{array}$

(b) $\begin{array}{r} 47 \\ -35 \\ \hline \end{array}$

(c) $\begin{array}{r} 429 \\ -318 \\ \hline \end{array}$

(d) $\begin{array}{r} 3927 \\ -2614 \\ \hline \end{array}$

(e) $\begin{array}{r} 5464 \\ -324 \\ \hline \end{array}$

(b) $18 - 13 = 5$
$18 = 13 + 5$

(c) $29 - 13 = 16$
$29 = 13 + 16$

Work Problem ❷ at the Side.

2   **Identify the minuend, subtrahend, and difference.**   In subtraction, as in addition, the numbers in a problem have names. For example, in the problem $8 - 5 = 3$, the number 8 is the **minuend,** 5 is the **subtrahend,** and 3 is the **difference** or answer.

$$8 \quad - \quad 5 \quad = 3 \leftarrow \text{Difference (answer)}$$
$$\quad\uparrow \qquad\qquad \uparrow$$
$$\text{Minuend} \qquad \text{Subtrahend}$$

$$\begin{array}{r} 8 \leftarrow \text{Minuend} \\ -5 \leftarrow \text{Subtrahend} \\ \hline 3 \leftarrow \text{Difference} \end{array}$$

3   **Subtract when no borrowing is needed.**   Subtract two numbers by lining up the numbers in columns so the digits in the ones place are in the same column. Next subtract by columns, starting at the right with the ones column.

Example 3   **Subtracting Two Numbers**

Subtract.

Ones digits are lined up in the same column.

(a) $\begin{array}{r} 53 \\ -21 \\ \hline 32 \end{array}$

$3 - 1 = 2$
$5 - 2 = 3$

Ones digits are lined up.

(b) $\begin{array}{r} 385 \\ -161 \\ \hline 224 \end{array} \leftarrow 5 - 1 = 4$

$8 - 6 = 2$
$3 - 1 = 2$

(c) $\begin{array}{r} 9431 \\ -210 \\ \hline 9221 \end{array} \leftarrow 1 - 0 = 1$

$3 - 1 = 2$
$4 - 2 = 2$
$9 - 0 = 9$

Work Problem ❸ at the Side.

4   **Check subtraction answers by adding.**   Use addition to check your answer to a subtraction problem. For example, check $8 - 3 = 5$ by *adding* 3 and 5:

$$3 + 5 = 8, \quad \text{so} \quad 8 - 3 = 5 \quad \text{is correct.}$$

Example 4 Checking Subtraction by Using Addition

Check each answer.

(a)   89
      $-$ 47
      42

Rewrite as an addition problem, as shown in Example 2.

Subtraction problem $\begin{cases} 89 \\ -47 \\ \overline{42} \\ \overline{89} \end{cases}$ Addition problem   47
                              $+$ 42
                              89

Because $47 + 42 = 89$, the subtraction was done correctly.

(b) $72 - 41 = 21$

Rewrite as an addition problem.

$$72 = 41 + 21$$

But, $41 + 21 = 62$, not 72, so the subtraction was done incorrectly. Rework the original subtraction to get the correct answer, 31.

(c)   374 ← — Match —┐
      $-$ 141   │
      233    $141 + 233 = 374$

The answer checks.

═══ Work Problem ❹ at the Side.

5 **Subtract by borrowing.**   When a digit in the minuend is less than the one directly below it we subtract by using **borrowing.**

Example 5 Subtracting with Borrowing

Subtract 19 from 57.

Write the problem.

      57
    $-$ 19

In the ones column, 7 is **less** than 9, so to subtract, we must **borrow a 10** from the 5 (which represents 5 tens, or 50).

$50 - 10 = 40 \rightarrow$ 4 17 $\leftarrow 10 + 7 = 17$
              5̸ 7̸
            $-$ 1 9

Now we can subtract $17 - 9$ in the ones column and then $4 - 1$ in the tens column.

      4 17
      5̸ 7̸
    $-$ 1 9
      3 8   Difference

Finally, $57 - 19 = 38$. Check by adding 19 and 38; you should get 57.

═══ Work Problem ❺ at the Side.

❹ Use addition to determine whether each answer is correct. If incorrect, what should it be?

(a)   76
      $-$ 45
      31

(b)   53
      $-$ 22
      21

(c)   374
      $-$ 251
      113

(d)   7531
      $-$ 4301
      3230

❺ Subtract.

(a)   58
      $-$ 19

(b)   86
      $-$ 38

(c)   41
      $-$ 27

(d)   863
      $-$ 47

(e)   762
      $-$ 157

❻ Subtract.

(a)
```
   536
 −  75
```

(b)
```
   348
 −  79
```

(c)
```
   477
 − 389
```

(d)
```
   1437
 −  988
```

(e)
```
   8739
 − 3892
```

ANSWERS
6. (a) 461   (b) 269   (c) 88
    (d) 449   (e) 4847

Example 6   **Subtracting with Borrowing**

Subtract by borrowing as necessary.

(a)
```
   7856
 −  137
```

There is no need to borrow, as 4 is greater than 3.

$10 + 6 = 16$

```
        4  16
   7 8 ⁵6̸ 6̸
 −   1 3  7
   7 7 1  9     Difference
```

(b)
```
   635
 − 546
```

$600 − 100 = 500$ $100 + 20 = 120 \ (12 \text{ tens} = 120)$ $10 + 5 = 15$

```
      5 12 15
      6̸  3̸  5̸
 −    5  4  6
         8  9     Difference
```

(c)
```
   3648
 − 1769
```

```
   2  15 13 18
   3̸  6̸  4̸  8̸
 − 1  7  6  9
   1  8  7  9
```

Work Problem ❻ at the Side.

Sometimes a minuend has 0s (zeros) in some of the positions. In such cases, borrowing may be a little more complicated than what we have shown so far.

Example 7   **Borrowing with Zeros**

Subtract.

$$\begin{array}{r} 4607 \\ -\ 3168 \end{array}$$

It is not possible to borrow from the tens position. Instead we must first borrow from the hundreds position.

$600 − 100 = 500$ $100 + 0 = 100$

```
       5 10
   4 6̸ 0̸ 7
 − 3 1 6 8
```

Now we may borrow from the tens position.

9 ←——— $100 − 10 = 90$
$5\ \ 10\ \ 17$ ←——— $10 + 7 = 17$

```
         9
   5 1̸0̸ 17
   4 6̸ 0̸ 7̸
 − 3 1 6 8
           9
```

Continued on Next Page

Complete the problem.

$$
\begin{array}{r}
\overset{9}{}\\
5\ \cancel{10}\ 17\\
4\ \cancel{6}\ \cancel{0}\ \cancel{7}\\
-\ 3\ 1\ 6\ 8\\
\hline
1\ 4\ 3\ 9
\end{array}\quad\text{Difference}
$$

Check by adding 1439 and 3168; you should get 4607.

══════ **Work Problem ❼ at the Side.**

Example 8 **Borrowing with Zeros**

Subtract.

(a)   708
       − 149

$$100 + 0 = 100 \longrightarrow\quad 100 - 10 = 90\ (9\ \text{tens} = 90)$$
$$700 - 100 = 600 \qquad\qquad 10 + 8 = 18$$

$$
\begin{array}{r}
9\\
6\ \cancel{10}\ 18\\
\cancel{7}\ \cancel{0}\ \cancel{8}\\
-\ 1\ 4\ 9\\
\hline
5\ 5\ 9
\end{array}
$$

(b)   380
       − 276

$$80 - 10 = 70\ (7\ \text{tens} = 70)\qquad 10 + 0 = 10$$
$$7\ 10$$

$$
\begin{array}{r}
7\ 10\\
3\ \cancel{8}\ \cancel{0}\\
-\ 2\ 7\ 6\\
\hline
1\ 0\ 4
\end{array}
$$

(c)   9000
       − 6999

$$
\begin{array}{r}
9\ \ 9\\
8\ \cancel{10}\ \cancel{10}\ 10\\
\cancel{9}\ \cancel{0}\ \cancel{0}\ \cancel{0}\\
-\ 6\ 9\ 9\ 9\\
\hline
2\ 0\ 0\ 1
\end{array}
$$

══════ **Work Problem ❽ at the Side.**

As we have seen, an answer to a subtraction problem can be checked by adding.

Example 9 **Checking Subtraction by Using Addition**

Using addition to check each answer.

                    Check
(a)   **613**   275
       − 275   *Match* + 338
       ─────   ─────
        338   **613**   Correct

══════ **Continued on Next Page**

❼ Subtract.

(a)   405
        − 363

(b)   304
        − 237

(c)   5073
        − 1632

❽ Subtract.

(a)   308
        − 159

(b)   480
        − 275

(c)   1570
        − 983

(d)   7001
        − 5193

(e)   4000
        − 1782

9 Use addition to check each answer. If the answer is incorrect, find the correct answer.

(a)
$$357$$
$$- 168$$
$$\overline{189}$$

(b)
$$570$$
$$- 328$$
$$\overline{252}$$

(c)
$$14{,}726$$
$$- 8\ 839$$
$$\overline{5\ 887}$$

(b)
$$1915$$
$$- 1635$$
$$\overline{280}$$

Match

Check
$$1635$$
$$+ 280$$
$$\overline{1915}$$ Correct

(c)
$$15{,}803$$
$$- 7\ 325$$
$$\overline{8\ 578}$$

No match

Check
$$7\ 325$$
$$+ 8\ 578$$
$$\overline{15{,}903}$$ Error

Rework the original problem to get the correct answer, 8478.

Work Problem 9 at the Side.

6 **Solve application problems with subtraction.** As shown in the next example, subtraction can be used to solve an application problem.

Example 10 **Applying Subtraction Skills**

Use the table to find how much more, on average, a person with an Associate of Arts degree earns each year than a high school graduate.

STUDENTS OF SUCCESS

The more education adults get, the higher their annual pay.

Education Level	Average Salary
Not a high-school graduate	$20,442
High-school graduate	$27,038
Some college	$31,128
Associate of Arts degree	$33,425
Bachelor's degree	$44,523
Master's degree	$55,384
Doctoral degree	$72,099
Professional degree	$98,197

Source: U.S. Bureau of the Census.

10 Use the table from Example 10 to find, on average,

(a) how much more each year a person with a Bachelor's degree earns than a person with an Associate of Arts degree.

(b) how much more each year a person with an Associate of Arts degree earns than a person who is not a high school graduate.

Approach The average salary for a person with an Associate of Arts degree is $33,425 each year and the average for a high school graduate is $27,038. Find how much more a college graduate earns by subtracting $27,038 from $33,425.

Solution
$$\$33{,}425$$ Associate of Arts degree
$$- 27{,}038$$ High school graduate
$$\overline{\$\ 6\ 387}$$ More earnings

On average, a person with an Associate of Arts degree earns $6387 more each year than a high school graduate.

Work Problem 10 at the Side.

ANSWERS
9. (a) correct
 (b) incorrect, should be 242
 (c) correct
10. (a) $11,098 (b) $12,983

1.3 **EXERCISES**

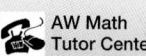
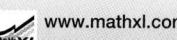

Work each subtraction problem. Use addition to check each answer. See Examples 3 and 4.

1. 54
− 23

2. 19
− 12

3. 86
− 53

4. 78
− 35

5. 77
− 60

6. 95
− 71

7. 335
− 122

8. 602
− 301

9. 552
− 451

10. 888
− 215

11. 6821
− 610

12. 4420
− 310

13. 5546
− 2134

14. 1875
− 1362

15. 6259
− 4148

16. 9654
− 4323

17. 24,392
− 11,232

18. 57,921
− 34,801

19. 46,253
− 5 143

20. 75,904
− 3 702

Use addition to check each subtraction problem. If an answer is not correct, find the correct answer. See Example 4.

21. 43
− 31
___
 12

22. 76
− 43
___
 33

23. 89
− 27
___
 63

24. 47
− 35
___
 13

25. 382
− 261
___
 131

26. 754
− 342
___
 412

27. 4683
− 3542
___
 1141

28. 5217
− 4105
___
 1132

29. 8643
− 1421
___
 7212

30. 9428
− 3124
___
 6324

Subtract by borrowing as necessary. See Examples 5–8.

31. 45
− 27

32. 86
− 28

33. 94
− 49

34. 98
− 69

35. 57
− 38

36. 83
− 55

37. 828
− 547

38. 916
− 618

39. 771
− 252

40. 973
− 788

41. 9861
  − 684

42. 6171
  − 1182

43. 9988
  − 2399

44. 3576
  − 1658

45. 38,335
  − 29,476

46. 61,278
  − 3 559

47. 40
  − 37

48. 80
  − 73

49. 60
  − 37

50. 70
  − 27

51. 308
  − 289

52. 600
  − 599

53. 4041
  − 1208

54. 4602
  − 2063

55. 9305
  − 1530

56. 7120
  − 6033

57. 1580
  − 1077

58. 3068
  − 2105

59. 2006
  − 1850

60. 8203
  − 5365

61. 8240
  − 6056

62. 7050
  − 6045

63. 8503
  − 2816

64. 16,004
  − 5 087

65. 80,705
  − 61,667

66. 81,000
  − 55,456

67. 66,000
  − 34,444

68. 77,000
  − 65,308

69. 20,080
  − 13,496

70. 80,056
  − 23,869

Use addition to check each subtraction problem. If an answer is incorrect, find the correct answer. See Example 9.

71. 5382
  − 4634
    748

72. 1671
  − 1325
   1346

73. 2548
  − 2278
    270

74. 5274
  − 1130
   4144

75. 76,326
  − 44,539
   31,787

76. 82,357
  − 14,396
   68,961

77. 36,778
  − 17,405
   19,373

78. 34,821
  − 17,735
   17,735

79. An addition problem can be changed to a subtraction problem and a subtraction problem can be changed to an addition problem. Give two examples of each to demonstrate this.

80. Can you use the commutative and the associative properties in subtraction? Explain.

Solve each application problem. See Example 10.

81. A man burns 103 calories during 30 minutes of bowling while a woman burns 88 calories during 30 minutes of bowling. How many fewer calories did the woman burn than the man?

82. Lillian Kim has $729 in her checking account. After she writes a check to the bookstore for $249, how much is remaining in her account?

83. Toronto's skyline is dominated by the CN Tower, which rises 1821 ft. The Sears Tower in Chicago is 1454 ft high. Find the difference in height between the two structures.

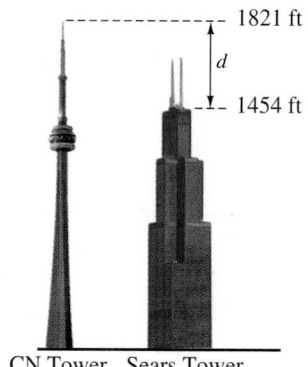

1821 ft

d

1454 ft

CN Tower   Sears Tower

84. The fastest animal in the world, the peregrine falcon, dives at 217 miles per hour (mph). A Boeing 747 cruises at 580 mph. How much faster is the plane?

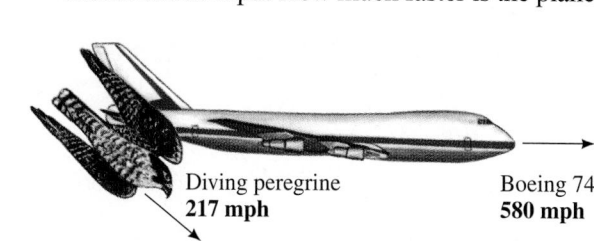

Diving peregrine
217 mph

Boeing 747
580 mph

85. An airplane is carrying 254 passengers. When it lands in Atlanta, 133 passengers get off. How many passengers are left on the plane?

86. On Tuesday, 5822 people went to a soccer game, and on Friday, 7994 people went. How many more people went to the game on Friday?

87. Last fall 12,625 students enrolled in classes. In the spring semester, 11,296 students enrolled. How many more students enrolled in the fall semester than in the spring?

88. In 1964, its first year on the market, the Ford Mustang sold for $2500. In 2000, the Ford Mustang sold for $24,870. Find the increase in price. (*Source:* Ford Motor Company.)

89. Patriot Flag Company manufactured 14,608 flags and sold 5069. How many flags remain unsold?

90. Eye exams have been given to 14,679 children in the school district. If there are 23,156 students in the school district, how many have not received eye exams?

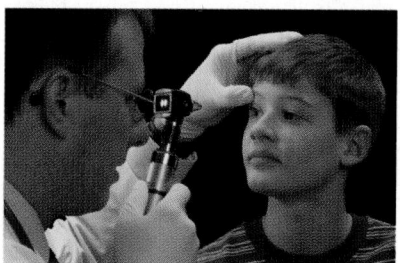

91. The Jordanos now pay rent of $650 per month. If they buy a house, their housing expense will be $913 per month. How much more will they pay per month if they buy a house?

92. A retired couple who used to receive a Social Security payment of $1479 per month now receive $1568 per month. Find the amount of the monthly increase.

93. On Monday, 11,594 people visited Arcade Amusement Park, and 12,352 people visited the park on Tuesday. How many more people visited the park on Tuesday?

94. Last month, Alice Blake earned $2382. This month she earned $2671. How much more did she earn this month than last month?

Solve each application problem. Add or subtract as necessary.

95. A survey of large hotels found that the average salary for a general manager of a deluxe spa and tennis resort is one hundred one thousand, five hundred dollars per year, while spa and tennis directors earn $44,000. How much more does a general manager earn than a spa and tennis director?

96. There are 24 million business enterprises in the United States. If only 7000 of these businesses are large businesses having 500 or more employees, how many are small and midsize businesses?

The table shows the deliveries made by Diana Lopez, a United Parcel Service driver. Use the table to answer Exercises 97–100.

PACKAGE DELIVERY (LOPEZ)

Day	Number of Deliveries
Monday	137
Tuesday	126
Wednesday	119
Thursday	89
Friday	147

97. How many more deliveries did Diana make on Monday than on Thursday?

98. How many more deliveries did Diana make on Friday than on Tuesday?

99. Find the total deliveries made on the two busiest days.

100. Find the total deliveries made on the two slowest days.

1.4 MULTIPLYING WHOLE NUMBERS

Suppose we want to know the total number of computers in a computer lab. The stations are arranged in four rows with three stations in each row. Adding the number 3 a total of 4 times gives 12.

$$3 + 3 + 3 + 3 = 12$$

This result can also be shown with a figure.

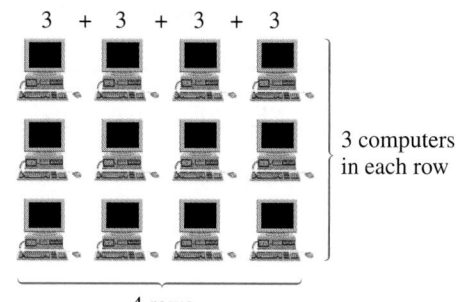

3 + 3 + 3 + 3

3 computers in each row

4 rows

1 **Identify the parts of a multiplication problem.** Multiplication is a shortcut for repeated addition. In the computer lab example, instead of *adding* $3 + 3 + 3 + 3$ to get 12, we can *multiply* 3 by 4 to get 12. The numbers being multiplied are called **factors.** The answer is called the **product.** For example, the product of 3 and 4 can be written with the symbol $\times$, a raised dot, or parentheses, as follows.

$$\begin{array}{r} 3 \\ \times\ 4 \\ \hline 12 \end{array}$$ ← Factor (also called *multiplicand*)
← Factor (also called *multiplier*)
← Product (answer)

$$3 \times 4 = 12 \quad or \quad 3 \cdot 4 = 12 \quad or \quad (3)(4) = 12$$

Work Problem 1 at the Side.

Commutative Property of Multiplication

By the **commutative property of multiplication,** the answer or product remains the same when the order of the factors is changed. For example,

$$3 \times 5 = 15 \quad and \quad 5 \times 3 = 15$$

CAUTION

Recall that addition also has a commutative property. Remember that $4 + 2$ is the same as $2 + 4$. Subtraction, however, is *not* commutative.

Example 1 **Multiplying Two Numbers**

Multiply. (Remember that a raised dot or parentheses means to multiply.)

(a) $3 \times 4 = 12$

(b) $6 \cdot 0 = 0$ (The product of any number and 0 is 0; if you give no money to each of 6 relatives, you give no money.)

(c) $(4)(8) = 32$

Work Problem 2 at the Side.

---

1 Identify the factors and the product in each multiplication problem.

(a) $2 \times 5 = 10$

(b) $6 \times 4 = 24$

(c) $7 \cdot 6 = 42$

(d) $(3)(9) = 27$

2 Multiply.

(a) 3×8

(b) 0×9

(c) $7 \cdot 5$

(d) $6 \cdot 5$

(e) $(3)(8)$

ANSWERS
1. **(a)** factors: 2, 5; product: 10
(b) factors: 6, 4; product: 24
(c) factors: 7, 6; product: 42
(d) factors: 3, 9; product: 27
2. **(a)** 24 **(b)** 0 **(c)** 35 **(d)** 30 **(e)** 24

❸ Multiply.

(a) $2 \times 5 \times 3$

2 **Do chain multiplications.** Some multiplications contain more than two factors.

Associative Property of Multiplication

By the **associative property of multiplication,** grouping the factors differently does not change the product.

Example 2 **Multiplying Three Numbers**

Multiply $2 \times 3 \times 5$.

$$(2 \times 3) \times 5 \qquad \text{Parentheses tell what to do first.}$$
$$6 \quad \times 5 = 30$$

Also,

$$2 \times (3 \times 5)$$
$$2 \times \quad 15 = 30$$

(b) $7 \cdot 1 \cdot 4$

Either grouping results in the same product.

Calculator Tip The calculator approach to Example 2 uses chain calculations.

$$2 \; ⊗ \; 3 \; ⊗ \; 5 \; ⊜ \; 30$$

A problem with more than two factors, such as the one in Example 2, is called a **chain multiplication.**

Work Problem ❸ at the Side.

3 **Multiply by single-digit numbers.** Carrying may be needed in multiplication problems with larger factors.

(c) $(6)(4)(0)$

Example 3 **Multiplying with Carrying**

Multiply.

(a) $\begin{array}{r} 53 \\ \times\ 4 \\ \hline \end{array}$

Start by multiplying in the ones column.

$$\begin{array}{r} 1 \\ 53 \\ \times\ 4 \qquad 4 \times 3 = 12 \\ \hline 2 \end{array} \qquad \begin{array}{l}\text{Carry the 1 to the tens column.} \\ \text{Write 2 in the ones column.}\end{array}$$

Next, multiply 4 ones and 5 tens.

$$\begin{array}{r} 1 \\ 53 \\ \times\ 4 \qquad 4 \times 5 = \textbf{20 } \text{tens} \\ \hline 2 \end{array}$$

Continued on Next Page

Add the 1 that was carried to the tens column.

$$\begin{array}{r} 1 \\ 53 \\ \times\ \ 4 \\ \hline 212 \end{array} \quad 20 + 1 = \mathbf{21}\ \text{tens}$$

(b) 724
$$\underline{\times\ \ 5}$$

Work as shown.

$$\begin{array}{r} 12 \\ 724 \\ \times\ \ 5 \\ \hline 3620 \end{array} \longleftarrow 5 \times 4 = \mathbf{20}\ \text{ones; write 0 ones and carry 2 tens.}$$

$5 \times 2 = \mathbf{10}$ tens; add the 2 tens to get 12 tens; write 2 tens and carry 1 hundred.

$5 \times 7 = \mathbf{35}$ hundreds; add the 1 hundred to get 36 hundreds.

=== **Work Problem ④ at the Side.**

4▬▬ **Use multiplication shortcuts for numbers ending in zeros.** The product of two whole number factors is also called a **multiple** of either factor. For example, since $4 \cdot 2 = 8$, the whole number 8 is a multiple of both 4 and 2. *Multiples of 10* are very useful when multiplying. A **multiple of 10** is a whole number that ends in 0, such as 10, 20, or 30; 100, 200, or 300; 1000, 2000, or 3000. There is a short way to multiply by these multiples of 10. Look at the following examples.

$$26 \times 1 = 26$$
$$26 \times 10 = 260$$
$$26 \times 100 = 2600$$
$$26 \times 1000 = 26{,}000$$

Do you see a pattern? These examples suggest the following rule.

Multiplying by Multiples of 10

To multiply a whole number by 10, 100, or 1000, attach one, two, or three 0s to the right of the whole number.

Example 4 **Using Multiples of 10 to Multiply**

Multiply.

(a) $59 \times 10 = 590$

Attach 0.

(b) $74 \times 100 = 7400$

Attach 00.

(c) $803 \times 1000 = 803{,}000$ ◄— Attach 000.

=== **Work Problem ⑤ at the Side.**

You can also find the product of other multiples of 10 by attaching 0s.

❹ Multiply.

(a)   62
$$\underline{\times\ \ 4}$$

(b)   98
$$\underline{\times\ \ 0}$$

(c)   758
$$\underline{\times\ \ \ 8}$$

(d)   2831
$$\underline{\times\ \ \ \ 7}$$

(e)   4714
$$\underline{\times\ \ \ \ 8}$$

❺ Multiply.

(a) 52×10

(b) 305×100

(c) 418×1000

Answers
4. (a) 248 **(b)** 0 **(c)** 6064
    (d) 19,817 **(e)** 37,712
5. (a) 520 **(b)** 30,500 **(c)** 418,000

❻ Multiply.

(a) 17×40

(b) 58×300

(c) $\begin{array}{r} 180 \\ \times\ \ 30 \\ \hline \end{array}$

(d) $\begin{array}{r} 4200 \\ \times\ \ \ \ 80 \\ \hline \end{array}$

(e) $\begin{array}{r} 700 \\ \times\ 400 \\ \hline \end{array}$

❼ Complete each multiplication.

(a) $\begin{array}{r} 35 \\ \times\ 54 \\ \hline 140 \\ 175\ \ \\ \hline \end{array}$

(b) $\begin{array}{r} 76 \\ \times\ 49 \\ \hline 684 \\ 304\ \ \\ \hline \end{array}$

Example 5   **Using Multiples of 10 to Multiply**

Multiply.

(a) 75×3000

Multiply 75 by 3, and then attach three 0s.

$$\begin{array}{r} 75 \\ \times\ \ \ 3 \\ \hline 225 \end{array} \qquad 75 \times 3000 = 225{,}000$$

Attach 000.

(b) 150×70

Multiply 15 by 7, and then attach two 0s.

$$\begin{array}{r} 15 \\ \times\ \ 7 \\ \hline 105 \end{array} \qquad 150 \times 70 = 10{,}500 \ \leftarrow \text{Attach 00.}$$

Work Problem ❻ at the Side.

5 **Multiply by numbers having more than one digit.**   The next example shows multiplication when both factors have more than one digit.

Example 6   **Multiplying with More Than One Digit**

Multiply 46 and 23.

First multiply 46 by 3.

$$\begin{array}{r} 1 \\ 46 \\ \times\ \ 3 \\ \hline 138 \end{array} \ \leftarrow 46 \times 3 = 138$$

Now multiply 46 by 20.

$$\begin{array}{r} 1 \\ 46 \\ \times\ 20 \\ \hline 920 \end{array} \ \leftarrow 46 \times 20 = 920$$

Add the results.

$$\begin{array}{r} 46 \\ \times\ 23 \\ \hline 138 \\ +\ 920 \\ \hline 1058 \end{array}$$

138 ← 46 × 3
+ 920 ← 46 × 20

Add.

Both 138 and 920 are called **partial products.** To save time, the 0 in 920 is usually not written.

$$\begin{array}{r} 46 \\ \times\ 23 \\ \hline 138 \\ 92 \\ \hline 1058 \end{array}$$

0 not written. Be very careful to place the 2 in the tens column.

Work Problem ❼ at the Side.

Example 7 **Using Partial Products**

Multiply.

(a)
```
      2 3 3
  ×   1 3 2
      4 6 6
      6 9 9     (Tens lined up)
    2 3 3       (Hundreds lined up)
  3 0,7 5 6     Product
```

(b)
```
    5 3 8
  ×  4 6
```

First multiply by 6.
```
        2 4
       5 3 8
     ×   4 6      Carrying is
       3 2 2 8    needed here.
```

Now multiply by 4, being careful to line up the tens.
```
       1 3
       2 4
      5 3 8
    ×   4 6
      3 2 2 8  ⎤
      2 1 5 2  ⎦ Finally, add the results.
    2 4,7 4 8
```

================ **Work Problem ❽ at the Side.**

When 0 appears in the multiplier, be sure to move the partial products to the left to account for the position held by the 0.

Example 8 **Multiplying with Zeros**

Multiply.

(a)
```
      1 3 7
  ×   3 0 6
      8 2 2
      0 0 0     (Tens lined up)
    4 1 1       (Hundreds lined up)
  4 1,9 2 2
```

(b)
```
      1 4 0 6                    1 4 0 6
  ×   2 0 0 1                ×   2 0 0 1
      1 4 0 6                    1 4 0 6
      0 0 0 0  ← (0s to line up tens)
      0 0 0 0  ← (0s to line up hundreds)  2 8 1 2 0 0  ← Zeros are
    2 8 1 2                                              written so this
  2,8 1 3,4 0 6                 2,8 1 3,4 0 6            partial product
                                                        starts in the
                                                        thousands
                                                        column.
```

❽ Multiply.

(a)
```
     46
  × 14
```

(b)
```
     41
  × 38
```

(c)
```
     75
  × 63
```

(d)
```
    234
  ×  73
```

(e)
```
    835
  × 189
```

9 Multiply.

(a)   36
      × 50

(b)   635
      × 40

(c)   562
      × 109

(d)   3526
      × 6002

10 Find the total cost of the following items.

(a) 289 redwood planters at $12 per planter

(b) 106 Microsoft Intelli-mouse Explorers at $69 each

(c) 12 delivery vans at $24,300 per van

NOTE

In Example 8(b) in the alternative method on the right, 0s were inserted so that thousands were placed in the thousands column. This is a commonly used shortcut.

Work Problem 9 at the Side.

6 Solve application problems with multiplication. The next example shows how multiplication can be used to solve an application problem.

Example 9 Applying Multiplication Skills

Find the total cost of 24 cellular phones priced at $54 each.

Approach To find the cost of all the cellular phones, multiply the number of phones (24) by the cost of one phone ($54).

Solution Multiply 24 by 54.

$$
\begin{array}{r}
24 \\
\times\ 54 \\
\hline
96 \\
120 \\
\hline
1296
\end{array}
$$

The total cost of the cellular phones is $1296.

Calculator Tip If you are using a calculator for Example 9, you will do this calculation.

24 ⊗ 54 ⊜ 1296

Work Problem 10 at the Side.

ANSWERS
9. (a) 1800   (b) 25,400
    (c) 61,258   (d) 21,163,052
10. (a) $3468   (b) $7314   (c) $291,600

| **1.4** | **EXERCISES** |

| FOR EXTRA HELP | Student's Solutions Manual | MyMathLab.com |  InterAct Math Tutorial Software | 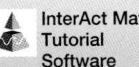 AW Math Tutor Center | 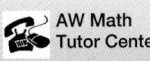 www.mathxl.com |  Digital Video Tutor CD 1 Videotape 2 |

Work each chain multiplication. See Example 2.

1. $4 \times 1 \times 4$ **2.** $3 \times 5 \times 3$ **3.** $8 \times 6 \times 1$ **4.** $2 \times 4 \times 5$

5. $7 \cdot 8 \cdot 0$ **6.** $9 \cdot 0 \cdot 5$ **7.** $4 \cdot 1 \cdot 6$ **8.** $1 \cdot 5 \cdot 7$

9. $(3)(2)(5)$ **10.** $(4)(1)(9)$ **11.** $(3)(0)(7)$ **12.** $(0)(9)(4)$

13. Explain in your own words the commutative property of multiplication. How do the commutative properties of addition and multiplication compare to each other?

14. Explain in your own words the associative property of multiplication. How do the associative properties of addition and multiplication compare to each other?

Multiply. See Example 3.

15. $\begin{array}{r} 25 \\ \times\ 6 \\ \hline \end{array}$ **16.** $\begin{array}{r} 62 \\ \times\ 8 \\ \hline \end{array}$ **17.** $\begin{array}{r} 34 \\ \times\ 7 \\ \hline \end{array}$ **18.** $\begin{array}{r} 76 \\ \times\ 5 \\ \hline \end{array}$

19. $\begin{array}{r} 642 \\ \times\ 5 \\ \hline \end{array}$ **20.** $\begin{array}{r} 472 \\ \times\ 4 \\ \hline \end{array}$ **21.** $\begin{array}{r} 624 \\ \times\ 3 \\ \hline \end{array}$ **22.** $\begin{array}{r} 852 \\ \times\ 7 \\ \hline \end{array}$

23. $\begin{array}{r} 2153 \\ \times\ 4 \\ \hline \end{array}$ **24.** $\begin{array}{r} 1137 \\ \times\ 3 \\ \hline \end{array}$ **25.** $\begin{array}{r} 2521 \\ \times\ 4 \\ \hline \end{array}$ **26.** $\begin{array}{r} 2544 \\ \times\ 3 \\ \hline \end{array}$

27. $\begin{array}{r} 2561 \\ \times\ 8 \\ \hline \end{array}$ **28.** $\begin{array}{r} 7326 \\ \times\ 5 \\ \hline \end{array}$ **29.** $\begin{array}{r} 36{,}921 \\ \times\ 7 \\ \hline \end{array}$ **30.** $\begin{array}{r} 28{,}116 \\ \times\ 4 \\ \hline \end{array}$

Multiply. See Examples 4 and 5.

31. $\begin{array}{r} 30 \\ \times\ 5 \\ \hline \end{array}$ **32.** $\begin{array}{r} 20 \\ \times\ 7 \\ \hline \end{array}$ **33.** $\begin{array}{r} 80 \\ \times\ 6 \\ \hline \end{array}$ **34.** $\begin{array}{r} 70 \\ \times\ 5 \\ \hline \end{array}$ **35.** $\begin{array}{r} 740 \\ \times\ 3 \\ \hline \end{array}$

36. $\begin{array}{r} 200 \\ \times\ 7 \\ \hline \end{array}$ **37.** $\begin{array}{r} 600 \\ \times\ 6 \\ \hline \end{array}$ **38.** $\begin{array}{r} 860 \\ \times\ 7 \\ \hline \end{array}$ **39.** $\begin{array}{r} 125 \\ \times\ 30 \\ \hline \end{array}$ **40.** $\begin{array}{r} 246 \\ \times\ 50 \\ \hline \end{array}$

41. $\begin{array}{r} 1485 \\ \times\ \ \ 30 \\ \hline \end{array}$

42. $\begin{array}{r} 8522 \\ \times\ \ \ 50 \\ \hline \end{array}$

43. $\begin{array}{r} 900 \\ \times\ \ 300 \\ \hline \end{array}$

44. $\begin{array}{r} 400 \\ \times\ \ 700 \\ \hline \end{array}$

45. $\begin{array}{r} 43{,}000 \\ \times\ \ \ 2\ 000 \\ \hline \end{array}$

46. $\begin{array}{r} 11{,}000 \\ \times\ \ \ 9\ 000 \\ \hline \end{array}$

47. $970 \cdot 50$

48. $730 \cdot 40$

49. $500 \cdot 700$

50. $850 \cdot 700$

51. $9700 \cdot 200$

52. $10{,}050 \cdot 300$

Multiply. See Examples 6–8.

53. $\begin{array}{r} 36 \\ \times 15 \\ \hline \end{array}$

54. $\begin{array}{r} 18 \\ \times 47 \\ \hline \end{array}$

55. $\begin{array}{r} 75 \\ \times 32 \\ \hline \end{array}$

56. $\begin{array}{r} 82 \\ \times 32 \\ \hline \end{array}$

57. $\begin{array}{r} 83 \\ \times 45 \\ \hline \end{array}$

58. $(62)(31)$

59. $(58)(41)$

60. $(82)(67)$

61. $(67)(92)$

62. $(26)(33)$

63. $(28)(564)$

64. $(58)(312)$

65. $(619)(35)$

66. $(681)(47)$

67. $(55)(286)$

68. $\begin{array}{r} 286 \\ \times\ \ 574 \\ \hline \end{array}$

69. $\begin{array}{r} 735 \\ \times\ \ 112 \\ \hline \end{array}$

70. $\begin{array}{r} 621 \\ \times\ \ 415 \\ \hline \end{array}$

71. $\begin{array}{r} 538 \\ \times\ \ 342 \\ \hline \end{array}$

72. $\begin{array}{r} 3228 \\ \times\ \ 751 \\ \hline \end{array}$

73. $\begin{array}{r} 9352 \\ \times\ \ \ 264 \\ \hline \end{array}$

74. $\begin{array}{r} 528 \\ \times\ \ 106 \\ \hline \end{array}$

75. $\begin{array}{r} 215 \\ \times\ \ 307 \\ \hline \end{array}$

76. $\begin{array}{r} 218 \\ \times\ \ 106 \\ \hline \end{array}$

77. $\begin{array}{r} 428 \\ \times\ \ 201 \\ \hline \end{array}$

78. $\begin{array}{r} 3706 \\ \times\ \ \ 208 \\ \hline \end{array}$

79. $\begin{array}{r} 6310 \\ \times\ \ 3078 \\ \hline \end{array}$

80. $\begin{array}{r} 3533 \\ \times\ \ 5001 \\ \hline \end{array}$

81. $\begin{array}{r} 2195 \\ \times\ \ 1038 \\ \hline \end{array}$

82. $\begin{array}{r} 1502 \\ \times\ \ 2009 \\ \hline \end{array}$

83. A classmate of yours is not clear on how to use a shortcut to multiply a whole number by 10, by 100, or by 1000. Write a short note explaining how this can be done.

84. Show two ways to multiply when a 0 is in the factor that is multiplying. Use the problem 291×307 to show this.

Solve each application problem. See Example 9.

85. There are 90 cartons of CDs loaded on a shipping pallet. If there are 200 loaded shipping pallets in the warehouse, how many cartons are in the warehouse?

86. A medical supply house has 30 bottles of vitamin C tablets, with each bottle containing 500 tablets. Find the total number of vitamin C tablets in the supply house.

87. There are 12 tomato plants to a flat. If a gardener buys 18 flats, find the total number of tomato plants he bought.

88. A hummingbird's wings beat about 65 times per second. How many times do the hummingbird's wings beat in 30 seconds?

89. A new Saturn automobile gets 38 miles per gallon on the highway. How many miles can it go on 11 gallons of gas?

90. Squid are being hauled out of the Santa Barbara Channel by the ton. They are then processed, renamed calamari, and exported. Last night 27 fishing boats each hauled out 40 tons of squid. What was the total catch for the night? (*Source: Santa Barbara News Press.*)

Find the total cost of the following items. See Examples 7–9.

91. 85 heater filters at $3 per filter

92. 38 employees at $64 per day

93. 65 rebuilt alternators at $24 per alternator

94. 76 flats of flowers at $22 per flat

95. 206 computers at $548 per computer

96. 520 printers at $219 per printer

 *Multiply.*

97. $21 \cdot 43 \cdot 56$

98. $(600)(8)(75)(40)$

Use addition, subtraction, or multiplication to solve each application problem.

99. In a forest-planting project, trees are planted 450 trees to an acre. Find the number of trees needed to plant 85 acres.

100. The largest living land mammal is the African elephant, and the largest mammal of all time is the blue whale. An African elephant weighs 15,225 pounds and a blue whale weighs 28 times that amount. Find the weight of the blue whale.

101. A large meal contains 1406 calories, while a small meal contains 348 calories. How many more calories are in the large meal than the small one?

102. As part of a Low Income Home Energy Assistance Program, the state of Florida will receive $1,100,000, while Vermont will receive $505,551. How much more will Florida receive than Vermont? (*Source:* U.S. Department of Energy.)

103. An insurance office purchased four computers at $680 each, four monitors at $295 each, and four printers at $230 each. Find the total cost of this equipment.

RELATING CONCEPTS (Exercises 104–113)   FOR INDIVIDUAL OR GROUP WORK

Work Exercises 104–113 in order.

104. Add.

   (a) $189 + 263$

   (b) $263 + 189$

105. Your answers to Exercise 104(a) and (b) should be the same. This shows that the order of numbers in an addition problem does not change the sum. This is known as the _____ property of addition.

106. Add. Recall that parentheses tell you what to do first.

   (a) $(65 + 81) + 135$

   (b) $65 + (81 + 135)$

107. Since the answers to Exercise 106(a) and (b) are the same, we see that grouping the addition of numbers in any order does not change the sum. This is known as the _____ property of addition.

108. Multiply.

   (a) 220×72

   (b) 72×220

109. Since the answers to Exercise 108(a) and (b) are the same, we see that the product remains the same when the order of the factors is changed. This is known as the _____ property of multiplication.

110. Multiply. Recall that parentheses tell you what to do first.

   (a) $(26 \times 18) \times 14$

   (b) $26(18 \times 14)$

111. Since the answers to Exercise 110(a) and (b) are the same, we see that the grouping of numbers in any order when multiplying gives the same product. This is known as the _____ property of multiplication.

112. Do the commutative and associative properties apply to subtraction? Explain your answer using several examples.

113. Do the commutative and associative properties apply to division? Explain your answer using several examples.

1.5   DIVIDING WHOLE NUMBERS

Suppose the cost of a fast-food lunch is $12 and is to be divided equally by three friends. Each person would pay $4, as shown here.

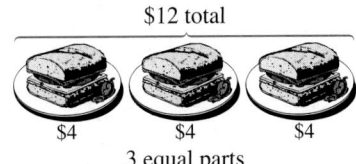

$12 total

$4 $4 $4

3 equal parts

1   **Write division problems in three ways.**   Just as $3 \cdot 4$, 3×4, and $(3)(4)$ are different ways of indicating the multiplication of 3 and 4, there are several ways to write 12 divided by 3.

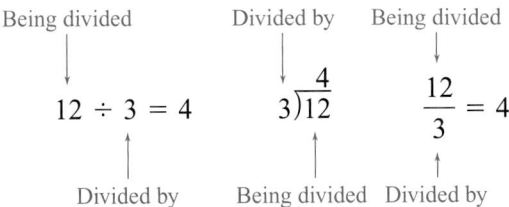

Being divided   Divided by   Being divided
                          ↓   ↓   ↓
$$12 \div 3 = 4 \qquad 3\overline{)12}^{\,4} \qquad \frac{12}{3} = 4$$
        ↑   ↑   ↑
  Divided by   Being divided   Divided by

We will use all three division symbols, $\div$, $\overline{)}\,$, and —. In courses such as algebra, a slash symbol, /, or a fraction bar, —, is most often used.

Example 1   **Using Division Symbols**

Write each division problem using two other symbols.

(a) $12 \div 4 = 3$
This division can also be written as
$$4\overline{)12}^{\,3} \quad \text{or} \quad \frac{12}{4} = 3.$$

(b) $\dfrac{15}{5} = 3$

$$15 \div 5 = 3 \quad \text{or} \quad 5\overline{)15}^{\,3}$$

(c) $5\overline{)20}^{\,4}$

$$20 \div 5 = 4 \quad \text{or} \quad \frac{20}{5} = 4$$

===================== **Work Problem ❶ at the Side.**

2   **Identify the parts of a division problem.**   In division, the number being divided is the **dividend,** the number divided by is the **divisor,** and the answer is the **quotient.**

$$\textbf{dividend} \div \textbf{divisor} = \textbf{quotient}$$

$$\textbf{divisor}\overline{)\textbf{dividend}}^{\,\textbf{quotient}} \qquad \frac{\textbf{dividend}}{\textbf{divisor}} = \textbf{quotient}$$

❶ Write each division problem using two other symbols.

(a) $48 \div 6 = 8$

(b) $24 \div 6 = 4$

(c) $9\overline{)36}^{\,4}$

(d) $\dfrac{42}{6} = 7$

ANSWERS

1. **(a)** $6\overline{)48}^{\,8}$ and $\dfrac{48}{6} = 8$

   (b) $6\overline{)24}^{\,4}$ and $\dfrac{24}{6} = 4$

   (c) $36 \div 9 = 4$ and $\dfrac{36}{9} = 4$

   (d) $6\overline{)42}^{\,7}$ and $42 \div 6 = 7$

❷ Identify the dividend, divisor, and quotient.

(a) $20 \div 5 = 4$

(b) $18 \div 6 = 3$

(c) $\dfrac{28}{7} = 4$

(d) $2\overline{)36}$ with 18 above

❸ Divide.

(a) $0 \div 10$

(b) $\dfrac{0}{6}$

(c) $\dfrac{0}{24}$

(d) $37\overline{)0}$

Example 2   **Identifying the Parts of a Division Problem**

Identify the dividend, divisor, and quotient.

(a) $35 \div 7 = 5$

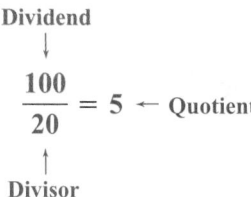

$$35 \div 7 = 5 \leftarrow \text{Quotient}$$

Dividend   Divisor

(b) $\dfrac{100}{20} = 5$

Dividend
↓
$$\dfrac{100}{20} = 5 \leftarrow \text{Quotient}$$
↑
Divisor

(c) $12\overline{)72}$ with 6 above

$6 \leftarrow$ Quotient
$12\overline{)72} \leftarrow$ Dividend
↑
Divisor

Work Problem ❷ at the Side.

3   **Divide 0 by a number.**   If no money, or \$0, is divided equally among five people, each person gets \$0. The general rule for dividing 0 follows.

Dividing 0 by a Number

The number **0** divided by any nonzero number is **0**.

Example 3   **Dividing 0 by a Number**

Divide.

(a) $0 \div 12 = 0$

(b) $0 \div 1728 = 0$

(c) $\dfrac{0}{375} = 0$

(d) $129\overline{)0}$ with 0 above

Work Problem ❸ at the Side.

Just as a subtraction such as $8 - 3 = 5$ can be written as the addition $8 = 3 + 5$, any division can be written as a multiplication. For example, $12 \div 3 = 4$ can be written as

$$3 \times 4 = 12 \quad \text{or} \quad 4 \times 3 = 12$$

> **Example 4** Changing Division Problems to Multiplication
>
> Change each division problem to a multiplication problem.
>
> **(a)** $\dfrac{20}{4} = 5$ becomes $4 \cdot 5 = 20$
>
> **(b)** $8\overline{)48}$ (quotient 6) becomes $8 \cdot 6 = 48$
>
> **(c)** $72 \div 9 = 8$ becomes $9 \cdot 8 = 72$

━━━━━━━━━━━━━ **Work Problem ④ at the Side.**

4 ▢ **Recognize that a number cannot be divided by 0.** Division by 0 cannot be done. To see why, try to find

$$9 \div 0 = \text{?}$$

As we have just seen, any division problem can be converted to a multiplication problem so that

divisor • quotient = dividend.

If you convert the preceding problem to its multiplication counterpart, it reads

$$0 \cdot \text{?} = 9.$$

You already know that 0 times any number must always be 0. Try any number you like to replace the "?" and you'll always get 0 instead of 9. Therefore, the division problem $9 \div 0$ cannot be done. Mathematicians say it is *undefined* and have agreed never to divide by 0. However, $0 \div 9$ *can* be done. Check by rewriting it as a multiplication problem.

$$0 \div 9 = 0 \quad \text{because} \quad 0 \cdot 9 = 0 \text{ is true.}$$

Dividing by 0

Since dividing by 0 cannot be done, we say that division by **0** is *undefined*. It is impossible to compute an answer.

> **Example 5** Dividing Numbers by 0
>
> All the following are undefined.
>
> **(a)** $\dfrac{6}{0}$ is undefined.
>
> **(b)** $0\overline{)8}$ is undefined.
>
> **(c)** $18 \div 0$ is undefined.
>
> **(d)** $\dfrac{0}{0}$ is undefined.

④ Write each division problem as a multiplication problem.

(a) $5\overline{)15}$ (quotient 3)

(b) $\dfrac{32}{4} = 8$

(c) $48 \div 8 = 6$

5 Divide. If the division is not possible, write "undefined."

(a) $\dfrac{6}{0}$

(b) $\dfrac{0}{6}$

(c) $0\overline{)28}$

(d) $28\overline{)0}$

(e) $100 \div 0$

(f) $0 \div 100$

6 Divide.

(a) $6 \div 6$

(b) $15\overline{)15}$

(c) $\dfrac{37}{37}$

Division Involving 0

$$0 \div \text{nonzero number} = 0 \quad \text{and} \quad \frac{0}{\text{nonzero number}} = 0$$

but

$$\frac{\text{nonzero number}}{0} \quad \text{and} \quad \text{nonzero number} \div 0 \text{ are undefined.}$$

CAUTION

When 0 is the divisor in a problem, you write "undefined" as the answer. Never divide by 0.

Work Problem 5 at the Side.

Calculator Tip   Try these two problems on your calculator. Jot down your answers.

$$9 \;\oplus\; 0 \;\ominus\; \underline{\hspace{1cm}} \qquad 0 \;\oslash\; 9 = \underline{\hspace{1cm}}$$

When you try to divide by 0, the calculator cannot do it, so it shows the word "Error" or the letter "E" (for error) in the display. But, when you divide 0 by 9 the calculator displays 0, which is the correct answer.

5 **Divide a number by itself.** What happens when a number is divided by itself? For example, what is $4 \div 4$ or $97 \div 97$?

Dividing a Number by Itself

Any nonzero number divided by itself is **1.**

Example 6   Dividing a Nonzero Number by Itself

Divide.

(a) $16 \div 16 = 1$

(b) $32\overline{)32}$ with quotient 1

(c) $\dfrac{57}{57} = 1$

Work Problem 6 at the Side.

6 **Divide a number by 1.** What happens when a number is divided by 1? For example, what is $5 \div 1$ or $86 \div 1$?

Dividing a Number by 1

Any number divided by 1 is itself.

Example 7 Dividing Numbers by 1

Divide.

(a) $8 \div 1 = 8$

(b) $1\overline{)26}$ with quotient 26

(c) $\dfrac{41}{1} = 41$

═══ Work Problem **7** at the Side.

7 ▭ **Use short division.** **Short division** is a method of dividing a number by a one-digit divisor.

Example 8 Using Short Division

Divide: $3\overline{)96}$.

First, divide 9 by 3.

$$\begin{array}{c} 3 \\ 3\overline{)96} \end{array} \leftarrow \dfrac{9}{3} = 3$$

Next, divide 6 by 3.

$$\begin{array}{c} 32 \\ 3\overline{)96} \end{array} \leftarrow \dfrac{6}{3} = 2$$

═══ Work Problem **8** at the Side.

When two numbers do not divide exactly, the leftover portion is called the **remainder.**

Example 9 Using Short Division with a Remainder

Divide 147 by 4.

Write the problem.

$$4\overline{)147}$$

Because 1 cannot be divided by 4, divide 14 by 4.

$$\begin{array}{c} 3 \\ 4\overline{)14\,^2 7} \end{array} \qquad \dfrac{14}{4} = 3 \text{ with 2 left over}$$

Next, divide 27 by 4. The final number left over is the remainder. Use R to indicate the remainder, and write the remainder to the side.

$$\begin{array}{c} 3\,6 \ \mathbf{R}3 \\ 4\overline{)14\,^2 7} \end{array} \qquad \dfrac{27}{4} = 6 \text{ with 3 left over}$$

═══ Work Problem **9** at the Side.

7 Divide.

(a) $6 \div 1$

(b) $1\overline{)18}$

(c) $\dfrac{36}{1}$

8 Divide.

(a) $2\overline{)18}$

(b) $3\overline{)39}$

(c) $4\overline{)88}$

(d) $2\overline{)462}$

9 Divide.

(a) $2\overline{)125}$

(b) $3\overline{)215}$

(c) $4\overline{)538}$

(d) $\dfrac{819}{5}$

ANSWERS
7. **(a)** 6 **(b)** 18 **(c)** 36
8. **(a)** 9 **(b)** 13 **(c)** 22 **(d)** 231
9. **(a)** 62 **R**1 **(b)** 71 **R**2 **(c)** 134 **R**2
    (d) 163 **R**4

10 Divide.

(a) $5\overline{)937}$

(b) $\dfrac{675}{7}$

(c) $3\overline{)1885}$

(d) $8\overline{)1135}$

Example 10 **Dividing with a Remainder**

Divide 1809 by 7.

Divide 7 into 18.

$$7\overline{)18^409}\quad\overset{2}{}\qquad \frac{18}{7} = 2 \text{ with 4 left over}$$

Divide 7 into 40.

$$7\overline{)18^40^59}\quad\overset{2\ 5}{}\qquad \frac{40}{7} = 5 \text{ with 5 left over}$$

Divide 7 into 59.

$$7\overline{)18^40^59}\quad\overset{2\ 5\ 8\ \mathbf{R3}}{}\qquad \frac{59}{7} = 8 \text{ with 3 left over}$$

Work Problem 10 at the Side.

NOTE

Short division takes practice but is useful in many situations.

8 Use multiplication to check the answer to a division problem. **Check** the answer to a division problem as follows.

Checking Division

(divisor × quotient) + remainder = dividend

Parentheses tell you what to do first: Multiply the divisor by the quotient, then add the remainder.

Example 11 **Checking Division by Using Multiplication**

Check each answer.

(a) $5\overline{)458}\quad\overset{91\ \mathbf{R3}}{}$

(divisor × quotient) + remainder = dividend

$$(5 \ \times \ 91) \ + \ 3$$

$$455 \quad + \quad 3 \quad = 458$$

Matches original dividend,
so the division was done correctly

Continued on Next Page

ANSWERS
10. (a) 187 **R2** **(b)** 96 **R3**
(c) 628 **R1** **(d)** 141 **R7**

(b) $6\overline{)1437}$ $\overset{239\ \textbf{R}4}{}$

(divisor × quotient) + remainder = dividend

$$(6 \quad \times \quad 239) \quad + \quad 4$$

$$1434 \quad + \quad 4 \quad = 1438$$

↑
Does not match original dividend.

The answer does not check. Rework the original problem to get the correct answer, 239 **R**3.

CAUTION

A common error when checking division is to forget to add the remainder. Be sure to add any remainder when checking a division problem.

Work Problem ⓫ at the Side.

9 ▭ **Use tests for divisibility.**   It is often important to know whether a number is *divisible* by another number. You will find this useful in Chapter 2 when writing fractions in lowest terms.

Divisibility

One whole number is **divisible** by another if the remainder is 0.

Use the following tests to decide whether one number is divisible by another number.

Tests for Divisibility

A number is divisible by

2 if it ends in 0, 2, 4, 6, or 8. These are the even numbers.
3 if the sum of its digits is divisible by 3.
4 if the last two digits make a number that is divisible by 4.
5 if it ends in 0 or 5.
6 if it is divisible by both 2 and 3.
7 has no simple test.
8 if the last three digits make a number that is divisible by 8.
9 if the sum of its digits is divisible by 9.
10 if it ends in 0.

The most commonly used tests are those for 2, 3, 5, and 10.

Divisibility by 2

A number is divisible by **2** if the number ends in 0, 2, 4, 6, or 8. All even numbers are divisible by 2.

⓫ Use multiplication to check each division. If an answer is incorrect, give the correct answer.

(a) $2\overline{)89}$ $\overset{44\ \textbf{R}1}{}$

(b) $8\overline{)739}$ $\overset{92\ \textbf{R}2}{}$

(c) $3\overline{)1223}$ $\overset{407\ \textbf{R}2}{}$

(d) $5\overline{)2383}$ $\overset{476\ \textbf{R}3}{}$

35. 24,040 ÷ 8

36. 8012 ÷ 4

37. 15,018 ÷ 3

38. 32,008 ÷ 8

39. 4867 ÷ 6

40. 5993 ÷ 7

41. 12,947 ÷ 5

42. 33,285 ÷ 9

43. 29,298 ÷ 4

44. 17,937 ÷ 6

45. 12,630 ÷ 4

46. 46,560 ÷ 7

47. $\dfrac{22,088}{4}$

48. $\dfrac{8199}{9}$

49. $\dfrac{74,751}{6}$

50. $\dfrac{72,543}{5}$

51. $\dfrac{71,776}{7}$

52. $\dfrac{77,621}{3}$

53. $\dfrac{128,645}{7}$

54. $\dfrac{172,255}{4}$

Use multiplication to check each answer. If an answer is incorrect, find the correct answer.
See Example 11.

55. $5\overline{)1877}$ 375 **R2**

56. $3\overline{)1282}$ 427 **R1**

57. $3\overline{)5725}$ 1908 **R2**

58. $5\overline{)2158}$ 432 **R3**

59. $7\overline{)4692}$ 650 **R2**

60. $9\overline{)5974}$ 663 **R5**

61. $6\overline{)21,409}$ 3 568 **R2**

62. $6\overline{)3192}$ 532

63. $8\overline{)16,019}$ 2 002 **R3**

64. $8\overline{)33,664}$ 4 208

65. $6\overline{)69,140}$ 11,523 **R2**

66. $3\overline{)82,598}$ 27,532 **R1**

67. $9\overline{)86,655}$ 9 628 **R7**

68. $7\overline{)50,809}$ 7 258 **R4**

69. $8\overline{)222,576}$ 27,822

70. $4\overline{)311,216}$ 77,804

71. Explain in your own words how to check a division problem using multiplication. Be sure to include what must be done if the quotient includes a remainder.

72. Describe the three divisibility rules that you feel might be most useful to you and tell why.

Solve each application problem.

73. A banquet caterer has 1165 dishes of the same pattern in inventory. If there are five dishes per place setting, how many place settings does the caterer have?

74. A school district will distribute 1620 new science books equally among 12 schools. How many books will each school receive?

75. Last year in Seattle, 56,000 people went to the circus over a 5-day period. If the same number of people attended each day, what was the daily attendance? (*Source:* Feld Entertainment, Vienna, Virginia.)

76. One gallon of orange juice will serve nine people. How many gallons are needed for 3483 people?

77. An estate of $127,400 is divided equally among seven family members. Find the amount received by each family member.

78. How many 5-lb bags of rice can be filled from 8750 lb of rice?

79. If 36 gallons of fertilizer are needed for each acre of land, find the number of acres that can be fertilized with 7380 gallons of fertilizer.

80. A roofing contractor has purchased 2268 squares (1 square measures 10 ft by 10 ft) of roofing material. If each home needs 21 squares of material, find the number of homes that can be roofed.

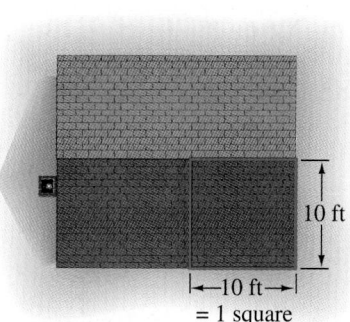

10 ft

|←10 ft→|
= 1 square

81. The Super Lotto payout of $8,100,000 will be divided equally by 36 people who purchased the winning ticket. Find the amount received by each person.

82. Ken Griffey Jr., signed a 9-year baseball contract for $117,000,000. Find his pay for each year. (*Source: USA Today.*)

83. Kaci Salmon, a supervisor at Albany Electric, earns $36,540 per year. Find the amount of her earnings in a 3-month period.

84. If Steven can assemble 168 light diffusers in an 8-hour shift, how many can he assemble in 3 hours?

Put a ✓ mark in the blank if the number at the left is divisible by the number at the top.
Put an X in the blank if the number is not divisible by the number at the top.
See Examples 12–15.

	2	3	5	10
85. 60	___	___	___	___
87. 92	___	___	___	___
89. 445	___	___	___	___
91. 903	___	___	___	___
93. 5166	___	___	___	___
95. 21,763	___	___	___	___

	2	3	5	10
86. 35	___	___	___	___
88. 96	___	___	___	___
90. 897	___	___	___	___
92. 500	___	___	___	___
94. 8302	___	___	___	___
96. 32,472	___	___	___	___

1.6 LONG DIVISION

If the total cost of 42 computer modems is $3066, we can find the cost of each modem using **long division.** Long division is used to divide by a number with more than one digit.

1 ▭ **Do long division.**   In long division, estimate the various numbers by using a **trial divisor,** which is used to get a **trial quotient.**

Example 1 ▶ **Using a Trial Divisor and a Trial Quotient**

Divide: $42\overline{)3066}$.

Because 42 is closer to 40 than to 50, use the first digit of the divisor as a trial divisor.

$$42$$

└────── Trial divisor

Try to divide the first digit of the dividend by 4. Since 3 cannot be divided by 4, use the first *two* digits, 30.

$$\frac{30}{4} = 7 \text{ with remainder 2}$$

$$\frac{7}{42\overline{)3066}} \leftarrow \text{Trial quotient}$$

└──── 7 goes over the 6, because $\frac{306}{42}$ is about 7.

Multiply 7 and 42 to get 294; next, subtract 294 from 306.

$$\begin{array}{r} 7 \\ 42\overline{)3066} \\ \underline{294} \leftarrow 7 \times 42 \\ 12 \leftarrow 306 - 294 \end{array}$$

Bring down the 6 at the right.

$$\begin{array}{r} 7 \\ 42\overline{)3066} \\ \underline{294}\downarrow \\ 126 \leftarrow 6 \text{ brought down} \end{array}$$

Use the trial divisor, 4.

First two digits of 126 ⟶ $\frac{12}{4} = 3$

$$\begin{array}{r} 73 \\ 42\overline{)3066} \\ \underline{294} \\ 126 \\ \underline{126} \leftarrow 3 \times 42 = 126 \\ 0 \end{array}$$

Check the answer by multiplying 42 and 73. The product should be 3066.

1 Divide.

(a) $14\overline{)1148}$

(b) $32\overline{)2048}$

(c) $61\overline{)4392}$

(d) $\dfrac{5394}{93}$

2 Divide.

(a) $48\overline{)2688}$

(b) $36\overline{)2236}$

(c) $65\overline{)5416}$

(d) $89\overline{)6649}$

Work Problem 1 at the Side.

Example 2 Dividing to Find a Trial Quotient

Divide: $58\overline{)2730}$.

Use 6 as a trial divisor, since 58 is closer to 60 than to 50.

First two digits of dividend ⟶ $\dfrac{27}{6} = 4$ with 3 left over

4 ⟵ Trial quotient

$$58\overline{)2730}$$
232 ⟵ $4 \times 58 = 232$
41 ⟵ $273 - 232 = 41$ (smaller than 58, the divisor)

Bring down the 0.

$$\begin{array}{r} 4 \\ 58\overline{)2730} \\ 232\downarrow \\ \hline 410 \end{array}$$ ⟵ 0 brought down

First two digits of 410 ⟶ $\dfrac{41}{6} = 6$ with 5 left over

46 ⟵ Trial quotient

$$58\overline{)2730}$$
232
410
348 ⟵ $6 \times 58 = 348$
62 ⟵ Greater than 58

The remainder, 62, is greater than the divisor, 58, so 7 should be used instead of 6.

$$\begin{array}{r} 47\ \textbf{R4} \\ 58\overline{)2730} \\ 232 \\ \hline 410 \\ 406 \\ \hline 4 \end{array}$$
⟵ $7 \times 58 = 406$
⟵ $410 - 406$

Work Problem 2 at the Side.

Sometimes it is necessary to insert a 0 in the quotient.

Example 3 Inserting 0s in the Quotient

Divide: 42$\overline{)8734}$.
Start as above.

$$\begin{array}{r} 2 \\ 42\overline{)8734} \\ 84 \\ \hline 3 \end{array}$$ ← 2 × 42 = 84
 ← 87 − 84 = 3

Bring down the 3.

$$\begin{array}{r} 2 \\ 42\overline{)8734} \\ 84\downarrow \\ \hline 33 \end{array}$$ ← 3 brought down

Since 33 cannot be divided by 42, place a 0 in the quotient as a placeholder.

$$\begin{array}{r} 20 \\ 42\overline{)8734} \\ 84 \\ \hline 33 \end{array}$$ ← 0 in quotient

Bring down the final digit, the 4.

$$\begin{array}{r} 20 \\ 42\overline{)8734} \\ 84\ \downarrow \\ \hline 334 \end{array}$$ ← 4 brought down

Complete the problem.

$$\begin{array}{r} 207\ \mathbf{R40} \\ 42\overline{)8734} \\ 84 \\ \hline 334 \\ 294 \\ \hline 40 \end{array}$$

The answer is 207 **R**40.

CAUTION

There **must be a digit** in the quotient (answer) above every digit in the dividend once the answer has begun. Notice in Example 3 that a **0** was used to assure an answer digit above every digit in the dividend.

Work Problem ❸ at the Side.

2 **Divide numbers ending in 0 by numbers ending in 0.** When the divisor and dividend both contain 0s at the far right, recall that these numbers are multiples of 10. As with multiplication, there is a short way to divide these multiples of 10. Look at the following examples.

$$26{,}000 \div 1 = 26{,}000$$
$$26{,}000 \div 10 = 2600$$
$$26{,}000 \div 100 = 260$$
$$26{,}000 \div 1000 = 26$$

Do you see a pattern? These examples suggest the following rule.

❸ Divide.

(a) 34$\overline{)3645}$

(b) 28$\overline{)5768}$

(c) 39$\overline{)15{,}933}$

(d) 78$\overline{)23{,}462}$

Real-Data Applications

Sharing Travel Expenses

Sharing costs for meals, entertainment, and travel can be a challenging mathematical problem, especially if the costs are incurred over several days and paid for by different people.

As an example, suppose that Donna and Jerry (couple), Peg and Richard (couple), and Jo (single) decide to spend a week touring Wales and Ireland. Donna has booked the accommodations, and each person will be responsible for their own meals and hotel expenses. However, to save on transportation costs, Richard (who lives in England) will use his personal car, and the group will share the costs for insurance, gasoline, parking, and the ferry. At the end of the trip, the costs were those listed in the table below. All the costs are rounded to the nearest U.S. dollar, although the travelers actually paid the amounts in British pounds and Irish punts.

Expense	Paid By	Costs (in U.S. dollars)
Supplemental car insurance	Richard	21
Ferry (prebooked)	Donna	220
Ferry (upgrade to fast boat)	Richard	40
Parking (Dublin, Ireland)	Richard	21
Gasoline (Wales)	Richard	46
Gasoline (Ireland)	Richard	13
Gasoline (Ireland)	Richard	22
Gasoline (Ireland)	Richard	24
Gasoline (Wales)	Richard	46
Gasoline (Wales)	Richard	21

1. What was the total amount spent on transportation?

2. To compute the shared costs, divide the total transportation costs by the number of travelers.

   (a) How many whole dollars does each traveler owe for transportation costs? (The division does not come out evenly; there is a remainder.)

   (b) How many dollars are left over (the remainder)? What should be done with the remainder costs to ensure a fair division of the costs?

   (c) If the total transportation costs are rounded up to the nearest multiple of 5, then how much would each traveler owe?

3. Using your answer from Problem 2(c), how much is Donna and Jerry's share of the costs as a couple? Do they owe money or are they owed money? If they owe money, how much and to whom should it be paid? If money is owed to them, who owes the money and how much is owed?

4. Using your answer from Problem 2(c), how much is Richard and Peg's share of the costs as a couple? Do they owe money or are they owed money? If they owe money, how much and to whom should it be paid? If money is owed to them, who owes the money and how much is owed?

5. Did each traveler pay the same amount toward transportation costs?

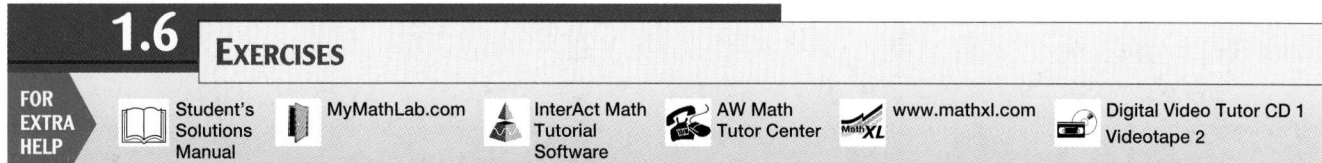

1.6   **EXERCISES**

FOR EXTRA HELP

📖 Student's Solutions Manual   📱 MyMathLab.com   🔺 InterAct Math Tutorial Software   📞 AW Math Tutor Center   MathXL www.mathxl.com   📼 Digital Video Tutor CD 1 Videotape 2

Decide where the first digit in the quotient would be located. Then without finishing the division, you can tell which of the three choices is the correct answer. Circle your choice. See Examples 1 and 2.

1. $25\overline{)550}$

   2   22   220

2. $14\overline{)476}$

   3   34   304

3. $18\overline{)4500}$

   2   25   250

4. $42\overline{)7560}$

   18   180   1800

5. $86\overline{)10,327}$

   12   120 R7   1200

6. $46\overline{)24,026}$

   5   52   522 R14

7. $52\overline{)68,025}$

   13   130 R1   1308 R9

8. $12\overline{)116,953}$

   974 R2   9746 R1   97,460

9. $21\overline{)149,826}$

   71   713   7134 R12

10. $64\overline{)208,138}$

   325 R2   3252 R10   32,521

11. $523\overline{)470,800}$

   9 R100   90 R100   900 R100

12. $230\overline{)253,230}$

   11   110   1101

Divide by using long division. Use multiplication to check each answer. See Examples 1–3.

13. $21\overline{)2272}$

14. $29\overline{)1827}$

15. $56\overline{)10,270}$

16. $83\overline{)39,692}$

17. $26\overline{)62,583}$

18. $28\overline{)84,249}$

19. $74\overline{)84,819}$

20. $238\overline{)186,948}$

21. $153\overline{)509,725}$

22. $308\overline{)26,796}$

23. $420\overline{)357,000}$

24. $900\overline{)153,000}$

Use multiplication to check each answer. If an answer in incorrect, find the correct answer. See Example 6.

          101 **R4**

25. 35)3549

          42 **R26**

26. 64)2712

          658 **R9**

27. 28)18,424

          239 **R121**

28. 145)34,776

          62 **R3**

29. 614)38,068

          174 **R368**

30. 557)97,286

31. Describe in your own words a shortcut you can use to divide multiples of 10 by 10, by 100, or by 1000. Write an example problem and solve it.

32. Suppose you have a division problem with a remainder in the answer. Explain how to check your answer by writing an example problem that has a remainder.

Solve each application problem by using addition, subtraction, multiplication, or division as needed. See Examples 3–5.

33. A private airplane flies 2304 miles to an air show in Oshkosh, Wisconsin, at 128 mph. How many hours did it take to get there?

34. The U.S. Government Printing Office uses 255,000 pounds of ink each year. If it does an equal amount of printing on each of 200 work days in a year, find the weight of the ink used each day. (*Source:* U.S. Government Printing Office.)

35. Don Gracey, the Mountain Timesmith, has serviced and repaired 636 clocks this year. He has worked on 272 wall clocks and 308 table clocks. The remainder were standing floor clocks. Find the number of floor clocks he worked on this year.

36. Two separated parents each share some of the $3718 education costs of their child. If one parent paid $1880, how much did the other pay?

37. Judy Martinez owes $3888 on a loan (including interest). Find her monthly payment if the loan is to be paid off in 36 months.

38. A consultant charged $13,050 for evaluating a school's compliance with the Americans with Disabilities Act. If the consultant worked 225 hours, find the rate charged per hour.

39. Clarence Hanks can assemble 42 circuit boards in 1 hour. How many circuit boards can he assemble in a 5-day workweek of 8 hours per day?

40. There are two conveyor lines in a factory, each of which packages 240 sacks of salt per hour. If the lines operate for 8 hours, find the total number of sacks of salt packaged by the two lines.

41. The average U.S. household of 2.5 people spent $2028 eating away from home last year. Find the average weekly cost of eating away from home. (*Hint:* 1 year equals 52 weeks.)(*Source:* U.S. Bureau of Labor Statistics consumer expenditure surveys.)

42. Former professional basketball player Junior Bridgeman now owns 120 Wendy's restaurants with 4080 employees. Find the average number of employees at each restaurant. (*Source:* National Basketball Retired Players Association.)

RELATING CONCEPTS (Exercises 43–50) | **FOR INDIVIDUAL OR GROUP WORK**

Knowing and using the rules of divisibility is necessary in problem solving.
Work Exercises 43–50 in order.

43. If you have $0 and you divide this amount among three people, how much will each receive?

44. When 0 is divided by any nonzero number, the result is _____ .

45. Divide.
$8 \div 0$

46. We say that division by 0 is *undefined* because it is _____ to compute the answer.
(possible/impossible)
Give an example involving cookies that will support your answer.

47. Divide.
(a) $14 \div 1$
(b) $1 \overline{)17}$
(c) $\dfrac{38}{1}$

48. Any number divided by 1 is the number itself. Is this also true when multiplying by 1? Give three examples that support your answer.

49. Divide.
(a) $32,000 \div 10$
(b) $32,000 \div 100$
(c) $32,000 \div 1000$

50. Write a rule that explains the shortcut for doing divisions like the ones in Exercise 49.

Real-Data Applications

Post Office Facts

The United States Postal Service posts a Web page on the Internet that gives a list of facts about their service. Some of those facts are given in the table below.

Resource/Service	Number
1. Mail collection boxes	312,000
2. Post offices	38,019
3. Delivery points	130 million
4. Pieces of First Class mail delivered each year	107 billion
5. Processing plants sorting and shipping the mail	331
6. Pounds of mail carried on commercial airline flights annually	2.7 billion
7. Commercial airline flights per day	15,000
8. Miles driven to move the mail annually	1.1 billion
9. Vehicles to pick up, transport, and deliver the mail	192,904
10. Customers a day who transact business at the post offices	7 million

Source: United States Postal Service, www.usps.com.

1. Write the number of delivery points in digits showing each period.

2. Write the number of miles driven annually to move the mail in digits showing each period.

3. Estimate the number of post offices, rounded to the nearest thousand.

4. Use the estimate found in Problem 3 to compute the average number of customers who transact business each day at post offices. Round the answer to the nearest person. (*Hint:* To find the average number of customers per day who transact business at post offices, divide the total number of customers per day by the number of post offices. Recall that you can use a short cut to simplify a division problem by crossing out the same number of 0s in the divisor and the dividend.)

5. Find the total number of commercial airline flights per year that carry postal freight. Write the result in words using period names. (*Hint:* Assume that planes fly 365 days per year.)

6. Find the average number of pounds of mail per commercial airline flight, rounded to the nearest 10 pounds. (*Hint:* Divide the total number of pounds of mail carried annually by the number of flights per year, calculated in Problem 5. Use the division shortcut described in Problem 4.)

7. Mail is sorted and shipped from processing plants to the post offices. Which of the following is a rough estimate of the number of post offices served by each processing plant: 10, 100, or 1000? Explain your choice.

1.7 ROUNDING WHOLE NUMBERS

One way to get a rough check on an answer is to *round* the numbers in the problem. **Rounding** a number means finding a number that is close to the original number, but easier to work with.

For example, the chancellor of a community college district might be discussing the need for more classrooms and lab facilities. In making her point, she probably would not need to say that the district has 61,832 students—she could probably just say there are 62,000 students, or even 60,000 students.

1 **Locate the place to which a number is to be rounded.** The first step in rounding a number is to locate the *place to which the number is to be rounded.*

Example 1 **Finding the Place to Which a Number Is to Be Rounded**

Locate and draw a line under the place to which each number is to be rounded.

(a) Round 83 to the nearest ten. Is 83 closer to 80 or to 90?

8̲3 is closer to 80.

Tens place ⎯⎯⎯⎯

(b) Round 54,702 to the nearest thousand. Is it closer to 54,000 or to 55,000?

5̲4,702 is closer to 55,000.

Thousands place ⎯⎯⎯⎯

(c) Round 2,806,124 to the nearest hundred thousand. Is it closer to 2,800,000 or to 2,900,000?

2,8̲06,124 is closer to 2,800,000.

⎯⎯⎯ Hundred thousands place

═════ **Work Problem ❶ at the Side.**

2 **Round numbers.** Use the following rules for rounding whole numbers.

Rounding Whole Numbers

Step 1	Locate the **place** to which the number is to be rounded. Draw a line under that place.
Step 2(a)	Look only at the next digit to the right of the one you underlined. If it is **5 or more,** increase the underlined digit by 1.
Step 2(b)	If the next digit to the right is **4 or less,** do not change the digit in the underlined place.
Step 3	**Change** all digits to the right of the underlined place to 0s.

Example 2 **Using Rounding Rules for 4 or Less**

Round 349 to the nearest hundred.

Step 1 Locate the place to which the number is being rounded. Draw a line under that place.

3̲49

⎯⎯⎯ Hundreds place

⎯⎯ **Continued on Next Page**

OBJECTIVES

1 Locate the place to which a number is to be rounded.

2 Round numbers.

3 Round numbers to estimate an answer.

4 Use front end rounding to estimate an answer.

❶ Locate and draw a line under the place to which each number is to be rounded. Then answer the question.

(a) 557 (nearest ten)

Is it closer to 550 or to 560? ⎯⎯⎯⎯

(b) 1482 (nearest thousand)

Is it closer to 1000 or to 2000? ⎯⎯⎯⎯

(c) 89,512 (nearest hundred)

Is it closer to 89,500 or 89,600? ⎯⎯⎯⎯

(d) 546,325 (nearest ten thousand)

Is it closer to 540,000 or 550,000? ⎯⎯⎯⎯

ANSWERS
1. **(a)** 5̲57 is closer to 560
   (b) 1̲482 is closer to 1000
   (c) 89,5̲12 is closer to 89,500
   (d) 54̲6,325 is closer to 550,000

❷ Round to the nearest ten.

(a) 43

(b) 92

(c) 164

(d) 6822

❸ Round to the nearest thousand.

(a) 2635

(b) 5508

(c) 43,766

(d) 74,803

Step 2 Because the next digit to the right of the underlined place is 4, which is 4 or less, do *not* change the digit in the underlined place.

Next digit is 4 or less.
349
3 remains 3.

Step 3 Change all digits to the right of the underlined place to 0s.

349 rounded to the nearest hundred is 300.

In other words, 349 is closer to 300 than to 400.

Work Problem ❷ at the Side.

Example 3 Using Rounding Rules for 5 or More

Round 36,833 to the nearest thousand.

Step 1 Find the place to which the number is to be rounded and draw a line under that place.

36,833
Thousands

Step 2 Because the next digit to the right of the underlined place is 8, which is 5 or more, add 1 to the underlined place.

Next digit is 5 or more.
36,833
Change 6 to 7.

Step 3 Change all digits to the right of the underlined place to 0s.

Change to 0.
36,833 rounded to the nearest thousand is 37,000.
Change 6 to 7.

In other words, 36,833 is closer to 37,000 than to 36,000.

Work Problem ❸ at the Side.

Example 4 Using Rounding Rules

(a) Round 2382 to the nearest ten.

Step 1 2382
Tens place

Step 2 The next digit to the right is 2, which is 4 or less.

Next digit is 4 or less.
2382
Leave 8 as 8.

Step 3 2382   Change to 0.

2382 rounded to the nearest ten is 2380. In other words, 2382 is closer to 2380 than to 2390.

Continued on Next Page

(b) Round 13,961 to the nearest hundred.

Step 1 13,961

↑————— Hundreds place

Step 2 The next digit to the right is 6.

┌— Next digit is 5 or more.
13,961

└— Change 9 to 10; write 0 and carry 1 into thousands place.
└— 3 + carried 1 = 4

┌— Change to 0.

Step 3 14,061

13,961 rounded to the nearest hundred is 14,000.

NOTE

In Step 2 of Example 4(a), notice that the first three digits increased from 139 to 140 when we added 1 to the hundreds place.

13,961 rounded to 14,000

Work Problem ❹ at the Side.

Example 5 **Rounding Large Numbers**

(a) Round 37,892 to the nearest ten thousand.

Step 1 37,892

↑——— Ten thousands place

Step 2 The next digit to the right is 7.

┌— Next digit is 5 or more.
37892

↑——— Change 3 to 4.

┌— Change to 0.

Step 3 47,892

37,892 rounded to the nearest ten thousand is 40,000.

(b) Round 528,498,675 to the nearest million.

Step 1 528,498,675

↑————— Millions place

┌— Next digit is 4 or less.

Step 2 528,498,675

└— Leave 8 as 8.

┌— Change to 0.

Step 3 528,498,675

528,498,675 rounded to the nearest million is 528,000,000.

Work Problem ❺ at the Side.

❹ Round each number as indicated.

(a) 7827 to the nearest ten

(b) 4342 to the nearest hundred

(c) 73,077 to the nearest hundred

(d) 56,961 to the nearest hundred

❺ Round each number as indicated.

(a) 13,586 to the nearest ten thousand

(b) 724,518,715 to the nearest million

6 Round each number to the nearest ten and then to the nearest hundred.

(a) 267

(b) 549

(c) 8709

Sometimes a number must be rounded to different places.

Example 6 **Rounding to Different Places**

Round 648 **(a)** to the nearest ten and **(b)** to the nearest hundred.

(a) to the nearest ten

```
                        ┌── Next digit is 5 or more.
                        ↓
              6 4̲ 8
                 └── Tens place (4 + 1 = 5)
```

648 to the nearest ten is 650.

(b) to the nearest hundred

```
                       ┌── Next digit is 4 or less.
                       ↓
              6̲ 4 8
               └── Hundreds place stays the same.
```

648 to the nearest hundred is 600.

Notice that 648 is rounded to the nearest ten is 650, and then 650 is rounded to the nearest hundred, the result is 700. If, however, 648 is rounded directly to the nearest hundred, the result is 600 (not 700).

CAUTION

Before rounding to a different place, always go back to the *original*, unrounded number.

Work Problem 6 at the Side.

Example 7 **Applying Rounding Rules**

Round each number to the nearest ten, nearest hundred, and nearest thousand.

(a) 4358
First round 4358 to the nearest ten.

```
                          ┌── Next digit is 5 or more.
                          ↓
              4 3 5̲ 8
                  └── Tens place (5 + 1 = 6)
```

4358 rounded to the nearest ten is 4360.

Now go back to 4358, the *original* number, before rounding to the nearest hundred.

```
                          ┌── Next digit is 5 or more.
                          ↓
              4 3̲ 5 8
                └── Hundreds place (3 + 1 = 4)
```

4358 rounded to the nearest hundred is 4400.

Again, go back to the *original* number before rounding to the nearest thousand.

```
                          ┌── Next digit is 4 or less.
                          ↓
              4̲ 3 5 8
               └── Thousands place stays the same.
```

4358 rounded to the nearest thousand is 4000.

Continued on Next Page

(b) 680,914

First, round to the nearest ten.

┌─ Next digit is 4 or less.

680,9<u>1</u>4

└─ Tens place stays the same.

680,914 rounded to the nearest ten is 680,910.
  Go back to 680,914, the *original* number, to round to the nearest hundred.

┌─ Next digit is 4 or less.

680,<u>9</u>14

└─ Hundreds place stays the same.

680,914 rounded to the nearest hundred is 680,900.
  Go back to the *original* number to round to the nearest thousand.

┌─ Next digit is 5 or more.

680,914

└─ Thousands place (0 + 1 = 1)

680,914 rounded to the nearest thousand is 681,000.

====== **Work Problem ❼ at the Side.**

❼ Round each number to the nearest ten, nearest hundred, and nearest thousand.

(a) 2087

(b) 46,364

(c) 268,328

3 ▭ **Round numbers to estimate an answer.** Numbers may be rounded to estimate an answer. An estimated answer is one that is close to the exact answer and may be used as a check when the exact answer is found. The "≈" sign is often used to show that an answer has been rounded or estimated and is almost equal to the exact answer. ≈ means "approximately equal to."

Example 8 Using Rounding to Estimate an Answer

Estimate each answer by rounding to the nearest ten.

(a) 76 ⟶ 80
     53 ⟶ 50   } Rounded to the nearest ten
     38 ⟶ 40
   + 91 ⟶ + 90
          260   Estimated answer

(b) 27   30   } Rounded to the nearest ten
      − 14   − 10
             20   Estimated answer

(c) 16   20   } Rounded to the nearest ten
      × 21   × 20
             400   Estimated answer

====== **Work Problem ❽ at the Side.**

❽ Estimate each answer by rounding to the nearest ten.

(a)   16
       74
       58
     + 31

(b)   53
       − 19

(c)   37
       × 84

Example 9 Using Rounding to Estimate an Answer

Estimate each answer by rounding to the nearest hundred.

(a) 152 ⟶ 200
     749 ⟶ 700   } Rounded to the nearest hundred
     576 ⟶ 600
   + 819 ⟶ + 800
          2300   Estimated answer

Continued on Next Page

❸ Simplify each expression.

(a) $2 + 6 + 3^2$

(b) $2^2 + 3^3$

(c) $4 \cdot 6 \div 12 - 2$

(d) $40 \div 5 \div 2$

(e) $8 + (14 \div 2) \cdot 6$

❹ Simplify each expression.

(a) $10 - 4 + 3^2$

(b) $3^2 + 4^2 - (7 \cdot 2)$

(c) $2 \cdot \sqrt{64} - 5 \cdot 3$

(d) $20 \div 2 + (7 - 5)$

(e) $15 \cdot \sqrt{9} - 8 \cdot \sqrt{4}$

Order of Operations

1. Do all operations inside **parentheses** or **other grouping symbols.**
2. Simplify any expressions with **exponents** and find any **square roots.**
3. **Multiply** or **divide,** proceeding from left to right.
4. **Add** or **subtract,** proceeding from left to right.

Example 3   Understanding the Order of Operations

Use the order of operations to simplify each expression.

(a) $8^2 + 5 + 2$

$$8^2 + 5 + 2$$
$$8 \cdot 8 + 5 + 2 \qquad \text{Evaluate exponent first; } 8^2 \text{ is } 8 \cdot 8.$$
$$64 + 5 + 2 \qquad \text{Add from left to right.}$$
$$69 + 2 = 71$$

(b) $35 \div 5 \cdot 6$ Divide first (start at left).
$$7 \cdot 6 = 42 \qquad \text{Multiply.}$$

(c) $9 + (20 - 4) \cdot 3$ Work inside parentheses first.
$$9 + 16 \cdot 3 \qquad \text{Multiply.}$$
$$9 + 48 = 57 \qquad \text{Add last.}$$

(d) $12 \cdot \sqrt{16} - 8 \cdot 4$ Find the square root first.
$$12 \cdot 4 - 8 \cdot 4 \qquad \text{Multiply from left to right.}$$
$$48 - 32 = 16 \qquad \text{Subtract last.}$$

Work Problem ❸ at the Side.

Example 4   Using the Order of Operations

Use the order of operations to simplify each expression.

(a) $15 - 4 + 2$ Subtract first (start at left).
$$11 + 2 = 13 \qquad \text{Add.}$$

(b) $8 + (7 - 3) \div 2$ Work inside parentheses first.
$$8 + 4 \div 2 \qquad \text{Divide.}$$
$$8 + 2 = 10 \qquad \text{Add last.}$$

(c) $4^2 \cdot 2^2 + (7 + 3) \cdot 2$ Parentheses first.
$$4^2 \cdot 2^2 + 10 \cdot 2 \qquad \text{Evaluate exponents.}$$
$$16 \cdot 4 + 10 \cdot 2 \qquad \text{Multiply from left to right.}$$
$$64 + 20 = 84 \qquad \text{Add last.}$$

(d) $4 \cdot \sqrt{25} - 7 \cdot 2$ Find the square root first.
$$4 \cdot 5 - 7 \cdot 2 \qquad \text{Multiply from left to right.}$$
$$20 - 14 = 6 \qquad \text{Subtract last.}$$

NOTE

Getting a correct answer always depends on correctly using the order of operations.

Work Problem ❹ at the Side.

1.8 EXERCISES

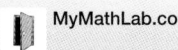
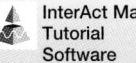
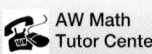

Identify the exponent and the base, and then simplify each expression. See Example 1.

1. 4^2

2. 2^3

3. 5^2

4. 3^2

5. 12^2

6. 10^3

7. 15^2

8. 11^3

Use the Perfect Squares Table on page 73 to find each square root. See Example 2.

9. $\sqrt{9}$

10. $\sqrt{25}$

11. $\sqrt{64}$

12. $\sqrt{36}$

13. $\sqrt{100}$

14. $\sqrt{121}$

15. $\sqrt{144}$

16. $\sqrt{225}$

Fill in each blank. See Example 2.

17. $8^2 = $ _____ so $\sqrt{} = 8$.

18. $9^2 = $ _____ so $\sqrt{} = 9$.

19. $20^2 = $ _____ so $\sqrt{} = 20$.

20. $30^2 = $ _____ so $\sqrt{} = 30$.

21. $35^2 = $ _____ so $\sqrt{} = 35$.

22. $38^2 = $ _____ so $\sqrt{} = 38$.

2 Use the bar graph to find the approximate number of fans who picked each sport as their favorite.

(a) Pro football

(b) Pro baseball

(c) Pro basketball

(d) College basketball

(e) Golf

3 Use the line graph to find the predicted population of the United States for each year.

(a) 2050

(b) 2075

(c) 2100

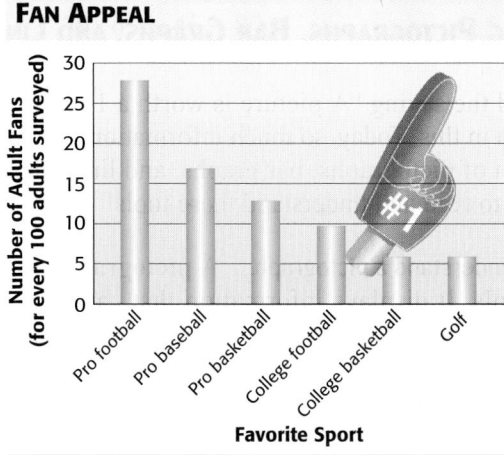

FAN APPEAL

Source: The Harris Poll.

Example 2 Using a Bar Graph

Use the bar graph to find the number of fans who picked college football as their favorite sport.

Use a ruler or straight edge to line up the top of the bar labeled "College Football," with the numbers on the left edge of the graph, labeled "Number of Adults Fans." We see that 10 out of 100 adult fans picked college football as their favorite sport.

Work Problem 2 at the Side.

3 **Read and understand a line graph.** A **line graph** is often used for showing a trend. The following line graph shows the U.S. Bureau of the Census predictions for U.S. population growth to the year 2100.

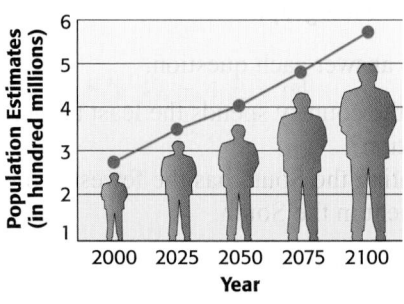

ANOTHER CENTURY OF GROWTH

Source: U.S. Bureau of the Census.

Example 3 Using a Line Graph

Use the line graph to answer each question.

(a) What trend or pattern is shown in the graph?
The population will continue to increase.

(b) What is the estimated population for 2025?
Use a ruler or straight edge to line up the dot above the year labeled 2025 on the horizontal line with the numbers along the left edge of the graph. Notice that the label on the left side says "in hundred millions." Since the 2025 dot is halfway between 3 and 4, the population in 2025 is halfway between 3 • 100,000,000 and 4 • 100,000,000 or 300,000,000 and 400,000,000. That means that the predicted population in 2025 is about 350,000,000 people.

Work Problem 3 at the Side.

The following pictograph shows the approximate number of video rental outlets in the eight countries with the greatest number of outlets in the world. Use the pictograph to answer Exercises 1–6. See Example 1.

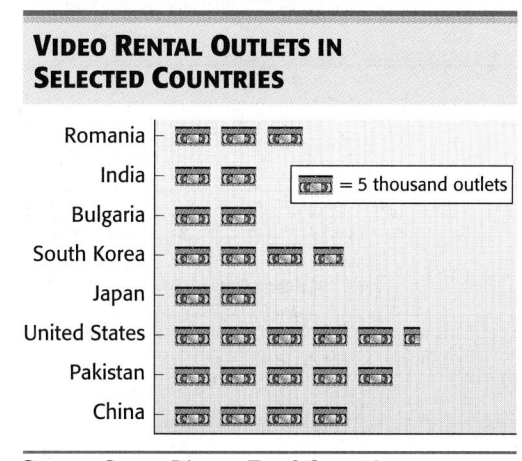

Source: Screen Digest—Top 8 Countries.

1. Find the number of video rental outlets in Romania.

2. How many video rental outlets are there in China?

3. Which country has the greatest number of outlets?

4. Which countries have the least number of outlets?

5. The United States has 28 thousand outlets. How many more is this than the second greatest number of outlets?

6. How many more outlets does China have than Bulgaria?

The following bar graph shows the results of a survey that was taken of 100 working adults to determine how they found their careers. Use the bar graph to answer Exercises 7–10. See Example 2.

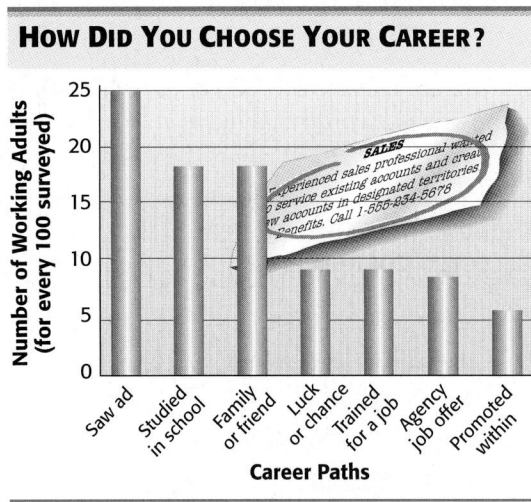

How Did You Choose Your Career?

Number of Working Adults (for every 100 surveyed)

Career Paths

Source: Market Facts/TeleNation for Career Education Corporation.

7. How many people found their careers as a result of training for a job?

8. How many people found their careers because they studied for the career in school?

9. (a) Which career path was taken by the greatest number of people?

(b) How many people used this path?

10. (a) Which career path was taken by the least number of people?

(b) How many people used this path?

11. How many more people found their careers as a result of "Studied in school" than "Luck or chance"?

12. Find the total number of people who found their careers as a result of either "Studied in school" or "Trained for a job."

A local real estate association collected home sales data for their area and prepared the following line graph. Remembering that the sales data is shown in thousands, use the line graph to answer Exercises 13–16. See Example 3.

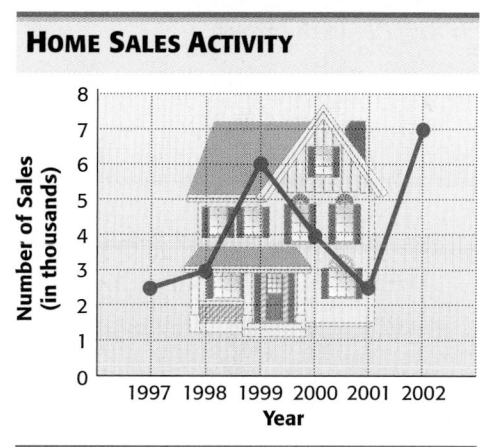

HOME SALES ACTIVITY

13. Which year had the greatest number of home sales? How many sales were there?

14. Which two years had the least number of home sales? How many sales were there in each of those years?

15. Find the increase in the number of home sales from 2001 to 2002.

16. Find the decrease in the number of home sales from 2000 to 2001.

17. Give three possible explanations for the decrease in home sales in 2000.

18. Give three possible explanations for the increase in home sales in 2002.

| RELATING CONCEPTS (Exercises 19–23) | FOR INDIVIDUAL OR GROUP WORK |

*Getting a correct answer in mathematics always depends on correctly using the order
of operations. Insert grouping symbols (parentheses) so that each given expression
evaluates to the given number.* **Work Exercises 19–23 in order.**

19. $5 + 1 \cdot 8 - 2$; evaluate to 46

20. $4 + 2 \cdot 5 + 1$; evaluate to 36

21. $36 \div 3 \cdot 3 \cdot 4$; evaluate to 16

22. $48 \div 2 \cdot 2 \cdot 2 + 2$; evaluate to 8

23. A fencing contractor is building 12 corrals like the one shown below for a dairy farmer.

    (a) Use the order of operations to write an expression for the amount of fencing
        needed for all 12 corrals.

    (b) How many feet of fencing are needed?

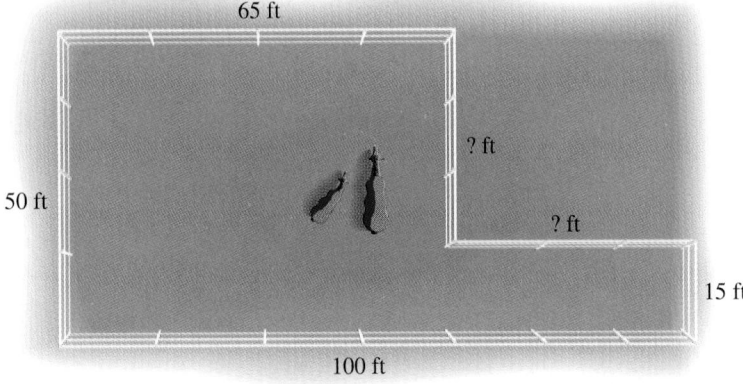

65 ft

? ft

50 ft

? ft

15 ft

100 ft

1.10 SOLVING APPLICATION PROBLEMS

Most problems involving applications of mathematics are written out in sentence form. You need to read the problem carefully to decide how to solve it.

OBJECTIVES

1 Find indicator words in application problems.

2 Solve application problems.

3 Estimate an answer.

1 **Find indicator words in application problems.** As you read an application problem, look for **indicator words** that help you determine the necessary operations—either addition, subtraction, multiplication, or division. Some of these indicator words are shown here.

Addition	Subtraction	Multiplication	Division	Equals
plus	less	product	divided by	is
more	subtract	double	divided into	the same as
more than	subtracted from	triple	quotient	equals
added to	difference	times	goes into	equal to
increased by	less than	of	divide	yields
sum	fewer	twice	divided equally	results in
total	decreased by	twice as much	per	are
sum of	loss of			
increase of	minus			
gain of	take away			

CAUTION

The word *and* does not always indicate addition, so it does not appear as an indicator word in the preceding table. Notice how the "and" shows the location of several different operation signs below.

The sum of 6 *and* 2 is 6 + 2.
The difference of 6 *and* 2 is 6 − 2.
The product of 6 *and* 2 is 6 • 2.
The quotient of 6 *and* 2 is 6 ÷ 2.

2 **Solve application problems.** Solve application problems by using the following six steps.

Solving an Application Problem

Step 1 **Read** the problem carefully and be certain you *understand* what the problem is asking. It may be necessary to read the problem several times.

Step 2 Before doing any calculations, **work out a plan** and try to visualize the problem. Draw a sketch if possible. Know which facts are given and which must be found. Use *indicator words* to help decide on the *plan* (whether you will need to add, subtract, multiply, or divide).

Step 3 **Estimate** a *reasonable answer* by using rounding.

Step 4 **Solve** the problem by using the facts given and your plan.

Step 5 **State the answer.**

Step 6 **Check** your work. If the answer does not seem reasonable, begin again by rereading the problem.

1 Pick the most reasonable answer for each problem.

   (a) A grocery clerk's hourly wage: $3; $11; $60

   (b) The total length of five sports utility vehicles: 8 ft; 18 ft; 80 ft; 800 ft

   (c) The cost of heart by-pass surgery: $500; $50,000; $5,000,000

2 Solve each problem.

   (a) On a recent geology field trip, 84 fossils were collected. If the fossils are divided equally among John, Sean, Jenn, and Kara, how many fossils will each receive?

   (b) There will be 384 people at an awards banquet. If each food server is assigned to 16 people, how many servers are needed?

CAUTION

Be careful not to begin solving a problem before you understand what the problem is asking. Be certain that you know what the problem is asking before you try to solve it.

3 **Estimate an answer.** The six problem-solving steps give a systematic approach for solving word problems. Each of the steps is important, but special emphasis should be placed on Step 3, estimating a *reasonable answer.* Many times an "answer" just does not fit the problem.

   What is a reasonable answer? Read the problem and try to determine the approximate size of the answer. Should the answer be part of a dollar, a few dollars, hundreds, thousands, or even millions of dollars? For example, if a problem asks for the cost of a man's shirt, would an answer of $20 be reasonable? $1000? $0.65? $65?

CAUTION

Always estimate the answer, then look at your final result to be sure it fits your estimate and is reasonable. This step will give greater success in problem solving.

Work Problem 1 at the Side.

Example 1 Applying Division

At a recent garage sale, the total sales were $584. If the money was divided equally among Tom, Rosetta, Maryann, and José, how much did each person get?

Step 1 **Read.** A reading of the problem shows that the four members in the group divided $584 equally.

Step 2 **Work out a plan.** The indicator words, **divided equally,** show that the amount each received can be found by dividing $584 by 4.

Step 3 **Estimate.** A reasonable answer would be a little less than $150 each, since $600 ÷ 4 = $150 ($584 rounded to $600).

Step 4 **Solve.** Find the actual answer by dividing $584 by 4.

$$\begin{array}{r} 146 \\ 4\overline{)584} \end{array}$$

Step 5 **State the answer.** Each person got $146.

Step 6 **Check.** The answer $146 is reasonable, as $146 is close to the estimated answer of $150. Is the answer $146 correct? Check by multiplying.

$$\begin{array}{r} \$146 \\ \times\ \ \ 4 \\ \hline \$584 \end{array}$$

   $146 Amount received by each person
   × 4 Number of people
   $584 Total sales; Matches number given in problem.

Work Problem 2 at the Side.

Example 2 Applying Addition

One week, Paula Crockett decided to total the sales at Starbucks. The daily sales figures were $2358 on Monday, $3056 on Tuesday, $2515 on Wednesday, $2475 on Thursday, $3378 on Friday, $3219 on Saturday, and $3008 on Sunday. Find the total sales for the week.

Step 1 **Read.** In this problem, the sales for each day are given and the total sales for the week must be found.

Step 2 **Work out a plan.** Add the daily sales to arrive at the weekly total.

Step 3 **Estimate.** Because the sales were about $3000 per day for a week of 7 days, a reasonable estimate would be around $21,000 ($7 \times \$3000 = \$21,000$).

Step 4 **Solve.** Find the actual answer by adding the sales for the 7 days.

$$
\begin{array}{r}
\underline{\$20,009} \quad \text{Check by adding up.} \\
\$\ 2\ 358 \\
3\ 056 \\
2\ 515 \\
2\ 475 \\
3\ 378 \\
3\ 219 \\
+\quad 3\ 008 \\
\hline
\$20,009 \quad \text{Sales for the week}
\end{array}
$$

Step 5 **State the answer.** Paula's total sales figure for the week was $20,009.

Step 6 **Check.** This answer of $20,009 is close to the estimate of $21,000, so it is reasonable. Add up the columns to check the exact answer.

🖩 **Calculator Tip**   The calculator solution to Example 2 uses chain calculations to get

2358 ⊕ 3056 ⊕ 2515 ⊕ 2475 ⊕ 3378 ⊕ 3219 ⊕ 3008 ⊜ 20009

Work Problem ❸ at the Side.

Step 4 **Solve.** Find the exact answer by subtracting 4084 from 21,382.

Example 3 Determining whether Subtraction Is Necessary

The number of students enrolled in Chabot College this year is 4084 fewer than the number enrolled last year. Enrollment last year was 21,382. Find the enrollment this year.

Step 1 **Read.** In this problem, the enrollment has decreased from last year to this year. The enrollment last year and the decrease in enrollment are given. This year's enrollment must be found.

Step 2 **Work out a plan.** The indicator word, *fewer*, shows that subtraction must be used to find the number of students enrolled this year.

Step 3 **Estimate.** Because the enrollment was about 21,000 students, and the decrease in enrollment is about 4000 students, a reasonable estimate would be 17,000 students ($21,000 - 4000 = 17,000$).

Continued on Next Page

❸ Solve each problem.

(a) During the semester, Cindy received the following points on examinations and quizzes: 92, 81, 83, 98, 15, 14, 15, and 12. Find her total points for the semester.

(b) Stephanie Dixon works at the telephone order desk of a catalog sales company. One week she had the following number of customer contacts: Monday, 78; Tuesday, 64; Wednesday, 118; Thursday, 102; and Friday, 196. How many customer contacts did she have that week?

4 Solve each problem.

(a) A home occupies 1450 square feet, while an apartment occupies 980 square feet. Find the difference in the number of square feet of the two living units.

$$\begin{array}{r} 21{,}382 \\ -\ 4\ 084 \\ \hline 17{,}298 \end{array}$$

Step 5 **State the answer.** The enrollment this year is 17,298 students.

Step 6 **Check.** The answer 17,298 is reasonable, as it is close to the estimate of 17,000. Check by adding.

17,298 ← Enrollment this year
+ 4 084 ← Decrease in enrollment
21,382 ← Enrollment last year; Matches number given in problem.

Work Problem ④ at the Side.

(b) Aim Electronics had 19,805 employees. After a layoff of 3980 employees, how many employees remain?

Example 4 Solving a Two-Step Problem

In May, a landlord received $720 from each of eight tenants. After paying $2180 in expenses, how much rent money did the landlord have left?

Step 1 **Read.** The problem asks for the amount of rent remaining after expenses have been paid.

Step 2 **Work out a plan.** The wording *from each of eight tenants* indicates that the eight rents must be totaled. Since the rents are all the same, use multiplication to find the total rent received. Finally, subtract expenses.

5 Solve each problem.

(a) Gwen is paid $315 for each car that she sells. If she sold five cars and had $280 in sales expense deducted, how much did she make?

Step 3 **Estimate.** The amount of rent is about $700, making the total rent received about $5600 ($700 × 8). The expenses are about $2000. A reasonable estimate of the amount remaining is $3600 ($5600 − $2000).

Step 4 **Solve.** Find the exact amount by first multiplying $720 by 8 (the number of tenants).

$$\begin{array}{r} \$\ 720 \\ \times\ \ \ \ 8 \\ \hline \$5760 \end{array}$$

Finally, subtract the $2180 in expenses from $5760.

$$\begin{array}{r} \$5760 \\ -\ \$2180 \\ \hline \$3580 \end{array}$$

(b) An Internet book company had sales of 12,628 books with a profit of $6 for each book sold. If 863 books are returned, how much profit remains?

Step 5 **State the answer.** The amount remaining is $3580.

Step 6 **Check.** The answer of $3580 is reasonable, since it is close to the estimated answer of $3600. Check by adding the expenses to the amount remaining and then dividing by 8.

$3580 + $2180 = $5760

$$\begin{array}{r} \$720 \\ 8\overline{)5760} \end{array}$$ ← Matches the rent amount given in the problem.

Work Problem ⑤ at the Side.

ANSWERS
4. **(a)** 470 square feet **(b)** 15,825 employees
5. **(a)** $1295 **(b)** $70,590

1.10   **EXERCISES**

| FOR EXTRA HELP | 📖 Student's Solutions Manual | 📙 MyMathLab.com | 🔺 InterAct Math Tutorial Software | ☎ AW Math Tutor Center | MathXL www.mathxl.com | 💿 Digital Video Tutor CD 1 Videotape 3 |

Solve each application problem. First use front end rounding to estimate the answer. Then find the exact answer. See Examples 1–4.

1. To train for a triathlon, Beth Andrews rode her bike 80 miles on Monday, 75 miles on Tuesday, 135 miles on Wednesday, 40 miles on Thursday, and 52 miles on Friday. How many miles did she ride in this 5-day period?

Estimate:

Exact:

2. During a recent week, Starbucks Coffee sold 325 lb of Estate Java Coffee, 75 lb of Encanta Blend Coffee, 137 lb of Ethiopia Sidamo Coffee, 495 lb of Starbucks House Blend Decaf Coffee, and 105 lb of New Guinea Peaberry Coffee. Find the total number of pounds of coffee sold.

Estimate:

Exact:

3. In a certain country, there were 70 ATM burglaries and attempted burglaries in 1992 and 200 in 1997. How many more of these crimes were there in 1997 than in 1992? (*Source: ATM Crime and Security Newsletter.*)

Estimate:

Exact:

4. The amount of cash in ATMs ranges from $15,000 in small machines to $250,000 in large bank machines. How much more money is there in the large machines than in the small machines? (*Source: Eric Sheppard of Security Corporation.*)

Estimate:

Exact:

5. A packing machine can package 236 first-aid kits each hour. At this rate, find the number of first-aid kits packaged in 24 hours.

Estimate:

Exact:

6. If 450 admission tickets to a classic car show are sold each day, how many tickets are sold in a 12-day period?

Estimate:

Exact:

7. Ted Slauson, coordinator of Toys for Tots, has collected 2628 toys. If his group can give the same number of toys to each of 657 children, how many toys will each child receive?

Estimate:

Exact:

8. If profits of $680,000 are divided evenly among a firm's 1000 employees, how much money will each employee receive?

Estimate:

Exact:

9. The number of boaters and campers at the lake was 8392 on Friday. If this was 4218 more than the number of people at the lake on Wednesday, how many were there on Wednesday?

Estimate:

Exact:

10. The community has raised $52,882 for the homeless shelter. If the total amount needed for the shelter is $75,650, find the additional amount needed.

Estimate:

Exact:

11. Turn down the thermostat in the winter and you can save money and energy. In the upper Midwest, setting back the thermostat from 68° to 55° at night can save $14 per month on fuel. Find the amount of money saved in five months.

Estimate:

Exact:

12. The cost of tuition and fees at a community college is $785 per quarter. If Gale Klein has five quarters remaining, find the total amount that she will need for tuition and fees.

Estimate:

Exact:

The table shows the average starting salaries for selected occupations. Refer to the table to answer Exercises 13–16. First use front end rounding to estimate the answer. Then find the exact answer.

13. How much more does an electrical engineer earn than an aircraft mechanic?

Estimate:

Exact:

ALL WALKS OF LIFE

Average starting salaries for selected occupations.

Secretary	$16,400
Flight attendant	$16,500
Photographer	$17,000
Journalist	$21,700*
Dental hygienist	$22,000
State police officer	$24,000
Aircraft mechanic	$29,000
Accountant	$29,800*
Electrical engineer	$38,700*

* requires a Bachelor's degree

Source: U.S. Department of Labor/U.S. Bureau of the Census.

14. How much more does an accountant earn than a journalist?

Estimate:

Exact:

15. Mr. Garrett is a journalist and Mrs. Garrett is an accountant. Mr. Harcos is a photographer and Mrs. Harcos is an electrical engineer.

(a) Which couple has higher earnings?

Estimate:

Exact:

(b) Find the difference in the earnings.

Estimate:

Exact:

16. Mr. Gonsalves is a dental hygienist and Mrs. Gonsalves is a secretary. Mr. Horton is a flight attendant and Mrs. Horton is a state police officer.

(a) Which couple has higher earnings?

Estimate:

Exact:

(b) Find the difference in the earnings.

Estimate:

Exact:

17. Dorene Cox decides to establish a budget. She will spend $450 for rent, $325 for food, $320 for child care, $182 for transportation, $150 for other expenses, and she will put the remainder in savings. If her monthly take-home pay is $1620, find her monthly savings.

Estimate:

Exact:

18. Jared Ueda had $2874 in his checking account. He wrote checks for $308 for auto repairs, $580 for child support, and $778 for an insurance payment. Find the amount remaining in his account.

Estimate:

Exact:

19. There are 43,560 square feet in one acre. How many square feet are there in 138 acres?

Estimate:

Exact:

20. The number of gallons of water polluted each day in an industrial area is 209,670. How many gallons of water are polluted each year? (Use a 365-day year.)

Estimate:

Exact:

The Internet was used to find the following minivan optional features and the price of each feature. Use this information to answer Exercises 21–24.

Safety and Security Options		Convenience and Comfort Options	
Option	**Cost**	**Option**	**Cost**
Seven-passenger seating with two child seats	$375	Driver's side sliding door	$595
		Roof rack	$215
Antilock brakes	$565	Air conditioning	$860
Rear window defroster	$195	Power windows	$340
Rear window wiper/washer	$115	Central locking system	$280
Full-size spare tire	$125		

Source: www.edmunds.com

21. Find the total cost of all Safety and Security Options listed.

Estimate:

Exact:

22. Find the total cost of all Convenience and Comfort Options listed.

Estimate:

Exact:

23. A new-car dealer offers an option value package that includes seven-passenger seating with two child seats, antilock brakes, a central locking system, and air conditioning at a cost of $1550. If Jill buys the value package instead of paying for each option separately, how much will she save?

Estimate:

Exact:

24. A new-car dealer offers an option package that includes seven-passenger seating with two child seats, antilock brakes, full-size spare tire, power windows, air conditioning, and a roof rack for a total of $1750. How much will Samuel save if he buys the option package instead of paying for each option separately?

Estimate:

Exact:

25. The Enabling Supply House purchased 6 wheelchairs at $1256 each and 15 speech compression recorder-players at $895 each. Find the total cost.

Estimate:

Exact:

26. A college bookstore buys 17 computers at $506 each and 13 printers at $482 each. Find the total cost.

Estimate:

Exact:

27. Being able to identify indicator words is helpful in determining how to solve an application problem. Write three indicators words for each of these operations: add, subtract, multiply, and divide. Write two indicator words that mean equals.

28. Identify and explain the six steps used to solve an application problem. You may refer to the text if you need help, but use your own words.

29. Write in your own words why it is important to estimate a reasonable answer. Give three examples of what might be a reasonable answer to a math problem from your daily activities.

30. First estimate by rounding to thousands, then find the exact answer to the following problem.

$$7438 + 6493 + 2380$$

Do the two answers vary by more than 1000? Why? Will estimated answers always vary from exact answers?

Estimate:

Exact:

Solve each application problem. See Examples 1–4.

31. A package of 3 undershirts costs $12, and a package of 6 pairs of socks costs $15. Find the total cost of 30 undershirts and 18 pairs of socks.

32. Brian earned $8 per hour for 38 hours of work. Maria earned $9 per hour for 39 hours of work. Find their total combined income.

33. A car weighs 2425 lb. If its 582-lb engine is removed and replaced with a 634-lb engine, what will the car weigh?

34. Barbara has $2324 in her preschool operating account. She spends $734 from this account, and then the class parents raise $568 in a rummage sale. Find the balance in the account after she deposits the money from the rummage sale.

35. In a recent survey of high-priced hotels, the cost per night at Harrah's in Reno was $59, while the cost at Harrah's Lake Tahoe was $159 per night. Find the amount saved on a 5-night stay at Harrah's in Reno instead of staying at Harrah's Lake Tahoe. (*Source: Harrah's Casinos and Hotels, weekday rates.*)

36. The most expensive hotel room in a recent study was the Ritz-Carlton at $375 per night, while the least expensive was Motel 6 at $32 per night. Find the amount saved in a 4-night stay at Motel 6 instead of staying at the Ritz-Carlton. (*Source: USA Today.*)

37. A youth soccer association raised $7588 through fund-raising projects. After expenses of $838 were paid, the balance of the money was divided evenly among the 18 teams. How much did each team receive?

38. Feather Farms Egg Ranch collected 3545 eggs in the morning and 2575 eggs in the afternoon. If the eggs are packed in flats containing 30 eggs each, find the number of flats needed for packing.

39. A theater owner wants to provide enough seating for 1250 people. The main floor has 30 rows of 25 seats in each row. If the balcony has 25 rows, how many seats must be in each balcony row to satisfy the owner's seating requirements?

40. Jennie makes 24 grapevine wreaths per week to sell to gift shops. She works 40 weeks a year and packages six wreaths per box. If she ships equal quantities to each of five shops, find the number of boxes each store will receive.

SUMMARY

 Study Skills Workbook
Activity 5

KEY TERMS

1.1	**whole numbers**	The whole numbers are 0, 1, 2, 3, 4, 5, 6, 7, 8, and so on.
	place value	The place value of each digit in a whole number is determined by its position in the whole number.
	table	A table is a display of facts in rows and columns.
1.2	**addition**	The process of finding the total is addition.
	addends	The numbers being added in an addition problem are addends.
	sum (total)	The answer in an addition problem is called the sum.
	commutative property of addition	The commutative property of addition states that the order of numbers in an addition problem can be changed without changing the sum.
	associative property of addition	The associative property of addition states that grouping the addition of numbers differently does not change the sum.
	carrying	The process of carrying is used in an addition problem when the sum of the digits in a column is greater than 9.
	perimeter	The perimeter is the distance around the outside edges of a figure.
1.3	**minuend**	The number from which another number (the subtrahend) is being subtracted is the minuend.
	subtrahend	The subtrahend is the number being subtracted in a subtraction problem.
	difference	The answer in a subtraction problem is called the difference.
	borrowing	The method of borrowing is used in subtraction if a digit is less than the one directly below.
1.4	**factors**	The numbers being multiplied are called factors. For example, in $3 \times 4 = 12$, both 3 and 4 are factors.
	product	The answer in a multiplication problem is called the product.
	commutative property of multiplication	The commutative property of multiplication states that the product in a multiplication problem remains the same when the order of the factors is changed.
	associative property of multiplication	The associative property of multiplication states that grouping the numbers differently does not change the product.
	multiple	The product of two whole number factors is a multiple of those numbers.
1.5	**dividend**	The number being divided by another number in a division problem is the dividend.
	divisor	The divisor is the number doing the dividing in a division problem.
	quotient	The answer in a division problem is called the quotient.
	short division	A method of dividing a number by a one-digit divisor is short division.
	remainder	The remainder is the number left over when two numbers do not divide exactly.
1.6	**long division**	The process of long division is used to divide by a number with more than one digit.
1.7	**rounding**	Rounding is used to find a number that is close to the original number, but easier to work with. Use the $\approx$ sign, which means "approximately equal to."
	estimate	An estimated answer is one that is close to the exact answer.
	front end rounding	Rounding to the highest possible place so that all the digits become 0s except the first one.
1.8	**square root**	The square root of a whole number is the number that can be multiplied by itself to produce the given number.
	perfect square	A number that is the square of a whole number is a perfect square.
	order of operations	For problems or expressions with more than one operation, the order of operations tells what to do first, second, and so on to get the correct answer. *(continued)*

1.9	**pictograph**	A graph that uses pictures or symbols to show data is a pictograph.
	bar graph	A graph that uses bars of various heights to show quantity is a bar graph.
	line graph	A graph that uses dots connected by lines to show trends is a line graph.
1.10	**indicator words**	Words in a problem that indicate the necessary operations—addition, subtraction, multiplication, or division—are indicator words.

TEST YOUR WORD POWER

See how well you have learned the vocabulary in this chapter. Answers follow the Quick Review.

1. When using **addends** you are performing
(a) division
(b) subtraction
(c) addition
(d) multiplication.

2. The **subtrahend** is the
(a) number being multiplied
(b) number being subtracted
(c) number being added
(d) answer in division.

3. A **factor** is
(a) the answer in an addition problem
(b) one of two or more numbers being added
(c) one of two or more numbers being multiplied
(d) one of two or more numbers being divided.

4. The **divisor** is
(a) the number being rounded
(b) the number being multiplied
(c) always the largest number
(d) the number doing the dividing.

5. We use **rounding** to
(a) avoid solving a problem
(b) purely guess at the answer
(c) help estimate a reasonable answer
(d) find the remainder.

6. A **perfect square** is
(a) the square of a whole number
(b) the same as square root
(c) similar to a perfect triangle
(d) used when rounding.

QUICK REVIEW

Concepts	*Examples*
1.1 Reading and Writing Whole Numbers Do not use the word *and* when writing a whole number. Commas help divide the periods or groups for ones, thousands, millions, and billions. A comma is not needed when a number has four digits or fewer.	795 is written *seven hundred ninety-five*. 9,768,002 is written *nine million, seven hundred sixty-eight thousand, two.*
1.2 Adding Whole Numbers Add from top to bottom, starting with the ones column and working left. To check, add from bottom to top.	
1.2 Commutative Property of Addition Changing the order of the addends in an addition problem does not change the sum.	$2 + 4 = 6$ $4 + 2 = 6$ By the commutative property, the sum is the same.
1.2 Associative Property of Addition Grouping the addends differently when adding does not change the sum.	$(2 + 3) + 4 = 9$ $2 + (3 + 4) = 9$ By the associative property, the sum is the same.
1.3 Subtracting Whole Numbers Subtract the subtrahend from the minuend to get the difference by borrowing when necessary. To check, add the difference to the subtrahend to get the minuend.	Problem Check 6 12 18 4 7 3 8 ← Minuend 4 0 8 9 − 6 4 9 Subtrahend + 6 4 9 4 0 8 9 Difference 4 7 3 8

Concepts	Examples

Concepts

1.4 Multiplying Whole Numbers

Use ×, • (a raised dot), or ( )( ) (two sets of parentheses) to indicate multiplication.

The numbers being multiplied are called *factors*. The multiplicand is being multiplied by the multiplier, giving the product. When the multiplier has more than one digit, partial products must be used and added to find the product.

Examples

$$\begin{array}{r} 78 \\ \times\ 24 \\ \hline 312 \\ 156 \\ \hline 1872 \end{array}$$

78   Multiplicand ⎱ Factors
× 24 Multiplier ⎰
312   Partial product
156   Partial product (move one position left)
1872   Product

1.4 Commutative Property of Multiplication

The product in a multiplication problem remains the same when the order of the factors is changed.

$$3 \times 4 = 12$$
$$4 \times 3 = 12$$

By the commutative property, the product is the same.

1.4 Associative Property of Multiplication

The grouping of factors differently when multiplying does not change the product.

$$(2 \times 3) \times 4 = 24$$
$$2 \times (3 \times 4) = 24$$

By the associative property, the product is the same.

1.5 Dividing Whole Numbers

÷ and ⟌ mean divide.
Also a —, as in $\frac{25}{5}$, means to divide the top number (dividend) by the bottom number (divisor).

Divisor → $4\overline{)88}$ ← Quotient ... Dividend

$$88 \div 4 = 22$$
Dividend | Quotient
Divisor

$$\frac{88}{4} = 22 \quad \leftarrow \text{Quotient}$$

1.7 Rounding Whole Numbers

Rules for Rounding:

Step 1 Locate the place to be rounded, and draw a line under it.

Step 2 If the next digit to the right is 5 or more, increase the underlined digit by 1. If the next digit is 4 or less, do not change the underlined digit.

Step 3 Change all digits to the right of the underlined place to 0s.

Round 726 to the nearest ten.

Next digit is 5 or more.

7<u>2</u>6

Tens place increases by 1 (2 + 1 = 3).

7<u>2</u>6 rounds to 7<u>3</u>0.

Round 1,498,586 to the nearest million.

Next digit is 4 or less.

<u>1</u>,498,586

Millions place does not change.

1,498,586 rounds to 1,000,000.

1.7 Front End Rounding

Front end rounding is rounding to the highest possible place so that all the digits become 0 except the first digit.

Round each number using front end rounding.

76 rounds to 80

348 rounds to 300

6512 rounds to 7000

23,751 rounds to 20,000

652,179 rounds to 700,000

[1.3] *Subtract.*

19. 52
−18

20. 38
−15

21. 375
−186

22. 573
−389

23. 6213
− 458

24. 5210
− 883

25. 2210
−1986

26. 99,704
−73,838

[1.4] *Multiply.*

27. 6
× 6

28. 9
× 0

29. 8×4

30. 8×8

31. $(7)(5)$

32. $(8)(7)$

33. $7 \cdot 8$

34. $9 \cdot 9$

Work each chain multiplication.

35. $2 \times 4 \times 6$

36. $9 \times 1 \times 5$

37. $4 \times 4 \times 3$

38. $2 \times 2 \times 2$

39. $(7)(0)(5)$

40. $(7)(1)(6)$

41. $6 \cdot 1 \cdot 8$

42. $7 \cdot 7 \cdot 0$

Multiply.

43. 32
× 5

44. 46
× 8

45. 58
× 9

46. 98
× 1

47. 589
× 7

48. 781
× 7

49. 1349
× 4

50. 9163
× 5

51. 8364
× 2

52. 5440
× 6

53. 93,105
× 5

54. 21,873
× 8

55. $\begin{array}{r} 45 \\ \times\ 15 \\ \hline \end{array}$

56. $\begin{array}{r} 63 \\ \times\ 28 \\ \hline \end{array}$

57. $\begin{array}{r} 98 \\ \times\ 12 \\ \hline \end{array}$

58. $\begin{array}{r} 68 \\ \times\ 75 \\ \hline \end{array}$

59. $\begin{array}{r} 472 \\ \times\ 33 \\ \hline \end{array}$

60. $\begin{array}{r} 392 \\ \times\ 77 \\ \hline \end{array}$

61. $\begin{array}{r} \times 4051 \\ \times\ 219 \\ \hline \end{array}$

62. $\begin{array}{r} 1527 \\ \times\ 328 \\ \hline \end{array}$

Find each total cost.

63. 20 CDs at $15 per CD

64. 76 subscribers at $14 per subscription

65. 318 drill bit sets at $64 per set

66. 168 welder's masks at $9 per mask

Multiply by using the shortcut for multiples of 10.

67. $\begin{array}{r} 280 \\ \times\ 50 \\ \hline \end{array}$

68. $\begin{array}{r} 340 \\ \times\ 70 \\ \hline \end{array}$

69. $\begin{array}{r} 517 \\ \times\ 400 \\ \hline \end{array}$

70. $\begin{array}{r} 637 \\ \times\ 500 \\ \hline \end{array}$

71. $\begin{array}{r} 16{,}000 \\ \times\ 8\,000 \\ \hline \end{array}$

72. $\begin{array}{r} 43{,}000 \\ \times\ 2\,100 \\ \hline \end{array}$

[1.5] *Divide. If the division is not possible, write "undefined."*

73. $15 \div 3$

74. $25 \div 5$

75. $42 \div 7$

76. $18 \div 9$

77. $\dfrac{54}{9}$

78. $\dfrac{36}{9}$

79. $\dfrac{49}{7}$

80. $\dfrac{0}{6}$

81. $\dfrac{148}{0}$

82. $\dfrac{0}{23}$

83. $\dfrac{64}{8}$

84. $\dfrac{81}{9}$

[1.5–1.6] *Divide.*

85. $3\overline{)276}$

86. $8\overline{)280}$

87. $6\overline{)26{,}532}$

88. $76\overline{)26{,}752}$

89. $2704 \div 18$

90. $15{,}525 \div 125$

[1.7] *Round as indicated.*

91. 476 to the nearest ten

92. 14,309 to the nearest hundred

93. 20,643 to the nearest thousand

94. 67,485 to the nearest ten thousand

Round each number to the nearest ten, nearest hundred, and nearest thousand. Remember to round from the original number.

	Ten	**Hundred**	**Thousand**
95. 3487	_____	_____	_____
96. 20,065	_____	_____	_____
97. 98,201	_____	_____	_____
98. 352,118	_____	_____	_____

[1.8] *Find each square root by using the Perfect Squares Table on page 73.*

99. $\sqrt{25}$　　　　**100.** $\sqrt{64}$　　　　**101.** $\sqrt{144}$　　　　**102.** $\sqrt{196}$

Identify the exponent and the base, and then simplify each expression.

103. 7^3　　　　**104.** 3^6　　　　**105.** 5^3　　　　**106.** 4^5

Simplify each expression by using the order of operations.

107. $8^2 - 10$　　　　**108.** $4^2 - 8$　　　　**109.** $2 \cdot 3^2 \div 2$

110. $9 \div 1 \cdot 2 \cdot 2 \div (11 - 2)$　　　　**111.** $\sqrt{9} + 2 \cdot 3$　　　　**112.** $6 \cdot \sqrt{16} - 6 \cdot \sqrt{9}$

[1.9] *The bar graph shows the number of parents out of 100 surveyed who nag their children about performing certain household chores.*

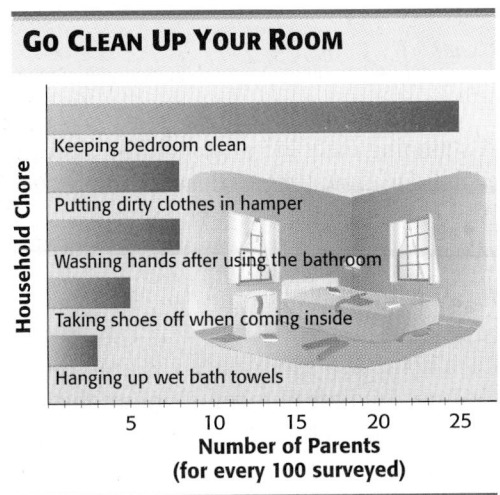

GO CLEAN UP YOUR ROOM

Household Chore (y-axis)

- Keeping bedroom clean
- Putting dirty clothes in hamper
- Washing hands after using the bathroom
- Taking shoes off when coming inside
- Hanging up wet bath towels

5   10   15   20   25

Number of Parents
(for every 100 surveyed)

Source: Opinion Research Corporation for the Soap and Detergent Association.

113. How many parents nagged their children about washing hands after using the bathroom?

114. Find the number of parents who nagged their children about taking shoes off when coming inside.

115. Which household chore was nagged about by the greatest number of parents?

116. Which household chore was nagged about by the least number of parents.

[1.10] *Solve each application problem. First use front end rounding to estimate the answer. Then find the exact answer.*

117. Find the cost of 75 picnic coolers at $10 per cooler.

Estimate:

Exact:

118. A pulley on an evaporative cooler turns 1400 revolutions per minute. How many revolutions will the pulley turn in 60 minutes?

Estimate:

Exact:

119. A McDonald's drink cooler contains 95 cups. Find the number of cups in four coolers.

Estimate:

Exact:

120. A drum contains 6000 brackets. How many brackets are in 30 drums?

Estimate:

Exact:

121. It takes 2000 hours of work to build one home. How many hours of work are needed to build 12 homes?

Estimate:

Exact:

122. A Japanese bullet train travels 80 miles in 1 hour. Find the number of miles traveled in 5 hours.

Estimate:

Exact:

123. The Houston Space Center charges $15 for each adult admission and $12 for each child. Find the total cost to admit a group of 18 adults and 26 children.

Estimate:

Exact:

124. A newspaper carrier has 62 customers who take the paper daily and 21 customers who take the paper on weekends only. A daily customer pays $16 per month and a weekend-only customer pays $7 per month. Find the total monthly collections.

Estimate:

Exact:

125. In a recent test of side-bagging lawn mowers, the most expensive model sold for $350 and the least expensive model sold for $170. Find the difference in price.

Estimate:

Exact:

126. Ruth Berry has $876 in her checking account. She writes checks for $408 and $216. How much does she have left in her account?

Estimate:

Exact:

127. A food canner uses 1 lb of pork for every 175 cans of pork and beans. How many pounds of pork are needed for 8750 cans?

Estimate:

Exact:

128. A stamping machine produces 986 license plates each hour. How long will it take to produce 32,538 license plates?

Estimate:

Exact:

129. Nitrogen sulfate is used in farming to enrich nitrogen-poor soil. If 625 lb of nitrogen sulfate are spread per acre, how many acres can be spread with 32,500 lb of nitrogen sulfate?

Estimate:

Exact:

130. Each home in a subdivision requires 180 ft of fencing. Find the number of homes that can be fenced with 5760 ft of fencing material.

Estimate:

Exact:

MIXED REVIEW EXERCISES*

Perform the indicated operations.

131.
$$\begin{array}{r} 56 \\ \times\ 5 \\ \hline \end{array}$$

132.
$$\begin{array}{r} 83 \\ \times\ 8 \\ \hline \end{array}$$

133.
$$\begin{array}{r} 207 \\ -\ 68 \\ \hline \end{array}$$

134.
$$\begin{array}{r} 835 \\ -\ 247 \\ \hline \end{array}$$

135.
$$\begin{array}{r} 662 \\ +\ 379 \\ \hline \end{array}$$

136.
$$\begin{array}{r} 789 \\ +\ 872 \\ \hline \end{array}$$

137.
$$\begin{array}{r} 38{,}140 \\ -\ 6\,078 \\ \hline \end{array}$$

138.
$$\begin{array}{r} 29{,}156 \\ -\ 4\,209 \\ \hline \end{array}$$

139. $21 \div 7$

140. $\dfrac{42}{6}$

141.
$$\begin{array}{r} 7\,218 \\ 3 \\ 18 \\ 1\,791 \\ 82{,}623 \\ +\ 1\,982 \\ \hline \end{array}$$

142.
$$\begin{array}{r} 3\,812 \\ 5 \\ 22 \\ 1\,836 \\ 75{,}134 \\ +\ 2\,369 \\ \hline \end{array}$$

143. $\dfrac{9}{0}$

144. $\dfrac{7}{1}$

145. $27{,}600 \div 4$

---

* The order of exercises in this final group does not correspond to the order in which topics occur in the chapter. This random ordering should help you prepare for the chapter test in yet another way.

146. 9220 ÷ 4   **147.** 8430
                                  × 128

148. 25,817
              ×   4

149. 34)3672   **150.** 68)14,076

151. Rewrite 376,853 in words.

152. Rewrite 408,610 in words.

153. Round 7549 to the nearest hundred.

154. Round 600,498 to the nearest thousand.

Find each square root.

155. $\sqrt{49}$

156. $\sqrt{81}$

Find each total cost.

157. 165 pairs of in-line skates at \$36 per pair

158. 84 dishwashers at \$370 per dishwasher

159. 208 baseball hats at \$11 per hat

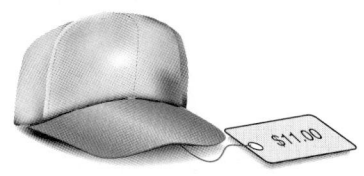

160. 607 boxes of avocados at \$26 per box

Solve each application problem.

161. There are 52 playing cards in a deck. How many cards are there in nine decks?

162. Your college bookstore receives textbooks packed 20 books per carton. How many textbooks are received in a delivery of 180 cartons?

163. "Push type" gasoline-powered lawn mowers cost \$100 less than self-propelled mowers that you walk behind. If a self-propelled mower costs \$380, find the cost of a "push-type" mower.

164. The Country Day School wants to raise \$218,450 to construct and equip a computer lab. If \$103,815 has already been raised, how much more must be raised to reach the goal?

American River Raft Rentals lists the following daily raft rental fees. Notice that there is an additional $2 launch fee payable to the park system for each raft rented. Use this information to solve Exercises 165 and 166.

AMERICAN RIVER RAFT RENTALS

Size	Rental Fee	Launch Fee
4-person	$28	$2
6-person	$38	$2
10-person	$70	$2
12-person	$75	$2
16-person	$85	$2

Source: American River Raft Rentals.

165. On a recent Tuesday the following rafts were rented: 6 4-person; 15 6-person; 10 10-person; 3 12-person; and 2 16-person. Find the total receipts, including the $2 per raft launch fee.

166. On the 4th of July the following rafts were rented: 38 4-person; 73 6-person; 58 10-person; 34 12-person; and 18 16-person. Find the total receipts, including the $2 per raft launch fee.

Chapter 1 TEST

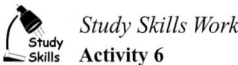 *Study Skills Workbook*
Activity 6

Write each number in words.

1. 6106

2. 85,055

3. Use digits to write four hundred twenty-six thousand, five.

Add.

4.
$$
\begin{array}{r}
762 \\
72 \\
5186 \\
+\ 2694 \\
\end{array}
$$

5.
$$
\begin{array}{r}
17,063 \\
7 \\
12 \\
1\ 505 \\
93,710 \\
+\ \ \ \ \ 333 \\
\end{array}
$$

Subtract.

6.
$$
\begin{array}{r}
6006 \\
-\ 4953 \\
\end{array}
$$

7.
$$
\begin{array}{r}
5062 \\
-\ 1978 \\
\end{array}
$$

Multiply.

8. $6 \times 8 \times 2$

9. $57 \cdot 3000$

10. $(95)(18)$

11.
$$
\begin{array}{r}
7381 \\
\times\ \ 603 \\
\end{array}
$$

Divide. If the division is not possible, write "undefined."

12. $16\overline{)112,752}$

13. $\dfrac{835}{0}$

14. $19,241 \div 42$

15. $280\overline{)44,800}$

Round as indicated.

16. 5238 to the nearest ten

17. 67,509 to the nearest thousand.

1. _____

2. _____

3. _____

4. _____

5. _____

6. _____

7. _____

8. _____

9. _____

10. _____

11. _____

12. _____

13. _____

14. _____

15. _____

16. _____

17. _____

Simplify each expression.

18. _____

18. $6^2 + 3 \times 4$ **19.** $7 \cdot \sqrt{64} - 14 \cdot 2$

19. _____

Solve each application problem. First use front end rounding to estimate the answer. Then find the exact answer.

20. *Estimate:* _____

Exact: _____

20. Amy collects the following monthly rents from the tenants in her four-plex: $485, $500, $515, and $425. After she pays expenses of $785, how much does she have left?

21. *Estimate:* _____

Exact: _____

21. Hewlett-Packard assembles 542 computer printers each day. Find the number of work days it would take to assemble 67,750 computer printers.

22. *Estimate:* _____

Exact: _____

22. Kenée Shadbourne paid $690 for tuition, $185 for books, and $68 for supplies. If this money was withdrawn from her checking account, which had a balance of $1108, find her new balance.

23. *Estimate:* _____

Exact: _____

23. An electronics manufacturer assembles 208 digital cameras each hour for 4 hours and 238 camcorders each hour for the next 4 hours. Find the number of both types of camera assembled in the 8-hour period.

24. _____

24. Explain in your own words the rules for rounding numbers. Give an example of rounding a number to the nearest ten thousand.

25. _____

25. List the six steps for solving application problems.

Multiplying and Dividing Fractions

2

Most recipes include ingredients that use common fractions and mixed numbers in their measurements. The recipe shown here will make 4 dozen (48) cookies, but suppose you wanted to make 2 dozen, 8 dozen, or even $3\frac{1}{2}$ dozen cookies? You would have to multiply or divide each of the ingredients to arrive at the proper amounts needed. In this chapter we discuss multiplication and division of fractions, which you need to know when cooking or baking. (See Exercises 25–28 in Section 2.8.)

M & M COOKIES	
$\frac{1}{3}$ c. shortening	Cream shortening, butter, sugar, and brown
$\frac{1}{3}$ c. butter, softened	sugar in mixer bowl until fluffy. Mix in egg and
$\frac{1}{2}$ c. sugar	
$\frac{1}{2}$ c. packed brown sugar	vanilla. Add sifted dry ingredients; mix well.
1 egg	Stir in nuts and candies. Drop by rounded tea-
1 tsp. vanilla extract	spoonfuls 2 inches apart onto ungreased cookie
$1\frac{3}{4}$ c. flour	sheet. Bake at 375 degrees for 8 to 10 minutes
$\frac{1}{2}$ tsp. soda	or until lightly browned. Cook slightly on
$\frac{1}{4}$ tsp. salt	
$\frac{3}{4}$ c. chopped nuts	cookie sheet. Remove to wire rack to cool com-
1 6oz. package M & M's	pletely. Yield: 4 dozen.

Source: *Recipes Are Naturally Good Eating.*

2.1 BASICS OF FRACTIONS

In Chapter 1 we discussed whole numbers. Many times, however, we find that parts of whole numbers are considered. One way to write parts of a whole is with **fractions.** Another way is with decimals, which is discussed in Chapter 4.

1 **Use a fraction to show which part of a whole is shaded.** The number $\frac{1}{8}$ is a fraction that represents 1 of 8 equal parts. Read $\frac{1}{8}$ as "one eighth."

Example 1 Identifying Fractions

Use fractions to represent the shaded portions and the unshaded portions of each figure.

(a) The figure on the left has 4 equal parts. The 1 shaded part is represented by the fraction $\frac{1}{4}$. The *un*shaded part is $\frac{3}{4}$.

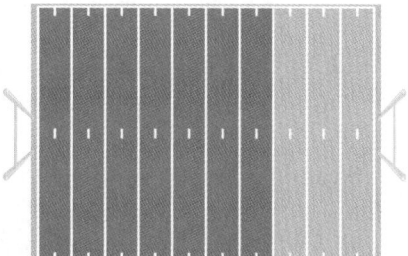

(b) The 3 shaded parts of the 10-part figure on the right are represented by the fraction $\frac{3}{10}$. The *un*shaded part is $\frac{7}{10}$.

Work Problem ① at the Side.

Fractions can be used to show more than one whole object.

① Write fractions for the shaded portions and the unshaded portions of each figure.

(a)

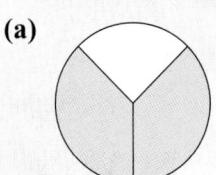

(b)

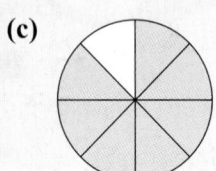

(c)

Example 2 Representing Fractions Greater Than 1

Use a fraction to represent the shaded part of each figure.

(a) $\frac{1}{4}$

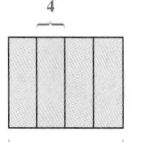

Whole object

An area equal to 5 of the $\frac{1}{4}$ parts is shaded. Write this as $\frac{5}{4}$.

(b)

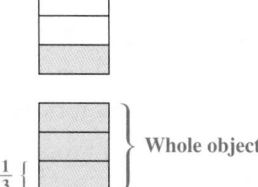

Whole object

$\frac{1}{3}$

An area equal to 4 of the $\frac{1}{3}$ parts is shaded, so $\frac{4}{3}$ is shaded.

Work Problem ② at the Side.

② Write fractions for the shaded portions of each figure.

(a) $\frac{1}{7}$

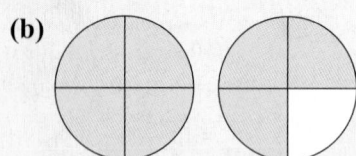

(b)

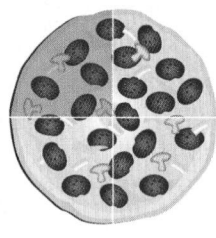

2 **Identify the numerator and denominator.** In the fraction $\frac{2}{3}$, the number 2 is the **numerator** and 3 is the **denominator.** The bar between the numerator and the denominator is the *fraction bar.*

Fraction bar → $\dfrac{2}{3}$ ← Numerator
← Denominator

Numerators and Denominators

The **denominator** of a fraction shows the number of equivalent parts in the whole, and the **numerator** shows how many parts are being considered.

NOTE

Remember that a bar, —, is one of the division symbols, and that division by 0 is undefined. A fraction with a denominator of 0 is also undefined.

Example 3 Identifying Numerators and Denominators

Identify the numerator and denominator in each fraction.

(a) $\dfrac{5}{8}$

(b) $\dfrac{9}{4}$

$\dfrac{5}{8}$ ← Numerator
← Denominator

$\dfrac{9}{4}$ ← Numerator
← Denominator

Work Problem **3** at the Side.

3 Identify proper and improper fractions. Fractions are sometimes called *proper* or *improper* fractions.

Proper and Improper Fractions

If the numerator of a fraction is *smaller* than the denominator, the fraction is a **proper fraction.**

If the numerator is *greater than or equal to* the denominator, the fraction is an **improper fraction.**

Proper Fractions	**Improper Fractions**
$\dfrac{1}{2},\ \dfrac{5}{11},\ \dfrac{35}{36}$	$\dfrac{9}{7},\ \dfrac{126}{125},\ \dfrac{7}{7}$

Example 4 Classifying Types of Fractions

(a) Identify all proper fractions in this list.

$$\frac{3}{4},\ \frac{5}{9},\ \frac{17}{5},\ \frac{9}{7},\ \frac{3}{3},\ \frac{12}{25},\ \frac{1}{9},\ \frac{5}{3}$$

Proper fractions have a numerator that is smaller than the denominator. The proper fractions are

$$\frac{3}{4}, \quad\text{← 3 is smaller than 4.}\qquad \frac{5}{9},\ \frac{12}{25},\ \text{and}\ \frac{1}{9}.$$

(b) Identify all improper fractions in the list in part (a).

Improper fractions have a numerator that is equal to or greater than the denominator. The improper fractions are

$$\frac{17}{5}, \quad\text{← 17 is greater than 5.}\qquad \frac{9}{7},\ \frac{3}{3},\ \text{and}\ \frac{5}{3}.$$

Work Problem **4** at the Side.

3 Identify the numerator and the denominator. Draw a picture with shaded parts to show each fraction. Your drawings may vary, but they should have the correct number of shaded parts.

(a) $\dfrac{2}{3}$

(b) $\dfrac{1}{4}$

(c) $\dfrac{8}{5}$

(d) $\dfrac{5}{2}$

4 From the following group of fractions:

$$\frac{2}{3},\ \frac{4}{3},\ \frac{3}{4},\ \frac{8}{8},\ \frac{3}{1},\ \frac{1}{3}$$

(a) list all proper fractions;

(b) list all improper fractions.

ANSWERS

3. (a) N: 2; D: 3

(b) N: 1; D: 4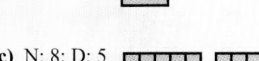

(c) N: 8; D: 5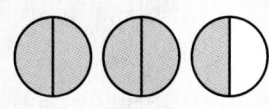

(d) N: 5; D: 2

4. (a) $\dfrac{2}{3},\dfrac{3}{4},\dfrac{1}{3}$ **(b)** $\dfrac{4}{3},\dfrac{8}{8},\dfrac{3}{1}$

❷ Write as improper fractions.

(a) $4\frac{1}{2}$

(b) $5\frac{3}{4}$

(c) $4\frac{7}{8}$

(d) $8\frac{5}{6}$

Example 1   Writing a Mixed Number as an Improper Fraction

Write $7\frac{2}{3}$ as an improper fraction (numerator greater than denominator).

Step 1 $7\frac{2}{3}$ $7 \cdot 3 = 21$ Multiply 7 and 3.

Step 2 $7\frac{2}{3}$ $21 + 2 = 23$ Add 2. The numerator is 23.

Step 3 $7\frac{2}{3} = \frac{23}{3}$ Use the same denominator.

Work Problem ❷ at the Side.

3▭ **Write improper fractions as mixed numbers.**   Write an improper fraction as a mixed number as follows.

Writing an Improper Fraction as a Mixed Number

Write an **improper fraction** as a mixed number by dividing the numerator by the denominator. The quotient is the whole number (of the mixed number), the remainder is the numerator of the fraction part, and the denominator remains unchanged.

Example 2   Writing Improper Fractions as Mixed Numbers

Write each improper fraction as a mixed number.

(a) $\frac{17}{5}$

Divide 17 by 5.

$$
\begin{array}{r}
3 \quad \leftarrow \text{Whole number part} \\
5\overline{)17} \\
\underline{15} \\
2 \quad \leftarrow \text{Remainder}
\end{array}
$$

The quotient **3** is the whole number part of the mixed number. The remainder **2** is the numerator of the fraction, and the denominator remains as **5.**

$$\frac{17}{5} = 3\frac{2}{5} \quad \begin{array}{l} \leftarrow \text{Remainder} \\ \text{Same denominator} \end{array}$$

We can verify this by using a diagram in which $\frac{17}{5}$ is shaded.

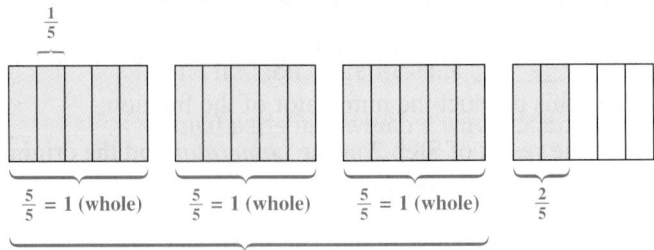

$\frac{5}{5} = 1$ (whole) $\frac{5}{5} = 1$ (whole) $\frac{5}{5} = 1$ (whole) $\frac{2}{5}$

3 wholes

Continued on Next Page

(b) $\dfrac{24}{4}$

Divide 24 by 4.

$$4\overline{)24} = 6 \quad \text{so} \quad \frac{24}{4} = 6$$

$$\begin{array}{r} 6 \\ 4\overline{)24} \\ \underline{24} \\ 0 \end{array} \leftarrow \text{No remainder}$$

NOTE

A proper fraction has a value that is less than 1, while an improper fraction has a value that is 1 or greater.

Work Problem 3 at the Side.

3 Write as whole or mixed numbers.

(a) $\dfrac{4}{3}$

(b) $\dfrac{9}{4}$

(c) $\dfrac{35}{5}$

(d) $\dfrac{85}{8}$

ANSWERS

3. (a) $1\frac{1}{3}$ **(b)** $2\frac{1}{4}$ **(c)** 7 **(d)** $10\frac{5}{8}$

55. Your classmate asks you how to change a mixed number to an improper fraction. Write a couple of sentences and give an example to show her how this is done.

56. Explain in a sentence or two how to change an improper fraction to a mixed number. Give an example to show how this is done.

Write each mixed number as an improper fraction.

57. $110\dfrac{3}{4}$

58. $218\dfrac{3}{5}$

59. $333\dfrac{1}{3}$

60. $401\dfrac{1}{2}$

61. $522\dfrac{3}{8}$

62. $622\dfrac{1}{4}$

Write each improper fraction as a whole or mixed number.

63. $\dfrac{837}{8}$

64. $\dfrac{219}{4}$

65. $\dfrac{2565}{15}$

66. $\dfrac{2915}{16}$

67. $\dfrac{3917}{32}$

68. $\dfrac{5632}{64}$

RELATING CONCEPTS (Exercises 69–74) **FOR INDIVIDUAL OR GROUP WORK**

Knowing the basics of fractions is necessary in problem solving. **Work Exercises 69–74 in order.**

69. Which of these fractions are proper fractions?

$$\frac{2}{3}, \frac{4}{5}, \frac{8}{5}, \frac{3}{4}, \frac{6}{6}, \frac{7}{10}$$

70. (a) The proper fractions in Exercise 69 are the ones where the _____ is smaller than the _____.

(b) Draw a picture with shaded parts to show each proper fraction in Exercise 69.

(c) The proper fractions in Exercise 69 are all _____ than 1.
(less/greater)

71. Which of these fractions are improper fractions?

$$\frac{5}{5}, \frac{3}{4}, \frac{10}{3}, \frac{2}{3}, \frac{5}{6}, \frac{6}{5}$$

72. (a) The improper fractions in Exercise 71 are the ones where the _____ is greater than or equal to the _____.

(b) Draw a picture with shaded parts to show each mixed number in Exercise 71.

(c) The improper fractions in Exercise 71 are all _____ than 1.
(less/greater)

73. Identify which of these fractions can be written as whole or mixed numbers and then write them as whole or mixed numbers.

$$\frac{5}{3}, \frac{7}{8}, \frac{7}{7}, \frac{11}{6}, \frac{4}{5}, \frac{15}{16}$$

74. (a) The fractions that can be written as whole or mixed numbers in Exercise 73 are _____ fractions, and their value is
(proper/improper)
always _____ 1.
(less than/greater than or equal to)

(b) Draw a picture with shaded parts to show each whole or mixed number in Exercise 73.

(c) Explain how to write an improper fraction as a whole or mixed number.

2.3 FACTORS

1〓 **Find factors of a number.**   You will recall that numbers multiplied to give a product are called **factors.** Because 2 • 5 = 10, both 2 and 5 are factors of 10. The numbers 1 and 10 are also factors of 10, because

$$1 \cdot 10 = 10.$$

The various tests for divisibility show that 1, 2, 5, and 10 are the only whole number factors of 10. The products 2 • 5 and 1 • 10 are called **factorizations** of 10.

> **NOTE**
>
> You might want to review the tests for divisibility in Section 1.5. The ones that you will want to remember are those for 2, 3, 5, and 10.

Example 1   **Using Factors**

Find all possible two-number factorizations of each number.

(a) 12

$$1 \cdot 12 = 12 \qquad 2 \cdot 6 = 12 \qquad 3 \cdot 4 = 12$$

The factors of 12 are 1, 2, 3, 4, 6, and 12.

(b) 60

$$1 \cdot 60 = 60 \qquad 2 \cdot 30 = 60$$
$$3 \cdot 20 = 60 \qquad 4 \cdot 15 = 60$$
$$5 \cdot 12 = 60 \qquad 6 \cdot 10 = 60$$

The factors of 60 are 1, 2, 3, 4, 5, 6, 10, 12, 15, 20, 30, and 60.

===================== **Work Problem ❶ at the Side.**

Composite Numbers

A number with a factor other than itself or 1 is called a **composite number.**

Example 2   **Identifying Composite Numbers**

Which of the following numbers are composite?

(a) 6
   Because 6 has factors of **2** and **3**, numbers other than 6 or 1, the number 6 is composite.

(b) 17
   The number 17 has only two factors, 17 and 1. It is not composite.

(c) 25
   A factor of 25 is **5**, so 25 is composite.

===================== **Work Problem ❷ at the Side.**

OBJECTIVES

1〓 Find factors of a number.

2〓 Identify prime numbers.

3〓 Find prime factorizations.

 Study Skills Workbook
Study Skills   **Activity 7**

❶ Find all the whole number factors of each number.

(a) 10

(b) 16

(c) 36

(d) 80

❷ Which of these numbers are composite?

2, 4, 5, 6, 8, 10, 11, 13, 19, 21, 27, 28, 33, 36, 42

❸ Which of the following are prime?

4, 7, 9, 13, 17, 19, 29, 33

2 ▭ **Identify prime numbers.** Whole numbers that are not composite are called **prime numbers,** except 0 and 1, which are neither prime nor composite.

Prime Numbers

A **prime number** is a whole number that has exactly *two different* factors, *itself* and *1.*

The number 3 is a prime number, since it can be divided evenly only by itself and 1. The number 8 is not a prime number (it is composite), since 8 can be divided evenly by 2 and 4, as well as by itself and 1.

CAUTION

A prime number has **only two** different factors, itself and 1. The number 1 is not a prime number because it does not have *two different* factors; the only factor of 1 is 1.

Example 3 **Finding Prime Numbers**

Which of the following numbers are prime?

2   5   11   15   27

The number 15 can be divided by 3 and 5, so it is not prime. Also, because 27 can be divided by 3 and 9, 27 is not prime. All the other numbers in the list are divisible by only themselves and 1, so they are prime.

Work Problem ❸ at the Side.

3 ▭ **Find prime factorizations.** For reference, here are the prime numbers smaller than 100.

2,	3,	5,	7,	11,
13,	17,	19,	23,	29,
31,	37,	41,	43,	47,
53,	59,	61,	67,	71,
73,	79,	83,	89,	97

CAUTION

All prime numbers are odd numbers except the number 2. Be careful though, because *all odd numbers are not prime numbers.* For example, 9, 15, and 21 are odd numbers, but are *not* prime numbers.

The **prime factorization** of a number can be especially useful when we are adding or subtracting fractions and need to find a common denominator or write a fraction in lowest terms.

Prime Factorization

A **prime factorization** of a number is a factorization in which every factor is a *prime number.*

Example 4   Determining the Prime Factorization

Find the prime factorization of 12.
  Try to divide 12 by the first prime, 2.

$$12 \div 2 = 6,$$

First prime

so

$$12 = 2 \cdot 6.$$

Try to divide 6 by the prime, 2.

$$6 \div 2 = 3,$$

so

$$12 = 2 \cdot \underbrace{2 \cdot 3}.$$

Factorization of 6

Because all factors are prime, the prime factorization of 12 is

$$2 \cdot 2 \cdot 3.$$

====================== **Work Problem ④ at the Side.**

④ Find the prime factorization of each number.

(a) 8

(b) 42

(c) 18

(d) 40

Example 5   Factoring by Using the Division Method

Find the prime factorization of 48.

$2\overline{)48}$ Divide 48 by 2 (first prime).

$2\overline{)24}$ Divide 24 by 2.

All
prime $2\overline{)12}$ Divide 12 by 2.
factors

$2\overline{)6}$ Divide 6 by 2.

$3\overline{)3}$ Divide 3 by 3.

1   Continue to divide until the quotient is 1.

Because all factors (divisors) are prime, the prime factorization of 48 is

$$2 \cdot 2 \cdot 2 \cdot 2 \cdot 3.$$

In Chapter 1, we wrote $2 \cdot 2 \cdot 2 \cdot 2$ as 2^4, so the prime factorization of 48 can be written, using exponents, as

$$2 \cdot 2 \cdot 2 \cdot 2 \cdot 3 = 2^4 \cdot 3.$$

====================== **Work Problem ⑤ at the Side.**

⑤ Find the prime factorization of each number. Write
the factorization with
exponents.

(a) 36

(b) 54

(c) 60

(d) 81

NOTE

When using the division method of factoring, the last quotient found is
1. The "1" is never used as a prime factor because 1 is neither prime nor
composite. Besides, 1 times any number is the number itself.

6 Write the prime factorization of each number by using exponents.

(a) 50

(b) 88

(c) 90

(d) 120

(e) 180

Example 6 Using Exponents with Prime Factorization

Find the prime factorization of 225.

All prime factors

$$3\overline{)225}$$ 225 is not divisible by 2; use 3.

$$3\overline{)75}$$ Divide 75 by 3.

$$5\overline{)25}$$ 25 is not divisible by 3; use 5.

$$5\overline{)5}$$ Divide by 5.

1   Continue to divide until the quotient is 1.

Write the prime factorization,

$$3 \cdot 3 \cdot 5 \cdot 5,$$

with exponents as

$$3^2 \cdot 5^2.$$

Work Problem 6 at the Side.

Another method of factoring is a *factor tree.*

Example 7 Factoring by Using a Factor Tree

Find the prime factorization of each number.

(a) 30
Try to divide by the first prime, 2. Write the factors under the 30. Circle the 2, since it is a prime.

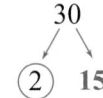

Since 15 cannot be divided evenly by 2, try the next prime, 3.

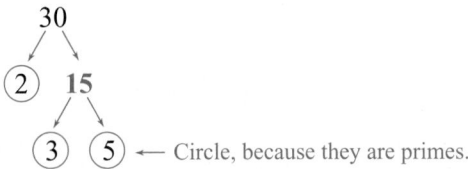

← Circle, because they are primes.

No uncircled factors remain, so the prime factorization (the circled factors) has been found.

$$30 = 2 \cdot 3 \cdot 5$$

(b) 24
Divide by 2.

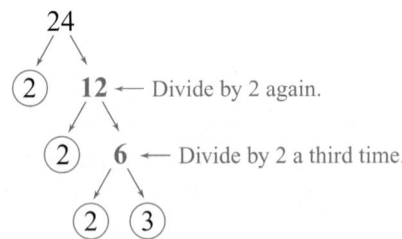

2 ← Divide by 2 again.

2   6 ← Divide by 2 a third time.

$$24 = 2 \cdot 2 \cdot 2 \cdot 3 \text{ or, using exponents, } 24 = 2^3 \cdot 3.$$

Continued on Next Page

(c) 45

Because 45 cannot be divided by 2, try 3.

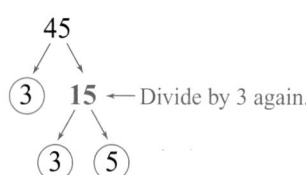

③ **15** ← Divide by 3 again.

③ ⑤

$45 = 3 \cdot 3 \cdot 5$ or, using exponents, $45 = 3^2 \cdot 5$.

NOTE

The diagrams used in Example 7 look like tree branches, and that is why this method is referred to as using a factor tree.

Work Problem 7 at the Side.

❼ Complete each factor tree and give the prime factorization.

(a) 28

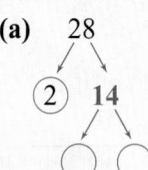

(b) 35

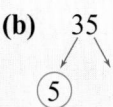

(c) 78

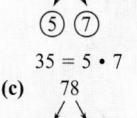

2.3 EXERCISES

 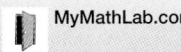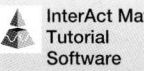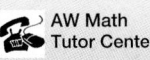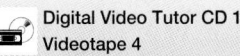

Find all the factors of each number. See Example 1.

1. 6

2. 15

3. 8

4. 28

5. 48

6. 30

7. 36

8. 20

9. 40

10. 60

11. 64

12. 84

Decide whether each number is prime *or* composite. *See Examples 2 and 3.*

13. 4

14. 6

15. 11

16. 7

17. 8

18. 9

19. 13

20. 65

21. 13

22. 17

23. 25

24. 26

25. 42

26. 47

27. 45

28. 53

Find the prime factorization of each number. Write answers with exponents when repeated factors appear. See Examples 4–7.

29. 10

30. 15

31. 20

32. 40

33. 25

34. 18

35. 36

36. 56

37. 68

38. 70

39. 72

40. 64

41. 44

42. 104

43. 100

44. 112

45. 125

46. 135

47. 180

48. 300

49. 320

50. 350

51. 360

52. 400

53. Give a definition in your own words of both a composite number and a prime number. Give three examples of each. Which whole numbers are neither prime nor composite?

54. With the exception of the number 2, all prime numbers are odd numbers. Nevertheless, all odd numbers are not prime numbers. Explain why these statements are true.

55. Explain the difference between finding all possible factors of 24 and finding the prime factorization of 24.

56. Use the division method to find the prime factorization of 36. Can you divide by 3s before you divide by 2s? Does the order of division change the answers?

Find the prime factorization of each number. Write answers using exponents.

57. 320   **58.** 640   **59.** 960   **60.** 1125

61. 1560   **62.** 2000   **63.** 1260   **64.** 2200

RELATING CONCEPTS (Exercises 65–70)   **FOR INDIVIDUAL OR GROUP WORK**

An understanding of factors and factorization will be needed to solve fraction problems.
Work Exercises 65–70 in order.

65. A prime number is a whole number that has exactly two different factors, itself and 1. List all prime numbers smaller than 50.

66. Explain how to determine which of the numbers in Exercise 65 are prime.

67. With the exception of the number 2, all prime numbers are odd numbers. Why is this true?

68. Can a multiple of a prime number (for example, 6, 9, 12, and 15 are multiples of 3) be prime? Explain.

69. Find the prime factorization of 2100. Do not use exponents in your answer.

70. Write the answer to Exercise 69 using exponents for repeated factors.

2.4 WRITING A FRACTION IN LOWEST TERMS

When working problems involving fractions, we must often compare two fractions to determine whether they represent the same portion of a whole. Consider the following diagrams.

$\frac{5}{6}$ is shaded. 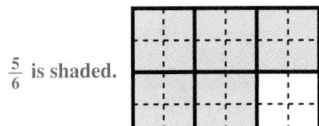    $\frac{20}{24}$ is shaded.

The figures show areas that are $\frac{5}{6}$ shaded and $\frac{20}{24}$ shaded. Because the shaded areas are equivalent (the same portion), the fractions $\frac{5}{6}$ and $\frac{20}{24}$ are **equivalent fractions.**

$$\frac{5}{6} = \frac{20}{24}$$

Because the numbers 20 and 24 both have 4 as a factor, 4 is called a **common factor** of the numbers. Other common factors of 20 and 24 are 1 and 2.

=== Work Problem ❶ at the Side.

1 ▭ **Tell whether a fraction is written in lowest terms.**   The fraction $\frac{5}{6}$ is written in *lowest terms* because the numerator and denominator have no common factor other than 1. However, the fraction $\frac{20}{24}$ is *not* in lowest terms because its numerator and denominator have common factors of 4, 2, and 1.

Writing a Fraction in Lowest Terms

A fraction is written in **lowest terms** when the numerator and denominator have no common factor other than 1.

Example 1   Understanding Lowest Terms

Are the following fractions in lowest terms?

(a) $\frac{3}{8}$

The numerator and denominator have no common factor other than 1, so the fraction is in lowest terms.

(b) $\frac{21}{36}$

The numerator and denominator have a common factor of 3, so the fraction is not in lowest terms.

=== Work Problem ❷ at the Side.

2 ▭ **Write a fraction in lowest terms using common factors.**   There are two common methods for writing a fraction in lowest terms. These methods are shown in the next examples. The first method works best when the numerator and denominator are small numbers.

❶ Decide whether the given factor is a common factor of both numbers.

(a) 8, 18; **2**

(b) 32, 64; **8**

(c) 16, 34; **16**

(d) 56, 73; **1**

❷ Are the following fractions in lowest terms?

(a) $\frac{3}{4}$

(b) $\frac{5}{15}$

(c) $\frac{9}{15}$

(d) $\frac{17}{46}$

ANSWERS
1. (a) yes   **(b)** yes   **(c)** no   **(d)** yes
2. (a) yes   **(b)** no   **(c)** no   **(d)** yes

❸ Write in lowest terms.

(a) $\dfrac{6}{12}$

(b) $\dfrac{6}{8}$

(c) $\dfrac{48}{60}$

(d) $\dfrac{30}{80}$

(e) $\dfrac{16}{40}$

Example 2 Writing Fractions in Lowest Terms

Write each fraction in lowest terms.

(a) $\dfrac{20}{24}$

The largest common factor of 20 and 24 is 4. Divide both numerator and denominator by 4.

$$\frac{20}{24} = \frac{20 \div 4}{24 \div 4} = \frac{5}{6}$$

(b) $\dfrac{30}{50} = \dfrac{30 \div 10}{50 \div 10} = \dfrac{3}{5}$ Divide both numerator and denominator by 10.

(c) $\dfrac{24}{42} = \dfrac{24 \div 6}{42 \div 6} = \dfrac{4}{7}$ Divide both numerator and denominator by 6.

(d) $\dfrac{60}{72}$

Suppose we made an error and thought 4 was the largest common factor of 60 and 72. Dividing by 4 would give

$$\frac{60}{72} = \frac{60 \div 4}{72 \div 4} = \frac{15}{18}.$$

But $\frac{15}{18}$ is not in lowest terms, because 15 and 18 have a common factor of 3. So we divide by 3.

$$\frac{15}{18} = \frac{15 \div 3}{18 \div 3} = \frac{5}{6}$$

The fraction $\frac{60}{72}$ could have been written in lowest terms in one step by dividing by 12, the largest common factor of 60 and 72.

$$\frac{60}{72} = \frac{60 \div 12}{72 \div 12} = \frac{5}{6}$$

NOTE

Dividing the numerator and denominator by the same number results in an equivalent fraction.

In Example 2, we wrote fractions in lowest terms by dividing by a common factor. This method is summarized in the following steps.

The Method of Dividing by a Common Factor

Step 1   Find the largest number that will divide evenly into both the numerator and denominator. This number is a *common factor*.

Step 2   *Divide* both numerator and denominator by the common factor.

Step 3   *Check* to see whether the new fraction has any common factors (besides 1). If it does, repeat Steps 2 and 3. If the only common factor is 1, the fraction is in lowest terms.

Work Problem ❸ at the Side.

ANSWERS

3. (a) $\dfrac{1}{2}$ (b) $\dfrac{3}{4}$ (c) $\dfrac{4}{5}$ (d) $\dfrac{3}{8}$ (e) $\dfrac{2}{5}$

3━━ **Write a fraction in lowest terms using prime factors.** The method of writing a fraction in lowest terms by division works well for fractions with small numerators and denominators. For larger numbers, it is common to use the method of *prime factors,* which is shown in the next example.

Example 3 **Using Prime Factors**

Write each fraction in lowest terms.

(a) $\dfrac{24}{42}$

Write the prime factorization of both numerator and denominator. See **Section 2.3** for help.

$$\frac{24}{42} = \frac{2 \cdot 2 \cdot 2 \cdot 3}{2 \cdot 3 \cdot 7}$$

Just as with the other method, divide both numerator and denominator by any common factors. Write a **1** by each factor that has been divided.

$$\frac{24}{42} = \frac{\overset{1}{\cancel{2}} \cdot 2 \cdot 2 \cdot \overset{1}{\cancel{3}}}{\underset{1}{\cancel{2}} \cdot \underset{1}{\cancel{3}} \cdot 7}$$

Multiply the remaining factors in both numerator and denominator.

$$\frac{24}{42} = \frac{1 \cdot 2 \cdot 2 \cdot 1}{1 \cdot 1 \cdot 7} = \frac{4}{7}$$

Finally, $\frac{24}{42}$ written in lowest terms is $\frac{4}{7}$.

(b) $\dfrac{162}{54}$

Write the prime factorization of both numerator and denominator.

$$\frac{162}{54} = \frac{2 \cdot 3 \cdot 3 \cdot 3 \cdot 3}{2 \cdot 3 \cdot 3 \cdot 3}$$

Now divide by the common factors. ***Do not forget to write the 1s.***

$$\frac{162}{54} = \frac{\overset{1}{\cancel{2}} \cdot \overset{1}{\cancel{3}} \cdot \overset{1}{\cancel{3}} \cdot \overset{1}{\cancel{3}} \cdot 3}{\underset{1}{\cancel{2}} \cdot \underset{1}{\cancel{3}} \cdot \underset{1}{\cancel{3}} \cdot \underset{1}{\cancel{3}}}$$

$$= \frac{1 \cdot 1 \cdot 1 \cdot 1 \cdot 3}{1 \cdot 1 \cdot 1 \cdot 1} = \frac{3}{1} = 3$$

(c) $\dfrac{18}{90}$

$$\frac{18}{90} = \frac{\overset{1}{\cancel{2}} \cdot \overset{1}{\cancel{3}} \cdot \overset{1}{\cancel{3}}}{\underset{1}{\cancel{2}} \cdot \underset{1}{\cancel{3}} \cdot \underset{1}{\cancel{3}} \cdot 5} = \frac{1 \cdot 1 \cdot 1}{1 \cdot 1 \cdot 1 \cdot 5} = \frac{1}{5}$$

CAUTION

In Example 3(c), all factors of the numerator were divided. But $1 \cdot 1 \cdot 1$ is still 1, so the final answer is $\frac{1}{5}$ (*not* 5).

❹ Use the method of prime factors to write each fraction in lowest terms.

(a) $\dfrac{12}{36}$

(b) $\dfrac{32}{56}$

(c) $\dfrac{74}{111}$

(d) $\dfrac{124}{340}$

❺ Is each pair of fractions equivalent?

(a) $\dfrac{24}{48}$ and $\dfrac{36}{72}$

(b) $\dfrac{45}{60}$ and $\dfrac{50}{75}$

(c) $\dfrac{20}{4}$ and $\dfrac{110}{22}$

(d) $\dfrac{120}{220}$ and $\dfrac{180}{320}$

4. (a) $\dfrac{1}{3}$ **(b)** $\dfrac{4}{7}$ **(c)** $\dfrac{2}{3}$ **(d)** $\dfrac{31}{85}$
5. (a) equivalent   **(b)** not equivalent
(c) equivalent   **(d)** not equivalent

In Example 3, we wrote fractions in lowest terms using prime factors. This method is summarized as follows.

The Method of Prime Factors

Step 1   Write the *prime factorization* of both numerator and denominator.

Step 2   *Divide* both numerator and denominator by any common factors.

Step 3   *Multiply* the remaining factors in the numerator and denominator.

Work Problem ❹ at the Side.

4 ▭   **Tell whether two fractions are equivalent.**   The next example shows how to tell whether two fractions are equivalent.

Example 4   **Determining whether Two Fractions Are Equivalent**

Determine whether each pair of fractions is equivalent.

(a) $\dfrac{16}{48}$ and $\dfrac{24}{72}$

Use the method of prime factors to write each fraction in lowest terms.

$$\frac{16}{48} = \frac{\overset{1}{\cancel{2}} \cdot \overset{1}{\cancel{2}} \cdot \overset{1}{\cancel{2}} \cdot \overset{1}{\cancel{2}}}{\underset{1}{\cancel{2}} \cdot \underset{1}{\cancel{2}} \cdot \underset{1}{\cancel{2}} \cdot \underset{1}{\cancel{2}} \cdot 3} = \frac{1 \cdot 1 \cdot 1 \cdot 1}{1 \cdot 1 \cdot 1 \cdot 1 \cdot 3} = \frac{1}{3}$$

$$\frac{24}{72} = \frac{\overset{1}{\cancel{2}} \cdot \overset{1}{\cancel{2}} \cdot \overset{1}{\cancel{2}} \cdot \overset{1}{\cancel{3}}}{\underset{1}{\cancel{2}} \cdot \underset{1}{\cancel{2}} \cdot \underset{1}{\cancel{2}} \cdot \underset{1}{\cancel{3}} \cdot 3} = \frac{1 \cdot 1 \cdot 1 \cdot 1}{1 \cdot 1 \cdot 1 \cdot 1 \cdot 3} = \frac{1}{3}$$

Equivalent $\left(\dfrac{1}{3} = \dfrac{1}{3}\right)$

(b) $\dfrac{32}{52}$ and $\dfrac{64}{112}$

$$\frac{32}{52} = \frac{\overset{1}{\cancel{2}} \cdot \overset{1}{\cancel{2}} \cdot 2 \cdot 2 \cdot 2}{\underset{1}{\cancel{2}} \cdot \underset{1}{\cancel{2}} \cdot 13} = \frac{2 \cdot 2 \cdot 2}{1 \cdot 1 \cdot 13} = \frac{8}{13}$$

$$\frac{64}{112} = \frac{\overset{1}{\cancel{2}} \cdot \overset{1}{\cancel{2}} \cdot \overset{1}{\cancel{2}} \cdot \overset{1}{\cancel{2}} \cdot 2 \cdot 2}{\underset{1}{\cancel{2}} \cdot \underset{1}{\cancel{2}} \cdot \underset{1}{\cancel{2}} \cdot \underset{1}{\cancel{2}} \cdot 7} = \frac{1 \cdot 1 \cdot 1 \cdot 1 \cdot 2 \cdot 2}{1 \cdot 1 \cdot 1 \cdot 1 \cdot 7} = \frac{4}{7}$$

Not equivalent $\left(\dfrac{8}{13} \neq \dfrac{4}{7}\right)$

(c) $\dfrac{75}{15}$ and $\dfrac{60}{12}$

$$\frac{75}{15} = \frac{\overset{1}{\cancel{3}} \cdot \overset{1}{\cancel{5}} \cdot 5}{\underset{1}{\cancel{3}} \cdot \underset{1}{\cancel{5}}} = \frac{1 \cdot 1 \cdot 5}{1 \cdot 1} = 5$$

$$\frac{60}{12} = \frac{\overset{1}{\cancel{2}} \cdot \overset{1}{\cancel{2}} \cdot \overset{1}{\cancel{3}} \cdot 5}{\underset{1}{\cancel{2}} \cdot \underset{1}{\cancel{2}} \cdot 3} = \frac{1 \cdot 1 \cdot 1 \cdot 5}{1 \cdot 1 \cdot 1} = 5$$

Equivalent $(5 = 5)$

Work Problem ❺ at the Side.

2.4 EXERCISES

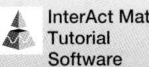

Put a ✓ mark in the blank if the number at the left is divisible by the number at the top. Put an X in the blank if the number is not divisible by the number at the top. (For help, see Section 1.5.)

	2	3	5	10			2	3	5	10
1. 30	___	___	___	___		**2.** 60	___	___	___	___
3. 48	___	___	___	___		**4.** 36	___	___	___	___
5. 160	___	___	___	___		**6.** 175	___	___	___	___
7. 138	___	___	___	___		**8.** 120	___	___	___	___

Write each fraction in lowest terms. See Example 2.

9. $\dfrac{9}{12}$
10. $\dfrac{5}{10}$
11. $\dfrac{16}{24}$
12. $\dfrac{4}{12}$

13. $\dfrac{25}{40}$
14. $\dfrac{32}{48}$
15. $\dfrac{36}{42}$
16. $\dfrac{22}{33}$

17. $\dfrac{63}{70}$
18. $\dfrac{21}{35}$
19. $\dfrac{180}{210}$
20. $\dfrac{72}{80}$

21. $\dfrac{36}{63}$
22. $\dfrac{73}{146}$
23. $\dfrac{12}{600}$
24. $\dfrac{8}{400}$

25. $\dfrac{96}{132}$
26. $\dfrac{165}{180}$
27. $\dfrac{60}{108}$
28. $\dfrac{112}{128}$

Write the numerator and denominator of each fraction as a product of prime factors and divide by the common factors. Then write the fraction in lowest terms. See Example 3.

29. $\dfrac{18}{24}$
30. $\dfrac{16}{64}$
31. $\dfrac{35}{40}$

32. $\dfrac{20}{32}$
33. $\dfrac{90}{180}$
34. $\dfrac{36}{48}$

35. $\dfrac{36}{12}$

36. $\dfrac{192}{48}$

37. $\dfrac{72}{225}$

38. $\dfrac{65}{234}$

Tell whether each pair of fractions is equivalent or not equivalent. See Example 4.

39. $\dfrac{2}{4}$ and $\dfrac{16}{32}$

40. $\dfrac{4}{10}$ and $\dfrac{20}{50}$

41. $\dfrac{10}{24}$ and $\dfrac{12}{30}$

42. $\dfrac{22}{32}$ and $\dfrac{32}{48}$

43. $\dfrac{15}{24}$ and $\dfrac{35}{52}$

44. $\dfrac{21}{33}$ and $\dfrac{9}{12}$

45. $\dfrac{14}{16}$ and $\dfrac{35}{40}$

46. $\dfrac{27}{90}$ and $\dfrac{24}{80}$

47. $\dfrac{48}{6}$ and $\dfrac{72}{8}$

48. $\dfrac{45}{15}$ and $\dfrac{96}{32}$

49. $\dfrac{25}{30}$ and $\dfrac{65}{78}$

50. $\dfrac{24}{72}$ and $\dfrac{30}{90}$

51. What does it mean when a fraction is expressed in lowest terms? Give three examples.

52. Explain what equivalent fractions are and give an example of a pair of equivalent fractions. Show that they are equivalent.

Write each fraction in lowest terms.

53. $\dfrac{224}{256}$

54. $\dfrac{363}{528}$

55. $\dfrac{356}{178}$

56. $\dfrac{525}{105}$

2.5 MULTIPLYING FRACTIONS

1 Multiply fractions. Suppose that you give $\frac{1}{2}$ of your Energy Bar to your kickboxing partner Jennifer. Then Jennifer gives $\frac{1}{2}$ of her share to Tony. How much of the Energy Bar does Tony get to eat?

Start with a sketch showing the Energy Bar cut in half (2 equal pieces).

$\frac{1}{2}$ to Jennifer

Next, take $\frac{1}{2}$ of the shaded area. (Here we are dividing $\frac{1}{2}$ into 2 equal parts and shading one darker than the other.)

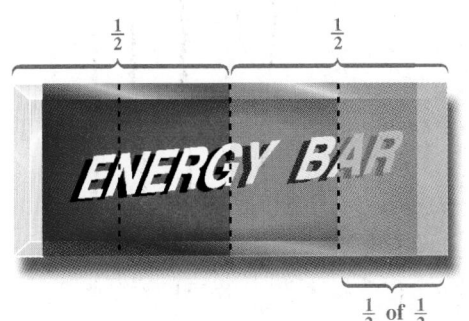

$\frac{1}{2}$ of $\frac{1}{2}$
to Tony

The sketch shows that Tony gets $\frac{1}{4}$ of the Energy Bar.

Tony gets $\frac{1}{2}$ of $\frac{1}{2}$ of the Energy Bar. When used between two fractions, the word **of** tells us to multiply.

$$\frac{1}{2} \text{ of } \frac{1}{2} \quad \text{means} \quad \frac{1}{2} \cdot \frac{1}{2}$$

Tony's share of the Energy Bar is

$$\frac{1}{2} \cdot \frac{1}{2} = \frac{1}{4}.$$

Work Problem ❶ at the Side.

The rule for multiplying fractions follows.

Multiplying Fractions

Multiply two fractions by multiplying the numerators and multiplying the denominators.

OBJECTIVES

1 Multiply fractions.

2 Use a multiplication shortcut.

3 Multiply a fraction and a whole number.

4 Find the area of a rectangle.

❶ Use these figures to find $\frac{1}{4}$ of $\frac{1}{2}$.

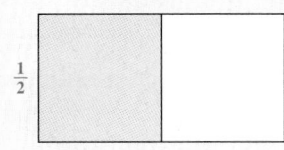

$\frac{1}{2}$

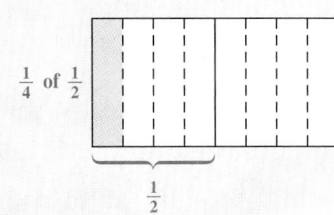

$\frac{1}{4}$ of $\frac{1}{2}$

$\frac{1}{2}$

❷ Multiply. Write answers in lowest terms.

(a) $\dfrac{3}{4} \cdot \dfrac{1}{2}$

(b) $\dfrac{1}{3} \cdot \dfrac{3}{5}$

(c) $\dfrac{1}{8} \cdot \dfrac{5}{6} \cdot \dfrac{1}{2}$

(d) $\dfrac{1}{2} \cdot \dfrac{3}{4} \cdot \dfrac{3}{8}$

Use this rule to find the product of $\frac{2}{3}$ and $\frac{1}{3}$ (multiply $\frac{2}{3}$ by $\frac{1}{3}$).

$$\dfrac{2}{3} \cdot \dfrac{1}{3} = \dfrac{2 \cdot 1}{3 \cdot 3} \quad \begin{array}{l}\longleftarrow \text{Multiply numerators.} \\ \longleftarrow \text{Multiply denominators.}\end{array}$$

$$= \dfrac{2}{9}$$

Finish multiplying.

$$\dfrac{2}{3} \cdot \dfrac{1}{3} = \dfrac{2 \cdot 1}{3 \cdot 3} = \dfrac{2}{9} \quad \begin{array}{l}\longleftarrow 2 \cdot 1 = 2 \\ \longleftarrow 3 \cdot 3 = 9\end{array}$$

Check that the final result is in lowest terms. $\frac{2}{9}$ is in lowest terms because 2 and 9 have no common factor other than 1.

Example 1 **Multiplying Fractions**

Multiply. Write answers in lowest terms.

(a) $\dfrac{5}{8} \cdot \dfrac{3}{4}$

Multiply the numerators and multiply the denominators.

$$\dfrac{5}{8} \cdot \dfrac{3}{4} = \dfrac{5 \cdot 3}{8 \cdot 4} = \dfrac{15}{32}$$

Notice that 15 and 32 have no common factors other than 1, so the answer is in lowest terms.

(b) $\dfrac{4}{7} \cdot \dfrac{2}{5}$

$$\dfrac{4}{7} \cdot \dfrac{2}{5} = \dfrac{4 \cdot 2}{7 \cdot 5} = \dfrac{8}{35}$$

(c) $\dfrac{5}{8} \cdot \dfrac{3}{4} \cdot \dfrac{1}{2}$

$$\dfrac{5}{8} \cdot \dfrac{3}{4} \cdot \dfrac{1}{2} = \dfrac{5 \cdot 3 \cdot 1}{8 \cdot 4 \cdot 2} = \dfrac{15}{64}$$

Work Problem ❷ at the Side.

2 ▭ **Use a multiplication shortcut.** A multiplication shortcut that can be used with fractions is shown in Example 2.

Example 2 **Using the Multiplication Shortcut**

Multiply $\frac{5}{6}$ and $\frac{9}{10}$. Write the answer in lowest terms.

$$\dfrac{5}{6} \cdot \dfrac{9}{10} = \dfrac{5 \cdot 9}{6 \cdot 10} = \dfrac{45}{60} \quad \text{Not in lowest terms}$$

The numerator and denominator have a common factor other than 1, so write the prime factorization of each number.

$$\dfrac{5}{6} \cdot \dfrac{9}{10} = \dfrac{5 \cdot 9}{6 \cdot 10} = \dfrac{5 \cdot 3 \cdot 3}{2 \cdot 3 \cdot 2 \cdot 5}$$

Continued on Next Page

Answers

2. (a) $\dfrac{3}{8}$ (b) $\dfrac{1}{5}$ (c) $\dfrac{5}{96}$ (d) $\dfrac{9}{64}$

Next, divide by the common factors of 5 and 3.

$$\frac{5}{6} \cdot \frac{9}{10} = \frac{5 \cdot 9}{6 \cdot 10} = \frac{\overset{1}{\cancel{5}} \cdot \overset{1}{\cancel{3}} \cdot 3}{2 \cdot \underset{1}{\cancel{3}} \cdot 2 \cdot \underset{1}{\cancel{5}}}$$

Finally, multiply the remaining factors in the numerator and in the denominator.

$$\frac{5}{6} \cdot \frac{9}{10} = \frac{1 \cdot 1 \cdot 3}{2 \cdot 1 \cdot 2 \cdot 1} = \frac{3}{4} \qquad \text{Lowest terms}$$

As a shortcut, instead of writing the prime factorization of each number, find the product of $\frac{5}{6}$ and $\frac{9}{10}$ as follows.

First, divide both 5 and 10 by 5.

$$\frac{\overset{1}{\cancel{5}}}{6} \cdot \frac{9}{\underset{2}{\cancel{10}}}$$

Next, divide both 6 and 9 by 3.

$$\frac{\overset{1}{\cancel{5}}}{\underset{2}{\cancel{6}}} \cdot \frac{\overset{3}{\cancel{9}}}{\underset{2}{\cancel{10}}}$$

Finally, multiply.

$$\frac{1 \cdot 3}{2 \cdot 2} \quad \frac{3}{4}$$

CAUTION

When using the multiplication shortcut, you are dividing a numerator and a denominator. Be certain that you divide a numerator and a denominator *by the same number.* If you do all possible divisions, your answer will be in lowest terms.

Example 3 **Using the Multiplication Shortcut**

Use the multiplication shortcut to find each product. Write the answers in lowest terms and as mixed numbers where possible.

(a) $\dfrac{6}{11} \cdot \dfrac{7}{8}$

Divide both 6 and 8 by 2. Next, multiply.

$$\frac{\overset{3}{\cancel{6}}}{11} \cdot \frac{7}{\underset{4}{\cancel{8}}} = \frac{3 \cdot 7}{11 \cdot 4} = \frac{21}{44} \qquad \text{Lowest terms}$$

(b) $\dfrac{7}{10} \cdot \dfrac{20}{21}$

Divide 7 and 21 by 7, and divide 10 and 20 by 10.

$$\frac{\overset{1}{\cancel{7}}}{\underset{1}{\cancel{10}}} \cdot \frac{\overset{2}{\cancel{20}}}{\underset{3}{\cancel{21}}} = \frac{1 \cdot 2}{1 \cdot 3} = \frac{2}{3} \qquad \text{Lowest terms}$$

Continued on Next Page

❸ Use the multiplication shortcut to find each product.

(a) $\dfrac{3}{4} \cdot \dfrac{2}{3}$

(b) $\dfrac{6}{11} \cdot \dfrac{33}{21}$

(c) $\dfrac{20}{4} \cdot \dfrac{3}{40} \cdot \dfrac{1}{3}$

(d) $\dfrac{18}{17} \cdot \dfrac{1}{36} \cdot \dfrac{2}{3}$

(c) $\dfrac{35}{12} \cdot \dfrac{32}{25}$

$$\dfrac{\overset{7}{\cancel{35}}}{\underset{3}{\cancel{12}}} \cdot \dfrac{\overset{8}{\cancel{32}}}{\underset{5}{\cancel{25}}} = \dfrac{7 \cdot 8}{3 \cdot 5} = \dfrac{56}{15} \quad \text{or} \quad 3\dfrac{11}{15} \quad \text{Mixed number}$$

(d) $\dfrac{2}{3} \cdot \dfrac{8}{15} \cdot \dfrac{3}{4}$

$$\dfrac{\overset{1}{\cancel{2}}}{\underset{1}{\cancel{3}}} \cdot \dfrac{\overset{4}{\cancel{8}}}{15} \cdot \dfrac{\overset{1}{\cancel{3}}}{\underset{1}{\cancel{4}}} = \dfrac{1 \cdot 4 \cdot 1}{1 \cdot 15 \cdot 1} = \dfrac{4}{15} \quad \text{Lowest terms}$$

This shortcut is especially helpful when the fractions involve large numbers.

NOTE

There is no specific order that must be used when dividing numerators and denominators as long as both numerator and denominator are divided by the same number.

Work Problem ❸ at the Side.

3 ▰ **Multiply a fraction and a whole number.** The rule for multiplying a fraction and a whole number follows.

Multiplying a Whole Number and a Fraction

Multiply a whole number and a fraction by writing the whole number as a fraction with a denominator of 1.

For example, write the whole numbers 8, 10, and 25 as follows.

$$8 = \dfrac{8}{1}, \quad 10 = \dfrac{10}{1}, \quad \text{and} \quad 25 = \dfrac{25}{1}$$

Example 4 Multiplying by Whole Numbers

Multiply. Write answers in lowest terms and as whole numbers where possible.

(a) $8 \cdot \dfrac{3}{4}$

Write 8 as $\frac{8}{1}$ and multiply.

$$8 \cdot \dfrac{3}{4} = \dfrac{\overset{2}{\cancel{8}}}{1} \cdot \dfrac{3}{\underset{1}{\cancel{4}}} = \dfrac{2 \cdot 3}{1 \cdot 1} = \dfrac{6}{1} = 6$$

Continued on Next Page

(b) $12 \cdot \dfrac{5}{6}$

$$12 \cdot \frac{5}{6} = \frac{\overset{2}{\cancel{12}}}{1} \cdot \frac{5}{\cancel{6}} = \frac{2 \cdot 5}{1 \cdot 1} = \frac{10}{1} = 10$$

Work Problem ❹ at the Side.

4 ▭ **Find the area of a rectangle.** To find the area of a rectangle (the amount of surface inside the rectangle), use the following formula.

Area of a Rectangle

The area of a rectangle is equal to the length multiplied by the width.

$$\textbf{area} = \textbf{length} \cdot \textbf{width}$$

For example, the rectangle shown here has an area of 12 square feet (ft²).

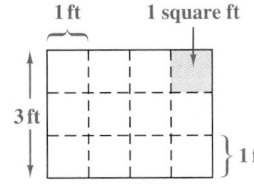

Area = length • width
Area = 4 ft • 3 ft
Area = 12 ft²

(See Section 8.3 for more information on area.)

Example 5 **Applying Fraction Skills**

Find the area of each rectangle.

(a) Find the area of this floor tile.

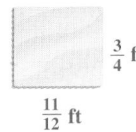

$\frac{3}{4}$ ft

$\frac{11}{12}$ ft

$$\text{Area} = \text{length} \cdot \text{width}$$
$$\text{Area} = \frac{11}{12} \cdot \frac{3}{4}$$
$$= \frac{11}{\underset{4}{\cancel{12}}} \cdot \frac{\overset{1}{\cancel{3}}}{4} \qquad \text{Divide numerator and denominator.}$$
$$= \frac{11}{16} \text{ ft}^2$$

Continued on Next Page

❹ Multiply. Write answers in lowest terms.

(a) $5 \cdot \dfrac{1}{5}$

(b) $\dfrac{5}{3} \cdot 12 \cdot \dfrac{3}{4}$

(c) $40 \cdot \dfrac{3}{10}$

(d) $\dfrac{3}{25} \cdot \dfrac{5}{11} \cdot 99$

ANSWERS

4. **(a)** 1 **(b)** 15 **(c)** 12 **(d)** $\dfrac{27}{5}$ or $5\dfrac{2}{5}$

❺ Find the area of each rectangle.

(a)

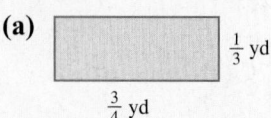

$\frac{1}{3}$ yd

$\frac{3}{4}$ yd

(b)

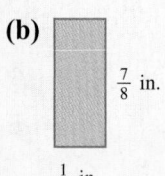

$\frac{7}{8}$ in.

$\frac{1}{3}$ in.

(c) a nature area that is $\frac{7}{5}$ miles by $\frac{5}{8}$ miles

$\frac{7}{5}$ mi

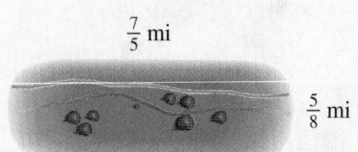

$\frac{5}{8}$ mi

(b) Find the area of this wallpaper sample.

$\frac{7}{9}$ yd

$\frac{3}{14}$ yd

Multiply the length and width.

$$\text{area} = \frac{7}{9} \cdot \frac{3}{14}$$

$$= \frac{\overset{1}{\cancel{7}}}{\underset{3}{\cancel{9}}} \cdot \frac{\overset{1}{\cancel{3}}}{\underset{2}{\cancel{14}}} \qquad \text{Divide numerator and denominator.}$$

$$= \frac{1}{6} \text{ square yard } \left(\text{yd}^2\right)$$

Work Problem ❺ at the Side.

2.5 EXERCISES

FOR EXTRA HELP

 Student's Solutions Manual

 MyMathLab.com

 InterAct Math Tutorial Software

 AW Math Tutor Center

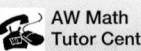 www.mathxl.com

 Digital Video Tutor CD 2 Videotape 5

Multiply. Write answers in lowest terms. See Examples 1–3.

1. $\dfrac{3}{4} \cdot \dfrac{1}{2}$

2. $\dfrac{1}{2} \cdot \dfrac{2}{3}$

3. $\dfrac{2}{3} \cdot \dfrac{5}{8}$

4. $\dfrac{2}{5} \cdot \dfrac{3}{4}$

5. $\dfrac{8}{5} \cdot \dfrac{15}{32}$

6. $\dfrac{4}{9} \cdot \dfrac{12}{7}$

7. $\dfrac{2}{3} \cdot \dfrac{7}{12} \cdot \dfrac{9}{14}$

8. $\dfrac{7}{8} \cdot \dfrac{16}{21} \cdot \dfrac{1}{2}$

9. $\dfrac{3}{4} \cdot \dfrac{5}{6} \cdot \dfrac{2}{3}$

10. $\dfrac{2}{5} \cdot \dfrac{3}{8} \cdot \dfrac{2}{3}$

11. $\dfrac{9}{22} \cdot \dfrac{11}{16}$

12. $\dfrac{5}{12} \cdot \dfrac{7}{10}$

13. $\dfrac{5}{8} \cdot \dfrac{16}{25}$

14. $\dfrac{6}{11} \cdot \dfrac{22}{15}$

15. $\dfrac{14}{25} \cdot \dfrac{65}{48} \cdot \dfrac{15}{28}$

16. $\dfrac{35}{64} \cdot \dfrac{32}{15} \cdot \dfrac{27}{72}$

17. $\dfrac{16}{25} \cdot \dfrac{35}{32} \cdot \dfrac{15}{64}$

18. $\dfrac{39}{42} \cdot \dfrac{7}{13} \cdot \dfrac{7}{24}$

Multiply. Write answers in lowest terms and as whole or mixed numbers where possible. See Example 4.

19. $6 \cdot \dfrac{5}{6}$

20. $40 \cdot \dfrac{3}{4}$

21. $\dfrac{5}{8} \cdot 64$

22. $\dfrac{3}{5} \cdot 45$

23. $28 \cdot \dfrac{3}{4}$

24. $30 \cdot \dfrac{3}{10}$

25. $36 \cdot \dfrac{5}{8} \cdot \dfrac{9}{15}$

26. $35 \cdot \dfrac{3}{5} \cdot \dfrac{1}{2}$

27. $100 \cdot \dfrac{21}{50} \cdot \dfrac{3}{4}$

28. $200 \cdot \dfrac{7}{8}$

29. $\dfrac{3}{5} \cdot 400$

30. $\dfrac{3}{7} \cdot 490$

31. $\dfrac{2}{3} \cdot 284$

32. $\dfrac{12}{25} \cdot 430$

33. $\dfrac{28}{21} \cdot 640 \cdot \dfrac{15}{32}$

34. $\dfrac{21}{13} \cdot 520 \cdot \dfrac{7}{20}$

35. $\dfrac{54}{38} \cdot 684 \cdot \dfrac{5}{6}$

36. $\dfrac{76}{43} \cdot 473 \cdot \dfrac{5}{19}$

Find the area of each rectangle. See Example 5.

37.

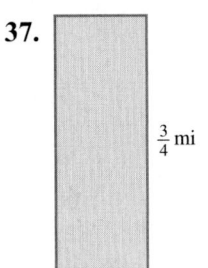

$\frac{3}{4}$ mi

$\frac{1}{3}$ mi

38.

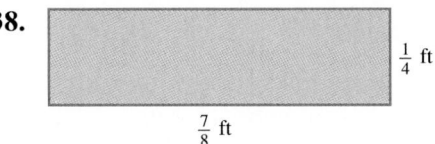

$\frac{1}{4}$ ft

$\frac{7}{8}$ ft

39.

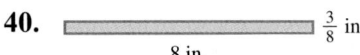

$\frac{3}{4}$ meter

12 meters

40.

$\frac{3}{8}$ in.

8 in.

41.

$\frac{3}{14}$ in.

$\frac{7}{5}$ in.

42.

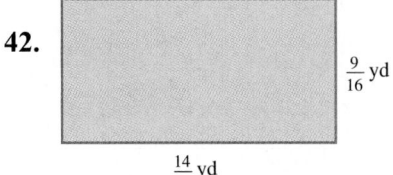

$\frac{9}{16}$ yd

$\frac{14}{15}$ yd

43. Write in your own words the rule for multiplying fractions. Make up an example problem to show how this works.

44. A useful shortcut when multiplying fractions is to divide a numerator and a denominator by the same number. Describe how this works and give an example.

Solve each application problem. Write answers in lowest terms and as whole or mixed numbers where possible. See Example 5.

45. Find the area of a heating-duct grill having a length of 2 yd and a width of $\frac{3}{4}$ yd.

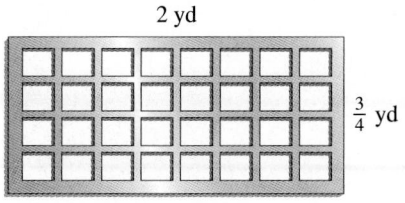

2 yd

$\frac{3}{4}$ yd

46. Find the area of the top of a computer desk having a length of 2 yd and a width of $\frac{7}{8}$ yd.

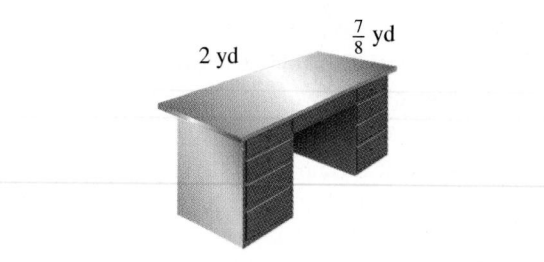

2 yd

$\frac{7}{8}$ yd

47. A parcel of land measures $\frac{1}{2}$ mile by 2 miles. Find the total area of the parcel.

48. A motorcycle race course is $\frac{3}{4}$ mile wide by 6 miles long. Find the area of the race course.

49. Parking lot A is $\frac{1}{4}$ mile long and $\frac{3}{16}$ mile wide, while parking lot B is $\frac{3}{8}$ mile long and $\frac{1}{8}$ mile wide. Which parking lot has the larger area?

50. The Rocking Horse Ranch is $\frac{3}{4}$ mile long and $\frac{1}{2}$ mile wide. The Silver Spur Ranch is $\frac{5}{8}$ mile long and $\frac{4}{5}$ mile wide. Which ranch has the larger area?

RELATING CONCEPTS (Exercises 51–56) For Individual or Group Work

Front end rounding can be used to estimate an answer when multiplying fractions. **Work Exercises 51–56 in order.**

The table shows the 10 best-selling vehicle models in the nation for the month of March.

MARCH 2000 TOP SELLERS
THE 10 BEST-SELLING VEHICLE MODELS

Vehicle	March Sales
Ford F-series pickup	90,522
Chevrolet C/K pickup	61,391
Ford Explorer	46,684
Ford Ranger	43,026
Ford Taurus	41,335
Dodge Caravan	37,457
Toyota Camry	36,076
Honda Accord	35,696
Dodge Ram	35,444
Ford Windstar	30,264

Source: *Autodata* April 4, 2000.

51. Use front end rounding to estimate the total sales for the 10 best-selling vehicles in the month of March.

52. Find the exact answer for total sales of the 10 best-selling vehicles sold in March using the exact numbers.

53. If $\frac{3}{4}$ of the Ford Explorers sold had 4-wheel drive, use front end rounding to estimate, and then find the exact number of Ford Explorers sold having 4-wheel drive.

54. If $\frac{3}{7}$ of the Dodge Rams sold had V-8 engines, use front end rounding to estimate, and then find the exact number of Dodge Rams sold having V-8 engines. Round to the nearest whole number.

Compare the estimated answers and the exact answers in Exercises 53 and 54. How can you get an estimated answer that is closer to the exact answer? Try rounding the number of vehicles to some multiple of the denominator in the fraction.

55. Refer to Exercise 53. Round the number of Ford Explorers to some multiple of the denominator in $\frac{3}{4}$. Now estimate the answer, showing your work.

56. Refer to Exercise 54. Round the number of Dodge Rams to some multiple of the denominator in $\frac{3}{7}$. Now estimate the answer, showing your work.

2.6   APPLICATIONS OF MULTIPLICATION

1 ▭   **Solve fraction application problems using multiplication.**   Many application problems are solved by multiplying fractions. Use the following indicator words for multiplication.

product
double
triple
times
of
twice
twice as much

Look for these indicator words in the following examples.

OBJECTIVE

1 ▭   Solve fraction application problems using multiplication.

❶ Solve each problem.

(a) Shafali Patel pays $\frac{1}{4}$ of her monthly salary as a house payment. If her monthly salary is $4320, find her monthly house payment.

Example 1   Applying Indicator Words

Lois Stevens gives $\frac{1}{10}$ of her income to her church. One month she earned $1980. How much did she give to the church that month?

Step 1   **Read** the problem. The problem asks us to find the amount of money given to the church.

Step 2   **Work out a plan.** The indicator word is *of:* Stevens gave $\frac{1}{10}$ *of* her income. The word *of* indicates multiplication, so find the amount given to the church by multiplying $\frac{1}{10}$ and $1980.

Step 3   **Estimate** a reasonable answer. Round the income of $1980 to $2000. Then divide $2000 by 10 to find $\frac{1}{10}$ of the income (one of 10 equal parts). Our estimate is $2000 ÷ 10 = $200. (Recall the shortcut for dividing by 10; drop one 0 from the dividend.)

Step 4   **Solve** the problem.

$$\text{amount} = \frac{1}{\overset{}{\underset{1}{10}}} \cdot \frac{\overset{198}{\cancel{1980}}}{1} = \frac{198}{1} = 198$$

Step 5   **State the answer.** Stevens gave $198 to her church that month.

Step 6   **Check.** The answer, $198, is close to our estimate of $198.

══════════ **Work Problem ❶ at the Side.**

(b) A retiring police officer will receive $\frac{5}{8}$ of her highest annual salary as retirement income. If her highest annual salary is $48,000, how much will she receive as retirement income?

Example 2   Solving a Fraction Application Problem

Of the 42 students in a biology class, $\frac{2}{3}$ went on a field trip. How many went on the trip?

Step 1   **Read** the problem. The problem asks us to find the number of students who went on the field trip.

Step 2   **Work out a plan.** Reword the problem to read

$\frac{2}{3}$ **of** the students went on a field trip.
  ↑
  Indicator word for multiplication

═══ **Continued on Next Page**

❷ At one pharmacy, $\frac{3}{16}$ of the prescriptions are paid by a third party (insurance company). If 2816 prescriptions are filled, find the number paid by a third party.

Use the six problem-solving steps.

❸ In a certain community, $\frac{1}{3}$ of the residents speak a foreign language. Of those speaking a foreign language, $\frac{3}{4}$ speak Spanish. What fraction of the residents speak Spanish?

Use the six problem-solving steps.

Step 3 **Estimate** a reasonable answer. Round the number of students in the class from 42 to 40. Then, $\frac{1}{2}$ of 40 is 20. Since $\frac{2}{3}$ is more than $\frac{1}{2}$, our estimate is that "more than 20 students" went on the trip.

Step 4 **Solve** the problem. Find the number who went on the trip by multiplying $\frac{2}{3}$ and 42.

$$\text{number who went} = \frac{2}{3} \cdot 42$$

$$= \frac{2}{\overset{}{\underset{1}{3}}} \cdot \frac{\overset{14}{42}}{1} = \frac{28}{1} = 28$$

Step 5 **State the answer.** 28 students went on the trip.

Step 6 **Check.** The answer, 28, fits our estimate of "more than 20."

Work Problem ❷ at the Side.

Example 3 **Finding a Fraction of a Fraction**

In her will, a woman divides her estate into 6 equal parts. Five of the 6 parts are given to relatives. Of the sixth part, $\frac{1}{3}$ goes to the Salvation Army. What fraction of her total estate goes to the Salvation Army?

Step 1 **Read** the problem. The problem asks for the fraction of an estate that goes to the Salvation Army.

Step 2 **Work out a plan.** Reword the problem to read the Salvation Army gets $\frac{1}{3}$ **of** $\frac{1}{6}$.

Indicator word for multiplication

Step 3 **Estimate** a reasonable answer. If the estate is divided into 6 equal parts and each of these parts was divided into 3 equal parts, we would have $6 \cdot 3 = 18$ equal parts. Our estimate is $\frac{1}{18}$.

Step 4 **Solve** the problem. The Salvation Army gets $\frac{1}{3}$ **of** $\frac{1}{6}$.

Indicator word

To find the fraction that the Salvation Army is to receive, multiply $\frac{1}{3}$ and $\frac{1}{6}$.

$$\text{fraction to Salvation Army} = \frac{1}{3} \cdot \frac{1}{6}$$

$$= \frac{1}{18}$$

Step 5 **State the answer.** The Salvation Army gets $\frac{1}{18}$ of the total estate.

Step 6 **Check.** The answer, $\frac{1}{18}$, matches our estimate.

Work Problem ❸ at the Side.

Example 4   Using Fractions with a Circle Graph

The circle graph, or pie chart, shows how people in a survey answered the question "Are unidentified flying objects (UFOs) real or imaginary?" If 1500 people were in the survey, find the number who answered "not sure."

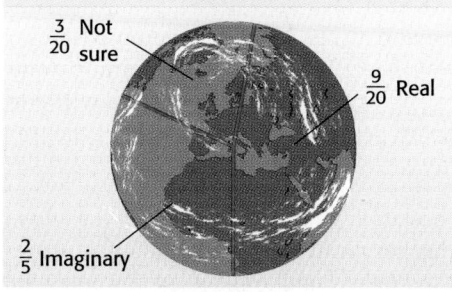

OUT OF THIS WORLD?
When asked in a survey if UFOs actually exist, respondents answered:

$\frac{3}{20}$ Not sure

$\frac{9}{20}$ Real

$\frac{2}{5}$ Imaginary

Source: Yankelovich Partners for *Life* magazine.

Step 1 **Read** the problem. The problem asks for the number of people who answered "not sure" in the survey.

Step 2 **Work out a plan.** Reword the problem to read

$$\frac{3}{20} \text{ of } 1500 \text{ people answered "not sure."}$$

↑
Indicator word for multiplication

Step 3 **Estimate** a reasonable answer. $\frac{1}{2}$ of 1500 people is 750 people. $\frac{3}{20}$ is less than $\frac{1}{2}$, so our estimate is "less than 750 people."

Step 4 **Solve** the problem. Find the number who were "not sure" by multiplying $\frac{3}{20}$ and 1500.

$$\text{number "not sure"} = \frac{3}{20} \cdot 1500$$

$$= \frac{3}{20} \cdot \frac{1500}{1}$$

$$= \frac{3}{\underset{2}{\cancel{20}}} \cdot \frac{\overset{75}{\cancel{\underset{\cancel{150}}{1500}}}}{1} \qquad \text{Divide both numerator and denominator.}$$

$$= \frac{225}{1} = 225$$

Step 5 **State the answer.** 225 people were "not sure."

Step 6 **Check.** The answer, 225 people, fits our estimate of "less than 750 people."

Work Problem ❹ at the Side.

❹ Solve each problem using the six problem-solving steps. Use the circle graph in Example 4.

(a) What fraction of the people answered that UFOs are real?

(b) What number of people answered that UFOs are real?

(c) What fraction of the people answered that UFOs are imaginary?

(d) What number of people answered that UFOs are imaginary?

ANSWERS

4. (a) $\frac{9}{20}$ **(b)** 675 people **(c)** $\frac{2}{5}$
(d) 600 people

Real-Data Applications

Heart-Rate Training Zone

Performing aerobic exercise is beneficial both for improving aerobic fitness and for burning fat. For best results, you should keep your heart rate within the training zone for a minimum of 12 minutes. If you train at the higher end of the training zone, you will burn glycogen and improve aerobic fitness. Training for longer periods at the lower end of the training zone results in your body using fat reserves for energy.

Example: The training zone (TZ) is based on your heart rate (HR) for one minute. To see if you are in the training zone, measure your heart rate for 15 seconds. Compare it to the 15-second training zone. Find the exact answer, and then round to the nearest whole number.

Instruction	Calculation	Example (age 22)
Calculate maximum heart rate (MHR)	$220 - $ your age	$220 - 22 = 198$
Calculate lower limit of training zone (TZ)	$\frac{3}{5} \times $ (MHR)	$\frac{3}{5} \times (198) = \frac{594}{5} = 118\frac{4}{5}$
Calculate upper limit of training zone (TZ)	$\frac{4}{5} \times $ (MHR)	$\frac{4}{5} \times (198) = \frac{792}{5} = 158\frac{2}{5}$
Calculate the exact 15-second training zone. Round the results to the nearest whole number.	$\left(\frac{1}{4} \times \text{lower TZ}, \frac{1}{4} \times \text{Upper TZ} \right)$	$\frac{1}{4} \times \frac{594}{5} = 29\frac{7}{10}; \frac{1}{4} \times \frac{792}{5} = 39\frac{3}{5}$ $29\frac{7}{10} < \text{HR} < 39\frac{3}{5}$ $30 < \text{HR} < 40$

Age	MHR	Lower Limit of TZ	Upper Limit of TZ	15-Second TZ (exact)	15-Second TZ (rounded)
18					
25					
30					
40					
50					
60					

1. Suppose you work in a physical fitness center and decide to design a poster to remind the clients of the training zone for their age. Compute the exact 15-second training zone for people of each of the following ages. Write fractions in lowest terms. Then round the answers to the nearest whole.

2. Explain why the lower and upper training zones (TZ) are multiplied by $\frac{1}{4}$.

3. Explain why the 15-second training zone is lower for a person aged 50 in comparison to a person aged 20.

4. Based on the chart, what would you tell a 45-year-old person about their 15-second training zone?

2.6 EXERCISES

Solve each application problem. Look for indicator words. See Examples 1–4.

1. A file cabinet top is $\frac{3}{4}$ yd by $\frac{2}{3}$ yd. Find its area.

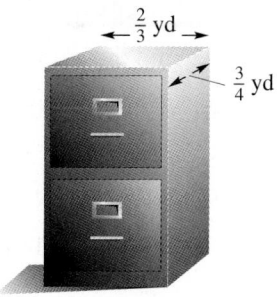

2. A dog bed is $\frac{7}{8}$ yd by $\frac{10}{9}$ yd. Find its area.

3. A cookie sheet is $\frac{4}{3}$ ft by $\frac{2}{3}$ ft. Find its area.

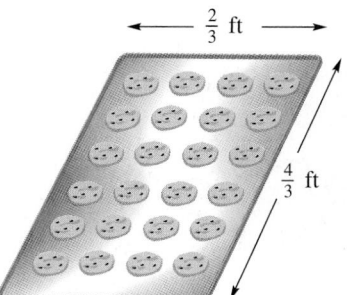

4. Al is helping Tim make a mahogany lamp table for Jill's birthday. Find the area of the top of the table if it is $\frac{4}{5}$ yd long by $\frac{3}{8}$ yd wide.

5. One-third of the players elected to the Baseball Hall of Fame were pitchers. If 183 players are in the Hall of Fame, how many were pitchers? (*Source: National Baseball Hall of Fame.*)

6. A minimarket sells 2500 items, of which $\frac{3}{25}$ are classified as junk food. How many of the items are junk food?

7. Dan Crump had expenses of $6848 during one semester of college. His part-time job provided $\frac{3}{8}$ of the amount he needed. How much did he earn on his job?

8. Erin Hernandez produces $5680 in profits for her employer. If her personal earnings are $\frac{2}{5}$ of these profits, find the amount of her earnings.

9. A school gives scholarships to $\frac{5}{24}$ of its 1800 freshmen. How many freshman students received scholarships?

10. Jason Todd estimates that it will cost him $8400, including living expenses, to attend college full time for one year. If he must earn $\frac{3}{4}$ of the cost and borrow the balance, find the amount that he must earn.

11. At the Garlic Festival Fun Run, $\frac{5}{12}$ of the runners are women. If there are 780 runners, how many are women?

12. A hotel has 408 rooms. Of these rooms, $\frac{9}{17}$ are for nonsmokers. How many rooms are for nonsmokers?

In a recent year, Americans purchased one billion books. The circle graph shows the purchase of these books by age group over a one-year period. Use this information to work Exercises 13–18.

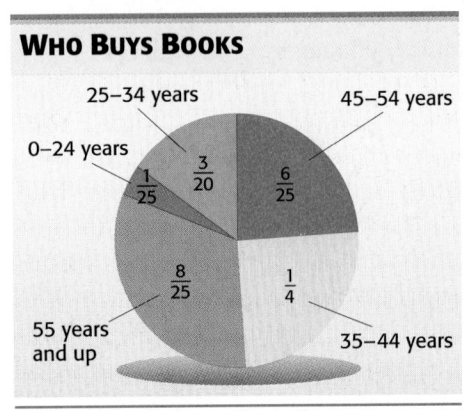

WHO BUYS BOOKS

Source: Book Industry Study Group.

13. Which age group purchased the least number of books? How many did this group purchase?

14. Which age group purchased the greatest number of books? How many did this group purchase?

15. Find the number of books purchased by those in the 35–44 year age group.

16. Find the number of books purchased by those in the 45–54 age group.

17. Without actually adding the fractions given for all the age groups, explain why their sum has to be 1.

18. Refer to Exercise 17. Suppose you added all the fractions for the age groups and did not get 1 as an answer. List some possible explanations.

The table shows the earnings for the Gomes family last year and the circle graph shows how they spent their earnings. Use this information to answer Exercises 19–24.

Month	Earnings	Month	Earnings
January	$3050	July	$3160
February	$2875	August	$2355
March	$3325	September	$2780
April	$3020	October	$3675
May	$2880	November	$3310
June	$3265	December	$4305

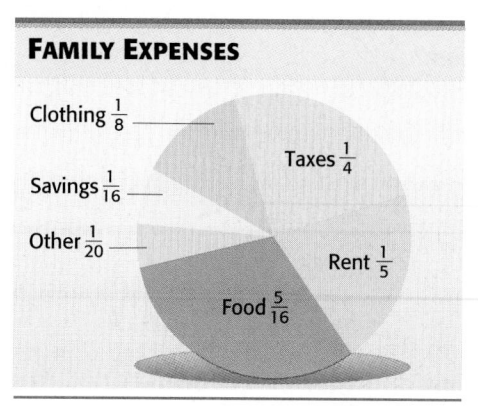

FAMILY EXPENSES

Clothing $\frac{1}{8}$

Savings $\frac{1}{16}$

Other $\frac{1}{20}$

Taxes $\frac{1}{4}$

Rent $\frac{1}{5}$

Food $\frac{5}{16}$

19. Find the Gomes family's total income for the year.

20. How much of their annual earnings went to taxes?

21. Find the amount of their rent for the year.

22. How much did they spend for food during the year?

23. Find their annual savings.

24. How much of their annual income was spent on clothing?

25. Here is how one student solved a multiplication problem. Find the error and solve the problem correctly.

$$\frac{9}{10} \times \frac{20}{21} = \frac{\overset{3}{\cancel{9}}}{\underset{1}{\cancel{10}}} \times \frac{\overset{2}{\cancel{20}}}{\underset{3}{\cancel{21}}} = \frac{6}{3} = 2$$

26. When two whole numbers are multiplied, the product is always larger than the numbers being multiplied. When two proper fractions are multiplied, the product is always smaller than the numbers being multiplied. Are these statements true? Why or why not?

Solve each application problem.

27. Pamela Denny is a waitress and earned $112 in one 8-hour day. How much money did she earn in 3 hours?

28. Ruth Berry jogged 40 miles in 8 hours. How far did she jog in 5 hours?

29. LaDonna Washington is running for city council. She needs to get $\frac{2}{3}$ of her votes from senior citizens and 27,000 votes in all to win. How many votes does she need from voters other than the senior citizens?

30. The start-up cost of a Subs and Sandwich Shop is $32,000. If the bank will loan you $\frac{9}{16}$ of the start up and you must pay the balance, how much more will you need to open a shop?

31. A will states that $\frac{7}{8}$ of an estate is to be divided among relatives. Of the remaining estate, $\frac{1}{4}$ goes to the American Cancer Society. What fraction of the estate goes to the American Cancer Society?

32. A couple has $\frac{1}{5}$ of their total investments in stocks. Of the remaining investment, $\frac{1}{8}$ is invested in bonds. What fraction of the total investment is invested in bonds?

2.7 DIVIDING FRACTIONS

1 ▭ **Find the reciprocal of a fraction.**   To divide fractions, we need to know how to find the **reciprocal** of a fraction.

Reciprocal of a Fraction

Two numbers are reciprocals of each other if their product is 1. To find the reciprocal of a fraction, interchange the numerator and denominator.

For example, the reciprocal of $\frac{3}{4}$ is $\frac{4}{3}$.

$$\text{Fraction} \quad \frac{3}{4} \diagdown\!\!\!\!\diagup \frac{4}{3} \quad \text{Reciprocal}$$

NOTE

Notice that you invert, or "flip" a fraction to find its reciprocal.

Example 1   **Finding Reciprocals**

Find the reciprocal of each fraction.

(a) The reciprocal of $\frac{1}{4}$ is $\frac{4}{1}$ because $\frac{1}{4} \cdot \frac{4}{1} = \frac{4}{4} = 1$.

(b) The reciprocal of $\frac{2}{3}$ is $\frac{3}{2}$ because $\frac{2}{3} \cdot \frac{3}{2} = \frac{6}{6} = 1$.

(c) The reciprocal of $\frac{3}{5}$ is $\frac{5}{3}$ because $\frac{3}{5} \cdot \frac{5}{3} = \frac{15}{15} = 1$.

(d) The reciprocal of 8 is $\frac{1}{8}$ because $\frac{8}{1} \cdot \frac{1}{8} = \frac{8}{8} = 1$.   Think of 8 as $\frac{8}{1}$.

═══ **Work Problem ❶ at the Side.**

NOTE

Every number has a reciprocal except 0. The number 0 has no reciprocal because there is no number that can be multiplied by 0 to get 1.

$$0 \cdot (\text{reciprocal}) = 1$$

There is no number to use here that will give an answer of 1. When you multiply by 0, you always get 0.

OBJECTIVES

1 ▭ Find the reciprocal of a fraction.

2 ▭ Divide fractions.

3 ▭ Solve application problems in which fractions are divided.

❶ Find the reciprocal of each fraction.

(a) $\frac{5}{8}$

(b) $\frac{2}{5}$

(c) $\frac{9}{4}$

(d) 5

In Chapter 1, we saw that the division problem $12 \div 3$ asks how many 3s are in 12. In the same way, the division problem $\frac{2}{3} \div \frac{1}{6}$ asks how many $\frac{1}{6}$s are in $\frac{2}{3}$. The figure illustrates $\frac{2}{3} \div \frac{1}{6}$.

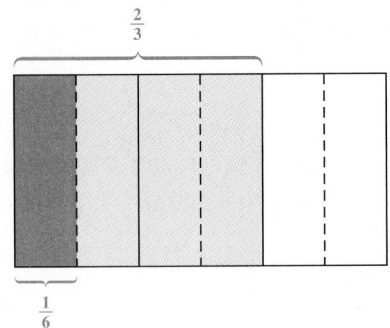

The figure shows that there are 4 of the $\frac{1}{6}$s in $\frac{2}{3}$, or

$$\frac{2}{3} \div \frac{1}{6} = 4.$$

2 ▭ **Divide fractions.** We will use reciprocals to divide fractions.

Dividing Fractions

To divide two fractions, multiply the first fraction by the reciprocal of the second fraction.

Example 2 Dividing One Fraction by Another

Divide. Write answers in lowest terms.

(a) $\dfrac{7}{8} \div \dfrac{15}{16}$

The reciprocal of $\dfrac{15}{16}$ is $\dfrac{16}{15}$.

Reciprocals

$$\frac{7}{8} \div \frac{15}{16} = \frac{7}{8} \cdot \frac{16}{15}$$

Change division to multiplication.

$$= \frac{7}{\overset{1}{\cancel{8}}} \cdot \frac{\overset{2}{\cancel{16}}}{15} \qquad \text{Divide the numerator and denominator by 8.}$$

$$= \frac{7 \cdot 2}{1 \cdot 15} \qquad \text{Multiply.}$$

$$= \frac{14}{15}$$

Continued on Next Page

(b) $\dfrac{\frac{4}{5}}{\frac{3}{10}}$

$$\frac{\frac{4}{5}}{\frac{3}{10}} = \frac{4}{5} \div \frac{3}{10} \qquad \text{Rewrite by using the } \div \text{ symbol for division.}$$

$$= \frac{4}{\overset{}{\underset{1}{\cancel{5}}}} \cdot \frac{\overset{2}{\cancel{10}}}{3} \qquad \begin{array}{l}\text{The reciprocal of } \frac{3}{10} \text{ is } \frac{10}{3}. \text{ Change "} \div \text{" to "} \bullet \text{", and} \\ \text{divide the numerator and denominator by 5.}\end{array}$$

$$= \frac{4 \cdot 2}{1 \cdot 3} \qquad \text{Multiply.}$$

$$= \frac{8}{3} = 2\frac{2}{3} \qquad \text{Mixed number}$$

CAUTION

Be certain that the divisor fraction is changed to its reciprocal *before* you divide numerators and denominators by common factors.

Work Problem ② at the Side.

Example 3   **Dividing with a Whole Number**

Divide. Write all answers in lowest terms and as whole or mixed numbers where possible.

(a) $5 \div \dfrac{1}{4}$

Write 5 as $\frac{5}{1}$. Next, use the reciprocal of $\frac{1}{4}$, which is $\frac{4}{1}$.

$$5 \div \frac{1}{4} = \frac{5}{1} \cdot \frac{4}{1} \qquad \text{Reciprocal of } \frac{1}{4} \text{ is } \frac{4}{1}.$$

Reciprocals

$$= \frac{5 \cdot 4}{1 \cdot 1} \qquad \text{Multiply.}$$

$$= \frac{20}{1} = 20 \qquad \text{Whole number}$$

Continued on Next Page

---

② Divide. Write answers in lowest terms.

(a) $\dfrac{1}{2} \div \dfrac{3}{4}$

(b) $\dfrac{5}{8} \div \dfrac{7}{8}$

(c) $\dfrac{\frac{2}{3}}{\frac{4}{5}}$

(d) $\dfrac{\frac{5}{6}}{\frac{7}{12}}$

❸ Divide. Write answers in lowest terms and as whole or mixed numbers where possible.

(a) $12 \div \dfrac{3}{4}$

(b) $8 \div \dfrac{8}{9}$

(c) $\dfrac{4}{5} \div 6$

(d) $\dfrac{3}{8} \div 4$

❹ Solve each problem using the six problem-solving steps.

(a) How many $\frac{3}{4}$-quart leaf-blower fuel tanks can be filled from 15 quarts of fuel?

(b) How many $\frac{2}{3}$-gallon garden sprayers can be filled from 36 gallons of insect spray?

(b) $\dfrac{2}{3} \div 6$

Write 6 as $\frac{6}{1}$. The reciprocal of $\frac{6}{1}$ is $\frac{1}{6}$.

$$\frac{2}{3} \div \frac{6}{1} = \frac{2}{3} \cdot \frac{1}{6}$$

Reciprocals

$$= \frac{\overset{1}{\cancel{2}}}{3} \cdot \frac{1}{\underset{3}{\cancel{6}}}$$ Divide the numerator and denominator by 2, then multiply.

$$= \frac{1 \cdot 1}{3 \cdot 3} = \frac{1}{9}$$

Work Problem ❸ at the Side.

3 ▭ **Solve application problems in which fractions are divided.** Many application problems require division of fractions. Recall that typical indicator words for division are **goes into, per, divide, divided by, divided equally,** and **divided into.**

Example 4 Applying Fraction Skills

Sophia, the manager of the Village Deli, must fill a 10-gallon dill pickle crock with salt brine. She has only a $\frac{2}{3}$-gallon container to use. How many times must she fill the $\frac{2}{3}$-gallon container and empty it into the 10-gallon crock?

Step 1 **Read** the problem. We need to find the number of times Sophia needs to use a $\frac{2}{3}$-gallon container in order to fill a 10-gallon crock.

Step 2 **Work out a plan.** We can solve the problem by finding the number of times 10 can be divided by $\frac{2}{3}$.

Step 3 **Estimate** a reasonable answer. Round $\frac{2}{3}$-gallon to 1-gallon. In order to fill the 10-gallon container, she would have to use the 1-gallon container 10 times, so our estimate is 10.

Step 4 **Solve** the problem.

Reciprocals

$$10 \div \frac{2}{3} = \frac{10}{1} \cdot \frac{3}{2}$$ Use the reciprocal, $\frac{3}{2}$, and change "÷" to "•".

$$= \frac{\overset{5}{\cancel{10}}}{1} \cdot \frac{3}{\underset{1}{\cancel{2}}}$$ Divide the numerator and denominator, and then multiply.

$$= \frac{15}{1} = 15$$

Step 5 **State the answer.** Sofia must fill the container 15 times.

Step 6 **Check.** The answer, 15 times, is close to our estimate of 10 times.

Work Problem ❹ at the Side.

Example 5 Applying Fraction Skills

At the Happi-Time Day Care Center, $\frac{6}{7}$ of the total operating fund goes to classroom operation. If there are 18 classrooms, what fraction of the classroom operating amount does each classroom receive?

Step 1 **Read** the problem. Since $\frac{6}{7}$ of the total operating fund must be split into 18 parts, we must find the fraction of the classroom operating amount received by each classroom.

Step 2 **Work out a plan.** We must divide the fraction of the total operating fund going to classroom operation $\left(\frac{6}{7}\right)$ by the number of classrooms (18).

Step 3 **Estimate** a reasonable answer. Round $\frac{6}{7}$ to 1. If all of the operating expenses (1 whole) were divided between 18 classrooms, each classroom would receive $\frac{1}{18}$ of the operating expenses, our estimate.

Step 4 **Solve** the problem. We solve by dividing $\frac{6}{7}$ by 18.

$$\frac{6}{7} \div 18 = \frac{6}{7} \div \frac{18}{1} \qquad \text{The reciprocal of } \frac{18}{1} \text{ is } \frac{1}{18}.$$

$$= \frac{\cancel{6}^{\,1}}{7} \cdot \frac{1}{\cancel{18}_{\,3}} \qquad \begin{array}{l}\text{Use the reciprocal, change "}\div\text{" to "}\bullet\text{", and}\\ \text{divide the numerator and denominator by 6.}\end{array}$$

$$= \frac{1}{21} \qquad \text{Multiply.}$$

Step 5 **State the answer.** Each classroom receives $\frac{1}{21}$ of the total operating fund.

Step 6 **Check.** The answer, $\frac{1}{21}$, is close to our estimate of $\frac{1}{18}$.

Work Problem ❺ at the Side.

❺ Solve each problem using the six problem-solving steps.

(a) The top 12 employees at Mayfield Manufacturing will divide $\frac{3}{4}$ of the annual bonus money. What fraction of the bonus money will each employee receive?

(b) The four top-performing students at Tulsa Community College will divide $\frac{1}{3}$ of the scholarship money awarded to students. What fraction of the scholarship money will each of these top students receive?

Real-Data Applications

Hotel Expenses

Mathematics teachers attending conferences in New Orleans, Louisiana, and San Jose, California, found the following information about hotel rates on the organizations' Internet Web sites.

Hotel	Single	Double	Triple	Quad	Suites
New Orleans (January 2001)					
Marriott	$116	$121	$121	$121	$603
Sheraton (Club)	$157	$169	$194	$219	$598
San Jose (February 2001)					
Crowne Plaza	$157	$157	$167	$177	—
Hilton Towers	$167	$187	$207	$227	—
Hyatt Sainte Claire	$156	$176	$196	$216	—

1. The double rate is for two people sharing a room. What fractional part does each person pay?

2. (a) Multiply the double rate at the Hilton Towers in San Jose by $\frac{1}{2}$. What is the result?

   (b) Divide the double rate at the Hilton Towers in San Jose by 2. What is the result?

   (c) Explain what happened. How much money would one person owe if he or she shared a double room at the Hilton Towers in San Jose?

3. The triple rate is for three people sharing a room. What fractional part does each person pay?

4. How much money would one person owe if he or she shared a triple room at the Hyatt Sainte Claire in San Jose? How much money would each person save if they could book a triple room at the Crowne Plaza instead of the Hyatt Sainte Claire? Find your answer using two different methods, based on your observations in Problem 2.

5. The quad rate is for four people sharing a room. What fractional part does each person pay?

6. How much money would one person owe if he or she shared a quad room at the Sheraton in New Orleans? Find your answer using two different methods, based on your observations in Problem 2.

7. How many people would have to share a suite at the Sheraton in New Orleans for the cost per person to be less than sharing a quad room at the same hotel? Round the answer to the nearest whole number. (*Hint:* Estimate the cost per person for a quad room first. Then estimate the number of people needed to share the cost of the suite. Check your work using actual values.)

8. Suppose you have a travel allotment of $500 that can be spent on transportation, hotel, food, and registration fees. You and a colleague decide to attend the 3-day New Orleans conference and plan to share a room at the Marriott. Registration costs $150; the flight costs $129 round-trip; the taxi ride to the hotel costs $20 per person each way; and you budget $35 per day for meals. How much out-of-pocket expense will you have to pay? How much would you save if you could recruit a third person to share the room?

2.7 EXERCISES

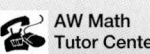

Find the reciprocal of each number. See Example 1.

1. $\dfrac{2}{3}$ **2.** $\dfrac{3}{4}$ **3.** $\dfrac{8}{5}$ **4.** $\dfrac{12}{7}$

5. $\dfrac{5}{6}$ **6.** $\dfrac{13}{20}$ **7.** 4 **8.** 10

Divide. Write answers in lowest terms and as whole or mixed numbers where possible. See Examples 2 and 3.

9. $\dfrac{1}{4} \div \dfrac{3}{4}$ **10.** $\dfrac{3}{8} \div \dfrac{5}{8}$ **11.** $\dfrac{7}{8} \div \dfrac{1}{3}$ **12.** $\dfrac{7}{8} \div \dfrac{3}{4}$

13. $\dfrac{3}{4} \div \dfrac{5}{3}$ **14.** $\dfrac{4}{5} \div \dfrac{9}{4}$ **15.** $\dfrac{7}{9} \div \dfrac{7}{36}$ **16.** $\dfrac{5}{8} \div \dfrac{5}{16}$

17. $\dfrac{15}{32} \div \dfrac{5}{64}$ **18.** $\dfrac{7}{12} \div \dfrac{14}{15}$ **19.** $\dfrac{\frac{13}{20}}{\frac{4}{5}}$ **20.** $\dfrac{\frac{9}{10}}{\frac{3}{5}}$

21. $\dfrac{\frac{5}{6}}{\frac{25}{24}}$ **22.** $\dfrac{\frac{28}{15}}{\frac{21}{5}}$ **23.** $12 \div \dfrac{2}{3}$ **24.** $7 \div \dfrac{1}{4}$

25. $\dfrac{\frac{15}{2}}{3}$ **26.** $\dfrac{9}{\frac{3}{4}}$ **27.** $\dfrac{\frac{4}{7}}{8}$ **28.** $\dfrac{\frac{7}{10}}{3}$

Solve each application problem by using division. See Examples 4 and 5.

29. Ms. Shaffer has a piece of property with an area that is $\frac{8}{9}$ acre. She wishes to divide it into 4 equal parts for her children. How many acres of land will each child get?

30. The Sweepstakes Lottery pays out $\frac{7}{8}$ of the total revenue to 14 top winners. What fraction of the total revenue does each winner receive?

31. Some college roommates want to make pancakes for their neighbors. They need 5 cups of flour, but have only a $\frac{1}{3}$-cup measuring cup. How many times will they need to fill their measuring cup?

32. Robert Cockrill has 10 quarts of lubricating oil. If each lubricating reservoir in a lathe holds $\frac{1}{3}$ quart of oil, how many reservoirs can be filled?

33. How many $\frac{1}{8}$-ounce eye drop dispensers can be filled with 11 ounces of eye drops?

34. It is estimated that each guest at a party will eat $\frac{5}{16}$ lb of peanuts. How many guests may be served with 10 lb of peanuts?

35. Pam Trizlia had a small pickup truck that could carry $\frac{2}{3}$-cord of firewood. Find the number of trips needed to deliver 40 cords of wood.

36. Manuel Servin has a 200-yd roll of weather stripping material. Find the number of pieces of weather stripping $\frac{5}{8}$ yd in length that may be cut from the roll.

37. A batch of double chocolate chip cookies requires $\frac{3}{4}$ lb of chocolate chips. If you have 9 lb of chocolate chips, how many batches of cookies can you make?

38. An upholsterer uses $\frac{7}{8}$ lb of brass tacks to reupholster one bar stool. How many bar stools can she reupholster with 21 lb of brass tacks?

39. Your classmate is confused on how to divide by a fraction. Write a short note telling him how this should be done.

40. If you multiply positive proper fractions, the product is smaller than the fractions multiplied. When you divide by a proper fraction, is the quotient smaller than the numbers in the problem? Prove your answer with examples.

41. Mike and Charlie have completed $\frac{4}{5}$ of their river rafting excursion down the Colorado River. If they have gone 304 miles so far, find the number of miles remaining in the excursion.

42. Sheila has been working on a job for 63 hours. The job is $\frac{7}{9}$ finished. How many *more* hours must she work to finish the job?

43. The Bridge Lighting Committee has raised $\frac{7}{8}$ of the funds necessary for their lighting project. If this amounts to $840,000, how much additional money must be raised?

44. The school bond committee has raised $\frac{9}{16}$ of the funds needed to promote the bond measure. If the amount raised so far is $45,000, how much additional money must be raised?

Many application problems are solved using multiplication and division of fractions.
Work Exercises 45–50 in order.

45. Perhaps the most common indicator word for multiplication is the word *of.* Circle the words in the list below that are also indicator words for multiplication.

more than	per
double	twice
times	product
less than	difference
equals	twice as much

46. Circle the words in the list below that are indicator words for division.

fewer	sum of
goes into	divide
per	quotient
equals	double
loss of	divided by

47. To divide two fractions, multiply the first fraction by the _____ of the second fraction.

48. Find the reciprocals for each number.

$$\frac{3}{4}; \quad \frac{7}{8}; \quad 5; \quad \frac{12}{19}$$

The size of a square postage stamp is shown here. Use this to answer Exercises 49 and 50.

$\frac{15}{16}$ in.

$\frac{15}{16}$ in.

49. Find the perimeter (distance around the outside edges) of the postage stamp. Explain how to find the perimeter of any 3-, 4-, 5-, or 6-sided figure.

50. Find the area of the postage stamp. Explain how to find the area of any square.

2.8 MULTIPLYING AND DIVIDING MIXED NUMBERS

In Section 2.2 we worked with mixed numbers—a whole number and a fraction written together. Many of the fraction problems you encounter in everyday life involve mixed numbers.

1   **Estimate the answer and multiply mixed numbers.**   When multiplying mixed numbers, it is a good idea to estimate the answer first. Then multiply the mixed numbers by using the following steps.

Multiplying Mixed Numbers

Step 1   *Change* each mixed number to an improper fraction.

Step 2   *Multiply* as fractions.

Step 3   Simplify the answer, which means to write it in *lowest terms*, and change it to a mixed number or whole number where possible.

To estimate the answer, round each mixed number to the nearest whole number. If the numerator is *half* of the denominator or *more,* round up the whole number part. If the numerator is *less* than half the denominator, leave the whole number as it is.

$$1\frac{5}{8} \quad \begin{array}{l} \leftarrow \text{5 is more than 4.} \\ \leftarrow \text{Half of 8 is 4.} \end{array} \Bigg\} \quad 1\frac{5}{8} \text{ rounds up to 2}$$

$$3\frac{2}{5} \quad \begin{array}{l} \leftarrow \text{2 is less than } 2\frac{1}{2}. \\ \leftarrow \text{Half of 5 is } 2\frac{1}{2}. \end{array} \Bigg\} \quad 3\frac{2}{5} \text{ rounds to 3}$$

Work Problem ❶ at the Side.

Example 1   Multiplying Mixed Numbers

First estimate the answer. Then multiply to get an exact answer. Simplify your answers.

(a) $2\frac{1}{2} \cdot 3\frac{1}{5}$

Estimate the answer by rounding the mixed numbers.

$$2\frac{1}{2} \text{ rounds to 3} \quad \text{and} \quad 3\frac{1}{5} \text{ rounds to 3}$$

$$3 \cdot 3 = 9 \quad \text{Estimated answer}$$

To find the exact answer, change each mixed number to an improper fraction.

$$\textit{Step 1} \quad 2\frac{1}{2} = \frac{5}{2} \quad \text{and} \quad 3\frac{1}{5} = \frac{16}{5}$$

Continued on Next Page

❶ Round each mixed number to the nearest whole number.

(a) $3\frac{2}{3}$

(b) $5\frac{2}{5}$

(c) $2\frac{3}{4}$

(d) $4\frac{7}{12}$

(e) $6\frac{1}{2}$

(f) $1\frac{4}{9}$

ANSWERS
1. (a) 4   **(b)** 5   **(c)** 3   **(d)** 5
 (e) 7   **(f)** 1

❷ First estimate the answer. Then multiply to find the exact answer. Write answers in lowest terms. Simplify your answers.

(a) $3\dfrac{1}{2}$ • $6\dfrac{1}{3}$

$= $ _____ *estimate*

(b) $4\dfrac{2}{3}$ • $2\dfrac{3}{4}$

$= $ _____ *estimate*

(c) $3\dfrac{3}{5}$ • $4\dfrac{4}{9}$

$= $ _____ *estimate*

(d) $5\dfrac{1}{4}$ • $3\dfrac{2}{5}$

$= $ _____ *estimate*

2. **(a)** *Estimate:* $4 \cdot 6 = 24$; *Exact:* $22\dfrac{1}{6}$

(b) *Estimate:* $5 \cdot 3 = 15$; *Exact:* $12\dfrac{5}{6}$

(c) *Estimate:* $4 \cdot 4 = 16$; *Exact:* 16

(d) *Estimate:* $5 \cdot 3 = 15$; *Exact:* $17\dfrac{17}{20}$

Next, multiply.

$$2\dfrac{1}{2} \cdot 3\dfrac{1}{5} = \dfrac{5}{2} \cdot \dfrac{16}{5} = \dfrac{\overset{1}{\cancel{5}}}{\underset{1}{\cancel{2}}} \cdot \dfrac{\overset{8}{\cancel{16}}}{\underset{1}{\cancel{5}}} = \dfrac{1 \cdot 8}{1 \cdot 1} = \dfrac{8}{1} = 8$$

The estimated answer is 9 and the exact answer is 8. The exact answer is reasonable.

(b) $3\dfrac{5}{8} \cdot 4\dfrac{4}{5}$

$3\dfrac{5}{8}$ rounds to 4   and $4\dfrac{4}{5}$ rounds to 5

$4 \cdot 5 = 20$ Estimated answer

Now find the exact answer.

$$3\dfrac{5}{8} \cdot 4\dfrac{4}{5} = \dfrac{29}{8} \cdot \dfrac{24}{5} = \dfrac{29}{\underset{1}{\cancel{8}}} \cdot \dfrac{\overset{3}{\cancel{24}}}{5} = \dfrac{29 \cdot 3}{1 \cdot 5} = \dfrac{87}{5}$$

As a mixed number,

$$\dfrac{87}{5} = 17\dfrac{2}{5}.$$ Simplified answer

The estimate was 20, so the exact answer is reasonable.

(c) $1\dfrac{3}{5} \cdot 3\dfrac{1}{3}$

$1\dfrac{3}{5}$ rounds to 2   and $3\dfrac{1}{3}$ rounds to 3

$2 \cdot 3 = 6$ Estimated answer

The exact answer is

$$1\dfrac{3}{5} \cdot 3\dfrac{1}{3} = \dfrac{8}{\underset{1}{\cancel{5}}} \cdot \dfrac{\overset{2}{\cancel{10}}}{3} = \dfrac{8 \cdot 2}{1 \cdot 3} = \dfrac{16}{3} = 5\dfrac{1}{3}.$$

The estimate was 6, so the exact answer is reasonable.

Work Problem ❷ at the Side.

2 ▭ **Estimate the answer and divide mixed numbers.** Just as you did when multiplying mixed numbers, it is also a good idea to estimate the answer when dividing mixed numbers. To divide mixed numbers, use the following steps.

Dividing Mixed Numbers

Step 1   *Change* each mixed number to an improper fraction.

Step 2   Use the *reciprocal* of the second fraction (divisor).

Step 3   *Multiply*.

Step 4   Simplify the answer, which means to write it in *lowest terms*, and change it to a mixed number or whole number where possible.

NOTE

Recall that the reciprocal of a fraction is found by interchanging the numerator and the denominator.

Example 2   Dividing Mixed Numbers

First estimate the answer. Then divide to find the exact answer. Simplify your answers.

(a) $2\frac{2}{5} \div 1\frac{1}{2}$

First estimate the answer by rounding each mixed number to the nearest whole number.

$$2\frac{2}{5} \quad \div \quad 1\frac{1}{2}$$

$$\big\downarrow \quad \text{Rounded} \quad \big\downarrow$$

$$2 \quad \div \quad 2 = 1 \qquad \text{Estimated answer}$$

To find the exact answer, first change each mixed number to an improper fraction.

$$\overset{\textit{Step 1}}{\overbrace{}}$$

$$2\frac{2}{5} \div 1\frac{1}{2} = \frac{12}{5} \div \frac{3}{2}$$

Next, use the reciprocal of the second fraction and multiply.

$$\frac{12}{5} \div \frac{3}{2} = \frac{\overset{4}{\cancel{12}}}{5} \cdot \frac{2}{\underset{1}{\cancel{3}}} = \frac{4 \cdot 2}{5 \cdot 1} = \frac{8}{5} = 1\frac{3}{5} \qquad \text{Simplified answer}$$

Reciprocals

The estimate was 1, so the exact answer is reasonable.

Continued on Next Page

❸ First estimate the answer.
Then divide to find the exact
answer. Simplify all answers.

(a) $3\dfrac{1}{8} \;\div\; 6\dfrac{1}{4}$

_____ $\div$ _____

= _____ _estimate_

(b) $5\dfrac{1}{3} \;\div\; 1\dfrac{1}{4}$

_____ $\div$ _____

= _____ _estimate_

(c) $8 \;\div\; 5\dfrac{1}{3}$

_____ $\div$ _____

= _____ _estimate_

(d) $13\dfrac{1}{2} \;\div\; 18$

_____ $\div$ _____

= _____ _estimate_

(b) $8 \div 3\dfrac{3}{5}$

$$8 \quad\div\quad 3\dfrac{3}{5}$$

Rounded

$$8 \quad\div\quad 4 = 2 \quad \text{Estimate}$$

Now find the exact answer.

Reciprocals

$$8 \div 3\dfrac{3}{5} = \dfrac{8}{1} \div \dfrac{18}{5} = \dfrac{\cancel{8}}{1} \cdot \dfrac{5}{\cancel{18}_{9}} = \dfrac{20}{9} = 2\dfrac{2}{9}$$

Write 8 as $\frac{8}{1}$.

The estimate was 2, so the exact answer is reasonable.

(c) $4\dfrac{3}{8} \div 5$

$$4\dfrac{3}{8} \quad\div\quad 5$$

Rounded

Reciprocals

$$4 \quad\div\quad 5 = \dfrac{4}{1} \div \dfrac{5}{1} = \dfrac{4}{1} \cdot \dfrac{1}{5} = \dfrac{4}{5} \quad \text{Estimate}$$

The exact answer is

Reciprocals

$$4\dfrac{3}{8} \div 5 = \dfrac{35}{8} \div \dfrac{5}{1} = \dfrac{35}{8} \cdot \dfrac{1}{\cancel{5}_{1}} = \dfrac{7}{8}.$$

Write 5 as $\frac{5}{1}$.

The estimate was $\frac{4}{5}$, so the exact answer is reasonable.

Work Problem ❸ at the Side.

3 ▓▓▓ **Solve application problems with mixed numbers.** The next two examples show how to solve application problems involving mixed numbers.

Example 3 Applying Multiplication Skills

The local Habitat for Humanity chapter is looking for 11 contractors who will each donate $3\frac{1}{4}$ days of labor to a community building project. How many days of labor will be donated in all?

Step 1 **Read** the problem. The problem asks for the total days of labor donated by the 11 contractors.

Step 2 **Work out a plan.** Multiply the number of contractors (11) and the amount of labor that each donates ($3\frac{1}{4}$ days).

Continued on Next Page

ANSWERS

3. **(a)** *Estimate:* $3 \div 6 = \dfrac{1}{2}$; *Exact:* $\dfrac{1}{2}$

(b) *Estimate:* $5 \div 1 = 5$; *Exact:* $4\dfrac{4}{15}$

(c) *Estimate:* $8 \div 5 = 1\dfrac{3}{5}$; *Exact:* $1\dfrac{1}{2}$

(d) *Estimate:* $14 \div 18 = \dfrac{7}{9}$; *Exact:* $\dfrac{3}{4}$

Step 3 **Estimate** a reasonable answer. Round $3\frac{1}{4}$ days to 3 days. Multiply 3 days by 11 contractors $(3 \cdot 11)$ to get an estimate of 33 days.

Step 4 **Solve** the problem. Find the exact answer.

$$11 \cdot 3\frac{1}{4} = 11 \cdot \frac{13}{4}$$

$$= \frac{11}{1} \cdot \frac{13}{4} = \frac{143}{4} = 35\frac{3}{4}$$

Step 5 **State the answer.** The community building project will receive $35\frac{3}{4}$ days of donated labor.

Step 6 **Check.** The answer, $35\frac{3}{4}$ days, is close to our estimate of 33 days.

===== **Work Problem ④ at the Side.**

Example 4 **Applying Division Skills**

A dome tent for backpacking requires $7\frac{1}{4}$ yd of nylon material. How many tents can be made from $65\frac{1}{4}$ yd of material?

Step 1 **Read** the problem. The problem asks how many tents can be made from $65\frac{1}{4}$ yd of material.

Step 2 **Work out a plan.** Divide the number of yards of cloth $\left(65\frac{1}{4} \text{ yd}\right)$ by the number of yards needed for one tent $\left(7\frac{1}{4} \text{ yd}\right)$.

Step 3 **Estimate** a reasonable answer.

$$65\frac{1}{4} \quad \div \quad 7\frac{1}{4}$$

Rounded

$$65 \quad \div \quad 7 \approx 9 \text{ tents} \quad \text{Estimate}$$

Step 4 **Solve** the problem. The exact answer is

$$65\frac{1}{4} \div 7\frac{1}{4} = \frac{261}{4} \div \frac{29}{4}$$

$$= \frac{\overset{9}{\cancel{261}}}{\underset{1}{\cancel{4}}} \cdot \frac{\overset{1}{\cancel{4}}}{\underset{1}{\cancel{29}}} = \frac{9}{1} = 9. \quad \text{Matches estimate}$$

Step 5 **State the answer.** 9 tents can be made from $65\frac{1}{4}$ yd of cloth.

Step 6 **Check.** The answer, 9, is close to our estimate.

===== **Work Problem ⑤ at the Side.**

NOTE

When rounding mixed numbers to estimate the answer to a problem, the estimated answer usually varies somewhat from the exact answer. However, the importance of the estimated answer is that it will show you whether your exact answer is reasonable or not.

④ Use the six problem-solving steps. Simplify all answers.

(a) If one hotel room requires $1\frac{7}{8}$ gallons of paint, find the number of gallons needed for 9 hotel rooms.

(b) Clare earns $\$9\frac{1}{4}$ per hour. How much would she earn in $6\frac{1}{2}$ hours? Write the answer as a mixed number.

⑤ Use the six problem-solving steps. Simplify all answers.

(a) The manufacture of one outboard engine propeller requires $4\frac{3}{4}$ lb of brass. How many propellers can be manu-factured from 57 lb of brass?

(b) Student help is paid $\$6\frac{1}{4}$ per hour. Find the number of hours of student help that can be paid for with $150.

Answers

4. (a) *Estimate:* $2 \cdot 9 = 18$; *Exact:* $16\frac{7}{8}$ gal

(b) *Estimate:* $9 \cdot 7 = 63$; *Exact:* $\$60\frac{1}{8}$

5. (a) *Estimate:* $57 \div 5 \approx 11$; *Exact:* 12 propellers

(b) *Estimate:* $150 \div 6 = 25$; *Exact:* 24 hr

Recipes

The side of the corn starch box shown has useful recipes for Fun-Time Dough and for Great Gravy. Suppose you have made the recipe before and know from experience that 1 pound makes enough Fun-Time Dough for three children.

ARGO

CORN STARCH

FAVORITE RECIPES

Fun-Time Dough

1½ cups Argo Corn Starch
½ cup flour
2 cups water
2 tsp cream of tartar
1 cup salt
1 T. vegetable oil

Mix all ingredients together in saucepan. Cook over medium heat, stirring constantly, until mixture gathers on the stirring spoon and forms dough. This will take about 6 minutes. Dump onto waxed paper until cool enough to handle and knead to form a pliable mass. Store in covered container or plastic bag. Food coloring may be added to make different colors.
Makes about 2 lbs. of Fun-Time Dough.

Great Gravy

3 T. bacon fat or meat drippings
2 T. Argo Corn Starch
1½ cups water
¾ tsp. salt
⅛ tsp. pepper

Blend fat and Argo Corn Starch over low heat until it is a rich brown color, stirring constantly. Gradually add water, salt, and pepper. Heat to boiling over direct heat and then boil gently 2 minutes, stirring constantly.
Makes 1½ cups.

—— **Satisfaction Guaranteed** ——

1. If you make the recipe as written, you will have enough Fun-Time Dough for how many children?

2. Suppose you work in a day care center. How many pounds of Fun-Time Dough will you need for nine children?

3. If you double the recipe, you would have 4 pounds of dough. By what fraction should you multiply each ingredient amount to make 3 pounds of dough?

4. Fill in the blanks with the ingredient amounts needed to make 3 pounds of Fun-Time Dough. Show your work.

   Corn starch: _____   Flour: _____

   Water: _____   Cream of tartar: _____

   Salt: _____   Vegetable oil: _____

You decide to make Great Gravy for Thanksgiving dinner.

5. How much gravy does the recipe make?

6. Suppose you decide to double the recipe. By what factor will you change the ingredient amounts?

7. You did not make enough! Suppose you decide to halve the recipe to make more gravy. Now, by what factor will you change the ingredient amounts?

8. If you need 4 cups of gravy, by what factor must you multiply each ingredient amount? Explain how you determined your answer.

2.8 **EXERCISES**

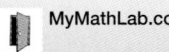
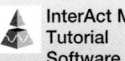

First estimate the answer. Then multiply to find the exact answer. Simplify all answers. See Example 1.

1. *Exact:*

$3\frac{1}{4} \cdot 2\frac{1}{2}$

Estimate:

____ • ____ = ____

2. *Exact:*

$3\frac{1}{2} \cdot 1\frac{1}{4}$

Estimate:

____ • ____ = ____

3. *Exact:*

$1\frac{2}{3} \cdot 2\frac{7}{10}$

Estimate:

____ • ____ = ____

4. *Exact:*

$4\frac{1}{2} \cdot 2\frac{1}{4}$

Estimate:

____ • ____ = ____

5. *Exact:*

$3\frac{1}{9} \cdot 1\frac{2}{7}$

Estimate:

____ • ____ = ____

6. *Exact:*

$6\frac{1}{4} \cdot 3\frac{1}{5}$

Estimate:

____ • ____ = ____

7. *Exact:*

$8 \cdot 6\frac{1}{4}$

Estimate:

____ • ____ = ____

8. *Exact:*

$6 \cdot 2\frac{1}{3}$

Estimate:

____ • ____ = ____

9. *Exact:*

$4\frac{1}{2} \cdot 2\frac{1}{5} \cdot 5$

Estimate:

____ • ____ • ____ = ____

10. *Exact:*

$5\frac{1}{2} \cdot 1\frac{1}{3} \cdot 2\frac{1}{4}$

Estimate:

____ • ____ • ____ = ____

11. *Exact:*

$3 \cdot 1\frac{1}{2} \cdot 2\frac{2}{3}$

Estimate:

____ • ____ • ____ = ____

12. *Exact:*

$\frac{2}{3} \cdot 3\frac{2}{3} \cdot \frac{6}{11}$

Estimate:

____ • ____ • ____ = ____

First estimate the answer. Then divide to find the exact answer. Simplify all answers. See Example 2.

13. *Exact:*

$2\frac{1}{2} \div 7\frac{1}{2}$

Estimate:

____ ÷ ____ = ____

14. *Exact:*

$1\frac{1}{8} \div 2\frac{1}{4}$

Estimate:

____ ÷ ____ = ____

15. *Exact:*

$2\frac{1}{2} \div 3$

Estimate:

____ ÷ ____ = ____

16. *Exact:*

$2\frac{3}{4} \div 2$

Estimate:

____ ÷ ____ = ____

17. *Exact:*

$9 \div 2\frac{1}{2}$

Estimate:

____ ÷ ____ = ____

18. *Exact:*

$5 \div 1\frac{7}{8}$

Estimate:

____ ÷ ____ = ____

19. *Exact:*

$$\frac{3}{4} \div 1\frac{3}{4}$$

Estimate:

_____ ÷ _____ = _____

20. *Exact:*

$$\frac{3}{4} \div 2\frac{1}{2}$$

Estimate:

_____ ÷ _____ = _____

21. *Exact:*

$$1\frac{7}{8} \div 6\frac{1}{4}$$

Estimate:

_____ ÷ _____ = _____

22. *Exact:*

$$8\frac{2}{5} \div 3\frac{1}{2}$$

Estimate:

_____ ÷ _____ = _____

23. *Exact:*

$$5\frac{2}{3} \div 6$$

Estimate:

_____ ÷ _____ = _____

24. *Exact:*

$$5\frac{3}{4} \div 2$$

Estimate:

_____ ÷ _____ = _____

For Exercises 25–40, first estimate the answer. Then solve each application problem by using the six problem-solving steps. Simplify all answers. See Examples 3 and 4.

Use the recipe for Quaker Choc Oat-Chip Cookies to work Exercises 25–28.

Quaker Choc Oat-Chip Cookies

1 cup (2 sticks) **margarine or butter,** softened
1¼ cups firmly packed **brown sugar**
½ cup **granulated sugar**
2 **eggs**
2 tablespoons **milk**
2 teaspoons **vanilla**
1⅝ cups **all purpose flour**
1 teaspoon **baking soda**

½ teaspoon **salt** (optional)
2½ cups **Quaker® Oats** (quick or old fashioned, uncooked)
One 12-ounce package (2 cups) **Nestle® Toll House® semi-sweet chocolate morsels**
1 cup coarsely chopped **nuts** (optional)

Heat oven to 375°F. **Beat** margarine and sugars until creamy.
Add eggs, milk and vanilla; beat well.
Add combined flour, baking soda and salt; mix well. **Stir** in oats, chocolate morsels and nuts; mix well.
Drop by rounded measuring tablespoonfuls onto ungreased cookie sheet.
Bake 9 to 10 minutes for a chewy cookie or 12 to 13 minutes for a crisp cookie.
Cool 1 minute on a cookie sheet; remove to wire rack. Cool completely.
ABOUT 5 DOZEN

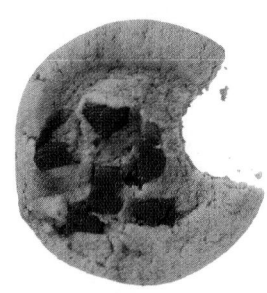

Source: Reprinted by permission of the Quaker Oats Company.

25. If the recipe is doubled, find the amount needed of each ingredient.

  (a) Quaker Oats

    Estimate:
    Exact:

  (b) Brown sugar

    Estimate:
    Exact:

  (c) Flour

    Estimate:
    Exact:

26. If the recipe is tripled, find the amount needed of each ingredient.

  (a) Flour

    Estimate:
    Exact:

  (b) Quaker Oats

    Estimate:
    Exact:

  (c) Granulated sugar

    Estimate:
    Exact:

27. How much of each ingredient is needed if you bake one-half of the recipe?

  (a) Brown sugar

    Estimate:
    Exact:

  (b) Granulated sugar

    Estimate:
    Exact:

  (c) Flour

    Estimate:
    Exact:

28. How much of each ingredient is needed if you bake one-third of the recipe?

  (a) Quaker Oats

    Estimate:
    Exact:

  (b) Salt

    Estimate:
    Exact:

  (c) Brown sugar?

    Estimate:
    Exact:

29. Each home in a housing development needs $109\frac{1}{2}$ yd of prefinished baseboard. How many homes can be fitted if there are 1314 yd of baseboard available?

Estimate:

Exact:

30. How many $1\frac{1}{8}$-ounce vials can be filled from 99 ounces of normal saline solution?

Estimate:

Exact:

31. An insect spray manufactured by Dutch Chemicals Incorporated is a mixture of $1\frac{3}{4}$ ounces of chemical per 1 gallon of water. How many ounces of chemical are needed for $12\frac{1}{2}$ gallons of water?

Estimate:

Exact:

32. To put roofing on the new homes in the Happy Trails subdivision, Duncan needs $37\frac{3}{4}$ lb of roofing nails per home. How many pounds of roofing nails will he need for 36 homes?

Estimate:

Exact:

33. Write the three steps for multiplying mixed numbers. Use your own words.

34. Refer to Exercise 33. In your own words, write the additional step that must be added to the rule for multiplying mixed numbers to make it the rule for dividing mixed numbers.

35. Manufacturing a fishing boat anchor requires $10\frac{3}{8}$ lb of steel. Find the number of anchors that can be manufactured with 25,730 lb of steel.

Estimate:

Exact:

36. To remodel the apartments in her complex, Joleen needs $62\frac{1}{2}$ square yards of carpet for each apartment unit. How many apartment units can she remodel with 6750 square yards of carpet?

Estimate:

Exact:

37. A manufacturer of bird feeders cuts spacers from a tube that is $9\frac{3}{4}$ in. long. How many spacers can be cut from the tube if each spacer must be $\frac{3}{4}$-in. thick?

Estimate:

Exact:

38. A building contractor must move 12 tons of sand. If his truck can carry $\frac{3}{4}$ ton of sand, how many trips must he make to move the sand?

Estimate:

Exact:

39. A fishing guide's boat motor uses $12\frac{3}{4}$ gallons of fuel on a full-day fishing trip and $7\frac{1}{8}$ gallons of fuel on a half-day fishing trip. Find the total number of gallons of fuel used in 28 full-day trips and 16 half-day trips.

Estimate:

Exact:

40. One necklace can be completed in $6\frac{1}{2}$ minutes, while a bracelet takes $3\frac{1}{8}$ minutes. Find the total time that it takes to complete 36 necklaces and 22 bracelets.

Estimate:

Exact:

Chapter 2

SUMMARY

 Study Skills Workbook
Activity 5

KEY TERMS

2.1	**numerator**	The number above the fraction bar in a fraction is called the numerator. It shows how many of the equivalent parts are being considered.
	denominator	The number below the fraction bar in a fraction is called the denominator. It shows the number of equal parts in a whole.
	proper fraction	In a proper fraction, the numerator is smaller than the denominator. The fraction is less than 1.
	improper fraction	In an improper fraction, the numerator is greater than or equal to the denominator. The fraction is equal to or greater than 1.
2.2	**mixed number**	A mixed number includes a fraction and a whole number written together.
2.3	**factors**	Numbers that are multiplied to give a product are factors.
	composite number	A composite number has at least one factor other than itself and 1.
	prime number	A prime number is a whole number other than 0 and 1 that has exactly two factors, itself and 1.
	factorizations	The numbers that can be multiplied to give a specific number (product) are factorizations of that number.
	prime factorization	In a prime factorization, every factor is a prime number.
2.4	**equivalent fractions**	Two fractions are equivalent when they represent the same portion of a whole.
	common factor	A common factor is a number that can be divided into two or more whole numbers.
	lowest terms	A fraction is written in lowest terms when its numerator and denominator have no common factor other than 1.
2.5	**multiplication shortcut**	When multiplying or dividing fractions, the process of dividing a numerator and denominator by a common factor can be used as a shortcut.
2.7	**reciprocal**	Two numbers are reciprocals of each other if their product is 1. To find the reciprocal of a fraction, interchange the numerator and the denominator.

NEW FORMULA

Area of a rectangle: Area = length • width

TEST YOUR WORD POWER

See how well you have learned the vocabulary in this chapter. Answers follow the Quick Review.

1. A **numerator** is
   (a) a number greater than 5
   (b) the number above the fraction bar in a fraction
   (c) any number
   (d) the number below the fraction bar in a fraction.

2. A **proper fraction**
   (a) has a value less than 1
   (b) has a whole number and a fraction
   (c) has a value greater than 1
   (d) is equal to 1.

3. A **mixed number** is
   (a) equal to 1
   (b) less than 1
   (c) a whole number and a fraction written together
   (d) a number multiplied by another number.

4. A **factor** is
   (a) one of two or more numbers that are added to get another number
   (b) the answer in division
   (c) one of two or more numbers that are multiplied to get another number
   (d) the answer in multiplication.

5. A whole number greater than 1 is **prime** if
   (a) it cannot be factored
   (b) it has just one factor
   (c) it has only itself and 1 as a factor
   (d) it has at least two different factors.

6. A **common factor** can
   (a) only be divided by itself and 1
   (b) be divided into two or more whole numbers
   (c) never be divided by 2
   (d) only be divided by the numbers 5 and 10.

7. A fraction is in **lowest terms** when
   (a) it cannot be divided
   (b) it is a common fraction
   (c) its numerator and denominator have no common factor other than 1
   (d) it has a value less than 1.

8. To find the **reciprocal** of a fraction
   (a) multiply it by itself
   (b) interchange the numerator and the denominator
   (c) change it to an improper fraction
   (d) change it to lowest terms.

QUICK REVIEW

Concepts	Examples
2.1 *Types of Fractions*	
Proper Numerator smaller than denominator.	$\dfrac{2}{3}, \dfrac{3}{4}, \dfrac{15}{16}, \dfrac{1}{8}$
Improper Numerator equal to or greater than denominator.	$\dfrac{17}{8}, \dfrac{19}{12}, \dfrac{11}{2}, \dfrac{5}{3}, \dfrac{7}{7}$
2.2 *Converting Fractions*	
Mixed to Improper Multiply denominator by whole number, add numerator, and place over denominator.	$7\dfrac{2}{3} = \dfrac{23}{3} \quad \leftarrow 3 \times 7 + 2$ Same denominator
Improper to Mixed Divide numerator by denominator and place remainder over denominator.	$\dfrac{17}{5} = 3\dfrac{2}{5}$ Same denominator

Concepts	Examples
2.3 Prime Numbers Determine whether a whole number is evenly divisible only by itself and 1. (By definition, 0 and 1 are not prime.)	The prime numbers less than 100 are 2, 3, 5, 7, 11, 13, 17, 19, 23, 29, 31, 37, 41, 43, 47, 53, 59, 61, 67, 71, 73, 79, 83, 89, and 97.
2.3 Finding the Prime Factorization of a Number Divide each factor by a prime number by using a diagram that forms the shape of tree branches.	Find the prime factorization of 30. Use a factor tree. 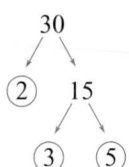 Prime factors are circled. $30 = 2 \cdot 3 \cdot 5$
2.4 Writing Fractions in Lowest Terms Divide the numerator and denominator by the greatest common factor.	Write $\frac{30}{42}$ in lowest terms. $$\frac{30}{42} = \frac{30 \div 6}{42 \div 6} = \frac{5}{7}$$
2.5 Multiplying Fractions 1. Multiply numerators by numerators and denominators by denominators. 2. Write answers in lowest terms if the multiplication shortcut was not used.	Multiply. $$\frac{6}{11} \cdot \frac{7}{8} = \frac{\overset{3}{\cancel{6}}}{11} \cdot \frac{7}{\underset{4}{\cancel{8}}} = \frac{3 \cdot 7}{11 \cdot 4} = \frac{21}{44}$$
2.7 Finding the Reciprocal To find the reciprocal of a fraction, interchange the numerator and denominator.	Find the reciprocal of each fraction. $\frac{3}{4}$ The reciprocal of $\frac{3}{4}$ is $\frac{4}{3}$. $\frac{8}{5}$ The reciprocal of $\frac{8}{5}$ is $\frac{5}{8}$. 9 The reciprocal of 9 is $\frac{1}{9}$.
2.7 Dividing Fractions Use the reciprocal of the second fraction (divisor) and multiply as fractions.	Divide. $$\frac{25}{36} \div \frac{15}{18} = \frac{\overset{5}{\cancel{25}}}{\underset{2}{\cancel{36}}} \cdot \frac{\overset{1}{\cancel{18}}}{\underset{3}{\cancel{15}}} = \frac{5 \cdot 1}{2 \cdot 3} = \frac{5}{6}$$ <center>Reciprocals</center>

Concepts	Examples

2.8 Multiplying Mixed Numbers

First estimate the answers. Then follow these steps.

Step 1 *Change* each mixed number to an improper fraction.

Step 2 *Multiply.*

Step 3 Simplify the answer which means to write it in *lowest terms*, and change it to a mixed number or whole number where possible.

First estimate the answer. Then multiply to get the exact answer.

Estimate: $\qquad$ *Exact:*

$$1\frac{3}{5} \quad \bullet \quad 3\frac{1}{3}$$

$$1\frac{3}{5} \bullet 3\frac{1}{3} = \frac{8}{\overset{}{\underset{1}{\cancel{5}}}} \bullet \frac{\overset{2}{\cancel{10}}}{3}$$

Rounded

$$2 \quad \bullet \quad 3 = 6$$

$$= \frac{8 \bullet 2}{1 \bullet 3}$$

$$= \frac{16}{3} = 5\frac{1}{3}$$

Close to estimate

2.8 Dividing Mixed Numbers

First estimate the answer. Then follow these steps.

Step 1 *Change* each mixed number to an improper fraction.

Step 2 Use the *reciprocal* of the second fraction (divisor).

Step 3 *Multiply.*

Step 4 Simplify the answer which means to write it in *lowest terms*, and change it to a mixed number or whole number where possible.

First estimate the answer. Then divide to get the exact answer.

Estimate: $\qquad$ *Exact:*

$$3\frac{5}{9} \quad \div \quad 2\frac{2}{5}$$

$$3\frac{5}{9} \div 2\frac{2}{5} = \frac{32}{9} \div \frac{12}{5}$$

Rounded

$$4 \quad \div \quad 2 = 2$$

$$= \frac{\overset{8}{\cancel{32}}}{9} \bullet \frac{5}{\underset{3}{\cancel{12}}} = \frac{40}{27}$$

$$= 1\frac{13}{27}$$

Close to estimate

ANSWERS TO TEST YOUR WORD POWER

1. (b) *Example:* In $\frac{3}{8}$, the numerator is 3. **2. (a)** *Example:* $\frac{1}{2}, \frac{3}{4},$ and $\frac{7}{8}$ are all proper fractions with a value less than 1. **3. (c)** *Example:* $2\frac{3}{8}$ and $5\frac{3}{4}$ are mixed numbers. **4. (c)** *Example:* Since $3 \bullet 5 = 15$, the numbers 3 and 5 are factors of 15. **5. (c)** *Example:* 3, 5, and 11 are prime numbers; 4, 8, and 12 are composite numbers. **6. (b)** *Example:* 3 is a common factor of both 6 and 9 because it can be evenly divided into each of them. **7. (c)** *Example:* $\frac{3}{8}, \frac{4}{5},$ and $\frac{5}{6}$ are in lowest terms but $\frac{6}{8}, \frac{3}{6},$ and $\frac{2}{4}$ are not.

8. (b) *Example:* The reciprocal of $\frac{3}{8}$ is $\frac{8}{3}$, the reciprocal of $\frac{25}{4}$ is $\frac{4}{25}$, and the reciprocal of 6 or $\frac{6}{1}$ is $\frac{1}{6}$.

Chapter 2 **REVIEW EXERCISES**

[2.1] *Write the fraction that represents each shaded portion.*

1.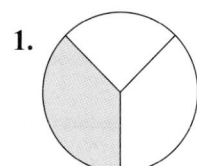

2.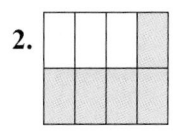

3.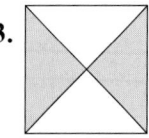

List the proper and improper fractions in each group.

	Proper	**Improper**

4. $\dfrac{1}{8}, \dfrac{4}{3}, \dfrac{5}{5}, \dfrac{3}{4}, \dfrac{2}{3}$ _____ _____

5. $\dfrac{6}{5}, \dfrac{15}{16}, \dfrac{16}{13}, \dfrac{1}{8}, \dfrac{5}{3}$ _____ _____

[2.2] *Write each mixed number as an improper fraction. Write each improper fraction as a mixed number.*

6. $5\dfrac{2}{3}$ 7. $10\dfrac{4}{5}$ 8. $\dfrac{17}{8}$ 9. $\dfrac{63}{5}$

[2.3] *Find all factors of each number.*

10. 6   11. 24   12. 55   13. 90

Write the prime factorization of each number by using exponents.

14. 27   15. 150   16. 168

Simplify each expression.

17. 6^2 18. $5^2 \cdot 2^3$ 19. $8^2 \cdot 3^3$ 20. $4^3 \cdot 2^5$

[2.4] *Write each fraction in lowest terms.*

21. $\dfrac{9}{12}$ 22. $\dfrac{30}{36}$ 23. $\dfrac{75}{80}$

Write the numerator and denominator of each fraction as a product of prime factors. Then, write the fraction in lowest terms.

24. $\dfrac{25}{60}$ 25. $\dfrac{384}{96}$

Decide whether each pair of fractions is equivalent or not equivalent, using the method of prime factors.

26. $\frac{3}{4}$ and $\frac{48}{64}$

27. $\frac{3}{4}$ and $\frac{42}{58}$

[2.5–2.8] *Multiply. Write answers in lowest terms, and as mixed numbers or whole numbers where possible.*

28. $\frac{2}{3} \cdot \frac{3}{4}$

29. $\frac{3}{10} \cdot \frac{5}{8}$

30. $\frac{70}{175} \cdot \frac{5}{14}$

31. $\frac{44}{63} \cdot \frac{3}{11}$

32. $\frac{5}{16} \cdot 48$

33. $\frac{5}{8} \cdot 1000$

Divide. Write answers in lowest terms, and as mixed numbers or whole numbers where possible.

34. $\frac{1}{2} \div \frac{3}{4}$

35. $\frac{5}{6} \div \frac{1}{2}$

36. $\dfrac{\frac{15}{18}}{\frac{10}{30}}$

37. $\dfrac{\frac{7}{8}}{\frac{7}{16}}$

38. $7 \div \frac{7}{8}$

39. $18 \div \frac{3}{4}$

40. $\frac{4}{5} \div 6$

41. $\frac{2}{3} \div 5$

42. $\dfrac{\frac{12}{13}}{3}$

Find the area of each rectangle.

43.

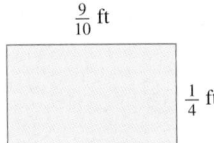
$\frac{9}{10}$ ft
$\frac{1}{4}$ ft

44.

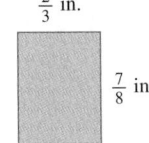
$\frac{2}{3}$ in.
$\frac{7}{8}$ in.

45. Find the area of a display shelf having a length of 10 ft and a width of $\frac{3}{4}$ ft.

46. Find the area of a rectangle having a length of 48 yd and a width of $\frac{3}{4}$ yd.

First estimate the answer. Then multiply or divide to find the exact answer. Simplify all answers.

47. *Exact:*

$5\frac{1}{2} \cdot 1\frac{1}{4}$

Estimate:

_____ • _____ = _____

48. *Exact:*

$2\frac{1}{4} \cdot 7\frac{1}{8} \cdot 1\frac{1}{3}$

Estimate:

_____ • _____ • _____ = _____

49. *Exact:*

$$15\frac{1}{2} \div 3$$

Estimate:

_____ ÷ _____ = _____

50. *Exact:*

$$4\frac{3}{4} \div 6\frac{1}{3}$$

Estimate:

_____ ÷ _____ = _____

Solve each application problem by using the six problem-solving steps.

51. How many $\frac{7}{8}$-lb boxes of sheetrock nails can be filled from 294 lb of sheetrock nails.

52. An estate is divided so that each of 5 children receives equal shares of $\frac{2}{3}$ of the estate. What fraction of the total estate will each receive?

53. How many window-blind pull cords can be made from $157\frac{1}{2}$ yd of cord if $4\frac{3}{8}$ yd of cord are needed for each blind? First estimate, and then find the exact answer.

Estimate:

Exact:

54. Working as a bookkeeper, Neta Fitzgerald is paid $\$8\frac{1}{2}$ per hour. Find her earnings for a week in which she works 38 hours. First estimate, and then find the exact answer.

Estimate:

Exact:

55. Ebony Wilson purchased 100 lb of rice at the food co-op. After selling $\frac{1}{4}$ of this to her neighbor, she gives $\frac{2}{3}$ of the remaining rice to her parents. How many pounds of rice does she have left?

56. Tiffany Crowder received her Social Security check for $1275. After paying $\frac{1}{3}$ of this amount for rent, she paid $\frac{2}{5}$ of the remaining amount for food, utilities, and transportation. How much money does she have left?

57. The Springvale Parish will divide $\frac{5}{8}$ of its library budget evenly among the 4 largest parish libraries. What fraction of the total library budget will each of these libraries receive?

58. In a morning of deep-sea fishing, 8 fishermen catch $\frac{2}{3}$ ton of halibut. If they divide the fish evenly, how much will each receive?

13. _____

13. Explain how to multiply fractions. What additional step must be taken when dividing fractions?

14. _____

Multiply or divide. Write answers in lowest terms, and as mixed numbers or whole numbers where possible.

14. $\dfrac{2}{5} \cdot \dfrac{3}{8}$

15. _____

15. $36 \cdot \dfrac{4}{9}$

16. _____

16. Find the area of a kitchen grill measuring $\dfrac{3}{4}$ yd by $\dfrac{1}{2}$ yd.

17. _____

17. There are 4224 people in a survey of licensed drivers. If $\dfrac{7}{8}$ of these drivers have had no traffic citations within the last 2 years, find the number of drivers who have had no citations.

18. _____

18. $\dfrac{3}{4} \div \dfrac{5}{6}$

19. $\dfrac{\dfrac{7}{4}}{9}$

19. _____

20. _____

20. The Lincoln Fun Run committee has acquired 120 quarts of Sports Drink to sell in $\dfrac{3}{5}$-quarts sports bottles at the race. Find the number of sports bottles that can be filled.

First estimate the answer. Then either multiply or divide to find the exact answer. Simplify all answers.

21. Estimate: _____

21. $4\dfrac{1}{8} \cdot 3\dfrac{1}{2}$

22. $1\dfrac{5}{6} \cdot 4\dfrac{1}{3}$

Exact: _____

22. Estimate: _____

Exact: _____

23. Estimate: _____

23. $9\dfrac{3}{5} \div 2\dfrac{1}{4}$

24. $\dfrac{8\dfrac{1}{2}}{1\dfrac{2}{3}}$

Exact: _____

24. Estimate: _____

Exact: _____

25. Estimate: _____

25. A new vaccine is synthesized at the rate of $2\dfrac{1}{2}$ grams per day. How many grams can be synthesized in $12\dfrac{1}{4}$ days?

Exact: _____

Name the digit that has the given place value in each number.

1. 579
   hundreds
   tens

2. 8,621,785
   millions
   ten thousands

Add, subtract, multiply, or divide as indicated.

3.
```
   43
   11
   85
 + 27
```

4.
```
   82,121
    5 468
      316
 +61,294
```

5.
```
   6537
 − 2085
```

6.
```
   4,819,604
 − 1,597,783
```

7.
```
   68
 ×  7
```

8. $8 \cdot 3 \cdot 4$

9.
```
   3784
 ×  573
```

10.
```
    563
 ×  800
```

11. $\dfrac{63}{7}$

12. $18\overline{)136,458}$

13. $33,886 \div 4$

14. $492\overline{)10,850}$

Round each number to the nearest ten, nearest hundred, and nearest thousand.

	Ten	Hundred	Thousand
15. 5737	_____	_____	_____
16. 76,271	_____	_____	_____

Simplify each expression by using the order of operations.

17. $3^3 − 8 \cdot 3$

18. $\sqrt{36} − 2 \cdot 3 + 5$

Solve each application problem using the six problem-solving steps.

19. The manager of Starbucks purchased 8 cases of small-size coffee cups for $46 per case and 12 cases of large-size coffee cups for $52 per case. Find the total cost of the 20 cases of coffee cups.

20. Home blood-pressure monitors sell for as little as $20 and as much as $150. If Scott bought the cheapest model and Jenn the most expensive model, how much more did Jenn pay than Scott?

21. A typical adult loses 100 hairs a day out of approximately 120,000 hairs. If the lost hairs were not replaced, find the number of hairs remaining after two years. (1 year = 365 days).

22. The Maxey family reunion will cost $4275. If the cost is to be divided evenly among 15 family members, find the cost for each family member.

23. The glass face for a picture frame measures $\frac{11}{12}$ ft by $\frac{3}{4}$ ft. Find its area.

24. The Municipal Utility District says that the cost of operating a hair dryer is $\frac{1}{5}$¢ per minute. Find the cost of operating the hair dryer for a half hour.

Write proper *or* improper *for each fraction.*

25. $\frac{2}{3}$

26. $\frac{6}{6}$

27. $\frac{9}{18}$

Write each mixed number as an improper fraction. Write each improper fraction as a whole or mixed number.

28. $3\frac{3}{8}$

29. $6\frac{2}{5}$

30. $\frac{14}{7}$

31. $\frac{103}{8}$

Find the prime factorization of each number. Write answers by using exponents.

32. 72

33. 126

34. 350

Simplify each expression.

35. $4^2 \cdot 2^2$

36. $2^3 \cdot 6^2$

37. $2^3 \cdot 4^2 \cdot 5$

Write each fraction in lowest terms.

38. $\frac{42}{48}$

39. $\frac{24}{36}$

40. $\frac{30}{54}$

Multiply or divide as indicated. Simplify all answers.

41. $\frac{1}{2} \cdot \frac{3}{4}$

42. $30 \cdot \frac{2}{3} \cdot \frac{3}{5}$

43. $7\frac{1}{2} \cdot 3\frac{1}{3}$

44. $\frac{3}{5} \div \frac{5}{8}$

45. $\frac{7}{8} \div 1\frac{1}{2}$

46. $3 \div 1\frac{1}{4}$

Adding and Subtracting Fractions

3

Most types of home improvement, such as carpentry, wallpapering a room, or hanging pictures on a wall, require the use of a tape measure. The inches on a tape measure are typically divided into smaller and smaller segments that represent $\frac{1}{2}$, $\frac{1}{4}$, $\frac{1}{8}$, and $\frac{1}{16}$ of an inch. In order to use a tape measure effectively, you need to be able to add and subtract common fractions and mixed numbers. (See Exercises 41 and 51 in Section 3.3 and Exercise 53 in Section 3.4.)

ADDISON - WESLEY
MyMathLab.com
You're Connected

3.1   ADDING AND SUBTRACTING LIKE FRACTIONS

In Chapter 2 we looked at the basics of fractions and then practiced with multiplication and division of common fractions and mixed numbers. In this chapter we will work with addition and subtraction of common fractions and mixed numbers.

1   **Define like and unlike fractions.**   Fractions with the same denominators are **like fractions.** Fractions with different denominators are **unlike fractions.**

1 Next to each pair of fractions write *like* or *unlike*.

(a) $\dfrac{2}{3}$ $\dfrac{1}{3}$ _____

Example 1   **Identifying Like and Unlike Fractions**

(a) $\dfrac{3}{4}, \dfrac{1}{4}, \dfrac{5}{4}, \dfrac{6}{4},$ and $\dfrac{4}{4}$ are **like** fractions.

  All denominators are the same.

(b) $\dfrac{7}{12}$ and $\dfrac{12}{7}$ are **unlike** fractions.

  All denominators are different.

Work Problem 1 at the Side.

(b) $\dfrac{3}{4}$ $\dfrac{3}{8}$ _____

NOTE

Like fractions have the *same* denominator.

2   **Add like fractions.**   The following figures show you how to add the fractions $\frac{2}{7}$ and $\frac{4}{7}$.

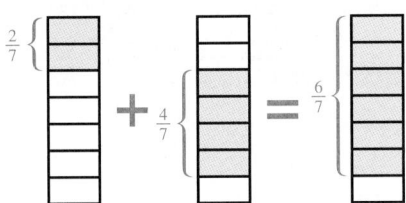

(c) $\dfrac{6}{15}$ $\dfrac{11}{15}$ _____

As the figures show,

$$\frac{2}{7} + \frac{4}{7} = \frac{6}{7}.$$

Add like fractions as follows.

(d) $\dfrac{5}{12}$ $\dfrac{5}{7}$ _____

Adding Like Fractions

Step 1   Add the numerators to find the numerator of the sum.

Step 2   Write the denominator of the like fractions as the denominator of the sum.

Step 3   Write the answer in lowest terms.

Example 2 Adding Like Fractions

Add and write the answer in lowest terms.

(a) $\dfrac{1}{5} + \dfrac{2}{5}$

$$\overbrace{\qquad}^{\text{Add numerators.}}$$

$$\frac{1}{5} + \frac{2}{5} = \frac{1 + 2}{5} = \frac{3}{5} \leftarrow \text{Same denominator}$$

(b) $\dfrac{1}{12} + \dfrac{7}{12} + \dfrac{1}{12}$

Step 1 $\qquad \dfrac{\overbrace{1 + 7 + 1}^{\text{Add numerators.}}}{12}$

Step 2 $\qquad = \dfrac{9}{12} \begin{array}{l} \leftarrow \text{Sum} \\ \leftarrow \text{Same denominator} \end{array}$

Step 3 $\qquad = \dfrac{9 \div 3}{12 \div 3} = \dfrac{3}{4} \quad \text{In lowest terms}$

CAUTION

Fractions may be added **only** if they have like denominators.

Work Problem ❷ at the Side.

3 Subtract like fractions. The figures show $\frac{7}{8}$ broken into $\frac{4}{8}$ and $\frac{3}{8}$.

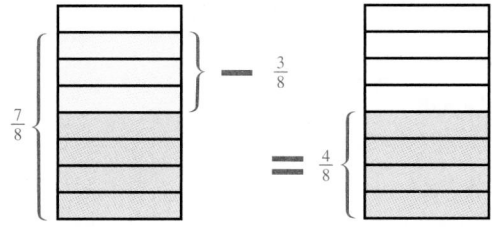

Subtracting $\frac{3}{8}$ from $\frac{7}{8}$ gives the answer $\frac{4}{8}$, or

$$\frac{7}{8} - \frac{3}{8} = \frac{4}{8}.$$

❷ Add and write the answers in lowest terms.

(a) $\dfrac{1}{5} + \dfrac{3}{5}$

(b) $\begin{array}{r} \dfrac{1}{9} \\ + \dfrac{5}{9} \\ \hline \end{array}$

(c) $\dfrac{5}{8} + \dfrac{1}{8}$

(d) $\dfrac{3}{10} + \dfrac{1}{10} + \dfrac{4}{10}$

3 Subtract and simplify.

(a) $\dfrac{7}{9} - \dfrac{5}{9}$

(b) $\dfrac{13}{16}$
$-\dfrac{5}{16}$

(c) $\dfrac{15}{3} - \dfrac{5}{3}$

(d) $\dfrac{25}{32}$
$-\dfrac{6}{32}$

Write $\frac{4}{8}$ in lowest terms.

$$\frac{7}{8} - \frac{3}{8} = \frac{4 \div 4}{8 \div 4} = \frac{1}{2}$$

The steps for subtracting like fractions are very similar to those for adding like fractions.

Subtracting Like Fractions

Step 1 Subtract the numerators to find the numerator of the difference.

Step 2 Write the denominator of the like fractions as the denominator of the difference.

Step 3 Write the answer in lowest terms.

Example 3 Subtracting Like Fractions

Subtract and simplify the answer.

(a) $\dfrac{15}{16} - \dfrac{3}{16}$

Step 1 $\dfrac{15}{16} - \dfrac{3}{16} = \dfrac{15 - 3}{16}$ ⏜ Subtract numerators.

Step 2 $= \dfrac{12}{16}$ ← Difference
  ← Same denominator

Step 3 $= \dfrac{12 \div 4}{16 \div 4} = \dfrac{3}{4}$ In lowest terms

(b) $\dfrac{13}{4} - \dfrac{6}{4}$

$\dfrac{13}{4} - \dfrac{6}{4} = \dfrac{13 - 6}{4}$ ⏜ Subtract numerators.
  ← Same denominator

$= \dfrac{7}{4}$

Write as a mixed number.

$$\frac{7}{4} = 1\frac{3}{4}$$

CAUTION

Fractions may be subtracted *only* if they have like denominators.

Work Problem 3 at the Side.

3.1 EXERCISES

FOR EXTRA HELP

 Student's Solutions Manual
 MyMathLab.com
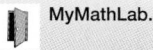 InterAct Math Tutorial Software
 AW Math Tutor Center
 www.mathxl.com
 Digital Video Tutor CD 2 Videotape 6

Add and simplify the answer. See Example 2.

1. $\dfrac{3}{5} + \dfrac{1}{5}$

2. $\dfrac{5}{8} + \dfrac{2}{8}$

3. $\dfrac{2}{6} + \dfrac{3}{6}$

4. $\dfrac{9}{11} + \dfrac{1}{11}$

5. $\dfrac{1}{6} + \dfrac{1}{6}$

6. $\dfrac{1}{12} + \dfrac{1}{12}$

7. $\begin{array}{r} \dfrac{9}{10} \\ + \dfrac{3}{10} \\ \hline \end{array}$

8. $\begin{array}{r} \dfrac{13}{12} \\ + \dfrac{5}{12} \\ \hline \end{array}$

9. $\begin{array}{r} \dfrac{2}{9} \\ + \dfrac{1}{9} \\ \hline \end{array}$

10. $\dfrac{3}{14} + \dfrac{5}{14}$

11. $\dfrac{6}{20} + \dfrac{4}{20} + \dfrac{3}{20}$

12. $\dfrac{1}{7} + \dfrac{2}{7} + \dfrac{3}{7}$

13. $\dfrac{4}{15} + \dfrac{2}{15} + \dfrac{5}{15}$

14. $\dfrac{5}{11} + \dfrac{1}{11} + \dfrac{4}{11}$

15. $\dfrac{3}{8} + \dfrac{7}{8} + \dfrac{2}{8}$

16. $\dfrac{4}{9} + \dfrac{1}{9} + \dfrac{7}{9}$

17. $\dfrac{2}{54} + \dfrac{8}{54} + \dfrac{12}{54}$

18. $\dfrac{7}{64} + \dfrac{15}{64} + \dfrac{20}{64}$

Subtract and simplify the answer. See Example 3.

19. $\dfrac{7}{8} - \dfrac{4}{8}$

20. $\dfrac{2}{3} - \dfrac{1}{3}$

21. $\dfrac{10}{11} - \dfrac{4}{11}$

22. $\dfrac{4}{5} - \dfrac{3}{5}$

23. $\dfrac{9}{10} - \dfrac{3}{10}$

24. $\dfrac{7}{8} - \dfrac{3}{8}$

25. $\begin{array}{r} \dfrac{31}{21} \\ - \dfrac{7}{21} \\ \hline \end{array}$

26. $\begin{array}{r} \dfrac{43}{24} \\ - \dfrac{13}{24} \\ \hline \end{array}$

27. $\begin{array}{r} \dfrac{27}{40} \\ - \dfrac{19}{40} \\ \hline \end{array}$

28. $\begin{array}{r} \dfrac{38}{55} \\ - \dfrac{16}{55} \\ \hline \end{array}$

29. $\dfrac{47}{36} - \dfrac{5}{36}$

30. $\dfrac{76}{45} - \dfrac{21}{45}$

31. $\dfrac{73}{60} - \dfrac{7}{60}$

32. $\dfrac{181}{100} - \dfrac{31}{100}$

33. In your own words, write an explanation of either how to add or subtract like fractions. Consider using three steps in your explanation.

34. Describe in your own words the difference between *like* fractions and *unlike* fractions. Give three examples of each type.

Solve each application problem. Write answers in lowest terms.

35. In June, Robert and Maria Hernandez had saved $\frac{3}{8}$ of the amount needed for a down payment on their first home. By November, they had saved another $\frac{3}{8}$ of the amount needed. What fraction of the amount needed have they saved?

ALMOST THERE...
How much have the Hernandezes saved, expressed as a fraction?

$\frac{3}{8}$
$\frac{3}{8}$?

36. After an initial payment to a lotto winner, the state lottery commission still owed the lotto winner $\frac{17}{20}$ of his total winnings. If the state pays the lotto winner another $\frac{3}{20}$ of the winnings, what fraction is still owed?

37. The Gerards owe $\frac{5}{9}$ of a loan for last year's vacation. If they pay $\frac{2}{9}$ of it this month, what fraction of the loan will they still owe?

38. Captain Sisko has been ordered to investigate the wreckage of a plane crash. Sisko leads a search-and-rescue team $\frac{5}{12}$ mile down a ravine and then $\frac{1}{12}$ mile along a creek bed to the crash site. How far does the search-and-rescue team travel?

39. An organic farmer purchased $\frac{9}{10}$ acre of land one year and $\frac{3}{10}$ acre the next year. She then planted carrots on $\frac{7}{10}$ acre of the land and squash on the remainder. How much land is planted in squash?

40. A forester planted $\frac{5}{12}$ acre in seedlings in the morning and $\frac{11}{12}$ acre in the afternoon. If $\frac{7}{12}$ acre of seedlings were destroyed by frost, how many acres remained?

3.2 LEAST COMMON MULTIPLES

Only *like* fractions can be added or subtracted. Because of this, we must rewrite *unlike* fractions as *like* fractions before we can add or subtract.

1 Find the least common multiple. We can rewrite unlike fractions as like fractions by finding the *least common multiple* of the denominators.

Least Common Multiple

The **least common multiple (LCM)** of two whole numbers is the smallest whole number divisible by both those numbers.

Example 1 Finding the Least Common Multiple

Find the least common multiple of 6 and 9.
    This list shows the multiples of 6.

$$\underbrace{6 \cdot 1}_{6,} \quad \underbrace{6 \cdot 2}_{12,} \quad \underbrace{6 \cdot 3}_{18,} \quad \underbrace{6 \cdot 4}_{24,} \quad \underbrace{6 \cdot 5}_{30,} \quad \underbrace{6 \cdot 6}_{36,} \quad \underbrace{6 \cdot 7}_{42,} \quad \underbrace{6 \cdot 8}_{48, \ldots}$$

(The three dots at the end of the list show that the list continues in the same pattern without stopping.) The next list shows multiples of 9.

$$\underbrace{9 \cdot 1}_{9,} \quad \underbrace{9 \cdot 2}_{18,} \quad \underbrace{9 \cdot 3}_{27,} \quad \underbrace{9 \cdot 4}_{36,} \quad \underbrace{9 \cdot 5}_{45,} \quad \underbrace{9 \cdot 6}_{54,} \quad \underbrace{9 \cdot 7}_{63,} \quad \underbrace{9 \cdot 8}_{72, \ldots}$$

The smallest number found in *both* lists is 18, so 18 is the **least common multiple** of 6 and 9; the number 18 is the smallest whole number divisible by both 6 and 9.

Multiples of 6:   6, 12, **18**, 24, 30, 36, 42, 48, . . .
Multiples of 9:   9, **18**, 27, 36, 45, 54, 63, 72, . . .

18 is the smallest number found in both lists. **18** is the least common multiple of 6 and 9.

━━━━━━━━━━━━━ **Work Problem ❶ at the Side.**

2 Find the least common multiple by using multiples of the largest number. There are several ways to find the least common multiple of a number. If the numbers are small, the least common multiple can often be found by inspection. Can you think of a number that can be divided evenly by both 3 and 4? What about 6 or 8, or perhaps 10 or 12? The number 12 will work; it is the least common multiple of the numbers 3 and 4. A method that works well to find the least common multiple is to write multiples of the larger number.
    In this case, write the multiples of 4:

$$4, 8, 12, 16, 20, \ldots$$

Now, check each multiple of 4 to see if it is divisible by 3.

4 is *not* divisible by 3.

8 is *not* divisible by 3.

12 *is* divisible by 3.

The first multiple of 4 that is divisible by 3 is 12, so 12 is the least common multiple of 3 and 4.

OBJECTIVES

1 Find the least common multiple.

2 Find the least common multiple by using multiples of the largest number.

3 Find the least common multiple by using prime factorization.

4 Find the least common multiple by using an alternative method.

5 Write a fraction with an indicated denominator.

❶ **(a)** List the multiples of 5.

5, ___, ___, ___,

___, ___, ___,

___, . . .

(b) List the multiples of 8.

8, ___, ___, ___,

___, ___, ___, . . .

(c) Find the least common multiple of 5 and 8.

ANSWERS
1. (a) 10, 15, 20, 25, 30, 35, 40, . . .
    (b) 16, 24, 32, 40, 48, 56, . . .
    (c) 40

2 Use multiples of the larger number to find the least common multiple in each set of numbers.

(a) 2 and 3

(b) 4 and 5

(c) 6 and 8

(d) 5 and 9

3 Use prime factorization to find the LCM for each pair of numbers.

(a) 15 and 18

(b) 12 and 20

Example 2 **Finding the Least Common Multiple**

Use multiples of the larger number to find the least common multiple of 6 and 9.

We start by writing the first few multiples of 9.

Multiples of 9
$$9, 18, 27, 36, 45, 54, \ldots$$

Now, we check each multiple of 9 to see if it is divisible by 6. The first multiple of 9 that is divisible by 6 is 18.

$$9, \mathbf{18}, 27, 36, 45, 54, \ldots$$

First multiple divisible by 6 ($18 \div 6 = 3$)

The least common multiple of the numbers 6 and 9 is 18.

Work Problem 2 at the Side.

3▭ **Find the least common multiple by using prime factorization.** Example 2 shows how to find the least common multiple of two numbers by making a list of the multiples of the *larger* number. Although this method works for smaller numbers, it is usually easier to find the least common multiple for larger numbers by using *prime factorization*, as shown in the next example.

Example 3 **Applying Prime Factorization Knowledge**

Use prime factorization to find the least common multiple of 9 and 12.
We start by finding the prime factorization of each number.

Factors of 9

$$9 = 3 \cdot 3$$
$$12 = 2 \cdot 2 \cdot 3$$

$$\text{LCM} = 3 \cdot 3 \cdot 2 \cdot 2 = 36$$

Factors of 12

Check to see that 36 is divisible by 9 (yes) and by 12 (yes). The smallest whole number divisible by both 9 and 12 is 36.

CAUTION

Notice that we did *not* have to repeat the factors that 9 and 12 have in common. In this case, the **3** in 2 • 2 • **3** = 12 was *not* used because 3 is already included in 3 • 3 = 9.

Work Problem 3 at the Side.

Example 4 **Using Prime Factorization**

Find the least common multiple of 12, 18, and 20.

Continued on Next Page

ANSWERS
2. (a) 6 **(b)** 20 **(c)** 24 **(d)** 45

3. (a) $15 = 3 \cdot 5$ LCM $= 2 \cdot 3 \cdot 3 \cdot 5 = 90$
    $18 = 2 \cdot 3 \cdot 3$

(b) $12 = 2 \cdot 2 \cdot 3$ LCM $= 2 \cdot 2 \cdot 3 \cdot 5 = 60$
    $20 = 2 \cdot 2 \cdot 5$

Find the prime factorization of each number. Then use the prime factors to build the LCM.

$$12 = 2 \cdot 2 \cdot 3$$
$$18 = 2 \cdot 3 \cdot 3$$
$$20 = 2 \cdot 2 \cdot 5$$

$$\text{LCM} = 2 \cdot 2 \cdot 3 \cdot 3 \cdot 5 = 180$$

Check to see that 180 is divisible by 12 (yes), and by 18 (yes) and by 20 (yes). The smallest whole number divisible by 12, 18, and 20 is 180.
Note: We did *not* repeat the factors that 12, 18, and 20 have in common.

================ **Work Problem ❹ at the Side.**

Example 5 **Finding the Least Common Multiple**

Find the least common multiple for each set of numbers.

(a) 5, 6, 35
Find the prime factorization for each number.

$$5 = 5$$
$$6 = 2 \cdot 3$$
$$35 = 5 \cdot 7$$

$$\text{LCM} = 2 \cdot 3 \cdot 5 \cdot 7 = 210$$

The least common multiple of 5, 6, and 35 is 210.

(b) 10, 20, 24
Find the prime factorization for each number.

$$10 = 2 \cdot 5$$
$$20 = 2 \cdot 2 \cdot 5$$
$$24 = 2 \cdot 2 \cdot 2 \cdot 3$$

$$\text{LCM} = 2 \cdot 2 \cdot 2 \cdot 3 \cdot 5 = 120$$

The least common multiple of 10, 20, and 24 is 120.

================ **Work Problem ❺ at the Side.**

---

4 Find the least common multiple by using an alternative method. Some people like the following *alternative method* for finding the least common multiple for larger numbers. Try both methods, and *use the one you prefer.* As a review, a list of the first few prime numbers follows.

$$2, 3, 5, 7, 11, 13, 17$$

---

❹ Find the least common multiple of the denominators in each set of fractions.

(a) $\dfrac{3}{8}$ and $\dfrac{6}{5}$

(b) $\dfrac{5}{6}$ and $\dfrac{1}{14}$

(c) $\dfrac{4}{9}$, $\dfrac{5}{18}$, and $\dfrac{7}{24}$

❺ Find the least common multiple for each set of numbers.

(a) 12, 15

(b) 8, 9, 12

(c) 18, 20, 30

(d) 15, 20, 30, 40

Answers

4. (a) $8 = 2 \cdot 2 \cdot 2$ $\text{LCM} = 2 \cdot 2 \cdot 2 \cdot 5 = 40$
   $5 = 1 \cdot 5$

(b) $6 = 2 \cdot 3$ $\text{LCM} = 2 \cdot 3 \cdot 7 = 42$
   $14 = 2 \cdot 7$

(c) $9 = 3 \cdot 3$ $\text{LCM} = 2 \cdot 2 \cdot 2 \cdot 3 \cdot 3 = 72$
   $18 = 2 \cdot 3 \cdot 3$
   $24 = 2 \cdot 2 \cdot 2 \cdot 3$

5. (a) 60 **(b)** 72 **(c)** 180 **(d)** 120

6 In the following problems, the divisions have already been worked out. Multiply the prime numbers on the left to find the least common multiple.

(a) 2 |6   1̸5̸
   3 |3   15
   5 |1̸   5
       1   1

(b) 2 |20   36
   2 |10   18
   3 |5̸   9
   3 |5̸   3
   5 |5   1̸
       1   1

Example 6 **Alternative Method for Finding the Least Common Multiple**

Find the least common multiple of each set of numbers.

(a) 14 and 21

Start by trying to divide 14 and 21 by the preceding list of prime numbers, 2, 3, 5, 7, 11, 13, and 17.
Use the following shortcut.
Divide by 2, the first prime

$$2 | 14 \quad 2̸1̸$$
$$\quad\quad 7 \quad 21$$

Because 21 cannot be divided evenly by 2, cross 21 out and bring it down.
Divide by 3, the second prime.

$$2 | 14 \quad 2̸1̸$$
$$3 | 7̸ \quad 21$$
$$\quad\quad 7 \quad 7$$

Since 7 cannot be divided evenly by the third prime, 5, skip 5 and divide by the next prime, 7.
Divide by 7, the fourth prime.

$$2 | 14 \quad 2̸1̸$$
$$3 | 7̸ \quad 21$$
$$7 | 7 \quad 7$$
$$\quad\quad 1 \quad 1 \quad \text{All quotients are 1.}$$

When all quotients are 1, multiply the prime numbers on the left side.

least common multiple = **2 • 3 • 7 = 42**

The least common multiple of 14 and 21 is 42.

(b) 6, 15, 18

Divide by 2.
$$2 | 6 \quad 1̸5̸ \quad 18$$
$$\quad 3 \quad 15 \quad 9 \quad \text{Cross out 15 and bring it down.}$$

Divide by 3.
$$2 | 6 \quad 1̸5̸ \quad 18$$
$$3 | 3 \quad 15 \quad 9$$
$$\quad 1 \quad 5 \quad 3$$

Divide by 3 again, since the remaining 3 can be divided.
$$2 | 6 \quad 1̸5̸ \quad 18$$
$$3 | 3 \quad 15 \quad 9$$
$$3 | 1̸ \quad 5̸ \quad 3$$
$$\quad 1 \quad 5 \quad 1$$

Finally, divide by 5.
$$2 | 6 \quad 1̸5̸ \quad 18$$
$$3 | 3 \quad 15 \quad 9$$
$$3 | 1̸ \quad 5̸ \quad 3$$
$$5 | 1̸ \quad 5 \quad 1̸$$
$$\quad 1 \quad 1 \quad 1 \quad \text{All quotients are 1.}$$

Multiply the prime numbers on the left side.

2 • 3 • 3 • 5 = 90 ← Least common multiple

Work Problem 6 at the Side.

5 **Write a fraction with an indicated denominator.** Before we can add or subtract unlike fractions, we must find the least common multiple, which is then used as the denominator of the fractions.

Example 7 Writing a Fraction with an Indicated Denominator

Write the fraction $\frac{2}{3}$ with a denominator of 15.
Find a numerator, so that

$$\frac{2}{3} = \frac{?}{15}.$$

To find the new numerator, first divide **15** by **3**.

$$\frac{2}{3} = \frac{?}{15} \qquad 15 \div 3 = 5$$

Multiply both numerator and denominator of the fraction $\frac{2}{3}$ by 5.

$$\frac{2}{3} = \frac{2 \cdot 5}{3 \cdot 5} = \frac{10}{15}$$

Work Problem 7 at the Side.

This process is just the opposite of writing a fraction in lowest terms. Check the answer by writing $\frac{10}{15}$ in lowest terms; you should get $\frac{2}{3}$.

Example 8 Writing Fractions with a New Denominator

Rewrite each fraction with the indicated denominator.

(a) $\frac{3}{8} = \frac{?}{48}$

Divide 48 by 8, getting 6. Now multiply both numerator and denominator of $\frac{3}{8}$ by 6.

$$\frac{3}{8} = \frac{3 \cdot 6}{8 \cdot 6} = \frac{18}{48} \qquad \text{Multiply numerator and denominator by 6.}$$

That is, $\frac{3}{8} = \frac{18}{48}$. As a check, write $\frac{18}{48}$ in lowest terms.

(b) $\frac{5}{6} = \frac{?}{42}$

Divide 42 by 6, getting 7. Next, multiply both numerator and denominator of $\frac{5}{6}$ by 7.

$$\frac{5}{6} = \frac{5 \cdot 7}{6 \cdot 7} = \frac{35}{42} \qquad \text{Multiply numerator and denominator by 7.}$$

This shows that $\frac{5}{6} = \frac{35}{42}$. As a check, write $\frac{35}{42}$ in lowest terms.

7 Find the least common multiple of each set of numbers.

(a) 4, 8, 12

(b) 25 and 30

(c) 9, 24

(d) 8, 21, 24

8 Rewrite each fraction with the indicated denominator.

(a) $\dfrac{1}{3} = \dfrac{?}{12}$

(b) $\dfrac{3}{4} = \dfrac{?}{8}$

(c) $\dfrac{4}{5} = \dfrac{?}{25}$

(d) $\dfrac{6}{11} = \dfrac{?}{33}$

NOTE

In Example 7, the fraction $\frac{2}{3}$ was multiplied by $\frac{5}{5}$. In Example 8, the fraction $\frac{3}{8}$ was multiplied by $\frac{6}{6}$ and the fraction $\frac{5}{6}$ was multiplied by $\frac{7}{7}$. The fractions $\frac{5}{5}$, $\frac{6}{6}$, and $\frac{7}{7}$ are all equal to 1.

$$\frac{5}{5} = 1 \qquad \frac{6}{6} = 1 \qquad \frac{7}{7} = 1$$

Recall that any number multiplied by 1 is the number itself.

Work Problem 8 at the Side.

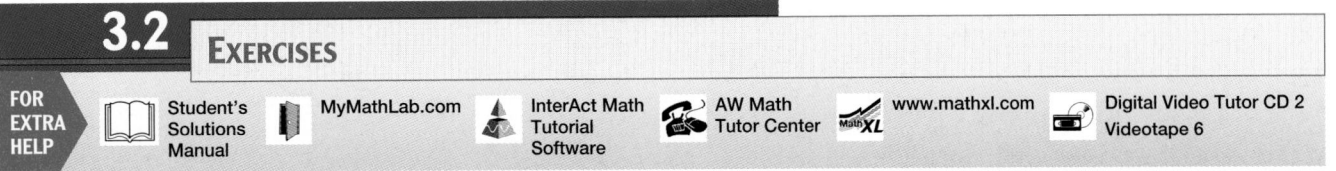

3.2 EXERCISES

FOR
EXTRA
HELP

Student's Solutions Manual   MyMathLab.com   InterAct Math Tutorial Software   AW Math Tutor Center   www.mathxl.com   Digital Video Tutor CD 2 Videotape 6

Use multiples of the larger number to find the least common multiple in each set of numbers. See Examples 1 and 2.

1. 2 and 4

2. 6 and 12

3. 3 and 5

4. 3 and 7

5. 4 and 9

6. 4 and 10

7. 2 and 7

8. 6 and 8

9. 6 and 10

10. 12 and 16

11. 20 and 50

12. 25 and 75

Find the least common multiple of each set of numbers. Use any method. See Examples 2–6.

13. 4, 10

14. 8, 10

15. 18, 24

16. 9 and 15

17. 6, 9, 12

18. 20, 24, 30

19. 4, 6, 8, 10

20. 8, 9, 12, 18

21. 12, 15, 18, 20

22. 6, 9, 27, 36

23. 8, 12, 16, 36

24. 5, 6, 25, 30

Rewrite each fraction with a denominator of 24. See Examples 7 and 8.

25. $\dfrac{1}{3} =$

26. $\dfrac{3}{8} =$

27. $\dfrac{3}{4} =$

28. $\dfrac{5}{12} =$

29. $\dfrac{5}{6} =$

30. $\dfrac{7}{8} =$

Rewrite each fraction with the indicated denominators.

31. $\dfrac{1}{4} = \dfrac{}{8}$

32. $\dfrac{2}{3} = \dfrac{}{12}$

33. $\dfrac{5}{8} = \dfrac{}{24}$

34. $\dfrac{7}{10} = \dfrac{}{30}$

35. $\dfrac{7}{8} = \dfrac{}{32}$

36. $\dfrac{5}{12} = \dfrac{}{48}$

37. $\dfrac{5}{6} = \dfrac{}{66}$

38. $\dfrac{7}{8} = \dfrac{}{96}$

39. $\dfrac{8}{5} = \dfrac{}{20}$

40. $\dfrac{7}{8} = \dfrac{}{40}$

41. $\dfrac{9}{7} = \dfrac{}{56}$

42. $\dfrac{3}{2} = \dfrac{}{64}$

43. $\dfrac{8}{3} = \dfrac{}{51}$

44. $\dfrac{7}{6} = \dfrac{}{120}$

45. $\dfrac{8}{11} = \dfrac{}{132}$

46. $\dfrac{4}{15} = \dfrac{}{165}$

47. $\dfrac{3}{16} = \dfrac{}{144}$

48. $\dfrac{7}{16} = \dfrac{}{112}$

49. There are several methods for finding the least common multiple (LCM). Do you prefer the method using multiples of the largest number or the method using prime factorizations? Why? Would you ever use the other method?

50. Explain in your own words how to write a fraction with an indicated denominator. As part of your explanation, show how to change $\dfrac{3}{4}$ to a fraction having 12 as a denominator.

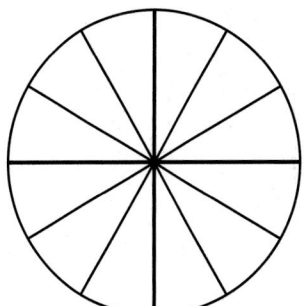

Find the least common multiple of the denominators of each pair of fractions.

51. $\dfrac{17}{800}, \dfrac{23}{3600}$

52. $\dfrac{53}{288}, \dfrac{115}{1568}$

53. $\dfrac{109}{1512}, \dfrac{23}{392}$

54. $\dfrac{61}{810}, \dfrac{37}{1170}$

RELATING CONCEPTS (Exercises 55–62)   FOR INDIVIDUAL OR GROUP WORK

Most people think that addition and subtraction of fractions is more difficult than mul-tiplication and division of fractions. This is probably because a common denominator must be used. **Work Exercises 55–62 in order.**

55. Fractions with the same denominators are _____ fractions and fractions with different denominators are _____ fractions.

56. To subtract like fractions, first find the numerator of the difference by subtracting the _____. Write the denominator of the like fractions as the _____ of the difference. The answer is then written in _____ terms.

57. The _____ common multiple (LCM) of two numbers is the _____ whole number (smallest/largest) divisible by both those numbers.

58. The following shows the common multiples for both 8 and 10. What is the least common multiple for these two numbers?

Multiples of 8:   8, 16, 24, 32, 40, 48, 56, 64, 72, 80, 88, . . .

Multiples of 10: 10, 20, 30, 40, 50, 60, 70, 80, 90, . . .

Find the least common multiple for each set of numbers.

59. 9, 8, 18, 12

60. 25, 18, 30, 5

61. Explain why the least common multiple for 8, 3, 5, 4, and 10 is not 240. Find the least common multiple.

62. Demonstrate that the least common multiple of 55 and 1760 is 1760.

3.3 ADDING AND SUBTRACTING UNLIKE FRACTIONS

1 **Add unlike fractions.**   In this section, we add and subtract unlike fractions. To add unlike fractions, we must first change them to like fractions (fractions with the same denominator). For example, the diagrams show $\frac{3}{8}$ and $\frac{1}{4}$.

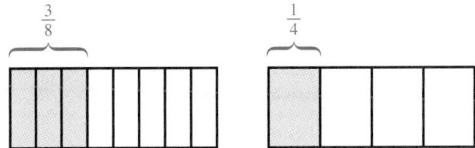

OBJECTIVES

1 Add unlike fractions.

2 Add fractions vertically.

3 Subtract unlike fractions.

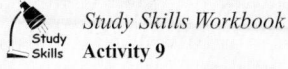 *Study Skills Workbook*
Activity 9

These fractions can be added by changing them to like fractions. Make like fractions by changing $\frac{1}{4}$ to the equivalent fraction $\frac{2}{8}$.

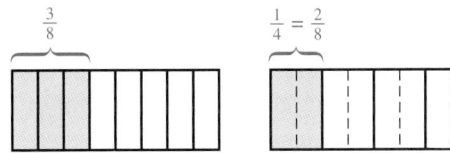

Next, add.

$$\overset{\text{Becomes}}{\frac{3}{8} + \frac{1}{4} = \frac{3}{8} + \frac{2}{8} = \frac{5}{8}}$$

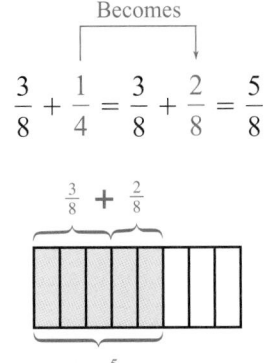

Use the following steps to add or subtract unlike fractions.

Adding or Subtracting Unlike Fractions

Step 1   Rewrite the *unlike fractions* as *like fractions* with the least common multiple as their new denominator. This new denominator is called the **least common denominator (LCD).**

Step 2   Add or subtract as with like fractions.

Step 3   Simplify the answer by writing in lowest terms and as a whole or mixed number where possible.

Example 1   Adding Unlike Fractions

Add $\frac{2}{3}$ and $\frac{1}{9}$.

The least common multiple of 3 and 9 is 9, so first write the fractions as like fractions with a denominator of 9. This is the *least common denominator* of 3 and 9.

Continued on Next Page

❶ Add.

(a) $\dfrac{1}{2} + \dfrac{3}{8}$

(b) $\dfrac{3}{4} + \dfrac{1}{8}$

(c) $\dfrac{3}{5} + \dfrac{3}{10}$

(d) $\dfrac{1}{12} + \dfrac{5}{6}$

❷ Add. Simplify all answers.

(a) $\dfrac{3}{10} + \dfrac{1}{5}$

(b) $\dfrac{5}{8} + \dfrac{1}{3}$

(c) $\dfrac{1}{10} + \dfrac{1}{3} + \dfrac{1}{6}$

Step 1 $\dfrac{2}{3} = \dfrac{?}{9}$

Divide 9 by 3, getting 3. Next, multiply numerator and denominator by 3.

$$\dfrac{2}{3} = \dfrac{2 \cdot 3}{3 \cdot 3} = \dfrac{6}{9}$$

Now, add the like fractions $\frac{6}{9}$ and $\frac{1}{9}$.

Becomes

Step 2 $\dfrac{2}{3} + \dfrac{1}{9} = \dfrac{6}{9} + \dfrac{1}{9} = \dfrac{6+1}{9} = \dfrac{7}{9}$

Step 3   Step 3 is not needed because $\frac{7}{9}$ is already in lowest terms.

Work Problem ❶ at the Side.

Example 2   Adding Fractions

Add the following fractions using the three steps. Simplify all answers.

(a) $\dfrac{1}{3} + \dfrac{1}{6}$

The least common multiple of 3 and 6 is 6. Rewrite both fractions as fractions with a least common denominator of 6.

Rewritten as like fractions

Step 1 $\dfrac{1}{3} + \dfrac{1}{6} = \dfrac{2}{6} + \dfrac{1}{6}$

Add numerators.

Step 2 $\dfrac{2}{6} + \dfrac{1}{6} = \dfrac{2+1}{6} = \dfrac{3}{6}$

Step 3 $\dfrac{3}{6} = \dfrac{1}{2}$ ⟵ In lowest terms

(b) $\dfrac{6}{15} + \dfrac{3}{10}$

The least common multiple of 15 and 10 is 30, so rewrite both fractions with a least common denominator of 30.

Rewritten as like fractions

Step 1 $\dfrac{6}{15} + \dfrac{3}{10} = \dfrac{12}{30} + \dfrac{9}{30}$

Add numerators.

Step 2 $\dfrac{12}{30} + \dfrac{9}{30} = \dfrac{12+9}{30} = \dfrac{21}{30}$

Step 3 $\dfrac{21}{30} = \dfrac{7}{10}$ ⟵ In lowest terms

Work Problem ❷ at the Side.

2 ▭ **Add fractions vertically.**   Fractions can also be added vertically (up and down).

Example 3   Vertical Addition of Fractions

Add the following fractions vertically.

(a)

┌───── Rewritten as like fractions

$$\frac{3}{8} = \frac{3 \cdot 3}{8 \cdot 3} = \frac{9}{24}$$

$$+\frac{7}{12} = \frac{7 \cdot 2}{12 \cdot 2} = \frac{14}{24}$$

$$\frac{23}{24} \longleftarrow \text{Add the numerators.}$$

(b)

┌───── Rewritten as like fractions

$$\frac{2}{9} = \frac{2 \cdot 4}{9 \cdot 4} = \frac{8}{36}$$

$$+\frac{1}{4} = \frac{1 \cdot 9}{4 \cdot 9} = \frac{9}{36}$$

$$\frac{17}{36} \longleftarrow \text{Add the numerators.}$$

═════════ **Work Problem ❸ at the Side.**

3 ▭ **Subtract unlike fractions.**   The next example shows subtraction of unlike fractions.

Example 4   Subtracting Unlike Fractions

Subtract. Simplify all answers.
   As with addition, rewrite unlike fractions with a least common denominator.

(a) $\dfrac{3}{4} - \dfrac{3}{8}$

┌──┬── Rewritten as like fractions

Step 1 $$\frac{3}{4} - \frac{3}{8} = \frac{6}{8} - \frac{3}{8}$$

Subtract numerators.

Step 2 $$\frac{6}{8} - \frac{3}{8} = \frac{6-3}{8} = \frac{3}{8}$$

Step 3   Not needed. $\left(\frac{3}{8} \text{ is in lowest terms.}\right)$

══ **Continued on Next Page**

---

❸ Add the following fractions vertically.

(a) $$\frac{3}{4}$$
$$+\frac{3}{16}$$

(b) $$\frac{5}{9}$$
$$+\frac{1}{3}$$

❹ Subtract. Simplify all answers.

(a) $\dfrac{5}{8} - \dfrac{1}{4}$

(b) $\dfrac{3}{4} - \dfrac{5}{9}$

Rewritten as like fractions

Step 1 $\qquad \dfrac{3}{4} - \dfrac{5}{9} = \dfrac{27}{36} - \dfrac{20}{36}$

Subtract numerators.

Step 2 $\qquad \dfrac{27}{36} - \dfrac{20}{36} = \dfrac{27 - 20}{36} = \dfrac{7}{36}$

Step 3 $\qquad$ Not needed. $\left(\dfrac{7}{36}\right.$ is in lowest terms.$\left.\right)$

Work Problem ❹ at the Side.

(b) $\dfrac{4}{5} - \dfrac{3}{4}$

(c) $\begin{array}{r} \dfrac{7}{8} \\[6pt] -\ \dfrac{2}{3} \\ \hline \end{array}$

3.3 EXERCISES

 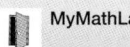
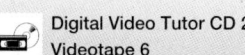

Add the following fractions. Simplify all answers. See Examples 1–3.

1. $\dfrac{1}{6} + \dfrac{2}{3}$

2. $\dfrac{1}{4} + \dfrac{1}{8}$

3. $\dfrac{2}{3} + \dfrac{2}{9}$

4. $\dfrac{3}{7} + \dfrac{1}{14}$

5. $\dfrac{9}{20} + \dfrac{3}{10}$

6. $\dfrac{5}{8} + \dfrac{1}{4}$

7. $\dfrac{3}{5} + \dfrac{3}{8}$

8. $\dfrac{5}{7} + \dfrac{3}{14}$

9. $\dfrac{2}{9} + \dfrac{5}{12}$

10. $\dfrac{5}{8} + \dfrac{1}{12}$

11. $\dfrac{1}{3} + \dfrac{3}{5}$

12. $\dfrac{2}{5} + \dfrac{3}{7}$

13. $\dfrac{1}{4} + \dfrac{2}{9} + \dfrac{1}{3}$

14. $\dfrac{3}{7} + \dfrac{2}{5} + \dfrac{1}{10}$

15. $\dfrac{3}{10} + \dfrac{2}{5} + \dfrac{3}{20}$

16. $\dfrac{1}{3} + \dfrac{3}{8} + \dfrac{1}{4}$

17. $\dfrac{4}{15} + \dfrac{1}{6} + \dfrac{1}{3}$

18. $\dfrac{5}{12} + \dfrac{2}{9} + \dfrac{1}{6}$

19. $\begin{array}{r} \dfrac{1}{3} \\ + \dfrac{1}{4} \\ \hline \end{array}$

20. $\begin{array}{r} \dfrac{7}{12} \\ + \dfrac{1}{8} \\ \hline \end{array}$

21. $\begin{array}{r} \dfrac{5}{12} \\ + \dfrac{1}{16} \\ \hline \end{array}$

22. $\begin{array}{r} \dfrac{3}{7} \\ + \dfrac{1}{3} \\ \hline \end{array}$

Subtract the following fractions. Simplify all answers. See Example 4.

23. $\dfrac{7}{8} - \dfrac{1}{2}$

24. $\dfrac{7}{8} - \dfrac{1}{4}$

25. $\dfrac{2}{3} - \dfrac{1}{6}$

26. $\dfrac{3}{4} - \dfrac{5}{8}$

27. $\dfrac{2}{3} - \dfrac{1}{5}$

28. $\dfrac{5}{6} - \dfrac{7}{9}$

29. $\dfrac{5}{12} - \dfrac{1}{4}$

30. $\dfrac{5}{7} - \dfrac{1}{3}$

31. $\dfrac{8}{9} - \dfrac{7}{15}$

32. $\begin{array}{r} \dfrac{4}{5} \\ -\dfrac{1}{3} \\ \hline \end{array}$

33. $\begin{array}{r} \dfrac{7}{8} \\ -\dfrac{4}{5} \\ \hline \end{array}$

34. $\begin{array}{r} \dfrac{5}{8} \\ -\dfrac{1}{3} \\ \hline \end{array}$

35. $\begin{array}{r} \dfrac{5}{12} \\ -\dfrac{1}{16} \\ \hline \end{array}$

36. $\begin{array}{r} \dfrac{7}{12} \\ -\dfrac{1}{3} \\ \hline \end{array}$

Solve each application problem.

37. To prepare her flower bed, Katie Lim ordered $\dfrac{1}{4}$ cubic yard of sand, $\dfrac{3}{8}$ cubic yard of mulch, and $\dfrac{1}{3}$ cubic yard of peat moss. How many cubic yards of material did she order?

38. Sergy Shibko has used his savings to repair his car. He spent $\dfrac{1}{4}$ of his savings for new tires, $\dfrac{1}{6}$ of it for brakes, $\dfrac{1}{10}$ of it for a tune-up, and $\dfrac{1}{12}$ of it for new belts and hoses. What fraction of his total savings has he spent?

39. The Weiner Works has $\frac{3}{4}$ acre of land. If $\frac{1}{6}$ acre must remain as a green belt and the remainder is buildable, find the amount of land that is buildable.

40. Della Daniel wants to open a day care center and has saved $\frac{2}{5}$ of the amount needed for start-up costs. If she saves another $\frac{1}{8}$ of the amount needed and then $\frac{1}{6}$ more, find the total portion of the start-up costs she has saved.

41. When installing cabinets, Cecil Feathers must be certain that the proper type and size of mounting screw is used. Find the total length of the screw shown.

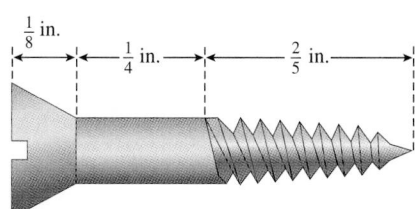

42. When installing a computer chassis, Carolyn Phelps must be certain that the proper type and size of bolt is used. Find the total length of the bolt shown.

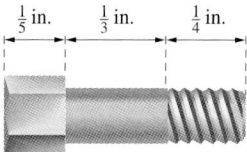

43. A hydraulic jack contains $\frac{7}{8}$ gallon of hydraulic fluid. A cracked seal resulted in a loss of $\frac{1}{6}$ gallon of fluid in the morning and another $\frac{1}{3}$ gallon in the afternoon. Find the amount of fluid remaining.

44. Adrian Ortega drives a tanker for the British Petroleum Company. He leaves the refinery with his tanker filled to $\frac{7}{8}$ of capacity. If he delivers $\frac{1}{4}$ of the tanker's capacity at the first stop and $\frac{1}{3}$ of the tanker's capacity at the second stop, find the fraction of the tanker's capacity remaining.

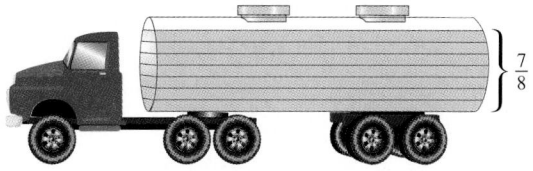

45. Step 1 in adding or subtracting unlike fractions is to rewrite the fractions so they have the least common multiple as a denominator. Explain in your own words why this is necessary.

46. Briefly list the three steps used for addition and subtraction of unlike fractions.

Refer to the circle graph to answer Exercises 47–52.

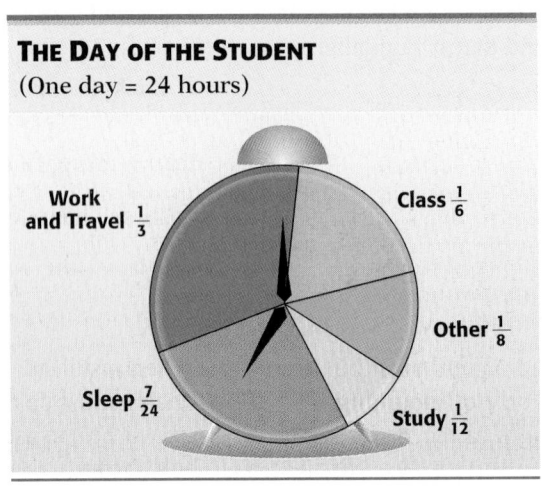

THE DAY OF THE STUDENT
(One day = 24 hours)

Work and Travel $\frac{1}{3}$

Class $\frac{1}{6}$

Other $\frac{1}{8}$

Sleep $\frac{7}{24}$

Study $\frac{1}{12}$

47. What fraction of the day was spent in class and study?

48. What fraction of the day was spent in work and travel and other?

49. In which activity was the greatest amount of time spent? How many hours did this activity take?

50. In which activity was the least amount of time spent? How many hours did this activity take?

51. Find the diameter of the hole in the mounting bracket shown. (The diameter is the distance across the center of the hole.)

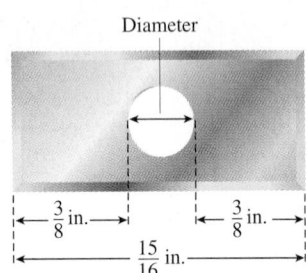

Diameter

$\frac{3}{8}$ in. $\frac{3}{8}$ in.

$\frac{15}{16}$ in.

52. Chakotay is fitting a turquoise stone into a bear claw pendant. Find the diameter of the hole in the pendant. (The diameter is the distance across the center of the hole.)

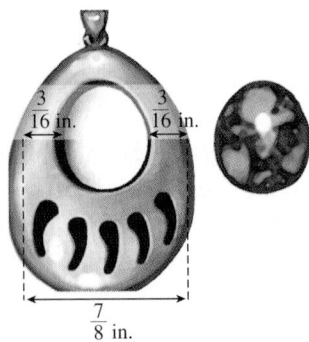

$\frac{3}{16}$ in. $\frac{3}{16}$ in.

$\frac{7}{8}$ in.

3.4 ADDING AND SUBTRACTING MIXED NUMBERS

Recall that a mixed number is the sum of a whole number and a fraction. For example,

$$3\frac{2}{5} \quad \text{means} \quad 3 + \frac{2}{5}.$$

1   **Estimate an answer, then add or subtract mixed numbers.**   Add or subtract mixed numbers by adding or subtracting the fraction parts and then the whole number parts. It is a good idea to estimate the answer first, as we did when multiplying and dividing mixed numbers in **Section 2.8.**

Work Problem ❶ at the Side.

Example 1   **Adding and Subtracting Mixed Numbers**

First estimate the answer. Then add or subtract to find the exact answer.

(a) $16\frac{1}{8} + 5\frac{5}{8}$

Estimate: *Exact:*

$$16 \xleftarrow{\text{Rounds to}} \left\{ 16\frac{1}{8} \right.$$

$$+ 6 \xleftarrow{\text{Rounds to}} \left\{ + 5\frac{5}{8} \right.$$

$$22 \qquad\qquad 21\frac{6}{8}$$

Sum of whole numbers ——┘   └—— Sum of fractions

In lowest terms $\frac{6}{8}$ *is* $\frac{3}{4}$, so the final answer is $21\frac{3}{4}$. This is reasonable because it is close to the estimate of 22.

(b) $8\frac{5}{8} - 3\frac{1}{12}$

Estimate: *Exact:*

$$9 \xleftarrow{\text{Rounds to}} \left\{ \; 8\frac{5}{8} = 8\frac{15}{24} \right.$$

$$- 3 \xleftarrow{\text{Rounds to}} \left\{ -3\frac{1}{12} = -3\frac{2}{24} \right.$$

Least common denominator

$$6 \qquad\qquad\quad 5\frac{13}{24}$$

Subtract whole numbers. ——┘   └—— Subtract fractions.

$5\frac{13}{24}$ is reasonable because it is close to the estimated answer of 6. Just as before, check by adding $5\frac{13}{24}$ and $3\frac{1}{12}$; the sum should be $8\frac{5}{8}$.

---

❶   As a review of mixed numbers, convert each mixed number to an improper fraction and each improper fraction to a mixed number.

(a) $\dfrac{7}{4}$

(b) $\dfrac{8}{3}$

(c) $5\dfrac{3}{5}$

(d) $3\dfrac{7}{8}$

❷ First estimate, and then add or subtract to find the exact answer.

(a) *Estimate:*   *Exact:*

$$7 \xleftarrow{\text{Rounds to}} \left\{ 6\frac{7}{8} \right.$$

$$+\, 2 \xleftarrow{\text{Rounds to}} \left\{ +\, 2\frac{1}{4} \right.$$

(b) $25\frac{3}{5} + 12\frac{3}{10}$

Estimate:

_____ + _____ = _____

Exact:

_____ + _____ = _____

(c) *Estimate:*   *Exact:*

$$\xleftarrow{\text{Rounds to}} \left\{ 4\frac{7}{9} \right.$$

$$- \xleftarrow{\text{Rounds to}} \left\{ -\, 2\frac{2}{3} \right.$$

❸ First estimate, and then add to find the exact answer.

(a) *Estimate:*   *Exact:*

$$\xleftarrow{\text{Rounds to}} \left\{ 9\frac{3}{4} \right.$$

$$+ \xleftarrow{\text{Rounds to}} \left\{ +\, 7\frac{1}{2} \right.$$

(b) *Estimate:*   *Exact:*

$$\xleftarrow{\text{Rounds to}} \left\{ 15\frac{4}{5} \right.$$

$$+ \xleftarrow{\text{Rounds to}} \left\{ +\, 12\frac{2}{3} \right.$$

ANSWERS

2. (a) $7 + 2 = 9$; $9\frac{1}{8}$

   (b) $26 + 12 = 38$; $37\frac{9}{10}$

   (c) $5 - 3 = 2$; $2\frac{1}{9}$

3. (a) $10 + 8 = 18$; $17\frac{1}{4}$

   (b) $16 + 13 = 29$; $28\frac{7}{15}$

NOTE

When estimating, if the numerator is *half* of the denominator or *more*, round up the whole number part. If the numerator is *less* than *half* the denominator, leave the whole number part as it is.

Work Problem ❷ at the Side.

When you add the fraction parts of mixed numbers, the sum may be greater than 1. If this happens, **carrying** from the fraction column to the whole number column is the best procedure. (You wouldn't want a whole number along with an improper fraction as the answer.)

Example 2   Carrying When Adding Mixed Numbers

First estimate, and then add $9\frac{5}{8} + 13\frac{7}{8}$.

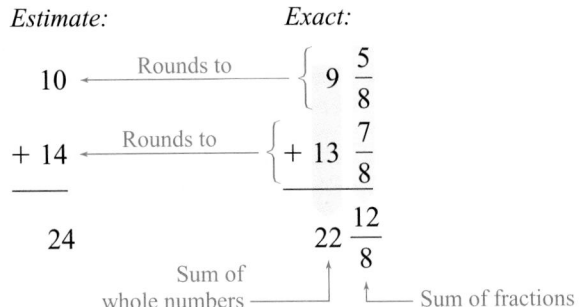

The improper fraction $\frac{12}{8}$ can be written in lowest terms as $\frac{3}{2}$. Because $\frac{3}{2} = 1\frac{1}{2}$, the sum is

$$22\frac{12}{8} = 22 + \frac{12}{8} = 22 + 1\frac{1}{2} = 23\frac{1}{2}.$$

The estimate was 24, so the exact answer is reasonable.

NOTE

When adding mixed numbers, first add the fraction parts, then add the whole number parts. Then combine the two answers simplifying the fraction part when necessary.

Work Problem ❸ at the Side.

2   Estimate an answer, then subtract mixed numbers by borrowing.   Borrowing is sometimes necessary when subtracting mixed numbers.

Example 3   Borrowing When Subtracting Mixed Numbers

First estimate, and then subtract.

(a) $7 - 2\frac{5}{6}$

Continued on Next Page

$$\begin{array}{r} \textit{Estimate:} \qquad \textit{Exact:} \\ 7 \xleftarrow{\text{Rounds to}} \left\{ 7 \right. \\ -\ 3 \xleftarrow{\text{Rounds to}} \left\{ -2\dfrac{5}{6} \right. \\ \hline 4 \end{array}$$

There is no fraction here from which to subtract $\frac{5}{6}$.

It is not possible to subtract $\frac{5}{6}$ without borrowing from the whole number 7 first.

Borrow 1.

$$7 = 6 + 1$$

$$1 = \frac{6}{6}$$

$$= 6 + \frac{6}{6}$$

$$= 6\frac{6}{6}$$

Now, subtract.

$$\begin{array}{r} 7 \ = \ 6\dfrac{6}{6} \\ -2\dfrac{5}{6} = -2\dfrac{5}{6} \\ \hline 4\dfrac{1}{6} \end{array}$$

The estimate was 4, so the exact answer is reasonable.

(b) $8\dfrac{1}{3} - 4\dfrac{3}{5}$

$$\begin{array}{r} \textit{Estimate:} \qquad\qquad \textit{Exact:} \\ 8 \xleftarrow{\text{Rounds to}} \left\{ 8\dfrac{1}{3} = \ 8\dfrac{5}{15} \right. \\ -\ 5 \xleftarrow{\text{Rounds to}} \left\{ -4\dfrac{3}{5} = -4\dfrac{9}{15} \right. \\ \hline 3 \end{array}$$

Least common denominator

It is not possible to subtract $\frac{9}{15}$ from $\frac{5}{15}$, so borrow from the whole number **8**.

Borrow 1.

$$8\frac{5}{15} = 8 + \frac{5}{15} = 7 + \ \mathbf{1} + \frac{5}{15}$$

$$1 = \tfrac{15}{15}.$$

$$= 7 + \frac{\mathbf{15}}{\mathbf{15}} + \frac{5}{15}$$

$$= 7 + \frac{\mathbf{20}}{\mathbf{15}} \quad\leftarrow\ \tfrac{15}{15} + \tfrac{5}{15}$$

$$= 7\frac{20}{15}$$

Continued on Next Page

④ First estimate and then subtract to find the exact answer.

(a) *Estimate:*   *Exact:*

$$\text{Rounds to} \longleftarrow \left\{ 7\dfrac{1}{3} \right.$$

$$- \qquad \text{Rounds to} \longleftarrow \left\{ - 4\dfrac{5}{6} \right.$$

(b) *Estimate:*   *Exact:*

$$\text{Rounds to} \longleftarrow \left\{ 4\dfrac{5}{8} \right.$$

$$- \qquad \text{Rounds to} \longleftarrow \left\{ - 2\dfrac{15}{16} \right.$$

(c) *Estimate:*   *Exact:*

$$\text{Rounds to} \longleftarrow \left\{ 15 \right.$$

$$- \qquad \text{Rounds to} \longleftarrow \left\{ - 6\dfrac{4}{9} \right.$$

⑤ Add or subtract by changing mixed numbers to improper fractions. Write answers as mixed numbers in lowest terms.

(a) $\quad 3\dfrac{3}{8}$ **(b)** $\quad 5\dfrac{3}{5}$

$\quad + 2\dfrac{1}{2}$ $+ 3\dfrac{1}{2}$

(c) $\quad 6\dfrac{3}{4}$ **(d)** $\quad 9\dfrac{7}{8}$

$\quad - 4\dfrac{2}{3}$ $- 4\dfrac{3}{4}$

Now, subtract.

$$8\dfrac{1}{3} = 8\dfrac{5}{15} = 7\dfrac{20}{15}$$
$$- 4\dfrac{3}{5} = 4\dfrac{9}{15} = 4\dfrac{9}{15}$$
$$\overline{ 3\dfrac{11}{15}}$$

The exact answer is $3\frac{11}{15}$ (lowest terms), which is reasonable because it is close to the estimate of 3.

Work Problem ④ at the Side.

3   **Add or subtract mixed numbers using an alternate method.** An alternate method for adding or subtracting mixed numbers is to first change the mixed numbers to improper fractions. Then rewrite the unlike fractions as like fractions. Finally, add or subtract the numerators and write the answer in lowest terms.

Example 4   **Adding or Subtracting Mixed Numbers**

Add or subtract.

(a) $\quad 2\dfrac{3}{8} = \dfrac{19}{8} = \dfrac{19}{8}$

$\quad\quad + 3\dfrac{3}{4} = \dfrac{15}{4} = \dfrac{30}{8}$ Least common denominator

$$\dfrac{49}{8} = 6\dfrac{1}{8} \quad \text{Answer as mixed number}$$

Change to improper fractions.

(b) $\quad 4\dfrac{2}{3} = \dfrac{14}{3} = \dfrac{70}{15}$

$\quad\quad - 2\dfrac{1}{5} = \dfrac{11}{5} = \dfrac{33}{15}$ Least common denominator

$$\dfrac{37}{15} = 2\dfrac{7}{15} \quad \text{Answer as mixed number}$$

Improper fractions

Work Problem ⑤ at the Side.

NOTE

The advantage of this alternate method of adding or subtracting mixed numbers is that it eliminates the need to carry when adding or to borrow when subtracting. However, if the mixed numbers are large, then the numerators of the improper fractions become so large that they are difficult to work with. In such cases, you may want to keep the numbers as mixed numbers.

3.4   EXERCISES

First estimate the answer. Then add to find the exact answer. Write answers as mixed numbers. See Examples 1 and 2.

1. *Estimate:*   *Exact:*

$\xleftarrow{\text{Rounds to}} \left\{ 6\dfrac{1}{3} \right.$

$+ \quad \xleftarrow{\text{Rounds to}} \left\{ + 2\dfrac{1}{2} \right.$

2. *Estimate:*   *Exact:*

$8\dfrac{3}{10}$

$+ \qquad\quad + 1\dfrac{3}{5}$

3. *Estimate:*   *Exact:*

$7\dfrac{1}{3}$

$+ \qquad\quad + 4\dfrac{1}{6}$

4. *Estimate:*   *Exact:*

$10\dfrac{1}{4}$

$+ \qquad\quad + 5\dfrac{5}{8}$

5. *Estimate:*   *Exact:*

$\dfrac{5}{8}$

$+ \qquad\quad + 3\dfrac{7}{12}$

6. *Estimate:*   *Exact:*

$12\dfrac{4}{5}$

$+ \qquad\quad + \dfrac{7}{10}$

7. *Estimate:*   *Exact:*

$24\dfrac{5}{6}$

$+ \qquad\quad + 18\dfrac{5}{6}$

8. *Estimate:*   *Exact:*

$14\dfrac{6}{7}$

$+ \qquad\quad + 15\dfrac{1}{2}$

9. *Estimate:*   *Exact:*

$33\dfrac{3}{5}$

$+ \qquad\quad + 18\dfrac{1}{2}$

10. *Estimate:*   *Exact:*

$18\dfrac{5}{8}$

$+ \qquad\quad + 6\dfrac{2}{3}$

11. *Estimate:*   *Exact:*

$22\dfrac{3}{4}$

$+ \qquad\quad + 15\dfrac{3}{7}$

12. *Estimate:*   *Exact:*

$7\dfrac{1}{4}$

$+ \qquad\quad + 25\dfrac{7}{8}$

13. *Estimate:*   *Exact:*

$$10\frac{3}{5}$$

$$17\frac{7}{10}$$

$+$ $+\ 15\frac{8}{15}$

14. *Estimate:*   *Exact:*

$$14\frac{9}{10}$$

$$8\frac{1}{4}$$

$+$ $+\ 13\frac{3}{5}$

First estimate the answer. Then subtract to find the exact answer. Simplify all answers.
See Examples 1 and 3.

15. *Estimate:*   *Exact:*

$$12\frac{3}{8}$$

$-$ $-\ 10\frac{1}{4}$

16. *Estimate:*   *Exact:*

$$14\frac{3}{4}$$

$-$ $-\ 11\frac{3}{8}$

17. *Estimate:*   *Exact:*

$$12\frac{2}{3}$$

$-$ $-\ 1\frac{1}{5}$

18. *Estimate:*   *Exact:*

$$11\frac{9}{20}$$

$-$ $-\ 4\frac{3}{5}$

19. *Estimate:*   *Exact:*

$$28\frac{3}{10}$$

$-$ $-\ 6\frac{1}{15}$

20. *Estimate:*   *Exact:*

$$15\frac{7}{20}$$

$-$ $-\ 6\frac{1}{8}$

21. *Estimate:*   *Exact:*

$$17$$

$-$ $-\ 6\frac{5}{8}$

22. *Estimate:*   *Exact:*

$$22$$

$-$ $-\ 4\frac{5}{8}$

23. *Estimate:*   *Exact:*

$$18\frac{3}{4}$$

$-$ $-\ 5\frac{4}{5}$

24. *Estimate:*   *Exact:*

$$17\dfrac{5}{8}$$

$$- \underline{\quad} \qquad -5\dfrac{2}{3}$$

25. *Estimate:*   *Exact:*

$$16\dfrac{3}{8}$$

$$- \underline{\quad} \qquad -10\dfrac{1}{2}$$

26. *Estimate:*   *Exact:*

$$20\dfrac{3}{5}$$

$$- \underline{\quad} \qquad -12\dfrac{7}{15}$$

Add or subtract by changing mixed numbers to improper fractions. Write answers as mixed numbers. See Example 4.

27. $3\dfrac{1}{2}$

$+ 5\dfrac{7}{8}$

28. $8\dfrac{3}{4}$

$+ 1\dfrac{5}{8}$

29. $4\dfrac{2}{3}$

$+ 6\dfrac{5}{6}$

30. $7\dfrac{5}{12}$

$+ 6\dfrac{2}{3}$

31. $2\dfrac{2}{3}$

$+ 1\dfrac{1}{6}$

32. $4\dfrac{1}{2}$

$+ 2\dfrac{3}{4}$

33. $3\dfrac{1}{4}$

$+ 3\dfrac{2}{3}$

34. $2\dfrac{4}{5}$

$+ 5\dfrac{1}{3}$

35. $1\dfrac{3}{8}$

$+ 6\dfrac{3}{4}$

36. $1\dfrac{5}{12}$

$+ 1\dfrac{7}{8}$

37. $3\dfrac{1}{2}$

$- 2\dfrac{2}{3}$

38. $4\dfrac{1}{4}$

$- 3\dfrac{7}{12}$

39. $5\frac{5}{8}$
$-2\frac{3}{4}$

40. $10\frac{1}{3}$
$-6\frac{5}{6}$

41. $7\frac{1}{4}$
$-4\frac{2}{3}$

42. $4\frac{1}{10}$
$-3\frac{7}{8}$

43. $9\frac{1}{5}$
$-3\frac{3}{4}$

44. $10\frac{2}{7}$
$-5\frac{5}{14}$

45. $6\frac{3}{7}$
$-2\frac{2}{3}$

46. $8\frac{2}{15}$
$-6\frac{1}{2}$

47. In your own words, explain the steps you would take to add two large mixed numbers.

48. When subtracting mixed numbers, explain when you need to borrow. Explain how to borrow using your own example.

First estimate the answer. Then solve each application problem.

49. The heaviest marine mammal in the world is the blue whale at $143\frac{3}{10}$ tons. The sixth heaviest is the humpback whale at $29\frac{1}{5}$ tons. Find the difference in their weight. (*Source: Top 10 of Everything, 2000.*)

Estimate:

Exact:

50. How much longer is the humpback whale, at $49\frac{1}{5}$ ft, than the southern elephant whale, at $21\frac{3}{10}$ ft? (*Source: Top 10 of Everything, 2000.*)

Estimate:

Exact:

51. The country with the lowest death rate is Qatar with $1\frac{3}{5}$ deaths per 1000 people each year. The death rate in the Marshall Islands is $3\frac{9}{10}$ per 1000 people. Find the difference in these death rates. (*Source:* United Nations.)

Estimate:

Exact:

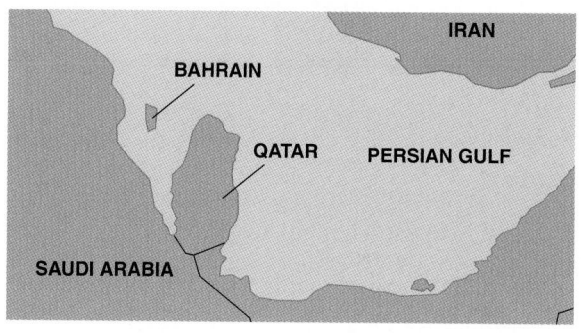

52. The death rate in the United States is $8\frac{4}{5}$ per 1000 people. The death rate in the United Arab Emirates is $2\frac{7}{10}$ per 1000 people. Find the difference in these death rates. (*Source:* United Nations.)

Estimate:

Exact:

53. A carpenter has two pieces of oak trim. One piece of trim is $12\frac{1}{2}$ ft long and the other is $8\frac{2}{3}$ ft long. How many feet of oak trim does he have in all?

Estimate:

Exact:

54. On Monday, $5\frac{3}{4}$ tons of cans were recycled, and $9\frac{3}{5}$ tons were recycled on Tuesday. How many tons were recycled on these two days?

Estimate:

Exact:

55. Andrea Abriani, a college student, works part-time at the Cyber Coffeehouse. She worked $3\frac{3}{8}$ hours on Monday, $5\frac{1}{2}$ hours on Tuesday, $4\frac{3}{4}$ hours on Wednesday, $3\frac{1}{4}$ hours on Thursday, and 6 hours on Friday. How many hours did she work altogether?

Estimate:

Exact:

56. On a recent vacation to Canada, Erin Gavin drove for $7\frac{3}{4}$ hours on the first day, $5\frac{1}{4}$ hours on the second day, $6\frac{1}{2}$ hours on the third day, and 9 hours on the fourth day. How many hours did she drive altogether?

Estimate:

Exact:

57. A craftsperson must attach a lead strip around all four sides of a stained glass window before it is installed. Find the length of lead stripping needed.

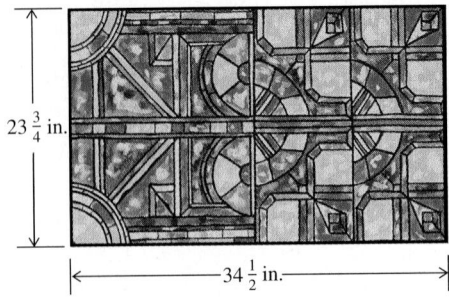

$23\frac{3}{4}$ in.

$34\frac{1}{2}$ in.

Estimate:

Exact:

59. A cement-truck driver has $8\frac{7}{8}$ cubic yards of concrete in a truck. If he unloads $2\frac{1}{2}$ cubic yards at the first stop, 3 cubic yards at the second stop, and $1\frac{3}{4}$ cubic yards at the third stop, how much concrete remains in the truck?

Estimate:

Exact:

61. The exercise yard at the correction center has four sides and is surrounded by $527\frac{1}{24}$ ft of security fencing. If three sides of the yard measure $107\frac{2}{3}$ ft, $150\frac{3}{4}$ ft, and $138\frac{5}{8}$ ft, find the length of the fourth side.

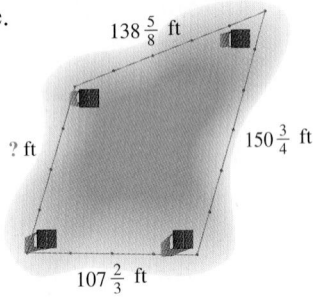

$138\frac{5}{8}$ ft

$150\frac{3}{4}$ ft

? ft

$107\frac{2}{3}$ ft

Estimate:

Exact:

58. To complete a custom order, Zak Morten of Home Depot must find the number of inches of brass trim needed to go around the four sides of the lamp base plate shown. Find the length of brass trim needed.

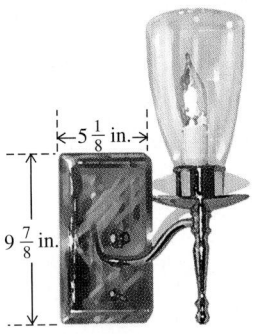

$5\frac{1}{8}$ in.

$9\frac{7}{8}$ in.

Estimate:

Exact:

60. Marv Levenson bought 15 yd of Italian silk fabric. He made two shirts with $3\frac{3}{4}$ yd of the material, a suit for his wife with $4\frac{1}{8}$ yd, and a jacket with $3\frac{7}{8}$ yd. Find the number of yards of material remaining.

Estimate:

Exact:

62. Three sides of a parking lot are $108\frac{1}{4}$ ft, $162\frac{3}{8}$ ft, and $143\frac{1}{2}$ ft. If the distance around the lot is $518\frac{3}{4}$ ft, find the length of the fourth side.

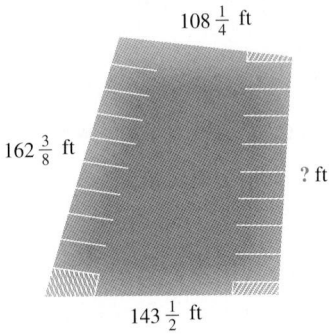

$108\frac{1}{4}$ ft

$162\frac{3}{8}$ ft

? ft

$143\frac{1}{2}$ ft

Estimate:

Exact:

63. A truck trailer is to be loaded with personal computers weighing $2\frac{5}{8}$ tons, computer monitors weighing $6\frac{1}{2}$ tons, computer printers weighing $1\frac{5}{6}$ tons, and assorted accessories weighing $3\frac{1}{4}$ tons. If the truck trailer weighs $7\frac{3}{8}$ tons empty, find the total weight after it has been loaded.

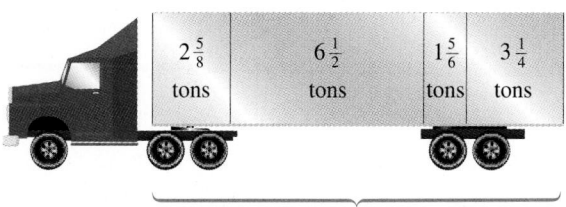

Weight of empty trailer = $7\frac{3}{8}$ tons

Estimate:

Exact:

64. Comet Auto Supply sold $16\frac{1}{2}$ cases of generic brand oil last week, $12\frac{1}{8}$ cases of Havoline oil, $8\frac{3}{4}$ cases of Valvoline oil, and $12\frac{5}{8}$ cases of Castrol oil. Find the total number of cases of oil sold during the week.

Estimate:

Exact:

Find the unknown length, labeled with a question mark, in each figure.

65.

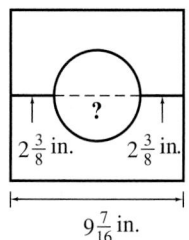

66.

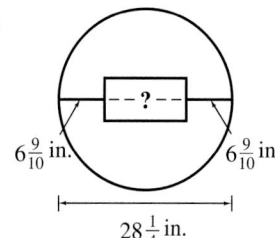

67.

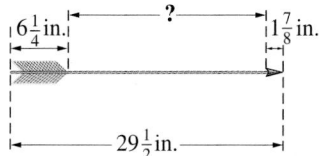

68.

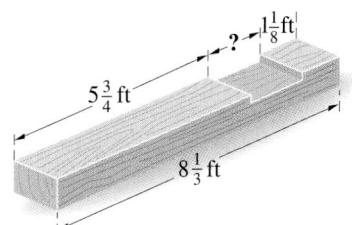

RELATING CONCEPTS (Exercises 69–74)   FOR INDIVIDUAL OR GROUP WORK

Most fraction problems include fractions with different denominators.
Work Exercises 69–74 in order.

69. To add or subtract fractions, we must first rewrite them as like fractions. Rewrite each fraction with the indicated denominator.

(a) $\dfrac{5}{9} = \dfrac{}{54}$ (b) $\dfrac{7}{12} = \dfrac{}{48}$

(c) $\dfrac{5}{8} = \dfrac{}{40}$ (d) $\dfrac{11}{5} = \dfrac{}{120}$

70. When rewriting unlike fractions as like fractions with the least common multiple as a denominator, the new denominator is called the _____ _____ _____, or LCD.

71. Add or subtract as indicated. Write answers in lowest terms.

(a) $\dfrac{5}{8} + \dfrac{1}{3}$ (b) $\dfrac{19}{20} - \dfrac{5}{12}$

(c) $\begin{array}{r} \dfrac{7}{12} \\[4pt] \dfrac{3}{16} \\[4pt] + \dfrac{3}{24} \\ \hline \end{array}$ (d) $\begin{array}{r} \dfrac{6}{7} \\[4pt] - \dfrac{2}{3} \\ \hline \end{array}$

72. A common method for adding or subtracting mixed numbers is to add or subtract the _____ _____ and then the whole number parts.

73. Another method for adding or subtracting mixed numbers is to first change the mixed numbers to _____ fractions. After adding or subtracting, write the answer as a mixed number in lowest terms.

74. Add or subtract the following fractions as indicated. First use the method where you add or subtract fraction parts and then whole number parts. Then use the method where you change each mixed number to an improper fraction, add or subtract, and then write the answer as a mixed number in lowest terms. Do you get the same answer using both methods? Which method do you prefer?

(a) $\begin{array}{r} 4\dfrac{5}{8} \\[4pt] + 3\dfrac{1}{4} \\ \hline \end{array}$ (b) $\begin{array}{r} 12\dfrac{2}{5} \\[4pt] - 8\dfrac{7}{8} \\ \hline \end{array}$

3.5   ORDER RELATIONS AND THE ORDER OF OPERATIONS

OBJECTIVES

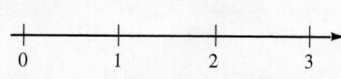

1   Identify the greater of two fractions.

2   Use exponents with fractions.

3   Use the order of operations.

There are times when we want to compare the size of two numbers. For example, we might want to know which is the greater amount, the larger size, or the longer distance.

Fractions, like whole numbers, can be located on a number line. For fractions, divide the space between whole numbers into equal parts.

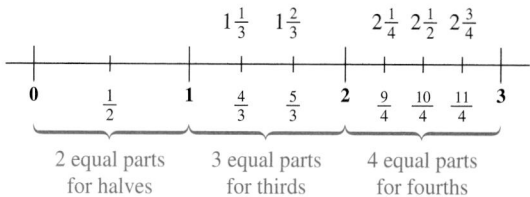

1   Identify the greater of two fractions.   To compare the size of two numbers, place the two numbers on a number line and use the following rule.

Comparing the Size of Two Numbers

The number farther to the left on the number line is always less and the number farther to the right on the number line is always greater.

For example, on the preceding number line, $\frac{1}{2}$ is to the *left* of $\frac{4}{3}$ $\left(1\frac{1}{3}\right)$, so $\frac{1}{2}$ is less than $\frac{4}{3}$ $\left(1\frac{1}{3}\right)$.

Work Problem 1 at the Side.

Write *order relations* by using the following symbols.

Symbols Used to Show Order Relations

$<$ is less than
$>$ is greater than

Example 1   Using Less-Than and Greater-Than Symbols

Rewrite the following using $<$ and $>$ symbols.

(a) $\frac{1}{2}$ is less than $\frac{4}{3}$.

$\frac{1}{2}$ **is less than** $\frac{4}{3}$ is written as $\frac{1}{2} < \frac{4}{3}$.

(b) $\frac{9}{4}$ is greater than 1.

$\frac{9}{4}$ **is greater than** 1 is written as $\frac{9}{4} > 1$.

(c) $\frac{5}{3}$ is less than $\frac{11}{4}$.

$\frac{5}{3}$ **is less than** $\frac{11}{4}$ is written as $\frac{5}{3} < \frac{11}{4}$.

1   Locate each fraction on the number line.

(a) $\frac{1}{4}$

(b) $\frac{2}{3}$

(c) $1\frac{1}{2}$

(d) $2\frac{3}{4}$

ANSWERS

1.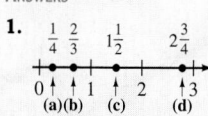

❷ Use the number line in the text to help you write < or > in each blank to make a true statement.

(a) 1 _____ $\dfrac{5}{4}$

(b) $\dfrac{8}{3}$ _____ $\dfrac{3}{2}$

(c) 0 _____ 1

(d) $\dfrac{17}{8}$ _____ $\dfrac{8}{4}$

❸ Write < or > in each blank to make a true statement.

(a) $\dfrac{2}{3}$ _____ $\dfrac{5}{8}$

(b) $\dfrac{5}{9}$ _____ $\dfrac{13}{8}$

(c) $\dfrac{3}{4}$ _____ $\dfrac{5}{6}$

(d) $\dfrac{9}{10}$ _____ $\dfrac{14}{15}$

NOTE

A number line is a very useful tool when working with order relations.

Work Problem ❷ at the Side.

The fraction $\frac{7}{8}$ represents 7 of 8 equivalent parts, while $\frac{3}{8}$ means 3 of 8 equivalent parts. Because $\frac{7}{8}$ represents more of the equivalent parts, $\frac{7}{8}$ is greater than $\frac{3}{8}$, or

$$\frac{7}{8} > \frac{3}{8}.$$

To identify the greater fraction, use the following steps.

Identifying the Greater Fraction

Step 1 Write the fractions as like fractions (same denominators).

Step 2 Compare the numerators. The fraction with the greater numerator is the greater fraction.

Example 2   Identifying the Greater Fraction

Determine which fraction in each pair is greater.

(a) $\dfrac{7}{8}, \dfrac{9}{10}$

First, write the fractions as like fractions. The least common multiple for 8 and 10 is 40, so

$$\frac{7}{8} = \frac{7 \cdot 5}{8 \cdot 5} = \frac{35}{40} \quad \text{and} \quad \frac{9}{10} = \frac{9 \cdot 4}{10 \cdot 4} = \frac{36}{40}.$$

Look at the numerators. Because 36 is greater than 35, $\frac{36}{40}$ is greater than $\frac{35}{40}$. Because $\frac{36}{40}$ is equivalent to $\frac{9}{10}$,

$$\frac{9}{10} > \frac{7}{8} \quad \text{or} \quad \frac{7}{8} < \frac{9}{10}.$$

The greater fraction is $\frac{9}{10}$.

(b) $\dfrac{8}{5}, \dfrac{23}{15}$

The least common multiple of 5 and 15 is 15.

$$\frac{8}{5} = \frac{8 \cdot 3}{5 \cdot 3} = \frac{24}{15} \quad \text{and} \quad \frac{23}{15} = \frac{23}{15}$$

This shows that $\frac{8}{5}$ is greater than $\frac{23}{15}$, or

$$\frac{8}{5} > \frac{23}{15}.$$

Work Problem ❸ at the Side.

ANSWERS

2. (a) < **(b)** > **(c)** < **(d)** >
3. (a) > **(b)** < **(c)** < **(d)** <

3.5 EXERCISES

FOR EXTRA HELP   Student's Solutions Manual   MyM

Locate each fraction in Exercise

1. $\dfrac{1}{2}$

2. $\dfrac{1}{4}$

7. $2\dfrac{1}{6}$

8. $3\dfrac{4}{5}$

Write < or > to make a true sta

13. $\dfrac{1}{2}$ ——— $\dfrac{3}{8}$ 14

17. $\dfrac{5}{12}$ ——— $\dfrac{3}{8}$ 1!

21. $\dfrac{11}{18}$ ——— $\dfrac{5}{9}$ 2

Simplify. See Example 3.

25. $\left(\dfrac{1}{2}\right)^2$

28. $\left(\dfrac{7}{8}\right)^2$

31. $\left(\dfrac{4}{5}\right)^3$

34. $\left(\dfrac{4}{3}\right)^4$

---

2 **Use exponents with fractions.**   Exponents were used in Chapter 1 to write repeated multiplication. For example,

Exponent

Exponent

$$3^2 = 3 \cdot 3 = 9 \quad \text{and} \quad 5^3 = 5 \cdot 5 \cdot 5 = 125.$$

Two factors of 3

Three factors of 5

The next example shows exponents used with fractions.

Example 3   Using Exponents with Fractions

Simplify.

(a) $\left(\dfrac{1}{2}\right)^3$

Three factors of $\frac{1}{2}$

$$\left(\dfrac{1}{2}\right)^3 = \dfrac{1}{2} \cdot \dfrac{1}{2} \cdot \dfrac{1}{2} = \dfrac{1}{8}$$

(b) $\left(\dfrac{5}{8}\right)^2$

Two factors of $\frac{5}{8}$

$$\left(\dfrac{5}{8}\right)^2 = \dfrac{5}{8} \cdot \dfrac{5}{8} = \dfrac{25}{64}$$

(c) $\left(\dfrac{3}{4}\right)^2 \cdot \left(\dfrac{2}{3}\right)^3$

$$\left(\dfrac{3}{4}\right)^2 \cdot \left(\dfrac{2}{3}\right)^3 = \left(\dfrac{3}{4} \cdot \dfrac{3}{4}\right) \cdot \left(\dfrac{2}{3} \cdot \dfrac{2}{3} \cdot \dfrac{2}{3}\right)$$

$$= \dfrac{\overset{1}{\cancel{3}} \cdot \overset{1}{\cancel{3}} \cdot \overset{1}{\cancel{2}} \cdot \overset{1}{\cancel{2}} \cdot \overset{1}{\cancel{2}}}{\underset{2}{\cancel{4}} \cdot \underset{2}{\cancel{4}} \cdot \underset{1}{\cancel{3}} \cdot \underset{1}{\cancel{3}} \cdot 3}$$ Divide out all the common factors.

$$= \dfrac{1}{6}$$

Work Problem ④ at the Side.

3 **Use the order of operations.**   Recall the *order of operations* from Chapter 1.

Order of Operations

1. Do all operations inside *parentheses or other grouping symbols*.
2. Simplify any expressions with *exponents* and find any *square roots*.
3. *Multiply* or *divide* proceeding from left to right.
4. *Add* or *subtract* proceeding from left to right.

---

④ Simplify.

(a) $\left(\dfrac{1}{4}\right)^2$

(b) $\left(\dfrac{3}{8}\right)^2$

(c) $\left(\dfrac{1}{2}\right)^3 \cdot \left(\dfrac{2}{3}\right)^2$

(d) $\left(\dfrac{1}{5}\right)^2 \cdot \left(\dfrac{5}{3}\right)^2$

ANSWERS

4. (a) $\dfrac{1}{16}$ (b) $\dfrac{9}{64}$ (c) $\dfrac{1}{18}$ (d) $\dfrac{1}{9}$

❺ Simplify by using the ord
of operations.

(a) $\dfrac{5}{9} - \dfrac{3}{4} \cdot \dfrac{2}{3}$

(b) $\dfrac{3}{5} \cdot \left(\dfrac{3}{4} - \dfrac{1}{3} \right)$

(c) $\dfrac{7}{8} \cdot \dfrac{2}{3} - \left(\dfrac{1}{2} \right)^2$

(d) $\dfrac{\left(\dfrac{5}{6} \right)^2}{\dfrac{4}{3}}$

5. (a) $\dfrac{1}{18}$ (b) $\dfrac{1}{4}$ (c) $\dfrac{1}{3}$ (d) $\dfrac{2}{4}$

RELATING CONCEPTS (Exercises 71–80) **FOR INDIVIDUAL OR GROUP WORK**

Problems involving order relations and order of operations commonly occur in problem solving. **Work Exercises 71–80 in order.**

71. When comparing the size of two numbers, we use the symbol _____ for **is less than** and the symbol _____ for **is greater than.**

72. (a) To identify the greater of two or more fractions, we must first write the fractions as _____ fractions and then compare the _____. The fraction with the greater _____ is the greater fraction.

(b) Write four pairs of fractions, all with different denominators, using the symbols for **less than** and **greater than.**

73. Fill in the blanks to complete the order of operations.

1. Do all operations inside _____ and other grouping symbols.

2. Simplify any expressions with _____ and find any _____ roots.

3. _____ or _____ proceeding from left to right.

4. _____ or _____ proceeding from left to right.

74. Use the order of operations to simplify the following.

$$\left(\dfrac{2}{3} \right)^2 - \left(\dfrac{4}{5} - \dfrac{3}{10} \right) \div \dfrac{5}{4}$$

Simplify, then place the results on the number line.

75. $\left(\dfrac{2}{3} \right)^2$

76. $\left(\dfrac{3}{8} \right)^2$

77. $\left(\dfrac{8}{7} \right)^3$

78. $\left(\dfrac{5}{4} \right)^4$

79. $4 + 2 - 2^2$

80. $\left(\dfrac{2}{3} \right)^2 - \left(\dfrac{4}{5} - \dfrac{3}{10} \right) \div \dfrac{5}{4}$

Summary Exercises on FRACTIONS

Write proper *or* improper *for each fraction.*

1. $\dfrac{7}{8}$

2. $\dfrac{6}{5}$

3. $\dfrac{8}{8}$

4. $\dfrac{9}{10}$

Write each fraction in lowest terms.

5. $\dfrac{24}{30}$

6. $\dfrac{175}{200}$

7. $\dfrac{15}{35}$

8. $\dfrac{115}{235}$

Add, subtract, multiply, or divide as indicated. Simplify all answers.

9. $\dfrac{3}{4} \cdot \dfrac{2}{3}$

10. $\dfrac{7}{12} \cdot \dfrac{9}{14}$

11. $56 \cdot \dfrac{5}{8}$

12. $\dfrac{5}{8} \div \dfrac{3}{4}$

13. $\dfrac{35}{45} \div \dfrac{10}{15}$

14. $21 \div \dfrac{3}{8}$

15. $\dfrac{7}{8} + \dfrac{2}{3}$

16. $\dfrac{5}{8} + \dfrac{3}{4} + \dfrac{7}{16}$

17. $\dfrac{7}{12} + \dfrac{5}{6} + \dfrac{2}{3}$

18. $\dfrac{5}{6} - \dfrac{3}{4}$

19. $\dfrac{7}{8} - \dfrac{5}{12}$

20. $\dfrac{4}{5} - \dfrac{2}{3}$

First estimate the answer. Then add or subtract to find the exact answer.

21. *Exact:*

$3\dfrac{1}{2} \cdot 2\dfrac{1}{4}$

Estimate:

___ • ___ = ___

22. *Exact:*

$5\dfrac{3}{8} \cdot 3\dfrac{1}{4}$

Estimate:

___ • ___ = ___

23. *Exact:*

$8 \cdot 5\dfrac{2}{3} \cdot 2\dfrac{3}{8}$

Estimate:

___ • ___ • ___ = ___

24. *Exact:*

$4\dfrac{3}{8} \div 3\dfrac{3}{4}$

Estimate:

___ ÷ ___ = ___

25. *Exact:*

$6\dfrac{7}{8} \div 2$

Estimate:

___ ÷ ___ = ___

26. *Exact:*

$4\dfrac{5}{8} \div \dfrac{3}{4}$

Estimate:

___ ÷ ___ = ___

27. *Estimate:*   *Exact:*

Rounds to $\left\{\ 4\dfrac{3}{8}\right.$

$+\ \underline{\qquad}$ Rounds to $\left\{\ +\ 3\dfrac{1}{2}\right.$

28. *Estimate:*   *Exact:*

$21\dfrac{3}{4}$

$+\ \underline{\qquad}$ $+\ 8\dfrac{5}{12}$

29. *Estimate:*   *Exact:*

$14\dfrac{3}{5}$

$+\ \underline{\qquad}$ $+\ 10\dfrac{2}{3}$

30. *Estimate:*   *Exact:*

$7\dfrac{1}{2}$

$-\ \underline{\qquad}$ $-\ 2\dfrac{3}{5}$

31. *Estimate:*   *Exact:*

14

$-\ \underline{\qquad}$ $-\ 7\dfrac{3}{8}$

32. *Estimate:*   *Exact:*

$31\dfrac{5}{6}$

$-\ \underline{\qquad}$ $-\ 22\dfrac{7}{12}$

Simplify by using the order of operations.

33. $\left(\dfrac{2}{3}-\dfrac{1}{4}\right)\cdot\dfrac{1}{5}$

34. $\dfrac{3}{4}\div\left(\dfrac{1}{2}+\dfrac{1}{3}\right)$

35. $\dfrac{2}{3}+\left(\dfrac{2}{3}\right)^{2}-\dfrac{5}{6}$

Find the least common multiple of each set of numbers.

36. 8, 10

37. 9, 18, 24

38. 4, 12, 21

Write each fraction by using the indicated denominator.

39. $\dfrac{5}{6}=\dfrac{}{42}$

40. $\dfrac{3}{7}=\dfrac{}{28}$

41. $\dfrac{11}{12}=\dfrac{}{60}$

Write < or > to make a true statement.

42. $\dfrac{7}{8}\ \underline{\qquad}\ \dfrac{15}{16}$

43. $\dfrac{16}{20}\ \underline{\qquad}\ \dfrac{23}{30}$

44. $\dfrac{11}{15}\ \underline{\qquad}\ \dfrac{7}{10}$

SUMMARY

3.1	**like fractions**	Fractions with the same denominator are called *like fractions*.
	unlike fractions	Fractions with different denominators are called *unlike fractions*.
3.2	**least common multiple**	Given two or more whole numbers, the least common multiple is the smallest whole number that is divisible by all the numbers.
	LCM	The abbreviation for *least common multiple* is LCM.
3.3	**least common denominator**	When unlike fractions are rewritten as like fractions having the least common multiple as the denominator, the new denominator is the least common denominator.
	LCD	The abbreviation for *least common denominator* is LCD.
3.4	**carrying**	Carrying is the method used when the sum of the fractions of mixed numbers is greater than 1. Carry from the fraction to the whole number.
	borrowing	Borrowing is the method used in subtracting mixed numbers when the fraction part of the minuend is less than the fraction part of the subtrahend.

TEST YOUR WORD POWER

See how well you have learned the vocabulary in this chapter. Answers follow the Quick Review.

1. **Like fractions** are
   (a) fractions that are equivalent
   (b) fractions that are not equivalent
   (c) fractions that have the same numerator
   (d) fractions that have the same denominator.

2. Two or more fractions are **unlike fractions** if
   (a) they are not equivalent
   (b) they have different numerators
   (c) they have different denominators
   (d) they are improper fractions.

3. The **least common multiple** is
   (a) the smallest whole number that is divisible by each of two or more numbers
   (b) the smallest numerator
   (c) the smallest denominator
   (d) the smallest whole number that is not divisible by a group of numbers.

4. The abbreviation **LCM** stands for
   (a) the largest common multiple
   (b) the longest common multiple
   (c) the most likely common multiplier
   (d) the least common multiple.

5. The **least common denominator** is
   (a) needed when multiplying fractions
   (b) needed when dividing fractions
   (c) the least common multiple of the denominators in a fraction problem
   (d) any denominator that is common to a group of fractions.

6. The abbreviation **LCD** stands for
   (a) the largest common denominator
   (b) the least common denominator
   (c) the least common divisor
   (d) the most likely common denominator.

QUICK REVIEW

Concepts	Examples

Concepts

Examples

3.1 *Adding Like Fractions*
Add numerators and write in lowest terms.

$$\frac{3}{4} + \frac{1}{4} + \frac{5}{4} = \frac{3 + 1 + 5}{4} = \frac{9}{4} = 2\frac{1}{4}$$

3.1 *Subtracting Like Fractions*
Subtract numerators and write in lowest terms.

$$\frac{7}{8} - \frac{5}{8} = \frac{7 - 5}{8} = \frac{2 \div 2}{8 \div 2} = \frac{1}{4}$$

3.2 *Finding the Least Common Multiple*
Method of using multiples of the larger number: List the first few multiples of the larger number. Check each one until you find the multiple that is divisible by the smaller number.

$$\frac{1}{3} + \frac{1}{4}$$

4, 8, 12, 16, . . . ⟵ Multiples of 4

First multiple divisible by 3
($12 \div 3 = 4$)

The least common multiple of the numbers 3 and 4 is 12.

3.2 *Finding the Least Common Multiple*
Method of prime numbers: First find the prime factorization of each number. Then use the prime factors to build the least common multiple.

Factors of 9

$$9 = 3 \cdot 3$$
$$\text{LCM} = 3 \cdot 3 \cdot 5 = 45$$
$$15 = 3 \cdot 5$$

Factors of 15

The least common multiple (LCM) is 45.

3.3 *Adding Unlike Fractions*
Step 1 Find the least common multiple (LCM).
Step 2 Rewrite fractions with the least common multiple as the denominator.
Step 3 Add numerators, placing the sum over the common denominator and simplify the answer.

$$\frac{1}{3} + \frac{1}{4} + \frac{1}{10} \qquad \text{LCM} = 60$$

$$\frac{1}{3} = \frac{20}{60}, \quad \frac{1}{4} = \frac{15}{60}, \quad \frac{1}{10} = \frac{6}{60}$$

$$\frac{20}{60} + \frac{15}{60} + \frac{6}{60} = \frac{41}{60}$$

3.3 *Subtracting Unlike Fractions*
Step 1 Find the least common multiple (LCM).
Step 2 Rewrite fractions with the least common multiple as the denominator.
Step 3 Subtract numerators, placing the difference over the common denominator and simplifying the answer.

$$\frac{5}{8} - \frac{1}{3} \qquad \text{LCM} = 24$$

$$\frac{5}{8} = \frac{15}{24}, \quad \frac{1}{3} = \frac{8}{24}$$

$$\frac{15}{24} - \frac{8}{24} = \frac{7}{24}$$

Concepts	Examples

3.4 Adding Mixed Numbers

Round the numbers and estimate the answer. Then find the exact answer.

1. Add fractions using a common denominator.
2. Add whole numbers.
3. Combine the sums of whole numbers and fractions, simplifying the fraction part when necessary.

Compare the exact answer to the estimate to see if it is reasonable.

Estimate: *Exact:*

$$10 \xleftarrow{\text{Rounds to}} \begin{cases} 9\dfrac{2}{3} = 9\dfrac{8}{12} \\[2mm] +\ 6\dfrac{3}{4} = 6\dfrac{9}{12} \end{cases}$$

$$+\ 7 \xleftarrow{\text{Rounds to}}$$

$$\overline{\rule{1cm}{0.4pt}}$$

$$17 \qquad\qquad 15\dfrac{17}{12} = 16\dfrac{5}{12}$$

The exact answer is reasonable because it is close to the estimate of 17.

3.4 Subtracting Mixed Numbers

Round the numbers and estimate the answer. Then find the exact answer.

1. Subtract fractions, using borrowing if necessary.
2. Subtract whole numbers.
3. Combine the differences of whole numbers and fractions, simplifying the fraction parts when necessary.

Compare the exact answer to the estimate to see if it is reasonable.

Estimate: *Exact:*

$$9 \xleftarrow{\text{Rounds to}} \begin{cases} 8\dfrac{5}{8} = 8\dfrac{15}{24} = 7\dfrac{39}{24} \\[2mm] -\ 3\dfrac{11}{12} = 3\dfrac{22}{24} = 3\dfrac{22}{24} \end{cases}$$

$$-\ 4 \xleftarrow{\text{Rounds to}}$$

$$\overline{\rule{1cm}{0.4pt}}$$

$$5 \qquad\qquad 4\dfrac{17}{24}$$

The exact answer is reasonable because it is close to the estimate of 5.

3.4 Adding or Subtracting Mixed Numbers Using an Alternate Method

1. Change the mixed numbers to improper fractions.
2. Rewrite the unlike fractions as like fractions.
3. Add or subtract the numerators and simplify the anwer.

Add.

$$2\dfrac{2}{3} = \dfrac{8}{3} = \dfrac{64}{24}$$

$$+\ 1\dfrac{3}{8} = \dfrac{11}{8} = +\ \dfrac{33}{24}$$

Least common denominator

$$\overline{\rule{2cm}{0.4pt}}$$

$$\dfrac{97}{24} = 4\dfrac{1}{24} \quad \text{Answer as mixed number}$$

Improper fractions

Subtract.

$$8\dfrac{2}{3} = \dfrac{26}{3} = \dfrac{104}{12}$$

$$-\ 5\dfrac{3}{4} = \dfrac{23}{4} = -\ \dfrac{69}{12}$$

Least common denominator

$$\overline{\rule{2cm}{0.4pt}}$$

$$\dfrac{35}{12} = 2\dfrac{11}{12} \quad \text{Answer as mixed number}$$

Improper fractions

Concepts	*Examples*

3.5 *Identifying the Larger of Two Fractions*
With unlike fractions, change to like fractions first. The fraction with the greater numerator is the greater fraction. Use these symbols:

$<$ is less than

$>$ is greater than

Identify the greater fraction.

$$\frac{7}{8}, \frac{9}{10}$$

$$\frac{7}{8} = \frac{7 \cdot 5}{8 \cdot 5} = \frac{35}{40}$$

$$\frac{9}{10} = \frac{9 \cdot 4}{10 \cdot 4} = \frac{36}{40}$$

$\frac{35}{40}$ is smaller than $\frac{36}{40}$, so $\frac{7}{8} < \frac{9}{10}$ or $\frac{9}{10} > \frac{7}{8}$.

$\frac{9}{10}$ is greater.

3.5 *Using the Order of Operations with Fractions*
Follow the order of operations.

1. Do all operations inside parentheses or other grouping symbols.
2. Simplify any expressions with exponents and find any square roots.
3. Multiply or divide from left to right.
4. Add or subtract from left to right.

Simplify by using the order of operations.

$$\frac{1}{2} \cdot \frac{2}{3} - \left(\frac{1}{4}\right)^2 \quad \text{Simplify fraction with exponent.}$$

$$= \frac{1}{2} \cdot \frac{2}{3} - \frac{1}{16} \quad \text{Next, multiply.}$$

$$= \frac{1}{3} - \frac{1}{16}$$

$$= \frac{16}{48} - \frac{3}{48} \quad \text{Change to common denominator and subtract.}$$

$$= \frac{13}{48}$$

ANSWERS TO TEST YOUR WORD POWER

1. **(d)** *Example:* Because the fractions $\frac{3}{8}$ and $\frac{10}{8}$ both have 8 as a denominator, they are like fractions.

2. **(c)** *Example:* The fractions $\frac{2}{3}$ and $\frac{3}{4}$ are unlike fractions because they have different denominators.

3. **(a)** *Example:* The least common multiple of 4 and 5 is 20 because 20 is the smallest number into which 4 and 5 will divide evenly.   4. **(d)** *Example:* LCM is the abbreviation for least common multiple.

5. **(c)** *Example:* The least common denominator of the fractions $\frac{2}{3}$ and $\frac{1}{2}$ is 6 because 6 is the least common multiple of 3 and 2. When written using the least common denominator, $\frac{2}{3}$ and $\frac{1}{2}$ become $\frac{4}{6}$ and $\frac{3}{6}$, respectively.

6. **(b)** *Example:* LCD is the abbreviation for least common denominator.

Chapter **3** **REVIEW EXERCISES**

[3.1] *Add or subtract. Write answers in lowest terms.*

1. $\dfrac{5}{9} + \dfrac{2}{9}$

2. $\dfrac{2}{7} + \dfrac{3}{7}$

3. $\dfrac{1}{8} + \dfrac{3}{8} + \dfrac{2}{8}$

4. $\dfrac{5}{16} - \dfrac{3}{16}$

5. $\dfrac{3}{10} - \dfrac{1}{10}$

6. $\dfrac{5}{12} - \dfrac{3}{12}$

7. $\dfrac{36}{62} - \dfrac{10}{62}$

8. $\dfrac{68}{75} - \dfrac{43}{75}$

Solve each application problem. Write answers in lowest terms.

9. Nurse Suzie Brasher screened $\dfrac{7}{16}$ of her patients in her first hour on duty and $\dfrac{5}{16}$ of her patients in the second hour. What fraction of her patients did she screen in the two hours?

10. The Koats for Kids committee members completed $\dfrac{5}{8}$ of their Web-page design in the morning and $\dfrac{3}{8}$ in the afternoon. How much less did they complete in the afternoon than in the morning?

[3.2] *Find the least common multiple of each set of numbers.*

11. 4, 3

12. 5, 4

13. 10, 12, 20

14. 3, 8, 4

15. 6, 8, 5, 15

16. 15, 9, 20

Rewrite each fraction using the indicated denominator.

17. $\dfrac{2}{3} = \dfrac{}{12}$

18. $\dfrac{3}{8} = \dfrac{}{56}$

19. $\dfrac{2}{5} = \dfrac{}{25}$

20. $\dfrac{5}{9} = \dfrac{}{81}$

21. $\dfrac{4}{5} = \dfrac{}{40}$

22. $\dfrac{5}{16} = \dfrac{}{64}$

[3.1–3.3] *Add or subtract. Write answers in lowest terms.*

23. $\dfrac{1}{2} + \dfrac{1}{3}$

24. $\dfrac{1}{5} + \dfrac{3}{10} + \dfrac{3}{8}$

25. $\begin{array}{r} \dfrac{5}{12} \\[2mm] + \dfrac{5}{24} \\ \hline \end{array}$

26. $\dfrac{2}{3} - \dfrac{1}{4}$

27. $\begin{array}{r} \dfrac{7}{8} \\[2mm] - \dfrac{1}{3} \\ \hline \end{array}$

28. $\begin{array}{r} \dfrac{11}{12} \\[2mm] - \dfrac{4}{9} \\ \hline \end{array}$

Solve each application problem.

29. The master gardener for a public garden used $\dfrac{3}{8}$ sack of fertilizer on the lawn, $\dfrac{1}{4}$ sack of fertilizer on the trees, and $\dfrac{1}{3}$ sack of fertilizer on the flower beds. How much of the fertilizer did he use?

30. Rachel is planning a birthday bash for Ross. Monica has raised $\dfrac{2}{5}$ of the amount needed through a bake sale; Joey has earned $\dfrac{1}{3}$ of the amount needed from an acting job, and Phoebe has raised another $\dfrac{1}{4}$ singing at the local coffeehouse. Find the portion of the total that has been raised.

[3.4] *First estimate the answer. Then add or subtract to find the exact answer. Write answers as mixed numbers.*

31. *Estimate:*   *Exact:*

$$\text{Rounds to} \longleftarrow \left\{ \; 18\frac{5}{8} \right.$$

$$+\underline{\quad\quad} \quad \underline{\text{Rounds to}} \left\{ \; +13\frac{3}{4} \right.$$

32. *Estimate:*   *Exact:*

$$22\frac{2}{3}$$

$$+\underline{\quad\quad} \quad +15\frac{4}{9}$$

33. *Estimate:*   *Exact:*

$$12\frac{3}{5}$$

$$8\frac{5}{8}$$

$$+\underline{\quad\quad} \quad +10\frac{5}{16}$$

34. *Estimate:*   *Exact:*

$$31\frac{3}{4}$$

$$-\underline{\quad\quad} \quad -14\frac{2}{3}$$

35. *Estimate:*   *Exact:*

$$34$$

$$-\underline{\quad\quad} \quad -15\frac{2}{3}$$

36. *Estimate:*   *Exact:*

$$215\frac{7}{16}$$

$$-\underline{\quad\quad} \quad -136$$

Add or subtract by changing mixed numbers to improper fractions. Simplify all answers.

37. $5\frac{2}{5}$

     $+3\frac{7}{10}$

38. $4\frac{3}{4}$

     $+5\frac{2}{3}$

39. 5

     $-1\frac{3}{4}$

40. $6\frac{1}{2}$

     $-4\frac{5}{6}$

41. $8\frac{1}{3}$

     $-2\frac{5}{6}$

42. $5\frac{5}{12}$

     $-2\frac{5}{8}$

First estimate the answer. Then solve each application problem.

43. A lab had $14\frac{2}{3}$ gallons of distilled water. If $5\frac{1}{2}$ gallons were used in the morning and $6\frac{3}{4}$ gallons were used in the afternoon, find the number of gallons remaining.

Estimate:

Exact:

44. The Boys and Girls Clubs of America collected $28\frac{2}{3}$ tons of newspapers on Saturday and $24\frac{3}{4}$ tons on Sunday. Find the total weight of the newspapers collected.

Estimate:

Exact:

45. In a recent bass-fishing derby, Matt Urban caught three largemouth bass weighing $8\frac{7}{8}$ lb, $9\frac{1}{3}$ lb, and $6\frac{3}{4}$ lb. Find their total weight.

Estimate:

Exact:

46. A developer wants to build a shopping center. She bought two parcels of land, one, $1\frac{11}{16}$ acres, and the other, $2\frac{3}{4}$ acres. If she needs a total of $8\frac{1}{2}$ acres for the center, how much additional land does she need to buy?

Estimate:

Exact:

[3.5] *Locate each fraction in Exercises 47–50 on the following number line.*

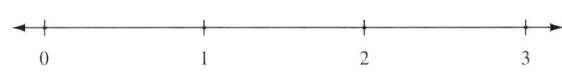

47. $\frac{3}{8}$ **48.** $\frac{7}{4}$ **49.** $\frac{8}{3}$ **50.** $2\frac{1}{5}$

Write $<$ or $>$ in each blank to make a true statement.

51. $\frac{2}{3}$ ___ $\frac{3}{4}$ **52.** $\frac{3}{4}$ ___ $\frac{7}{8}$ **53.** $\frac{1}{2}$ ___ $\frac{7}{15}$ **54.** $\frac{7}{10}$ ___ $\frac{8}{15}$

55. $\frac{9}{16}$ ___ $\frac{5}{8}$ **56.** $\frac{7}{20}$ ___ $\frac{8}{25}$ **57.** $\frac{19}{36}$ ___ $\frac{29}{54}$ **58.** $\frac{19}{132}$ ___ $\frac{7}{55}$

Simplify each expression.

59. $\left(\frac{1}{2}\right)^2$ **60.** $\left(\frac{2}{3}\right)^2$ **61.** $\left(\frac{3}{10}\right)^3$ **62.** $\left(\frac{3}{8}\right)^4$

Simplify by using the order of operations.

63. $6 \cdot \left(\dfrac{1}{3}\right)^2$

64. $\left(\dfrac{2}{3}\right)^2 \cdot 15$

65. $\left(\dfrac{3}{4}\right)^2 \cdot \left(\dfrac{8}{9}\right)^2$

66. $\dfrac{7}{8} \div \left(\dfrac{1}{8} + \dfrac{3}{4}\right)$

67. $\left(\dfrac{1}{2}\right)^2 \cdot \left(\dfrac{1}{4} + \dfrac{1}{2}\right)$

68. $\left(\dfrac{1}{4}\right)^3 + \left(\dfrac{5}{8} + \dfrac{3}{4}\right)$

MIXED REVIEW EXERCISES

Simplify by using the order of operations as necessary. Write answers in lowest terms and as whole or as mixed numbers when possible.

69. $\dfrac{5}{6} - \dfrac{1}{6}$

70. $\dfrac{7}{8} - \dfrac{3}{4}$

71. $\dfrac{29}{32} - \dfrac{5}{16}$

72. $\dfrac{1}{4} + \dfrac{1}{8} + \dfrac{5}{16}$

73. $\begin{array}{r} 6\frac{2}{3} \\ -\ 4\frac{1}{2} \\ \hline \end{array}$

74. $\begin{array}{r} 9\frac{1}{2} \\ +\ 16\frac{3}{4} \\ \hline \end{array}$

75. $\begin{array}{r} 7 \\ -\ 1\frac{5}{8} \\ \hline \end{array}$

76. $\begin{array}{r} 2\frac{3}{5} \\ 8\frac{5}{8} \\ +\ \frac{5}{16} \\ \hline \end{array}$

77. $\begin{array}{r} 32\frac{5}{12} \\ -\ 17 \\ \hline \end{array}$

78. $\dfrac{7}{22} + \dfrac{3}{22} + \dfrac{3}{11}$

79. $\left(\dfrac{1}{4}\right)^2 \cdot \left(\dfrac{2}{5}\right)^3$

80. $\dfrac{3}{8} \div \left(\dfrac{1}{2} + \dfrac{1}{4}\right)$

81. $\left(\dfrac{2}{3}\right)^2 \cdot \left(\dfrac{1}{3} + \dfrac{1}{6}\right)$

82. $\left(\dfrac{2}{3}\right)^3 + \left(\dfrac{2}{3} - \dfrac{5}{9}\right)$

Write < or > in each blank to make a true statement.

83. $\dfrac{2}{3}$ _____ $\dfrac{7}{12}$

84. $\dfrac{8}{9}$ _____ $\dfrac{15}{8}$

85. $\dfrac{17}{30}$ _____ $\dfrac{36}{60}$

86. $\dfrac{5}{8}$ _____ $\dfrac{17}{30}$

Find the least common multiple of each set of numbers.

87. 12, 18

88. 6, 8, 10, 12

89. 9, 14, 21

Rewrite each fraction using the indicated denominator.

90. $\dfrac{2}{3} = \dfrac{}{27}$

91. $\dfrac{9}{12} = \dfrac{}{144}$

92. $\dfrac{4}{5} = \dfrac{}{75}$

First estimate the answer. Then solve each application problem.

93. A cement contractor needs $13\dfrac{1}{2}$ ft of wire mesh for a concrete walkway and $22\dfrac{3}{8}$ ft of wire mesh for a driveway. If the contractor has a roll of wire from which he is cutting that is $92\dfrac{3}{4}$ ft long, find the number of feet of wire remaining after the two jobs have been completed.

Estimate:

Exact:

94. The Horticulture Club used $3\dfrac{5}{8}$ gallons of insecticide on one crop of plants and $8\dfrac{1}{2}$ gallons on another crop. If their original supply of insecticide was 15 gallons, find the number of gallons remaining.

Estimate:

Exact:

Chapter 3 TEST

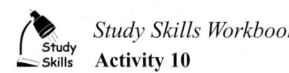
Study Skills Workbook
Activity 10

Add or subtract. Write answers in lowest terms.

1. $\dfrac{3}{8} + \dfrac{3}{8}$

2. $\dfrac{3}{16} + \dfrac{5}{16}$

3. $\dfrac{7}{10} - \dfrac{3}{10}$

4. $\dfrac{7}{12} - \dfrac{5}{12}$

Find the least common multiple of each set of numbers.

5. 2, 3, 4

6. 6, 3, 5, 15

7. 6, 9, 27, 36

Add or subtract. Write answers in lowest terms.

8. $\dfrac{3}{8} + \dfrac{1}{4}$

9. $\dfrac{2}{9} + \dfrac{5}{12}$

10. $\dfrac{7}{8} - \dfrac{2}{3}$

11. $\dfrac{2}{5} - \dfrac{3}{8}$

First estimate the answer. Then add or subtract to find the exact answer; simplify exact answers.

12. $7\dfrac{2}{3} + 4\dfrac{5}{6}$

13. $16\dfrac{2}{5} - 11\dfrac{2}{3}$

14. $18\dfrac{3}{4} + 9\dfrac{2}{5} + 12\dfrac{1}{3}$

15. $24 - 18\dfrac{3}{8}$

1. _____

2. _____

3. _____

4. _____

5. _____

6. _____

7. _____

8. _____

9. _____

10. _____

11. _____

12. *Estimate:* _____

   Exact: _____

13. *Estimate:* _____

   Exact: _____

14. *Estimate:* _____

   Exact: _____

15. *Estimate:* _____

   Exact: _____

16. _____

16. Most students say that "addition and subtraction of fractions is more difficult than multiplication and division of fractions." Why do you think they say this? Do you agree with these students?

17. _____

17. Devise and explain a method of estimating an answer to addition and subtraction problems involving mixed numbers. Might your estimated answer vary from the exact answer? If it did, what would the estimation accomplish?

First estimate the answer. Then solve each application problem.

18. *Estimate:* _____

Exact: _____

18. A professional football player trains with body building, jogging, and wind sprints. He trains $5\frac{2}{3}$ hours on Monday, $4\frac{3}{4}$ hours on Tuesday, $3\frac{5}{6}$ hours on Wednesday, $7\frac{1}{3}$ hours on Thursday, and $5\frac{1}{6}$ hours on Friday. Find the total number of hours that he trained.

19. *Estimate:* _____

Exact: _____

19. A painting contractor arrived at a 6-unit apartment complex with $147\frac{1}{2}$ gallons of exterior paint. If his crew sprayed $68\frac{1}{2}$ gallons on the wood siding, rolled $37\frac{3}{8}$ gallons on the masonry exterior, and brushed $5\frac{3}{4}$ gallons on the trim, find the number of gallons of paint remaining.

Write < or > to make a true statement.

20. $\frac{3}{4}$ ____ $\frac{17}{24}$

21. $\frac{19}{24}$ ____ $\frac{17}{36}$

20. _____

21. _____

Simplify. Use the order of operations as needed.

22. $\left(\frac{1}{3}\right)^3 \cdot 54$

23. $\left(\frac{3}{4}\right)^2 - \left(\frac{7}{8} \cdot \frac{1}{3}\right)$

22. _____

23. _____

24. $\left(\frac{7}{8} - \frac{7}{16}\right) \cdot 4$

25. $\frac{5}{6} + \frac{4}{3} \cdot \frac{3}{8}$

24. _____

25. _____

For each number, name the digit that has the given place value.

1. 871

hundreds

ones

2. 5,629,428

millions

thousands

Round each number to the nearest ten, nearest hundred, and nearest thousand.

	Ten	**Hundred**	**Thousand**
3. 1438	_____	_____	_____
4. 59,803	_____	_____	_____

Use front end rounding to estimate each answer. Then add, subtract, multiply, or divide to find the exact answer.

5. *Estimate:*

Exact:

Rounds to → 2361
Rounds to → 386
Rounds to → 47,304
+ _____ Rounds to → + 29,728

6. *Estimate:*

Exact:

24,276
− 9 887
− _____

7. *Estimate:*

Exact:

4468
× _____   × 280

8. *Estimate:*

Exact:

$\overline{)}$

$35\overline{)112,385}$

Add, subtract, multiply, or divide as indicated.

9.
4
8
5
+ 9

10.
375,899
521,742
+ 357,968

11.
1687
− 1096

12.
3,896,502
− 1,094,807

13. $3 \times 8 \times 5$

14. $6 \times 3 \times 7$

15. $5 \times 8 \times 4$

16.
57
× 8

17.
962
× 384

18.
340
× 50

19. $8\overline{)1080}$

20. $13,467 \div 5$

21. $506\overline{)16,358}$

Use front end rounding to estimate the answer to each application problem. Then find the exact answer.

22. The Americans with Disabilities Act provides the single parking space design shown. Find the perimeter of (distance around) this parking space, including the accessible aisle.

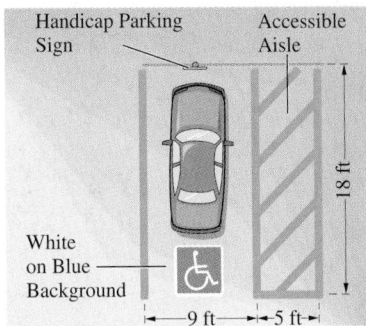

Handicap Parking Sign

Accessible Aisle

18 ft

White on Blue Background

9 ft 5 ft

Estimate:

Exact:

23. The single parking space design in Exercise 22 measures 18 ft by 14 ft. Find its area.

Estimate:

Exact:

24. Find the number of 32 ounce (1 quart) cartons of orange juice that can be filled with 40,320 ounces of orange juice.

Estimate:

Exact:

25. A dentist's drill makes 3600 revolutions each minute. How many revolutions would it make in 60 minutes of operation?

Estimate:

Exact:

Round the mixed number in each problem to the nearest whole number and estimate the answer. Then find the exact answer.

26. The top of a rectangular pool table is $1\frac{3}{4}$ yd by $2\frac{2}{3}$ yd. Find its area.

Estimate:

Exact:

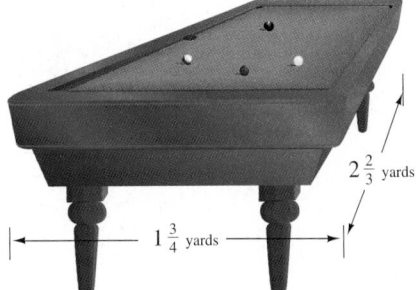

$2\frac{2}{3}$ yards

$1\frac{3}{4}$ yards

27. A wilderness park is $4\frac{5}{8}$ miles wide by $6\frac{2}{3}$ miles long. How many square miles are in the wilderness park?

Estimate:

Exact:

$6\frac{2}{3}$ miles long

$4\frac{5}{8}$ miles wide

28. Larry Foxworthy cuts, splits, and delivers firewood. If his truck, when fully loaded, holds $5\frac{1}{4}$ cords of firewood, find the number of cords he could deliver in $3\frac{1}{2}$ loads.

Estimate:

Exact:

29. The Sears Tower in Chicago is 110 stories tall. The total height of the building is $1536\frac{7}{8}$ ft, including a flagpole at the top of the building that is $82\frac{1}{2}$ ft tall. Find the height of the building itself. (*Source: Top 10 of Everything, 1999.*)

Estimate:

Exact:

Find the prime factorization of each number. Write answers by using exponents.

30. 18

31. 200

32. 1225

Simplify.

33. $2^4 \cdot 3^2$

34. $3^3 \cdot 2^2$

35. $6^2 \cdot 3^3$

Find each square root.

36. $\sqrt{36}$

37. $\sqrt{81}$

38. $\sqrt{144}$

Simplify by using the order of operations.

39. $5^2 - 2 \cdot 8$

40. $\sqrt{25} + 5 \cdot 9 - 6$

41. $\left(\dfrac{4}{5} - \dfrac{2}{3}\right) \cdot \dfrac{2}{3}$

42. $\dfrac{3}{4} \div \left(\dfrac{1}{3} + \dfrac{1}{2}\right)$

43. $\dfrac{7}{8} + \left(\dfrac{3}{4}\right)^2 - \dfrac{3}{8}$

Write proper *or* improper *for each fraction.*

44. $\dfrac{5}{6}$

45. $\dfrac{6}{6}$

46. $\dfrac{11}{10}$

Write each fraction in lowest terms.

47. $\dfrac{25}{60}$

48. $\dfrac{84}{96}$

49. $\dfrac{63}{70}$

Add, subtract, multiply, or divide as indicated. Simplify all answers.

50. $\dfrac{3}{4} \cdot \dfrac{2}{3}$

51. $\dfrac{3}{8} \cdot \dfrac{5}{6}$

52. $42 \cdot \dfrac{7}{8}$

53. $\dfrac{5}{8} \div \dfrac{3}{8}$

54. $\dfrac{25}{40} \div \dfrac{10}{35}$

55. $9 \div \dfrac{2}{3}$

56. $\dfrac{3}{4} + \dfrac{1}{7}$

57. $\dfrac{3}{8} + \dfrac{3}{16} + \dfrac{1}{4}$

58. $\dfrac{11}{18} - \dfrac{5}{12}$

First estimate the answer. Then add or subtract to find the exact answer. Write exact answers as mixed numbers.

59. *Estimate:*   *Exact:*

$$\xleftarrow{\text{Rounds to}} \begin{cases} 3\dfrac{3}{8} \\ \\ +\ 4\dfrac{1}{2} \end{cases}$$

$+ \underline{\quad}$ $\xleftarrow{\text{Rounds to}}$

60. *Estimate:*   *Exact:*

$$21\dfrac{7}{8}$$
$$+\ 4\dfrac{5}{12}$$

$+ \underline{\quad}$

61. *Estimate:*   *Exact:*

$$5$$
$$-2\dfrac{3}{8}$$

$- \underline{\quad}$

Find the least common multiple of each set of numbers.

62. 8, 12

63. 3, 8, 15

64. 12, 16, 18

Rewrite each fraction using the indicated denominator.

65. $\dfrac{4}{5} = \dfrac{\ }{45}$

66. $\dfrac{7}{9} = \dfrac{\ }{72}$

67. $\dfrac{9}{15} = \dfrac{\ }{135}$

68. $\dfrac{5}{7} = \dfrac{\ }{84}$

Locate each fraction in Exercises 69–72 on the following number line.

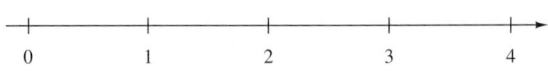

69. $\dfrac{3}{4}$

70. $\dfrac{1}{9}$

71. $\dfrac{5}{3}$

72. $\dfrac{10}{3}$

Write $<$ or $>$ in each blank to make a true statement.

73. $\dfrac{3}{5} \underline{\quad} \dfrac{5}{8}$

74. $\dfrac{17}{20} \underline{\quad} \dfrac{3}{4}$

75. $\dfrac{7}{12} \underline{\quad} \dfrac{11}{18}$

Decimals

4

Freshwater fishing is America's third most popular recreational activity, and saltwater fishing is ninth. (*Source:* Sporting Goods Manufacturers Association.) In Section 4.3, Exercises 47–50, this father will use decimal numbers when he pays for his daughter's equipment. But will decimals help him and his daughter to catch their limit? (See also Section 4.1, Exercises 59–62, and Section 4.6, Exercises 65–68.)

ADDISON - WESLEY
MyMathLab.com
You're Connected

4.1 READING AND WRITING DECIMALS

OBJECTIVES

1 Write parts of a whole using decimals.

2 Identify the place value of a digit.

3 Read and write decimals in words.

4 Write decimals as fractions or mixed numbers.

❶ There are 10 dimes in one dollar. Each dime is $\frac{1}{10}$ of a dollar. Write the yellow shaded portion of each dollar as a fraction, as a decimal, and in words.

(a)

(b)

(c)

Fractions are used to represent parts of a whole. In this chapter, **decimals** are used as another way to show parts of a whole. For example, our money system is based on decimals. One dollar is divided into 100 equivalent parts. One cent ($0.01) is one of the parts, and a dime ($0.10) is 10 of the parts. Metric measurement (see Chapter 7) is also based on decimals.

1 **Write parts of a whole using decimals.**   Decimals are used when a whole is divided into 10 equivalent parts, or into 100 or 1000 or 10,000 equivalent parts. In other words, decimals are fractions with denominators that are a power of 10. For example, the square below is cut into 10 equivalent parts. Written as a fraction, each part is $\frac{1}{10}$ of the whole. Written as a decimal, each part is 0.1. Both $\frac{1}{10}$ and 0.1 are read as "*one tenth.*"

$\frac{1}{10}$ | 0.1

One-tenth of the square is shaded.

The dot in 0.1 is called the **decimal point.**

$$0.1$$
$$\uparrow$$
Decimal point

The square at the right has 7 of its 10 parts shaded.

Written as a *fraction*, $\frac{7}{10}$ of the square is shaded.

Written as a *decimal*, 0.7 of the square is shaded.

Both $\frac{7}{10}$ and 0.7 are read as "*seven tenths.*"

Work Problem ❶ at the Side.

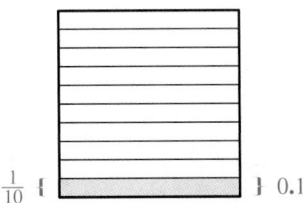

$\frac{7}{10}$

0.7

Seven-tenths of the square is shaded.

The square below is cut into 100 equivalent parts. Written as a *fraction,* each part is $\dfrac{1}{100}$ of the whole.

Written as a decimal, each part is **0.01** of the whole. Both $\frac{1}{100}$ and 0.01 are read as "one hundredth."

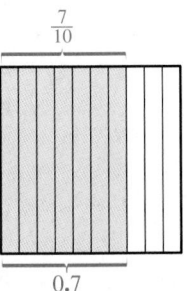

$\frac{1}{100}$ 0.01

The square above has 87 parts shaded.

Written as a fraction, $\dfrac{87}{100}$ of the total area is shaded.

Written as a decimal, **0.87** of the total area is shaded.
Both $\frac{87}{100}$ and 0.87 are read as "*eighty-seven hundredths.*"

Work Problem ❷ at the Side.

Example 1 shows several numbers written as both fractions and decimals.

> **Example 1** **Using the Decimal Forms of Fractions**
>
> | | Fraction | Decimal | Read As |
> |---|---|---|---|
> | **(a)** | $\frac{4}{10}$ | 0.4 | four tenths |
> | **(b)** | $\frac{9}{100}$ | 0.09 | nine hundredths |
> | **(c)** | $\frac{71}{100}$ | 0.71 | seventy-one hundredths |
> | **(d)** | $\frac{8}{1000}$ | 0.008 | eight thousandths |
> | **(e)** | $\frac{45}{1000}$ | 0.045 | forty-five thousandths |
> | **(f)** | $\frac{832}{1000}$ | 0.832 | eight hundred thirty-two thousandths |

=== **Work Problem ❸ at the Side.**

2 **Identify the place value of a digit.** The decimal point separates the *whole number part* from the *fractional part* in a decimal number. In the chart below, you see that the **place value names** for fractional parts are similar to those on the whole number side, but end in "*ths*."

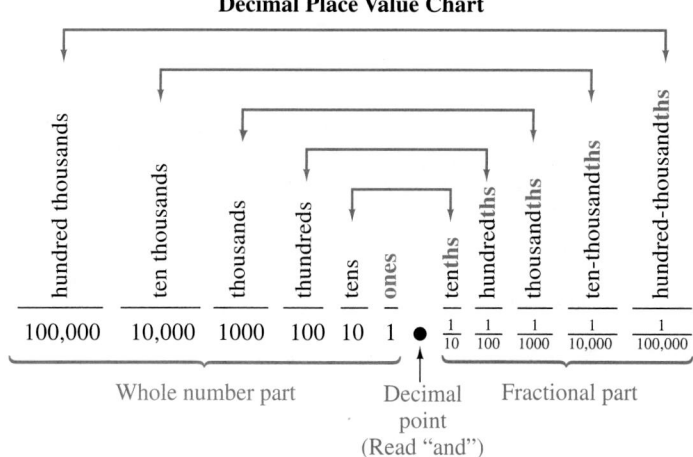

Decimal Place Value Chart

Notice that the **ones** place is at the center. (There is no "oneths" place.) Also notice that each place is 10 times the value of the place to its right.

❷ Write the portion of each square that is shaded as a fraction, as a decimal, and in words.

(a)

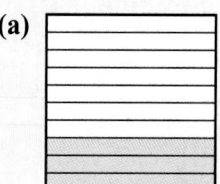

(b)

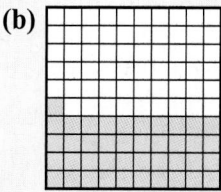

❸ Write each decimal as a fraction.

(a) 0.7

(b) 0.2

(c) 0.03

(d) 0.69

(e) 0.047

(f) 0.351

ANSWERS

2. (a) $\frac{3}{10}$; 0.3; three tenths

  (b) $\frac{41}{100}$; 0.41; forty-one hundredths

3. (a) $\frac{7}{10}$ **(b)** $\frac{2}{10}$ **(c)** $\frac{3}{100}$

  (d) $\frac{69}{100}$ **(e)** $\frac{47}{1000}$ **(f)** $\frac{351}{1000}$

④ Identify the place value of each digit.

(a) 971.54

(b) 0.4

(c) 5.60

(d) 0.0835

⑤ Tell how to read each decimal in words.

(a) 0.6

(b) 0.46

(c) 0.05

(d) 0.409

(e) 0.0003

(f) 0.0703

(g) 0.088

CAUTION

In this chapter, if a number does *not* have a decimal point, it is a *whole number*. A whole number has no fractional part. If you want to show the decimal point in a whole number, it is just to the *right* of the digit in the ones place. For example:

$$8 = 8. \qquad\qquad 306 = 306.$$

↑ Decimal point　　　　↑ Decimal point

Example 2 Identifying the Place Value of a Digit

Identify the place value of each digit.

(a) 178.36　　　　　　(b) 0.00935

hundreds	tens	ones	tenths	hundredths
1	7	8 .	3	6

ones	tenths	hundredths	thousandths	ten-thousandths	hundred-thousandths
0 .	0	0	9	3	5

Notice in Example 2(b) that we do *not* use commas on the right side of the decimal point.

Work Problem ④ at the Side.

3 Read and write decimals in words. A decimal is read according to its form as a fraction. We read 0.9 as "nine tenths" because 0.9 is the same as $\frac{9}{10}$. Notice that 0.9 ends in the tenths place.

ones	tenths
0.	9

We read 0.02 as "two hundredths" because 0.02 is the same as $\frac{2}{100}$. Notice that 0.02 ends in the hundredths place.

ones	tenths	hundredths
0.	0	2

Example 3 Reading Decimal Numbers

Tell how to read each decimal in words.

(a) 0.3

Because $0.3 = \dfrac{3}{10}$, read the decimal as three <u>tenths</u>.

(b) 0.49　　Read it as:　forty-nine <u>hundredths</u>.

(c) 0.08　　Read it as:　eight <u>hundredths</u>.

(d) 0.918　　Read it as:　nine hundred eighteen <u>thousandths</u>.

(e) 0.0106　　Read it as:　one hundred six <u>ten-thousandths</u>.

Work Problem ⑤ at the Side.

ANSWERS

4. (a)
hundreds	tens	ones	tenths	hundredths
9	7	1 .	5	4

(b)
ones	tenths
0 .	4

(c)
ones	tenths	hundredths
5 .	6	0

(d)
ones	tenths	hundredths	thousandths	ten-thousandths
0 .	0	8	3	5

5. (a) six tenths
(b) forty-six hundredths
(c) five hundredths
(d) four hundred nine thousandths
(e) three ten-thousandths
(f) seven hundred three ten-thousandths
(g) eighty-eight thousandths

Reading Decimal Numbers

Step 1   Read any whole number part to the *left* of the decimal point as you normally would.

Step 2   Read the decimal point as "*and.*"

Step 3   Read the part of the number to the *right* of the decimal point as if it were an ordinary whole number.

Step 4   Finish with the place value name of the rightmost digit; these names all end in "*ths.*"

NOTE

If there is *no whole number part,* you will use only Steps 3 and 4.

Example 4   **Reading Decimals**

Read each decimal.

(a)

9 is in tenths place.

16.9

sixteen **and** nine **tenths**

16.9 is read "sixteen and nine tenths."

(b)

5 is in hundredths place.

482.35

four hundred eighty-two **and** thirty-five **hundredths**

482.35 is read "four hundred eighty-two and thirty-five hundredths."

3 is in thousandths place.

(c)   0.063 is "sixty-three **thousandths.**" (No whole number part.)

(d)   11.1085 is "eleven **and** one thousand eighty-five **ten-thousandths.**"

CAUTION

Use "and" *only* when reading a decimal point. A common mistake is to read the whole number 405 as "four hundred *and* five." But there is *no decimal point* shown in 405, so it is read "four hundred five."

Work Problem ❻ at the Side.

4   **Write decimals as fractions or mixed numbers.**   Knowing how to read decimals will help you when writing decimals as fractions.

Writing Decimals as Fractions or Mixed Numbers

Step 1   The digits to the right of the decimal point are the numerator of the fraction.

Step 2   The denominator is 10 for tenths, 100 for hundredths, 1000 for thousandths, 10,000 for ten-thousandths, and so on.

Step 3   If the decimal has a whole number part, the fraction will be a mixed number with the same whole number part.

❻ Tell how to read each decimal in words.

(a) 3.8

(b) 15.001

(c) 0.0073

(d) 64.309

7 Write each decimal as a fraction or mixed number.

(a) 0.7

(b) 12.21

(c) 0.101

(d) 0.007

(e) 1.3717

8 Write each decimal as a fraction or mixed number in lowest terms.

(a) 0.5

(b) 12.6

(c) 0.85

(d) 3.05

(e) 0.225

(f) 420.0802

Example 5 Writing Decimals as Fractions or Mixed Numbers

Write each decimal as a fraction or mixed number.

(a) 0.19

The digits to the right of the decimal point, 19, are the numerator of the fraction. The denominator is 100 for hundredths because the right-most digit is in the hundredths place.

$$0.19 = \frac{19}{100} \leftarrow 100 \text{ for hundredths}$$

↑ Hundredths place

(b) 0.863

$$0.863 = \frac{863}{1000} \leftarrow 1000 \text{ for thousandths}$$

↑ Thousandths place

(c) 4.0099

The whole number part stays the same.

$$4.0099 = 4\frac{99}{10,000} \leftarrow 10,000 \text{ for ten-thousandths}$$

↑ Ten-thousandths place

Work Problem 7 at the Side.

CAUTION

After you write a decimal as a fraction or a mixed number, make sure the fraction is in lowest terms.

Example 6 Writing Decimals as Fractions or Mixed Numbers in Lowest Terms

Write each decimal as a fraction or mixed number in lowest terms.

(a) $0.4 = \dfrac{4}{10} \leftarrow 10$ for tenths

Write $\dfrac{4}{10}$ in lowest terms. $\dfrac{4}{10} = \dfrac{4 \div 2}{10 \div 2} = \dfrac{2}{5} \leftarrow$ Lowest terms

(b) $0.75 = \dfrac{75}{100} = \dfrac{75 \div 25}{100 \div 25} = \dfrac{3}{4} \leftarrow$ Lowest terms

(c) $18.105 = 18\dfrac{105}{1000} = 18\dfrac{105 \div 5}{1000 \div 5} = 18\dfrac{21}{200} \leftarrow$ Lowest terms

(d) $42.8085 = 42\dfrac{8085}{10,000} = 42\dfrac{8085 \div 5}{10,000 \div 5} = 42\dfrac{1617}{2000} \leftarrow$ Lowest terms

Work Problem 8 at the Side.

Calculator Tip In this book you'll notice that we use a 0 in the ones place for decimal fractions. We write **0.45** instead of just **.45**, to emphasize that there is no whole number. Your scientific (not graphing) calculator shows these 0s also. Enter ⊙ ④ ⑤. Notice that the display automatically shows 0.45 even though you did not press 0. For comparison, enter the whole number 45 by pressing ④ ⑤ ⊕ and notice where the decimal point is shown in the display. (It automatically appears to the *right* of the 5.)

ANSWERS

7. (a) $\dfrac{7}{10}$ (b) $12\dfrac{21}{100}$ (c) $\dfrac{101}{1000}$
   (d) $\dfrac{7}{1000}$ (e) $1\dfrac{3717}{10,000}$

8. (a) $\dfrac{1}{2}$ (b) $12\dfrac{3}{5}$ (c) $\dfrac{17}{20}$ (d) $3\dfrac{1}{20}$
   (e) $\dfrac{9}{40}$ (f) $420\dfrac{401}{5000}$

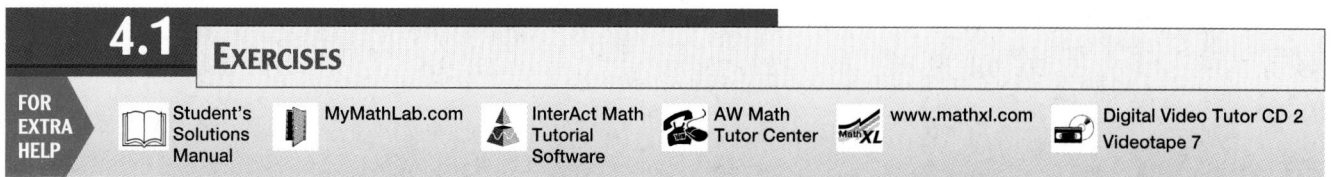

4.1 EXERCISES

Identify the digit that has the given place value. See Example 2.

1. 70.489
   tens
   ones
   tenths

2. 135.296
   ones
   tenths
   tens

3. 0.2518
   hundredths
   thousandths
   ten-thousandths

4. 0.9347
   hundredths
   thousandths
   ten-thousandths

5. 93.01472
   thousandths
   ten-thousandths
   tenths

6. 0.51968
   tenths
   ten-thousandths
   hundredths

7. 314.658
   tens
   tenths
   hundreds

8. 51.325
   tens
   tenths
   hundredths

9. 149.0832
   hundreds
   hundredths
   ones

10. 3458.712
   hundreds
   hundredths
   tenths

11. 6285.7125
   thousands
   thousandths
   hundredths

12. 5417.6832
   thousands
   thousandths
   ones

Write the decimal number that has the specified place values. See Example 2.

13. 0 ones, 5 hundredths, 1 ten, 4 hundreds, 2 tenths

14. 7 tens, 9 tenths, 3 ones, 6 hundredths, 8 hundreds

15. 3 thousandths, 4 hundredths, 6 ones, 2 ten-thousandths, 5 tenths

16. 8 ten-thousandths, 4 hundredths, 0 ones, 2 tenths, 6 thousandths

17. 4 hundredths, 4 hundreds, 0 tens, 0 tenths, 5 thousandths, 5 thousands, 6 ones

18. 7 tens, 7 tenths, 6 thousands, 6 thousandths, 3 hundreds, 3 hundredths, 2 ones

Write each decimal as a fraction or mixed number in lowest terms. See Examples 1, 5, and 6.

19. 0.7   **20.** 0.1   **21.** 13.4   **22.** 9.8   **23.** 0.25

24. 0.55   **25.** 0.66   **26.** 0.33   **27.** 10.17   **28.** 31.99

29. 0.06   **30.** 0.08   **31.** 0.205   **32.** 0.805

33. 5.002   **34.** 4.008   **35.** 0.686   **36.** 0.492

Tell how to read each decimal in words. See Examples 3 and 4.

37. 0.5   **38.** 0.2

39. 0.78   **40.** 0.55

41. 0.105   **42.** 0.609

43. 12.04   **44.** 86.09

45. 1.075   **46.** 4.025

Write each decimal in numbers. See Examples 3 and 4.

47. six and seven tenths

48. eight and twelve hundredths

49. thirty-two hundredths

50. one hundred eleven thousandths

51. four hundred twenty and eight thousandths

52. two hundred and twenty-four thousandths

53. seven hundred three ten-thousandths

54. eight hundred and six hundredths

55. seventy-five and thirty thousandths

56. sixty and fifty hundredths

57. Anne read the number 4302 as "four thousand three hundred and two." Explain what is wrong with the way Anne read the number.

58. Jerry read the number 9.0106 as "nine and one hundred and six ten-thousandths." Explain the error he made.

The dad on the first page of this chapter needs to select the correct fishing line for his daughter's reel. Fishing line is sold according to how many pounds of "pull" the line can withstand before breaking. Use the table to answer Exercises 59–62. Write all fractions in lowest terms. (Note: The diameter of the fishing line is its thickness.)

RELATING FISHING LINE DIAMETER TO TEST STRENGTH

Test Strength (pounds)	Average Diameter (inches)
4	0.008
8	0.010
12	0.013
14	0.014
17	0.015
20	0.016

Source: Berkley Outdoor Technologies Group.

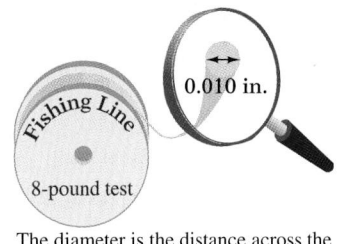

The diameter is the distance across the end of the line, or its thickness.

59. Write the diameter of 8-pound test line in words and as a fraction.

60. Write the diameter of 17-pound test line in words and as a fraction.

61. What is the test strength of the line with a diameter of $\frac{13}{1000}$ inch?

62. What is the test strength of the line with a diameter of sixteen thousandths inch?

Suppose your job is to take phone orders for precision parts. Use the table, and in Exercises 63–68, write the correct part number that matches what you hear the customer say over the phone. In Exercises 67–68, write the words you would say to the customer.

Part Number	Size in Centimeters
3-A	0.06
3-B	0.26
3-C	0.6
3-D	0.86
4-A	1.006
4-B	1.026
4-C	1.06
4-D	1.6
4-E	1.602

63. "Please send the six-tenths centimeter bolt."

Part number _____.

64. "The part missing from our order was the one and six hundredths size."

Part number _____.

65. "The size we need is one and six thousandths centimeters."

Part number _____.

66. "Do you still stock the twenty-six hundredths centimeter bolt?"

Part number _____.

67. "What size is part number 4-E?" Write your answer in words.

68. "What size is part number 4-B?" Write your answer in words.

RELATING CONCEPTS (Exercises 69–76) **FOR INDIVIDUAL OR GROUP WORK**

*Use your knowledge of place value to **work Exercises 69–76 in order.***

69. Look back at the decimal place value chart on page 249. What do you think would be the names of the next four places to the *right* of hundred-thousandths? What information did you use to come up with these names?

70. A common mistake is to think that the first place to the right of the decimal point is "oneths" and the second place is "tenths." Why might someone make that mistake? How would you explain why there is no "oneths" place?

71. Use your answer to Exercise 69 to write 0.72436955 in words.

72. Use your answer to Exercise 69 to write 0.000678554 in words.

73. Write 8006.500001 in words.

74. Write 20,060.000505 in words.

75. Write this decimal in numbers:

Three hundred two thousand forty ten-millionths.

76. Write this decimal in numbers:

nine billion, eight hundred seventy-six million, five hundred forty-three thousand, two hundred ten and one hundred million two hundred thousand three hundred billionths.

4.2 ROUNDING DECIMALS

Section 1.7 showed how to round whole numbers. For example, 89 rounded to the nearest ten is 90, and 8512 rounded to the nearest hundred is 8500.

1   **Learn the rules for rounding decimals.**   It is also important to be able to **round** decimals. For example, a store is selling 2 candy mints for $0.75 but you want only one mint. The price of each mint is $0.75 ÷ 2, which is $0.375, but you cannot pay part of a cent. Is $0.375 closer to $0.37 or to $0.38? Actually, it's exactly halfway between. When this happens in everyday situations, the rule is to round *up*. The store will charge you $0.38 for the mint.

OBJECTIVES

1   Learn the rules for rounding decimals.

2   Round decimals to any given place.

3   Round money amounts to the nearest cent or nearest dollar.

Rounding Decimals

Step 1   Find the place to which the rounding is being done. Draw a "cut-off" line *after* that place to show that you are cutting off and dropping the rest of the digits.

Step 2   Look *only* at the *first* digit you are cutting off.

Step 3(a)   If this digit is **4 or less,** the part of the number you are keeping **stays the same.**

Step 3(b)   If this digit is **5 or more,** you must **round up** the part of the number you are keeping.

Step 4   You can use the "≈" sign to indicate that the rounded number is now an approximation (close, but not exact). "≈" means "is approximately equal to."

CAUTION

Do *not* move the decimal point when rounding.

2   **Round decimals to any given place.**   The following examples show you how to round decimals.

Example 1   Rounding a Decimal Number

Round 14.39652 to the nearest thousandth. Is it closer to 14.396 or to 14.397?

Step 1   Draw a "cut-off" line after the thousandths place.

$$1\ 4\ .\ 3\ 9\ 6 \;|\; 5\ 2$$

Thousandths ⎯⎯↑

You are cutting off the 5 and 2. They will be dropped.

Step 2   Look *only* at the *first* digit you are cutting off. Ignore the other digits you are cutting off.

$$1\ 4\ .\ 3\ 9\ 6 \;|\; 5\ 2$$

⎯ Look only at the 5.
Ignore the 2.

Continued on Next Page

❸ Round each money amount to the nearest cent.

(a) $14.595

(b) $578.0663

(c) $0.849

(d) $0.0548

Example 3 Rounding to the Nearest Cent

Round each money amount to the nearest cent.

(a) $2.4238 (Is it closer to $2.42 or to $2.43?)

First digit cut is 4 or less, so the part you are keeping stays the same.

$2.42|38

$2.42 ← You pay

(b) $0.695 (Is it closer to $0.69 or to $0.70?)

5 or more; round up

$0.69|5

$0.69
+ $0.01 ← To round up, add 1 hundredth (1 cent).
─────────
$0.70 ← You pay

Work Problem ❸ at the Side.

It is also common to round money amounts to the nearest dollar. For example, you can do that on your federal and state income tax returns to make the calculations easier.

Example 4 Rounding to the Nearest Dollar

Round to the nearest dollar.

(a) $48.69 (Is it closer to $48 or to $49?)

First digit cut is 5 or more, so round up by adding $1.

$48.|69

$48
+ 1
─────
$49

CAUTION

$48.69 rounded to the nearest dollar is $49. Write the answer as $49 to show that the rounding is to the *nearest dollar.* Writing $49.00 would show rounding to the nearest *cent.*

(b) $594.36 (Is it closer to $594 or to $595?)

First digit cut is 4 or less, so the part you keep stays the same.

$594.|36

$594

$594.36 rounded to the nearest dollar is $594.

(c) $349.88 (Is it closer to $349 or to $350?)

5 or more, so round up by adding $1.

$349.|88

$349
+   1
──────
$350

$349.88 rounded to the nearest dollar is $350.

Continued on Next Page

(d) $2689.50 rounded to the nearest dollar is $2690.

(e) $0.61 rounded to the nearest dollar is $1.

📅 **Calculator Tip**   Accountants and other people who work with money amounts often set their calculators to automatically round to two decimal places (nearest cent) or to round to zero decimal places (nearest dollar). Your calculator may have this feature.

Work Problem ④ at the Side.

④ Round to the nearest dollar.

(a) $29.10

(b) $136.49

(c) $990.91

(d) $5949.88

(e) $49.60

(f) $0.55

(g) $1.08

Focus on Real-Data Applications

Lawn Fertilizer

Gotta Be Green

A lot's being said about personal responsibility these days, and the idea seems to be ending up on the front lawn—literally! Each spring, homeowners across the country gear up to green up their lawns, and the increased use of fertilizer has a lot of environmentalists concerned about the potential effects of chemical runoff into nearby rivers and streams.

Each year, according to a study conducted by the University of Minnesota's Department of Agriculture, each household in the Minneapolis/St. Paul metro area uses an average of 36 pounds of lawn fertilizer. That adds up to 25,529,295 pounds, or 12,765 tons. Add to that another 193,000 pounds of weed killer and you're looking at the total picture for keeping it green in the Twin Cities.

Source: Minneapolis Star Tribune.

1. According to the article,
   (a) How many pounds of lawn fertilizer are used each year in the *entire metro area?*
   (b) Do a division on your calculator to find the number of *households* in the metro area.
   (c) Would it make sense to round your answer to part (b)? If so, how would you round it?

   (d) How many pounds of *weed killer* are used each year in the entire metro area*?*
   (e) Do a division on your calculator to find the number of pounds of *weed killer* used by each household in the metro area. Round your answer to the nearest hundredth.

2. There are 2000 pounds in one ton.
   (a) Find the number of tons equivalent to 25,529,295 pounds of fertilizer.
   (b) Does your answer match the figure given in the article? If not, what did the author of the article do to get 12,765 tons?

   (c) Is the author's figure accurate? Why or why not?

   (d) Find the number of tons equivalent to 193,000 pounds of weed killer.
   (e) How can you write the division problem $193,000 \div 2,000$ in a simpler way? Explain what you did.

3. According to the article,
   (a) "each household in the Minneapolis/St. Paul metro area uses an average of 36 pounds of lawn fertilizer" each year. What mathematical operation do you do to find an average?

   (b) When the calculations were done to find the average, the answer was probably not *exactly* 36 pounds. List six different values that are *less than* 36 that would round to 36. List two values with one decimal place; two values with two decimal places, and two values with three decimal places.

   (c) List six different values that are *greater than* 36 that would round to 36. List two values each with one, two, and three decimal places.

   (d) What is the *smallest* number that can be rounded to 36? What is the *largest* number?

4.2 EXERCISES

Round each number to the place indicated. See Examples 1 and 2.

1. 16.8974 to the nearest tenth

2. 193.845 to the nearest hundredth

3. 0.95647 to the nearest thousandth

4. 96.81584 to the nearest ten-thousandth

5. 0.799 to the nearest hundredth

6. 0.952 to the nearest tenth

7. 3.66062 to the nearest thousandth

8. 1.5074 to the nearest hundredth

9. 793.988 to the nearest tenth

10. 476.1196 to the nearest thousandth

11. 0.09804 to the nearest ten-thousandth

12. 176.004 to the nearest tenth

13. 48.512 to the nearest one

14. 3.385 to the nearest one

15. 9.0906 to the nearest hundredth

16. 30.1290 to the nearest thousandth

17. 82.000151 to the nearest ten-thousandth

18. 0.400594 to the nearest ten-thousandth

Nardos is grocery shopping. The store will round the amount she pays for each item to the nearest cent. Write the rounded amounts. See Example 3.

19. Soup is three cans for $2.45, so one can is $0.81666. Nardos pays _____.

20. Orange juice is two cartons for $2.69, so one carton is $1.345. Nardos pays _____.

21. Facial tissue is four boxes for $4.89, so one box is $1.2225. Nardos pays _____.

22. Muffin mix is three packages for $1.75, so one package is $0.58333. Nardos pays _____.

23. Candy bars are six for $2.99, so one bar is $0.4983. Nardos pays _____.

24. Spaghetti is four boxes for $3.59, so one box is $0.8975. Nardos pays _____.

As she gets ready to do her income tax return, Ms. Chen rounds each amount to the nearest dollar. Write the rounded amounts. See Example 4.

25. Income from job, $48,649.60

26. Income from interest on bank account, $69.58

27. Union dues, $310.08

28. Federal withholding, $6064.49

29. Donations to charity, $848.91

30. Medical expenses, $609.38

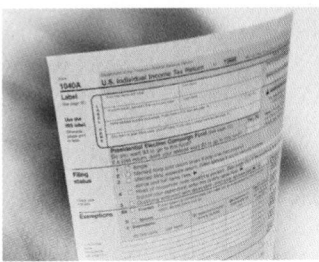

Round each money amount as indicated.

31. $499.98 to the nearest dollar.

32. $9899.59 to the nearest dollar.

33. $0.996 to the nearest cent.

34. $0.09929 to the nearest cent.

35. $999.73 to the nearest dollar.

36. $9999.80 to the nearest dollar.

The table lists speed records for various types of transportation. Use the table to answer Exercises 37–40.

Record	Speed (miles per hour)
Land speed record (specially built car)	763.04
Motorcycle speed record (specially adapted motorcycle)	322.15
Fastest train (regular passenger service)	162.7
Fastest X-15 (military jet)	4520
Boeing 737-300 airplane (regular passenger service)	495
Indianapolis 500 auto race (fastest average winning speed)	185.984
Daytona 500 auto race (fastest average winning speed)	177.602

Source: The Top 10 of Everything 2000.

37. Round these speed records to the nearest tenth.

(a) Indianapolis 500 average winning speed

(b) Land speed record

38. Round these speed records to the nearest hundredth.

(a) Daytona 500 average winning speed

(b) Indianapolis 500 average winning speed

39. Round these speed records to the nearest whole number.

(a) Motorcycle   (b) Train

40. Round these speed records to the nearest hundred.

(a) X-15 military jet   (b) Boeing 737-300 airplane

RELATING CONCEPTS (Exercises 41–44)   **FOR INDIVIDUAL OR GROUP WORK**

Use your knowledge about rounding money amounts to **work Exercises 41–44 in order.**

41. Explain what happens when you round $0.499 to the nearest dollar. Why does this happen?

42. Look again at Exercise 41. How else could you round $0.499 that would be more helpful? What kind of guideline does this suggest about rounding to the nearest dollar?

43. Explain what happens when you round $0.0015 to the nearest cent. Why does this happen?

44. Suppose you want to know which of these amounts is less, so you round them both to the nearest cent.

$0.5968 $0.6014

Explain what happens. Describe what you could do instead of rounding to the nearest cent.

4.3 ADDING AND SUBTRACTING DECIMALS

1 **Add decimals.** When adding or subtracting *whole* numbers (**Sections 1.2** and **1.3**), you lined up the numbers in columns so that you were adding ones to ones, tens to tens, and so on. A similar idea applies to adding or subtracting *decimal* numbers. With decimals, you line up the decimal points to make sure you are adding tenths to tenths, hundredths to hundredths, and so on.

Adding and Subtracting Decimals

Step 1 Write the numbers in columns with the decimal points lined up.

Step 2 If necessary, write in 0s so both numbers have the same number of decimal places. Then add or subtract as if they were whole numbers.

Step 3 Line up the decimal point in the answer directly below the decimal points in the problem.

Example 1 Adding Decimal Numbers

Add.

(a) 16.92 and 48.34
Step 1 Write the numbers in columns with the decimal points lined up.

$$\begin{array}{r} 16.92 \\ + 48.34 \end{array}$$

Decimal points are lined up.

Step 2 Add as if these were whole numbers.

$$\begin{array}{r} 1\ 1 \\ 16.92 \\ + 48.34 \\ \hline 65.26 \end{array}$$

Step 3 Decimal point in answer is lined up under decimal points in problem.

(b) 5.897 + 4.632 + 12.174
Write the numbers vertically with decimal points lined up. Then add.

$$\begin{array}{r} 11\ 21 \\ 5.897 \\ 4.632 \\ + 12.174 \\ \hline 22.703 \end{array}$$

Decimal points are lined up.

Work Problem ❶ at the Side.

In Example 1(a), both numbers had *two* decimal places (two digits to the right of the decimal point). In Example 1(b), all the numbers had *three decimal places* (three digits to the right of the decimal point). That made it easy to add tenths to tenths, hundredths to hundredths, and so on.

❶ Find each sum.

(a) 2.86 + 7.09

(b) 13.761 + 8.325

(c) 0.319 + 56.007 + 8.252

(d) 39.4 + 0.4 + 177.2

ANSWERS
1. (a) 9.95 **(b)** 22.086 **(c)** 64.578 **(d)** 217.0

❷ Find each sum.

(a) $6.54 + 9.8$

If the number of decimal places does *not* match, you can write in 0s as placeholders to make them match. This is shown in Example 2.

Example 2 Writing 0s as Placeholders before Adding

Add.

(a) $7.3 + 0.85$

There are two decimal places in 0.85 (tenths and hundredths), so write a 0 in the hundredths place in 7.3 so that it has two decimal places also.

$$\begin{array}{r} 7.30 \\ + 0.85 \\ \hline 8.15 \end{array}$$ ← One 0 is written in.

7.30 is equivalent to 7.3 because

$7\dfrac{30}{100}$ in lowest terms is $7\dfrac{3}{10}$

(b) $0.831 + 222.2 + 10$

(b) $6.42 + 9 + 2.576$

Write in 0s so that all the addends have three decimal places. Notice how the whole number 9 is written with the decimal point at the *far right* side. (If you put the decimal point on the *left* side of the 9, you would turn it into the decimal fraction 0.9.)

$$\begin{array}{r} 6.4\,2\,0 \\ 9.0\,0\,0 \\ + 2.5\,7\,6 \\ \hline 17.9\,9\,6 \end{array}$$

6.4 2 0 ← One 0 is written in.
9.0 0 0 ← 9 is a whole number; decimal point and three 0s are written in.
+ 2.5 7 6 ← No 0s are needed.

Decimal points are lined up.

(c) $8.64 + 39.115 + 3.0076$

NOTE

Writing 0s to the right of a *decimal* number does *not* change the value of the number.

Work Problem ❷ at the Side.

❷ **Subtract decimals.** Subtraction of decimals is done in much the same way as addition of decimals. You can check the answers to subtraction problems using addition, as you did with whole numbers (see **Section 1.3**).

(d) $5 + 429.823 + 0.76$

Example 3 Subtracting Decimal Numbers

Subtract. Check your answers using addition.

(a) 15.82 from 28.93

Step 1

$$\begin{array}{r} 28.93 \\ - 15.82 \end{array}$$

Line up decimal points. Then you will be subtracting hundredths from hundredths and tenths from tenths.

Step 2

$$\begin{array}{r} 28.93 \\ - 15.82 \\ \hline 13\ 11 \end{array}$$

Both numbers have two decimal places; no need to write in 0s.

Subtract as if they were whole numbers.

Continued on Next Page

Step 3

$$\begin{array}{r} 28.\overset{\cdot}{9}3 \\ -\ 15.82 \\ \hline 13.11 \end{array}$$

└──── Decimal point in answer is lined up.

Check the answer by adding 13.11 and 15.82. If the subtraction is done correctly, the sum will be 28.93.

(b) 146.35 minus 58.98
Borrowing is needed here.

$$\begin{array}{r} {\scriptstyle 0\ \ 13\ 15\ \ \ \ 12\ 15} \\ \not{1}\,\not{4}\,\not{6}\,.\,\not{3}\,\not{5} \\ -\ \ \ 5\ 8\,.\,9\ 8 \\ \hline 8\ 7\,.\,3\ 7 \end{array}$$

└──── Line up decimal points.

Check the answer by adding 87.37 and 58.98. If you did the subtraction correctly, the sum will be 146.35. (If it *isn't,* you need to rework the problem.)

═════ **Work Problem ❸ at the Side.**

Example 4 **Writing 0s as Placeholders before Subtracting**

Subtract.

(a) 16.5 from 28.362
Use the same steps as in Example 3 above, remembering to write in 0s so both numbers have three decimal places.

┌──── Line up decimal points.

$$\begin{array}{r} 28.\overset{\cdot}{3}62 \\ -\ 16.500 \\ \hline 11.862 \end{array}$$ ← Write two 0s.
← Subtract as usual.

Check the answer by adding.

$$\begin{array}{r} 16.500 \\ +\ 11.862 \\ \hline 28.362 \end{array}$$ ← Matches minuend in original problem.

(b) 59.7 − 38.914

$$\begin{array}{r} 59.700 \\ -\ 38.914 \\ \hline 20.786 \end{array}$$ ← Write two 0s
← Subtract as usual.

(c) 12 less 5.83

$$\begin{array}{r} 12.00 \\ -\ 5.83 \\ \hline 6.17 \end{array}$$ ← Write a decimal point and two 0s.
← Subtract as usual.

═════ **Work Problem ❹ at the Side.**

3 ▭ **Estimate the answer when adding or subtracting decimals.** A common error in working decimal problems by hand is to misplace the decimal point in the answer. Or, when using a calculator, you may accidentally press the wrong key. **Estimating** the answer will help you avoid these mistakes. Start by using *front end rounding* on each number (as you did in **Section 1.7**). Here are several examples. Notice that in the rounded numbers only the leftmost digit is something other than 0.

3.25	rounds to	3		6.812	rounds to	7
532.6	rounds to	500		26.397	rounds to	30
7094.2	rounds to	7000		351.24	rounds to	400

❸ Subtract. Check your answers using addition.

(a) 22.7 from 72.9

(b) 6.425 from 11.813

(c) 20.15 − 19.67

❹ Subtract. Check your answers using addition.

(a) 18.651 from 25.3

(b) 5.816 − 4.98

(c) 40 less 3.66

(d) 1 − 0.325

⑤ First, use front end rounding and estimate each answer. Then add or subtract to find the exact answer.

(a) 2.83 + 5.009 + 76.1

(b) 11.365 from 58

(c) 398.81 + 47.658 + 4158.7

(d) Find the difference between 12.837 meters and 46.091 meters.

(e) $19.28 plus $1.53

> **Example 5   Estimating Decimal Answers**
>
> Use front end rounding to round each number. Then add or subtract the rounded numbers to get an estimated answer. Finally, find the exact answer.
>
> (a) Add 194.2 and 6.825.
>
> | | *Estimate:* | | *Exact:* |
> |---|---|---|---|
> | | 200 | ← Rounds to | 194.200 |
> | + | 7 | ← Rounds to | + 6.825 |
> | | 207 | | 201.025 |
>
> The estimate goes out to the hundreds place (three places to the *left* of the decimal point), and so does the exact answer. Therefore, the decimal point is probably in the correct place in the exact answer.
>
> (b) $69.42 + $13.78
>
> | | *Estimate:* | | *Exact:* |
> |---|---|---|---|
> | | $70 | ← Rounds to | $69.42 |
> | + | 10 | ← Rounds to | + 13.78 |
> | | $80 | | $83.20 ← Answer is close to estimate, so the problem is probably set up correctly. |
>
> (c) Find the difference between 0.92 ft and 8 ft.
>   Use subtraction to find the difference between two numbers. The larger number, 8, is written on top.
>
> | | *Estimate:* | | *Exact:* |
> |---|---|---|---|
> | | 8 | ← Rounds to | 8.00 ← Write a decimal point and two 0s. |
> | − | 1 | ← Rounds to | − 0.92 |
> | | 7 | | 7.08 ft ← Answer is close to estimate. |
>
> (d) Subtract 1.8614 from 7.3
>
> | | *Estimate:* | | *Exact:* |
> |---|---|---|---|
> | | 7 | ← Rounds to | 7.3000 ← Write three 0s. |
> | − | 2 | ← Rounds to | − 1.8614 |
> | | 5 | | 5.4386 ← Answer is close to estimate. |

Work Problem ⑤ at the Side.

▦ Calculator Tip   If you are *adding* numbers, you can enter them in any order on your calculator. Try these; jot down the answers.

9.82 ⊕ 1.86 ⊜ _____   1.86 ⊕ 9.82 ⊜ _____

The answers are the same because addition is *commutative*. (See **Section 1.2.**) But subtraction is *not* commutative. It *does* matter which number you enter first. Try these:

9.82 ⊖ 1.86 ⊜ _____   1.86 ⊖ 9.82 ⊜ _____

The second answer has a negative sign (−) next to it. A negative number is less than 0. If it's in your checkbook, you'd be "in the hole" by $7.96. (See **Section 9.1** for more about negative numbers.)

ANSWERS
5. (a) 3 + 5 + 80 = 88; 83.939
   (b) 60 − 10 = 50; 46.635
   (c) 400 + 50 + 4000 = 4450; 4605.168
   (d) 50 − 10 = 40; 33.254 meters
   (e) $20 + $2 = $22; $20.81

4.3   EXERCISES

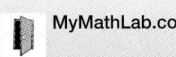
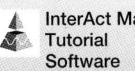
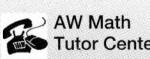

Find each sum or difference. See Examples 1–4.

1. $5.69 + 11.79$

2. $372.1 - 33.7$

3. $24.008 - 0.995$

4. $0.7759 + 9.8883$

5. $8.263 - 0.5$

6. $47.658 - 20.9$

7. $76.5 + 0.506$

8. $1.87 + 9.749$

9. $21 - 0.896$

10. $9 - 1.183$

11. $0.4 - 0.291$

12. $0.35 - 0.088$

13. $39.76005 + 182 + 4.799 + 98.31 + 5.9999$

14. $489.76 + 0.9993 + 38 + 8.55087 + 80.697$

This drawing of a human skeleton shows the average length of the longest bones, in inches. Use the drawing to answer Exercises 15–18.

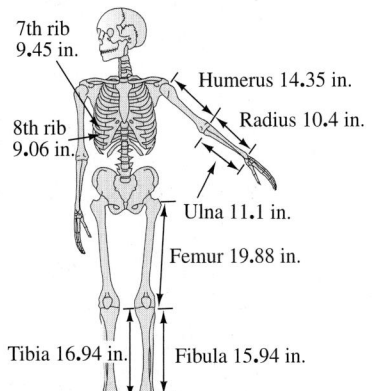

7th rib 9.45 in.
8th rib 9.06 in.
Humerus 14.35 in.
Radius 10.4 in.
Ulna 11.1 in.
Femur 19.88 in.
Tibia 16.94 in.
Fibula 15.94 in.

Source: *The Top 10 of Everything 2000.*

15. (a) What is the combined length of the humerus and radius bones?

(b) What is the difference in the lengths of these two bones?

16. (a) What is the total length of the femur and tibia bones?

(b) How much longer is the femur than the tibia?

17. (a) Find the sum of the lengths of the humerus, ulna, femur, and tibia.

(b) How much shorter is the 8th rib than the 7th rib?

18. (a) What is the difference in the lengths of the two bones in the lower arm?

(b) What is the difference in the lengths of the two bones in the lower leg?

19. Explain and correct
the error that a student
made when he added
$0.72 + 6 + 39.5$ this way:

$$
\begin{array}{r}
0.72 \\
6 \\
+\ 39.50 \\
\hline
40.28
\end{array}
$$

20. Explain the difference between saying
"subtract 2.9 from 8" and saying
"2.9 minus 8."

*Use front end rounding to round each number. Then add or subtract the rounded
numbers to get an estimated answer. Finally, find the exact answer. See Example 5.*

21. *Estimate:*   *Exact:*

$$
\begin{array}{r}
\$\ \ \ \ \ \ \ \ \ \$19.74 \\
-\ \ \ \ \ \ \ \ \ \ -\ \ 6.58 \\
\hline
\$\ \ \ \ \ \ \ \ \ \$
\end{array}
$$

22. *Estimate:*   *Exact:*

$$
\begin{array}{r}
\$\ \ \ \ \ \ \ \ \ \$27.96 \\
-\ \ \ \ \ \ \ \ \ \ -\ \ 8.39 \\
\hline
\$\ \ \ \ \ \ \ \ \ \$
\end{array}
$$

23. *Estimate:*   *Exact:*

$$
\begin{array}{r}
392.7 \\
0.865 \\
+\ \ \ \ \ \ \ \ \ +\ \ 21.08 \\
\hline
\end{array}
$$

24. *Estimate:*   *Exact:*

$$
\begin{array}{r}
38.55 \\
7.716 \\
+\ \ \ \ \ \ \ \ \ +\ \ 0.6 \\
\hline
\end{array}
$$

25. What is 8.6 less 3.751?

   Estimate: *Exact:*

26. What is 31.7 less 4.271?

   Estimate: *Exact:*

27. *Estimate:*   *Exact:*

$$
\begin{array}{r}
62.8173 \\
539.99 \\
+\ \ \ \ \ \ \ \ \ +\ \ 5.629 \\
\hline
\end{array}
$$

28. *Estimate:*   *Exact:*

$$
\begin{array}{r}
332.607 \\
12.5 \\
+\ \ \ \ \ \ \ \ \ +\ 823.3949 \\
\hline
\end{array}
$$

*Use your estimation skills to pick the most reasonable answer for each example.
Do **not** solve the problems. Circle your choice.*

29. $12 - 11.725$

   2.75   0.275   27.5

30. $20 - 1.37$

   0.1863   1.863   18.63

31. $6.5 + 0.007$

   6.507   0.6507   65.07

32. $9.67 + 0.09$

   0.976   9.76   0.00976

33. $456.71 - 454.9$

   18.1   181   1.81

34. $803.25 - 0.6$

   802.65   0.80265   8.0265

35. $6004.003 + 52.7172$

   60.567202   605.67202   6056.7202

36. $128.35 + 97.0093$

   2253.593   225.3593   0.2253593

First use front end rounding to round each number and estimate the answer. Then find the exact answer.

37. The cost of Julie's tennis racket, with tax, is $41.09. She gave the clerk two $20 bills and a $10 bill. What amount of change did Julie receive?

Estimate:

Exact:

38. The tallest known land mammal is a prehistoric ancestor of the rhino measuring 6.4 meters. Find the combined heights of these NBA basketball stars: Charles Barkley at 1.98 meters, Karl Malone at 2.06 meters, and David Robinson at 2.16 meters. Is their combined height greater or less than the prehistoric rhino? (*Source:* Harper's Index and NBA.)

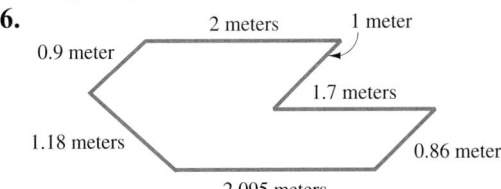

6.4 meters

Estimate:

Exact:

39. Find the difference between 1.981 in. and 2 in.

Estimate:

Exact:

40. Find the difference between 13.582 meters and 28 meters.

Estimate:

Exact:

41. The U.S. population in 2000 was approximately 281.42 million. The U.S. Bureau of the Census estimates that it will be 393.9 million in the year 2050. The increase in population during that 50-year period is how many millions of people? (*Source:* U.S. Bureau of the Census.)

Estimate:

Exact:

42. At a bakery, Sue Chee bought $7.42 worth of muffins and $10.09 worth of croissants for a staff party and a $0.69 cookie for herself. How much money did she spend altogether?

Estimate:

Exact:

43. Namiko is comparing two boxes of chicken nuggets. One box weighs 9.85 ounces and the other weighs 10.5 ounces. What is the difference in the weight of the two boxes?

Estimate:

Exact:

44. Sammy works in a veterinarian's office. He weighed two newborn kittens. One was 3.9 ounces and the other was 4.05 ounces. What was the difference in the weight of the two kittens?

Estimate:

Exact:

Find the perimeter of (distance around) each figure by adding the lengths of the sides.

45.

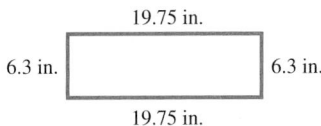

19.75 in.

6.3 in.   6.3 in.

19.75 in.

Estimate:

Exact:

46.

2 meters   1 meter

0.9 meter

1.7 meters

1.18 meters

0.86 meter

2.095 meters

Estimate:

Exact:

The dad and daughter buying fishing equipment on the first page of this chapter brought along the store's sale insert from the Sunday paper. Use the information on sale prices to answer Exercises 47–50.

Fishing Opener Sale
Catch your limit of savings!

Bobbers 3 for 87¢

4-, 6-, or 8-pound test line
330 yd for $5.32
110 yd for $2.72
No-See Line

Environmentally safe tin split shot
$1.57

Small tackle boxes
One tray $6.38
Two trays $7.96

Leaded split shot
94¢

Spinning reels: $8.96, $13.47, $17.96, $29.46
Light weight rods: $6.96, $12.97, $19.94

Source: Wal-Mart.

47. What is the difference in price between the most expensive and least expensive spinning reel?

Estimate:

Exact:

48. How much more would 330 yd of line cost than three bobbers?

Estimate:

Exact:

49. What is the total cost of the middle-priced rod, the second most expensive reel, a one-tray tackle box, 110 yd of line, and a package of environmentally safe split shot?

Estimate:

Exact:

50. Dad also bought his daughter a cap for $8.49, SPF45 sunscreen for $6.97, and a child-size flotation vest for $19.99. How much did he spend on these items?

Estimate:

Exact:

Olivia Sanchez kept track of her expenses for one month. Use her list to answer Exercises 51–56.

Monthly Expenses	
Rent	$994
Car payment	$190.78
Car repairs, gas	$205
Cable TV	$39.95
Internet access	$19.95
Electricity	$40.80
Telephone	$57.32
Groceries	$186.81
Entertainment	$97.75
Clothing, laundry	$107

51. What were Olivia's total expenses for the month?

52. How much did Olivia pay for telephone, cable TV, and Internet access?

53. What was the difference in the amounts spent for groceries and for the car payment?

54. Compare the amount Olivia spent on entertainment to the amount spent on car repairs and gas. What is the difference?

55. How much more did Olivia spend on rent than on all her car expenses?

56. How much less did Olivia spend on clothing and laundry than on all her car expenses?

Find the length of the dashed line in each rectangle or circle.

57.

0.91 cm   0.7 cm b

3 cm

58.

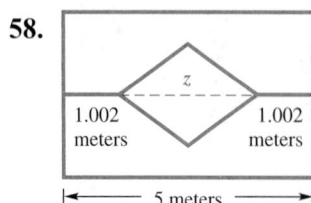

z

1.002 meters   1.002 meters

5 meters

59.

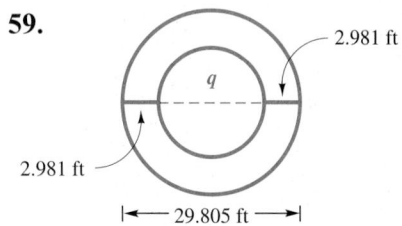

2.981 ft

q

2.981 ft

29.805 ft

4.4 MULTIPLYING DECIMALS

1   **Multiply decimals.** The decimals 0.3 and 0.07 can be multiplied by writing them as fractions.

$$0.\underset{\uparrow}{3} \times 0.\underset{\vee}{07} = \frac{3}{10} \times \frac{7}{100} = \frac{3 \times 7}{10 \times 100} = \frac{21}{1000} = 0.021$$

1 decimal place + 2 decimal places = 3 decimal places

Can you see a way to multiply decimals without writing them as fractions? Try these steps. Remember that each number in a multiplication problem is called a *factor,* and the answer is called the *product.*

Multiplying Decimals

Step 1   Multiply the numbers (the factors) as if they were whole numbers.

Step 2   Find the *total* number of decimal places in *both* factors.

Step 3   Write the decimal point in the product (the answer) so it has the same number of decimal places as the total from Step 2. You may need to write in extra 0s on the left side of the product to get the correct number of decimal places.

NOTE

When multiplying decimals, you do ***not*** need to line up decimal points. (You ***do*** need to line up decimal points when adding or subtracting decimals.)

Example 1   Multiplying Decimal Numbers

Multiply: 8.34 times 4.2.
Step 1   Multiply the numbers as if they were whole numbers.

```
      8.3 4
   ×    4.2
      1 6 6 8
    3 3 3 6
    3 5 0 2 8
```

Step 2   Count the total number of decimal places in both factors.

```
      8.3 4  ← 2 decimal places
   ×    4.2  ← 1 decimal place
      1 6 6 8    3 total decimal places
    3 3 3 6
    3 5 0 2 8
```

Step 3   Count over 3 places in the product and write the decimal point. Count from *right to left.*

```
      8.3 4  ← 2 decimal places
   ×    4.2  ← 1 decimal place
      1 6 6 8    3 total decimal places
    3 3 3 6
    3 5.0 2 8  ← 3 decimal places in product
```
Count over 3 places from right to left to position the decimal point.

Work Problem 1 at the Side.

OBJECTIVES

1 Multiply decimals.
2 Estimate the answer when multiplying decimals.

1 Multiply.

(a)   2.6
   × 0.4

(b)   45.2
   × 0.25

(c)   0.104 ← 3 decimal places
   ×   7 ← 0 decimal places
        ← 3 decimal places in the product

(d)   3.18
   × 2.23

(e)   611
   × 3.7

ANSWERS
1. (a) 1.04   (b) 11.300   (c) 0.728
(d) 7.0914   (e) 2260.7

❷ Multiply.

(a) 0.04×0.09

(b) $0.2 \cdot 0.008$

(c) $(0.063)(0.04)$

(d) $0.0081 \cdot 0.003$

(e) $(0.11)(0.0005)$

❸ First use front end rounding and estimate the answer. Then find the exact answer.

(a) $(11.62)(4.01)$

(b) $(5.986)(33)$

(c) $8.31 \cdot 4.2$

(d) 58.6×17.4

Example 2   Writing 0s as Placeholders in the Product

Multiply 0.042 by 0.03.

Start by multiplying, then count decimal places.

$$
\begin{array}{r}
0.0\,4\,2 \leftarrow \text{3 decimal places} \\
\times \quad 0.0\,3 \leftarrow \text{2 decimal places} \\
\hline
1\,2\,6 \leftarrow \text{5 decimal places needed in product}
\end{array}
$$

After multiplying, the answer has only three decimal places, but five are needed. So write two 0s on the *left* side of the answer.

$$
\begin{array}{r}
0.0\,4\,2 \\
\times \quad 0.0\,3 \\
\hline
0\,0\,1\,2\,6
\end{array}
\qquad
\begin{array}{r}
0.0\,4\,2 \leftarrow \text{3 decimal places} \\
\times \quad 0.0\,3 \leftarrow \text{2 decimal places} \\
\hline
.0\,0\,1\,2\,6 \leftarrow \text{5 decimal places}
\end{array}
$$

Write two 0s on *left* side of answer.   Now count over 5 places and write in the decimal point.

The final product is 0.00126, which has five decimal places.

Work Problem ❷ at the Side.

2 ▮▮▮ **Estimate the answer when multiplying decimals.** If you are doing multiplication problems by hand, estimating the answer helps you check that the decimal point is in the right place. When you are using a calculator, estimating helps you catch an error like pressing the ⊝ key instead of the ⊗ key.

Example 3   Estimating before Multiplying

First estimate $(76.34)(12.5)$ using front end rounding to round each number. Then find the exact answer.

Estimate: *Exact:*

$$
\begin{array}{r}
80 \\
\times \; 10 \\
\hline
800
\end{array}
\qquad
\begin{array}{r}
76.3\,4 \leftarrow \text{2 decimal places} \\
\times \quad 1\,2.5 \leftarrow \text{1 decimal place} \\
\hline
3\,8\,1\,7\,0 \quad \text{3 decimal places needed} \\
1\,5\,2\,6\,8 \qquad \text{in product.} \\
7\,6\,3\,4 \\
\hline
9\,5\,4.2\,5\,0
\end{array}
$$

Both the estimate and the exact answer go out to the hundreds place, so the decimal point in 954.250 is probably in the correct place.

Work Problem ❸ at the Side.

▦ Calculator Tip   When working with money amounts, you may need to write a 0 in your answer. For example, try multiplying $\$3.54 \times 5$ on your calculator. Write down the result.

$$3.54 \; ⊗ \; 5 \; ⊜ \; \underline{\qquad}$$

Notice that the result is 17.7, which is *not* the way to write a money amount. You have to write the 0 in the hundredths place: $\$17.70$ is correct. The calculator does not show the "extra" 0 because:

$$17.70 \text{ or } 17\frac{70}{100} \quad \text{reduces to} \quad 17\frac{7}{10} \text{ or } 17.7.$$

So keep an eye on your calculator—it doesn't know when you're working with money amounts.

ANSWERS
2. (a) 0.0036 **(b)** 0.0016 **(c)** 0.00252
(d) 0.0000243 **(e)** 0.000055
3. (a) $(10)(4) = 40$; 46.5962
(b) $(6)(30) = 180$; 197.538
(c) $8 \cdot 4 = 32$; 34.902
(d) $60 \times 20 = 1200$; 1019.64

4.4   EXERCISES

 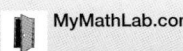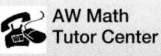

Multiply. See Example 1.

1.   0.042
    × 3.2

2.   0.571
    × 2.9

3.   21.5
    × 7.4

4.   85.4
    × 3.5

5.   23.4
    × 0.666

6.   0.896
    × 0.799

7. $51.88
    × 665

8. $736.75
    × 118

Use the fact that $72 \times 6 = 432$ *to help you answer Exercises 9–16 by simply counting decimal places. See Examples 1 and 2.*

9. $72 \times 0.6 =$ 4 3 2

10. $7.2 \times 6 =$ 4 3 2

11. $(7.2)(0.06) =$ 4 3 2

12. $(0.72)(0.6) =$ 4 3 2

13. $0.72 \times 0.06 =$ 4 3 2

14. $72 \times 0.0006 =$ 4 3 2

15. $0.0072 \times 0.6 =$ 4 3 2

16. $0.072 \times 0.006 =$ 4 3 2

Multiply. See Example 2.

17. $(0.006)(0.0052)$

18. $(0.0052)(0.009)$

19. $0.003 \cdot 0.002$

20. $0.0079 \cdot 0.006$

RELATING CONCEPTS (Exercises 21–22)   FOR INDIVIDUAL OR GROUP WORK

Look for patterns in the multiplications as you **work Exercises 21 and 22 in order.**

21. Do these multiplications:

$(5.96)(10)$ $(3.2)(10)$

$(0.476)(10)$ $(80.35)(10)$

$(722.6)(10)$ $(0.9)(10)$

What pattern do you see? Write a "rule" for multiplying by 10. What do you think the rule is for multiplying by 100? by 1000? Write the rules and try them out on the numbers above.

22. Do these multiplications:

$(59.6)(0.1)$ $(3.2)(0.1)$

$(0.476)(0.1)$ $(80.35)(0.1)$

$(65)(0.1)$ $(523)(0.1)$

What pattern do you see? Write a "rule" for multiplying by 0.1. What do you think the rule is for multiplying by 0.01? by 0.001? Write the rules and try them out on the numbers above.

First use front end rounding to round each number and estimate the answer. Then find the exact answer. See Example 3.

23. *Estimate:*   *Exact:*

Rounds to ⟵
Rounds to ⟵   39.6
× ____   × 4.8

24. *Estimate:*   *Exact:*

               18.7
× ____   × 2.3

25. *Estimate:*   *Exact:*

               37.1
× ____   ×   42

26. *Estimate:*   *Exact:*

               5.08
× ____   ×   71

27. *Estimate:*   *Exact:*

               6.53
× ____   × 4.6

28. *Estimate:*   *Exact:*

               7.51
× ____   × 8.2

29. *Estimate:*   *Exact:*

               2.809
× ____   × 6.85

30. *Estimate:*   *Exact:*

               73.52
× ____   × 22.34

Even with most of the problem missing, you can tell whether or not these answers are reasonable. Circle reasonable *or* unreasonable. *If the answer is unreasonable, move the decimal point, or insert a decimal point, to make the answer reasonable.*

31. How much was his car payment? $18.90

    reasonable

    unreasonable, should be _____

32. How many hours did she work today? 25 hours

    reasonable

    unreasonable, should be _____

33. How tall is her son? 60.5 in.

    reasonable

    unreasonable, should be _____

34. How much does he pay for rent now? $6.92

    reasonable

    unreasonable, should be _____

35. What is the price of one gallon of milk? $319

    reasonable

    unreasonable, should be _____

36. How long is the living room? 16.8 feet

    reasonable

    unreasonable, should be _____

37. How much did the baby weigh? 0.095 pounds

    reasonable

    unreasonable, should be _____

38. What was the sale price of the jacket? $1.49

    reasonable

    unreasonable, should be _____

Solve each application problem. If the problem involves money, round to the nearest cent, when necessary.

39. LaTasha worked 50.5 hours over the last two weeks. She earns $18.73 per hour. How much did she make?

40. Michael's time card shows 42.2 hours at $10.03 per hour. What are his gross earnings?

41. Sid needs 0.6 meter of canvas material to make a carry-all bag that fits on his wheelchair. If canvas is $4.09 per meter, how much will Sid spend? (*Note:* $4.09 *per* meter means $4.09 for *one* meter.)

42. How much will Mrs. Nguyen pay for 3.5 yards of lace trim that costs $0.87 per yard?

43. Michelle filled the tank of her pickup truck with regular unleaded gas. Use the information shown on the pump to find how much she paid for gas.

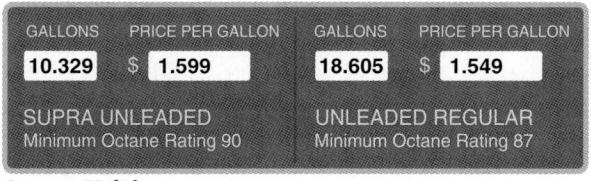

GALLONS	PRICE PER GALLON	GALLONS	PRICE PER GALLON
10.329	$ 1.599	18.605	$ 1.549
SUPRA UNLEADED Minimum Octane Rating 90		UNLEADED REGULAR Minimum Octane Rating 87	

Source: Holiday.

44. Ground beef and spicy chicken wings are on sale. Juma bought 1.7 pounds of wings. Use the information in the ad to find the amount she paid.

BIG ONE FOODS Sale

Ground Beef $1.89 per pound   Spicy Wings $0.98 per pound

PRICES GOOD THROUGH SUNDAY!

45. Ms. Rolack is a real estate broker who helps people sell their homes. Her fee is 0.07 times the price of the home. What was her fee for selling a $175,300 home?

46. Alex Rodriguez, shortstop for the Seattle Mariners, had a batting average of 0.316 in the 2000 season. If he went to bat 554 times, how many hits did he make? (*Hint:* Multiply his batting average by the number of times at bat.) Round to the nearest whole number. (*Source: World Almanac*, 2001.)

47. Judy Lewis pays $28.96 per month for basic cable TV. How much will she pay for cable over one year? How much would she pay in a year for the deluxe cable package that costs $59.95 per month?

48. Chuck's car payment is $220.27 per month for three years. How much will he pay altogether?

49. Paper for the copy machine at the library costs $0.015 per sheet. How much will the library pay for 5100 sheets?

50. A student group collected 2200 pounds of plastic as a fund-raiser. How much will they make if the recycling center pays $0.142 per pound?

51. The National Aquarium in Baltimore charges $11.95 for adults, $10.50 for seniors, and $7.50 for children. How much will a mother with four children spend for her family and three senior relatives? (*Source:* Lyon Group.)

52. (Complete Exercise 51 first.) How much *less* would the same family spend at the Texas State Aquarium, which charges $8 for adults, $5.75 for seniors, and $4.50 for children? (*Source:* Lyon Group.)

53. Ms. Sanchez paid $29.95 a day to rent a car, plus $0.29 per mile. Find the cost of her rental for a four-day trip of 926 miles.

54. The Bell family rented a motor home for $375 per week plus $0.35 per mile. What was the rental cost for their three-week vacation trip of 2650 miles?

55. Barry bought 16.5 meters of rope at $0.47 per meter and three meters of wire at $1.05 per meter. How much change did he get from three $5 bills?

56. Susan bought a VCR that cost $229.88. She paid $45 down and $37.98 per month for six months. How much could she have saved by paying cash?

Use the information from the Look Smart mail order catalog to answer Exercises 57–60.

Knit Shirt Ordering Information		
43–2A	Short sleeve, solid colors	$14.75 each
43–2B	Short sleeve, stripes	$16.75 each
43–3A	Long sleeve, solid colors	$18.95 each
43–3B	Long sleeve, stripes	$21.95 each
Extra-large size, add $2 per shirt.		
Monogram, $4.95 each. Gift box, $5 each.		

Total Price of All Items (excluding monograms and gift boxes)	Shipping, Packing, and Handling
$0–25.00	$3.50
$25.01–75.00	$5.95
$75.01–125.00	$7.95
$125.01+	$9.95
Shipping to each additional address add $4.25.	

57. Find the total cost of ordering four long-sleeve, solid-color shirts and two short-sleeve, striped shirts, all in the extra-large size, and all shipped to your home.

58. What is the total cost of eight long-sleeve shirts, five in solid colors and three striped? Include the cost of shipping the solid shirts to your home and the striped shirts to your brother's home.

59. (a) What is the total cost, including shipping, of sending three short-sleeve, solid-color shirts, with monograms, in a gift box to your aunt for her birthday?

  (b) How much did the monograms, gift box, and shipping add to the cost of your gift?

60. (a) Suppose you order one of each type of shirt for yourself, adding a monogram on each of the solid-color shirts. At the same time, you order three long-sleeved striped shirts, in the extra-large size, shipped to your dad in a gift box. Find the total cost of your order.

  (b) What is the difference in total cost (excluding shipping) between the shirts for yourself and the gift for your dad?

4.5 DIVIDING DECIMALS

There are two kinds of decimal division problems; those in which a decimal is divided by a whole number, and those in which a decimal is divided by a decimal. First recall the parts of a division problem from **Section 1.5.**

$$\begin{array}{r} 8 \leftarrow \text{Quotient} \\ \text{Divisor} \rightarrow 4\overline{)33} \leftarrow \text{Dividend} \\ \underline{32} \\ 1 \leftarrow \text{Remainder} \end{array}$$

1 **Divide a decimal by a whole number.**   When the divisor is a whole number, use these steps.

Dividing Decimals by Whole Numbers

Step 1   Write the decimal point in the quotient (answer) directly above the decimal point in the dividend.

Step 2   Divide as if both numbers were whole numbers.

Example 1   Dividing Decimals by Whole Numbers

Divide.

(a) 21.93 by 3

Dividend — Divisor

Rewrite the division problem. $3\overline{)21.93}$

Step 1   Write the decimal point in the quotient directly above the decimal point in the dividend. $3\overline{)21.93}$ — Decimal points lined up

Step 2   Divide as if the numbers were whole numbers.
$$3\overline{)21.93}^{\,7.31}$$

Check by multiplying the quotient times the divisor.
$$\begin{array}{r} 7.31 \\ \times\ \ \ 3 \\ \hline 21.93 \end{array}$$ Matches, so 7.31 is correct.

The quotient (answer) is 7.31.

(b) $9\overline{)470.7}$

Divisor — Dividend

Write the decimal point in the quotient above the decimal point in the dividend. Then divide as if the numbers were whole numbers.

Decimal points lined up
$$\begin{array}{r} 52.3 \\ 9\overline{)470.7} \\ \underline{45} \\ 20 \\ \underline{18} \\ 27 \\ \underline{27} \\ 0 \end{array}$$

Check:
$$\begin{array}{r} 52.3 \\ \times\ \ \ 9 \\ \hline 470.7 \end{array}$$ Matches

The quotient is 52.3.

OBJECTIVES

1 Divide a decimal by a whole number.
2 Divide a decimal by a decimal.
3 Estimate the answer when dividing decimals.
4 Use the order of operations with decimals.

1 Divide. Check your answers by multiplying.

(a) $4\overline{)93.6}$

(b) $6\overline{)6.804}$

(c) $11\overline{)278.3}$

(d) $0.51835 \div 5$

(e) $213.45 \div 15$

ANSWERS
1. (a) 23.4; (23.4)(4) = 93.6
 (b) 1.134; (1.134)(6) = 6.804
 (c) 25.3; (25.3)(11) = 278.3
 (d) 0.10367; (0.10367)(5) = 0.51835
 (e) 14.23; (14.23)(15) = 213.45

Work Problem ❶ at the Side.

❷ Divide. Check your answers by multiplying.

(a) $5\overline{)6.4}$

(b) $30.87 \div 14$

(c) $\dfrac{259.5}{30}$

(d) $0.3 \div 8$

Example 2 Writing Extra 0s to Complete a Division

Divide 1.5 by 8.

Keep dividing until the remainder is 0, or until the digits in the quotient begin to repeat in a pattern. In Example 1(b), you ended up with a remainder of 0. But sometimes you run out of digits in the dividend before that happens. If so, write extra 0s on the right side of the dividend so you can continue dividing.

$$
\begin{array}{r}
0.1 \\
8\overline{)1.5} \\
\underline{8} \\
7
\end{array}
$$
← All digits have been used.
← Remainder is not yet 0.

Write a 0 after the 5 in the dividend so you can continue dividing. Keep writing more 0s in the dividend if needed. Recall that writing 0s to the *right* of a decimal number does ***not*** change its value.

$$
\begin{array}{r}
0.1\,8\,7\,5 \\
8\overline{)1.5\,0\,0\,0} \\
\underline{8} \\
7\,0 \\
\underline{6\,4} \\
6\,0 \\
\underline{5\,6} \\
4\,0 \\
\underline{4\,0} \\
0
\end{array}
$$
← Three 0s needed to complete the division.
← Stop dividing when the remainder is 0.

Check:

$$
\begin{array}{r}
0.1875 \\
\times \qquad 8 \\
\hline
1.5000
\end{array}
$$
Matches dividend, so 0.1875 is correct.

📟 **Calculator Tip** When *multiplying* numbers, you can enter them in any order because multiplication is commutative (see **Section 1.4**). But division is *not* commutative. It *does* matter which number you enter first. Try Example 2 both ways; jot down your answers.

1.5 ⊘ 8 ⊜ _____   8 ⊘ 1.5 ⊜ _____

Notice that the first answer, 0.1875, matches the result from Example 2. But the second answer is much different: 5.333333333. Be careful to enter the dividend first.

CAUTION

When dividing decimals, notice that the dividend may *not* be the larger number, as it was in whole numbers. In Example 2 the dividend is 1.5, which is *smaller* than 8.

Work Problem ❷ at the Side.

The next example shows a quotient (answer) that must be rounded because you will never get a remainder of 0.

Example 3 Rounding a Decimal Quotient

Divide 4.7 by 3. Round the quotient to the nearest thousandth.
 Write extra 0s in the dividend so you can continue dividing.

$$
\begin{array}{r}
1.5\,6\,6\,6 \\
3\overline{)4.7\,0\,0\,0} \quad \leftarrow \text{Three 0s added so far} \\
\underline{3} \\
1\,7 \\
\underline{1\,5} \\
2\,0 \\
\underline{1\,8} \\
2\,0 \\
\underline{1\,8} \\
2\,0 \\
\underline{1\,8} \\
2 \quad \leftarrow \text{Remainder is still not 0.}
\end{array}
$$

Notice that the digit 6 in the answer is repeating. It will continue to do so. The remainder will *never be 0*. There are two ways to show that the answer is a **repeating decimal** that goes on forever. You can write three dots after the answer, or you can write a bar above the digits that repeat (in this case, the 6).

$$
\underbrace{1.5666\ldots}_{\text{Three dots}} \quad \text{or} \quad 1.5\overline{6} \quad \begin{array}{l} \leftarrow \text{Bar above} \\ \text{repeating digit} \end{array}
$$

When repeating decimals occur, round the answer according to the directions in the problem. In this example, to round to thousandths, divide out one *more* place, to ten-thousandths.

$$4.7 \div 3 = 1.5666\ldots \quad \text{rounds to} \quad 1.567$$

Check the answer by multiplying 1.567 by 3. Because 1.567 is a rounded answer, the check will not give exactly 4.7, but it should be very close.

$$(1.567)(3) = 4.701 \quad \begin{array}{l} \leftarrow \text{Does not equal exactly 4.7} \\ \text{because 1.567 was rounded.} \end{array}$$

CAUTION

When checking answers that you've rounded, the check will *not* match the dividend exactly, but it should be very close.

Work Problem ❸ at the Side.

2   **Divide a decimal by a decimal.**   To divide by a *decimal* divisor, first change the divisor to a whole number. Then divide as before. To see how this is done, write the problem in fraction form. Here is an example.

$$1.2\overline{)6.36} \quad \text{can be written} \quad \frac{6.36}{1.2}$$

In **Section 3.2** you learned that multiplying the numerator and denominator by the same number gives an equivalent fraction. We want the divisor (1.2) to be a whole number. Multiplying by 10 will accomplish that.

$$\underset{\substack{\text{Decimal} \\ \text{divisor}}}{\underbrace{\frac{6.36}{1.2}}} = \frac{6.36 \cdot 10}{1.2 \cdot 10} = \underset{\substack{\text{Whole number} \\ \text{divisor}}}{\underbrace{\frac{63.6}{12}}}$$

❸ Divide. Round answers to the nearest thousandth. If it is a repeating decimal, also write the answer using a bar. Check your answers by multiplying.

(a) $13\overline{)267.01}$

(b) $6\overline{)20.5}$

(c) $\dfrac{10.22}{9}$

(d) $16.15 \div 3$

(e) $116.3 \div 7$

ANSWERS
3. (a) 20.539 (rounded); no repeating digits visible on calculator;
   (20.539)(13) = 267.007
 (b) 3.417 (rounded); 3.41$\overline{6}$;
   (3.417)(6) = 20.502
 (c) 1.136 (rounded); 1.13$\overline{5}$;
   (1.136)(9) = 10.224
 (d) 5.383 (rounded); 5.38$\overline{3}$;
   (5.383)(3) = 16.149
 (e) 16.614 (rounded); starts repeating in eighth decimal place as 16.6$\overline{142857}$;
   (16.614)(7) = 116.298

4 Divide. If the quotient does not come out even, round to the nearest hundredth.

(a) $0.2\overline{)1.04}$

(b) $0.06\overline{)1.8072}$

(c) $0.005\overline{)32}$

(d) $8.1 \div 0.025$

(e) $\dfrac{7}{1.3}$

(f) $5.3091 \div 6.2$

The short way to multiply by 10 is to move the decimal point *one place* to the *right* in both the divisor and the dividend.

$$1.2\,\overline{)6.3\,6} \quad \text{is equivalent to} \quad 12\overline{)63.6}$$

NOTE

Moving the decimal points the *same* number of places in *both* the divisor and dividend will *not* change the answer.

Dividing by Decimals

Step 1 Count the number of decimal places in the divisor and move the decimal point that many places to the *right*. (This changes the divisor to a whole number.)

Step 2 Move the decimal point in the dividend the *same* number of places to the *right*. (Write in extra 0s if needed.)

Step 3 Write the decimal point in the quotient directly above the decimal point in the dividend. Then divide as usual.

Example 4 Dividing by Decimals

(a) $0.003\overline{)27.69}$

Move the decimal point in the divisor *three* places to the *right* so 0.003 becomes the whole number 3. To move the decimal point in the dividend the same number of places, write in an extra 0.

$$0.003\overline{)27.690}$$

Move decimal points in divisor and dividend. Then line up decimal point in answer.

Moving decimal point three places is the same as multiplying by 1000. ⟶

$$3\overline{)27690.}\quad{}^{9230.}$$ Divide as usual.

(b) Divide 5 by 4.2. Round to the nearest hundredth.

Move the decimal point in the divisor one place to the right so 4.2 becomes the whole number 42. The decimal point in the dividend starts on the right side of 5 and is also moved one place to the right.

```
            1.1 9 0  ← To round to hundredths,
  4.2 )5.0 0 0 0        divide out one more
      4 2               place, to thousandths.
      ───
        8 0
        4 2
        ───
        3 8 0
        3 7 8
        ─────
            2 0
```

Round the quotient. It is 1.19 (rounded to the nearest hundredth).

Work Problem 4 at the Side.

3 ▭ **Estimate the answer when dividing decimals.** Estimating the answer to a division problem helps you catch errors. Compare the estimate to your exact answer. If they are very different, do the division again.

> **Example 5** Estimating before Dividing

First use front end rounding to round each number and estimate the answer. Then divide to find the exact answer.

$$580.44 \div 2.8$$

Here is how one student solved this problem. She rounded 580.44 to 600 and rounded 2.8 to 3 to estimate the answer.

Estimate: *Exact:*

$$\begin{array}{r} 200 \\ 3\overline{)600} \end{array} \qquad \begin{array}{r} 2\,7.3 \\ 2.8\overline{)5\,8\,0.4\,4} \\ \underline{5\,6} \\ 2\,0\,4 \\ \underline{1\,9\,6} \\ 8\,4 \\ \underline{8\,4} \\ 0 \end{array}$$

Very different; need to rework the problem.

Notice that the estimate, which is in the hundreds, is very different from the exact answer, which is only in the tens. This tells the student that she needs to rework the problem. Can you find the error? (The exact answer should be 207.3, which fits with the estimate of 200.)

Work Problem **5** at the Side.

4 ▭ **Use the order of operations with decimals.** Use the order of operations when a decimal problem involves more than one operation, as you did with whole numbers in **Section 1.8.**

Order of Operations

1. Do all operations inside *parentheses* or *other grouping symbols.*
2. Simplify any expressions with *exponents* and find any *square roots.*
3. *Multiply* or *divide,* proceeding from left to right.
4. *Add* or *subtract,* proceeding from left to right.

> **Example 6** Using the Order of Operations

Use the order of operations to simplify each expression.

(a) $2.5 + 6.3^2 + 9.62$ Use exponent first.

$2.5 + 39.69 + 9.62$ Add from left to right.

$ 42.19 + 9.62$

$ 51.81$

(b) $1.82 + (6.7 - 5.2) \cdot 5.8$ Work inside parentheses.

$1.82 + 1.5 \cdot 5.8$ Multiply next.

$1.82 + 8.7$ Add last.

$ 10.52$

── **Continued on Next Page**

5 Decide whether each answer is reasonable by rounding the numbers and estimating the answer. If the exact answer is *not* reasonable, find and correct the error.

(a) $42.75 \div 3.8 = 1.125$

Estimate:

(b) $807.1 \div 1.76 = 458.580$
to nearest thousandth

Estimate:

(c) $48.63 \div 52 = 93.519$
to nearest thousandth

Estimate:

(d) $9.0584 \div 2.68 = 0.338$

Estimate:

ANSWERS
5. (a) Estimate is $40 \div 4 = 10$;
answer not reasonable,
should be 11.25
(b) Estimate is $800 \div 2 = 400$;
answer is reasonable.
(c) Estimate is $50 \div 50 = 1$;
answer is not reasonable,
should be 0.935.
(d) Estimate is $9 \div 3 = 3$;
answer is not reasonable;
should be 3.38.

6 Use the order of operations to simplify each expression.

(a) $4.6 - 0.79 + 1.5^2$

(b) $3.64 \div 1.3 \cdot 3.6$

(c) $0.08 + 0.6 \cdot (3 - 2.99)$

(d) $10.85 - 2.3 \times 5.2 \div 3.2$

(c) $\underbrace{3.7^2} - 1.8 \times 5.1 \div 1.5$ Use exponent first.

$13.69 - \underbrace{1.8 \times 5.1} \div 1.5$ Multiply and divide from left to right.

$13.69 - \underbrace{9.18 \div 1.5}$

$13.69 - 6.12$ Subtract last.

$ 7.57$

Work Problem 6 at the Side.

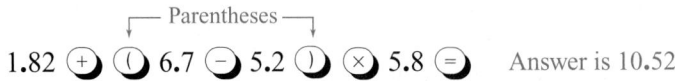

Calculator Tip   Most scientific calculators that have parentheses keys ( ) can handle calculations like those in Example 6 just by entering the numbers in the order given. For example, the keystrokes for Example 6(b) are:

┌── Parentheses ──┐
1.82 (+) (6.7 (−) 5.2) (×) 5.8 (=)   Answer is 10.52.

Standard, four-function calculators generally do not have parentheses keys and will *not* give the correct answer if you simply enter the numbers in the order given.

Check the instruction manual that came with your calculator for information on "order of calculations" to see if your machine has the rules for order of operations built into it. For a quick check, try entering this problem:

2 (+) 2 (×) 2 (=)

If the result is 6, the calculator follows the order of operations. If the result is 8, it does *not* have the rules built into it. To see why this test works, do the calculations by hand.

Use order of operations.	Work from left to right.
$2 + \underbrace{2 \times 2}$	$\underbrace{2 + 2} \times 2$
$2 + 4$	4×2
6 ←— Correct	8 ←— Incorrect

4.5 EXERCISES

 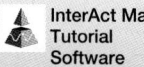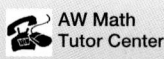
Divide. See Examples 1 and 4.

1. $7\overline{)27.3}$

2. $8\overline{)50.4}$

3. $\dfrac{4.23}{9}$

4. $\dfrac{1.62}{6}$

5. $0.05\overline{)20.01}$

6. $0.08\overline{)16.04}$

7. $1.5\overline{)54}$

8. $2.4\overline{)132}$

Use the fact that $108 \div 18 = 6$ to work Exercises 9–12 simply by moving decimal points. See Examples 2 and 4.

9. $0.108 \div 1.8$

10. $10.8 \div 18$

11. $0.018\overline{)108}$

12. $0.18\overline{)1.08}$

Divide. Round quotients to the nearest hundredth if necessary. See Examples 3 and 4.

13. $4.6\overline{)116.38}$

14. $2.6\overline{)4.992}$

15. $\dfrac{3.1}{0.006}$

16. $\dfrac{1.7}{0.09}$

Divide. Round quotients to the nearest thousandth. See Example 4.

17. $240 \div 9.88$

18. $7643 \div 5.36$

19. $0.034\overline{)342.81}$

20. $0.043\overline{)1748.4}$

RELATING CONCEPTS (Exercises 21–22)   **FOR INDIVIDUAL OR GROUP WORK**

*Look back at your work in Exercises 21 and 22 in Section 4.4. Then **work Exercises 21 and 22 in order.***

21. Do these division problems:

$3.77 \div 10$	$9.1 \div 10$
$0.886 \div 10$	$30.19 \div 10$
$406.5 \div 10$	$6625.7 \div 10$

What pattern do you see? Write a "rule" for dividing by 10. What do you think the rule is for dividing by 100? by 1000? Write the rules and try them out on the numbers above.

22. Do these division problems:

$40.2 \div 0.1$	$7.1 \div 0.1$
$0.339 \div 0.1$	$15.77 \div 0.1$
$46 \div 0.1$	$873 \div 0.1$

What pattern do you see? Write a "rule" for dividing by 0.1. What do you think the rule is for dividing by 0.01? by 0.001? Write the rules and try them out on the numbers above.

Decide whether each answer is reasonable or unreasonable by rounding the numbers and estimating the answer. If the exact answer is not reasonable, find the correct answer. See Example 5.

23. $37.8 \div 8 = 47.25$

Estimate:

24. $345.6 \div 3 = 11.52$

Estimate:

25. $54.6 \div 48.1 = 1.135$

Estimate:

26. $2428.8 \div 4.8 = 50.6$

Estimate:

27. $307.02 \div 5.1 = 6.2$

Estimate:

28. $395.415 \div 5.05 = 78.3$

Estimate:

29. $9.3 \div 1.25 = 0.744$

Estimate:

30. $78 \div 14.2 = 0.182$

Estimate:

Solve each application problem. Round money answers to the nearest cent, if necessary.

31. Alfred has discovered that Batman's favorite brand of superhero tights are on sale. He's been told to buy only one pair for Robin. How much will he pay for one pair?

Special Purchase!
Tights
6 pairs for $23.98
Stock up now!

32. The bookstore has a special price on notepads. How much did Randall pay for one notepad?

Notepads 4 for $1.69

33. It will take 21 equal monthly payments for Aimee to pay off her charge account balance of $408.66. How much is she paying each month?

34. Marcella Anderson bought 2.6 meters of suede fabric for $18.19. How much did she pay per meter?

35. Adrian Webb bought 619 bricks to build a barbecue pit, paying $185.70. Find the cost per brick. (*Hint:* Cost *per* brick means the cost for *one* brick.)

36. Lupe Wilson is a newspaper distributor. Last week she paid the newspaper $130.51 for 842 copies. Find the cost per copy.

37. Darren Jackson earned $356.80 for 40 hours of work. Find his earnings per hour.

38. At a record manufacturing company, 400 records cost $289. Find the cost per record.

39. It took 16.35 gallons of gas to fill Kim's car gas tank. She had driven 346.2 miles since her last fill-up. How many miles per gallon did her car get? Round to the nearest tenth.

40. Mr. Rodriquez pays $53.19 each month to House-hold Finance. How many months will it take him to pay off $1436.13?

Use the table of longest long jumps (through the year 2000) to answer Exercises 41–46. To find an average, add up the values you are interested in and then divide the sum by the number of values. Round your answer to the nearest hundredth. Some of the other exercises may require subtraction or multiplication.

Athlete	Country	Year	Length (meters)
M. Powell	U.S.	1991	8.95
B. Beamon	U.S.	1968	8.90
C. Lewis	U.S.	1991	8.87
R. Emmiyan	USSR	1987	8.86
L. Myricks	U.S.	1988	8.74
E. Walder	U.S.	1994	8.74
I. Pedroso	Cuba	1995	8.71
K. Streete-Thompson	U.S.	1994	8.63
J. Beckford	Jamaica	1997	8.62

Source: www.Olympics.com

41. Find the average length of the long jumps made by U.S. athletes.

42. Find the average length of all the long jumps listed in the table.

43. How much longer was the second-place jump than the third-place jump?

44. If the first-place athlete made six jumps of the same length, what was the total distance jumped?

45. What was the total length jumped by the top three athletes?

46. How much less was the last-place jump than the next-to-last-place jump?

17. _____

17. Write $2\dfrac{5}{8}$ as a decimal. Round to the nearest thousandth, if necessary.

Arrange in order from smallest to largest.

18. _____

18. 0.44, 0.451, $\dfrac{9}{20}$, 0.4506

Use the order of operations to simplify this expression.

19. _____

19. $6.3^2 - 5.9 + 3.4 \cdot 0.5$

Solve each application problem.

20. _____

20. Jennifer had $71.15 in her checking account. Yesterday her account earned $0.95 interest for the month, and she deposited a paycheck for $390.77. The bank charged her $16 for new checks. What is the new balance in her account?

21. _____

21. During her career, Jackie Joyner-Kersee won three Olympic medals in the long jump event. In 1988 she jumped 7.4 meters in Seoul. In 1992 she jumped 7.07 meters in Barcelona, and in 1996 she jumped 7 meters in Atlanta. In which year did she have the longest jump? How much longer was it than the second best jump? (*Source:* Associated Press.)

22. _____

22. Mr. Yamamoto bought 1.85 pounds of cheese at $2.89 per pound. How much did he pay for the cheese, to the nearest cent?

23. _____

23. Loren's baby had a temperature of 102.7 degrees. Later in the day it was 99.9 degrees. How much had the baby's temperature dropped?

24. _____

24. Pat bought 3.4 meters of fabric. She paid $15.47. What was the cost per meter?

25. _____

25. Write your own application problem using decimals. Make it different from problems 20–24. Then show how to solve your problem.

Name the digit that has the given place value.

1. 19,076,542
hundreds
millions
ones

2. 83.0754
tenths
thousandths
tens

Round each number as indicated.

3. 499,501 to the nearest thousand

4. 602.4937 to the nearest hundredth

5. $709.60 to the nearest dollar

6. $0.0528 to the nearest cent

First use front end rounding to round each number and estimate the answer. Then find the exact answer.

7. *Estimate:*   *Exact:*

$$\begin{array}{r} 3672 \\ 589 \\ +\ 9078 \\ \hline \end{array}$$

+ ____

8. *Estimate:*   *Exact:*

$$\begin{array}{r} 4.06 \\ 15.7 \\ +\ 0.923 \\ \hline \end{array}$$

+ ____

9. *Estimate:*   *Exact:*

$$\begin{array}{r} 5018 \\ -\ 1809 \\ \hline \end{array}$$

– ____

10. *Estimate:*   *Exact:*

$$\begin{array}{r} 51.6 \\ -\ 7.094 \\ \hline \end{array}$$

– ____

11. *Estimate:*   *Exact:*

$$\begin{array}{r} 3317 \\ \times\ 166 \\ \hline \end{array}$$

× ____

12. *Estimate:*   *Exact:*

$$\begin{array}{r} 6.82 \\ \times\ 7.3 \\ \hline \end{array}$$

× ____

13. *Estimate:*   *Exact:*

____) _____ $46\overline{)123{,}740}$

14. *Estimate:*   *Exact:*

____) _____ $8.4\overline{)37.8}$

15. *Estimate:*   *Exact:*

____ • ____ = ____ $1\dfrac{9}{10} \cdot 3\dfrac{3}{4}$

16. *Estimate:*   *Exact:*

____ ÷ ____ = ____ $2\dfrac{1}{3} \div \dfrac{5}{6}$

17. *Estimate:*   *Exact:*

____ + ____ = ____ $1\dfrac{4}{5} + 1\dfrac{2}{3}$

18. *Estimate:*   *Exact:*

____ – ____ = ____ $4\dfrac{1}{2} - 1\dfrac{7}{8}$

Add, subtract, multiply, or divide as indicated.

19. $10 - 0.329$

20. $2\frac{3}{5} \cdot \frac{5}{9}$

21. $9 + 72,417 + 799$

22. $11\frac{1}{5} \div 8$

23. $5006 - 92$

24. $0.7 + 85 + 7.903$

25. Write your answer using R for the remainder.

$7\overline{)2831}$

26. $\frac{5}{6} + \frac{7}{8}$

27. 332×704

28. $(0.006)(5.44)$

29. 3.2×2.5

30. $25.2 \div 0.56$

31. $\frac{2}{3} \div 5\frac{1}{6}$

32. $5\frac{1}{4} - 4\frac{7}{12}$

33. $4.7 \div 9.3$
Round to nearest hundredth.

Use the order of operations to simplify each expression.

34. $10 - 4 \div 2 \cdot 3$

35. $\sqrt{36} + 3 \cdot 8 - 4^2$

36. $\frac{2}{3} \cdot \left(\frac{7}{8} - \frac{1}{2} \right)$

37. $0.9^2 + 10.6 \div 0.53$

38. $4^3 \cdot 3^2$

39. $\sqrt{196}$

40. Find the prime factorization of 200. Write your answer using exponents.

41. Write 40.035 in words.

42. Write three hundred six ten-thousandths in numbers.

Write each decimal as a fraction or mixed number in lowest terms.

43. 0.125

44. 3.08

Write each fraction or mixed number as a decimal. Round to the nearest thousandth if necessary.

45. $2\dfrac{3}{5}$

46. $\dfrac{7}{11}$

47. Write $<$ or $>$ in the blank to make a true statement: $\dfrac{5}{8}$ _____ $\dfrac{4}{9}$.

Arrange each group of numbers in order, from smallest to largest.

48. 7.005, 7.5005, 7.5, 7.505

49. $\dfrac{7}{8}$, 0.8, $\dfrac{21}{25}$, 0.8015

Use the information in the table to answer Exercises 50–53. All measurements are in inches.

Children's Hats	Head Size	Order Size:
Measure around head above eyebrow ridges.	16½"–18"	XXS
	18¼"–19"	XS
	19¼"–20"	S
	20¼"–21⅛"	M
	21½"–22¼"	L

Source: Lands' End.

50. If the distance around your child's head is $21\dfrac{1}{16}$ in., which hat size should you order?

51. Find the difference between the smaller and larger measurements for each of the hat sizes.

52. Change the measurements for the medium (M) size to decimals. Then find the difference in the measurements. Show why this answer is equivalent to your answer from Exercise 51.

53. Did you prefer doing the subtraction using fractions (as in Exercise 51) or using decimals (as in Exercise 52)? Explain your reasoning.

First round the numbers and estimate the answer to each application problem. Then find the exact answer.

54. Lameck had two $10 bills. He spent $7.96 on gasoline and $0.87 for a candy bar at the convenience store. How much money does he have left?

Estimate:

Exact:

55. Manuela's daughter is 50 in. tall. Last year she was $46\frac{5}{8}$ in. tall. How much has she grown?

(When estimating, round to the nearest whole number.)

Estimate:

Exact:

56. Sharon records textbooks on tape for students who are blind. Her hourly wage is $11.63. How much did she earn working 16.5 hours last week, to the nearest cent?

Estimate:

Exact:

57. The Farnsworth Elementary School has eight classrooms with 22 students in each one and 12 classrooms with 26 students in each one. How many students attend the school?

Estimate:

Exact:

58. Toshihiro bought $2\frac{1}{3}$ yd of cotton fabric and $3\frac{7}{8}$ yd of wool fabric. How many yards did he buy in all?

Estimate:

Exact:

59. Kimberly had $29.44 in her checking account. She wrote a check for $40 and deposited a $220.06 paycheck into her account, but not in time to prevent an $18 overdraft charge. What is the new balance in her account?

Estimate:

Exact:

60. Paulette bought 2.7 pounds of grapes for $2.56. What was the cost per pound, to the nearest cent?

Estimate:

Exact:

61. Carter Community College received a $78,000 grant from a local computer company to help students pay tuition for computer classes. How much money could be given to each of 107 students? Round to the nearest dollar.

Estimate:

Exact:

Use the information in the table to answer Exercises 62–64.

Animal	Average Weight of Animal (ounces)	Average Weight of Food Eaten Each Day (ounces)
Hamster	3.5	0.4
Queen bee	0.004	?
Hummingbird	?	0.07

Source: NCTM News Bulletin.

 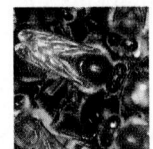 

62. (a) In how many days will a hamster eat enough food to equal its body weight? Round to the nearest whole number of days.

(b) If a 140-pound woman ate her body weight of food in the same number of days as the hamster, how much would she eat each day? Round to the nearest tenth.

63. While laying eggs, a queen bee eats eighty times her weight each day. Use this information to fill in one of the missing values in the table.

64. A hummingbird's body weight is about $1\frac{3}{5}$ times the weight of its daily food intake. Find its body weight using decimal numbers and using fractions. Then prove that the two answers are equivalent.

Ratio and Proportion

5

Nearly $\frac{1}{3}$ of the people in the United States own a cell phone. (*Source: Scientific American.*) Everyone likes to talk, but no one likes to pay the bills! Now you can get the best possible deal on cellular phone service by finding unit rates. (See Section 5.2, Exercises 47–50.)

ADDISON · WESLEY
MyMathLab.com
You're Connected

5.1 RATIOS

OBJECTIVES

1 Write ratios as fractions.

2 Solve ratio problems involving decimals or mixed numbers.

3 Solve ratio problems after converting units.

A **ratio** compares two quantities. You can compare two numbers, such as 8 and 4, or two measurements that have the same type of units, such as 3 days and 12 days.

Ratios can help you see important relationships. For example, if the ratio of your monthly expenses to your monthly income is 10 to 9, then you are spending $10 for every $9 you earn, going deeper into debt.

1 **Write ratios as fractions.**   A ratio can be written in three ways.

Writing a Ratio

The ratio of $7 **to** $3 can be written:

$$7 \text{ to } 3 \quad \text{or} \quad 7{:}3 \quad \text{or} \quad \frac{7}{3} \;\leftarrow \text{Fraction bar indicates "to."}$$

":" indicates "**to**."

Writing a ratio as a fraction is the most common method, and the one we will use here. All three ways are read, "the ratio of 7 **to** 3." The word **to** separates the quantities being compared.

Writing a Ratio as a Fraction

Order is important when writing a ratio. The quantity mentioned **first** is the **numerator**. The quantity mentioned **second** is the **denominator**. For example:

The ratio of **5** to **12**   is written $\dfrac{5}{12}$.

Example 1 Writing Ratios

The Anasazi, ancestors of the Pueblo Indians, built multistory apartment towns in New Mexico about 1100 years ago. A room might measure 14 ft long, 11 ft wide, and 15 ft high.

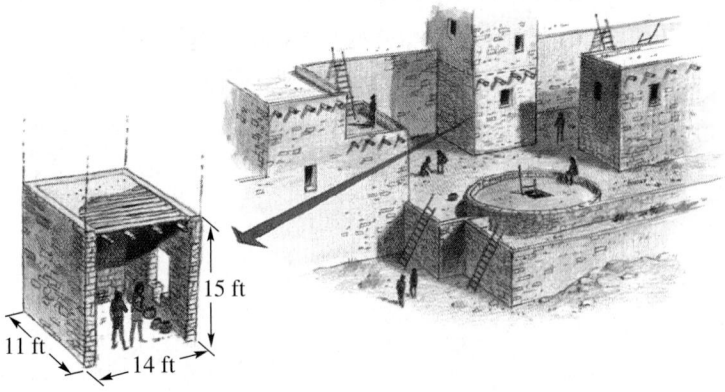

15 ft

11 ft   14 ft

Continued on Next Page

Write each ratio as a fraction, using these room measurements.

(a) Ratio of length to width

The ratio of **length to width** is $\dfrac{14\ \cancel{ft}}{11\ \cancel{ft}} = \dfrac{14}{11}$.

Numerator (mentioned first) Denominator (mentioned second)

You can divide out common *units* just like you divided out common *factors* when writing fractions in lowest terms. (See **Section 2.4.**)

(b) Ratio of width to height

The ratio of **width to height** is $\dfrac{11\ \cancel{ft}}{15\ \cancel{ft}} = \dfrac{11}{15}$.

CAUTION

Remember, the *order* of the numbers is important in a ratio. Look for the words "ratio of <u>a</u> to <u>b</u>." Write the ratio as $\dfrac{a}{b}$, ***not*** $\dfrac{b}{a}$. The quantity mentioned first is the numerator.

Work Problem ❶ at the Side.

Any ratio can be written as a fraction. Therefore, you can write a ratio in *lowest terms,* just as you do with any fraction.

Example 2 **Writing Ratios in Lowest Terms**

Write each ratio in lowest terms.

(a) 60 days to 20 days

The ratio is $\frac{60}{20}$. Write this ratio in lowest terms by dividing the numerator and denominator by 20.

$$\frac{60}{20} = \frac{60 \div 20}{20 \div 20} = \frac{3}{1} \leftarrow \left\{ \begin{array}{l} \text{Ratio in} \\ \text{lowest terms} \end{array} \right.$$

CAUTION

In the fractions chapters you would have rewritten $\frac{3}{1}$ as 3. But a *ratio* compares *two* quantities, so you need to keep both parts of the ratio and write it as $\frac{3}{1}$.

(b) 50 ounces of medicine to 120 ounces of medicine

The ratio is $\frac{50}{120}$. Divide the numerator and denominator by 10.

$$\frac{50}{120} = \frac{50 \div 10}{120 \div 10} = \frac{5}{12} \leftarrow \left\{ \begin{array}{l} \text{Ratio in} \\ \text{lowest terms} \end{array} \right.$$

Continued on Next Page

---

❶ Shane spent $14 on meat, $5 on milk, and $7 on fresh fruit. Write each ratio as a fraction.

(a) The ratio of amount spent on fruit to amount spent on milk.

(b) The ratio of amount spent on milk to amount spent on meat.

(c) The ratio of amount spent on meat to amount spent on milk.

ANSWERS

1. (a) $\dfrac{7}{5}$ **(b)** $\dfrac{5}{14}$ **(c)** $\dfrac{14}{5}$

2 Write each ratio as a fraction in lowest terms.

(a) 9 hours to 12 hours

(b) 100 meters to 50 meters

(c) Write the ratio of width to length for this rectangle.

Length
48 ft

Width
24 ft

3 Write each ratio as a ratio of whole numbers in lowest terms.

(a) The price of Tamar's favorite brand of lipstick increased from $5.50 to $7.00. Find the ratio of the increase in price to the original price.

(b) Last week, Lance worked 4.5 hours each day. This week he cut back to 3 hours each day. Find the ratio of the decrease in hours to the original number of hours.

(c) 15 people in a large van to 6 people in a small van

$$\text{The ratio is } \frac{15}{6} = \frac{15 \div 3}{6 \div 3} = \frac{5}{2} \leftarrow \left\{ \begin{array}{l} \text{Ratio in} \\ \text{lowest terms} \end{array} \right.$$

NOTE

Although $\frac{5}{2} = 2\frac{1}{2}$, ratios are *not* written as mixed numbers. Nevertheless, in Example 2(c), the ratio $\frac{5}{2}$ does mean the large van holds $2\frac{1}{2}$ times as many people as the small van.

Work Problem ② at the Side.

2 ▃▃ **Solve ratio problems involving decimals or mixed numbers.** Sometimes a ratio compares two decimal numbers or two fractions. It is easier to understand if we rewrite the ratio as a ratio of two whole numbers.

Example 3 **Using Decimal Numbers in a Ratio**

The price of a Sunday newspaper increased from $1.50 to $1.75. Find the ratio of the increase in price to the original price.

The words increase in price are mentioned first, so the increase will be the numerator. How much did the price go up? Use subtraction.

$$\begin{array}{ccc} \text{new price} - \text{original price} &=& \text{increase} \\ \$1.75 \quad - \quad \$1.50 &=& \$0.25 \end{array}$$

The words the original price are mentioned second, so the original price of $1.50 is the denominator.

The ratio of increase in price to original price is

$$\frac{0.25}{1.50} \begin{array}{l} \leftarrow \text{increase} \\ \leftarrow \text{original price} \end{array}$$

Now rewrite the ratio as a ratio of whole numbers. Recall that if you multiply both the numerator and denominator of a fraction by the same number, you get an equivalent fraction. The decimals in this example are hundredths, so multiply by 100 to get whole numbers. (If the decimals are tenths, multiply by 10. If thousandths, multiply by 1000.) Then write the ratio in lowest terms.

$$\frac{0.25}{1.50} = \frac{0.25 \cdot 100}{1.50 \cdot 100} = \frac{25}{150} = \frac{25 \div 25}{150 \div 25} = \frac{1}{6} \leftarrow \left\{ \begin{array}{l} \text{Ratio in} \\ \text{lowest terms} \end{array} \right.$$

Ratio as two whole numbers

Work Problem ③ at the Side.

Example 4 **Using Mixed Numbers in Ratios**

Write each ratio as a comparison of whole numbers in lowest terms.

(a) 2 days to $2\frac{1}{4}$ days

Write the ratio as follows. Divide out the common units.

$$\frac{2 \text{ days}}{2\frac{1}{4} \text{ days}} = \frac{2}{2\frac{1}{4}}$$

Continued on Next Page

ANSWERS

2. (a) $\frac{3}{4}$ **(b)** $\frac{2}{1}$ **(c)** $\frac{1}{2}$

3. (a) $\frac{1.50 \times 100}{5.50 \times 100} = \frac{150 \div 50}{550 \div 50} = \frac{3}{11}$

(b) $\frac{1.5 \times 10}{4.5 \times 10} = \frac{15 \div 15}{45 \div 15} = \frac{1}{3}$

Next, write 2 as $\frac{2}{1}$ and $2\frac{1}{4}$ as the improper fraction $\frac{9}{4}$.

$$\frac{2}{2\frac{1}{4}} = \frac{\frac{2}{1}}{\frac{9}{4}}$$

Now rewrite the problem in horizontal format, using the "÷" symbol for division. Finally, multiply by the reciprocal of the divisor, as you did in **Section 2.7.**

Reciprocal

$$\frac{\frac{2}{1}}{\frac{9}{4}} = \frac{2}{1} \div \frac{9}{4} = \frac{2}{1} \cdot \frac{4}{9} = \frac{8}{9}$$

The ratio, in lowest terms, is $\frac{8}{9}$.

(b) $3\frac{1}{4}$ to $1\frac{1}{2}$

Write the ratio as $\dfrac{3\frac{1}{4}}{1\frac{1}{2}}$. Then write $3\frac{1}{4}$ and $1\frac{1}{2}$ as improper fractions.

$$3\frac{1}{4} = \frac{13}{4} \quad \text{and} \quad 1\frac{1}{2} = \frac{3}{2}$$

The ratio is shown below.

$$\frac{3\frac{1}{4}}{1\frac{1}{2}} = \frac{\frac{13}{4}}{\frac{3}{2}}$$

Rewrite as a division problem in horizontal format, using the "÷" symbol. Then multiply by the reciprocal of the divisor.

$$\frac{13}{4} \div \frac{3}{2} = \frac{13}{\overset{2}{\cancel{4}}} \cdot \frac{\overset{1}{\cancel{2}}}{3} = \frac{13}{6} \leftarrow \left\{ \begin{array}{l} \text{Ratio in} \\ \text{lowest terms} \end{array} \right.$$

════ **Work Problem ④ at the Side.**

3▭ **Solve ratio problems after converting units.** When a ratio compares measurements, both measurements must be in the *same* units. For example, *feet* must be compared to *feet, hours* to *hours, pints* to *pints,* and *inches* to *inches.*

╭─ **Example 5** **Ratio Applications Using Measurement**

(a) Write the ratio of the length of the board on the left to the length of the board on the right. Compare in inches.

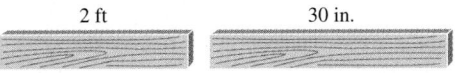

2 ft   30 in.

First, express 2 ft in inches. Because 1 ft has 12 in., 2 ft is

$$2 \cdot 12 \text{ in.} = 24 \text{ in.}$$

──── **Continued on Next Page**

④ Write each ratio as a ratio of whole numbers in lowest terms.

(a) $3\frac{1}{2}$ to 4

(b) $5\frac{5}{8}$ pounds to $3\frac{3}{4}$ pounds

(c) $3\frac{1}{2}$ in. to $\frac{7}{8}$ in.

5 Write each ratio as a fraction in lowest terms. (*Hint:* Recall that it is usually easier to write the ratio using the smaller measurement unit.)

(a) 9 in. to 6 ft

(b) 2 days to 8 hours

(c) 7 yd to 14 ft

(d) 3 quarts to 3 gallons

(e) 25 minutes to 2 hours

(f) 4 pounds to 12 ounces

The length of the board on the left is 24 in., so the ratio of the lengths is

$$\frac{24 \text{ in.}}{30 \text{ in.}} = \frac{24}{30}$$

Write the ratio in lowest terms

$$\frac{24}{30} = \frac{24 \div 6}{30 \div 6} = \frac{4}{5} \leftarrow \left\{ \begin{array}{l} \text{Ratio in} \\ \text{lowest terms} \end{array} \right.$$

The shorter board on the left is $\frac{4}{5}$ the length of the longer board on the right.

NOTE

Notice that we wrote the ratio using the smaller unit (inches are smaller than feet). Using the smaller unit will help you avoid working with fractions. If we wrote the ratio using feet, then

$$30 \text{ in.} = 2\frac{1}{2} \text{ ft.}$$

So the ratio in feet is shown below.

$$\frac{2 \text{ ft}}{2\frac{1}{2} \text{ ft}} = \frac{2}{1} \div \frac{5}{2} = \frac{2}{1} \cdot \frac{2}{5} = \frac{4}{5} \leftarrow \text{Same result.}$$

The ratio is the same, but it takes more steps to get the answer. Using the smaller unit is usually easier.

(b) Write the ratio of 28 days to 3 weeks.

Since it is easier to write the ratio using the smaller measurement unit, compare in *days* because days are shorter than weeks.

First express 3 weeks in days. Because 1 week has 7 days, 3 weeks is

$$3 \cdot 7 \text{ days} = 21 \text{ days}.$$

So the ratio in days is shown below.

$$\frac{28 \text{ days}}{21 \text{ days}} = \frac{28}{21} = \frac{28 \div 7}{21 \div 7} = \frac{4}{3} \leftarrow \text{Lowest terms}$$

The following table will help you set up ratios that compare measurements. You will work with these measurements again in Chapter 7.

Measurement Comparisons

Length	Capacity (Volume)
1 foot = 12 inches	1 pint = 2 cups
1 yard = 3 feet	1 quart = 2 pints
1 mile = 5280 feet	1 gallon = 4 quarts
Weight	**Time**
1 pound = 16 ounces	1 minute = 60 seconds
1 ton = 2000 pounds	1 hour = 60 minutes
	1 day = 24 hours
	1 week = 7 days

Work Problem 5 at the Side.

5.2

A *ratio*
9 **feet**
we make

This typ

1 \
rate at w

In a rat
compari

CAUTION

When
lars, a
divide

Example

Write eac

(a) 5 gall

(b) $1500

(c) 2225

2 Fin
unit rate.
1 hour of

Or, you dri

Use **per** or

5.1 **EXERCISES**

FOR EXTRA HELP

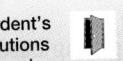 Student's Solutions Manual   MyMathLab.com    InterAct Math Tutorial Software    AW Math Tutor Center    www.mathxl.com   Digital Video Tutor CD 3 Videotape 9

Write each ratio as a fraction in lowest terms. See Examples 1 and 2.

1. 8 to 9

2. 11 to 15

3. $100 to $50

4. 35¢ to 7¢

5. 30 minutes to 90 minutes

6. 9 pounds to 36 pounds

7. 80 miles to 50 miles

8. 300 people to 450 people

9. 6 hours to 16 hours

10. 45 books to 35 books

Write each ratio as a ratio of whole numbers in lowest terms. See Examples 3 and 4.

11. $4.50 to $3.50

12. $0.08 to $0.06

13. 15 to $2\frac{1}{2}$

14. 5 to $1\frac{1}{4}$

15. $1\frac{1}{4}$ to $1\frac{1}{2}$

16. $2\frac{1}{3}$ to $2\frac{2}{3}$

Write each ratio as a fraction in lowest terms. For help, use the table of measurement relationships on page 316. See Example 5.

17. 4 ft to 30 in.

18. 8 ft to 4 yd

19. 5 minutes to 1 hour

20. 8 quarts to 5 pints

21. 15 hours to 2 days

22. 3 pounds to 6 ounces

23. 5 gallons to 5 quarts

24. 3 cups to 3 pints

Writ

43.

45.

 t

 s

 8

 t

Use y

47. In
 lo
 ot
 sti

❹ Solve each problem.

(a) Some batteries claim to last longer than others. If you believe these claims, which brand is the best buy?

Four-pack of AA-size batteries for $2.79

One AA-size battery for $1.19; lasts twice as long

(b) Which tube of toothpaste is the better buy? You have a coupon for 85¢ off Brand C and a coupon for 20¢ off Brand D.

Brand C is $3.89 for 6 ounces.

Brand D is $1.59 for 2.5 ounces.

Answers

4. (a) One battery that lasts twice as long (like getting two) is the better buy. The cost per unit is $0.595 per battery. The four-pack is $0.698 per battery (rounded).

(b) Brand C with the 85¢ coupon is the better buy at $0.507 per ounce (rounded). Brand D with the 20¢ coupon is $0.556 per ounce.

Example 4   **Solving Best Buy Applications**

Solve each application problem.

(a) There are many brands of liquid laundry detergent. If you feel they all do a good job of cleaning your clothes, you can base your purchase on cost per unit. But some brands are "concentrated" so you can use less detergent for each load of clothes. Which of the choices shown below is the best buy?

To find Sudzy's unit cost, divide $3.99 by 64 ounces, not 50 ounces. You're getting as many clothes washed as if you bought 64 ounces. Similarly, to find White-O's unit cost, divide $9.89 by 256 ounces (twice 128 ounces, or 2 • 128 ounces = 256 ounces).

$$\text{Sudzy} \quad \frac{\$3.99}{64 \text{ ounces}} \approx \$0.062 \text{ per ounce}$$

$$\text{White-O} \quad \frac{\$9.89}{256 \text{ ounces}} \approx \$0.039 \text{ per ounce}$$

White-O has the lower cost per ounce and is the better buy. (However, if you try it and it really doesn't get out all the stains, Sudzy may be worth the extra cost.)

(b) "Cents-off" coupons also affect the best buy. Suppose you are looking at these choices for "extra strength" aspirin.

Brand X is $2.29 for 50 tablets.

Brand Y is $10.75 for 200 tablets.

You have a 40¢ coupon for Brand X and a 75¢ coupon for Brand Y. Which choice is the best buy?

To find the better buy, first subtract the coupon amounts, then divide to find the lower cost per ounce.

Brand X costs $2.29 − $0.40 = $1.89

$$\frac{\$1.89}{50 \text{ tablets}} \approx \$0.038 \text{ per tablet}$$

Brand Y costs $10.75 − $0.75 = $10.00

$$\frac{\$10.00}{200 \text{ tablets}} = \$0.05 \text{ per tablet}$$

Brand X has the lower cost per tablet and is the better buy.

Work Problem ❹ at the Side.

5.2 EXERCISES

  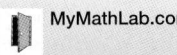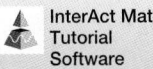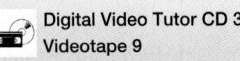

Write each rate as a fraction in lowest terms. See Example 1.

1. 10 cups for 6 people

2. $12 for 30 pens

3. 15 feet in 35 seconds

4. 100 miles in 30 hours

5. 14 people for 28 dresses

6. 12 wagons for 48 horses

7. 25 letters in 5 minutes

8. 68 pills for 17 people

9. $63 for 6 visits

10. 25 doctors for 310 patients

11. 72 miles on 4 gallons

12. 132 miles on 8 gallons

Find each unit rate. See Example 2.

13. $60 in 5 hours

14. $2500 in 20 days

15. 50 eggs from 10 chickens

16. 36 children from 12 families

17. 7.5 pounds for 6 people

18. 44 bushels from 8 trees

19. $413.20 for 4 days

20. $74.25 for 9 hours

Earl kept the following record of the gas he bought for his car. For each entry, find the number of miles he traveled and the unit rate. Round your answers to the nearest tenth.

	Date	Odometer at Start	Odometer at End	Miles Traveled	Gallons Purchased	Miles per Gallon
21.	2/4	27,432.3	27,758.2		15.5	
22.	2/9	27,758.2	28,058.1		13.4	
23.	2/16	28,058.1	28,396.7		16.2	
24.	2/20	28,396.7	28,704.5		13.3	

Source: Author's car records.

Find the best buy (based on the cost per unit) for each item. See Example 3. (Source: Piggly Wiggly.)

25. Black pepper

26. Shampoo

27. Cereal
   13 ounces for $2.80
   15 ounces for $3.15
   18 ounces for $3.98

28. Soup
   2 cans for $0.95
   3 cans for $1.45
   5 cans for $2.29

29. Chunky peanut butter
   12 ounces for $1.29
   18 ounces for $1.79
   28 ounces for $3.39
   40 ounces for $4.39

30. Pork and beans
   8 ounces for $0.37
   16 ounces for $0.77
   21 ounces for $0.99
   31 ounces for $1.50

31. Suppose you are choosing between two brands of chicken noodle soup. Brand A is $0.48 per can and Brand B is $0.58 per can. But Brand B has more chunks of chicken in it. Which soup is the better buy? Explain your choice.

32. A small bag of potatoes costs $0.19 per pound. A large bag costs $0.15 per pound. But there are only two people in your family, so half the large bag would probably rot before you use it up. Which bag is the better buy? Explain.

Solve each application problem. See Examples 2–4.

33. Makesha lost 10.5 pounds in six weeks. What was her rate of loss in pounds per week?

34. Enrique's taco recipe uses four pounds of meat to feed 10 people. Give the rate in pounds per person.

35. Russ works 7 hours to earn $85.82. What is his pay rate per hour?

36. Find the cost of 1 gallon of gas if 18 gallons cost $26.28.

37. Ms. Keskinen bought 150 shares of stock for $1725. Find the cost of one share.

38. A company pays $6450 in dividends for the 2500 shares of its stock. Find the dividend per share.

39. In the 2000 Olympics, Michael Johnson ran the 400-meter event in a record time of approximately 44 seconds (actually 43.84 seconds). Give his rate in seconds per meter and in meters per second. Use 44 seconds as the time. (*Source:* www. Olympics.com)

40. Sofia can clean and adjust five hearing aids in four hours. Give her rate in hearing aids per hour and in hours per hearing aid.

41. A long-distance phone service advertised that all calls were 5¢ a minute, with a 50¢ minimum per completed call. (*Source:* VarTech Telecom, Inc.) Find the actual cost per minute for: **(a)** a three-minute call, **(b)** a four-minute call, and **(c)** a six-minute call. Round your answers to the nearest tenth of a cent.

42. Another long-distance phone service advertised a rate of 7¢ a minute, with no minimum per call, plus a $5.95 monthly fee. (*Source:* AT&T.) Find the actual cost per minute during one month if you make long-distance calls totaling **(a)** 10 minutes, **(b)** 30 minutes, or **(c)** 60 minutes. Round your answers to the nearest cent.

43. If you believe the claims that some batteries last longer, which is the better buy?

 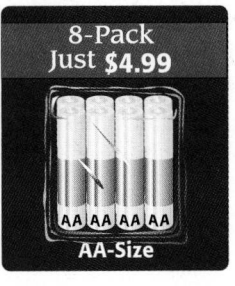

44. Which is the better buy, assuming these laundry detergents both clean equally well?

45. Three brands of cornflakes are available. Brand G is priced at $2.39 for 10 ounces. Brand K is $3.99 for 20.3 ounces and Brand P is $3.39 for 16.5 ounces. You have a coupon for 50¢ off Brand P and a coupon for 60¢ off Brand G. Which cereal is the best buy based on cost per unit?

46. Two brands of facial tissue are available. Brand K is on special at three boxes of 175 tissues each for $5. Brand S is priced at $1.29 per box of 125 tissues. You have a coupon for 20¢ off one box of Brand S and a coupon for 45¢ off one box of Brand K. How can you get the best buy on one box of tissue?

RELATING CONCEPTS (Exercises 47–50) **FOR INDIVIDUAL OR GROUP WORK**

On the first page of this chapter, we said that unit rates can help you get the best deal on cell phone service. Use the information in the table to **work Exercises 47–50 in order.**

Company	Anytime Minutes	Weekend Minutes	Monthly Charge	Other Costs
Verizon	500	1000	$35	$6.95 monthly access fee
Qwest	400	1000	$39.99	$25 one-time activation fee
VoiceStream	75	250	$19.99	
Sprint PCS	250	250	$29.95	

Source: Advertisements appearing in *Minneapolis Star Tribune.*

Notes:

1. All companies require that you sign up for 12 months of service and charge a $150 fee if you quit early.

2. Weekend minutes can only be used from midnight Friday to midnight Sunday.

3. Unused minutes cannot be carried over to the next month.

47. If you sign up for one year of service with Qwest, how much will the activation fee increase your average monthly charge?

48. Find the actual average cost per minute for each company, including "other costs." Assume you use all the minutes and no more. Decide how to round your answers so you can find the best buy.

49. How many hours is 1000 minutes? Over the course of a year, what is the average number of "weekend days" per month? How many hours would you have to talk on each "weekend day" to use up 1000 minutes per month? (Round answers to nearest tenth.)

50. Suppose that after two months you canceled your service because you found that you only used 100 minutes per month. Under those conditions, find the actual cost per minute for each company, to the nearest cent.

5.3 PROPORTIONS

1 ⬛ **Write proportions.** A **proportion** states that two ratios (or rates) are equivalent. For example,

$$\frac{\$20}{4 \text{ hours}} = \frac{\$40}{8 \text{ hours}}$$

is a proportion that says the rate $\dfrac{\$20}{4 \text{ hours}}$ is equivalent to the rate $\dfrac{\$40}{8 \text{ hours}}$.

As the amount of money doubles, the number of hours also doubles. This proportion is read:

20 dollars **is to** 4 hours **as** 40 dollars **is to** 8 hours.

Example 1 **Writing Proportions**

Write each proportion.

(a) 6 ft is to 11 ft **as** 18 ft is to 33 ft.

$$\frac{6 \text{ ft}}{11 \text{ ft}} = \frac{18 \text{ ft}}{33 \text{ ft}} \quad \text{so} \quad \frac{6}{11} = \frac{18}{33} \qquad \begin{array}{l}\text{The common units (ft) divide} \\ \text{out and are not written.}\end{array}$$

(b) \$9 is to 6 liters **as** \$3 is to 2 liters.

$$\frac{\$9}{6 \text{ liters}} = \frac{\$3}{2 \text{ liters}} \qquad \text{Units must be written.}$$

══════════ **Work Problem ❶ at the Side.**

2 ⬛ **Determine whether proportions are true or false.** There are two ways to see whether a proportion is true. One way is to *write both of the ratios in lowest terms.*

Example 2 **Writing Both Ratios in Lowest Terms**

Are the following proportions true?

(a) $\dfrac{5}{9} = \dfrac{18}{27}$

Write each ratio in lowest terms.

$$\frac{5}{9} \leftarrow \begin{array}{l}\text{Already in} \\ \text{lowest terms}\end{array} \qquad \frac{18 \div 9}{27 \div 9} = \frac{2}{3} \leftarrow \begin{array}{l}\text{Lowest} \\ \text{terms}\end{array}$$

Because $\frac{5}{9}$ is *not* equivalent to $\frac{2}{3}$, the proportion is *false*.

(b) $\dfrac{16}{12} = \dfrac{28}{21}$

Write each ratio in lowest terms.

$$\frac{16 \div 4}{12 \div 4} = \frac{4}{3} \quad \text{and} \quad \frac{28 \div 7}{21 \div 7} = \frac{4}{3}$$

Both ratios are equivalent to $\frac{4}{3}$, so the proportion is *true*.

══════════ **Work Problem ❷ at the Side.**

❶ Write each proportion.

(a) \$7 is to 3 cans as \$28 is to 12 cans

(b) 9 meters is to 16 meters as 18 meters is to 32 meters

(c) 5 is to 7 as 35 is to 49

(d) 10 is to 30 as 60 is to 180

❷ Determine whether each proportion is true or false by writing both ratios in lowest terms.

(a) $\dfrac{6}{12} = \dfrac{15}{30}$

(b) $\dfrac{20}{24} = \dfrac{3}{4}$

(c) $\dfrac{25}{40} = \dfrac{30}{48}$

(d) $\dfrac{35}{45} = \dfrac{12}{18}$

(e) $\dfrac{21}{45} = \dfrac{56}{120}$

ANSWERS

1. (a) $\dfrac{\$7}{3 \text{ cans}} = \dfrac{\$28}{12 \text{ cans}}$ **(b)** $\dfrac{9}{16} = \dfrac{18}{32}$

(c) $\dfrac{5}{7} = \dfrac{35}{49}$ **(d)** $\dfrac{10}{30} = \dfrac{60}{180}$

2. (a) true **(b)** false **(c)** true
(d) false **(e)** true

3 ▭ **Find cross products.** Another way to test whether the ratios in a proportion are equivalent is to compare *cross products.*

Using Cross Products to Determine Whether a Proportion Is True

To see whether a proportion is true, first multiply along one diagonal, then multiply along the other diagonal, as shown here.

$$
\begin{array}{c}
5 \cdot 4 = 20 \\[4pt]
\dfrac{2}{5} \diagdown\!\!\!\!\diagup \dfrac{4}{10} \\[4pt]
2 \cdot 10 = 20
\end{array}
\quad
\begin{array}{l}
\text{Cross products} \\
\text{are equal.}
\end{array}
$$

In this case the **cross products** are both 20. When cross products are *equal,* the proportion is *true.* If the cross products are *unequal,* the proportion is *false.*

The cross products test is based on rewriting both fractions with the common denominator of $5 \cdot 10$, or 50.

$$
\frac{2 \cdot 10}{5 \cdot 10} = \frac{20}{50} \quad \text{and} \quad \frac{4 \cdot 5}{10 \cdot 5} = \frac{20}{50}
$$

We see that $\frac{2}{5}$ and $\frac{4}{10}$ are equivalent because both can be rewritten as $\frac{20}{50}$. The cross product test takes a shortcut by comparing only the two numerators $(20 = 20)$.

Example 3 Using Cross Products

Use cross products to see whether each proportion is true or false.

(a) $\dfrac{3}{5} = \dfrac{12}{20}$

Multiply along one diagonal and then along the other diagonal.

$$
\begin{array}{c}
5 \cdot 12 = 60 \\[4pt]
\dfrac{3}{5} = \dfrac{12}{20} \\[4pt]
3 \cdot 20 = 60
\end{array}
\quad \text{Equal}
$$

The cross products are *equal,* so the proportion is *true.*

Continued on Next Page

(b) $\dfrac{2\frac{1}{3}}{3\frac{1}{3}} = \dfrac{9}{16}$

Cross multiply.

Changed to improper fractions

$$\frac{2\frac{1}{3}}{3\frac{1}{3}} = \frac{9}{16}$$

$$3\frac{1}{3} \cdot 9 = \frac{10}{\cancel{3}} \cdot \frac{\cancel{9}^{3}}{1} = \frac{30}{1} = 30$$

$$2\frac{1}{3} \cdot 16 = \frac{7}{3} \cdot \frac{16}{1} = \frac{112}{3} = 37\frac{1}{3}$$

Unequal

The cross products are *unequal,* so the proportion is *false.*

NOTE

The numbers in a proportion do *not* have to be whole numbers. They can be fractions, mixed numbers, decimal numbers, and so on.

Work Problem ❸ at the Side.

❸ Cross multiply to see whether each proportion is true or false.

(a) $\dfrac{5}{9} = \dfrac{10}{18}$

(b) $\dfrac{32}{15} = \dfrac{16}{8}$

(c) $\dfrac{10}{17} = \dfrac{20}{34}$

(d) $\dfrac{2.4}{6} = \dfrac{5}{12}$ $6 \cdot 5 =$

$2.4 \cdot 12 =$

(e) $\dfrac{3}{4.25} = \dfrac{24}{34}$

(f) $\dfrac{1\frac{1}{6}}{2\frac{1}{3}} = \dfrac{4}{8}$

Tour of the West

Visits to Mount Rushmore, Devil's Tower, Yellowstone National Park, the Grand Tetons, Bryce Canyon, Zion National Park, the Painted Desert, and the Grand Canyon are highlights of a tour advertised to British citizens. (*Source:* www.archersdirect.co.uk) The itinerary is shown on the map to the right. (*Source:* www.mapquest.com)

The travel distances and times between the daily stopping points are estimates, based on information from the Web site www.mapquest.com. The table below gives the daily route, the travel distances, and the travel times between the locations.

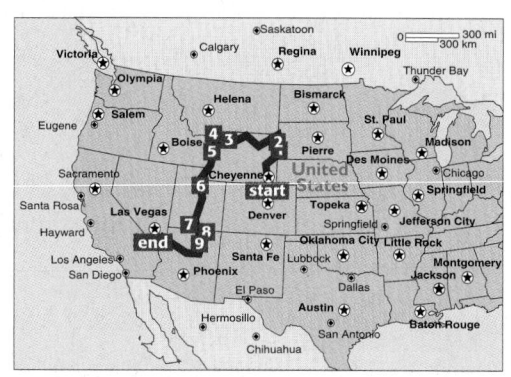

Source: ©2000 MapQuest.com, Inc.; ©2000 AND Data Solutions B.V.

1. Calculate the average speed in miles per hour (mph) for each segment of the trip, rounded to the nearest whole number. Notice that you are working with rates that compare distance to time (miles to hours). (*Hint:* Divide the distance traveled by the time elapsed. You must first rewrite the time in decimal form. For example, on Day 4, 1 hr 40 min, is $1 + \frac{40}{60}$ or 1.6666666, which rounds to 1.67. The average speed is $56 \div 1.67 \approx 33.5$ or 34 mph.)

Day	Location	Distance	Time	Average Speed (mph)
1	London, England to Denver, Colorado (CO)	4693 mi	14 hr flight	
2	Denver, CO			
3	Denver, CO to Custer, South Dakota (SD)	365 mi	8 hr	
4	Custer, SD to Lead, SD	56 mi	1 hr, 40 min	34 mph
5	Lead, SD to Cody, Wyoming (WY)	361 mi	7 hr, 50 min	
6	Cody, WY to Yellowstone National Park, WY	110 mi	3 hr, 10 min	
7	Yellowstone, WY to Jackson, WY	131 mi	3 hr, 45 min	
8	Jackson, WY to Salt Lake City, Utah (UT)	269 mi	7 hr	
9	Salt Lake City, UT			
10	Salt Lake City, UT to Cedar City, UT	251 mi	4 hr, 20 min	
11	Cedar City, UT to Page, Arizona (AZ)	155 mi	4 hr, 30 min	
12	Page, AZ to Grand Canyon, AZ	138 mi	4 hr	
13	Grand Canyon, AZ to Las Vegas, Nevada (NV)	279 mi	6 hr, 30 min	
14	Las Vegas, NV			
15	Las Vegas, NV to San Francisco, CA San Francisco, CA to London, England	420 mi 5376 mi	1 hr 40 min flight 17 hr flight	
	Totals (driving portion of the tour)			

2. Calculate the total driving distance and total driving time during this tour. Find the overall average speed.

3. Why do you think the average speeds for this trip are not closer to 55 mph?

4. If a friend from Scotland asked your opinion about how interesting and feasible these tourist attractions would be, what advice would you give?

5.3 EXERCISES

FOR EXTRA HELP

 Student's Solutions Manual

 MyMathLab.com

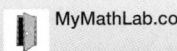 InterAct Math Tutorial Software

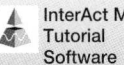 AW Math Tutor Center

 www.mathxl.com

 Digital Video Tutor CD 3 Videotape 9

Write each proportion. See Example 1.

1. $9 is to 12 cans as $18 is to 24 cans.

2. 28 people is to 7 cars as 16 people is to 4 cars.

3. 200 adults is to 450 children as 4 adults is to 9 children.

4. 150 trees is to 1 acre as 1500 trees is to 10 acres.

5. 120 ft is to 150 ft as 8 ft is to 10 ft.

6. $6 is to $9 as $10 is to $15.

Determine whether each proportion is true or false by writing the ratios in lowest terms. See Example 2.

7. $\dfrac{6}{10} = \dfrac{3}{5}$

8. $\dfrac{1}{4} = \dfrac{9}{36}$

9. $\dfrac{5}{8} = \dfrac{25}{40}$

10. $\dfrac{2}{3} = \dfrac{20}{27}$

11. $\dfrac{150}{200} = \dfrac{200}{300}$

12. $\dfrac{100}{120} = \dfrac{75}{100}$

13. $\dfrac{42}{15} = \dfrac{28}{10}$

14. $\dfrac{18}{16} = \dfrac{36}{32}$

15. $\dfrac{32}{18} = \dfrac{48}{27}$

16. $\dfrac{15}{48} = \dfrac{10}{24}$

17. $\dfrac{7}{6} = \dfrac{54}{48}$

18. $\dfrac{28}{21} = \dfrac{44}{33}$

Use cross multiplication to determine whether each proportion is true or false. Circle the correct answer. See Example 3.

19. $\dfrac{2}{9} = \dfrac{6}{27}$

True   False

20. $\dfrac{20}{25} = \dfrac{4}{5}$

True   False

21. $\dfrac{20}{28} = \dfrac{12}{16}$

True   False

22. $\dfrac{16}{40} = \dfrac{22}{55}$

True   False

23. $\dfrac{110}{18} = \dfrac{160}{27}$

True   False

24. $\dfrac{600}{420} = \dfrac{20}{14}$

True   False

25. $\dfrac{3.5}{4} = \dfrac{7}{8}$

True   False

26. $\dfrac{36}{23} = \dfrac{9}{5.75}$

True   False

27. $\dfrac{18}{16} = \dfrac{2.8}{2.5}$

True   False

28. $\dfrac{0.26}{0.39} = \dfrac{1.3}{1.9}$

True   False

29. $\dfrac{6}{3\frac{2}{3}} = \dfrac{18}{11}$

True   False

30. $\dfrac{16}{13} = \dfrac{2}{1\frac{5}{8}}$

True   False

31. $\dfrac{2\frac{5}{8}}{3\frac{1}{4}} = \dfrac{21}{26}$

True   False

32. $\dfrac{28}{17} = \dfrac{9\frac{1}{3}}{5\frac{2}{3}}$

True   False

33. $\dfrac{\frac{2}{3}}{2} = \dfrac{2.7}{8}$

True   False

34. $\dfrac{3.75}{1\frac{1}{4}} = \dfrac{7.5}{2\frac{1}{2}}$

True   False

35. $\dfrac{2\frac{3}{10}}{8.05} = \dfrac{\frac{1}{4}}{0.9}$

True   False

36. $\dfrac{3}{\frac{5}{6}} = \dfrac{1.5}{\frac{7}{12}}$

True   False

37. Suppose Jerome Walton of the Atlanta Braves had 16 hits in 50 times at bat and Mariano Duncan of the New York Yankees was at bat 400 times and got 128 hits. Paul is trying to convince Jamie that the two men hit equally well. Show how you could use a proportion and cross products to see whether Paul is correct.

38. Jay worked 3.5 hours and packed 91 cartons. Craig packed 126 cartons in 5.25 hours. To see whether the men worked equally fast, Barry set up this proportion:

$$\frac{3.5}{91} = \frac{126}{5.25}$$

Explain what is wrong with Barry's proportion and write a correct one. Is the correct proportion true or false?

5.4 SOLVING PROPORTIONS

1☐ **Find the unknown number in a proportion.** Four numbers are used in a proportion. If any three of these numbers are known, the fourth can be found. For example, find the unknown number that will make this proportion true.

$$\frac{3}{5} = \frac{x}{40}$$

The x represents the unknown number. Start by finding the cross products.

$$\frac{3}{5} = \frac{x}{40}$$

$5 \cdot x$

$3 \cdot 40$

Cross products

To make the proportion true, the cross products must be equal.

$$5 \cdot x = 3 \cdot 40$$
$$5 \cdot x = 120$$

The equal sign says that $5 \cdot x$ and 120 are equivalent. If $5 \cdot x$ and 120 are both divided by 5, the results will still be equivalent.

$$\frac{5 \cdot x}{5} = \frac{120}{5} \quad \leftarrow \text{Divide each side by 5.}$$

Divide out 5 in numerator and denominator.

$$\frac{\overset{1}{\cancel{5}} \cdot x}{\underset{1}{\cancel{5}}} = 24 \qquad \text{On the right side, divide} \\ 120 \text{ by 5 to get 24.}$$

Multiplying by 1 does *not* change a number, so in the numerator on the left side, $1 \cdot x$ is the same as x.

$$\frac{x}{1} = 24$$

Dividing by 1 does *not* change a number, so on the left side, $\frac{x}{1}$ is the same as x.

$$x = 24$$

The unknown number in the proportion is 24. The complete proportion is shown below.

$$\frac{3}{5} = \frac{24}{40} \quad \leftarrow x \text{ is 24.}$$

Check by finding the cross products. If they are equal, you solved the problem correctly. If they are unequal, rework the problem.

$$\frac{3}{5} = \frac{24}{40}$$

$5 \cdot 24 = 120$

$3 \cdot 40 = 120$

Equal; proportion is true.

The cross products are equal, so the solution, $x = 24$, is correct.

CAUTION

The solution is 24, which is the unknown number in the proportion. 120 is *not* the solution; it is the cross product you get when *checking* the solution.

Solve a proportion for an unknown number by using the following steps.

Finding an Unknown Number in a Proportion

Step 1   Find the cross products.

Step 2   Show that the cross products are equivalent.

Step 3   Divide both products by the number multiplied by x (the number next to x).

Step 4   Check by writing the solution in the proportion and finding the cross products.

Example 1   Solving for Unknown Numbers

Find the unknown number in each proportion. Round to hundredths, if necessary.

(a) $\dfrac{16}{x} = \dfrac{32}{20}$

Recall that ratios can be rewritten in lowest terms. If desired, you can do that before finding the cross products. In this example, write $\frac{32}{20}$ in lowest terms $\left(\frac{8}{5}\right)$ to get $\frac{16}{x} = \frac{8}{5}$.

Step 1
$$\frac{16}{x} = \frac{8}{5}$$
$x \cdot 8$
$16 \cdot 5$
Find the cross products.

Step 2 $x \cdot 8 = \underline{16 \cdot 5}$ ← Show that cross products are equivalent.
$x \cdot 8 = \quad 80$

Step 3 $\dfrac{x \cdot \cancel{8}}{\cancel{8}} = \dfrac{80}{8}$ ——— Divide each side by 8.

$x = 10$ ——— Find x. (No rounding necessary.)

Step 4   Write the solution in the proportion and check by finding cross products.

$$\frac{16}{10} = \frac{8}{5}$$
x is 10. →
$10 \cdot 8 = 80$
$16 \cdot 5 = 80$
Equal; proportion is true.

The cross products are equal, so 10 is the correct solution.

NOTE

It is not necessary to write the ratios in lowest terms before solving. However, if you do, you will have smaller numbers to work with.

Continued on Next Page

(b) $\dfrac{7}{12} = \dfrac{15}{x}$

$$12 \cdot 15 = 180$$

$$\dfrac{7}{12} \bowtie \dfrac{15}{x}$$ Find the cross products.

$$7 \cdot x$$

Show that cross products are equivalent.

$$7 \cdot x = 180$$

Divide each side by 7.

$$\dfrac{\overset{1}{\cancel{7}} \cdot x}{\underset{1}{\cancel{7}}} = \dfrac{180}{7}$$

$$x \approx 25.71 \quad \leftarrow \text{Rounded to nearest hundredth}$$

When the division does not come out even, check for directions on how to round your answer. Divide out one more place, then round.

$$\begin{array}{r} 25.714 \quad \leftarrow \text{Divide out to thousandths.} \\ 7\overline{)180.000} \quad \text{Round to hundredths.} \end{array}$$

Write the solution in the proportion and check by finding the cross products.

$$12 \cdot 15 = \mathbf{180}$$

$$\dfrac{7}{12} \bowtie \dfrac{15}{25.71}$$ Very close, but not equal

$$7 \cdot 25.71 = \mathbf{179.97}$$

The cross products are slightly different because you rounded the value of x. However, they are close enough to see that the problem was done correctly and 25.71 is the solution.

═══ **Work Problem ① at the Side.**

2 Find the unknown number in a proportion with mixed numbers or decimals. The following examples show how to work with mixed numbers or decimals in a proportion.

Example 2 Solving Proportions with Mixed Numbers and Decimals

Find the unknown number in each proportion.

(a) $\dfrac{2\frac{1}{5}}{6} = \dfrac{x}{10}$

$$6 \cdot x$$

$$\dfrac{2\frac{1}{5}}{6} \bowtie \dfrac{x}{10} \quad 2\dfrac{1}{5} \cdot 10$$ Find the cross products.

Find $2\frac{1}{5} \cdot 10$.

$$2\frac{1}{5} \cdot 10 = \frac{11}{5} \cdot \frac{10}{1} = \frac{11}{\cancel{5}} \cdot \frac{\overset{2}{\cancel{10}}}{1} = \frac{22}{1} = 22$$

   Changed to improper fraction

═══ **Continued on Next Page**

---

① Find the unknown numbers. Round to hundredths, if necessary. Check your answers by finding the cross products.

(a) $\dfrac{1}{2} = \dfrac{x}{12}$

(b) $\dfrac{6}{10} = \dfrac{15}{x}$

(c) $\dfrac{28}{x} = \dfrac{21}{9}$

(d) $\dfrac{x}{8} = \dfrac{3}{5}$

(e) $\dfrac{14}{11} = \dfrac{x}{3}$

ANSWERS
1. **(a)** $x = 6$ **(b)** $x = 25$
   (c) $x = 12$ **(d)** $x = 4.8$
   (e) $x \approx 3.82$ (rounded to nearest hundredth)

❷ Find the unknown numbers. Round to hundredths on the decimal problems, if necessary. Check your answers by finding the cross products.

(a) $\dfrac{3\frac{1}{4}}{2} = \dfrac{x}{8}$

(b) $\dfrac{x}{3} = \dfrac{1\frac{2}{3}}{5}$

(c) $\dfrac{0.06}{x} = \dfrac{0.3}{0.4}$

(d) $\dfrac{2.2}{5} = \dfrac{13}{x}$

(e) $\dfrac{x}{6} = \dfrac{0.5}{1.2}$

(f) $\dfrac{0}{2} = \dfrac{x}{7.092}$

Show that the cross products are equivalent.
$$6 \cdot x = 22$$
Divide each side by 6.
$$\dfrac{\overset{1}{\cancel{6}} \cdot x}{\underset{1}{\cancel{6}}} = \dfrac{22}{6}$$

Write the answer as a mixed number in lowest terms.
$$x = \dfrac{22 \div 2}{6 \div 2} = \dfrac{11}{3} = 3\frac{2}{3}$$

Write the solution in the proportion and check by finding the cross products.

$$6 \cdot 3\frac{2}{3} = \dfrac{\overset{2}{\cancel{6}}}{1} \cdot \dfrac{11}{\underset{1}{\cancel{3}}} = \dfrac{22}{1} = 22$$

$$\dfrac{2\frac{1}{5}}{6} \; \overset{\times}{=} \; \dfrac{3\frac{2}{3}}{10}$$

$$2\frac{1}{5} \cdot 10 = \dfrac{11}{\underset{1}{\cancel{5}}} \cdot \dfrac{\overset{2}{\cancel{10}}}{1} = \dfrac{22}{1} = 22$$

Equal

The cross products are equal, so $3\frac{2}{3}$ is the correct solution.

(b) $\dfrac{1.5}{0.6} = \dfrac{2}{x}$

Show that cross products are equivalent.
$$1.5 \cdot x = \underbrace{0.6 \cdot 2}$$
$$1.5 \cdot x = \quad 1.2$$

Divide each side by 1.5.
$$\dfrac{\overset{1}{\cancel{1.5}} \cdot x}{\underset{1}{\cancel{1.5}}} = \dfrac{1.2}{1.5}$$

$$x = \dfrac{1.2}{1.5}$$

Complete the division.
$$1.5\overline{)1.20}^{\;.8}$$

So the unknown number is 0.8. Check by finding the cross products.

$$0.6 \cdot 2 \quad = 1.2$$

$$\dfrac{1.5}{0.6} \; \overset{\times}{=} \; \dfrac{2}{0.8}$$

$$1.5 \cdot 0.8 = 1.2$$

Equal

The cross products are equal, so 0.8 is the correct solution.

Work Problem ❷ at the Side.

5.4 EXERCISES

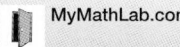
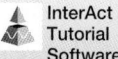

Find the unknown number in each proportion. Round your answers to hundredths, if necessary. Check your answers by finding the cross products. See Examples 1 and 2.

1. $\dfrac{1}{3} = \dfrac{x}{12}$

2. $\dfrac{x}{6} = \dfrac{15}{18}$

3. $\dfrac{15}{10} = \dfrac{3}{x}$

4. $\dfrac{5}{x} = \dfrac{20}{8}$

5. $\dfrac{x}{11} = \dfrac{32}{4}$

6. $\dfrac{12}{9} = \dfrac{8}{x}$

7. $\dfrac{42}{x} = \dfrac{18}{39}$

8. $\dfrac{49}{x} = \dfrac{14}{18}$

9. $\dfrac{x}{25} = \dfrac{4}{20}$

10. $\dfrac{6}{x} = \dfrac{4}{8}$

11. $\dfrac{8}{x} = \dfrac{24}{30}$

12. $\dfrac{32}{5} = \dfrac{x}{10}$

13. $\dfrac{99}{55} = \dfrac{44}{x}$

14. $\dfrac{x}{12} = \dfrac{101}{147}$

15. $\dfrac{0.7}{9.8} = \dfrac{3.6}{x}$

16. $\dfrac{x}{3.6} = \dfrac{4.5}{6}$

17. $\dfrac{250}{24.8} = \dfrac{x}{1.75}$

18. $\dfrac{4.75}{17} = \dfrac{43}{x}$

Find the unknown number in each proportion. Write your answers as whole or mixed numbers when possible. See Example 2.

19. $\dfrac{15}{1\frac{2}{3}} = \dfrac{9}{x}$

20. $\dfrac{x}{\frac{3}{10}} = \dfrac{2\frac{2}{9}}{1}$

21. $\dfrac{2\frac{1}{3}}{1\frac{1}{2}} = \dfrac{x}{2\frac{1}{4}}$

22. $\dfrac{1\frac{5}{6}}{x} = \dfrac{\frac{3}{14}}{\frac{6}{7}}$

Solve each proportion two different ways. First change all the numbers to decimal form and solve. Then change all the numbers to fraction form and solve; write your answers in lowest terms.

23. $\dfrac{\frac{1}{2}}{x} = \dfrac{2}{0.8}$

24. $\dfrac{\frac{3}{20}}{0.1} = \dfrac{0.03}{x}$

25. $\dfrac{x}{\frac{3}{50}} = \dfrac{0.15}{1\frac{4}{5}}$

26. $\dfrac{8\frac{4}{5}}{1\frac{1}{10}} = \dfrac{x}{0.4}$

RELATING CONCEPTS (Exercises 27–28) **FOR INDIVIDUAL OR GROUP WORK**

Work Exercises 27–28 in order. *First prove that the proportions are **not** true. Then create four true proportions for each exercise by changing one number at a time.*

27. $\dfrac{10}{4} = \dfrac{5}{3}$

28. $\dfrac{6}{8} = \dfrac{24}{30}$

5.5 SOLVING APPLICATION PROBLEMS WITH PROPORTIONS

1 Use proportions to solve application problems. Proportions can be used to solve a wide variety of problems. Watch for problems in which you are given a ratio or rate and then asked to find part of a corresponding ratio or rate. Remember that a ratio or rate compares two quantities and often includes one of the following indicator words.

OBJECTIVE

1 Use proportions to solve application problems.

in   for   on   per   from   to

Use the six problem-solving steps you learned in **Section 1.10**.

Step 1 **Read** the problem.   *Step 4* **Solve** the problem.

Step 2 **Work out a plan.**   *Step 5* **State the answer.**

Step 3 **Estimate** a reasonable answer.   *Step 6* **Check** your work.

Example 1 Solving a Proportion Application

Mike's car can travel 163 **miles on** 6.4 **gallons** of gas. How far can it travel on a full tank of 14 **gallons** of gas? Round to the nearest whole mile.

Step 1 **Read** the problem. The problem asks for the number of miles the car can travel on 14 gallons of gas.

Step 2 **Work out a plan.** Decide what is being compared. This example compares **miles** to **gallons**. Write a proportion using the two rates. Be sure that *both* rates compare miles to gallons in the same order. In other words, miles is in both numerators and gallons is in both denominators. Use a letter to represent the unknown number.

Step 3 **Estimate** a reasonable answer. To estimate the answer, notice that 14 gallons is a little more than *twice as much* as 6.4 gallons, so the car should travel a little more than *twice as far*; 2 • 163 miles = 326 miles.

Step 4 **Solve** the problem. With the proportion set up correctly, solve for the unknown number.

Matching units

$$\text{This rate compares miles to gallons.} \left\{ \rightarrow \frac{163 \text{ miles}}{6.4 \text{ gallons}} = \frac{x \text{ miles}}{14 \text{ gallons}} \leftarrow \right\{ \text{This rate compares miles to gallons.}$$

Matching units

Ignore the units while finding the cross products and dividing each side by 6.4.

$$6.4 \cdot x = \underbrace{163 \cdot 14}_{} \quad \text{Show that cross products are equivalent.}$$

$$6.4 \cdot x = 2282$$

$$\frac{\cancel{6.4} \cdot x}{\cancel{6.4}} = \frac{2282}{6.4} \quad \text{Divide each side by 6.4.}$$

$$x = 356.5625$$

Continued on Next Page

➊ Set up and solve a proportion for each problem.

(a) If 2 pounds of fertilizer will cover 50 square feet of garden, how many pounds are needed for 225 square feet?

(b) A U.S. map has a scale of 1 inch to 75 miles. Lake Superior is 4.75 inches long on the map. What is the lake's actual length in miles?

(c) Cough syrup is to be given at the rate of 30 milliliters for each 100 pounds of body weight. How much should be given to a 34-pound child? Round to the nearest whole milliliter.

Step 5 **State the answer.** Rounded to the nearest mile, the car can travel 357 miles on a full tank of gas.

Step 6 **Check** your work. The answer, 357 miles, is a little more than the estimate of 326 miles, so it is reasonable.

Work Problem ➊ at the Side.

Example 2 **Solving a Proportion Application**

A newspaper report says that 7 out of 10 people surveyed watch the news on TV. At that rate, how many of the 3200 people in town would you expect to watch the news?

Step 1 **Read** the problem. The problem asks how many of the 3200 people in town would be expected to watch TV news.

Step 2 **Work out a plan.** You are comparing people who watch the news to people surveyed. Set up a proportion using the two rates described in the example. Be sure that both rates make the same comparison. "People who watch the news" is mentioned first, so it should be in the numerator of *both* rates.

Step 3 **Estimate** a reasonable answer. To estimate the answer, notice that 7 out of 10 people is more than half the people, but less than all the people. Half of 3200 people is $3200 \div 2 = 1600$, so our estimate is between 1600 and 3200 people.

Step 4 **Solve** the problem. Solve for the unknown number in the proportion.

$$\text{People who watch news} \rightarrow \frac{7}{10} = \frac{x}{3200} \leftarrow \text{People who watch news}$$
$$\text{Total group} \rightarrow \quad \quad \quad \leftarrow \text{Total group}$$
$$\text{(people surveyed)} \quad \quad \quad \text{(people in town)}$$

$$10 \cdot x = \underbrace{7 \cdot 3200} \quad \text{Show that cross products are equivalent.}$$

$$10 \cdot x = 22{,}400$$

$$\frac{\overset{1}{\cancel{10}} \cdot x}{\underset{1}{\cancel{10}}} = \frac{22{,}400}{10} \quad \text{Divide each side by 10.}$$

$$x = 2240$$

Step 5 **State the answer.** You would expect 2240 people in town to watch the news on TV.

Step 6 **Check** your work. The answer, 2240 people, is between 1600 and 3200, as called for in the estimate.

Continued on Next Page

CAUTION

Always check that your answer is reasonable. If it is not, look at the way your proportion is set up. Be sure you have matching units in the numerators and matching units in the denominators.

For example, suppose you had set up the last proportion *incorrectly* as shown here.

$$\frac{7}{10} = \frac{3200}{x} \leftarrow \text{Incorrect setup}$$

$$7 \cdot x = 10 \cdot 3200$$

$$\frac{\overset{1}{\cancel{7}} \cdot x}{\underset{1}{\cancel{7}}} = \frac{32,000}{7}$$

$$x \approx 4571 \text{ people} \leftarrow \text{Unreasonable answer}$$

This answer is *unreasonable* because there are only 3200 people in the town; it is *not* possible for 4571 people to watch the news.

═══ **Work Problem ❷ at the Side.**

❷ Solve each problem to find a reasonable answer. Then flip one side of your proportion to see what answer you get with an *incorrect* setup. Explain why the second answer is *unreasonable*.

(a) A survey showed that 2 out of 3 people would like to lose weight. At this rate, how many people in a group of 150 want to lose weight?

(b) In one state, 3 out of 5 college students receive financial aid. At this rate, how many of the 4500 students at Central Community College receive financial aid?

(c) An advertisement says that 9 out of 10 dentists recommend sugarless gum. If the ad is true, how many of the 60 dentists in our city would recommend sugarless gum?

ANSWERS

2. (a) 100 people (reasonable); incorrect setup gives 225 people (only 150 people in the group).
(b) 2700 students (reasonable); incorrect setup gives 7500 students (only 4500 students at the college).
(c) 54 dentists (reasonable); incorrect setup gives about 67 dentists (only 60 dentists in the city).

Feeding Hummingbirds

A recipe can be used to make as much of a mixture as you might need as long as the ingredients are kept proportional. Use the recipe for a homemade mixture of sugar water for hummingbird feeders to answer these problems.

1. What is the ratio of sugar to water in the recipe?

What is the ratio of water to sugar in the recipe?

2. Complete each table.

Sugar	Water
1 cup	4 cups
	5 cups
	6 cups
	7 cups
2 cups	8 cups

Sugar	Water
1 cup	4 cups
	3 cups
	2 cups
	1 cup

3. How much water would you need

(a) if you wanted to use 3 cups of sugar?

(b) if you wanted to use 4 cups of sugar?

(c) if you wanted to use $\frac{1}{3}$ cup of sugar?

4. One cup of sugar weighs about 200 grams and one cup of water weighs about 235 grams. If you mix 1 cup of sugar and 4 cups of water, what is the approximate weight of the resulting mixture?

5. The article says that the nectar from wildflowers visited by hummingbirds has an average sugar concentration of 21 percent. That represents a ratio of 21:100. If you wanted to mix a sugar solution in the same proportion as the wildflower nectar, how much water should you use for 1 cup of sugar? What proportion did you set up to solve this problem?

Feeding Hummingbirds

After getting a hummingbird feeder, the next step is to fill it! You have two choices at this point: you can either buy one of the commercial mixtures or you can make your own solution:

> **Recipe for Homemade Mixture:**
> 1 part sugar (not honey)
> 4 parts water
> Boil for 1 to 2 minutes. Cool.
> Store extra in refrigerator.

The concentration of the sugar is important. A 1 to 4 ratio of sugar to water is recommended because it approximates the ratio of sugar to water found in the nectar of many hummingbird flowers. A recent study of 21 native California wildflowers visited by hummingbirds showed that their nectar had an average sugar concentration of 21 percent. This is sweet enough to attract the hummers without being too sweet. If you increase the concentration of sugar, it may be harder for the birds to digest; if you decrease the concentration, they may lose interest.

Boiling the solution helps retard fermentation. Sugar-and-water solutions are subject to rapid spoiling, especially in hot weather.

Source: The Hummingbird Book.

6. As you change the amounts of water and sugar, should you change the length of time that you boil the mixture? Explain your answer.

7. Will the length of time it takes to get the water hot enough to start boiling change? Explain your answer.

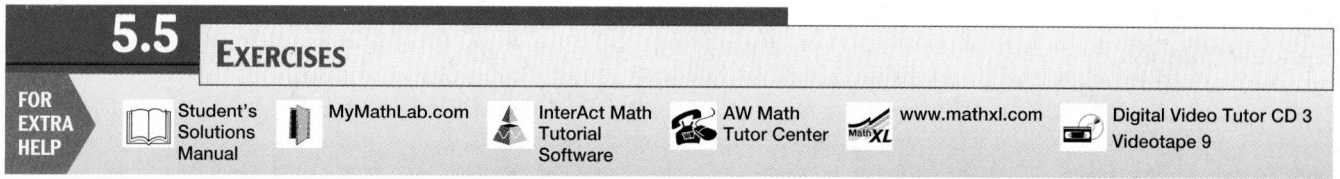

5.5   EXERCISES

FOR EXTRA HELP

📖 Student's Solutions Manual   📗 MyMathLab.com   🔺 InterAct Math Tutorial Software   ☎ AW Math Tutor Center   MathXL www.mathxl.com   💿 Digital Video Tutor CD 3 Videotape 9

Set up and solve a proportion for each application problem. See Example 1.

1. Caroline can sketch four cartoon strips in five hours. How long will it take her to sketch 18 strips?

2. The Cosmic Toads recorded eight songs on their first CD in 26 hours. How long will it take them to record 14 songs for their second CD?

3. Sixty newspapers cost $27. Find the cost of 16 newspapers.

4. Twenty-two guitar lessons cost $396. Find the cost of 12 lessons.

5. If three pounds of fescue grass seed cover about 350 square feet of ground, how many pounds are needed for 4900 square feet?

6. Anna earns $1242.08 in 14 days. How much does she earn in 260 days?

7. Tom makes $455.75 in 5 days. How much does he make in 3 days?

8. If 5 ounces of a medicine must be mixed with 8 ounces of water, how many ounces of medicine would be mixed with 20 ounces of water?

9. The bag of rice noodles shown below makes 7 servings. (*Source:* Everfresh Foods.) At that rate, how many ounces of noodles do you need for 12 servings, to the nearest ounce?

10. This can of sweet potatoes is enough for 4 servings. (*Source:* Moody Dunbar, Inc.) How many ounces are needed for 9 servings, to the nearest ounce?

11. Three quarts of a latex enamel paint will cover about 270 square feet of wall surface. (*Source:* Thompson and Formby, Inc.) How many quarts will you need to cover 350 square feet of wall surface in your kitchen and 100 square feet of wall surface in your bathroom?

12. One gallon of clear gloss wood finish covers about 550 square feet of surface. (*Source:* The Flecto Company.) If you need to apply three coats of finish to 400 square feet of surface, how many gallons do you need, to the nearest tenth?

Use the floor plan shown to complete Exercises 13–16. On the plan, one inch represents four feet.

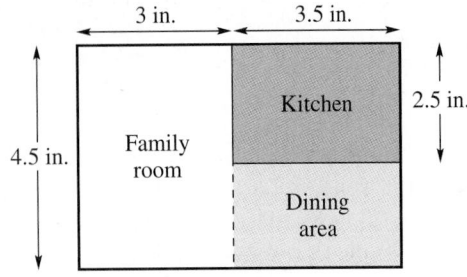

13. What is the actual length and width of the kitchen?

14. What is the actual length and width of the family room?

15. What is the actual length and width of the dining area?

16. What is the actual length and width of the entire floor plan?

17. Vince Carter scored 18 points during the 28 minutes that he played in the Olympic basketball game between the United States and Lithuania. (*Source:* Associated Press.) At that rate, how many points would he make if he played the entire game (40 minutes), to the nearest whole point?

18. In the U.S.–France gold medal Olympic basketball game, Alonzo Mourning played 26 minutes and made 7 rebounds. (*Source:* Associated Press.) How many rebounds would you expect him to make in 40 minutes (a complete Olympic game), to the nearest whole number?

Set up a proportion to solve each problem. Check to see whether your answer is reasonable. Then flip one side of your proportion to see what answer you get with an incorrect setup. Explain why the second answer is unreasonable. See Example 2.

19. About 7 out of 10 people entering a community college need to take a refresher math course. If there are 2950 entering students, how many will probably need refresher math? (*Source:* Minneapolis Community and Technical College.)

20. In a survey, only 3 out of 100 people like their eggs poached. At that rate, how many of the 60 customers who ordered eggs at Soon-Won's restaurant this morning asked to have them poached? Round to the nearest whole person.

21. Nearly 4 out of 5 people choose vanilla as their favorite ice cream flavor. (*Source:* Baskin-Robbins.) If 238 people attend an ice cream social, how many would you expect to choose vanilla? Round to the nearest whole person.

22. In a test of 200 sewing machines, only one had a defect. At that rate, how many of the 5600 machines shipped from the factory have defects?

23. About 9 out of 10 U.S. households have TV remote controls, according to a survey. There were 102,500,000 U.S. households in 1998. If the survey is accurate, how many U.S. households have TV remote controls? (*Source:* Magnavox, U.S. Bureau of the Census.)

24. In a survey, 3 out of 100 dog owners washed their pets by having the dogs go into the shower with them. If the survey is accurate, how many of the 31,200,000 dog owners in the United States use this method? (*Source:* Teledyne Water Pik, American Veterinary Medical Association.)

Set up and solve a proportion for each problem.

25. The stock market report says that 5 stocks went up for every 6 stocks that went down. If 750 stocks went down yesterday, how many went up?

26. The human body contains 90 pounds of water for every 100 pounds of body weight. How many pounds of water are in a child who weighs 80 pounds?

27. The ratio of the length of an airplane wing to its width is 8 to 1. If the length of a wing is 32.5 meters, how wide must it be? Round to the nearest hundredth.

28. The Rosebud School District wants a student-to-teacher ratio of 19 to 1. How many teachers are needed for 1850 students? Round to the nearest whole number.

29. The number of calories you burn depends on your weight. A 150-pound person burns 222 calories during 30 minutes of tennis. How many calories would a 210-pound person burn, to the nearest whole number? (*Source: Wellness Encyclopedia.*)

30. (Complete Exercise 29 first.) A 150-pound person burns 189 calories during 45 minutes of grocery shopping. How many calories would a 115-pound person burn, to the nearest whole number? (*Source: Wellness Encyclopedia.*)

31. At 3 P.M., Coretta's shadow is 1.05 meters long. Her height is 1.68 meters. At the same time, a tree's shadow is 6.58 meters long. How tall is the tree? Round to the nearest hundredth.

32. Refer to Exercise 31. Later in the day, Coretta's shadow was 2.95 meters long. How long a shadow did the tree have at that time? Round to the nearest hundredth.

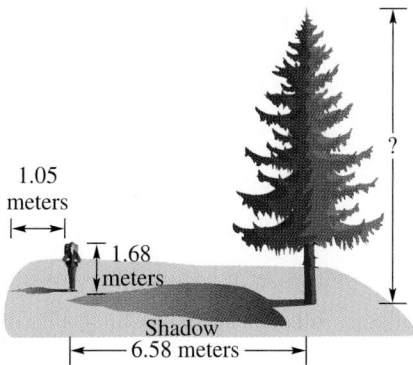

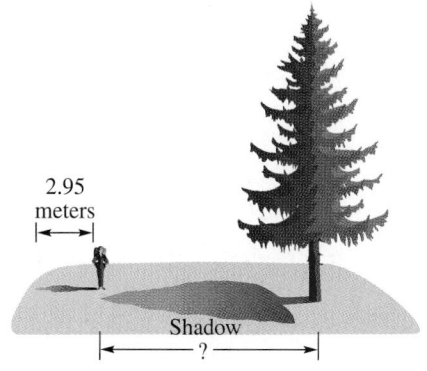

33. Can you set up a proportion to solve this problem? Explain why or why not. Jim is 25 years old and weighs 180 pounds. How much will he weigh when he is 50 years old?

34. Write your own application problem that can be solved by setting up a proportion. Also show the proportion and the steps needed to solve your problem.

35. A survey of college students shows that 4 out of 5 drink coffee. Of the students who drink coffee, 1 out of 8 adds cream to it. How many of the 38,000 students at the University of Minnesota would be expected to use cream in their coffee?

36. About 9 out of 10 adults think it is a good idea to exercise regularly. But of the ones who think it is a good idea, only 1 in 6 actually exercise at least three times a week. At this rate, how many of the 300 employees in our company exercise regularly?

37. The nutrition information on a bran cereal box says that a $\frac{1}{3}$-cup serving provides 80 calories and 8 grams of dietary fiber. (*Source:* Kraft Foods, Inc.) At that rate, how many calories and grams of fiber are in a $\frac{1}{2}$-cup serving?

38. A $\frac{2}{3}$-cup serving of penne pasta has 210 calories and 2 grams of dietary fiber. (*Source:* Borden Foods.) How many calories and grams of fiber would be in a 1-cup serving?

RELATING CONCEPTS (Exercises 39–42) **FOR INDIVIDUAL OR GROUP WORK**

*A box of instant mashed potatoes has the list of ingredients shown in the table. Use this information to **work Exercises 39–42 in order.***

Ingredient	For 12 Servings
Water	$3\frac{1}{2}$ cups
Margarine	6 Tbsp
Milk	$1\frac{1}{2}$ cups
Potato flakes	4 cups

Source: General Mills.

39. Find the amount of each ingredient for 6 servings. Show *two* different methods for finding the amounts. One method should use proportions.

40. Find the amount of each ingredient for 18 servings. Show *two* different methods for finding the amounts, one using proportions and one using your answers from Exercise 39.

41. Find the amount of each ingredient for 3 servings, using your answers from either Exercise 39 or Exercise 40.

42. Find the amount of each ingredient for 9 servings, using your answers from either Exercise 40 or Exercise 41.

SUMMARY

5.1	**ratio**	A ratio compares two quantities having the same type of units. For example, the ratio of 6 apples to 11 apples is written in fraction form as $\frac{6}{11}$. The common units (apples) divide out.
5.2	**rate**	A rate compares two measurements with different types of units. Examples are 96 dollars for 8 hours, or 450 miles on 18 gallons.
	unit rate	A unit rate has 1 in the denominator.
	cost per unit	Cost per unit is a rate that tells how much you pay for one item or one unit. The lowest cost per unit is the best buy.
5.3	**proportion**	A proportion states that two ratios or rates are equivalent.
	cross products	Cross multiply to get the cross products of a proportion. If the cross products are equal, the proportion is true.

TEST YOUR WORD POWER

See how well you have learned the vocabulary in this chapter. Answers follow the Quick Review.

1. A **ratio**
   (a) can be written only as a fraction
   (b) compares two quantities that have the same type of units
   (c) compares two quantities that have different types of units
   (d) is the reciprocal of a rate.

2. A **rate**
   (a) can be written only as a decimal
   (b) compares two quantities that have the same type of units
   (c) compares two quantities that have different types of units
   (d) is the reciprocal of a ratio.

3. A **unit rate**
   (a) has a numerator of 1
   (b) has a denominator of 1
   (c) is found by cross multiplying
   (d) is usually written in fraction form.

4. **Cost per unit** is
   (a) the best buy
   (b) a ratio written in lowest terms
   (c) found by comparing cross products
   (d) the price of one item or one unit.

5. A **proportion**
   (a) shows that two ratios or rates are equivalent
   (b) contains only whole numbers or decimals
   (c) always has one unknown number
   (d) states that two improper fractions are equivalent.

6. **Cross products** are
   (a) used only with ratios, not with rates
   (b) equal when a proportion is false
   (c) used to find the best buy
   (d) equal when a proportion is true.

QUICK REVIEW

Concepts

5.1 *Writing a Ratio*
A ratio compares two quantities. A ratio is usually written as a fraction with the number that is mentioned first in the numerator. The common units divide out and are not written in the answer. Check that the fraction is in lowest terms.

Examples

Write this ratio as a fraction in lowest terms.

60 ounces of medicine **to** 160 ounces of medicine

$$\frac{60 \text{ ounces}}{160 \text{ ounces}} = \frac{60 \div 20}{160 \div 20} = \frac{3}{8} \leftarrow \text{Lowest terms}$$

↑
Divide out common units.

Concepts	Examples

Concepts

Examples

5.1 Using Mixed Numbers in a Ratio

If a ratio has mixed numbers, change the mixed numbers to improper fractions. Rewrite the problem in horizontal format using the "÷" symbol for division. Finally, multiply by the reciprocal of the divisor.

Write as a ratio of whole numbers in lowest terms.

$$2\frac{1}{2} \quad \text{to} \quad 3\frac{3}{4}$$

$$\frac{2\frac{1}{2}}{3\frac{3}{4}} = \frac{\frac{5}{2}}{\frac{15}{4}}$$

Reciprocal

$$= \frac{5}{2} \div \frac{15}{4} = \frac{5}{2} \cdot \frac{4}{15}$$

$$= \frac{\overset{1}{\cancel{5}}}{\underset{1}{\cancel{2}}} \cdot \frac{\overset{2}{\cancel{4}}}{\underset{3}{\cancel{15}}} = \frac{2}{3} \quad \leftarrow \text{Ratio in lowest terms}$$

5.1 Using Measurements in Ratios

When a ratio compares measurements, both measurements must be in the *same* units. It is usually easier to compare the measurements using the smaller unit, for example, inches instead of feet.

Write as a ratio in lowest terms.

$$8 \text{ in. to } 6 \text{ ft}$$

Compare using the smaller unit, inches. Because 1 ft has 12 in., 6 ft is

$$6 \cdot \mathbf{12 \text{ in.}} = 72 \text{ in.}$$

The ratio is shown below.

$$\frac{8 \cancel{\text{ in.}}}{72 \cancel{\text{ in.}}} = \frac{8 \div 8}{72 \div 8} = \frac{1}{9}$$

↑
Divide out common units.

5.2 Writing Rates

A rate compares two measurements with different types of units. The units do *not* divide out, so you must write them as part of the rate.

Write the rate as a fraction in lowest terms.

$$475 \text{ miles in } 10 \text{ hours}$$

$$\frac{475 \text{ miles} \div 5}{10 \text{ hours} \div 5} = \frac{95 \text{ miles}}{2 \text{ hours}} \quad \rceil \text{Must write units: miles and hours}$$

5.2 Finding a Unit Rate

A unit rate has 1 in the denominator. To find the unit rate, divide the numerator by the denominator. Write unit rates using the word **per** or a / mark.

Write as a unit rate: $1278 in 9 days.

$$\frac{\$1278}{9 \text{ days}} \quad \leftarrow \text{The fraction bar indicates division.}$$

$$9\overline{)1278} \atop 142 \quad \text{so} \quad \frac{\$1278 \div 9}{9 \text{ days} \div 9} = \frac{\$142}{1 \text{ day}}$$

Write the answer as $142 **per** day or $142/day.

Concepts	Examples
5.2 *Finding the Best Buy* The best buy is the item with the lowest cost per unit. Divide the price by the number of units. Round to thousandths, if necessary. Then compare to find the lowest cost per unit.	Find the best buy on cheese. You have a coupon for 50¢ off on 2 pounds or 75¢ off on 3 pounds. 2 pounds for $2.75 3 pounds for $4.15 Find the cost per unit (cost per pound) after subtracting the coupon. 2 pounds cost $2.75 − $0.50 = $2.25. $$\frac{\$2.25}{2} = \$1.125 \text{ per pound}$$ 3 pounds cost $4.15 − $0.75 = $3.40. $$\frac{\$3.40}{3} \approx \$1.133 \text{ per pound}$$ The lower cost per pound is $1.125, so 2 pounds is the better buy.

Concepts	Examples
5.3 *Writing Proportions* A proportion states that two ratios or rates are equivalent. The proportion "5 is to 6 as 25 is to 30" is written as shown below. $$\frac{5}{6} = \frac{25}{30}$$ To see whether a proportion is true or false, cross multiply one way, then cross multiply the other way. If the two cross products are equal, the proportion is true. If the two cross products are unequal, the proportion is false.	Write as a proportion: 8 is to 40 as 32 is to 160. $$\frac{8}{40} = \frac{32}{160}$$ Is this proportion true or false? $$\frac{6}{8\frac{1}{2}} = \frac{24}{34}$$ Cross multiply. $$8\frac{1}{2} \cdot 24 = \frac{17}{2} \cdot \frac{\overset{12}{\cancel{24}}}{1} = 204$$ $$6 \cdot 34 = 204 \quad \text{Equal}$$ The cross products are equal, so the proportion is true.

Concepts	Examples
5.4 *Solving Proportions* Solve for an unknown number in a proportion by using these steps. *Step 1* Find the cross products. (If desired, you can rewrite the ratios in lowest terms before finding the cross products.)	Find the unknown number. $$\frac{12}{x} = \frac{6}{8}$$ $$\frac{12}{x} = \frac{3}{4} \quad \text{Lowest terms}$$ $$\frac{12}{x} = \frac{3}{4} \quad \begin{array}{l} x \cdot 3 \\ \text{Find cross} \\ \text{products.} \\ 12 \cdot 4 \end{array}$$ *(continued)*

Concepts	*Examples*
Step 2 Show that the cross products are equivalent.	$x \cdot 3 = \underline{12 \cdot 4}$ Show that cross products are $x \cdot 3 = \quad 48$ equivalent.
Step 3 Divide both products by the number multiplied by x (the number next to x).	$\dfrac{x \cdot \overset{1}{\cancel{3}}}{\underset{1}{\cancel{3}}} = \dfrac{48}{3}$ Divide each side by 3. $x = 16$
Step 4 Check by writing the solution in the proportion and finding the cross products.	x is 16. → $\dfrac{12}{16} = \dfrac{6}{8}$ $\begin{array}{l}16 \cdot 6 = 96\\ 12 \cdot 8 = 96\end{array}$ Equal
	The cross products are equal, so 16 is the correct solution.

5.5 *Solving Application Problems with Proportions*

Decide what is being compared. Set up and solve a proportion using the two rates described in the problem. Be sure that *both* rates compare things in the *same order*. Use a letter, like x, to represent the unknown number. Use the six problem-solving steps.	If 3 pounds of grass seed cover 450 square feet of lawn, how much seed is needed for 1500 square feet of lawn?
Step 1 **Read** the problem carefully.	The problem asks for the pounds of grass seed needed for 1500 square feet of lawn.
Step 2 **Work out a plan.**	Pounds of seed is compared to square feet of lawn. Set up and solve a proportion using the two given rates. Be sure that pounds of seed is in both numerators and square feet of lawn is in both denominators.
Step 3 **Estimate** a reasonable answer.	Notice that 1500 square feet is about three times as much lawn as 450 square feet. So about three times as much seed will be needed, and $3 \cdot 3$ pounds $= 9$ pounds is a reasonable estimate.
Step 4 **Solve** the problem.	With the proportion set up correctly, solve for the unknown number.
	Matching units $\dfrac{3 \text{ pounds}}{450 \text{ square feet}} = \dfrac{x \text{ pounds}}{1500 \text{ square feet}}$ Matching units
	Both sides compare pounds to square feet. Ignore the units while finding cross products.
	$450 \cdot x = \underline{3 \cdot 1500}$ Show that cross $450 \cdot x = \quad 4500$ products are equivalent.
	$\dfrac{\overset{1}{\cancel{450}} \cdot x}{\underset{1}{\cancel{450}}} = \dfrac{4500}{450}$ Divide each side by 450. $x = 10$
Step 5 **State the answer.**	10 pounds of grass seed are needed.
Step 6 **Check** your work.	The answer, 10 pounds of seed, is close to our estimate of 9 pounds, so it is reasonable.

ANSWERS TO TEST YOUR WORD POWER

1. **(b)** *Example:* The ratio of 3 miles to 4 miles is $\frac{3}{4}$; the common units (miles) divide out. 2. **(c)** *Example:*

$4.50 for 3 pounds is a rate comparing dollars to pounds. 3. **(b)** *Example:* $\frac{\$1.79}{1 \text{ pound}}$ is a unit rate.

We write it as $1.79 per pound or $1.79/pound. 4. **(d)** *Example:* $1.65 per gallon tells the price of one gallon

(one unit). 5. **(a)** *Example:* $\frac{5}{6} = \frac{25}{30}$ is a proportion because $\frac{5}{6}$ is equivalent to $\frac{25}{30}$. 6. **(d)** *Example:* The cross

products for $\frac{5}{6} = \frac{25}{30}$ are $6 \cdot 25 = 150$ and $5 \cdot 30 = 150$.

[5.2] *Write each rate as a fraction in lowest terms.*

15. $88 for 8 dozen

16. 96 children in 40 families

17. In his keyboarding class, Patrick can type four pages in 20 minutes. Give his rate in pages per minute and minutes per page.

18. Elena made $24 in three hours. Give her earnings in dollars per hour and hours per dollar.

Find the best buy.

19. Minced onion

  13 ounces for $2.29

  8 ounces for $1.45

  3 ounces for $0.95

20. Dog food; you have a coupon for $1 off on 25 pounds or more.

  50 pounds for $19.95

  25 pounds for $10.40

  8 pounds for $3.40

[5.3] *Use either the method of writing in lowest terms or of cross multiplication to decide whether each proportion is true or false.*

21. $\dfrac{6}{10} = \dfrac{9}{15}$

22. $\dfrac{6}{48} = \dfrac{9}{36}$

23. $\dfrac{47}{10} = \dfrac{98}{20}$

24. $\dfrac{64}{36} = \dfrac{96}{54}$

25. $\dfrac{1.5}{2.4} = \dfrac{2}{3.2}$

26. $\dfrac{3\frac{1}{2}}{2\frac{1}{3}} = \dfrac{6}{4}$

[5.4] *Find the unknown number in each proportion. Round answers to the nearest hundredth, if necessary.*

27. $\dfrac{4}{42} = \dfrac{150}{x}$

28. $\dfrac{16}{x} = \dfrac{12}{15}$

29. $\dfrac{100}{14} = \dfrac{x}{56}$

30. $\dfrac{5}{8} = \dfrac{x}{20}$

31. $\dfrac{x}{24} = \dfrac{11}{18}$

32. $\dfrac{7}{x} = \dfrac{18}{21}$

33. $\dfrac{x}{3.6} = \dfrac{9.8}{0.7}$

34. $\dfrac{13.5}{1.7} = \dfrac{4.5}{x}$

35. $\dfrac{0.82}{1.89} = \dfrac{x}{5.7}$

[5.5] *Set up and solve a proportion for each application problem.*

36. The ratio of cats to dogs at the animal shelter is 3 to 5. If there are 45 dogs, how many cats are there?

37. Danielle had 8 hits in 28 times at bat during last week's games. If she continues to hit at the same rate, how many hits will she get in 161 times at bat?

38. If 3.5 pounds of ground beef cost $9.77, what will 5.6 pounds cost? Round to the nearest cent.

39. About 4 out of 10 students are expected to vote in campus elections. There are 8247 students. How many are expected to vote? Round to the nearest whole number.

40. The scale on Brian's model railroad is 1 in. to 16 ft. One of the scale model boxcars is 4.25 in. long. What is the length of a real boxcar in feet?

41. In the hospital pharmacy, Michiko sees that a certain medicine is to be given at the rate of 3.5 milligrams for every 50 pounds of body weight. How much medicine should be given to a patient who weighs 210 pounds?

42. A 180-pound person burns 284 calories playing basketball for 25 minutes. How many calories would the person burn in 45 minutes, to the nearest whole number? (*Source: Wellness Encyclopedia.*)

43. Marvette makes necklaces to sell at a local gift shop. She made 2 dozen necklaces in $16\frac{1}{2}$ hours. How long will it take her to make 40 necklaces?

MIXED REVIEW EXERCISES

Find the unknown number in each proportion. Round answers to the nearest hundredth, if necessary.

44. $\dfrac{x}{45} = \dfrac{70}{30}$

45. $\dfrac{x}{52} = \dfrac{0}{20}$

46. $\dfrac{64}{10} = \dfrac{x}{20}$

47. $\dfrac{15}{x} = \dfrac{65}{100}$

48. $\dfrac{7.8}{3.9} = \dfrac{13}{x}$

49. $\dfrac{34.1}{x} = \dfrac{0.77}{2.65}$

Use cross multiplication to decide whether each proportion is true or false. Circle the correct answer.

50. $\dfrac{55}{18} = \dfrac{80}{27}$

51. $\dfrac{5.6}{0.6} = \dfrac{18}{1.94}$

52. $\dfrac{\frac{1}{5}}{2} = \dfrac{1\frac{1}{6}}{11\frac{2}{3}}$

   True   False

   True   False

   True   False

Write each ratio as a fraction in lowest terms. Change to the same units when necessary.

53. 4 dollars to 10 quarters

54. $4\frac{1}{8}$ in. to 10 in.

55. 10 yd to 8 ft

56. $3.60 to $0.90

57. 12 eggs to 15 eggs

58. 37 meters to 7 meters

59. 3 pints to 4 quarts

60. 15 minutes to 3 hours

61. $4\frac{1}{2}$ miles to $1\frac{3}{10}$ miles

62. Nearly 7 out of 8 fans buy something to drink at rock concerts. How many of the 28,500 fans at today's concert would be expected to buy a beverage? Round to the nearest hundred fans.

63. Emily spent $150 on car repairs and $400 on car insurance. What is the ratio of the amount spent on insurance to the amount spent on repairs?

64. Antonio is choosing among three packages of plastic wrap. Is the best buy 25 ft for $0.78; 75 ft for $1.99; or 100 ft for $2.59? He has a coupon for 50¢ off on either of the larger two packages.

65. On this scale drawing of a backyard patio, 0.5 in. represents 6 ft. If the patio measures 1.25 in. long on the drawing, what will the actual length of the patio be when it is built?

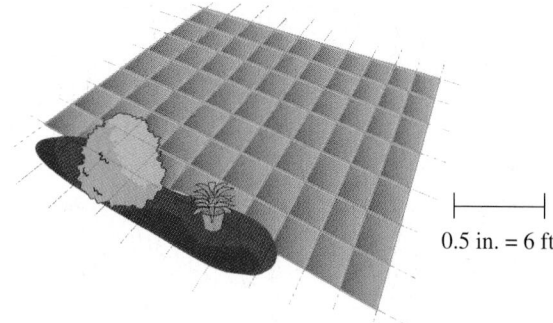

0.5 in. = 6 ft

66. A lawn mower uses 0.8 gallon of gas every 3 hours. The gas tank holds 2 gallons. How long can the mower run on a full tank?

67. An antibiotic is to be given at the rate of $1\frac{1}{2}$ teaspoons for every 24 pounds of body weight. How much should be given to an infant who weighs 8 pounds?

68. Charles made 251 points during 169 minutes of playing time last year. If he plays 14 minutes in tonight's game, how many points would you expect him to make? Round to the nearest whole number.

69. Refer to Exercise 67. Explain each step you took in solving the problem. Be sure to tell how you decided which way to set up the proportion and how you checked your answer.

70. A vitamin supplement for cats is to be given at the rate of 1000 milligrams for a 5-pound cat.
(a) How much should be given to a 7-pound cat?

(b) How much should be given to an 8-ounce kitten? (*Source:* St. Jon Pet Care Products.)

Chapter 5 TEST

Write each rate or ratio as a fraction in lowest terms. Change to the same units when necessary.

1. 16 fish to 20 fish

1. _____

2. 300 miles on 15 gallons

2. _____

3. $15 for 75 minutes

3. _____

4. The little theater has 320 seats. The auditorium has 1200 seats. Find the ratio of auditorium seats to theater seats.

4. _____

5. 3 quarts to 60 gallons

5. _____

6. 3 hours to 40 minutes

6. _____

7. Find the best buy on spaghetti sauce. You have a coupon for 75¢ off Brand X and a coupon for 25¢ off Brand Y.
   28 ounces of Brand X for $3.89
   18 ounces of Brand Y for $1.89
   13 ounces of Brand Z for $1.29

7. _____

8. Suppose the ratio of your income last year to your income this year is 3 to 2. Explain what this means. Give an example of the dollars earned last year and this year that fits the 3 to 2 ratio.

8. _____

9. _____

Determine whether each proportion is true or false.

9. $\dfrac{6}{14} = \dfrac{18}{45}$

10. $\dfrac{8.4}{2.8} = \dfrac{2.1}{0.7}$

10. _____

Find the unknown number in each proportion. Round the answers to the nearest hundredth, if necessary.

11. $\dfrac{5}{9} = \dfrac{x}{45}$

12. $\dfrac{3}{1} = \dfrac{8}{x}$

13. $\dfrac{x}{20} = \dfrac{6.5}{0.4}$

14. $\dfrac{2\frac{1}{3}}{x} = \dfrac{\frac{8}{9}}{4}$

Set up and solve a proportion for each application problem.

15. Pedro types 240 words in five minutes. At that rate, how many words can he type in 12 minutes?

16. Just 0.8 ounce of wildflower seeds is enough for 50 square feet of ground. What weight of seeds is needed for a garden with 225 square feet? (*Source:* White Swan Ltd.)

17. About 2 out of every 15 people are left-handed. How many of the 650 students in our school would you expect to be left-handed? Round to the nearest whole number.

18. A student set up the proportion for Exercise 17 this way and arrived at an answer of 4875.

$$\dfrac{2}{15} = \dfrac{650}{x} \qquad \text{Check:} \qquad \dfrac{2}{15} = \dfrac{650}{4875}$$

$$15 \cdot 650 = 9750$$
$$2 \cdot 4875 = 9750$$

Because the cross products are equal, the student said the answer is correct. Is the student right? Explain why or why not.

19. A medication is given at the rate of 8.2 grams for every 50 pounds of body weight. How much should be given to a 145-pound person? Round to the nearest tenth.

20. On a scale model, 1 in. represents 8 ft. If a building in the model is 7.5 in. tall, what is the actual height of the building in feet?

11. _____

12. _____

13. _____

14. _____

15. _____

16. _____

17. _____

18. _____

19. _____

20. _____

Name the digit that has the given place value.

1. 216,475,038

thousands
tens
millions
hundred thousands

2. 340.6915

hundredths
ones
ten-thousandths
hundreds

Round each number as indicated.

3. 9903 to the nearest hundred

4. 617.0519 to the nearest tenth

5. $99.81 to the nearest dollar

6. $3.0555 to the nearest cent

First use front end rounding to round each number and estimate the answer. Then find the exact answer.

7. *Estimate:*   *Exact:*
$$\begin{array}{r} 28 \\ 5206 \\ +\ 351 \\ \hline \end{array}$$
$+$ ___

8. *Estimate:*   *Exact:*
$$\begin{array}{r} 63.1 \\ -\ 5.692 \\ \hline \end{array}$$
$-$ ___

9. *Estimate:*   *Exact:*
$$\begin{array}{r} 4716 \\ \times\ 804 \\ \hline \end{array}$$
$\times$ ___

10. *Estimate:*   *Exact:*
$$\begin{array}{r} 0.982 \\ \times\ 17.8 \\ \hline \end{array}$$
$\times$ ___

11. *Estimate:*   *Exact:*

___$)$___ $53\overline{)48{,}071}$

12. *Estimate:*   *Exact:*

___$)$___ $4.5\overline{)1638}$

13. *Estimate:*   *Exact:*

___ $\cdot$ ___ $=$ ___ $1\frac{5}{6} \cdot 3\frac{3}{5}$

14. *Estimate:*   *Exact:*

___ $\div$ ___ $=$ ___ $5\frac{1}{4} \div \frac{7}{8}$

15. *Estimate:*   *Exact:*

___ $-$ ___ $=$ ___ $2\frac{4}{5} - 1\frac{5}{6}$

16. *Estimate:*   *Exact:*

___ $+$ ___ $=$ ___ $2\frac{9}{10} + 10\frac{1}{2}$

Add, subtract, multiply, or divide as indicated.

17. $988 + 373{,}422 + 6$

18. $30 - 0.66$

19. Write your answer using R for the remainder.

$$33\overline{)20{,}157}$$

20. $(1.9)(0.004)$

21. $3020 - 708$

22. $0.401 + 62.98 + 5$

23. $1.39 \div 0.025$

24. $(6392)(5609)$

Use the order of operations to simplify each expression.

25. $36 + 18 \div 6$

26. $8 \div 4 + (10 - 3^2) \cdot 4^2$

27. $88 \div \sqrt{121} \cdot 2^3$

28. $(16.2 - 5.85) - 2.35 \cdot 4$

29. Write 0.0105 in words.

30. Write sixty and seventy-one thousandths in numbers.

Nest boxes for birds are made with different sizes of entry holes. The right size opening allows only certain types of birds to use the nest box and helps prevent entry by predators. Use the information in the table to answer Exercises 31–34. The diameter is the distance across the opening at its widest point.

Eastern bluebird nest box with 1.5 in. opening.

Bird	Diameter of Opening
Eastern bluebird	1.5 in.
Western bluebird	$1\frac{9}{16}$ in.
Chickadee	1.25 in.
Swallow	$1\frac{3}{8}$ in.
Wren	$1\frac{1}{8}$ in.

Source: Duncraft.

31. Write each mixed number measurement in the table as a decimal. Round to the nearest thousandth, if necessary.

32. List the nest box openings in order from smallest to largest.

33. Find the difference in diameter between the largest and smallest openings. Write your answer as a fraction or mixed number in lowest terms.

34. What is the difference in diameter between the openings for an eastern bluebird and a western bluebird? Write your answer as a fraction and as a decimal.

Write each rate or ratio as a fraction in lowest terms. Change to the same units when necessary. Use the circle graph for Exercises 35–37.

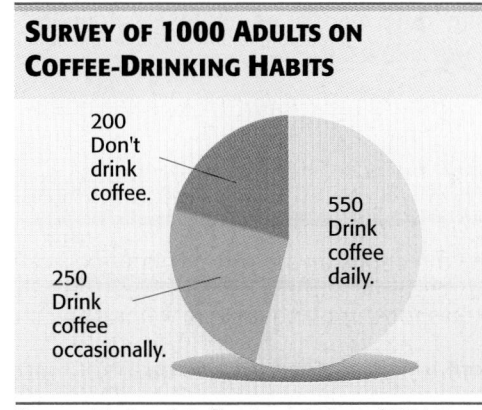

SURVEY OF 1000 ADULTS ON COFFEE-DRINKING HABITS

200 Don't drink coffee.

550 Drink coffee daily.

250 Drink coffee occasionally.

Source: National Coffee Association of USA Inc.

35. Write the ratio of those who drink coffee daily to those who drink it occasionally.

36. What is the ratio of those who do not drink coffee to the total number of people in the survey?

37. Find the ratio of all the adults who drink coffee to those who never drink it.

38. Write the ratio of 20 minutes to 4 hours.

39. Find the ratio of 2 ft to 8 in.

40. Find the best buy on instant mashed potatoes. You have a coupon for 50¢ off on either the 36-serving or 48-serving box.

A box that makes 20 servings for $1.59

A box that makes 36 servings for $3.24

A box that makes 48 servings for $4.99

Find the unknown number in each proportion. Round your answers to the nearest hundredth, if necessary.

41. $\dfrac{9}{12} = \dfrac{x}{28}$

42. $\dfrac{7}{12} = \dfrac{10}{x}$

43. $\dfrac{x}{\frac{3}{4}} = \dfrac{2\frac{1}{2}}{\frac{1}{6}}$

44. $\dfrac{6.7}{x} = \dfrac{62.8}{9.15}$

Solve each application problem.

45. The college honor society has a goal of collecting 1500 pounds of food to fill Thanksgiving baskets. So far they've collected $\dfrac{5}{6}$ of their goal. How many more pounds do they need?

46. Tara has a photo that is 10 centimeters wide by 15 centimeters long. If the photo is enlarged to a length of 40 centimeters, find the new width, to the nearest tenth.

47. The distance around Dunning Pond is $1\frac{1}{10}$ miles. Norma ran around the pond four times in the morning and $2\frac{1}{2}$ times in the afternoon. How far did she run in all?

48. Rodney bought 49.8 gallons of gas for his minivan while driving 896.5 miles on a vacation. How many miles per gallon did he get, rounded to the nearest tenth?

49. In a survey, 5 out of 8 apartment residents said they are sometimes bothered by noise from their neighbors. How many of the 224 residents at Harris Towers would you expect to be bothered by noise?

50. The directions on a bottle of plant food call for $\frac{1}{2}$ teaspoon in two quarts of water. How much plant food is needed for five quarts? (*Source: Schultz Company.*)

Use the bar graph of average yearly earnings to answer Exercises 51–56.

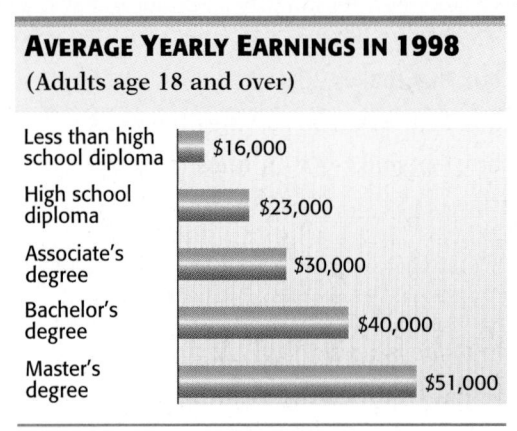

AVERAGE YEARLY EARNINGS IN 1998
(Adults age 18 and over)

Less than high school diploma — $16,000
High school diploma — $23,000
Associate's degree — $30,000
Bachelor's degree — $40,000
Master's degree — $51,000

Source: U.S. Bureau of the Census.

51. Write the ratio of bachelor's degree earnings to "less than high school diploma" earnings as a fraction in lowest terms.

52. Explain what the ratio in Exercise 51 is saying about the yearly earnings of the two groups.

53. What is the average monthly salary for a person with a high school diploma, to the nearest dollar?

54. What is the average monthly salary for a person with a master's degree, to the nearest dollar?

55. If an average work year is 2000 hours of work time, find the difference in the hourly wage of a person with an associate's degree and a person without a high school diploma.

56. Compare the lifetime earnings of a person with a high school diploma who works from age 18 to 65 to a person with a bachelor's degree who works from age 22 to 65.

Percent

6

Percents are a large part of our everyday lives. For example, interest rates on automobile loans, home loans, and other installment loans are always given as percents. Sales tax, commission rates, and discounts on sale items are other everyday applications that show the importance of understanding percent. For example, knowing how to calculate the sales tax on any item you purchase allows you to know the true cost of anything you buy. (See Examples 1 and 2, Section 6.6.)

You're Connected

6 Fill in the blanks.

(a) 50% of 100 windows is

_____.

(b) 50% of 64 e-mails is

_____.

(c) 10% of 3850 members

is _____.

(d) 1% of 240 ft

is _____.

Example 6 Finding 50%, 10%, and 1% of a Number

Fill in the blanks.

(a) 50% of 2420 hours is _____.

50% is half of the hours. So, 50% of 2420 hours is 1210 hours .

(b) 10% of 280 pages is _____.

10% is $\frac{1}{10}$ of the pages. Move the decimal point one place to the left. So, 10% of 280. pages is 28 pages .

(c) 1% of $540 is _____.

1% is $\frac{1}{100}$ of the money. Move the decimal point two places to the left. So, 1% of $540. is $5.40 .

Work Problem 6 at the Side.

ANSWERS
6. (a) 50 windows (b) 32 e-mails
(c) 385 members (d) 2.4 ft

Write each percent as a decimal. See Examples 2 and 3.

1. 15%   **2.** 41%   **3.** 60%   **4.** 40%

5. 25%   **6.** 35%   **7.** 140%   **8.** 250%

9. 5.5%   **10.** 6.7%   **11.** 100%   **12.** 600%

13. 0.5%   **14.** 0.25%   **15.** 0.35%   **16.** 0.75%

Write each decimal as a percent. See Example 4.

17. 0.8   **18.** 0.4   **19.** 0.58   **20.** 0.25

21. 0.01   **22.** 0.07   **23.** 0.125   **24.** 0.875

25. 0.375   **26.** 0.625   **27.** 2   **28.** 5

29. 3.7   **30.** 2.2   **31.** 0.0312   **32.** 0.0625

33. 4.162   **34.** 8.715   **35.** 0.0028   **36.** 0.0064

37. Fractions, decimals, and percents are all used to describe a part of something. The use of percents is much more common than fractions and decimals. Why do you suppose this is true?

38. List five uses of percent that are or will be part of your life. Consider the activities of working, shopping, saving, and planning for the future.

Write each percent as a decimal and each decimal as a percent. See Examples 2–4.

39. In Lincoln, 45% of the refuse is recycled.

40. At College of DuPage, 82% of the students work part-time.

41. At Bett's Boutique, 18% of the items sold are returned.

42. There was a 43.2% voter turnout at the election.

43. The property tax rate in Alpine County is 0.035.

44. A church building fund has 0.49 of the money needed.

45. The number of people successfully completing CPR training this session is 2 times that of the last session.

46. The number of newspaper subscribers was 4 times as great as last quarter.

47. Only 0.005 of the total population has this genetic defect.

48. The return rate of defective keyboards is 0.0075 of total output.

49. The patient's blood pressure was 153.6% of normal.

50. Success with the diet was 248.7% greater than anticipated.

Fill in the blanks. Remember that 100% is all of something, 200% is two times as many, and 300% is three times as many. See Example 5.

51. There are 20 children in the preschool class. 100% of the children are served breakfast and lunch. How many children are served both meals?

_____

52. The company owns 345 vans. 100% of the vans are painted white with blue lettering. How many vans are painted white with blue lettering?

_____

53. Last year we had 210 employees. This year we have 200% of that number. How many employees do we have this year? _____

54. This week's expenses are 200% of last week's $380. This week's expenses are _____.

55. If we need 300% of the 90 chairs that we needed for last week's meeting, we will need _____.

56. Jim's new car gets 300% of the 12 miles per gallon that his old car got. His new car gets

_____.

Fill in the blanks. Remember that 50% is half of something; 10% is found by moving the decimal point one place to the left, and 1% is found by moving the decimal point two places to the left. See Example 6.

57. John owes $285 for tuition. Financial aid will pay 50% of the cost. Financial aid will pay

_____.

58. The Animal Humane Society took in 20,000 animals last year. About 50% of them were dogs. The number of dogs taken in was _____.

59. Only 10% of 8200 commuters are carpooling to work. How many commuters carpool?

_____

60. Sarah knows that she will not be able to sell 10% of the 240 dozen plants in her greenhouse. The number of unsold plants will be _____.

61. The naturalist said that 1% of the 2600 plants in the park are poisonous. How many plants are poisonous? _____

62. Of the 4800 accidents, only 1% was caused by mechanical failure. How many accidents were caused by mechanical failure? _____

63. (a) Describe a shortcut method of finding 100% of a number.

(b) Show an example using your shortcut.

64. (a) Describe a shortcut method of finding 50% of a number.

(b) Show an example using your shortcut.

65. (a) Describe a shortcut method of finding 200% of a number.

(b) Show an example using your shortcut.

67. (a) Describe a shortcut method of finding 10% of a number.

(b) Show an example using your shortcut.

66. (a) Describe a shortcut method of finding 300% of a number.

(b) Show an example using your shortcut.

68. (a) Describe a shortcut method of finding 1% of a number.

(b) Show an example using your shortcut.

College students were asked to rank the most important issues they would like presidential candidates to address. The bar graph shows the ranking of these issues and the percent of students selecting each issue. Use this graph to answer Exercises 69–72. Write each answer as a percent and as a decimal.

69. What portion of the students selected education as a top issue?

70. What portion of the students selected environmental issues as a top issue?

71. (a) What was the second most important issue?

(b) Write the portion of the students who selected that issue.

72. (a) What was the third most important issue?

(b) Write the portion of the students who selected that issue?

EDUCATION IS TOPS
Education ranks as the most important issue that college students would like presidential candidates to address.

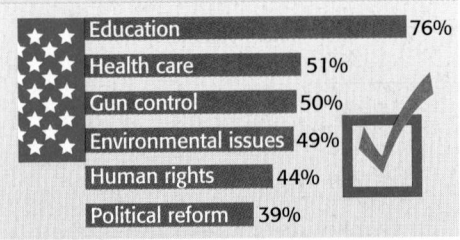

Education	76%
Health care	51%
Gun control	50%
Environmental issues	49%
Human rights	44%
Political reform	39%

Source: Greenfield Online, YouthStream: Pulsefinder On-Campus Market Study.

In the United States, 12.7% of the population is 65 years old or older. The bar graph shows the countries with the highest percent of people 65 years old or older. Use this graph to answer Exercises 73–76. Write each answer as a percent and as a decimal.

73. What portion of the population of Spain is 65 or older?

74. What portion of the population of Greece is 65 or older?

75. (a) Which two countries have the lowest portion of the population 65 or older?

(b) What portion is this?

65 AND UP
Countries with the highest percentage of seniors:

Country	Percent
Italy	18.2%
Sweden	17.2%
Greece	17.2%
Belgium	17.1%
Japan	17%
Spain	16.8%
Germany	16.5%
Bulgaria	16.5%

Source: U.S. Bureau of the Census
International Programs Center.

76. (a) Which country has the highest portion of the population 65 or older?

(b) What portion is this?

Write a percent for both the shaded and unshaded part of each figure.

77.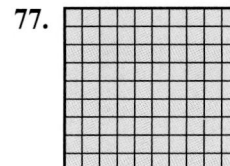

78.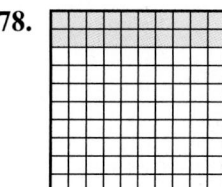

79.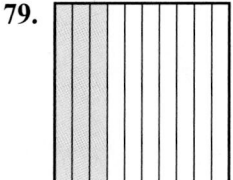

80.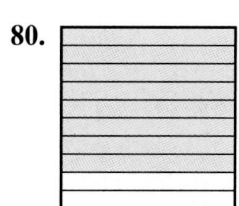

81.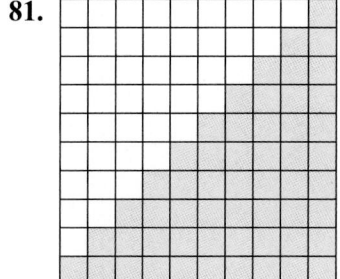

82.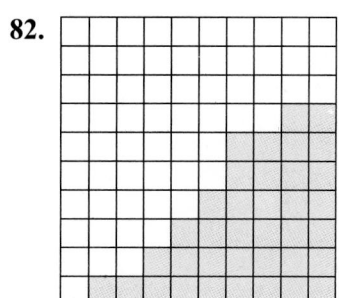

6.3   USING THE PERCENT PROPORTION AND IDENTIFYING THE COMPONENTS IN A PERCENT PROBLEM

There are two ways to solve percent problems. One method uses proportions and is discussed in this and the next section. The other method uses the percent equation and is explained in **Section 6.5.**

1 **Learn the percent proportion.**   We have seen that a statement of two equivalent ratios is called a proportion.

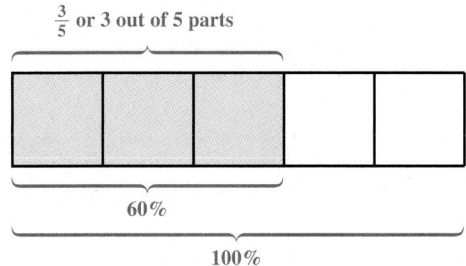

$\frac{3}{5}$ or 3 out of 5 parts

60%

100%

For example, the fraction $\frac{3}{5}$ is the same as the ratio 3 to 5, and 60% is the same as the ratio 60 to 100. As the figure shows, these two ratios are equivalent and make a proportion.

Work Problem ① at the Side.

The **percent proportion** can be used to solve percent problems.

Percent Proportion

Part is to *whole* as percent is to 100.

$$\frac{\text{part}}{\text{whole}} = \frac{\text{percent}}{100} \quad \leftarrow \text{Always 100 because percent means per 100}$$

In the figure at the top of the page, the **whole** is 5 (the entire quantity), the **part** is 3 (the part of the whole), and the **percent** is 60. Write the percent proportion as follows.

$$\frac{\text{part} \rightarrow 3}{\text{whole} \rightarrow 5} = \frac{60 \leftarrow \text{percent}}{100 \leftarrow 100}$$

2 **Solve for an unknown value in a proportion.**   As shown in **Section 5.4,** if any two of the three values (part, whole, or percent) in the percent proportion are known, the third can be found by solving the proportion.

Example 1   Using the Percent Proportion

Use the percent proportion and solve for the unknown value. Let x represent the unknown value.

(a) part = 12, percent = 25; find the whole.

$$\frac{\text{part}}{\text{whole}} = \frac{\text{percent}}{100} \quad \text{Percent proportion}$$

$$\text{Part} \rightarrow \frac{12}{x} = \frac{25}{100} \quad \text{or} \quad \frac{12}{x} = \frac{1}{4} \quad \text{Lowest terms}$$
Whole (unknown) →

— **Continued on Next Page**

① As a review of proportions, use the method of cross products to decide whether each proportion is *true* or *false.*

(a) $\dfrac{1}{2} = \dfrac{25}{50}$

(b) $\dfrac{3}{5} = \dfrac{75}{125}$

(c) $\dfrac{7}{8} = \dfrac{180}{200}$

(d) $\dfrac{32}{53} = \dfrac{160}{265}$

(e) $\dfrac{112}{41} = \dfrac{332}{123}$

ANSWERS
1. (a) true   **(b)** true   **(c)** false
(d) true   **(e)** false

❷ Use the percent proportion $\left(\dfrac{\text{part}}{\text{whole}} = \dfrac{\text{percent}}{100}\right)$ and solve for the unknown value.

(a) part = 12, percent = 16

(b) part = 30, whole = 120

(c) whole = 210, percent = 20

(d) whole = 4000, percent = 32

(e) part = 74, whole = 185

ANSWERS

2. (a) whole = 75   **(b)** percent = 25
(so, the percent is 25%)   **(c)** part = 42
(d) part = 1280   **(e)** percent = 40
(so, the percent is 40%)

Find the cross products to solve this proportion.

$$\dfrac{12}{x} = \dfrac{1}{4}$$

Show that the cross products are equivalent.

$$x \cdot 1 = 12 \cdot 4$$
$$x = 48$$

The whole is 48.

CAUTION

You cannot divide out common factors from a numerator and denominator in a proportion. This can be done **only** when the fractions are being multiplied.

(b) part = 30, whole = 50; find the percent.
   Use the percent proportion.

$$\text{Part} \to \dfrac{30}{50} = \dfrac{x}{100} \gets \text{Percent (unknown)} \qquad \text{Percent proportion}$$

$$\dfrac{3}{5} = \dfrac{x}{100} \qquad \text{Lowest terms}$$

$$5 \cdot x = 3 \cdot 100 \qquad \text{Cross products}$$

$$5 \cdot x = 300$$

$$\dfrac{\cancel{5} \cdot x}{\cancel{5}} = \dfrac{300}{5} \qquad \text{Divide each side by 5.}$$

$$x = 60$$

The percent is 60, written as 60%.

(c) whole = 150, percent = 18; find the part.

$$\text{Part (unknown)} \to \dfrac{x}{150} = \dfrac{18}{100} \quad \text{or} \quad \dfrac{x}{150} = \dfrac{9}{50} \qquad \text{Lowest terms}$$

$$x \cdot 50 = 150 \cdot 9 \qquad \text{Cross products}$$

$$x \cdot 50 = 1350$$

$$\dfrac{x \cdot \cancel{50}}{\cancel{50}} = \dfrac{1350}{50} \qquad \text{Divide each side by 50.}$$

$$x = 27$$

The part is 27.

Work Problem ❷ at the Side.

As a help in solving percent problems, keep in mind this basic idea.

Percent Problems

All percent problems involve a comparison between a part of something and the whole.

Solving these problems requires identifying the three components of a percent proportion: part, whole, and percent.

3▭ **Identify the percent.** Look for the percent first. It is the easiest to identify.

Percent

The **percent** is the ratio of a part to a whole, with 100 as the denominator. In a problem, the percent appears with the word *percent* or with the symbol "**%**" after it.

> **Example 2** **Finding the Percent in Percent Problems**
>
> Find the percent in the following.
>
> **(a)** 32% of the 900 men were retired.
>
> ↓ Percent
>
> The percent is 32. The number 32 appears with the symbol %.
>
> **(b)** $150 is 25 percent of what number?
>
> ↓ Percent
>
> The percent is 25 because 25 appears with the word *percent*.
>
> **(c)** What percent of the 350 women will go?
>
> ↓ Percent (unknown)
>
> The word *percent* has no number with it, so the percent is the unknown part of the problem.
>
> ══════════════ **Work Problem ❸ at the Side.**

4▭ **Identify the whole.** Next, look for the whole.

Whole

The **whole** is the entire quantity. In a percent problem, the whole often appears after the word **of**.

❸ Identify the percent.

(a) Of the $1800, 18% will be spent on window coverings.

(b) Of the 620 children, 10% will have birthdays this month.

(c) Find the amount of sales tax by multiplying $590 and $6\frac{1}{2}$ percent.

(d) 105 is 3% of what number?

(e) What percent of the 380 guests will return this year?

4 Identify the whole.

(a) Of the $1800, 18% will be spent on window coverings.

(b) Of the 620 children, 10% will have birthdays this month.

(c) Find the amount of sales tax by multiplying sales of $590 and $6\frac{1}{2}$ percent.

(d) $105 is 3% of what number?

(e) What percent of the 380 guests will return this year?

5 Identify the part.

(a) Of the $1800, 18%, or $324 will be spent on window coverings.

(b) Of the 620 children, 10%, or 62 children, will have birthdays this month.

(c) Find the sales tax by multiplying $590 and $6\frac{1}{2}$ percent.

(d) $105 is 3% of what number?

(e) 80% of the 380 guests will return this year.

Example 3 Finding the Whole in Percent Problems

Identify the whole in the following.

(a) 32% **of** the 900 men were too large for the imported car.
↓
Whole

The whole is 900. The number 900 appears after the word *of*.

(b) $150 is 25 percent **of** what number?
↓
Whole The whole is the unknown part of the problem.

(c) 85% **of** 7000 is what number?
↓
Whole

Work Problem **at the Side.**

5 **Identify the part.** Finally, look for the part.

Part

The **part** is the portion being compared with the whole.

NOTE

If you have trouble identifying the part, find the whole and percent first. The remaining number is the part.

Example 4 Finding the Part in Percent Problems

Identify the part in the following.

(a) 54% **of** 700 students is 378 students.
First find the percent and the whole.

54% **of** 700 students is 378 students.
↓   ↓
Percent; with % sign   Whole; follows "of"

The remaining number is the part.

54% **of** 700 students is 378 students.
↓   ↓   ↓
Percent   Whole   Part

The part is 378.

(b) $150 is 25% **of** what number?
↓   ↓
Percent   Whole (unknown)

$150 is the remaining number, so the part is $150.

(c) 85% **of** $7000 is what number?
↓   ↓   ↓
Percent   Whole   Part (unknown)

Work Problem **5** **at the Side.**

6.3 EXERCISES

FOR EXTRA HELP

 Student's Solutions Manual    MyMathLab.com   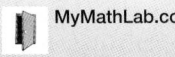 InterAct Math Tutorial Software   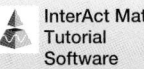 AW Math Tutor Center    www.mathxl.com    Digital Video Tutor CD 4 Videotape 10

Find the unknown value in the percent proportion $\dfrac{part}{whole} = \dfrac{percent}{100}$. *Round to the nearest tenth if necessary. If the answer is a percent, be sure to include a percent sign (%). See Example 1.*

1. part = 5, percent = 10

2. part = 20, percent = 25

3. part = 30, percent = 20

4. part = 25, percent = 25

5. part = 28, percent = 40

6. part = 11, percent = 5

7. part = 15, whole = 60

8. part = 105, whole = 35

9. part = 36, whole = 24

10. part = 1.5, whole = 4.5

11. part = 9.25, whole = 27.75

12. part = 12.8, whole = 9.6

13. whole = 52, percent = 50

14. whole = 160, percent = 35

15. whole = 72, percent = 30

16. whole = 115, percent = 38

17. whole = 94.4, part = 25

18. whole = 89.6, part = 50

Solve each problem. If the answer is a percent, be sure to include a percent sign (%). See Examples 2–4.

19. Find the whole if the part is 46 and the percent is 40.

20. The percent is 45 and the whole is 160. Find the part.

21. The whole is 5000 and the part is 20. Find the percent.

22. Suppose the part is 15 and the whole is 2500. Find the percent.

23. Find the percent if the whole is 4300 and the part is $107\frac{1}{2}$.

24. What is the part, if the percent is $12\frac{3}{4}$ and the whole is 5600?

25. The whole is 6480 and the part is 19.44. Find the percent.

26. Suppose the part is 281.25 and the percent is $1\frac{1}{4}$. Find the whole.

Identify the percent, whole, and part in the following. Do not try to solve for any unknowns. See Examples 2–4.

	Percent	Whole	Part
27. 10% of how many bicycles is 60 bicycles?	_____	_____	_____
28. 58% of how many preschoolers is 203 preschoolers?	_____	_____	_____
29. 75% of $800 is $600.	_____	_____	_____
30. 93% of $1500 is $1395.	_____	_____	_____
31. What is 25% of $970?	_____	_____	_____
32. What is 61% of 830 homes?	_____	_____	_____
33. 12 injections is 20% of what number of injections?	_____	_____	_____
34. 92 servings is 26% of what number of servings?	_____	_____	_____
35. 34 trophies is 50% of 68 trophies.	_____	_____	_____
36. 410 pallets is $33\frac{1}{3}$% of 1230 pallets.	_____	_____	_____
37. What percent of $296 is $177.60?	_____	_____	_____
38. What percent of $120.80 is $30.20?	_____	_____	_____
39. 54.34 is 3.25% of what number?	_____	_____	_____
40. 16.74 is 11.9% of what number?	_____	_____	_____
41. 0.68% of $487 is what amount?	_____	_____	_____
42. What amount is 6.21% of $704.35?	_____	_____	_____

43. Identify the three components in a percent problem. In your own words, write one sentence telling how you will identify each of these three components.

44. Write one short sentence or statement using numbers and words. The statement should include a percent, a whole, and a part. Identify each of these three components.

Find the percent, whole, and part in each application problem. Do not try to solve for any unknowns.

45. In a tree-planting project, 640 of the 810 trees planted were still living 1 year later. What percent of the trees planted were still living?

46. Ivory Soap is $99\frac{44}{100}\%$ pure. If a bar of Ivory Soap weighs 9 ounces, how many ounces are pure? (*Source:* Procter & Gamble.)

47. Of the 142 people attending a movie theater, 86 bought buttered popcorn. What percent bought buttered popcorn?

48. On her first check from the Pizza Hut Restaurant, 15% is withheld from Maria's total earnings of $225. What amount is withheld?

49. Of the lunch and dinner customers at the Heston Grill, 23% prefer a fat-free salad dressing. If the total number of customers is 610, find the number who prefer fat-free dressing.

50. There are 680 1-gigahertz computer chips in a secured storage area designed to hold 2000 computer chips. What percent of the storage area is filled?

51. Of the total candy bars contained in a vending machine, 240 bars have been sold. If 25% of the bars have been sold, find the total number of candy bars that were in the machine.

52. There have been 36 cups of coffee served from a banquet-sized coffee pot. If this is 30% of the capacity of the pot, find the capacity of the pot.

36 cups

53. In a recent survey of 480 adults, 55% said that they would prefer to have their wedding at a religious site. How many said they would prefer the religious site? (*Source:* National Family Opinion Research.)

54. Sue Ann needs 64 credits to graduate. If she has completed 48 of the credits needed, what percent of the credits has she already completed?

55. In a poll of 822 people, 49.5% said that they get their news from television. Find the number of people who said they get their news from television. (*Source: Brills Content.*)

56. The sales tax on a new car is $820. If the sales tax rate is 5%, find the price of the car before the sales tax is added.

57. A medical clinic found that 16.8% of the patients were late for their appointments. The number of patients who were late was 504. Find the total number of patients.

58. The state troopers tested 924 cars for safety. There were 231 cars that failed the safety test for one or more reasons. Find the percent of cars that failed the test.

6.4 Using Proportions to Solve Percent Problems

This is the percent proportion.

$$\frac{\text{part}}{\text{whole}} = \frac{\text{percent}}{100}$$

As discussed in Section 6.3, if any two of the three values are known, the third can be found by solving the percent proportion.

1 **Use the percent proportion to find the part.** The first example shows how to use the percent proportion to find the part.

> **Example 1** **Finding the Part with the Percent Proportion**
>
> Find 15% of $160.
>   Here the percent is 15 and the whole is 160. (Recall that the whole often comes after the word *of*.) Now find the part. Let x represent the unknown part.
>
> $$\frac{\text{part}}{\text{whole}} = \frac{\text{percent}}{100} \quad \text{so} \quad \frac{x}{160} = \frac{15}{100} \quad \text{or} \quad \frac{x}{160} = \frac{3}{20} \quad \text{Lowest terms}$$
>
> Find the cross products in the proportion.
>
> $$x \cdot 20 = 160 \cdot 3 \quad \text{Cross products}$$
> $$x \cdot 20 = 480$$
>
> $$\frac{x \cdot \overset{1}{\cancel{20}}}{\cancel{20}} = \frac{480}{20} \quad \text{Divide each side by 20.}$$
>
> $$x = 24 \quad \text{Part}$$
>
> 15% of $160 is **$24**.

─────────── **Work Problem ❶ at the Side.**

Just as with the application problems given earlier, the word *of* is an indicator word meaning **multiply.** For example:

$$15\% \text{ of } 160$$
$$\downarrow$$
$$15\% \ \cdot \ 160.$$

Because of this, there is another way to find the part.

Finding the Part Using Multiplication

To find the part:

Step 1 Identify the percent. Write the percent as a decimal.

Step 2 Multiply this decimal by the whole.

❶ Use the percent proportion to find the part.

(a) 10% of 1250 sailboats

(b) 15% of $3220

(c) 7% of 2700 miles

(d) 39% of 1220 meters

❷ Use multiplication to find the part.

(a) 55% of 10,000 injections

(b) 16% of 120 miles

(c) 135% of 60 dosages

(d) 0.5% of $238

> ### Example 2   Finding the Part by Using Multiplication
>
> Use multiplication to find the part.
>
> **(a)** Find 42% of 830 yards.
>
> > *Step 1*   Here, the percent is 42. Write 42% as the decimal 0.42.
> >
> > *Step 2*   Multiply 0.42 and the whole, which is 830.
> >
> > $$\text{part} = 0.42 \cdot 830$$
> > $$= 348.6 \text{ yd}$$
>
> It is a good idea to estimate the answer, to make sure no mistakes were made with decimal points. Round 42% to 40% or 0.4, and round 830 as 800. Next, 40% of 800 is
>
> $$0.4 \cdot 800 = 320, \leftarrow \text{Estimate.}$$
>
> so 348.6 is a reasonable answer.
>
> **(b)** Find 25% of 1680 cars.
> Identify the percent as 25. Write 25% in decimal form as 0.25. Now, multiply 0.25 and 1680.
>
> $$\text{part} = 0.25 \cdot 1680 = 420 \text{ cars} \quad \text{Multiply.}$$
>
> You can use a shortcut to estimate the answer. Since 25% means 25 parts out of 100 parts, this is the same as $\frac{1}{4}$ of the whole ($\frac{25}{100} = \frac{1}{4}$). Do you see a shortcut here? You can find $\frac{1}{4}$ of a number by dividing the number by 4. So, this shortcut gives us the exact answer, $1680 \div 4 = 420$.
>
> **(c)** Find 140% of 60 miles.
> In this problem, the percent is 140. Write 140% as the decimal 1.40. Next, multiply 1.40 and 60.
>
> $$\text{part} = 1.40 \cdot 60 = 84 \text{ miles} \quad \text{Multiply.}$$
>
> You can estimate the answer by realizing that 140% is close to 150% (which is $1\frac{1}{2}$) and $1\frac{1}{2}$ times 60 is 90. So, 84 miles is a reasonable answer.
>
> **(d)** Find 0.4% of 50 kilometers.
>
> $$\text{part} = 0.004 \cdot 50 = 0.2 \text{ kilometers} \quad \text{Multiply.}$$
>
> Write 0.4% as a decimal.
>
> Estimate the answer. 0.4% is less than 1%.
>
> $$1\% \text{ of } 50 \text{ miles} = 50. = 0.5 \text{ miles}$$
>
> So our answer should be *less than* 0.5 miles, and 0.2 miles fits this requirement.
>
> **Work Problem ❷ at the Side.**

> ### Example 3   Solving for the Part in an Application Problem
>
> Raley's Markets has 850 employees. Of these employees, 28% are students. How many of the employees are students?
>
> **Continued on Next Page**

Step 1   **Read** the problem. The problem asks us to find the number of employees who are students.

Step 2   **Work out a plan.** Look for the word *of* as an indicator word for multiplication.

28% **of** the employees are students.

└── Indicator word

   The total number of employees is 850, so the whole is 850. The percent is 28. To find the number of students, find the part.

Step 3   **Estimate** a reasonable answer. You can estimate the answer by rounding 28% to 25% and 850 to 900. Remember 25% is 25 parts out of 100, which is equivalent to $\frac{1}{4}$. So divide 900 by 4.

$900 \div 4 = 225$ students   ← Estimate

Step 4   **Solve** the problem.

part $= 0.28 \cdot 850 = 238$ Multiply.

└── Write 28% as a decimal.

Step 5   **State the answer.** Raley's Markets has 238 student employees.

Step 6   **Check.** The answer, 238 students, is close to our estimate of 225 students.

Work Problem ❸ at the Side.

⊞ Calculator Tip   If you are using a calculator, you could solve Example 3 like this.

0.28 ⓧ 850 ⊜ 238

Or, you can use this alternate approach on calculators with a % key.

850 ⓧ 28 ⓟ ⊜ 238

2⃞   **Fiind the whole using the percent proportion.**   The next example shows how to use the percent proportion to find the whole.

NOTE

⌐ Remember, the whole is the entire quantity.

Example 4   **Finding the Whole with the Percent Proportion**

(a) 8 tables is 4% of what number of tables?
   Here the percent is 4, the whole is unknown, and the part is 8. Use the percent proportion to find the whole. Let *x* represent the unknown whole.

$\dfrac{\text{part}}{\text{whole}} = \dfrac{\text{percent}}{100}$ so $\dfrac{8}{x} = \dfrac{4}{100}$ or $\dfrac{8}{x} = \dfrac{1}{25}$ Lowest terms

Find the cross products.

$x \cdot 1 = 8 \cdot 25$

$x = 200$

8 tables is 4% of **200 tables**.

──── **Continued on Next Page**

---

❸ Use the six problem-solving steps to solve each problem.

(a) One day on Jacob's mail route there were 2920 pieces of mail. If 45% of those were advertising pieces, find the number of advertising pieces.

(b) There are 8550 students attending a college. If 28% of the students drink coffee, find the number of coffee drinkers.

ANSWERS
3. (a) 1314 advertising pieces
    (b) 2394 coffee drinkers

④ Use the percent proportion to find the unknown whole.

(a) 150 tickets is 25% of what number of tickets?

(b) 28 antiques is 35% of what number of antiques?

(c) 387 customers is 36% of what number of customers?

(d) 292.5 miles is 37.5% of what number of miles?

(b) 135 tourists is 15% of what number of tourists?
The percent is 15 and part is 135, so

Part → $\dfrac{135}{x} = \dfrac{15}{100}$ ← Percent

$$\dfrac{135}{x} = \dfrac{3}{20} \qquad \text{Lowest terms}$$

$$x \cdot 3 = 135 \cdot 20 \qquad \text{Cross products}$$

$$x \cdot 3 = 2700$$

$$\dfrac{x \cdot \cancel{3}}{\cancel{3}} = \dfrac{2700}{3} \qquad \text{Divide each side by 3.}$$

$$x = 900.$$

135 tourists is 15% of **900 tourists**.

Work Problem ④ at the Side.

Example 5 **Applying the Percent Proportion**

At Newark Salt Works, 78 employees are absent because of illness. If this is 5% of the total number of employees, how many employees does the company have?

Step 1 **Read** the problem. The problem asks for the total number of employees.

Step 2 **Work out a plan.** From the information in the problem, the percent is 5 and the part of the total number of employees is 78. The total number of employees or entire quantity, which is the whole, is the unknown.

Step 3 **Estimate** a reasonable answer. Round the number of employees from 78 to 80. Then, 5% is equivalent to the fraction $\frac{1}{20}$. Since 80 is $\frac{1}{20}$ of the total number of employees,

$$80 \cdot 20 = 1600 \text{ employees} \leftarrow \text{Estimate}$$

Step 4 **Solve** the problem. Use the percent proportion to find the whole (the total number of employees).

Part → $\dfrac{78}{x} = \dfrac{5}{100}$ ← Percent

$$\dfrac{78}{x} = \dfrac{1}{20} \qquad \text{Lowest terms}$$

Find the cross products.

$$x \cdot 1 = 78 \cdot 20$$

$$x = 1560$$

Step 5 **State the answer.** The company has **1560 employees**.

Step 6 **Check.** The answer, 1560 employees, is close to our estimate of 1600 employees.

NOTE

To estimate the answer to Example 5, the 5% was changed to its fraction equivalent $\frac{1}{20}$. Because 80 (rounded) is $\frac{1}{20}$ of the total employees, 80 was multiplied by 20 to get 1600, the estimated answer.

Work Problem ❺ at the Side.

3 ▭ **Find the percent using the percent proportion.** Finally, if the part and the whole are known, the percent proportion can be used to find the percent.

Example 6 **Using the Percent Proportion to Find the Percent**

(a) 13 roofs is what percent of 52 roofs?
The whole is 52 (follows *of*) and the part is 13. Next, find the percent.

$$\frac{\text{part}}{\text{whole}} = \frac{\text{percent}}{100}$$

$$\text{Part} \rightarrow \frac{13}{52} = \frac{x}{100} \leftarrow \text{Percent (unknown)}$$
$$\text{Whole} \rightarrow$$

$$\frac{1}{4} = \frac{x}{100} \quad \text{Lowest terms}$$

Find the cross products.

$$4 \cdot x = 1 \cdot 100$$

$$\frac{\cancel{4} \cdot x}{\cancel{4}} = \frac{100}{4} \quad \text{Divide each side by 4.}$$

$$x = 25$$

13 roofs is **25%** of 52 roofs.

(b) What percent of $500 is $100?
The whole is 500 (follows *of*) and the part is 100, so

$$\frac{100}{500} = \frac{x}{100} \leftarrow \text{Percent (unknown)}$$

$$\frac{1}{5} = \frac{x}{100} \quad \text{Lowest terms}$$

$$5 \cdot x = 1 \cdot 100 \quad \text{Cross products}$$
$$5 \cdot x = 100$$

$$\frac{\cancel{5} \cdot x}{\cancel{5}} = \frac{100}{5} \quad \text{Divide each side by 5.}$$

$$x = 20$$

20% of $500 is $100.

❺ Use the six problem-solving steps and the percent proportion to solve each problem.

(a) A freeze resulted in a loss of 52% of an avocado crop. If the loss was 182 tons, find the total number of tons in the crop.

(b) A metal alloy contains 450 pounds of zinc, which is 8% of the alloy. Find the total weight of the alloy.

6 Use the percent proportion to solve each problem.

(a) $21 is what percent of $105?

(b) What percent of 320 Internet companies is 48 Internet companies?

(c) What percent of 2280 court trials is 1026 trials?

(d) 432 snowboarders is what percent of 108 snowboarders?

7 Solve each problem.

(a) The bid price on an auction item is $289 while the minimum acceptable price is $425. What percent of the minimum acceptable price is the bid price?

(b) A late-model domestic car gets 38 miles per gallon on the highway and 32.3 miles per gallon around town. What percent of the highway mileage does the car get around town?

CAUTION

When finding the percent, be sure to label your answer with the percent symbol (%).

Work Problem 6 at the Side.

Example 7 Applying the Percent Proportion

A roof is expected to last 20 years before needing replacement. If the roof is now 15 years old, what percent of the roof's life has been used?

Step 1 **Read** the problem. The problem asks for the percent of the roof's life that is already used.

Step 2 **Work out a plan.** The expected life of the roof is the entire quantity or whole, which is 20. The part of the roof's life that is already used is 15. Use the percent proportion to find the percent of the roof's life used.

Step 3 **Estimate** a reasonable answer. Since the roof is 15 years old, it is $\frac{15}{20}$ or $\frac{3}{4}$ used. Remember that $\frac{3}{4}$ is equivalent to 75%, our estimate.

Step 4 **Solve** the problem. Let x represent the unknown percent.

$$\text{Part} \to \frac{15}{20} = \frac{x}{100} \quad \text{or} \quad \frac{3}{4} = \frac{x}{100} \quad \text{Lowest terms}$$
$$\text{Whole} \to$$

Find the cross products.

$$4 \cdot x = 3 \cdot 100$$
$$4 \cdot x = 300$$

$$\frac{\cancel{4} \cdot x}{\cancel{4}} = \frac{300}{4} \quad \text{Divide each side by 4.}$$

$$x = 75$$

Step 5 **State the answer.** **75%** of the roof's life has been used.

Step 6 **Check.** The answer, 75%, matches our estimate of 75%.

Work Problem 7 at the Side.

Example 8 Applying the Percent Proportion

Rainfall this year was 33 in., while normal rainfall is only 30 in. What percent of normal rainfall is this year's rainfall?

Step 1 **Read** the problem. The problem asks us to find what percent this year's rainfall is of normal rainfall.

Step 2 **Work out a plan.** The normal rainfall is the whole, which is 30. This year's rainfall is all of normal rainfall and more, or 33 (part = 33). You need to find the percent that this year's rainfall is of normal rainfall.

Continued on Next Page

Step 3 **Estimate** a reasonable answer. The increase in rainfall is 3 inches and the whole is 30 inches. The increase is $\frac{3}{30} = \frac{1}{10}$ or 10%. The whole is 100%, so 100% + 10% = 110%, our estimate.

Step 4 **Solve** the problem. Let x represent the unknown percent.

$$\frac{33}{30} = \frac{x}{100} \quad \text{or} \quad \frac{11}{10} = \frac{x}{100} \quad \text{Lowest terms}$$

Find the cross products.

$$10 \cdot x = 11 \cdot 100$$
$$10 \cdot x = 1100$$

$$\frac{\overset{1}{\cancel{10}} \cdot x}{\underset{1}{\cancel{10}}} = \frac{1100}{10}$$

$$x = 110$$

Step 5 **State the answer.** This year's rainfall is **110%** of normal rainfall.

Step 6 **Check.** The answer, 110%, matches our estimate of 110%.

═══════════ **Work Problem ❽ at the Side.**

❽ Solve each problem.

(a) The number of students who usually take this class is 300. If 450 students are taking this class now, find the percent of the usual number who are now taking the class.

(b) The service department set a goal of 360 service calls this week. If they made 504 service calls, find the percent of their goal that they completed.

❷ Find the rate of sales tax.

    (a) The tax on a $190 table is $15.20.

Example 2   **Finding the Sales Tax Rate**

The sales tax on a $14,800 Honda Civic is $962. Find the rate of the sales tax.

Step 1   **Read** the problem. This problem asks us to find the sales tax rate.

Step 2   **Work out a plan.** Use the sales tax formula.

$$\text{sales tax} = \text{rate of tax} \cdot \text{cost of item}$$

    Solve for the rate of tax, which is the percent. The cost of the Honda Civic (the whole) is $14,800, and the amount of sales tax (the part) is $962. Use r for the *rate* of tax (the percent).

Step 3   **Estimate** a reasonable answer. Round $14,800 to $15,000 and round $962 to $1000. The sales tax is $\frac{1000}{15,000}$ or $\frac{1}{15}$ of the cost of the car. So divide 1 by 15 to find the percent (rate) of sales tax.

$$\frac{1}{15} = .06\overline{6} = 7\% \text{ Rounded estimate}$$

(b) The tax on a $24 sweatshirt is $1.56.

Step 4   **Solve** the problem.

$$\text{sales tax} = \text{rate of tax} \cdot \text{cost of item}$$
$$\$962 = \quad r \quad \cdot \quad \$14,800$$

$$\frac{962}{14,800} = \frac{\overset{1}{\cancel{14,800}} \cdot r}{\underset{1}{\cancel{14,800}}} \qquad \text{Divide each side by } 14,800.$$

$$0.065 = r$$
$$0.065 \text{ is } 6.5\% \qquad \text{Write the decimal as a percent.}$$

Step 5   **State the answer.** The sales tax rate is 6.5% or $6\frac{1}{2}\%$.

Step 6   **Check.** The answer, $6\frac{1}{2}\%$, is close to our estimate of 7%.

(c) The tax on $14,820 worth of light fixtures is $592.80.

NOTE

You can use the sales tax formula to find the amount of sales tax, the cost of an item, or the rate of sales tax (the percent).

Work Problem ❷ at the Side.

3   **Find commissions.**   Many salespeople are paid by *commission* rather than an hourly wage. If you are paid by **commission,** you are paid a certain percent of the total sales dollars. The following formula for finding the commission is based on the percent equation.

Commission Formula

Part	=	Percent	·	Whole
↓		↓		↓

amount of commission = rate of commission · amount of sales

Example 3 Determining the Amount of Commission

Scott Samuels had pharmaceutical sales of $42,500 last month. If his commission rate is 9%, find the amount of his commission.

Step 1 **Read** the problem. The problem asks for the amount of commission that Scott earned.

Step 2 **Work out a plan.** Use the commission formula. Write the rate of commission (9%) as a decimal (0.09). The amount of sales ($42,500) is the whole. Use c for the *amount* of commission.

Step 3 **Estimate** a reasonable answer. Round the commission rate of 9% to 10%. Round the amount of sales from $42,500 to $40,000. Since 10% is equivalent to $\frac{1}{10}$, divide $40,000 by 10 to find the amount of commission.

$$\$40,000 \div 10 = \$4000 \text{ Our estimate}$$

Step 4 **Solve** the problem.

amount of commission = rate of commission • amount of sales

$$c = 9\% \cdot \$42,500$$
$$c = 0.09 \cdot \$42,500$$
$$c = \$3825 \quad \text{Amount of commission}$$

Step 5 **State the answer.** Samuels earned a commission of $3825 for selling the pharmaceuticals.

Step 6 **Check.** The answer, $3825, is close to our estimate of $4000.

Work Problem ❸ at the Side.

Example 4 Finding the Rate of Commission

Chris Knudson earned a commission of $510 for selling $17,000 worth of shipping supplies. Find the rate of commission.

Step 1 **Read** the problem. In this problem, we must find the rate (percent) of commission.

Step 2 **Work out a plan.** You could use the commission formula. Another approach is to use the percent proportion. The *whole* is $17,000, the *part* is $510, and the *percent* is unknown. (The rate of commission is the percent.)

Step 3 **Estimate** a reasonable answer. Round the commission, $510, to $500. The commission in fraction form is $\frac{\$500}{\$20,000}$, which divides out to $\frac{5}{200}$. Changing $\frac{5}{200}$ to a percent gives $2\frac{1}{2}\%$ (rounded), our estimate.

Step 4 **Solve** the problem.

$$\frac{\text{part}}{\text{whole}} = \frac{x}{100} \longleftarrow \text{Percent (unknown)}$$

$$\frac{510}{17,000} = \frac{x}{100}$$

$$17,000 \cdot x = 510 \cdot 100 \quad \text{Cross multiply.}$$

$$\frac{\overset{1}{\cancel{17,000}} \cdot x}{\underset{1}{\cancel{17,000}}} = \frac{51,000}{17,000} \quad \text{Divide each side by 17,000.}$$

$$x = 3$$

Continued on Next Page

❸ Find the amount of commission.

(a) Jill Buteo sells dental equipment at a commission rate of 12% and has sales for the month of $28,750.

(b) Last month Pam Prentice sold $71,800 worth of furniture with a commission rate of 4%.

ANSWERS
3. (a) $3450 **(b)** $2872

④ Find the rate of commission.

(a) A commission of $450 is earned on one sale of computer products worth $22,500.

(b) Jamal Story earns $2898 for selling office furniture worth $32,200.

⑤ Find the amount of the discount and the sale price.

(a) An Easy-Boy leather recliner originally priced at $950 is offered at a 42% discount.

(b) WAL-MART has women's sweater sets on sale at 35% off. One sweater set was originally priced at $30.

Step 5   **State the answer.** The rate of commission is 3%.

Step 6   **Check.** The answer, 3%, is the same as our estimate of $2\frac{1}{2}$%.

Work Problem ④ at the Side.

3   **Find the discount and sale price.**   Most of us prefer buying things when they are on sale. A store will reduce prices, or **discount,** to attract additional customers. Use the following formula to find the discount and the sale price.

Discount Formula and Sale Price Formula

amount of discount = rate (or percent) of discount • original price

sale price = original price − amount of discount

Example 5   **Finding a Sale Price**

The Oak Mill Furniture Store has a home entertainment center with an original price of $840 on sale at 15% off. Find the sale price of the entertainment center.

Step 1   **Read** the problem. This problem asks for the price of an entertainment center after a discount of 15%.

Step 2   **Work out a plan.** This problem is solved in two steps. First, find the amount of the discount, that is, the amount that will be "taken off" (subtracted), by multiplying the original price ($840) by the rate of the discount (15%). The second step is to subtract the amount of discount from the original price. This gives you the sale price, which is what you will actually pay for the entertainment center.

Step 3   **Estimate** a reasonable answer. Round the original price from $840 to $800, and the rate of discount from 15% to 20%. Since 20% is equivalent to $\frac{1}{5}$, the discount is $800 ÷ 5 = $160. The sale price is $800 − $160 = $640, our estimate.

Step 4   **Solve** the problem. First find the amount of the discount.

amount of discount = rate of discount • original price

$a = 0.15 • \$840$ Write 15% as a decimal.

$a = \$126$ Amount of discount

Now find the sale price of the entertainment center by subtracting the amount of the discount ($126) from the original price.

sale price = original price − amount of discount

$= \$840 − \$126$

$= \$714$ Sale price

Step 5   **State the answer.** The sale price of the entertainment center is $714.

Step 6   **Check.** The answer, $714, is close to our estimate of $640.

Work Problem ⑤ at the Side.

 **Calculator Tip** In Example 5, you can use a calculator to find the amount of discount and subtract the discount from the original price directly.

$$840 \ominus .15 \otimes 840 \ominus 714$$

Original price · Amount of discount · Sale price

A scientific calculator observes the order of operations, so it will automatically do the multiplication before the subtraction.

4 ▭ **Find the percent of change.** We are often interested in looking at increases or decreases in sales, production, population, and many other areas. This type of problem involves finding the *percent of change*. Use the following steps to find the **percent of increase.**

Finding the Percent of Increase

Step 1 Use subtraction to find the amount of increase.

Step 2 Use the percent proportion to find the percent of increase.

$$\frac{\textbf{amount of increase (part)}}{\textbf{original value (whole)}} = \frac{\textbf{percent}}{\textbf{100}}$$

Example 6 Finding the Percent of Increase

Attendance at county parks climbed from 18,300 last month to 56,730 this month. Find the percent of increase.

Step 1 **Read** the problem. The problem asks for the percent of increase.

Step 2 **Work out a plan.** Subtract the attendance last month (18,300) from the attendance this month (56,730) to find the amount of increase in attendance. Next, use the percent proportion. The whole is 18,300 (last month's original attendance), the part is 38,430 (amount of increase in attendance), the percent is unknown.

Step 3 **Estimate** a reasonable answer. Round 18,300 to 20,000 and 56,730 to 60,000. The amount of increase is $60,000 - 20,000 = 40,000$. Since 40,000 (the increase) is *twice* as large as the original amount, the percent of increase is 200%, our estimate.

Step 4 **Solve** the problem.

$$56,730 - 18,300 = 38,430 \quad \text{Amount of increase in attendance}$$

$$\frac{38,430}{18,300} = \frac{x}{100} \quad \text{Percent proportion}$$

Solve this proportion to find that $x = 210$.

Step 5 **State the answer.** The percent of increase is 210%.

Step 6 **Check.** The answer, 210%, is close to our estimate of 200%.

═══ **Work Problem 6 at the Side.**

6 Find the percent of increase.

(a) A manufacturer of snowboards increased production from 14,100 units last year to 19,035 this year.

(b) The number of flu cases rose from 496 cases last week to 620 this week.

7 Find the percent of decrease.

Use the following steps to find the **percent of decrease.**

(a) The number of new trainees fell from 760 last month to 570 this month.

Finding the Percent of Decrease

Step 1 Use subtraction to find the amount of decrease.

Step 2 Use the percent proportion to find the percent of decrease.

$$\frac{\text{amount of decrease (part)}}{\text{original value (whole)}} = \frac{\text{percent}}{100}$$

Example 7 Finding the Percent of Decrease

The number of production employees this week fell to 1406 people from 1480 people last week. Find the percent of decrease.

Step 1 **Read** the problem. The problem asks for the percent of decrease.

Step 2 **Work out a plan.** Subtract the number of employees this week (1406) from the number of employees last week (1480) to find the amount of decrease. Next, use the percent proportion. The whole is 1480 (last week's original number of employees), the part is 74 (decrease in employees), and the percent is unknown.

Step 3 **Estimate** a reasonable answer. Estimate the answer by rounding 1406 to 1400 and 1480 to 1500. The decrease is $1500 - 1400 = 100$. Since 100 is $\frac{1}{15}$ of 1500 our estimate is $1 \div 15 \approx .07$ or 7%.

Step 4 **Solve** the problem.

$$1480 - 1406 = 74 \qquad \text{Decrease in number of employees}$$

$$\frac{74}{1480} = \frac{x}{100} \qquad \text{Percent proportion}$$

Solve this proportion to find that $x = 5$.

Step 5 **State the answer.** The percent of decrease is 5%.

Step 6 **Check.** The answer, 5%, is close to our estimate of 7%.

(b) The number of workers applying for unemployment fell from 4850 last month to 3977 this month.

CAUTION

When solving for percent of increase or decrease, the *whole is always the original value* or *value before the change occurred.* The part is the change in values, that is, how much something went up or went down.

Work Problem 7 at the Side.

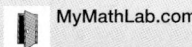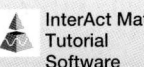
Find the amount of sales tax or the tax rate and the total cost (amount of sale + amount of tax = total cost). Round money answers to the nearest cent if necessary. See Examples 1 and 2.

Amount of Sale	Tax Rate	Amount of Tax	Total Cost
1. $8	5%	_____	_____
2. $15	4%	_____	_____
3. $425	_____	$12.75	_____
4. $322	_____	$19.32	_____
5. $284	_____	$14.20	_____
6. $84	_____	$5.88	_____
7. $12,229	$5\frac{1}{2}\%$	_____	_____
8. $11,789	$7\frac{1}{2}\%$	_____	_____

Find the commission earned or the rate of commission. Round money answers to the nearest cent if necessary. See Examples 3 and 4.

Sales	Rate of Commission	Commission
9. $400	10%	_____
10. $325	4%	_____
11. $3000	_____	$600
12. $7800	_____	$1170
13. $6183.50	3%	_____
14. $4416.70	7%	_____
15. $73,500	9%	_____
16. $55,800	6%	_____

Find the amount or rate of discount and the sale price after the discount. Round money answers to the nearest cent if necessary. See Example 5.

	Original Price	Rate of Discount	Amount of Discount	Sale Price
17.	$199.99	10%	_____	_____
18.	$29.95	15%	_____	_____
19.	$180	_____	$54	_____
20.	$38	_____	$9.50	_____
21.	$17.50	25%	_____	_____
22.	$76	60%	_____	_____
23.	$58.40	15%	_____	_____
24.	$99.80	30%	_____	_____

25. You are trying to decide between Company A paying a 10% commission and Company B paying an 8% commission. For which company would you prefer to work? Are there considerations other than commission rate that would be important to you? What would they be?

26. Give four examples of where you might use the percent of increase or the percent of decrease in your own personal activities. Think in terms of work, school, home, hobbies, and sports. Write an increase or a decrease problem about one of these four examples, then show how to solve it.

Solve each application problem. Round money answers to the nearest cent and rates to the nearest tenth of a percent if necessary. See Examples 1–7.

27. If the sales tax rate is 4% and today's sales at the Dollar Store are $1453, find the amount of sales tax collected.

28. An Exer-Cycle Machine sells for $590 plus 7% sales tax. Find the amount of sales tax.

29. Diamonds at Discounts sells a diamond engagement ring at 40% off the regular price. Find the sale price of a $\frac{1}{2}$-carat diamond ring normally priced at $1950.

30. During a January inventory sale, Heather Hall purchased a new car at 12% below sticker price. If the sticker price was $18,350, what was her cost?

31. An Anderson wood frame French door is priced at $1980 with a sales tax of $99. Find the rate of sales tax.

32. Textbooks for three classes cost $165 plus sales tax of $11.55. Find the sales tax rate.

33. Harley-Davidson, the only major U.S.-based motorcycle manufacturer, says that it expects to build 145,000 motorcycles this year, up from 131,000 last year. Find the percent of increase in production. (*Source:* Harley-Davidson, Inc.)

34. Americans are eating more fish. This year the average American will eat 15.5 pounds compared to only 12.5 pounds per year a decade ago. Find the percent of increase. (*Source: Consumer Reports.*)

35. The number of industrial accidents this month fell to 989 accidents from 1276 accidents last month. Find the percent of decrease.

36. The average number of hours worked in manufacturing jobs last week fell from 41.1 to 40.9. Find the percent of decrease.

37. A "60% off sale" begins today at Wanda's Women's Wear. What is the sale price of women's wool coats normally priced at $335?

38. What is the sale price of a $769 Kenmore washer/dryer set with a discount of 25%?

The weekly sales record of the top four sales people at the Family Shoe Store is shown in the table. Use this information to answer Exercises 39–42.

Employee	Sales	Rate of Commission	Commission
McKee, J.	$9480	3%	$78.40
Brown, D.	$10,730	3%	$321.90
Poznick, M.	$8840		$353.60
Washington, R.	$11,522		$576.10

39. Find the commission for McKee.

40. Find the commission for Brown.

41. What is the rate of commission for Poznick?

42. What is the rate of commission for Washington?

43. An 8-millimeter camcorder normally priced at $590 is on sale at 18% off. Find the discount and the sale price.

44. A Dodge Durango is offered at 17% off the manufacturer's suggested retail price. Find the discount and the sale price of this Durango, originally priced at $28,700.

45. The price per share of Toys Я Us stock fell from $35.50 to $33.50. Find the percent of decrease in price.

46. In the past five years, the cost of generating electricity from the sun has been brought down from 24 cents per kilowatt hour to 8 cents (less than the newest nuclear power plants). Find the percent of decrease.

47. College students are offered a 6% discount on a dictionary that sells for $18.50. If the sales tax is 6%, find the cost of the dictionary including the sales tax.

48. A personal computer and monitor priced at $698 is marked down 8%. If the sales tax is also 8%, find the cost of the computer and monitor including sales tax.

49. A real estate agent sells a house for $129,605. A sales commission of 6% is charged. The agent gets 55% of this commission. How much money does the agent get?

50. The local real estate agents' association collects a fee of 2% on all money received by its members. The members charge 6% of the selling price of a property as their fee. How much does the association get, if its members sell property worth a total of $8,680,000?

51. What is the total price of a ski boat with an original price of $10,214, if it is sold at a 15% discount? The sales tax rate is $3\frac{3}{4}$%.

52. A commercial security alarm system originally priced at $10,800 is discounted 22%. Find the total price of the system if the sales tax rate is $7\frac{1}{4}$%.

RELATING CONCEPTS (Exercises 53–58) **FOR INDIVIDUAL OR GROUP WORK**

Knowing how to use the percent equation is important when solving application problems involving sales tax. **Work Exercises 53–58 in order.**

53. The percent equation is

part = _____ • _____

54. The formula used to find sales tax is an application of the percent equation. The sales tax formula is

sales tax = _____ • _____

In the United States there are certain items on which an excise tax is charged in addition to a sales tax. A table of federal excise taxes is shown here. Use this table to answer Exercises 55–58. (Excise tax is calculated on the amount of the sale before sales tax is added.) Round answers to the nearest cent.

FEDERAL EXCISE TAXES

Product or Service	Rate	Product or Service	Rate
Telephone service	3%	Tires (by weight)	
Teletypewriter service	3%	Under 40 pounds	No tax
Air transportation	7.5%	40–69 pounds	15¢/pound over 40 pounds
International air travel	$12.20/person	70–89 pounds	$4.50 plus 30¢/pound over 70 pounds
Air freight	6.25%		
Coal		90 pounds and more	$10.50 plus 50¢/pound over 90 pounds
Underground (lower amount)	$1.10/ton or 4.4%		
Surface (lower amount)	55¢/ton or 4.4%	Truck and trailer, chassis and bodies	12%
Fishing rods	10%	Inland waterways fuel	24.4¢/gallon
Bows and arrows	12.4%	Ship passenger tax	$3/passenger
Gasoline	18.4¢/gallon	Luxury cars (amount over $36,000)	6%
Diesel fuel	24.4¢/gallon		
Aviation fuel	21.9¢/gallon		

Source: Publication 510, I.R.S., Excise Taxes for 1999.

55. Some archery equipment (bows and arrows) are priced at $123. Use the federal excise tax table and a sales tax rate of $6\frac{1}{2}$% to find the cost of the equipment, including both taxes. (Round to the nearest cent.)

56. Refer to Exercise 55. Calculate the two taxes separately and then add them together. Now, add the two tax rates together and then find the tax. Are your answers the same? Why or why not? (*Hint:* Recall the commutative and associative properties of multiplication.)

57. The price of an international airline ticket is $1248. Use the federal excise tax table and a sales tax rate of $7\frac{3}{4}$% to find the total cost of one ticket. (Excise tax is not charged on sales tax.)

58. Refer to Exercise 57. Can the federal excise tax be added to the sales tax rate to find the total tax? Why or why not?

6.7 SIMPLE INTEREST

When we open a savings account, we are actually lending money to the bank or credit union. It will in turn lend this money to individuals and businesses. These people then become borrowers. The bank or credit union pays a fee to the savings account holders and charges a higher fee to its borrowers. These fees are called *interest.*

Interest is a fee paid or a charge made for lending or borrowing money. The amount of money borrowed is called the **principal.** The charge for interest is often given as a percent, called the interest rate or **rate of interest.** The rate of interest is assumed to be *per year,* unless stated otherwise.

1 ▭ **Find the simple interest on a loan.** In most cases, interest on a loan is computed on the *original principal* and is called **simple interest.** We use the following **interest formula** to find simple interest.

Formula for Simple Interest

$$\text{interest} = \text{principal} \cdot \text{rate} \cdot \text{time}$$

The formula is usually written in letters.

$$I = p \cdot r \cdot t$$

NOTE

Simple interest is used for most short-term business loans, most real estate loans, and many automobile and consumer loans.

Example 1 **Finding Interest for a Year**

Find the interest on $5000 at 6% for 1 year.

The amount borrowed, or principal (p), is $5000. The interest rate (r) is 6%, which is 0.06 as a decimal, and the time of the loan (t) is 1 year. Use the formula.

$$I = p \cdot r \cdot t$$
$$I = 5000 \cdot (0.06) \cdot 1$$
$$I = 300$$

The interest is $300.

═══════ **Work Problem ❶ at the Side.**

Example 2 **Finding Interest for More Than a Year**

Find the interest on $4200 at 8% for three and a half years.

The principal (p) is $4200. The rate ($r$) is 8%, or 0.08 as a decimal, and the time (t) is $3\frac{1}{2}$ or 3.5 years. Use the formula.

$$I = p \cdot r \cdot t$$
$$I = 4200 \cdot (0.08) \cdot (3.5)$$
$$I = 1176$$

The interest is $1176.

═══════ **Work Problem ❷ at the Side.**

OBJECTIVES

1 ▭ Find the simple interest on a loan.
2 ▭ Find the total amount due on a loan.

❶ Find the interest.

(a) $1000 at 5% for 1 year

(b) $3650 at 4% for 1 year

❷ Find the interest.

(a) $820 at 6% for $3\frac{1}{2}$ years

(b) $4850 at 8% for $2\frac{1}{2}$ years

(c) $16,800 at 5% for $2\frac{3}{4}$ years

ANSWERS
1. (a) $50 (b) $146
2. (a) $172.20 (b) $970 (c) $2310

❸ Find the interest.

(a) $1800 at 8% for 4 months

(b) $28,000 at $9\frac{1}{2}$% for 3 months

❹ Find the total amount due on each loan.

(a) $3800 at $6\frac{1}{2}$% for 6 months

(b) $12,400 at 7% for 5 years

(c) $2400 at 11% for $2\frac{3}{4}$ years

CAUTION

It is best to change fractions of percents or fractions of years into their decimal form. In Example 2, the $3\frac{1}{2}$ years becomes **3.5** years.

Interest rates are given *per year.* For loan periods of less than one year, be careful to express time as a fraction of a year.

If time is given in months, for example, use a denominator of 12, because there are 12 months in a year. A loan for 9 months would be for $\frac{9}{12}$ of a year.

Example 3 Finding Interest for Less Than 1 Year

Find the interest on $840 at $8\frac{1}{2}$% for 9 months.

The principal is $840. The rate is $8\frac{1}{2}$% or 0.085, and the time is $\frac{9}{12}$ of a year. Use the formula $I = p \cdot r \cdot t$.

$$I = \underbrace{840 \cdot (0.085)} \cdot \frac{9}{12} \qquad \text{9 months} = \tfrac{9}{12} \text{ of a year.}$$

$$= 71.4 \quad \cdot \frac{3}{4} \qquad \tfrac{9}{12} \text{ in lowest terms is } \tfrac{3}{4}.$$

$$= \frac{(71.4) \cdot 3}{4}$$

$$= \frac{214.2}{4} = 53.55$$

The interest is $53.55.

Calculator Tip The calculator solution to Example 3 uses chain calculations.

840 ⊗ 0.085 ⊗ 9 ⊘ 12 ⊜ 53.55

Work Problem ❸ at the Side.

2 ▭ **Find the total amount due on a loan.** When a loan is repaid, the interest is added to the original principal to find the total amount due.

Formula for Amount Due

amount due = principal + interest

Example 4 Calculating the Total Amount Due

A loan of $3240 has been made at 12% for 3 months. Find the total amount due.

First find the interest, then add the principal and the interest to find the total amount due.

$$I = 3240 \cdot (0.12) \cdot \frac{3}{12} \qquad \text{3 months} = \tfrac{3}{12} \text{ of a year.}$$

$$I = \$97.20$$

The interest is $97.20.

amount due = principal + interest
= $3240 + $97.20 = $3337.20

The total amount due is $3337.20.

Work Problem ❹ at the Side.

6.7 EXERCISES

 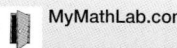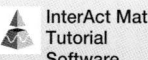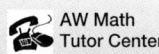
Find the interest. See Examples 1 and 2.

	Principal	Rate	Time in Years	Interest
1.	$100	6%	1	_____
2.	$200	3%	1	_____
3.	$700	5%	3	_____
4.	$900	4%	4	_____
5.	$240	4%	3	_____
6.	$190	3%	2	_____
7.	$2300	$8\frac{1}{2}\%$	$2\frac{1}{2}$	_____
8.	$4700	$5\frac{1}{2}\%$	$1\frac{1}{2}$	_____
9.	$10,800	$7\frac{1}{2}\%$	$2\frac{3}{4}$	_____
10.	$12,400	$6\frac{1}{2}\%$	$3\frac{3}{4}$	_____

Find the interest. Round to the nearest cent if necessary. See Example 3.

	Principal	Rate	Time in Months	Interest
11.	$400	5%	6	_____
12.	$600	7%	5	_____
13.	$820	6%	12	_____

Principal	Rate	Time in Months	Interest
14. $780	8%	24	_____
15. $940	3%	18	_____
16. $178	4%	12	_____
17. $1225	$5\frac{1}{2}\%$	3	_____
18. $2660	$7\frac{1}{2}\%$	3	_____
19. $15,300	$7\frac{1}{4}\%$	7	_____
20. $13,700	$3\frac{3}{4}\%$	11	_____

Find the total amount due on the following loans. Round to the nearest cent if necessary. See Example 4.

Principal	Rate	Time	Total Amount Due
21. $200	5%	1 year	_____
22. $400	4%	6 months	_____
23. $740	6%	9 months	_____
24. $1180	3%	2 years	_____

	Principal	Rate	Time	Total Amount Due
25.	$1800	9%	18 months	_____
26.	$9000	6%	7 months	_____
27.	$3250	10%	6 months	_____
28.	$7600	5%	1 year	_____
29.	$16,850	$7\frac{1}{2}\%$	9 months	_____
30.	$19,450	$5\frac{1}{2}\%$	6 months	_____

31. The amount of interest paid on savings accounts and charged on loans can vary from one institution to another. However, when the amount of interest is calculated, three factors are used in the calculation. Name these three factors and describe them in your own words.

32. Interest rates are usually given as a rate per year (annual rate). Explain what must be done when time is given in months. Write your own problem where time is given in months and then show how to solve it.

Solve each application problem. Round to the nearest cent if necessary.

33. Reann Chang deposits $825 at 5% for 1 year. How much interest will she earn?

34. The Jidobu family invests $18,000 at 9% for 6 months. What amount of interest will the family earn?

35. The Bank of Boston loans $150,000 to a business at 10% for 18 months. How much interest will the bank earn?

36. Pat Martin, a retiree, deposits $80,000 at 7% for 3 years. How much interest will he earn?

4 Find the compound amount and the interest.

(a) $4000 at 8% for 10 years

(b) $12,600 at $5\frac{1}{2}$% for

8 years

(c) $32,700 at 6% for 12 years

5 Find the compound amount and the amount of compound interest. Find the compound amount and interest as follows.

Finding the Compound Amount and the Interest

Compound Amount
Find the compound amount for any amount of principal by multiplying the principal by the compound amount for $1.

Interest
Find the amount of interest earned on a deposit by subtracting the amount originally deposited from the compound amount.

Example 4 Finding Compound Amount and Interest

Find the compound amount and the interest.

(a) $1000 at $5\frac{1}{2}$% interest for 12 years
Look in the table for $5\frac{1}{2}$% (5.50%) and 12 periods; find the number 1.9012 but do *not* round it. Multiply this number and the principal of $1000.

$$\$1000 \cdot (1.9012) = \$1901.20$$

The account will contain $1901.20 after 12 years.
Find the amount of interest by subtracting the original deposit from the compound amount.

Compound amount ——— Original amount ——— Amount of interest

$$\$1901.20 - \$1000 = \$901.20$$

(b) $6400 at 8% for 7 years
Look in the table for 8% and 7 periods, finding 1.7138. Multiply.

$$\$6400 \cdot (1.7138) = \$10,968.32 \quad \text{Compound amount}$$

Subtract the original deposit from the compound amount.

$$\$10,968.32 - \$6400 = \$4568.32 \quad \text{Interest}$$

A total of $4568.32 in interest was earned.

Work Problem 4 at the Side.

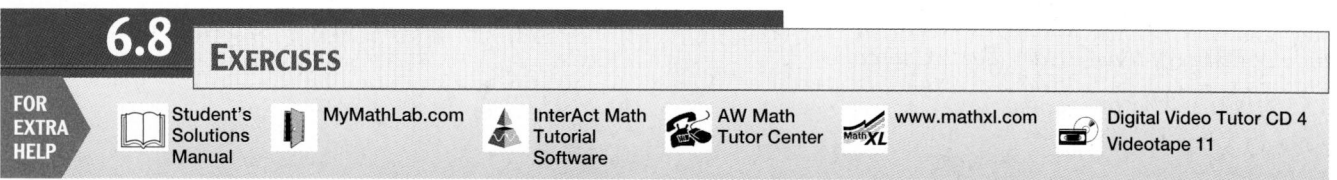

| FOR EXTRA HELP | | Student's Solutions Manual | | MyMathLab.com | | InterAct Math Tutorial Software | | AW Math Tutor Center | MathXL | www.mathxl.com | | Digital Video Tutor CD 4 Videotape 11 |

Find the compound amount given the following deposits. Calculate the interest each year, then add it to the previous year's amount. See Example 1.

1. $500 at 4% for 2 years

2. $1000 at 5% for 3 years

3. $1800 at 3% for 3 years

4. $2000 at 8% for 3 years

5. $3500 at 7% for 4 years

6. $5500 at 6% for 4 years

Find each compound amount by multiplying the original amount deposited by 100% plus the compound rate in the following. See Example 2. Round to the nearest cent if necessary.

7. $1000 at 5% for 2 years

8. $500 at 4% for 3 years

9. $1400 at 6% for 5 years

10. $1800 at 8% for 4 years

11. $1180 at 7% for 8 years

12. $12,800 at 6% for 7 years

13. $10,940 at 8% for 6 years

14. $15,710 at 10% for 8 years

Use the table on page 439 to find the compound amount and the interest. Interest is compounded annually. Round to the nearest cent if necessary. See Example 3.

15. $1000 at 4% for 5 years

16. $10,000 at 3% for 4 years

17. $8000 at 6% for 10 years

18. $7800 at 5% for 8 years

19. $8428.17 at $4\frac{1}{2}$% for 6 years

20. $10,472.88 at $5\frac{1}{2}$% for 12 years

21. Write a definition for compound interest. Describe in your own words what compound interest means to you.

22. What is the difference between the compound amount and compound interest?

Use the table on page 439 to solve each application problem. Round to the nearest cent if necessary. See Examples 3 and 4.

23. Jane Chavez deposited $8450 in an account that pays 6% interest compounded annually. Find the amount she will have (compound amount) at the end of 8 years.

24. Rob Diamond borrowed $22,500 from his uncle to open Bookkeeping and More. He will repay the loan at the end of 5 years at 8% interest compounded annually. Find the amount he will repay.

25. Al Granard lends $38,000 to the owner of Rick's Limousine Service. He will be repaid at the end of 9 years at 6% interest compounded annually. Find **(a)** the total amount that he should be repaid and **(b)** the amount of interest earned.

26. Sadie Simms has $28,500 in an Individual Retirement Account (IRA) that pays 5% interest compounded annually. Find **(a)** the total amount she will have at the end of 5 years and **(b)** the amount of interest earned.

27. Jennifer Barrister deposits $30,000 at 6% interest compounded annually. Two years after she makes the first deposit, she adds another $40,000, also at 6% compounded annually.

(a) What total amount will she have 5 years after her first deposit?

(b) What amount of interest will she have earned?

28. Kara Ivee invests $25,000 at 8% interest compounded annually. Three years after she makes the first deposit, she adds another $25,000, also at 8% compounded annually.

(a) What total amount will she have 5 years after her first deposit?

(b) What amount of interest will she have earned?

| RELATING CONCEPTS (Exercises 29–34) | FOR INDIVIDUAL OR GROUP WORK |

Knowing how to solve interest problems is important to business people and consumers alike.
Work Exercises 29–34 in order.

29. Simple interest calculation is used for most short-term business loans, most real estate loans, and many automobile and consumer loans. The formula for simple interest is

   Interest = _____ • _____ • _____

   or $I =$ _____ • _____ • _____ .

30. When a loan is repaid, the interest is added to the original principal. The formula for amount due is

   amount due = _____ + _____

31. Compound interest is paid on most savings accounts and many other types of investments. Compound interest is interest calculated on _____ plus past _____ .

32. The compound amount is the total amount in an account at the end of a period of time. Compound amount is the original _____ + compound _____ .

33. Dick Gonsalves has two choices. He can invest $4350 at 6% simple interest for 6 years, or he can invest the same amount at 6% compounded annually for 6 years.

   (a) Find the difference in the amount of interest earned in these two accounts.

   (b) If the length of time is doubled from 6 to 12 years, will the difference in the amount of interest earned also double?

   (c) Use your own examples to determine that what you found in part (b) is true with other interest rates and other lengths of time.

34. One account is opened with $20,000 at 8% simple interest for 12 years. Another account is opened with $16,000 at 8% compounded annually for 12 years.

   (a) At the end of the 12 years, which account has the higher balance and by how much?

   (b) What does this tell you about compound interest? Why?

Chapter 6

SUMMARY

KEY TERMS

6.1	**percent**	Percent means per one hundred. A percent is a ratio with a denominator of 100.
6.3	**percent proportion**	The proportion $\dfrac{\text{part}}{\text{whole}} = \dfrac{\text{percent}}{100}$ is used to solve percent problems.
	whole	The whole in a percent problem is the entire quantity, the total, or the base.
	part	The part in a percent problem is the portion being compared with the whole.
6.5	**percent equation**	The percent equation is part = percent • whole. It is another way to solve percent problems.
6.6	**sales tax**	Sales tax is a percent of the total sales charged as a tax.
	commission	Commission is a percent of the dollar value of total sales paid to a salesperson.
	discount	Discount is often expressed as a percent of the original price; it is then deducted from the original price, resulting in the sale price.
	percent of increase or decrease	Percent of increase or decrease is the amount of increase or decrease expressed as a percent of the original amount.
6.7	**interest**	Interest is a fee paid or a charge for lending or borrowing money.
	interest formula	The interest formula is used to calculate interest. It is interest = principal • rate • time or $I = p \bullet r \bullet t$.
	simple interest	Interest that is computed on the original principal is simple interest.
	principal	Principal is the amount of money on which interest is earned.
	rate of interest	Often referred to as "rate," it is the charge for interest and is given as a percent.
6.8	**compound interest**	Compound interest is interest paid on past interest as well as on principal.
	compound amount	The total amount in an account including compound interest and the original principal.
	compounding	Interest that is compounded once each year is compounded annually.

NEW FORMULAS

To write percents as decimals: $p\% = p \div 100$

To write percents as fractions: $p\% = \dfrac{p}{100}$

Percent proportion: $\dfrac{\text{part}}{\text{whole}} = \dfrac{\text{percent}}{100}$

Percent equation: part = percent • whole

Amount of sales tax: amount of sales tax = rate of tax • cost of item

Amount of commission: amount of commission = rate of commission • amount of sales

Amount of discount: amount of discount = rate of discount • original price

Sale price: sale price = original price − amount of discount

Percent of increase: $\dfrac{\text{percent}}{100} = \dfrac{\text{amount of increase}}{\text{original value}}$

Percent of decrease: $\dfrac{\text{percent}}{100} = \dfrac{\text{amount of decrease}}{\text{original value}}$

Interest: $I = p \bullet r \bullet t$ (principal • rate • time)

Amount due: amount due = principal + interest

TEST YOUR WORD POWER

See how well you have learned the vocabulary in this chapter. Answers follow the Quick Review.

1. To write a **percent as a decimal,** you drop the percent sign
   (a) after finding the decimal point
   (b) after removing the decimal point
   (c) after moving the decimal point two places to the right
   (d) after moving the decimal point two places to the left.

2. To write a **decimal as a percent,** you add the percent sign
   (a) after finding the decimal point
   (b) after removing the decimal point
   (c) after moving the decimal point two places to the right
   (d) after moving the decimal point two places to the left.

3. **Percent** means
   (a) the same as interest
   (b) per every ten
   (c) per one thousand
   (d) per one hundred.

4. When you use
   $$\frac{\text{part}}{\text{whole}} = \frac{\text{percent}}{100}$$ to solve
   percent problems, you are using the
   (a) simple interest formula
   (b) percent proportion
   (c) percent equation
   (d) percent sales tax formula.

5. The **percent equation** is
   (a) part = percent • whole
   (b) $I = p \cdot r \cdot t$
   (c) $p\% = \dfrac{p}{100}$
   (d) amount due = principal + interest.

6. In a **percent of increase or decrease** problem, the increase or decrease is a percent of
   (a) the largest amount
   (b) the original amount
   (c) the new or most recent amount
   (d) all the amounts.

7. In the formula $I = p \cdot r \cdot t$, the p stands for
   (a) proportion
   (b) product
   (c) principal
   (d) percent.

8. The term **rate** in an interest problem represents the
   (a) whole
   (b) percent
   (c) part
   (d) amount of interest.

QUICK REVIEW

Concepts	Examples
6.1 Basics of Percent	
Writing a Percent as a Decimal To write a percent as a decimal, move the decimal point two places to the left and drop the % sign.	50% (.50%) = 0.50 or just 0.5 3% (.03%) = 0.03
Writing a Decimal as a Percent To write a decimal as a percent, move the decimal point two places to the right and attach a % sign.	0.75 (0.75) = 75% 3.6 (3.60) = 360%
6.2 Writing a Fraction as a Percent Use a proportion and solve for p to change a fraction to percent.	$\dfrac{2}{5} = \dfrac{p}{100}$ Proportion $5 \cdot p = 2 \cdot 100$ Cross products $5 \cdot p = 200$ $\dfrac{\overset{1}{\cancel{5}} \cdot p}{\underset{1}{\cancel{5}}} = \dfrac{200}{5}$ Divide each side by 5. $p = 40$ $\dfrac{2}{5} = 40\%$ Attach % sign.

Concepts	Examples
6.3 *Learning the Percent Proportion* Part is to whole as percent is to 100. $$\frac{\text{part}}{\text{whole}} = \frac{\text{percent}}{100}$$	Use the percent proportion to solve for the unknown value. part = 30, whole = 50; find the percent. Percent (unknown) $$\frac{30}{50} = \frac{x}{100} \qquad \text{Percent proportion}$$ $$\frac{3}{5} = \frac{x}{100} \qquad \text{Lowest terms}$$ $$5 \cdot x = 3 \cdot 100 \qquad \text{Cross products}$$ $$5 \cdot x = 300$$ $$\frac{\overset{1}{\cancel{5}} \cdot x}{\underset{1}{\cancel{5}}} = \frac{300}{5} \qquad \text{Divide each side by 5.}$$ $$x = 60$$ The percent is 60, which is 60%.
6.3 *Identifying Percent, Whole, and Part in a Percent Problem* The percent appears with the word **percent** or with the symbol %. The whole often appears after the word **of**. The whole is the entire quantity or total. The part is the portion of the total. If the percent and the whole are found first, the remaining number is the part.	Find the percent, whole, and part in the following. 10% of the 500 pies is how many pies? Percent / Whole / Part (unknown) 20 cats is 5% of what number of cats? Part / Percent / Whole (unknown) What percent of \$220 is \$33? Percent (unknown) / Whole / Part
6.4 *Applying the Percent Proportion* Read the problem and identify the percent, whole, and part. Use the percent proportion to solve for the unknown quantity (whole).	A tank contains 35% distilled water. If 28 gallons of distilled water are in the tank when it is full, find the capacity of the tank. $$\text{percent} = 35 \quad \text{and} \quad \text{part} = 28$$ Use the percent proportion to find the whole. Whole (unknown) → $$\frac{\text{part}}{x} = \frac{\text{percent}}{100}$$ $$\frac{28}{x} = \frac{35}{100}$$ $$\frac{28}{x} = \frac{7}{20} \qquad \text{Lowest terms}$$ $$x \cdot 7 = 560 \qquad \text{Cross products}$$ $$\frac{x \cdot \overset{1}{\cancel{7}}}{\underset{1}{\cancel{7}}} = \frac{560}{7} \qquad \text{Divide each side by 7.}$$ $$x = 80$$ The capacity of the tank is 80 gallons.

Concepts	*Examples*

6.5 *Using the Percent Equation*

The percent equation is part = percent • whole. Identify the percent, whole, and part and solve for the unknown quantity. Always write the percent as a decimal before using the equation.

Solve each problem.

(a) Find 20% of 220 applicants.

$$\text{part (unknown)} = \text{percent} \cdot \text{whole}$$
$$x = 0.2 \cdot 220$$
$$x = 44$$

20% of 220 applicants is 44 applicants.

(b) 8 balls is 4% of what number of balls?

$$\text{part} = \text{percent} \cdot \text{whole (unknown)}$$
$$8 = 0.04 \cdot x$$
$$\frac{8}{0.04} = \frac{\overset{1}{\cancel{0.04}} \cdot x}{\underset{1}{\cancel{0.04}}}$$
$$x = 200$$

8 balls is 4% of 200 balls.

(c) $13 is what percent of $52?

$$\text{part} = \text{percent (unknown)} \cdot \text{whole}$$
$$13 = x \cdot 52$$
$$\frac{13}{52} = \frac{x \cdot \overset{1}{\cancel{52}}}{\underset{1}{\cancel{52}}}$$
$$x = 0.25 = 25\%$$

$13 is 25% of $52.

6.6 *Solving Application Problems with Proportions*

To solve for **sales tax**, use the formula

amount of sales tax = rate of tax • cost of item.

The cost of a 35-inch television is $699, and the sales tax is 5%. Find the sales tax.

amount of sales tax = 5% • $699
$$= 0.05 \cdot \$699 = \$34.95$$

To find **commissions**, use the formula

amount of commission = rate of commission • amount of sales.

The sales are $92,000 with a commission rate of 3%. Find the commission.

amount of commission = 3% • $92,000
$$= 0.03 \cdot \$92,000$$
$$= \$2760$$

To find the **discount** and the **sale price**, use these formulas

amount of discount = rate of discount • original price

sale price = original price − amount of discount.

A gas oven originally priced at $480 is offered at a 25% discount. Find the amount of the discount and the sale price.

discount = 0.25 • $480 = $120

sale price = $480 − $120 = $360

Concepts	Examples
6.6 *Solving Application Problems with Proportions* (*continued*) To find the **percent of change,** subtract to find the amount of change (increase or decrease), which is the part. The whole is the original value or value before the change.	The number of parking violations rose from 1980 violations to 2277. Find the percent of increase. $$2277 - 1980 = 297 \quad \text{Increase}$$ $$\frac{297}{1980} = \frac{\text{percent}}{100}$$ Solve the proportion to find that the percent = 15, so the percent of increase is 15%.

6.7 *Finding Simple Interest*

Use the formula $I = p \cdot r \cdot t$.

$$\textbf{interest} = \textbf{principal} \cdot \textbf{rate} \cdot \textbf{time}$$

Time (t) is in years. When the time is given in months, use a fraction with 12 in the denominator because there are 12 months in a year.

$2800 is deposited at 8% for 3 months. Find the amount of interest.

$$I = \quad p \quad \cdot \quad r \quad \cdot \quad t$$

$$= \underbrace{2800 \cdot 0.08} \cdot \frac{3}{12}$$

$$= \quad 224 \quad \cdot \frac{1}{4} = \frac{224 \cdot 1}{4} = \$56$$

6.8 *Finding Compound Amount and Compound Interest*

There are three methods for finding the compound amount.

1. Calculate the interest for each compound interest period, then add it back to the principal.

2. Multiply the original deposit by 100% plus the compound interest rate.

3. Use the table to find the interest on $1. Then, multiply the table value by the principal.

The compound interest is found with the formula

$$\begin{matrix} \textbf{compound} \\ \textbf{interest} \end{matrix} = \begin{matrix} \textbf{compound} \\ \textbf{amount} \end{matrix} - \begin{matrix} \textbf{original} \\ \textbf{deposit} \end{matrix}$$

Find the compound amount and interest if $1500 is deposited at 5% interest for 3 years.

		Compound Interest	Compound Amount
1.	Year 1	$1500 • (0.05) • 1 = $75	
			$1500 + $75 = $1575
	Year 2	$1575 • (0.05) • 1 = $78.75	
			$1575 + $78.75 = $1653.75
	Year 3	$1653.75 • (0.05) ≈ $82.69	
			$1653.75 + $82.69 = $1736.44

2. $1500 • $\underbrace{(1.05) \cdot (1.05) \cdot (1.05)}$ ≈ $1736.44

  ↑ Original deposit   ↑ Compound amount

       100% + 5% = 105% = 1.05

3. Locate 5% across the top of the table and 3 periods at the left. The table value is 1.1576.

compound amount = $1500 • (1.1576) = $1736.40*

interest = $1736.40 − $1500 = $236.40

* The difference in the compound amount results from rounding in the table.

ANSWERS TO TEST YOUR WORD POWER

1. **(d)** *Example:* 50% written as a decimal is 0.50 or 0.5. **2.** **(c)** *Example:* 0.25 written as a percent is 0.25% or 25%. **3.** **(d)** *Example:* 8% means 8 per 100. **4.** **(b)** *Example:* Part = 4, and whole = 25. To find the percent, $\frac{4}{25} = \frac{x}{100}$; $x = 16$ or 16% (percent). **5.** **(a)** *Example:* Percent = 25, and whole = 300. To find the part, $0.25 \times 300 = 75$ (part). **6.** **(b)** *Example:* Original = 200, and increase or decrease = 40. To find the percent of increase or decrease, $\frac{40}{200} = 0.2 = 20\%$ (percent of increase or decrease). **7.** **(c)** *Example:* Principal (p) = \$800, rate (r) = 5%, and time (t) = 1 year. Then, $I = p \cdot r \cdot t$ so $I = \$800 \cdot 0.05 \cdot 1 = \$40$ (interest). **8.** **(b)** *Example:* Principal (p) = \$1650, rate (r) = 4%, and time (t) = $\frac{1}{2}$ year. To find the interest (I),

$$I = \$1650 \cdot 0.04 \cdot \frac{1}{2} = \$33 \text{ (interest)}.$$

Chapter 6 REVIEW EXERCISES

[6.1] *Write each percent as a decimal and each decimal as a percent.*

1. 35% **2.** 150% **3.** 99.44% **4.** 0.085%

5. 3.15 **6.** 0.02 **7.** 0.875 **8.** 0.002

[6.2] *Write each percent as a fraction or mixed number in lowest terms and each fraction as a percent.*

9. 15% **10.** 37.5% **11.** 175% **12.** 0.25%

13. $\dfrac{3}{4}$ **14.** $\dfrac{5}{8}$ **15.** $3\dfrac{1}{4}$ **16.** $\dfrac{1}{200}$

Complete this chart.

Fraction	Decimal	Percent
$\dfrac{1}{8}$	**17.** _____	**18.** _____
19. _____	0.25	**20.** _____
21. _____	**22.** _____	180%

[6.3] *Find the unknown value in the percent proportion* $\dfrac{part}{whole} = \dfrac{percent}{100}$.

23. part = 25, percent = 10

24. whole = 480, percent = 5

Identify each percent, whole, and part. Do not try to solve.

25. 35% of 820 mailboxes is 287 mailboxes.

26. 73 brooms is what percent of 90 brooms?

27. Find 14% of 160 bicycles.

28. 418 curtains is 16% of what number of curtains?

29. A golfer lost three of his eight golf balls. What percent were lost?

30. Only 88% of the door keys cut will operate properly. If there are 1280 keys cut, find the number of keys that will operate properly.

[6.4] *Find the part using the percent proportion or the multiplication shortcut.*

31. 18% of 950 programs

32. 60% of 1450 reference books

33. 0.6% of 5200 acres

34. 0.2% of 1400 kilograms

Find the whole using the percent proportion.

35. 105 crates is 14% of what number of crates?

36. 348 test tubes is 15% of what number of test tubes?

37. 677.6 miles is 140% of what number of miles?

38. 2.5% of what number of cases is 425 cases?

Find the percent using the percent proportion. Round percent answers to the nearest tenth if necessary.

39. 649 tulip bulbs is what percent of 1180 tulip bulbs?

40. What percent of 1620 dinner rolls is 85 dinner rolls?

41. What percent of 380 pairs of socks is 36 pairs?

42. What percent of 650 soup cans is 200 soup cans?

[6.1–6.4] *Solve each application problem. Round percent answers to the nearest tenth if necessary.*

43. Eight years ago there were 112,000 high school math teachers in the United States. Since that time the number of high school math teachers has increased by 25%. How many high school math teachers are there today? (*Source: USA Today.*)

44. Scientists tell us that there are 9600 species of birds and that 1000 of these species are in danger of extinction. What percent of the bird species are in danger of extinction?

[6.5] *Use the percent equation to answer each question.*

45. 32% of $454 is what amount?

46. 155% of 120 trucks is how many trucks?

47. 0.128 ounces is what percent of 32 ounces?

48. 304.5 meters is what percent of 174 meters?

49. 33.6 miles is 28% of what number of miles?

50. $92 is 16% of what number?

[6.6] *Find the amount of sales tax or the tax rate and the total cost. Round to the nearest cent if necessary.*

Amount of Sale	Tax Rate	Amount of Tax	Total Cost
51. $630	5%	_____	_____
52. $780	_____	$58.50	_____

Find the commission earned or the rate of commission.

Sales	Rate of Commission	Commission
53. $3450	8%	_____
54. $65,300	_____	$3265

Find the amount or rate of discount and the amount paid after the discount. Round to the nearest cent if necessary.

Original Price	Rate of Discount	Amount of Discount	Sale Price
55. $112.50	30%	_____	_____
56. $252	_____	$63	_____

[6.7] *Find the simple interest due on each loan.*

	Principal	Rate	Time in Years	Interest
57.	$200	4%	1	_____
58.	$1080	5%	$1\frac{1}{4}$	_____

Find the simple interest paid on each investment.

	Principal	Rate	Time in Months	Interest
59.	$400	7%	3	_____
60.	$1560	$6\frac{1}{2}\%$	18	_____

Find the total amount due on the following simple interest loans.

	Principal	Rate	Time	Total Amount Due
61.	$750	$5\frac{1}{2}\%$	2 years	_____
62.	$1530	6%	9 months	_____

[6.8] *Find the compound amount and compound interest in the following. Interest is compounded annually. You may use the table on page 439. Round to the nearest cent if necessary.*

	Principal	Rate	Time in Years	Compound Amount	Compound Interest
63.	$2000	5%	10	_____	_____
64.	$1870	4%	4	_____	_____
65.	$3600	8%	3	_____	_____
66.	$12,500	$5\frac{1}{2}\%$	5	_____	_____

MIXED REVIEW EXERCISES

Find the unknown value in the percent proportion $\dfrac{part}{whole} = \dfrac{percent}{100}$.

67. whole = 80, percent = 15

68. part = 738, percent = 45

Use the percent proportion or equation to answer each question.

69. 12% of 194 meters is how many meters?

70. 327 cars is what percent of 218 cars?

71. 0.6% of $85 is what amount?

72. 396 employees is 20% of what number of employees?

73. 76 chickens is what percent of 190 chickens?

74. 214.484 liters is 43% of what number of liters?

Write each percent as a decimal and each decimal as a percent.

75. 55%

76. 300%

77. 5

78. 4.71

79. 8.6%

80. 0.621

81. 0.375%

82. 0.0006

Write each percent as a fraction in lowest terms and each fraction as a percent.

83. $\dfrac{3}{4}$

84. 42%

85. 87.5%

86. $\dfrac{3}{8}$

87. $32\dfrac{1}{2}\%$

88. $\dfrac{3}{5}$

89. 0.25%

90. $3\dfrac{3}{4}$

Solve each application problem. Round percent answers to the nearest tenth and money amounts to the nearest cent if necessary.

91. Jim Bralley invests $16,850 at $10\dfrac{1}{2}\%$ for 9 months. Find the amount of interest earned.

92. Karl Schmidt borrows $14,750 at 12% for 18 months to buy a small Lionel train collection. Find the total amount due.

93. A Hotpoint refrigerator has a capacity of 11.5 cubic feet in the refrigerator and 5.5 cubic feet in the freezer. What percent of the total capacity is the capacity of the freezer?

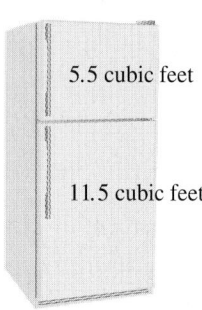

5.5 cubic feet

11.5 cubic feet

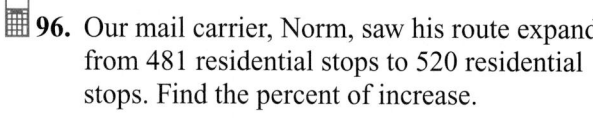 **94.** Tommy Downs invests the money he inherited from his aunt at 6% compounded annually for 4 years. If the amount of money invested is $12,500, find

    (a) the compound amount at the end of 4 years and

    (b) the amount of interest that he earned. Do not use the table.

95. Tom Dugally, a real estate agent, sold two properties, one for $125,000 and the other for $290,000. After all of his expenses, he receives a commission of $1\frac{1}{2}$ % of total sales. Find the commission that he earned.

96. Our mail carrier, Norm, saw his route expand from 481 residential stops to 520 residential stops. Find the percent of increase.

 **97.** A digital camera priced at $598 is marked down 7% to promote the new model. If the sales tax is also 7%, find the cost of the digital camera including sales tax.

98. The DaimlerChrysler Company hopes to sell 35,000 of the new Chrysler PT Cruisers each year in foreign markets. If this amounts to 18.9% of the total annual PT Cruiser sales, find the total predicted annual sales. Round to the nearest whole number. (*Source:* Daimler-Chrysler.)

99. Stephen and Heather Hall established a budget allowing 25% for rent, 30% for food, 8% for clothing, 20% for travel and recreation, and the remainder for savings. Stephen takes home $2075 per month, and Heather takes home $32,500 per year. How much money will the couple save in a year?

100. The mileage on a car dropped from 32.8 miles per gallon to 28.5 miles per gallon. Find the percent of decrease.

Chapter 6 TEST

 Study Skills Workbook
Activity 12

Write each percent as a decimal and each decimal as a percent.

1. 65%   **2.** 0.8

3. 1.75   **4.** 0.875

5. 300%   **6.** 0.05%

Write each decimal as a fraction in lowest terms.

7. 12.5%   **8.** 0.25%

Write each fraction or mixed number as a percent.

9. $\dfrac{3}{5}$ **10.** $\dfrac{5}{8}$

11. $2\dfrac{1}{2}$

Solve each problem.

12. 32 sacks is 4% of what number of sacks?

13. $680 is what percent of $3400?

14. Ann Barnes has saved 65% of the amount needed for a down payment on a home. If she has saved $12,025, find the total down payment needed.

15. The price of a copy machine is $2680 plus sales tax of $6\dfrac{1}{2}$%. Find the total cost of the copy machine including sales tax.

16. An insurance company pays its salespeople a commission of 8% on all sales. Find the commission earned on insurance sales of $7850.

1. _____

2. _____

3. _____

4. _____

5. _____

6. _____

7. _____

8. _____

9. _____

10. _____

11. _____

12. _____

13. _____

14. _____

15. _____

16. _____

17. _____

17. Attendance at the homecoming game increased from 4320 fans last year to 5616 fans this year. Find the percent of increase.

18. _____

18. A problem includes last year's salary, this year's salary, and asks for the percent of increase. Explain how you would identify the part, the whole, and the percent in the problem. Show the percent proportion that you would use.

19. _____

19. Write the formula used to find interest. Explain the difference in what to do if the time is expressed in months or in years. Write a problem that involves finding interest for 9 months and another problem that involves finding interest for $2\frac{1}{2}$ years. Use your own numbers for the principal and the rate. Show how to solve your problems.

Find the amount of discount and the sale price. Round answers to the nearest cent.

	Original Price	**Rate of Discount**
20.	$96	12%
21.	$280	32.5%

20. _____

21. _____

Find the simple interest on each loan.

	Principal	**Rate**	**Time**
22.	$4200	6%	$1\frac{1}{2}$ years
23.	$6400	5%	9 months

22. _____

23. _____

24. _____

24. A parent borrows $5300 to help her son start college. The loan is for 9 months at 9% interest. Find the total amount due on the loan.

25. (a) _____

   (b) _____

25. The River City School PTA Emergency Fund deposited $4000 at 6% compounded annually. Two years after the first deposit, they add another $5000, also at 6% compounded annually. Use the compound interest table.

   (a) What total amount will they have 4 years after their first deposit? Round to the nearest dollar.

   (b) What amount of interest will they have earned?

Round the numbers in each problem using front end rounding. Then add, subtract, multiply, or divide the rounded numbers, as indicated, to estimate the answer. Finally, solve for the exact answer.

1. *Estimate:*   *Exact:*
$$5608$$
$$94$$
$$+ \quad \qquad + 739$$

2. *Estimate:*   *Exact:*
$$0.56$$
$$49.614$$
$$+ \quad \qquad + 8.4$$

3. *Estimate:*   *Exact:*
$$75{,}078$$
$$- \quad \qquad - 46{,}090$$

4. *Estimate:*   *Exact:*
$$7.8$$
$$- \quad \qquad - 3.5029$$

5. *Estimate:*   *Exact:*
$$6538$$
$$\times \quad \qquad \times 708$$

6. *Estimate:*   *Exact:*
$$65.3$$
$$\times \quad \qquad \times 8.7$$

7. *Estimate:*   *Exact:*
$$\overline{) \qquad \quad} \qquad 43\overline{)38{,}786}$$

8. *Estimate:*   *Exact:*
$$\overline{) \qquad \quad} \qquad 7.6\overline{)2432}$$

9. *Estimate:*   *Exact:*
$$\overline{) \qquad \quad} \qquad 0.8\overline{)6.76}$$

Use the order of operations to simplify each expression.

10. $6^2 - 3 \cdot 6$

11. $\sqrt{49} + 5 \cdot 4 - 8$

12. $9 + 6 \div 3 + 7 \cdot 4$

Round each number to the place shown.

13. 4677 to the nearest ten

14. 7,583,281 to the nearest hundred thousand

15. $513.499 to the nearest dollar

16. $362.735 to the nearest cent

Add, subtract, multiply, or divide as indicated. Write answers in lowest terms and as whole or mixed numbers when possible.

17. $\dfrac{3}{4} + \dfrac{5}{8}$

18. $\dfrac{1}{3} + \dfrac{3}{4}$

19. $\begin{array}{r} 5\dfrac{3}{4} \\[2mm] + 7\dfrac{5}{8} \\ \hline \end{array}$

20. $\dfrac{7}{8} - \dfrac{3}{4}$

21. $\begin{array}{r} 8\dfrac{3}{8} \\[2mm] - 4\dfrac{1}{2} \\ \hline \end{array}$

22. $\begin{array}{r} 26\dfrac{1}{3} \\[2mm] - 17\dfrac{4}{5} \\ \hline \end{array}$

23. $\dfrac{7}{8} \cdot \dfrac{2}{3}$

24. $7\dfrac{3}{4} \cdot 3\dfrac{3}{8}$

25. $36 \cdot \dfrac{4}{5}$

26. $\dfrac{5}{8} \div \dfrac{5}{7}$

27. $12 \div \dfrac{3}{4}$

28. $2\dfrac{3}{4} \div 7\dfrac{1}{2}$

Write < or > to make a true statement.

29. $\dfrac{5}{8}$ _____ $\dfrac{2}{3}$

30. $\dfrac{8}{15}$ _____ $\dfrac{11}{20}$

31. $\dfrac{2}{3}$ _____ $\dfrac{7}{12}$

Simplify each expression. Use the order of operations as needed.

32. $\left(\dfrac{7}{8} - \dfrac{1}{2}\right) \cdot \dfrac{2}{3}$

33. $\dfrac{7}{8} \div \left(\dfrac{3}{4} + \dfrac{1}{8}\right)$

34. $\left(\dfrac{5}{6} - \dfrac{5}{12}\right) - \left(\dfrac{1}{2}\right)^2 \cdot \dfrac{2}{3}$

Write each fraction as a decimal. Round to the nearest thousandth if necessary.

35. $\dfrac{3}{5}$

36. $\dfrac{7}{8}$

37. $\dfrac{7}{12}$

38. $\dfrac{11}{20}$

Write each ratio as a fraction in lowest terms. Be sure to make all necessary conversions.

39. 2 hours to 40 minutes

40. If there are 27 people and 36 life jackets, what is the ratio of life jackets to people?

41. $1\dfrac{5}{8}$ to 13

Use cross multiplication to decide whether each proportion is true *or* false. *Circle the correct answer.*

42. $\dfrac{5}{8} = \dfrac{45}{72}$

    True   False

43. $\dfrac{64}{144} = \dfrac{48}{108}$

    True   False

Find the unknown value in each proportion.

44. $\dfrac{1}{6} = \dfrac{x}{36}$
45. $\dfrac{315}{45} = \dfrac{21}{x}$
46. $\dfrac{8}{x} = \dfrac{72}{144}$
47. $\dfrac{x}{120} = \dfrac{7.5}{30}$

Write each percent as a decimal. Write each decimal as a percent.

48. 78%
49. 3%
50. 200%
51. 0.5%

52. 0.87
53. 3.8
54. 0.023

Write each percent as a fraction or mixed number in lowest terms. Write each fraction as a percent.

55. 8%
56. 62.5%
57. 175%

58. $\dfrac{7}{8}$
59. $\dfrac{3}{10}$
60. $4\dfrac{1}{5}$

Solve each percent problem.

61. 35% of 1400 watches is how many watches?

62. $5\dfrac{1}{2}$% of $720 is how much?

63. 72 tires is 40% of what number of tires?

64. $4\dfrac{1}{2}$% of what number of miles is 76.5 miles?

65. What percent of 656 books is 328 books?

66. 72 hours is what percent of 180 hours?

Find the amount of sales tax or the tax rate and the total cost. Round to the nearest cent if necessary.

	Amount of Sale	Tax Rate	Amount of Tax	Total Cost
67.	$108.95	5%	_____	_____
68.	$460	_____	$29.90	_____

Find the commission earned or the rate of commission.

	Sales	Rate of Commission	Commission
69.	$12,538	8%	_____
70.	$225,300	_____	$5632.50

Find the amount or rate of discount and the amount paid after the discount. Round to the nearest cent if necessary.

	Original Price	Rate of Discount	Amount of Discount	Sale Price
71.	$456	45%	_____	_____
72.	$1085	_____	$162.75	_____

Find the total amount due on each simple interest loan. Round to the nearest cent if necessary.

	Principal	Rate	Time	Total Amount To Be Repaid
73.	$970	10%	$1\frac{1}{2}$ years	_____
74.	$18,350	11%	9 months	_____

Set up and solve a proportion for each problem.

75. Carol can test 11 cars for emissions in 4 hours. Find the number of cars that she can test in 12 hours.

76. If 3.5 ounces of weed killer is needed to make 6 gallons of spray, how much weed killer is needed for 102 gallons of spray?

Solve each application problem. Use the table to answer Exercises 77–80. Round answers to the nearest tenth of a percent if necessary.

EXISTING HOME SALES

Region	Last Year	This Year
Northeast	32,000	36,000
Midwest	65,000	66,300
South	82,000	77,500
West	54,000	49,600

77. Find the percent of increase in sales in the northeastern region.

78. Find the percent of increase in sales in the midwestern region.

79. What is the percent of decrease in sales in the southern region?

80. What is the percent of decrease in sales in the western region?

81. Joan Ong has $26,880 invested in her home. If this amount is 25% of her total assets, find her total assets.

82. Pat Ueda deposits $15,000 in a savings account that pays 5.5% interest compounded annually. Find
(a) the amount that he will have in the account at the end of 5 years, and

(b) the amount of interest earned. Do not use the table.

Measurement

7

At the 2000 Sydney Olympics, Michael Johnson won a pair of gold medals in track events to go with his gold shoes. The shoes, made especially for Johnson by 3M Company, contained 5 grams of pure gold. (*Source:* 3M Company.) But did the gold weigh Johnson down, slowing his speed? To find out, read about grams in Section 7.3. Then find out how to convert grams to ounces in Section 7.5. (See Exercise 13 in Section 7.5.)

You're Connected

7.1   PROBLEM SOLVING WITH ENGLISH MEASUREMENT

OBJECTIVES

1 Learn the basic measurement units in the English system.

2 Convert among measurement units using multiplication or division.

3 Convert among measurement units using unit fractions.

4 Solve application problems using English measurement.

❶ After memorizing the measurement conversions, answer these questions.

(a) 1 c = _____ fl oz

(b) _____ qt = 1 gal

(c) 1 wk = _____ days

(d) _____ ft = 1 yd

(e) 1 ft = _____ in.

(f) _____ oz = 1 lb

(g) 1 T = _____ lb

(h) _____ min = 1 hr

(i) 1 pt = _____ c

(j) _____ hr = 1 day

(k) 1 min = _____ sec

(l) 1 qt = _____ pt

(m) _____ ft = 1 mi

We measure things all the time: the distance traveled on vacation, the floor area we want to cover with carpet, the amount of milk in a recipe, the weight of the bananas we buy at the store, the number of hours we work, and many more.

In the United States we still use the **English system** of measurement for many everyday activities. Examples of English units are inches, feet, quarts, ounces, and pounds. However, the fields of science, medicine, sports, and manufacturing use the **metric system** (meters, liters, and grams). And, because the rest of the world uses only the metric system, U.S. businesses are beginning to change to the metric system in order to compete internationally.

1 Learn the basic measurement units in the English system.   Until the switch to the metric system is complete, we still need to know how to use the English system of measurement. The table below lists the relationships you should memorize. The time relationships are used in both the English and metric systems.

English Measurement Relationships

Length	Weight
1 foot (ft) = 12 inches (in.)	1 pound (lb) = 16 ounces (oz)
1 yard (yd) = 3 feet (ft)	1 ton (T) = 2000 pounds (lb)
1 mile (mi) = 5280 feet (ft)	

Capacity	Time
1 cup (c) = 8 fluid ounces (fl oz)	1 week (wk) = 7 days
1 pint (pt) = 2 cups (c)	1 day = 24 hours (hr)
1 quart (qt) = 2 pints (pt)	1 hour (hr) = 60 minutes (min)
1 gallon (gal) = 4 quarts (qt)	1 minute (min) = 60 seconds (sec)

As you can see, there is no simple or "natural" way to convert among these various measures. The units evolved over hundreds of years and were based on a variety of "standards." For example, one yard was the distance from the tip of a king's nose to his thumb when his arm was outstretched. An inch was three dried barleycorns laid end to end.

Example 1   Knowing English Measurement Units

Memorize the English measurement conversions. Then answer these questions.

(a) 24 hr = _____ day   Answer: 1 day

(b) 1 yd = _____ ft   Answer: 3 ft

Work Problem ❶ at the Side.

2 Convert among measurement units using multiplication or division. You often need to convert from one unit of measure to another. Two methods of converting measurements are shown here. Study each way and use the method you prefer. The first method involves deciding whether to multiply or divide.

Converting among Measurement Units

1. *Multiply* when converting from a larger unit to a smaller unit.
2. *Divide* when converting from a smaller unit to a larger unit.

ANSWERS
1. **(a)** 8 **(b)** 4 **(c)** 7 **(d)** 3 **(e)** 12
(f) 16 **(g)** 2000 **(h)** 60 **(i)** 2 **(j)** 24
(k) 60 **(l)** 2 **(m)** 5280

Example 2 **Converting from One Unit of Measure to Another**

Convert each measurement.

(a) 7 ft to inches

You are converting from a *larger* unit to a *smaller* unit (feet to inches), so multiply.

Because *1 ft = 12 in.,* multiply by 12.

$$7 \text{ ft} = 7 \cdot 12 = 84 \text{ in.}$$

(b) $3\frac{1}{2}$ lb to ounces

You are converting from a *larger* unit to a *smaller* unit (pounds to ounces), so multiply.

Because *1 lb = 16 oz,* multiply by 16.

$$3\frac{1}{2} \text{ lb} = 3\frac{1}{2} \cdot 16 = \frac{7}{2} \cdot \frac{\overset{8}{\cancel{16}}}{1} = \frac{56}{1} = 56 \text{ oz}$$

(c) 20 qt to gallons

You are converting from a *smaller* unit to a *larger* unit (quarts to gallons), so divide.

Because *4 qt = 1 gal,* divide by 4.

$$20 \text{ qt} = \frac{20}{4} = 5 \text{ gal}$$

Divide by 4.

(d) 45 min to hours

You are converting from a *smaller* unit to a *larger* unit (minutes to hours), so divide.

Because *60 min = 1 hr,* divide by 60 and write the fraction in lowest terms.

$$45 \text{ min} = \frac{45}{60} = \frac{45 \div 15}{60 \div 15} = \frac{3}{4} \text{ hr} \leftarrow \text{Lowest terms}$$

Divide by 60.

Work Problem **2** at the Side.

3 **Convert among measurement units using unit fractions.** If you have trouble deciding whether to multiply or divide when converting measurements, use *unit fractions* to solve the problem. You'll also find this method useful in science classes. A **unit fraction** is equivalent to 1. Here is an example.

$$\frac{12 \text{ in.}}{12 \text{ in.}} = \frac{\overset{1}{\cancel{12 \text{ in.}}}}{\underset{1}{\cancel{12 \text{ in.}}}} = 1$$

Use the table of measurement relationships on the first page of this section to find that 12 in. is the same as 1 ft. So you can substitute 1 ft for 12 in. in the numerator, or you can substitute 1 ft for 12 in. in the denominator. This makes two useful unit fractions.

$$\frac{1 \text{ ft}}{12 \text{ in.}} = 1 \quad \text{or} \quad \frac{12 \text{ in.}}{1 \text{ ft}} = 1$$

To convert from one measurement unit to another, just multiply by the appropriate unit fraction. Remember, a unit fraction is equivalent to 1. Multiplying something by 1 does *not* change its value.

2 Convert each measurement using multiplication or division.

(a) $5\frac{1}{2}$ ft to inches

(b) 64 oz to pounds

(c) 6 yd to feet

(d) 2 T to pounds

(e) 35 pt to quarts

(f) 20 min to hours

(g) 4 wk to days

ANSWERS
2. (a) 66 in. **(b)** 4 lb **(c)** 18 ft
(d) 4000 lb **(e)** $17\frac{1}{2}$ qt **(f)** $\frac{1}{3}$ hr
(g) 28 days

❸ First write the unit fraction needed to make each conversion. Then complete the conversion.

(a) 36 in. to feet

$$\left.\begin{array}{l}\text{unit}\\\text{fraction}\end{array}\right\}\ \dfrac{1\ \text{ft}}{12\ \text{in.}}$$

(b) 14 ft to inches

$$\left.\begin{array}{l}\text{unit}\\\text{fraction}\end{array}\right\}\ \dfrac{\text{in.}}{\text{ft}}$$

(c) 60 in. to feet

$$\left.\begin{array}{l}\text{unit}\\\text{fraction}\end{array}\right\}\ \underline{\quad\quad}$$

(d) 4 yd to feet

$$\left.\begin{array}{l}\text{unit}\\\text{fraction}\end{array}\right\}\ \underline{\quad\quad}$$

(e) 39 ft to yards

$$\left.\begin{array}{l}\text{unit}\\\text{fraction}\end{array}\right\}\ \underline{\quad\quad}$$

(f) 2 mi to feet

$$\left.\begin{array}{l}\text{unit}\\\text{fraction}\end{array}\right\}\ \underline{\quad\quad}$$

Use these guidelines to choose the correct unit fraction.

Choosing a Unit Fraction

The *numerator* should use the measurement unit you want in the *answer*.

The *denominator* should use the measurement unit you want to *change*.

Example 3 Using Unit Fractions with Length Measurements

(a) Convert 60 in. to feet.

Use a unit fraction with feet (the unit for your answer) in the numerator, and inches (the unit being changed) in the denominator. Because *1 ft = 12 in.*, the necessary unit fraction is

$$\dfrac{1\ \text{ft}}{12\ \text{in.}}\quad\begin{array}{l}\leftarrow\text{ Unit for your answer is feet.}\\\leftarrow\text{ Unit being changed is inches.}\end{array}$$

Next, multiply 60 in. times this unit fraction. Write 60 in. as the fraction $\dfrac{60\ \text{in.}}{1}$. Then divide out common units and factors wherever possible.

$$60\ \text{in.}\cdot\dfrac{1\ \text{ft}}{12\ \text{in.}}=\dfrac{\overset{5}{\cancel{60}\ \cancel{\text{in.}}}}{1}\cdot\dfrac{1\ \text{ft}}{\underset{1}{\cancel{12}\ \cancel{\text{in.}}}}=\dfrac{5\cdot1\ \text{ft}}{1}=5\ \text{ft}$$

These units should match.

Divide out inches.
Divide 60 and 12 by 12.

(b) Convert 9 ft to inches.

Select the correct unit fraction to change 9 ft to inches.

$$\dfrac{12\ \text{in.}}{1\ \text{ft}}\quad\begin{array}{l}\leftarrow\text{ Unit for your answer is inches.}\\\leftarrow\text{ Unit being changed is feet.}\end{array}$$

Multiply 9 ft times the unit fraction.

$$9\ \text{ft}\cdot\dfrac{12\ \text{in.}}{1\ \text{ft}}=\dfrac{9\ \cancel{\text{ft}}}{1}\cdot\dfrac{12\ \text{in.}}{1\ \cancel{\text{ft}}}=\dfrac{9\cdot12\ \text{in.}}{1}=108\ \text{in.}$$

These units should match.

Divide out feet.

CAUTION

If no units will divide out, you made a mistake in choosing the unit fraction.

Work Problem ❸ at the Side.

Example 4 Using Unit Fractions with Capacity and Weight Measurements

(a) Convert 9 pt to quarts.

First select the correct unit fraction.

$$\dfrac{1\ \text{qt}}{2\ \text{pt}}\quad\begin{array}{l}\leftarrow\text{ Unit for your answer is quarts.}\\\leftarrow\text{ Unit being changed is pints.}\end{array}$$

Continued on Next Page

ANSWERS

3. (a) 3 ft **(b)** $\dfrac{12\ \text{in.}}{1\ \text{ft}}$; 168 in.

(c) $\dfrac{1\ \text{ft}}{12\ \text{in.}}$; 5 ft **(d)** $\dfrac{3\ \text{ft}}{1\ \text{yd}}$; 12 ft

(e) $\dfrac{1\ \text{yd}}{3\ \text{ft}}$; 13 yd **(f)** $\dfrac{5280\ \text{ft}}{1\ \text{mi}}$; 10,560 ft

Next multiply.

Write as mixed number.

$$9 \text{ pt} \cdot \frac{1 \text{ qt}}{2 \text{ pt}} = \frac{9 \text{ pt}}{1} \cdot \frac{1 \text{ qt}}{2 \text{ pt}} = \frac{9}{2} \text{ qt} = 4\frac{1}{2} \text{ qt}$$

These units   Divide out pints.
should match.

(b) Convert $7\frac{1}{2}$ gal to quarts.

Write as an improper fraction.

$$\frac{7\frac{1}{2} \text{ gal}}{1} \cdot \frac{4 \text{ qt}}{1 \text{ gal}} = \frac{15}{2} \cdot \frac{4}{1} \text{ qt}$$

$$= \frac{15}{2} \cdot \frac{\overset{2}{\cancel{4}}}{1} \text{ qt}$$

$$= 30 \text{ qt}$$

(c) Convert 36 oz to pounds.

$$\frac{\overset{9}{\cancel{36}} \text{ oz}}{1} \cdot \frac{1 \text{ lb}}{\underset{4}{\cancel{16}} \text{ oz}} = \frac{9}{4} \text{ lb} = 2\frac{1}{4} \text{ lb}$$

NOTE

In Example 4(c) you get $\frac{9}{4}$ lb. Recall that $\frac{9}{4}$ means $9 \div 4$. If you do $9 \div 4$ on your calculator, you get 2.25 lb. English measurements usually use fractions or mixed numbers, like $2\frac{1}{4}$ lb. However, 2.25 lb is also correct and is the way grocery stores often show weights of produce, meat, and cheese.

Work Problem ④ at the Side.

Example 5 **Using Several Unit Fractions**

Sometimes you may need to use two or three unit fractions in problems like these.

(a) Convert 63 in. to yards.

Use the unit fraction $\dfrac{1 \text{ ft}}{12 \text{ in.}}$ to change inches to feet and the unit fraction $\dfrac{1 \text{ yd}}{3 \text{ ft}}$ to change feet to yards. Notice how all the units divide out except yards, which is the unit you want in the answer.

$$\frac{63 \text{ in.}}{1} \cdot \frac{1 \text{ ft}}{12 \text{ in.}} \cdot \frac{1 \text{ yd}}{3 \text{ ft}} = \frac{63}{36} \text{ yd} = 1\frac{3}{4} \text{ yd}$$

Continued on Next Page

④ Convert using unit fractions.

(a) 16 qt to gallons

(b) 3 c to pints

(c) $3\frac{1}{2}$ T to pounds

(d) $1\frac{3}{4}$ lb to ounces

(e) 4 oz to pounds

ANSWERS

4. **(a)** 4 gal **(b)** $1\frac{1}{2}$ pt or 1.5 pt **(c)** 7000 lb

(d) 28 oz **(e)** $\frac{1}{4}$ lb or 0.25 lb

5 Convert using two or three unit fractions.

(a) 4 T to ounces

(b) 3 mi to inches

(c) 36 pt to gallons

(d) 2 wk to minutes

You can also divide out common factors in the numbers.

$$\frac{\overset{7}{\cancel{\underset{}{63}}}}{1} \cdot \frac{1}{\underset{4}{\cancel{12}}} \cdot \frac{1}{\underset{1}{\cancel{3}}} = \frac{7}{4} = 1\frac{3}{4} \text{ yd}$$

Instead of changing $\frac{7}{4}$ to $1\frac{3}{4}$, you can enter $7 \div 4$ on your calculator to get 1.75 yd. Both answers are correct because 1.75 is equivalent to $1\frac{3}{4}$.

(b) Convert 2 days to seconds.

Use three unit fractions. The first one changes days to hours, the second one changes hours to minutes, and the third one changes minutes to seconds. All the units divide out except seconds, which is what you want in your answer.

$$\frac{2 \text{ d\cancel{ays}}}{1} \cdot \frac{24 \text{ \cancel{hr}}}{1 \text{ d\cancel{ay}}} \cdot \frac{60 \text{ m\cancel{in}}}{1 \text{ \cancel{hr}}} \cdot \frac{60 \text{ seconds}}{1 \text{ m\cancel{in}}} = 172,800 \text{ seconds}$$

Divide out **days**.
Divide out **hr.**
Divide out **min.**

Work Problem 5 at the Side.

4 Solve application problems using English measurement. To solve application problems, we will use the steps you learned in **Section 1.10.** Those steps are summarized here.

Step 1 **Read** the problem carefully.

Step 2 **Work out a plan.**

Step 3 **Estimate** a reasonable answer.

Step 4 **Solve** the problem.

Step 5 **State the answer.**

Step 6 **Check** your work.

Example 6 Solving English Measurement Applications

(a) A 36-oz can of coffee is on sale at Cub Foods for $4.98. What is the cost per pound, to the nearest cent? (*Source:* Cub Foods.)

Step 1 **Read** the problem. The problem asks for the cost per pound of coffee.

Step 2 **Work out a plan.** The word *per* indicates division. You need to divide the cost by the number of pounds. Since the weight of the coffee is given in ounces, convert ounces to pounds.

Step 3 **Estimate** a reasonable answer. To estimate, round $4.98 to $5. Then, there are 16 oz in a pound, so 36 oz as a little more than 2 lb. So, $5 ÷ 2 = $2.50 per pound as our estimate.

Continued on Next Page

ANSWERS
5. (a) 128,000 oz **(b)** 190,080 in.

(c) $4\frac{1}{2}$ gal or 4.5 gal **(d)** 20,160 min

Step 4   **Solve** the problem. Use a unit fraction to convert 36 oz to pounds.

$$\frac{36 \; \cancel{oz}^{\,9}}{1} \cdot \frac{1 \; lb}{\cancel{16} \; \cancel{oz}_{4}} = \frac{9}{4} lb = 2.25 \; lb$$

Then divide.

$$\frac{\$4.98}{2.25 \; lb} = 2.21\overline{3} \approx 2.21 \quad \text{(rounded)}$$

Step 5   **State the answer.** The coffee costs $2.21 per pound (to the nearest cent).

Step 6   **Check** your work. The answer, $2.21, is close to our estimate of $2.50.

(b) Bilal's favorite dessert recipe uses $1\frac{2}{3}$ c of milk. If he makes six desserts for a bake sale at his son's school, how many quarts of milk will he need?

Step 1   **Read** the problem. The problem asks for the number of quarts of milk needed for six desserts.

Step 2   **Work out a plan.** Multiply to find the cups of milk for six desserts. Then convert cups to quarts.

Step 3   **Estimate** a reasonable answer. To estimate, round $1\frac{2}{3}$ c to 2 c. Then, 2 c times 6 = 12 c. There are 4 c in a quart, so 12 c ÷ 4 = 3 qt as our estimate.

Step 4   **Solve** the problem. First multiply. Then use unit fractions to convert.

$$1\frac{2}{3} \cdot 6 = \frac{5}{\cancel{3}_{1}} \cdot \frac{\cancel{6}^{\,2}}{1} = \frac{10}{1} = 10 \; c$$

$$\frac{\cancel{10}^{\,5} \; \cancel{c}}{1} \cdot \frac{1 \; \cancel{pt}}{\cancel{2} \; \cancel{c}_{1}} \cdot \frac{1 \; qt}{2 \; \cancel{pt}} = \frac{5}{2} qt = 2\frac{1}{2} \; qt$$

Step 5   **State the answer.** Bilal needs $2\frac{1}{2}$ qt (or 2.5 qt) of milk.

Step 6   **Check** your work. The answer, $2\frac{1}{2}$ qt, is close to our estimate of 3 qt.

══════════════ **Work Problem ❻ at the Side.**

❻ Solve each application problem using the six problem-solving steps.

(a) Kristin paid $3.29 for 12 oz of extra sharp cheddar cheese. What is the price per pound, to the nearest cent?

(b) A moving company estimates 11,000 lb of furnishings for an average 3-bedroom house. (*Source:* North American Van Lines.) If the company made five such moves last week, how many tons of furnishings did they move?

Growing Sunflowers

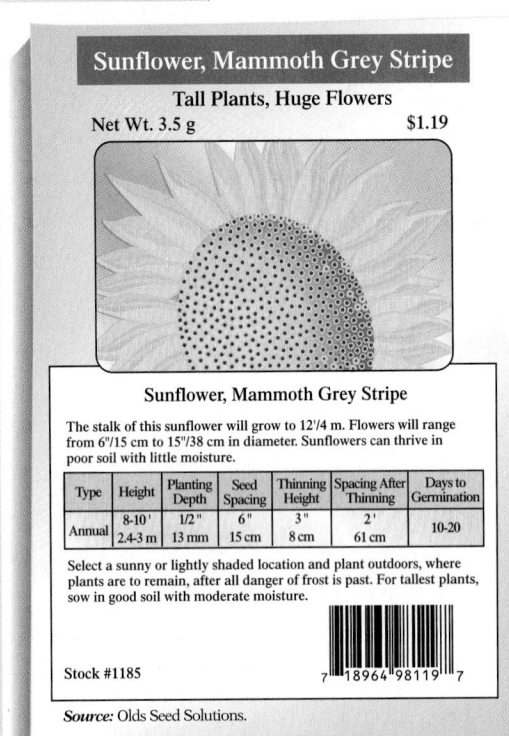

Sunflower, Mammoth Grey Stripe

Tall Plants, Huge Flowers

Net Wt. 3.5 g $1.19

Sunflower, Mammoth Grey Stripe

The stalk of this sunflower will grow to 12'/4 m. Flowers will range from 6"/15 cm to 15"/38 cm in diameter. Sunflowers can thrive in poor soil with little moisture.

Type	Height	Planting Depth	Seed Spacing	Thinning Height	Spacing After Thinning	Days to Germination
Annual	8-10' 2.4-3 m	1/2 " 13 mm	6" 15 cm	3" 8 cm	2' 61 cm	10-20

Select a sunny or lightly shaded location and plant outdoors, where plants are to remain, after all danger of frost is past. For tallest plants, sow in good soil with moderate moisture.

Stock #1185

7 18964 98119 7

Source: Olds Seed Solutions.

The front and back of a seed packet for sunflowers are shown. Look at the front of the packet first.

1. There were 42 seeds in the packet. If 40 of the seeds sprouted, what was the cost per sprout, to the nearest cent?

2. If vegetable and flower seeds were on sale at 30% off, what was the cost per sprout, to the nearest cent?

3. What percent of the seeds sprouted, to the nearest whole percent?

4. How many seeds would weigh 1 gram?

5. The table on the back of the packet uses an apostrophe (') which is a symbol for feet, and " which is a symbol for inches.

   (a) How tall will the plants grow, in feet?

   (b) How tall will they grow in inches?

   (c) How tall will they grow in yards?

6. If you plant all 42 seeds in one long row, using the spacing given on the package, how long will your row be in feet?

7. How many inches tall should the plants be when you thin them (remove less vigorous plants to give others room to grow)? How tall is that in feet?

8. What is the range in the diameter of the flowers, in inches, and in feet? Diameter is the distance across the circular flower.

9. (a) Using the information in the article, how many gum wrappers are needed to make 1 foot of chain?

   (b) To make 1 inch of chain?

   (c) How many inches of chain, to the nearest hundredth, are made from one wrapper?

10. Is the article correct in saying that 125 miles is 34 million gum wrappers?

Sometimes a person's life's work makes it into a museum. So it figures that Michael Knutson's 128-foot chain of gum wrappers now resides in the Yellow Medicine County Museum in Granite Falls, Minnesota.

According to the *Redwood Gazette,* the wrapper chain started in 1974 when Knutson, now a 37-year-old woodworker, became bored during study hall. "I've never been much of a gum chewer, but I found most of the wrappers on the streets of the city," he said. (FYI: It takes 6602 wrappers to make a 128-foot chain.)

Granite Falls is about 125 miles—or 34 million gum wrappers—west of the Twin Cities.

Source: Minneapolis Star Tribune.

7.1 EXERCISES

 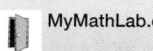

Fill in the blanks with the measurement relationships you have memorized. See Example 1.

1. 1 yd = _____ ft

2. 1 ft = _____ in.

3. _____ fl oz = 1 c

4. _____ qt = 1 gal

5. 1 mi = _____ ft

6. 1 wk = _____ days

7. _____ lb = 1 T

8. _____ oz = 1 lb

9. 1 min = _____ sec

10. 1 day = _____ hr

Convert each measurement using unit fractions. See Examples 3 and 4.

11. 120 sec = _____ min

12. 180 min = _____ hr

13. 8 qt = _____ gal

14. 6 gal = _____ qt

15. An adult sperm whale may weigh 38 to 40 tons. How many pounds could it weigh? (*Source: Grolier Multimedia Encyclopedia.*)

16. Recent fossil finds in Argentina indicate that the largest meat-eating dinosaur may have been 45 ft long. How many yards long was this dinosaur? (*Source: Washington Post.*)

45 ft

17. 9 yd = _____ ft

18. 20,000 lb = _____ T

19. 7 lb = _____ oz

20. 96 oz = _____ lb

21. 5 qt = _____ pt

22. 26 pt = _____ qt

23. 90 min = _____ hr

24. 45 sec = _____ min

25. 3 in. = _____ ft

26. 30 in. = _____ ft

27. 24 oz = _____ lb

28. 36 oz = _____ lb

29. 5 c = _____ pt

30. 15 qt = _____ gal

31. Mr. Kashpaws and his son worked for 12 hours doing traditional harvesting of wild rice. What part of a day did they work?

32. Michelle prepares 4-oz hamburgers at a fast-food restaurant. Each hamburger is what part of a pound?

33. $2\frac{1}{2}$ T = _____ lb

34. $4\frac{1}{2}$ pt = _____ c

35. $4\frac{1}{4}$ gal = _____ qt

36. $2\frac{1}{4}$ hr = _____ min

37. Our premature baby weighed $2\frac{3}{4}$ lb at birth. How many ounces did our baby weigh?

38. Gheorge Muresan, the tallest person playing basketball in the NBA, is $7\frac{2}{3}$ ft tall. What is his height in inches? (*Source:* NBA.)

Use two or three unit fractions to convert the following. See Example 5.

39. 6 yd = _____ in.

40. 2 T = _____ oz

41. 112 c = _____ qt

42. 336 hr = _____ wk

43. 6 days = _____ sec

44. 5 gal = _____ c

45. $1\frac{1}{2}$ T = _____ oz

46. $3\frac{1}{3}$ yd = _____ in.

47. The statement 8 = 2 is *not* true. But with appropriate measurement units, it *is* true.

$$8 \text{ } quarts = 2 \text{ } gallons$$

Attach measurement units to these numbers to make the statements true.

(a) 1 _____ = 16 _____

(b) 10 _____ = 20 _____

(c) 120 _____ = 2 _____

(d) 2 _____ = 24 _____

(e) 6000 _____ = 3 _____

(f) 35 _____ = 5 _____

48. Explain in your own words why you can add 2 feet + 12 inches to get 3 feet, but you cannot add 2 feet + 12 pounds.

Convert the following. See Example 5.

49. $2\frac{3}{4}$ mi = _____ in.

50. $5\frac{3}{4}$ tons = _____ oz

51. $6\frac{1}{4}$ gal = _____ fl oz

52. $3\frac{1}{2}$ days = _____ sec

53. 24,000 oz = _____ T

54. 57,024 in. = _____ mi

Solve each application problem. See Example 6.

55. Geralyn bought a 20-oz box of strawberries for $2.29. What was the price per pound for the strawberries, to the nearest cent?

56. Zach paid $0.79 for a 1.6-oz candy bar. (*Source: Byerly's Foods.*) What was the cost per pound?

57. Dan orders supplies for the science labs. Each of the 24 stations in the chemistry lab needs 2 ft of rubber tubing. If rubber tubing sells for $8.75 per yard, how much will it cost to equip all the stations?

58. In 1998, Marquette, Michigan, had 136 in. of snowfall, while Detroit, Michigan, had 14 in. (*Source: World Almanac,* 2000.) What was the difference in snowfall between the two cities, in feet? Round to the nearest tenth.

59. At the day care center, each of the 15 toddlers drinks about $\frac{2}{3}$ cup of milk with lunch. The center is open 5 days a week.

   (a) How many quarts of milk will the center need for one week of lunches?

   (b) If the center buys milk in gallon jugs, how many jugs should be ordered for one week?

60. A snail moves at an average speed of 2 feet every 3 minutes. (*Source: Beakman and Jax.*) At that rate, how long would it take the snail to travel one mile? Give your answer

   (a) in hours and

   (b) in days.

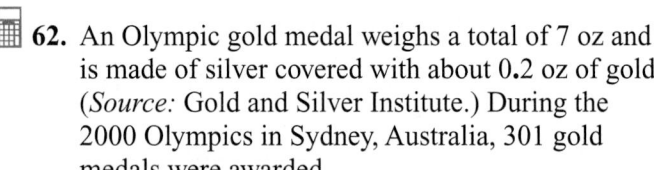

61. Bob's Candies in Albany, Georgia, makes 135,000 pounds of candy canes each day. (*Source:* Bob's Candies, Inc.)

   (a) How many tons of candy canes are produced during a 5-day workweek?

   (b) The plant operates 24 hours per day. How many tons of candy canes are produced each hour, to the nearest tenth?

62. An Olympic gold medal weighs a total of 7 oz and is made of silver covered with about 0.2 oz of gold (*Source:* Gold and Silver Institute.) During the 2000 Olympics in Sydney, Australia, 301 gold medals were awarded.

   (a) How many pounds of gold were used to make all the medals, to the nearest tenth?

   (b) How many pounds of silver were used to make all the medals, to the nearest tenth?

RELATING CONCEPTS (Exercises 63–66)   **FOR INDIVIDUAL OR GROUP WORK**

People often complain about the price of a gallon of gasoline. In response, AutoWeek *magazine asked readers to look at the price per gallon of other liquids. See what they discovered as you* **work Exercises 63–66 in order.**

63. If 16 fl oz of Ocean Spray cranberry juice costs $1.25, how could you find the cost per gallon? Here is one way.

   Step 1   Use unit fractions to find the number of fluid ounces in one gallon.

   Step 2   Set up and solve a proportion. (*Hint:* One side of the proportion should compare 16 fl oz to $1.25.)

64. Find the cost per gallon for Evian water if an 11.2 fl oz bottle sells for $1.49. Use the same method as in Exercise 63.

65. If you pay $3.85 for a bottle of Pepto Bismol containing 4 fl oz, what is the cost per gallon? Use a different method to find the answer than you used in Exercises 63 and 64.

66. Does it make sense to compare the price per gallon of gasoline to the price per gallon of cranberry juice or Pepto Bismol? Explain why or why not.

7.2 THE METRIC SYSTEM—LENGTH

Around 1790, a group of French scientists developed the metric system of measurement. It is an organized system based on multiples of 10, like our number system and our money. After you are familiar with metric units, you will see that they are easier to use than the hodgepodge of English measurement relationships you used in **Section 7.1.**

1 **Learn the basic metric units of length.**   The basic unit of length in the metric system is the **meter** (also spelled *metre*). Use the symbol **m** for meter; do not put a period after it. If you put five of the pages from this textbook side by side, they would measure about 1 meter. Or, look at a yardstick—a meter is just a little longer. A yard is 36 inches long; a meter is about 39 inches long.

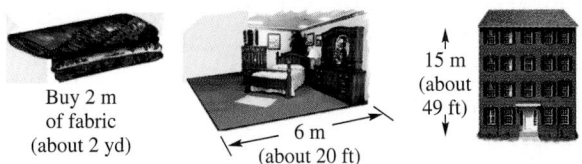

In the metric system you use meters for things like buying fabric for sewing projects, measuring the length of your living room, talking about heights of buildings, or describing track and field athletic events.

Buy 2 m of fabric (about 2 yd)

6 m (about 20 ft)

15 m (about 49 ft)

Work Problem 1 at the Side.

To make longer or shorter length units in the metric system, **prefixes** are written in front of the word *meter.* For example, the prefix *kilo* means 1000, so a *kilo*meter is 1000 meters. The table below shows how to use the prefixes for length measurements. It is helpful to memorize the prefixes because they are also used with weight and capacity measurements. The blue boxes are the units you will use most often in daily life.

Prefix	kilo-meter	hecto-meter	deka-meter	meter	deci-meter	centi-meter	milli-meter
Meaning	1000 meters	100 meters	10 meters	1 meter	$\frac{1}{10}$ of a meter	$\frac{1}{100}$ of a meter	$\frac{1}{1000}$ of a meter
Symbol	km	hm	dam	m	dm	cm	mm

Here are some comparisons to help you get acquainted with the commonly used length units: km, m, cm, mm.

*Kilo*meters are used instead of miles. A kilometer is 1000 meters. It is about 0.6 mile (a little more than half a mile) or about 5 to 6 city blocks. If you participate in a 10 km run, you'll run about 6 miles.

1 Circle the items that measure about 1 meter.

Length of a pencil

Length of a baseball bat

Height of doorknob from the floor

Height of a house

Basketball player's arm length

Length of a paper clip

ANSWERS

1. baseball bat, height of doorknob, basketball player's arm length.

❷ Write the most reasonable metric unit in each blank. Choose from km, m, cm, and mm.

(a) The woman's height is

168 _____.

(b) The man's waist is

90 _____ around.

(c) Louise ran the 100 _____

dash in the track meet.

(d) A postage stamp is

22 _____ wide.

(e) Michael paddled his

canoe 2 _____ down

the river.

(f) The pencil lead is

1 _____ thick.

(g) A stick of gum is

7 _____ long.

(h) The highway speed limit

is 90 _____ per hour.

(i) The classroom was

12 _____ long.

(j) A penny is about

18 _____ across.

A meter is divided into 100 smaller pieces called *centi*meters. Each centimeter is $\frac{1}{100}$ of a meter. Centimeters are used instead of inches. A centimeter is a little shorter than $\frac{1}{2}$ inch. The cover of this textbook is about 21 cm wide. A nickel is about 2 cm across. Measure the width and length of your little finger on this centimeter ruler. The width of your little finger is probably about 1 cm.

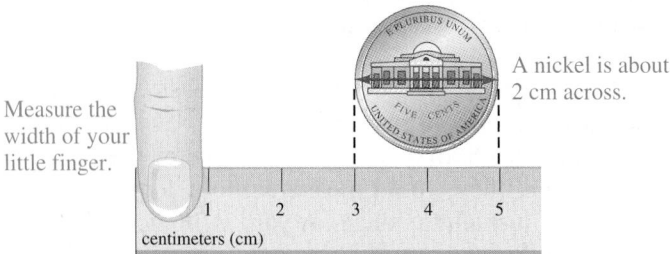

A meter is divided into 1000 smaller pieces called *milli*meters. Each millimeter is $\frac{1}{1000}$ of a meter. It takes 10 mm to equal 1 cm, so it is a very small length. The thickness of a dime is about 1 mm. Measure the width of your pen or pencil and the width of your little finger on this millimeter ruler.

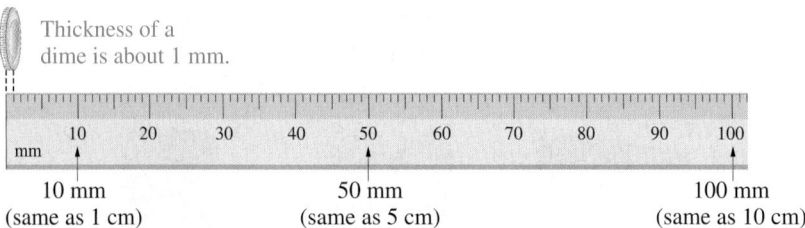

Example 1 Using Metric Length Units

Write the most reasonable metric unit in each blank. Choose from km, m, cm, and mm.

(a) The distance from home to work is 20 _____.

20 <u>km</u> because kilometers are used instead of miles.

20 km is about 12 miles.

(b) My wedding ring is 4 _____ wide.

4 <u>mm</u> because the width of a ring is very small.

(c) The newborn baby is 50 _____ long.

50 <u>cm</u>, which is half of a meter; a meter is about 39 inches so half a meter is around 20 inches.

Work Problem ❷ at the Side.

2 ▓▓▓ **Use unit fractions to convert among units.** You can convert among metric length units using unit fractions. Keep these relationships in mind when setting up the unit fractions.

Metric Length Relationships

1 km = 1000 m so the unit fractions are:	**1 m = 1000 mm** so the unit fractions are:
$\dfrac{1 \text{ km}}{1000 \text{ m}}$ or $\dfrac{1000 \text{ m}}{1 \text{ km}}$	$\dfrac{1 \text{ m}}{1000 \text{ mm}}$ or $\dfrac{1000 \text{ mm}}{1 \text{ m}}$
1 m = 100 cm so the unit fractions are:	**1 cm = 10 mm** so the unit fractions are:
$\dfrac{1 \text{ m}}{100 \text{ cm}}$ or $\dfrac{100 \text{ cm}}{1 \text{ m}}$	$\dfrac{1 \text{ cm}}{10 \text{ mm}}$ or $\dfrac{10 \text{ mm}}{1 \text{ cm}}$

Example 2 ▶ **Using Unit Fractions to Convert Length Measurements**

Convert the following.

(a) 5 km to m

Put the unit for the answer (meters) in the numerator of the unit fraction; put the unit you want to change (km) in the denominator.

Unit fraction equivalent to 1 $\left\{ \dfrac{1000 \text{ m}}{1 \text{ km}} \right.$ ← Unit for answer ← Unit being changed

Multiply. Divide out common units where possible.

$$5 \text{ km} \cdot \dfrac{1000 \text{ m}}{1 \text{ km}} = \dfrac{5 \text{ km}}{1} \cdot \dfrac{1000 \text{ m}}{1 \text{ km}} = \dfrac{5 \cdot 1000 \text{ m}}{1} = 5000 \text{ m}$$

These units should match.

The answer makes sense because a kilometer is much longer than a meter, so 5 km will contain many meters.

(b) 18.6 cm to m

Multiply by a unit fraction that allows you to divide out centimeters.

Unit fraction

$$\dfrac{18.6 \text{ cm}}{1} \cdot \dfrac{1 \text{ m}}{100 \text{ cm}} = \dfrac{18.6}{100} \text{ m} = 0.186 \text{ m}$$

There are 100 cm in a meter, so 18.6 cm will be a small part of a meter. The answer makes sense.

══════ **Work Problem ❸ at the Side.**

---

❸ First write the unit fraction needed to make each conversion. Then complete the conversion.

(a) 3.67 m to cm

unit fraction $\left. \right\}$ $\dfrac{100 \text{ cm}}{1 \text{ m}}$

(b) 92 cm to m

unit $\left. \right\}$ $\dfrac{\text{m}}{\text{cm}}$

(c) 432.7 cm to m

unit fraction $\left. \right\}$ _____

(d) 65 mm to cm

unit fraction $\left. \right\}$ _____

(e) 0.9 m to mm

unit fraction $\left. \right\}$ _____

(f) 2.5 cm to mm

unit fraction $\left. \right\}$ _____

Answers

3. (a) 367 cm **(b)** $\dfrac{1 \text{ m}}{100 \text{ cm}}$; 0.92 m

(c) $\dfrac{1 \text{ m}}{100 \text{ cm}}$; 4.327 m

(d) $\dfrac{1 \text{ cm}}{10 \text{ mm}}$; 6.5 cm

(e) $\dfrac{1000 \text{ mm}}{1 \text{ m}}$; 900 mm

(f) $\dfrac{10 \text{ mm}}{1 \text{ cm}}$; 25 mm

❹ Do each multiplication or division by hand or on a calculator. Compare your answer to the one you get by moving the decimal point.

(a) 43.5 • 10 = _____

43.5 gives 435.

(b) 43.5 ÷ 10 = _____

43.5 gives _____.

(c) 28 • 100 = _____

28.00 gives _____.

(d) 28 ÷ 100 = _____

28. gives _____.

(e) 0.7 • 1000 = _____

0.700 gives _____.

(f) 0.7 ÷ 1000 = _____

000.7 gives _____.

3 ▬▬▬ **Move the decimal point to convert among units.** By now you have probably noticed that conversions among metric units are made by multiplying or dividing by 10, by 100, or by 1000. A quick way to *multiply* by 10 is to move the decimal point one place to the *right*. Move it two places to the right to multiply by 100, three places to multiply by 1000. *Dividing* is done by moving the decimal point to the *left* in the same manner.

Work Problem ❹ at the Side.

An alternate conversion method to unit fractions is moving the decimal point using this **metric conversion line.**

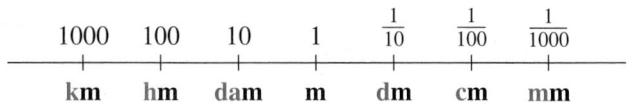

Here are the steps for using the conversion line.

Using the Metric Conversion Line

Step 1 Find the unit you are given on the metric conversion line.

Step 2 Count the number of places to get from the unit you are given to the unit you want in the answer.

Step 3 Move the decimal point the **same number of places** and in the **same direction** as you did on the conversion line.

Example 3 **Using the Metric Conversion Line**

Use the metric conversion line to make the following conversions.

(a) 5.702 km to m

Find **km** on the metric conversion line. To get to **m**, you move *three places* to the *right*. So move the decimal point in 5.702 *three places* to the *right*.

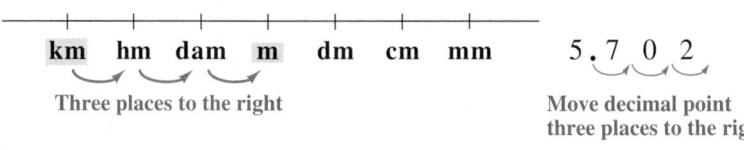

5.702 km = 5702 m

(b) 69.5 cm to m

Find **cm** on the conversion line. To get to **m**, move *two places* to the *left*.

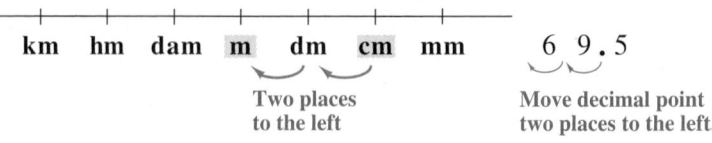

69.5 cm = 0.695 m

Continued on Next Page

(c) 8.1 cm to mm

From **cm** to **mm** is *one place* to the *right*.

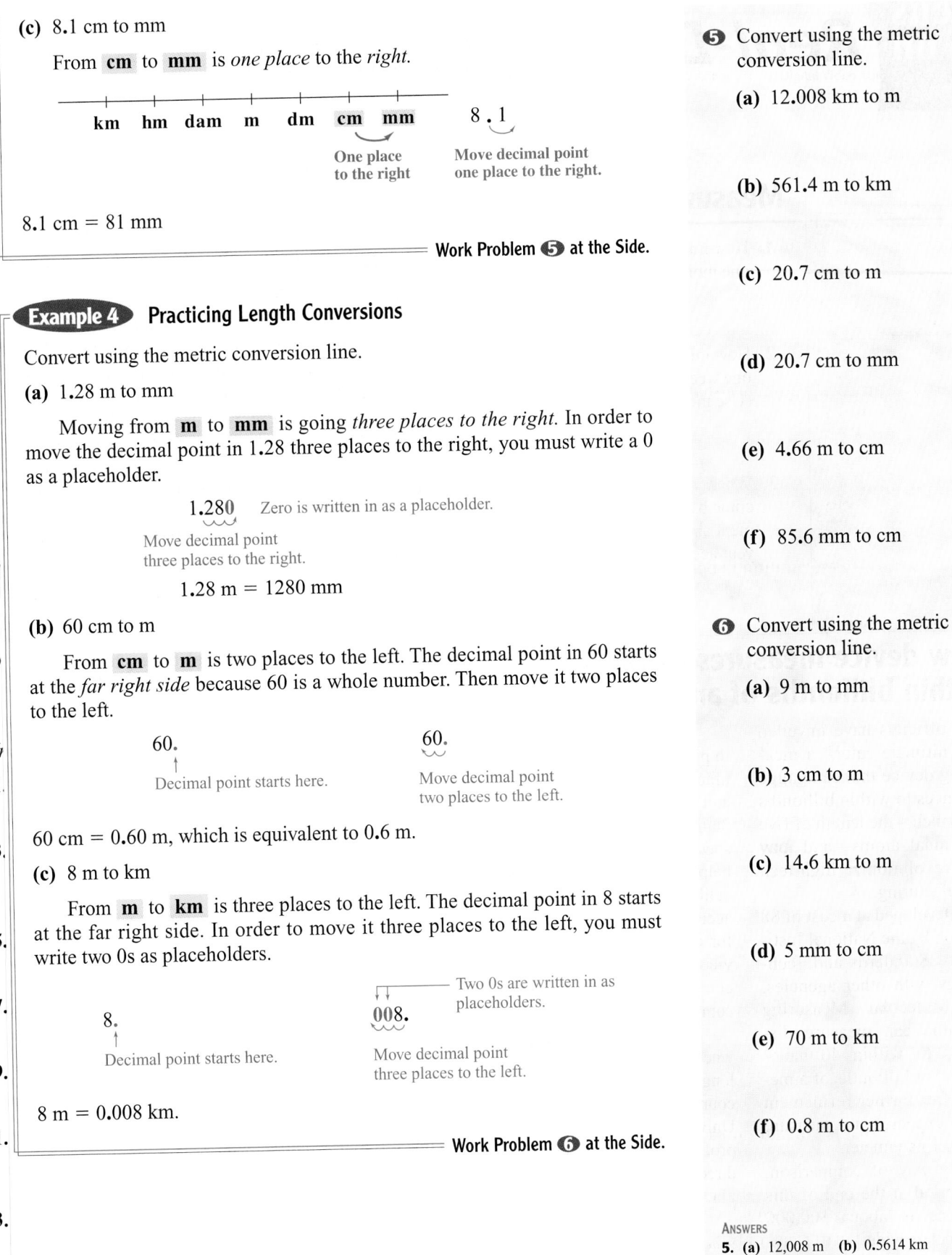

km   hm   dam   m   dm   **cm**   **mm**   8 . 1

One place   Move decimal point
to the right   one place to the right.

8.1 cm = 81 mm

═══ **Work Problem 5 at the Side.**

Example 4 **Practicing Length Conversions**

Convert using the metric conversion line.

(a) 1.28 m to mm

Moving from **m** to **mm** is going *three places to the right*. In order to move the decimal point in 1.28 three places to the right, you must write a 0 as a placeholder.

  1.280   Zero is written in as a placeholder.

Move decimal point
three places to the right.

  1.28 m = 1280 mm

(b) 60 cm to m

From **cm** to **m** is two places to the left. The decimal point in 60 starts at the *far right side* because 60 is a whole number. Then move it two places to the left.

  60.   60.

Decimal point starts here.   Move decimal point
                               two places to the left.

60 cm = 0.60 m, which is equivalent to 0.6 m.

(c) 8 m to km

From **m** to **km** is three places to the left. The decimal point in 8 starts at the far right side. In order to move it three places to the left, you must write two 0s as placeholders.

                              ┌── Two 0s are written in as
                              │   placeholders.
  8.   008.

Decimal point starts here.   Move decimal point
                              three places to the left.

8 m = 0.008 km.

═══ **Work Problem 6 at the Side.**

---

5 Convert using the metric conversion line.

(a) 12.008 km to m

(b) 561.4 m to km

(c) 20.7 cm to m

(d) 20.7 cm to mm

(e) 4.66 m to cm

(f) 85.6 mm to cm

6 Convert using the metric conversion line.

(a) 9 m to mm

(b) 3 cm to m

(c) 14.6 km to m

(d) 5 mm to cm

(e) 70 m to km

(f) 0.8 m to cm

ANSWERS
5. (a) 12,008 m **(b)** 0.5614 km
  (c) 0.207 m **(d)** 207 mm
  (e) 466 cm **(f)** 8.56 cm
6. (a) 9000 mm **(b)** 0.03 m
  (c) 14,600 m **(d)** 0.5 cm
  (e) 0.07 km **(f)** 80 cm

(left margin, partially visible): U, 1, 4, U:, 7, 9, W_, 11., 13., 15., 17., 19., 21., 23.

❷ Write the most reasonable metric unit in each blank. Choose from L and mL.

(a) I bought 8 _____ of milk at the store.

(b) The nurse gave me 10 _____ of cough syrup.

(c) This is a 100 _____ garbage can.

(d) It took 10 _____ of paint to cover the bedroom walls.

(e) My car's gas tank holds 50 _____.

(f) I added 15 _____ of oil to the pancake mix.

(g) The can of orange soda holds 350 _____.

(h) My friend gave me a 30 _____ bottle of expensive perfume.

Answers
2. (a) L (b) mL (c) L
(d) L (e) L (f) mL
(g) mL (h) mL

The capacity units you will use most often in daily life are liters (L) and *milli*liters (mL). A tiny box that measures 1 cm on every side holds exactly one milliliter. (In medicine, this small amount is also called 1 cubic centimeter, or 1 cc for short.) It takes 1000 mL to make 1 L. Here are some other useful comparisons.

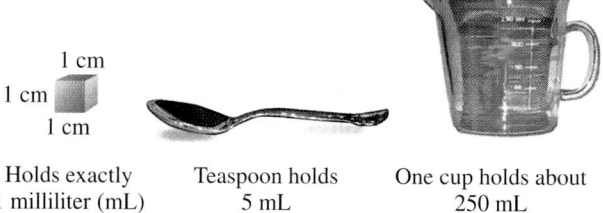

1 cm × 1 cm × 1 cm Holds exactly 1 milliliter (mL)
Teaspoon holds 5 mL
One cup holds about 250 mL

Example 1 Using Metric Capacity Units

Write the most reasonable metric unit in each blank. Choose from L and mL.

(a) The bottle of shampoo held 500 _____.
500 mL because 500 L would be about 500 quarts, which is too much.

(b) I bought a 2 _____ carton of orange juice.
2 L because 2 mL would be less than a teaspoon.

Work Problem ❷ at the Side.

2 Convert among metric capacity units. Just as with length units, you can convert between milliliters and liters using unit fractions.

Metric Capacity Relationships

1L = 1000 mL, so the unit fractions are:

$$\frac{1\ L}{1000\ mL} \quad or \quad \frac{1000\ mL}{1\ L}$$

Or you can use a metric conversion line to decide how to move the decimal point.

1000   100   10   1 $\frac{1}{10}$ $\frac{1}{100}$ $\frac{1}{1000}$
kL   hL   daL L   dL   cL   mL

Example 2 Converting among Metric Capacity Units

Convert using the metric conversion line or unit fractions.

(a) 2.5 L to mL
Using the metric conversion line:
From **L** to **mL** is *three places* to the *right*.

2.500   Write two 0s as placeholders.

2.5 L = 2500 mL

Using unit fractions:

Multiply by a unit fraction that allows you to divide out liters.

$$\frac{2.5\ \cancel{L}}{1} \cdot \frac{1000\ mL}{1\ \cancel{L}} = 2500\ mL$$

Continued on Next Page

(b) 80 mL to L

Using the metric conversion line:

From **mL** to **L** is *three places* to the *left*.

80.　　　　　080.

↑　　　　　　⌣⌣⌣
Decimal point　Move three
starts here.　　places left.

80 mL = 0.080 L or 0.08 L

Using unit fractions:

Multiply by a unit fraction that allows you to divide out mL.

$$\frac{80 \text{ mL}}{1} \cdot \frac{1 \text{ L}}{1000 \text{ mL}}$$

$$= \frac{80}{1000} \text{ L} = 0.08 \text{ L}$$

=== **Work Problem ❸ at the Side.**

3 ▮▮ **Learn the basic metric units of weight (mass).** The **gram** is the basic metric unit for *mass*. Although we often call it "weight," there is a difference. Weight is a measure of the pull of gravity; the farther you are from the center of the earth, the less you weigh. In outer space you become weightless, but your mass, the amount of matter in your body, stays the same regardless of where you are. In science courses, it will be important to distinguish between the weight of an object and its mass. But for everyday purposes, we will use the word *weight*.

The gram is related to metric length in this way: The weight of the water in a box measuring 1 cm on every side is 1 gram. This is a very tiny amount of water (1 mL) and a very small weight. One gram is also the weight of a dollar bill or a single raisin. A nickel weighs 5 grams. A regular hamburger weighs from 175 to 200 grams.

The 1 mL of water in this box weighs 1 gram.

A nickel weighs 5 grams.

(Now look back at the information about Michael Johnson's golden shoes at the start of this chapter.)

A dollar bill weighs 1 gram.

A hamburger weighs 175 to 200 grams.

Work Problem ❹ at the Side.

❸ Convert.

(a) 9 L to mL

(b) 0.75 L to mL

(c) 500 mL to L

(d) 5 mL to L

(e) 2.07 L to mL

(f) 3275 mL to L

❹ Which things would weigh about 1 gram?

A small paperclip

A pair of scissors

One playing card from a deck of cards

A calculator

An average-sized apple

The check you wrote at the grocery store

ANSWERS
3. (a) 9000 mL **(b)** 750 mL **(c)** 0.5 L
(d) 0.005 L **(e)** 2070 mL **(f)** 3.275 L
4. paperclip, playing card, check

❺ Write the most reasonable metric unit in each blank. Choose from kg, g, and mg.

(a) A thumbtack weighs

800 _____.

(b) A teenager weighs

50 _____.

(c) This large cast-iron

frying pan weighs

1 _____.

(d) Jerry's basketball

weighed 600 _____.

(e) Tamlyn takes a

500 _____ calcium

tablet every morning.

(f) On his diet, Greg can eat

90 _____ of meat for

lunch.

(g) One strand of hair weighs

2 _____.

(h) One banana might weigh

150 _____.

To make larger or smaller weight units, we use the same **prefixes** as we did with length and capacity units. For example, *kilo* means 1000 so a *kilo*meter is 1000 meters, a *kilo*liter is 1000 liters, and a *kilo*gram is 1000 grams.

Prefix	kilo-gram	hecto-gram	deka-gram	gram	deci-gram	centi-gram	milli-gram
Meaning	1000 grams	100 grams	10 grams	1 gram	$\frac{1}{10}$ of a gram	$\frac{1}{100}$ of a gram	$\frac{1}{1000}$ of a gram
Symbol	kg	hg	dag	g	dg	cg	mg

The units you will use most often in daily life are kilograms (kg), grams (g), and milligrams (mg). *Kilo*grams are used instead of pounds. A kilogram is 1000 grams. It is about 2.2 pounds. Two packages of butter plus one stick of butter weigh about 1 kg. An average newborn baby weighs 3 to 4 kg; a college football player might weigh 100 to 130 kg.

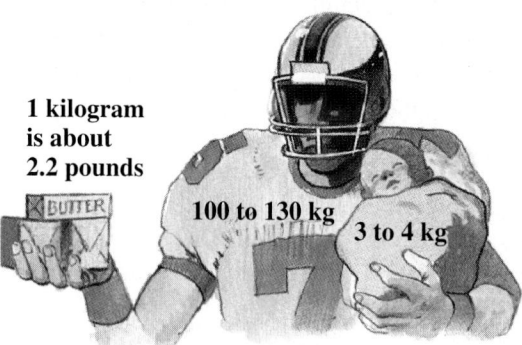

1 kilogram is about 2.2 pounds

BUTTER

100 to 130 kg

3 to 4 kg

Extremely small weights are measured in *milli*grams. It takes 1000 mg to make 1 g. Recall that a dollar bill weighs about 1 g. Imagine cutting it into 1000 pieces; the weight of one tiny piece would be 1 mg. Dosages of medicine and vitamins are given in milligrams. You will also use milligrams in science classes.

Cut a dollar bill into 1000 pieces. One tiny piece weighs 1 milligram.

Example 3 Using Metric Weight Units

Write the most reasonable metric unit in each blank. Choose from kg, g, and mg.

(a) Ramon's suitcase weighed 20 _____.
20 kg because kilograms are used instead of pounds.
20 kg is about 44 pounds.

(b) LeTia took a 350 _____ aspirin tablet.
350 mg because 350 g would be more than the weight of a hamburger, which is too much.

(c) Jenny mailed a letter that weighed 30 _____.
30 g because 30 kg would be much too heavy and 30 mg is less than the weight of a dollar bill.

Work Problem ❺ at the Side.

4 ▭ **Convert among metric weight (mass) units.** As with length and capacity, you can convert among metric weight units by using unit fractions. The unit fractions you need are shown here.

Metric Weight (Mass) Relationships

1 kg = 1000 g so the
unit fractions are:

$$\frac{1 \text{ kg}}{1000 \text{ g}} \quad \text{or} \quad \frac{1000 \text{ g}}{1 \text{ kg}}$$

1 g = 1000 mg so the
unit fractions are:

$$\frac{1 \text{ g}}{1000 \text{ mg}} \quad \text{or} \quad \frac{1000 \text{ mg}}{1 \text{ g}}$$

Or you can use a metric conversion line to decide how to move the decimal point.

```
   1000   100    10     1    1/10   1/100  1/1000
    +------+------+------+------+------+------+
    kg     hg    dag     g     dg     cg     mg
```

Example 4 Converting among Metric Weight Units

Convert using the metric conversion line or unit fractions.

(a) 7 mg to g
Using the metric conversion line:
From **mg** to **g** is *three places* to the *left*.

7. → 007.

Decimal point starts here. Move three places left.

7 mg = 0.007 g

Using unit fractions:
Multiply by a unit fraction that allows you to divide out mg.

$$\frac{7 \text{ mg}}{1} \cdot \frac{1 \text{ g}}{1000 \text{ mg}} = \frac{7}{1000} \text{ g}$$
$$= 0.007 \text{ g}$$

(b) 13.72 kg to g
Using the metric conversion line:
From **kg** to **g** is *three* places to the *right*.

13.720 Decimal point moves three places to the right.

13.72 kg = 13,720 g
↑
A comma
(not a decimal point)

Using unit fractions:
Multiply by a unit fraction that allows you to divide out kg.

$$\frac{13.72 \text{ kg}}{1} \cdot \frac{1000 \text{ g}}{1 \text{ kg}} = 13,720 \text{ g}$$
↑
A comma
(not a decimal point)

Work Problem ⑥ at the Side.

⑥ Convert.

(a) 10 kg to g

(b) 45 mg to g

(c) 6.3 kg to g

(d) 0.077 g to mg

(e) 5630 g to kg

(f) 90 g to kg

7 First decide which type of units are needed: length, capacity, or weight. Then write the most appropriate unit in the blank. Choose from km, m, cm, mm, L, mL, kg, g, and mg.

(a) Gail bought a 4 _____ can of paint.

Use _____ units.

(b) The bag of chips weighed 450 _____.

Use _____ units.

(c) Give the child 5 _____ of liquid aspirin.

Use _____ units.

(d) The width of the window is 55 _____.

Use _____ units.

(e) Akbar drives 18 _____ to work.

Use _____ units.

(f) Each computer weighs 5 _____.

Use _____ units.

(g) A credit card is 55 _____ wide.

Use _____ units.

5 ▮▮▮▮ **Distinguish among basic metric units of length, capacity, and weight (mass).** As you encounter things to be measured at home, on the job, or in your classes at school, be careful to use the correct type of measurement unit.

Use *length units* (kilometers, meters, centimeters, millimeters) to measure:

how long	how high	how far away
how wide	how tall	how far around (perimeter)
how deep	distance	

Use *capacity units* (liters, milliliters) to measure liquids (things that can be poured) such as:

water	shampoo	gasoline
milk	perfume	oil
soft drinks	cough syrup	paint

Also use liters and milliliters to describe how much liquid something can hold, such as an eyedropper, measuring cup, pail, or bathtub.

Use *weight units* (kilograms, grams, milligrams) to measure:

the weight of something   how heavy something is

In **Chapter 8** you will use square units (such as square meters) to measure area, and cubic units (such as cubic centimeters) to measure volume.

Example 5   **Using a Variety of Metric Units**

First decide which type of units are needed: length, capacity, or weight. Then write the most appropriate metric unit in the blank. Choose from km, m, cm, mm, L, mL, kg, g, and mg.

(a) The letter needs another stamp because it weighs 40 _____.

Use _____ units.

The letter weighs 40 **grams** because 40 mg is less than the weight of a dollar bill and 40 kg would be about 88 pounds.
Use **weight** units because of the word "weighs."

(b) The swimming pool is 3 _____ deep at the deep end.

Use _____ units.

The pool is 3 **meters** deep because 3 cm is only about an inch and 3 km is about 1.8 miles.
Use **length** units because of the word "deep."

(c) This is a 340 _____ can of juice.

Use _____ units.

It is a 340 **milliliter** can because 340 liters would be more than 340 quarts.

Use **capacity** units because juice is a liquid.

Work Problem 7 at the Side.

7.4 PROBLEM SOLVING WITH METRIC MEASUREMENT

1 ▦ Solve application problems involving metric measurements. One advantage of the metric system is the ease of comparing measurements in application situations. Just be sure that you are comparing similar units: mg to mg, km to km, and so on.

Once again, we will use the six problem-solving steps you learned in **Section 1.10.**

❶ Solve this problem using the six steps.

Satin ribbon is on sale at $0.89 per meter. How much will 75 cm cost, to the nearest cent?

Example 1   Solving a Metric Application

Cheddar cheese is on sale at $8.99 per kilogram. Jake bought 350 g of the cheese. How much did he pay, to the nearest cent?

Step 1   **Read** the problem. The problem asks for the cost of 350 g of cheese.

Step 2   **Work out a plan.** The price is $8.99 per *kilogram,* but the amount Jake bought is given in *grams.* Convert grams to kilograms (the unit in the price). Then multiply the weight times the cost per kilogram.

Step 3   **Estimate** a reasonable answer. Round the cost of 1 kg from $8.99 to $9. There are 1000 g in a kilogram, so 350 g is about $\frac{1}{3}$ of a kilogram. Jake is buying about $\frac{1}{3}$ of a kilogram, so $\frac{1}{3}$ of $9 = $3 as our estimate.

Step 4   **Solve** the problem. Use a unit fraction to convert 350 g to kilograms.

$$\frac{350 \cancel{g}}{1} \cdot \frac{1 \text{ kg}}{1000 \cancel{g}} = \frac{350}{1000} \text{ kg} = 0.35 \text{ kg}$$

Now multiply 0.35 kg times the cost per kilogram.

$$\frac{\$8.99}{1 \cancel{kg}} \cdot \frac{0.35 \cancel{kg}}{1} = \$3.1465 \approx \$3.15 \quad \text{(rounded)}$$

Step 5   **State the answer.** Jake paid $3.15, rounded to the nearest cent.

Step 6   **Check** your work. The answer, $3.15, is close to our estimate of $3.

━━━━━━━━━━━━ **Work Problem ❶ at the Side.**

Example 2   Solving a Metric Application

Olivia has 2.6 m of lace. How many centimeters of lace can she use to trim each of six hair ornaments? Round to the nearest tenth of a centimeter.

Step 1   **Read** the problem. The problem asks for the number of centimeters of lace for each of six hair ornaments.

Step 2   **Work out a plan.** The given amount of lace is in *meters,* but the answer must be in *centimeters.* Convert meters to centimeters, then divide by 6 (the number of hair ornaments).

━━━━ **Continued on Next Page**

❷ Lucinda's doctor wants her to take 1.2 g of medication each day in three equal doses. How many milligrams should be in each dose? Use the six problem-solving steps.

Step 3 **Estimate** a reasonable answer. To estimate, round 2.6 m of lace to 3 m. Then, 3 m = 300 cm, and 300 ÷ 6 = 50 cm as our estimate.

Step 4 **Solve** the problem. On the metric conversion line, moving from **m** to **cm** is two places to the right, so move the decimal point in 2.6 m two places to the right. Then divide by 6.

$$2.60 \text{ m} = 260 \text{ cm} \qquad \frac{260 \text{ cm}}{6 \text{ ornaments}} \approx 43.3 \text{ cm per ornament}$$

Step 5 **State the answer.** Olivia can use about 43.3 cm of lace on each ornament.

Step 6 **Check** your work. The answer, 43.3 cm, is close to our estimate of 50 cm.

Work Problem ❷ at the Side.

NOTE

In Example 1 we used a unit fraction to convert the measurement, and in Example 2 we moved the decimal point. Use whichever method you prefer. Also, there is more than one way to solve an application problem. Another way to solve Example 2 is to divide 2.6 m by 6 to get 0.4333 m of lace for each ornament. Then convert 0.4333 m to 43.3 cm (rounded to the nearest tenth).

Example 3 Solving a Metric Application

Rubin measured a board and found that the length was 3 m plus an additional 5 cm. He cut off a piece measuring 1 m 40 cm for a shelf. Find the length in meters of the remaining piece of board.

❸ Andrea has two pieces of fabric. One measures 2 m 35 cm and the other measures 1 m 85 cm. How many meters of fabric does she have in all? Use the six problem-solving steps.

Step 1 **Read** the problem. Part of a board is cut off. The problem asks what length of board, in meters, is left over.

Step 2 **Work out a plan.** The lengths involve two units, m and cm. Rewrite both lengths in meters (the unit called for in the answer), and then subtract.

Step 3 **Estimate** a reasonable answer. To estimate, 3 m 5 cm can be rounded to 3 m, because 5 cm is less than half of a meter (less than 50 cm). Round 1 m 40 cm down to 1 m. Then, 3 m − 1 m = 2 m as our estimate.

Step 4 **Solve** the problem. Rewrite the lengths in meters. Then subtract.

$$
\begin{array}{llll}
3\text{m} \rightarrow & 3.00 \text{ m} & 1\text{ m} \rightarrow & 1.0 \text{ m} \\
\text{plus 5 cm} \rightarrow & \underline{+\ 0.05 \text{ m}} & \text{plus 40 cm} \rightarrow & \underline{+\ 0.4 \text{ m}} \\
& 3.05 \text{ m} & & 1.4 \text{ m}
\end{array}
$$

$$
\begin{array}{ll}
\text{Subtract to find} & 3.05 \text{ m} \leftarrow \text{Board} \\
\text{leftover length.} & \underline{-\ 1.40 \text{ m}} \leftarrow \text{Shelf} \\
& 1.65 \text{ m} \leftarrow \text{Leftover piece}
\end{array}
$$

Step 5 **State the answer.** The length of the remaining piece of board is 1.65 m.

Step 6 **Check** your work. The answer, 1.65 m, is close to our estimate of 2 m.

Work Problem ❸ at the Side.

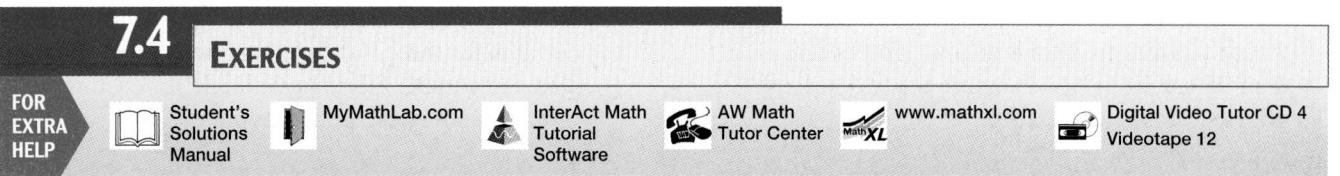

7.4 EXERCISES

| FOR EXTRA HELP | | Student's Solutions Manual | | MyMathLab.com | | InterAct Math Tutorial Software | | AW Math Tutor Center | | www.mathxl.com | | Digital Video Tutor CD 4 Videotape 12 |

Solve each application problem. Round money answers to the nearest cent. See Examples 1–3.

1. Bulk rice is on special at $0.65 per kilogram. Pam scooped some rice into a bag and put it on the scale. How much will she pay for 2 kg 50 g of rice?

2. Lanh is buying a piece of plastic tubing measuring 3 m 15 cm for the science lab. The price is $4.75 per meter. How much will Lanh pay?

3. A miniature Yorkshire terrier, one of the smallest dogs, may weigh only 500 g. But a St. Bernard, the heaviest dog, could easily weigh 90 kg. What is the difference in the weights of the two dogs, in kilograms? (*Source: Big Book of Knowledge.*)

4. The world's longest insect is the giant stick insect of Indonesia, measuring 33 cm. The fairy fly, the smallest insect, is just 0.2 mm long. How much longer is the giant stick insect, in millimeters? (*Source: Big Book of Knowledge.*)

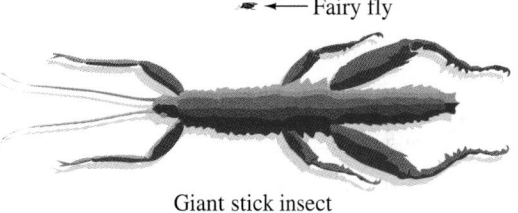

Fairy fly

Giant stick insect

5. An adult human body contains about 5 L of blood. If each beat of the heart pumps 70 mL of blood, how many times must the heart beat to pass all the blood through the heart? Round to the nearest whole number of beats. (*Source: Harper's Index.*)

6. A floor tile measures 30 cm by 30 cm and weighs 185 g. How many kilograms would a stack of 24 tiles weigh? How much would five stacks of tile weigh? (*Source: The Tile Shop.*)

7. Each piece of lead for a mechanical pencil has a thickness of 0.5 mm and is 60 mm long. Find the total length in centimeters of the lead in a package with 30 pieces. (*Source: Pentel.*) If the price of the package is $3.29, find the cost per centimeter for the lead.

60 mm

8. The apartment building caretaker puts 750 mL of chlorine into the swimming pool every day. How many liters should he order to have a one-month (30-day) supply on hand? If chlorine is sold in containers that hold 4 L, how many containers should be ordered for one month? How much chlorine will be left at the end of the month?

9. Rosa is building a bookcase. She has one board that is 2 m 8 cm long and another that is 2 m 95 cm long. How long are the two boards together, in meters?

10. Janet has 10 m 30 cm of fabric. She wants to make curtains for three windows that are all the same size. How much fabric is available for each window, to the nearest tenth of a meter?

11. In chemistry class, each of the 45 students needs 85 mL of acid. How many one-liter bottles of acid need to be ordered?

12. James needs 3 m 80 cm of wood molding to frame a picture. The price is $5.89 per meter plus a 7% sales tax. How much will James pay?

13. As a fund raiser, the PTA bought 40 kg of nuts for $113.50. They sold the nuts in 250 g bags for $2.95 each. Find the amount of profit.

14. Which case of shampoo is the better buy: a $16 case that holds 12 1-liter bottles or an $18 case that holds 36 400-mL bottles?

RELATING CONCEPTS (Exercises 15–18)   **FOR INDIVIDUAL OR GROUP WORK**

It is difficult to weigh very light objects, such as a single sheet of paper or a single staple (unless you have a very expensive scientific scale). One way around this problem is to weigh a large number of the items and then divide to find the weight of one item. Of course, before dividing, you must subtract the weight of the box or wrapper that the items are packaged in to find the net weight. **Work Exercises 15–18 in order,** *to complete the table.*

	Item	Total Weight	Weight of Packaging	Net Weight	Weight of One Item in Grams	Weight of One Item in Milligrams
15.	Box of 50 envelopes	255 g	40 g	_____	_____	_____
16.	Box of 1000 staples	350 g	20 g	_____	_____	_____
17.	Ream of paper (500 sheets)	_____	50 g	_____	_____	3000 mg
18.	Box of 100 small paper clips	_____	5 g	_____	_____	500 mg

7.5   METRIC–ENGLISH CONVERSIONS AND TEMPERATURE

1 Use unit fractions to convert between metric and English units.   Until the United States has switched completely from the English system to the metric system, it will be necessary to make conversions from one system to the other. *Approximate* conversions can be made with the help of the following table, in which the values have been rounded to the nearest hundredth or thousandth. (The only value that is exact, not rounded, is 1 inch = 2.54 cm.)

Metric to English		English to Metric	
1 kilometer	≈ 0.62 mile	1 mile	≈ 1.61 kilometers
1 meter	≈ 1.09 yards	1 yard	≈ 0.91 meter
1 meter	≈ 3.28 feet	1 foot	≈ 0.30 meter
1 centimeter	≈ 0.39 inch	1 inch	= 2.54 centimeters
1 liter	≈ 0.26 gallon	1 gallon	≈ 3.78 liters
1 liter	≈ 1.06 quarts	1 quart	≈ 0.95 liter
1 kilogram	≈ 2.20 pounds	1 pound	≈ 0.45 kilogram
1 gram	≈ 0.035 ounce	1 ounce	≈ 28.35 grams

Example 1   **Converting between Metric and English Length Units**

Convert 10 m to yards using unit fractions. Round your answer to the nearest tenth if necessary.

We're changing from a metric unit to an English unit. In the "Metric to English" side of the table, you see that 1 meter ≈ 1.09 yards. Two unit fractions can be written using that information.

$$\frac{1\text{ m}}{1.09\text{ yd}} \quad \text{or} \quad \frac{1.09\text{ yd}}{1\text{ m}}$$

Multiply by the unit fraction that allows you to divide out meters (that is, meters is in the denominator).

$$10\text{ m} \cdot \frac{1.09\text{ yd}}{1\text{ m}} = \frac{10\text{ m}}{1} \cdot \frac{1.09\text{ yd}}{1\text{ m}} = \frac{(10)(1.09\text{ yd})}{1} = 10.9\text{ yd}$$

These units should match.

10 m ≈ 10.9 yd

NOTE

You could also use the other numbers from the table involving meters and yards: 1 yard ≈ 0.91 meter.

$$\frac{10\text{ m}}{1} \cdot \frac{1\text{ yd}}{0.91\text{ m}} = \frac{10}{0.91}\text{ yd} ≈ 10.99\text{ yd}$$

The answer is slightly different because the values in the table are approximate. Also, you have to divide instead of multiply, which is usually more difficult to do without a calculator. We will use the first method in this chapter.

Work Problem ❶ at the Side.

OBJECTIVES

1 Use unit fractions to convert between metric and English units.

2 Learn common temperatures on the Celsius scale.

3 Convert temperatures by using the order of operations.

❶ Convert using unit fractions. Round your answers to the nearest tenth.

(a) 23 m to yards

(b) 40 cm to inches

(c) 5 mi to kilometers (Look at the "English to Metric" side of the table.)

(d) 12 in. to centimeters

ANSWERS
1. **(a)** 23 m ≈ 25.1 yd   **(b)** 40 cm ≈ 15.6 in.
(c) 5 mi ≈ 8.1 km   **(d)** 12 in. ≈ 30.5 cm

❷ Convert. Use the values from the table on the previous page to make unit fractions. Round answers to the nearest tenth.

(a) 17 kg to pounds

(b) 5 L to quarts

(c) 90 g to ounces

(d) 3.5 gal to liters

(e) 145 lb to kilograms

(f) 8 oz to grams

Example 2 **Converting between Metric and English Weight and Capacity Units**

Convert using unit fractions. Round your answers to the nearest tenth.

(a) 3.5 kg to pounds

Look in the "Metric to English" side of the table on the previous page to see that 1 kilogram ≈ 2.20 pounds. Use this information to write a unit fraction that allows you to divide out kilograms.

$$\frac{3.5 \text{ kg}}{1} \cdot \frac{2.20 \text{ lb}}{1 \text{ kg}} = \frac{(3.5)(2.20 \text{ lb})}{1} = 7.7 \text{ lb}$$

3.5 kg ≈ 7.7 lb

(b) 18 gal to liters

Look in the "English to Metric" side of the table to see that 1 gallon ≈ 3.78 liters. Write a unit fraction that will allow you to divide out gallons.

$$\frac{18 \text{ gal}}{1} \cdot \frac{3.78 \text{ L}}{1 \text{ gal}} = \frac{(18)(3.78 \text{ L})}{1} = 68.04 \text{ L}$$

68.04 rounded to the nearest tenth is 68.0.
18 gal ≈ 68.0 L

(c) 300 g to ounces

In the "Metric to English" side of the table, 1 gram ≈ 0.035 ounce.

$$\frac{300 \text{ g}}{1} \cdot \frac{0.035 \text{ oz}}{1 \text{ g}} = 10.5 \text{ oz}$$

300 g ≈ 10.5 oz

CAUTION

Because the metric and English systems were developed independently, almost all comparisons are approximate. Your answers should be written with the "≈" symbol to show they are approximate.

Work Problem ❷ at the Side.

2 Learn common temperatures on the Celsius scale. In the metric system, temperature is measured on the **Celsius scale.** On the Celsius scale, water freezes at 0 °C and boils at 100 °C. The small raised circle stands for "degrees" and capital **C** is for Celsius. Read the temperatures like this:

Water freezes at 0 degrees Celsius (0 °C).

Water boils at 100 degrees Celsius (100 °C).

The English temperature system, used only in the United States, is measured on the **Fahrenheit scale.** On this scale:

Water freezes at 32 degrees Fahrenheit (32 °F).

Water boils at 212 degrees Fahrenheit (212 °F).

ANSWERS
2. (a) 17 kg ≈ 37.4 lb **(b)** 5 L ≈ 5.3 qt
(c) 90 g ≈ 3.2 oz **(d)** 3.5 gal ≈ 13.2 L
(e) 145 lb ≈ 65.3 kg **(f)** 8 oz ≈ 226.8 g

The thermometer below shows some typical temperatures in both Celsius and Fahrenheit. For example, comfortable room temperature is about 20 °C or 68 °F, and normal body temperature is about 37 °C or 98.6 °F.

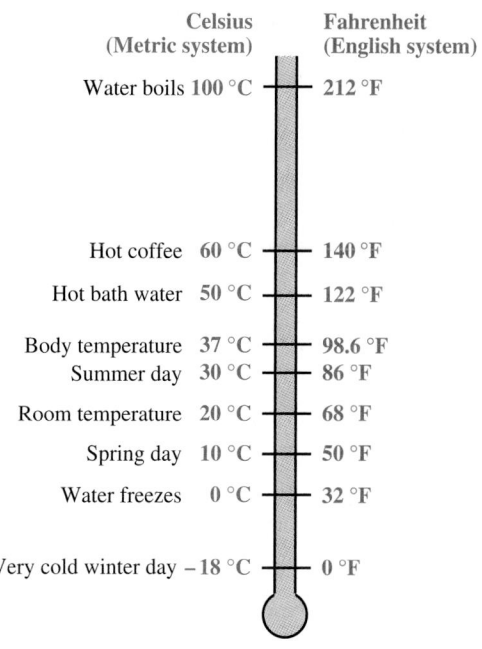

	Celsius (Metric system)	Fahrenheit (English system)
Water boils	100 °C	212 °F
Hot coffee	60 °C	140 °F
Hot bath water	50 °C	122 °F
Body temperature	37 °C	98.6 °F
Summer day	30 °C	86 °F
Room temperature	20 °C	68 °F
Spring day	10 °C	50 °F
Water freezes	0 °C	32 °F
Very cold winter day	−18 °C	0 °F

NOTE

The freezing and boiling temperatures are exact. The other temperatures are approximate. Even normal body temperature varies slightly from person to person.

Example 3 Using Celsius Temperatures

Circle the Celsius temperature that is most reasonable for each situation.

(a) Warm summer day   29 °C   64 °C   90 °C

29 °C is reasonable. 64 °C and 90 °C are too hot; they're both above the temperature of hot bath water (above 122 °F).

(b) Inside a freezer   −10 °C   3 °C   25 °C

−10 °C is the reasonable temperature because it is the only one below the freezing point of water (0 °C). Your frozen foods would start thawing at 3 °C or 25 °C.

Work Problem ❸ at the Side.

3 Convert temperatures by using the order of operations. You can use these formulas to convert between Celsius and Fahrenheit temperatures.

Celsius–Fahrenheit Conversion Formulas

Converting from Fahrenheit (F) to Celsius (C)

$$C = \frac{5(F - 32)}{9}$$

Converting from Celsius (C) to Fahrenheit (F)

$$F = \frac{9 \cdot C}{5} + 32$$

---

❸ Circle the Celsius temperature that is *most* reasonable for each situation.

(a) Set the living room thermostat at:
11 °C   21 °C   71 °C

(b) The baby has a fever of:
29 °C   39 °C   49 °C

(c) Wear a sweater outside because it's:
15 °C   25 °C   50 °C

(d) My iced tea is:
−5 °C   5 °C   30 °C

(e) Time to go swimming! It's:
95 °C   65 °C   35 °C

(f) Inside a refrigerator (not the freezer) it's:
−15 °C   0 °C   3 °C

(g) There's a blizzard outside. It's:
10 °C   0 °C   −20 °C

(h) I need hot water to get these clothes clean. It should be:
55 °C   105 °C   200 °C

ANSWERS
3. (a) 21 °C **(b)** 39 °C **(c)** 15 °C
(d) 5 °C **(e)** 35 °C **(f)** 3 °C
(g) −20 °C **(h)** 55 °C

④ Convert to Celsius.

(a) 59 °F

(b) 41 °F

(c) 212 °F

(d) 98.6 °F

⑤ Convert to Fahrenheit.

(a) 100 °C

(b) 25 °C

(c) 80 °C

(d) 5 °C

As you use these formulas, be sure to follow the order of operations from **Section 1.8.**

Order of Operations

1. Do all operations inside *parentheses* or *other grouping symbols.*
2. Simplify any expressions with *exponents* and find any *square roots.*
3. *Multiply* or *divide,* proceeding from left to right.
4. *Add* or *subtract,* proceeding from left to right.

Example 4 **Converting Fahrenheit to Celsius**

Convert 68 °F to Celsius.

Use the formula and follow the order of operations.

$$C = \frac{5(F - 32)}{9}$$

$$= \frac{5(68 - 32)}{9} \qquad \text{Work inside parentheses first.}$$

$$= \frac{5(36)}{9}$$

$$= \frac{5(\overset{4}{\cancel{36}})}{\underset{1}{\cancel{9}}} = 20 \qquad \begin{array}{l}\text{Divide out common factors.}\\ \text{Multiply.}\end{array}$$

Thus, 68 °F = 20 °C.

Work Problem ④ at the Side.

Example 5 **Converting Celsius to Fahrenheit**

Convert 15 °C to Fahrenheit.

Use the formula and follow the order of operations.

$$F = \frac{9 \cdot C}{5} + 32$$

$$= \frac{9 \cdot 15}{5} + 32$$

$$= \frac{9 \cdot \overset{3}{\cancel{15}}}{\underset{1}{\cancel{5}}} + 32 \qquad \begin{array}{l}\text{Divide out common factors.}\\ \text{Multiply.}\end{array}$$

$$= 27 + 32 \qquad \text{Add.}$$

$$= 59$$

Thus, 15 °C = 59 °F.

Work Problem ⑤ at the Side.

7.5 EXERCISES

Use the table on the first page of this section and unit fractions to make approximate conversions from metric to English or English to metric. Round your answers to the nearest tenth. See Examples 1 and 2.

1. 20 m to yards

2. 8 km to miles

3. 80 m to feet

4. 85 cm to inches

5. 16 ft to meters

6. 3.2 yd to meters

7. 150 g to ounces

8. 2.5 oz to grams

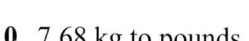 **9.** 248 lb to kilograms

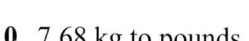 **10.** 7.68 kg to pounds

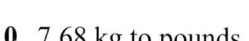 **11.** 28.6 L to quarts

 **12.** 15.75 L to gallons

13. On the first page of this chapter, we said that the 3M Company used 5 g of pure gold to coat Michael Johnson's Olympic track shoes.

    (a) How many ounces of gold were used, to the nearest tenth?

    (b) Was this enough extra weight to slow him down?

14. The fastest nerve signals in the human body travel 120 m per second. (*Source: Big Book of Knowledge.*) How many feet per second do the signals travel?

15. The heavy duty wash cycle in a dishwasher uses 8.4 gal of water. How many liters does it use, to the nearest tenth? (*Source:* Frigidaire.)

16. The rinse-and-hold cycle in a dishwasher uses only 4.5 L of water. How many gallons does it use, to the nearest tenth? (*Source:* Frigidaire.)

17. The smallest pet fish are dwarf gobies, which are half an inch long. (*Source: Big Book of Knowledge.*) How many centimeters long is a dwarf gobie, to the nearest tenth? (*Hint:* Write half an inch in decimal form.)

18. A Toshiba Protégé laptop computer weighs 4.4 lb. How many kilograms does it weigh, to the nearest tenth? (*Source:* Toshiba.)

Circle the more reasonable temperature for each situation. See Example 3.

19. A snowy day

    28 °C   28 °F

20. Brewing coffee

    80 °C   80 °F

21. A high fever

    40 °C   40 °F

22. Swimming pool water

    78 °C   78° F

23. Oven temperature

    150 °C   150 °F

24. Light jacket weather

    10 °C   10 °F

25. Would a drop in temperature of 20 Celsius degrees be more or less than a drop of 20 Fahrenheit degrees? Explain your answer.

26. Describe one advantage of switching from the Fahrenheit temperature scale to the Celsius scale. Describe one disadvantage.

Use the conversion formulas from this section and the order of operations to convert Fahrenheit temperatures to Celsius and Celsius temperatures to Fahrenheit. Round your answers to the nearest degree if necessary. See Examples 4 and 5.

27. 60 °F

28. 80 °F

29. 104 °F

30. 36 °F

31. 8 °C

32. 18 °C

33. 35 °C

34. 0 °C

Solve each application problem. Round your answers to the nearest degree if necessary.

35. The highest temperature ever recorded on Earth was 136 °F at Aziza, Libya, in 1922. Convert this temperature to Celsius. (*Source: World Almanac, 2000.*)

36. Hummingbirds have a normal body temperature of 107 °F. But on cold nights they go into a state of torpor where their body temperature drops to 39 °F. What are these temperatures in Celsius? (*Source: Wildbird.*)

37. (a) Here is the tag on a pair of Sorrel boots. In what kind of weather would you wear these boots?

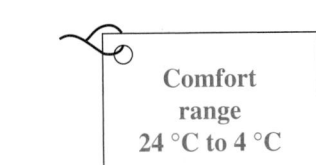

Comfort range 24 °C to 4 °C

Source: Sorrel.

(b) For what Fahrenheit temperatures are the boots designed?

(c) What range of metric temperatures would you have in January where you live?

38. (a) What are the picture directions on this tea bag package telling you to do?

for a perfect cup of tea

100 °C   4 MIN

Source: Pickwick Teas.

(b) What Fahrenheit temperature would give the same result?

RELATING CONCEPTS (Exercises 39–42) **FOR INDIVIDUAL OR GROUP WORK**

After years of discussion, the National Collegiate Athletic Association (NCAA) decided to "go metric" in 2000 by changing the lengths of swimming events from yards to meters. This will help American swimmers prepare for international competitions, such as the Olympics, where all events are in meters. (Source: NCAA Swimming and Diving Committee.) **Work Exercises 39–42 in order,** *rounding answers to the nearest tenth.*

39. All 25-yd swimming races will now be 25 m. Is the new distance longer or shorter than the old distance? By how many yards? (Use the table on the first page of this section. Round your answer to the nearest tenth.)

40. All 50-yd races will now be 50 m and all 100-yd races will be 100 m. Use your rounded answer from Exercise 39 (not the table) to find how much longer or shorter each new race will be, in yards.

41. Now use the table to find how much longer or shorter the 50 m and 100 m races will be. Why are your answers different from the ones you got in Exercise 40?

42. The 500-yd race will now be 400 m. Is the new distance longer or shorter than the old distance? By how many yards?

SUMMARY

KEY TERMS

7.1	**English system**	The English system of measurement (American system of units) is the system used for many daily activities in the United States. Common units in this system include quarts, pounds, feet, miles, and degrees Fahrenheit.
	metric system	The metric system of measurement is an international system of measurement used in manufacturing, science, medicine, sports, and other fields. Common units in this system include meters, liters, grams, and degrees Celsius.
	unit fraction	A unit fraction involves measurement units and is equivalent to 1. Unit fractions are used to convert among different measurements.
7.2	**meter**	The meter is the basic unit of length in the metric system. The symbol **m** is used for meter. One meter is a little longer than a yard.
	prefixes	Attaching a prefix to meter, liter, or gram produces larger or smaller units. For example, the prefix *kilo* means 1000 so a *kilo*meter is 1000 meters.
	metric conversion line	The metric conversion line is a line showing the various metric measurement prefixes and their size relationship to each other.
7.3	**liter**	The liter is the basic unit of capacity in the metric system. The symbl **L** is used for liter. One liter is a little more than one quart.
	gram	The gram is the basic unit of weight (mass) in the metric system. The symbol **g** is used for gram. One gram is the weight of 1 milliliter of water or one dollar bill.
7.5	**Celsius**	The Celsius scale is the scale used to measure temperature in the metric system. Water boils at 100 °C and freezes at 0 °C.
	Fahrenheit	The Fahrenheit scale is the scale used to measure temperature in the English system. Water boils at 212 °F and freezes at 32 °F.

NEW FORMULAS

Converting from Celsius to Fahrenheit:

$$F = \frac{9 \cdot C}{5} + 32$$

Converting from Fahrenheit to Celsius:

$$C = \frac{5(F - 32)}{9}$$

TEST YOUR WORD POWER

See how well you have learned the vocabulary in this chapter. Answers follow the Quick Review.

1. The **metric system**
   (a) uses meters, liters, and degrees Fahrenheit
   (b) is based on multiples of 10
   (c) is used only in the United States
   (d) has evolved over centuries.

2. The **English system** of measurement
   (a) is used throughout the world
   (b) is based on multiples of 12
   (c) uses feet, inches, quarts, and pounds
   (d) was developed by a group of scientists in 1790.

3. A **unit fraction**
   (a) has the unit you want to change in the numerator
   (b) has a denominator of 1
   (c) must be written in lowest terms
   (d) is equivalent to 1.

4. A **gram** is
   (a) the weight of 1 mL of water
   (b) abbreviated gm
   (c) equivalent to 1000 kg
   (d) approximately equal to 2.2 pounds.

5. A **meter** is
   (a) equivalent to 1000 cm
   (b) approximately equal to $\frac{1}{2}$ inch
   (c) abbreviated m with no period after it
   (d) the basic unit of capacity in the metric system.

6. The **Celsius** temperature scale
   (a) shows water freezing at 32°
   (b) is used in the English system of measurement
   (c) shows water boiling at 100°
   (d) cannot be converted to the Fahrenheit temperature scale.

QUICK REVIEW

Concepts	*Examples*

7.1 The English System of Measurement
Memorize the basic measurement relationships. Then, to convert units, multiply when changing from a larger unit to a smaller unit; divide when changing from a smaller unit to a larger unit.

Convert each measurement.
(a) 5 ft to inches

$$5 \text{ ft} = 5 \cdot \mathbf{12} = 60 \text{ in.}$$

(b) 3 lb to ounces

$$3 \text{ lb} = 3 \cdot \mathbf{16} = 48 \text{ oz}$$

(c) 15 qt to gallons

$$15 \text{ qt} = \frac{15}{4} = 3\frac{3}{4} \text{ gal}$$

7.1 Using Unit Fractions
Another, more useful, conversion method is multiplying by a unit fraction. The unit you want in the answer should be in the numerator. The unit you want to change should be in the denominator.

Convert 32 oz to pounds.

$$\left.\frac{32 \text{ oz}}{1} \cdot \frac{1 \text{ lb}}{16 \text{ oz}}\right\} \text{ Unit fraction}$$

$$= \frac{\overset{2}{\cancel{32} \cancel{oz}}}{1} \cdot \frac{1 \text{ lb}}{\underset{1}{\cancel{16} \cancel{oz}}} \quad \begin{array}{l}\text{Divide out ounces.}\\ \text{Divide out common factors.}\end{array}$$

$$= 2 \text{ lb}$$

7.1 Solving English Measurement Application Problems
To solve application problems, use the six problem-solving steps.

Use the six steps to solve this problem.

Mr. Green has 10 yd of rope. He is cutting it into eight pieces so his sailing class can practice knot tying. How many feet of rope will each of his eight students get?

Step 1   **Read** the problem carefully.

The problem asks how many feet of rope can be given to each of eight students.

Step 2   **Work out a plan.**

Convert 10 yd to feet (the unit required in the answer). Then divide by eight students.

Step 3   **Estimate** a reasonable answer.

There are 3 ft in one yard, so there are 30 ft in 10 yd. Then 30 ft ÷ 8 ≈ 4 ft as our estimate.

Step 4   **Solve** the problem.

Use a unit fraction to convert 10 yd to feet, then divide.

$$\frac{10 \text{ yd}}{1} \cdot \frac{3 \text{ ft}}{1 \text{ yd}} = 30 \text{ ft}$$

$$\frac{30 \text{ ft}}{8 \text{ students}} = 3\frac{3}{4} \text{ ft or } 3.75 \text{ ft per student}$$

Step 5   **State the answer.**

Each student gets $3\frac{3}{4}$ or 3.75 ft.

Step 6   **Check** your work.

The answer, 3.75 ft, is close to our estimate of 4 ft.

Concepts	Examples

7.2 Basic Metric Length Units
Use approximate comparisons to judge which units are appropriate:
 	
      1 mm is the thickness of a dime.
      1 cm is about $\frac{1}{2}$ inch.
      1 m is a little more than 1 yard.
      1 km is about 0.6 mile.
	Write the most reasonable metric unit in each blank. Choose from km, m, cm, and mm.
	
	The room is 6 __m__ long.
	
	A paper clip is 30 __mm__ long.
	
	He drove 20 __km__ to work.

7.2 and 7.3 *Converting within the Metric System*

Using Unit Fractions

One conversion method is to multiply by a unit fraction. Use a fraction with the unit you want in the answer in the numerator and the unit you want to change in the denominator.

Convert 9 g to kg.

$$\frac{9\ \cancel{g}}{1} \cdot \frac{1\ kg}{1000\ \cancel{g}} = \frac{9}{1000}\,kg = 0.009\ kg$$

Convert 3.6 m to cm

$$\frac{3.6\ \cancel{m}}{1} \cdot \frac{100\ cm}{1\ \cancel{m}} = 360\ cm$$

Using the Metric Conversion Line

Another conversion method is to find the unit you are given on the metric conversion line. Count the number of places to get from the unit you are given to the unit you want. Move the decimal point the same number of places and in the same direction.

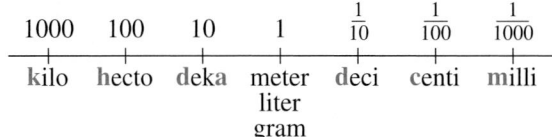

Convert.

(a) 68.2 kg to g

From **kg** to **g** is three places to the right.

   6 8.2 0 0   Decimal point is moved three places to the right.

68.2 kg = 68,200 g

(b) 300 mL to L

From **mL** to **L** is three places to the left.

   3 0 0.   Decimal point is moved three places to the left.

300 mL = 0.3 L

(c) 825 cm to m

From **cm** to **m** is two places to the left.

   8 2 5.   Decimal point is moved two places to the left.

825 cm = 8.25 m

Concepts	Examples
7.3 Basic Metric Capacity Units Use approximate comparisons to judge which units are appropriate:	Write the most appropriate metric unit in each blank. Choose from L and mL.
1 L is a little more than 1 quart.	The pail holds 12 ___L___ .
1 mL is the amount of water in a cube 1 cm on each side.	The milk carton from the vending machine holds 250 ___mL___.
5 mL is about one teaspoon.	
250 mL is about one cup.	

7.3 Basic Metric Weight (Mass) Units
Use approximate comparisons to judge which units are appropriate:

> 1 kg is about 2.2 pounds.
>
> 1 g is the weight of 1 mL of water or one dollar bill.
>
> 1 mg is $\dfrac{1}{1000}$ of a gram; very tiny!

Write the most appropriate metric unit in each blank. Choose from kg, g, and mg.

The wrestler weighed 95 ___kg___ .

She took a 500 ___mg___ aspirin tablet.

One banana weighs 150 ___g___ .

7.4 Solving Metric Application Problems
Convert units so you are comparing kg to kg, cm to cm, and so on. When a measurement involves two units, such as 6 m 20 cm, write it in terms of the unit called for in the answer (6.2 m or 620 cm).

Use the six problem-solving steps.

Step 1 **Read** the problem carefully.

Step 2 **Work out a plan.**

Step 3 **Estimate** a reasonable answer.

Step 4 **Solve** the problem.

Step 5 **State the answer.**

Step 6 **Check** your work.

Use the six steps to solve this problem.

George cut 1 m 35 cm off of a 3 m board. How long was the leftover piece, in meters?

The problem asks for the length of the leftover piece in meters.

Convert the cut-off length to meters (the unit required in the answer), and then subtract to find the "leftover."

To estimate, round 1 m 35 cm to 1 m, because 35 cm is less than half a meter. Then, 3 m − 1 m = 2 m as our estimate.

Convert the cut-off measurement to meters, then subtract.

$$
\begin{array}{ccc}
1\text{ m} \rightarrow & 1.00\text{ m} & 3.00\text{ m} \leftarrow \text{Board} \\
\text{plus } 35\text{ cm} \rightarrow & +\,0.35\text{ m} & -\,1.35\text{ m} \leftarrow \text{Cut off} \\
\hline
 & 1.35\text{ m} & 1.65\text{ m} \leftarrow \text{Left}
\end{array}
$$

The leftover piece is 1.65 m long.

The answer, 1.65 m, is close to our estimate of 2 m.

Concepts	Examples

Concepts

7.5 *Converting between Metric and English Units*
Use the values in the table of conversion factors to write a unit fraction. Because the values in the table are rounded, your answers will be approximate.

Examples

Convert. Round answers to the nearest tenth.

(a) 23 m to yards
From the table, 1 meter ≈ 1.09 yards.

$$\frac{23 \ \cancel{m}}{1} \cdot \frac{1.09 \ yd}{1 \ \cancel{m}} = 25.07 \ yd$$

25.07 rounds to 25.1, so 23 m ≈ 25.1 yd.

(b) 4 oz to grams

$$\frac{4 \ \cancel{oz}}{1} \cdot \frac{28.35 \ g}{1 \ \cancel{oz}} = 113.4 \ g$$

So 4 oz ≈ 113.4 g.

7.5 *Common Celsius Temperatures*
Use approximate and exact comparisons to judge which temperatures are appropriate.

Exact comparisons:
0 °C is the freezing point of water (32 °F).

100 °C is the boiling point of water (212 °F).

Approximate comparisons:
10 °C for a spring day (50 °F)

20 °C for room temperature (68 °F)

30 °C for summer day (86 °F)

37 °C for body temperature (98.6 °F)

Circle the Celsius temperature that is most reasonable.

(a) Hot summer day:
   ⟨35 °C⟩   90 °C   110 °C

(b) The first snowy day in winter:
   −20 °C   ⟨0 °C⟩   15 °C

7.5 *Converting between Fahrenheit and Celsius Temperatures*
Use these formulas.

$$C = \frac{5(F - 32)}{9}$$

Convert 176 °F to Celsius.

$$C = \frac{5(176 - 32)}{9}$$

$$= \frac{5(\overset{16}{\cancel{144}})}{\underset{1}{\cancel{9}}} \qquad \text{Divide out common factors.}$$
$$\text{Then multiply.}$$

$$= 80$$

176 °F = 80 °C

$$F = \frac{9 \cdot C}{5} + 32$$

Convert 80 °C to Fahrenheit.

$$F = \frac{9 \cdot 80}{5} + 32$$

$$= \frac{9 \cdot \overset{16}{\cancel{80}}}{\underset{1}{\cancel{5}}} + 32 \qquad \text{Divide out common factors.}$$
$$\text{Then multiply.}$$

$$= 144 + 32 \qquad \text{Add.}$$

$$= 176$$

80 °C = 176 °F

ANSWERS TO TEST YOUR WORD POWER

1. (b) *Examples:* 10 meters = 1 dekameter; 100 meters = 1 hectometer; 1000 meters = 1 kilometer.
2. (c) *Examples:* feet and inches are used to measure length, quarts to measure capacity, pounds to measure weight.   **3. (d)** *Example:* Because 12 in. = 1 ft, the unit fraction $\dfrac{12 \text{ in.}}{1 \text{ ft}}$ is equivalent to $\dfrac{12 \text{ in.}}{12 \text{ in.}} = 1$.

4. (a) *Example:* A small box measuring 1 cm on every edge holds exactly 1 mL of water, and the water weighs 1 g.   **5. (c)** *Example:* A measurement of 16 meters is written 16 m (without a period).
6. (c) *Example:* In the metric system, water freezes at 0 °C and boils at 100 °C. The English system uses the Fahrenheit temperature scale where water freezes at 32 °F and boils at 212 °F.

Chapter 7

REVIEW EXERCISES

[7.1] *Fill in the blanks with the measurement relationships you have memorized.*

1. 1 lb = _____ oz

2. _____ ft = 1 yd

3. 1 T = _____ lb

4. _____ qt = 1 gal

5. 1 hr = _____ min

6. 1 c = _____ fl oz

7. _____ sec = 1 min

8. _____ ft = 1 mi

9. _____ in. = 1 ft

Convert using unit fractions.

10. 4 ft = _____ in.

11. 6000 lb = _____ T

12. 64 oz = _____ lb

13. 18 hr = _____ day

14. 150 min = _____ hr

15. $1\frac{3}{4}$ lb = _____ oz

16. $6\frac{1}{2}$ ft = _____ in.

17. 7 gal = _____ c

18. 4 days = _____ sec

19. The average depth of the world's oceans is 12,460 ft. (*Source: Handy Ocean Answer Book.*)

   (a) What is the average depth in yards?

   (b) What is the average depth in miles, to the nearest tenth?

20. During the first year of a program to recycle office paper, a company recycled 123,260 pounds of paper. The company received $40 per ton for the paper. (*Source: I. C. System.*) How much money did the company make? Use the six problem-solving steps.

[7.2] *Write the most reasonable metric length unit in each blank. Choose from km, m, cm, and mm.*

21. My thumb is 20 _____ wide.

22. Her waist measurement is 66 _____.

23. The two towns are 40 _____ apart.

24. A basketball court is 30 _____ long.

25. The height of the picnic bench is 45 _____.

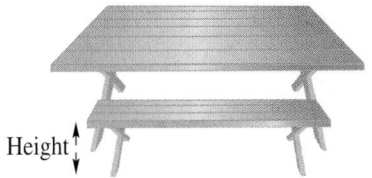

Height

26. The eraser on the end of my pencil is 5 _____ long.

Convert using unit fractions or the metric conversion line.

27. 5 m to cm

28. 8.5 km to m

29. 85 mm to cm

30. 370 cm to m   **31.** 70 m to km   **32.** 0.93 m to mm

[7.3] *Write the most reasonable metric unit in each blank. Choose from L, mL, kg, g, and mg.*

33. The eyedropper holds 1 _____.   **34.** I can heat 3 _____ of water in this pan.

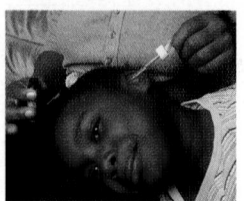   

35. Loretta's hammer weighed 650 _____.   **36.** Yongshu's suitcase weighed 20 _____ when it was packed.

37. My fish tank holds 80 _____ of water.   **38.** I'll buy the 500 _____ bottle of mouthwash.

39. Mara took a 200 _____ antibiotic pill.   **40.** This piece of chicken weighs 100 _____.

Convert using unit fractions or the metric conversion line.

41. 5000 mL to L   **42.** 8 L to mL   **43.** 4.58 g to mg

44. 0.7 kg to g   **45.** 6 mg to g   **46.** 35 mL to L

[7.4] *Solve each application problem. Use the six problem-solving steps.*

47. Each serving of punch at the wedding reception will be 180 mL. How many liters of punch are needed for 175 guests?

48. Jason is serving a 10 kg turkey to 28 people. How many grams of meat is he allowing for each person? Round to the nearest whole gram.

49. Yerald weighed 92 kg. Then he lost 4 kg 750 g. What is his weight now, in kilograms?

50. Young-Mi bought 2 kg 20 g of onions. The price was $1.49 per kilogram. How much did she pay, to the nearest cent?

[7.5] *Use the table on the first page of **Section 7.5** and unit fractions to make approximate conversions. Round your answers to the nearest tenth if necessary.*

51. 6 m to yards   **52.** 30 cm to inches

53. 108 km to miles

54. 800 mi to kilometers

55. 23 qt to liters

56. 41.5 L to quarts

*Write the appropriate **metric** temperature in each blank.*

57. Water freezes at _____.

58. Water boils at _____.

59. Normal body temperature is about _____.

60. Comfortable room temperature is about _____.

*Use the conversion formulas in **Section 7.5** to convert each temperature to Fahrenheit or to Celsius. Round to the nearest degree if necessary.*

61. 77 °F

62. 92 °F

63. 6 °C

64. Water coming into a dishwasher should be at least 49 °C to clean the dishes properly. What Fahrenheit temperature is that? (*Source:* Frigidaire.)

MIXED REVIEW EXERCISES

Write the most reasonable metric unit in each blank. Choose from km, m, cm, mm, L, mL, kg, g, and mg.

65. I added 1 _____ of oil to my car.

66. The box of books weighed 15 _____.

67. Larry's shoe is 30 _____ long.

68. Jan used 15 _____ of shampoo on her hair.

69. My fingernail is 10 _____ wide.

70. I walked 2 _____ to school.

71. The tiny bird weighed 15 _____.

72. The new library building is 18 _____ wide.

73. The cookie recipe uses 250 _____ of milk.

74. Renee's pet mouse weighs 30 _____.

75. One postage stamp weighs 90 _____.

76. I bought 30 _____ of gas for my car.

Convert the following using unit fractions, the metric conversion line, or the temperature conversion formulas.

77. 10.5 cm to millimeters

78. 45 min to hours

79. 90 in. to feet

80. 1.3 m to centimeters

81. 25 °C to Fahrenheit

82. $3\frac{1}{2}$ gal to quarts

83. 700 mg to grams

84. 0.81 L to milliliters

85. 5 lb to ounces

86. 60 kg to grams

87. 1.8 L to milliliters

88. 86 °F to Celsius

89. 0.36 m to centimeters

90. 55 mL to liters

Solve each application problem. Use the six problem-solving steps in Exercises 91–92.

91. Peggy had a board measuring 2 m 4 cm. She cut off 78 cm. How long is the board now, in meters?

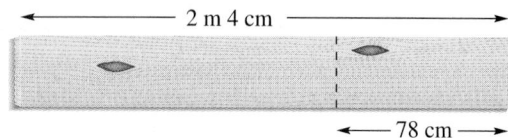

92. During the 12-day Minnesota State Fair, one of the biggest in the United States, Sweet Martha's booth sells an average of 3000 pounds of cookies per day. How many tons of cookies are sold in all? (*Source: Minneapolis Star Tribune.*)

93. Olivia is sending a recipe to her mother in Mexico. Among other things, the recipe calls for 4 oz of rice and a baking temperature of 350 °F. Convert these measurements to metric, rounding to the nearest gram and nearest degree.

94. While on vacation in Canada, Jalo became ill and went to a health clinic. They said he weighed 80.9 kg and was 1.83 m tall. Find his weight in pounds and height in feet. Round to the nearest tenth.

Fill in the blank spaces in this table. Then use the information in the table to answer Exercises 99–101.

		Weight		Battery Life	
	Cell Phone	ounces (nearest tenth)	grams (nearest whole)	Talking	Standby
95.	Audiovox CDM 9000	4.8 oz		180 min	170 hr
96.	Motorola V2282		150 g	3.5 hr	135 hr
97.	NeoPoint 1000		180 g	2.5 hr	$1\frac{2}{3}$ days
98.	Samsung SCH-3500	5.5 oz		168 min	$6\frac{1}{4}$ days

Source: Access.

99. Suppose you want a lightweight cell phone. List the phones in order from the lightest to the heaviest.

100. (a) List the phones in order from the one with the longest battery life for talking to the one with the shortest.

(b) What is the difference in length of talking battery life between the longest and shortest?

101. Which cell phone would you choose, based on the table? Explain your answer.

Convert the following measurements.

1. 9 gal = _____ qt

2. 45 ft = _____ yd

3. 135 min = _____ hr

4. 9 in. = _____ ft

5. $3\frac{1}{2}$ lb = _____ oz

6. 5 days = _____ min

Write the most reasonable metric unit in each blank. Choose from km, m, cm, mm, L, mL, kg, g, and mg.

7. My husband weighs 75 _____.

8. I hiked 5 _____ this morning.

9. She bought 125 _____ of cough syrup.

10. This apple weighs 180 _____.

11. This page is about 21 _____ wide.

12. My watch band is 10 _____ wide.

13. I bought 10 _____ of soda for the picnic.

14. The bracelet is 16 _____ long.

Convert the following measurements.

15. 250 cm to meters

16. 4.6 km to meters

17. 5 mm to centimeters

18. 325 mg to grams

19. 16 L to milliliters

20. 0.4 kg to grams

21. 10.55 m to centimeters

22. 95 mL to liters

1. _____

2. _____

3. _____

4. _____

5. _____

6. _____

7. _____

8. _____

9. _____

10. _____

11. _____

12. _____

13. _____

14. _____

15. _____

16. _____

17. _____

18. _____

19. _____

20. _____

21. _____

22. _____

23. _____

23. The rainiest place in the world is Mount Waialeale in Hawaii, which receives 460 in. of rain each year. (*Source:* National Geographic Society.) What is the average rainfall per month, in feet, to the nearest tenth?

24. (a) _____

(b) _____

24. A 6-inch Subway "Vegie Delite" sandwich has 590 mg of sodium. A 6-inch "Super Subway Melt" sandwich has 2.9 g of sodium. (*Source:* Subway.)

(a) How much more sodium is in the "Super Melt" sandwich, in milligrams, than in the "Vegie Delite"?

(b) The recommended amount of sodium is less than 2400 mg daily. How much more or less sodium does the "Super Melt" have than the recommended daily amount?

Pick the metric temperature that is most appropriate in each situation.

25. _____

25. The water is almost boiling.
210 °C   155 °C   95 °C

26. The tomato plants may freeze tonight.
30 °C   20 °C   0 °C

26. _____

Use the table from Section 7.5 and unit fractions to convert the following measurements. Round your answers to the nearest tenth if necessary.

27. _____

27. 6 ft to meters

28. 125 lb to kilograms

28. _____

29. _____

29. 50 L to gallons

30. 8.1 km to miles

30. _____

Use the conversion formulas to convert each temperature. Round your answers to the nearest degree if necessary.

31. _____

31. 74 °F to Celsius

32. 2 °C to Fahrenheit

32. _____

Solve this application problem.

33. _____

33. Denise is making five matching pillows. She needs 1 m 20 cm of braid to trim each pillow. If the braid costs $3.98 per yard, how much will she spend to trim the pillows, to the nearest cent? (First find the number of meters of braid Denise needs.)

34. _____

34. Describe two benefits the United States would achieve by switching entirely to the metric system.

First use front end rounding to round each number and estimate the answer. Then find the exact answer.

1. *Estimate:*   *Exact:*

$$107.5$$
$$2.548$$
$$+ \underline{} \qquad + 68.79$$

2. *Estimate:*   *Exact:*

$$31{,}007$$
$$- \underline{} \qquad - 829$$

3. *Estimate:*   *Exact:*

$$92{,}075$$
$$\times \underline{} \qquad \times 183$$

4. *Estimate:*   *Exact:*

$$56.52$$
$$\times \underline{} \qquad \times 4.7$$

5. *Estimate:*   *Exact:*

$$\underline{}\big) \qquad 37\overline{)19{,}610}$$

6. *Estimate:*   *Exact:*

$$\underline{}\big) \qquad 8.3\overline{)38.18}$$

7. *Estimate:*   *Exact:*

$$1\frac{7}{10}$$
$$+ \underline{} \qquad + 3\frac{4}{5}$$

8. *Estimate:*   *Exact:*

$$5\frac{1}{2}$$
$$- \underline{} \qquad - 1\frac{2}{7}$$

9. *Exact:*

$$3\frac{1}{6} \cdot 4\frac{2}{3} =$$

Estimate:

$$\underline{} \cdot \underline{} = \underline{}$$

10. *Exact:*

$$2\frac{1}{4} \div \frac{9}{10} =$$

Estimate:

$$\underline{} \div \underline{} = \underline{}$$

Add, subtract, multiply, or divide as indicated. Write answers to fraction problems in lowest terms and as whole or mixed numbers when possible.

11. $3 - 2\frac{5}{16}$

12. $7 + 484{,}099 + 3939$

13. $12 \cdot 2\frac{2}{9}$

14. $0.86 \div 0.066$
Round to nearest tenth.

15. $\frac{3}{8} + \frac{5}{6}$

16. $8 - 0.9207$

17. $3\frac{3}{4} \div 6$

18. Write your answer using R for the remainder.

$$47\overline{)14{,}467}$$

19. $(2.54)(0.003)$

Use the order of operations to simplify each expression.

20. $24 - 12 \div 6 \cdot 8 + (25 - 25)$

21. $3^2 + 2^5 \cdot \sqrt{64}$

22. Write 307.19 in words.

23. Write eighty-two ten-thousandths in numbers.

24. Arrange in order from smallest to largest.

0.67   0.067   0.6   0.6007

25. Find the best buy on extra large disposable diapers.

Package of 16 for $3.87
Package of 22 for $5.96
Package of 36 for $11.69

Complete the table.

Fraction or Mixed Number	Decimal	Percent
$\frac{7}{8}$	**26.** _____	**27.** _____
28. _____	0.05	**29.** _____
30. _____	**31.** _____	350%

The bar graph shows the amount of sales where people purchased items from businesses on the Internet. Use the graph to answer Exercises 32–34. Write each ratio as a fraction in lowest terms.

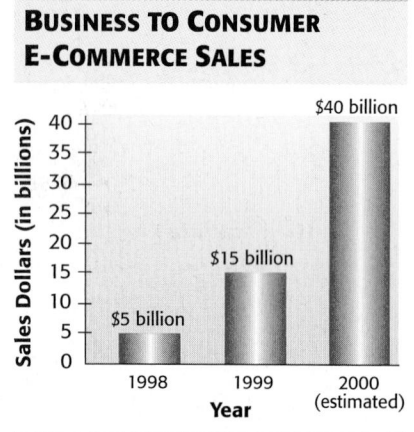

BUSINESS TO CONSUMER E-COMMERCE SALES

Sales Dollars (in billions)

$40 billion
$15 billion
$5 billion

1998   1999   2000 (estimated)
Year

Source: BizRate.com as seen in the Industry Standard.

32. What is the ratio of 1998 sales to 1999 sales?

33. Find the ratio of 2000 sales to 1998 sales.

34. Write the ratio of 2000 sales to 1999 sales.

Find the unknown number in each proportion. Round your answers to hundredths if necessary.

35. $\dfrac{x}{16} = \dfrac{3}{4}$

36. $\dfrac{0.9}{0.75} = \dfrac{2}{x}$

Solve each percent problem.

37. $4 is what percent of $80?

38. 36 hours is 120% of what number of hours?

*Convert the following measurements. Use the table in **Section 7.5** and the temperature conversion formulas when necessary.*

39. $2\dfrac{1}{2}$ ft to inches

40. 105 sec to minutes

41. 2.8 m to centimeters

42. 65 mg to grams

43. 198 km to miles

44. 50 °F to Celsius

Write the most reasonable metric unit in each blank. Choose from km, m, cm, mm, L, mL, kg, g, and mg.

45. Ron bought the tube of toothpaste weighing

    100 _____.

46. The teacher's desk is 140 _____ long.

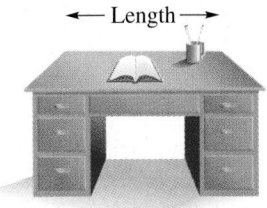

←— Length —→

47. The hallway is 3 _____ wide.

48. Joe's hammer weighed 1 _____.

49. Anne took a 500 _____ tablet of vitamin C.

50. Tia added 125 _____ of milk to her cereal.

Circle the metric temperature that is most appropriate in each situation.

51. John has a slight fever.

    38 °C   70 °C   99 °C

52. You'll need a light jacket outside.

    0 °C   12 °C   45 °C

Solve each application problem.

53. Calbert works at a WAL-MART store and used his employee discount to buy a $189.94 DVD player at 10% off. Find the amount of his discount, to the nearest cent, and the sale price. (*Source:* WAL-MART.)

54. Danielle had a roll of 35 mm film developed. She received 24 prints for $10.35 plus $6\dfrac{1}{2}$% sales tax. What was the cost per print, to the nearest cent?

55. Bags of slivered almonds weigh 4 oz each. They are packed in a carton that weighs 12 oz. How many pounds would a carton containing 48 bags weigh?

56. On the Illinois map, one centimeter represents 12 km. The center of Springfield is 7.8 cm from the center of Bloomington on the map.

(a) What is the actual distance in kilometers?

(b) What is the actual distance in miles, to the nearest tenth?

57. Dimitri took out a $3\frac{1}{2}$ year car loan for $8750 at 9% simple interest. Find the interest and the total amount due on the loan.

58. On a 35-problem math test, Juana solved 31 problems correctly. What percent of the problems were correct? Round to the nearest tenth of a percent.

59. Mark bought 650 g of maple sugar candy on his vacation in Montreal. The candy is priced at $14.98 per kilogram. How much did Mark pay, to the nearest cent?

60. The Jackson family is making three kinds of holiday cookies that require brown sugar. The recipes call for $2\frac{1}{4}$ cups, $1\frac{1}{2}$ cups, and $\frac{3}{4}$ cup, respectively. They bought two packages of brown sugar, each holding $2\frac{1}{3}$ cups. The amount bought is how much more or less than the amount needed?

61. Akuba is knitting a scarf. Six rows of knitting result in 5 cm of scarf. At that rate, how many rows will she knit to make a 100 cm scarf?

62. A survey of the 5600 students on our campus found that $\frac{3}{8}$ of the students work 20 hours or more per week. How many students work 20 hours or more?

Use the table to answer Exercises 63–65.

OLYMPIC TELEVISION COVERAGE

Year	Olympic Host	Network	Hours	Cost (in millions)
1964	Tokyo	NBC	14.5	$1.5
1976	Montreal	ABC	76.5	$25
1984	Los Angeles	ABC	180	$225
1992	Barcelona	NBC	161	$401
2000	Sydney	NBC	171.5	$715

Source: USA Today.

63. (a) What is the increase in the hours of television coverage from 1964 to 1984?

(b) The hours of coverage in 1964 are what percent of the hours of coverage in 1984, to the nearest tenth of a percent?

64. (a) What is the decrease in the hours of coverage from 1984 to 2000?
(b) What is the percent of decrease, to the nearest tenth?

65. What was the cost per hour of coverage in
(a) 1984;
(b) 1992;
(c) 2000?
Round to the nearest tenth.

Geometry

8

In May 1998, an F-1 tornado (wind speeds of 73 to 112 mph) blew down many large trees in Shoreview, MN. Now, forestry major Bill Masterson is making an inventory of Shoreview's trees to help the city plan replacement plantings. He measures the circumference of each tree at chest height. Then, using the formulas in Section 8.6, he calculates the diameter. This information helps him analyze growth patterns and tree age. (See Exercise 29 in Section 8.6.) (*Source: Shoreview Press.*)

ADDISON - WESLEY
MyMathLab.com
You're Connected

8.1 BASIC GEOMETRIC TERMS

Geometry developed centuries ago when people needed a way to measure land. The name *geometry* comes from the Greek words *ge,* meaning earth, and *metron,* meaning measure. Today we still use geometry to measure farmland. It is also important in architecture, construction, navigation, art and design, physics, chemistry, and astronomy. You can use it at home when you buy carpet or wallpaper, hang a picture, or build a fence. This chapter discusses the basic terms of geometry and the common geometric shapes that are all around us.

Geometry starts with the idea of a point. A **point** is a location in space. It has no length or width. A point is represented by a dot and is named by writing a capital letter next to the dot.

Point *P*

1 **Identify lines, line segments, and rays.** A **line** is a straight row of points that goes on forever in both directions. A line is drawn by using arrowheads to show that it never ends. The line is named by using the letters of any two points on the line.

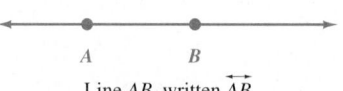

Line *AB*, written $\overleftrightarrow{AB}$

A piece of a line that has two endpoints is called a **line segment.** A line segment is named using its endpoints. The segment with endpoints *P* and *Q* is shown below. It can be named $\overline{PQ}$ or $\overline{QP}$.

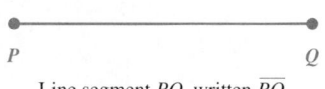

Line segment *PQ*, written $\overline{PQ}$

A **ray** is a part of a line that has only one endpoint and goes on forever in one direction. A ray is named by using the endpoint and some other point on the ray. The endpoint is always mentioned first.

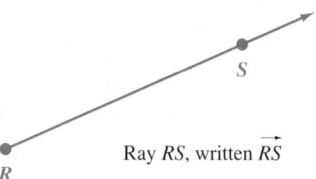

Ray *RS*, written $\overrightarrow{RS}$

Example 1 **Identifying Lines, Rays, and Line Segments**

Identify each figure as a line, line segment, or ray.

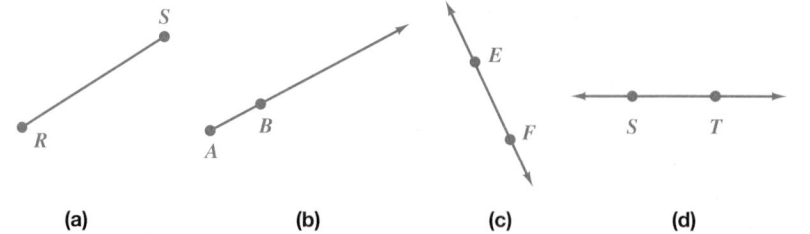

(a)   (b)   (c)   (d)

Figure **(a)** has two endpoints, so it is a line segment.
Figure **(b)** starts at point *A* and goes on forever in one direction, so it is a ray.
Figures **(c)** and **(d)** go on forever in both directions, so they are lines.

Work Problem ❶ at the Side.

❶ Identify each figure as a line, line segment, or ray.

(a)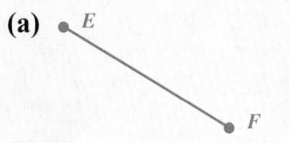

(b)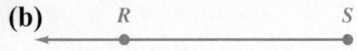

(c)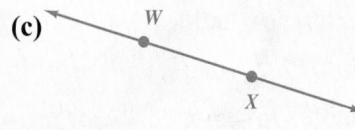

(d)

2 ▧ **Identify parallel and intersecting lines.**   A *plane* is an infinitely large flat surface. A floor or a wall is a part of a plane. Lines that are in the same plane, but that never intersect (never cross), are called **parallel lines,** while lines that cross are called **intersecting lines.** (Think of an intersection, where two streets cross each other.)

Example 2   **Identifying Parallel and Intersecting Lines**

Label each pair of lines as parallel or intersecting.

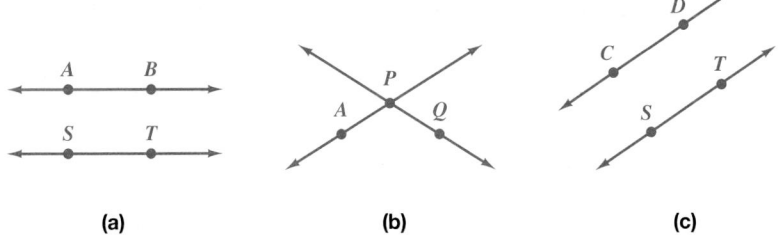

(a)   (b)   (c)

The lines in Figures **(a)** and **(c)** never intersect. They are parallel lines. The lines in Figure **(b)** cross at *P*, so they are intersecting lines.

═══ **Work Problem ❷ at the Side.**

3 ▧ **Identify and name angles.**   An **angle** is made up of two rays that start at a common endpoint. This common endpoint is called the *vertex.*

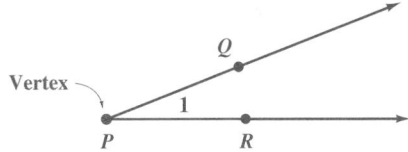

The rays *PQ* and *PR* are called *sides.* The angle can be named in four different ways, as shown below.

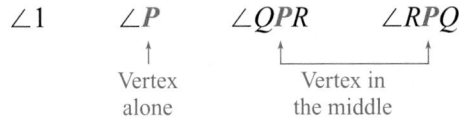

Naming an Angle

When naming an angle, the vertex is written alone or it is written in the middle of two other points. If two or more angles have the *same vertex,* as in Example 3 below, do *not* use the vertex alone to name an angle.

Example 3   **Identifying and Naming an Angle**

Name the highlighted angle.

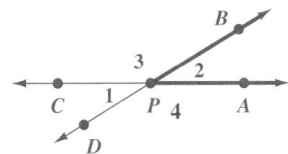

The angle can be named ∠*BPA*, ∠*APB*, or ∠2. It cannot be named ∠*P*, using the vertex alone, because four different angles have *P* as their vertex.

═══ **Work Problem ❸ at the Side.**

❷ Label each pair of lines as parallel or intersecting.

(a)
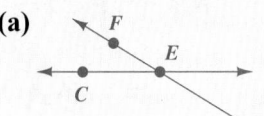

(b)
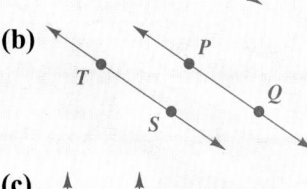

(c)
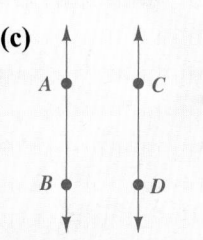

❸ **(a)** Name the highlighted angle in three different ways.

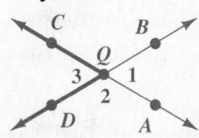

(b) Darken the rays that make up ∠*ZTW*.

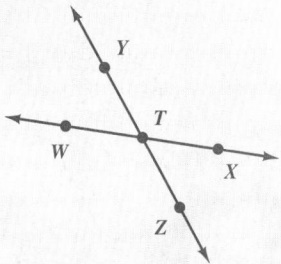

(c) Name this angle in four different ways.

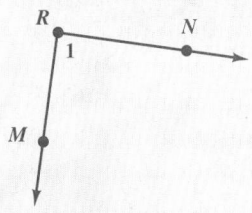

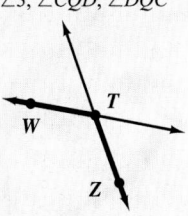

4 ▦   Classify angles as right, acute, straight, or obtuse.   Angles can be measured in **degrees.** The symbol for degrees is a small, raised circle °. Think of the minute hand on a clock as a ray of an angle. Suppose it is at 12:00. During one hour of time, the minute hand moves around in a complete circle. It moves 360 *degrees*, or 360°. In half an hour, at 12:30, the minute hand has moved half way around the circle, or 180°. An angle of 180° is called a **straight angle.** When two rays go in opposite directions, the rays form a straight angle.

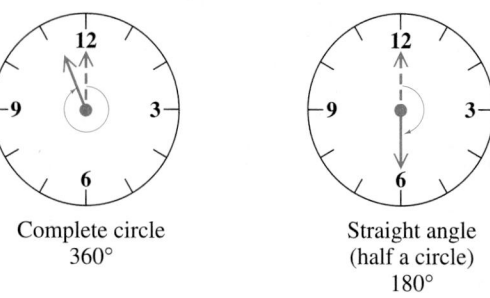

Complete circle
360°

Straight angle
(half a circle)
180°

In a quarter of an hour, at 12:15, the minute hand has moved $\frac{1}{4}$ of the way around the circle, or 90°. An angle of 90° is called a **right angle.** Sometimes you hear it called a *square angle.* The minute hands at 12:00 and 12:15 form one corner of a square. So, to show that an angle is a **right angle**, we draw a **small square** at the vertex.

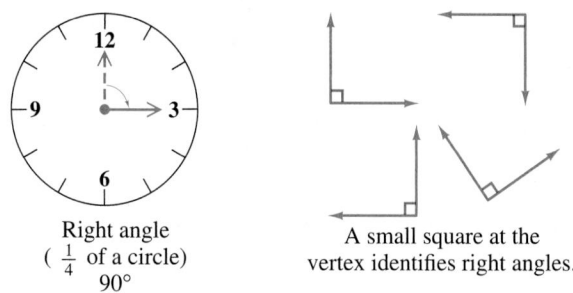

Right angle
($\frac{1}{4}$ of a circle)
90°

A small square at the
vertex identifies right angles.

In one minute, the minute hand moves 6°. From this you can tell that an angle of 1° is very small.

1° angle

Some other terms used to describe angles are shown below.

Acute angles measure less than 90°.

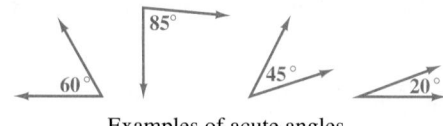

Examples of acute angles

Obtuse angles measure more than 90° but less than 180°.

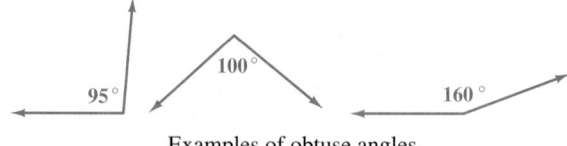

Examples of obtuse angles

Section 10.1 shows you how to use a tool called a *protractor* to measure the number of degrees in an angle.

Classifying Angles

Acute angles measure less than 90°.
Right angles measure exactly 90°.
Obtuse angles measure more than 90° but less than 180°.
Straight angles measure exactly 180°.

NOTE
Angles can also be measured in radians, which you will learn about in a later math course.

Example 4   **Classifying Angles**

Label each angle as acute, right, obtuse, or straight.

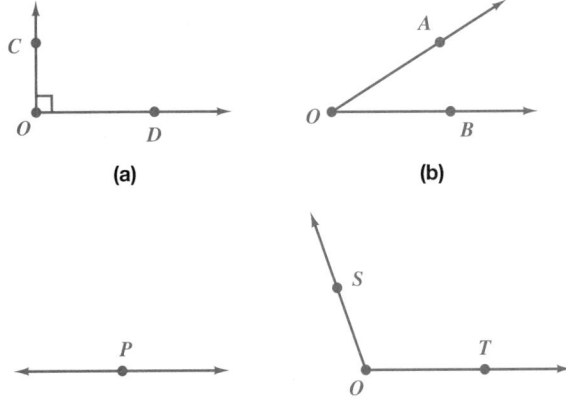

Figure **(a)** shows a *right angle* (exactly 90° and identified by a small square at the vertex).
Figure **(b)** shows an *acute angle* (less than 90°).
Figure **(c)** shows a *straight angle* (exactly 180°).
Figure **(d)** shows an *obtuse angle* (more than 90° but less than 180°).

======================== **Work Problem ❹ at the Side.**

5▭ **Identify perpendicular lines.**   Two lines are called **perpendicular lines** if they intersect to form a right angle.

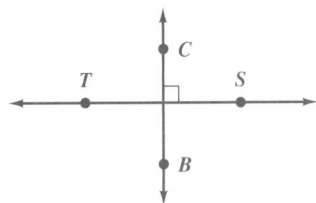

Lines *CB* and *ST* are **perpendicular** because they intersect at right angles.
This can be written in the following way: $\overleftrightarrow{CB} \perp \overleftrightarrow{ST}$.

❹ Label each figure as an acute, right, obtuse, or straight angle.

(a)
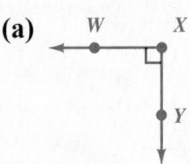

(b)
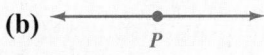

(c)
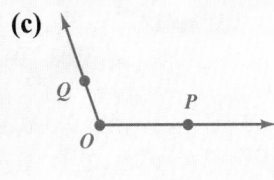

(d)
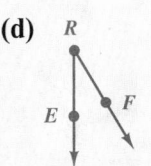

5 Which pair of lines is perpendicular? How can you describe the other pair of lines?

(a)

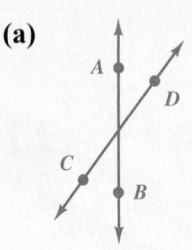

(b)

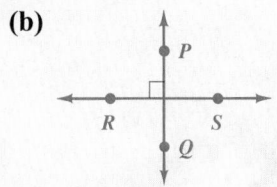

Example 5 Identifying Perpendicular Lines

Which pairs of lines are perpendicular?

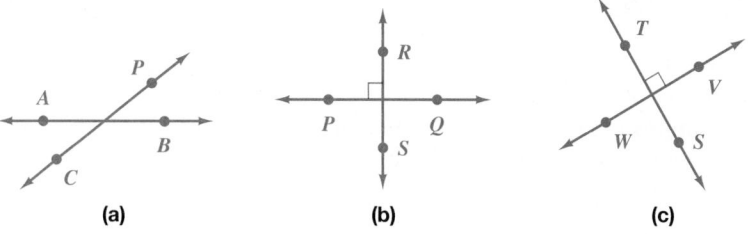

(a)　　　　(b)　　　　(c)

The lines in Figures **(b)** and **(c)** are perpendicular to each other because they intersect at right angles.

The lines in Figure **(a)** are intersecting lines, but they are not perpendicular because they do not form a right angle.

Work Problem 5 at the Side.

8.1 EXERCISES

 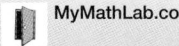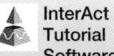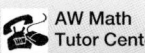

Name each line, line segment, or ray using the appropriate symbol. See Example 1.

1.

2.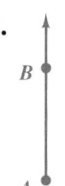

3.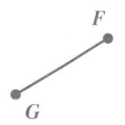

4.

5.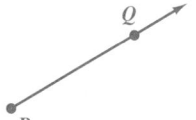

6.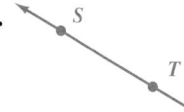

Label each pair of lines as parallel, perpendicular, or intersecting. See Examples 2 and 5.

7.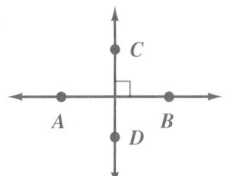

8.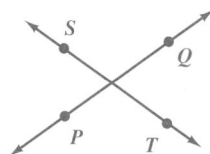

9.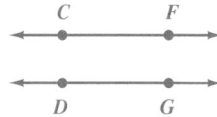

10.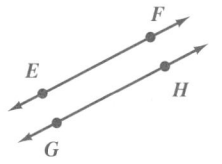

11.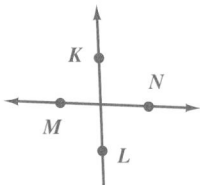

12.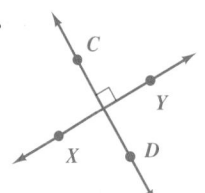

Name each highlighted angle by using the three-letter form of identification. See Example 3.

13.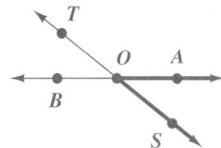

14.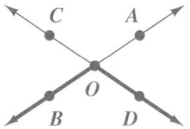

15.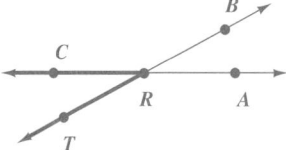

16.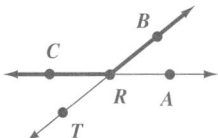

17.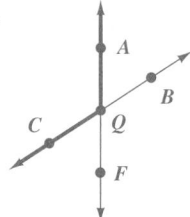

18.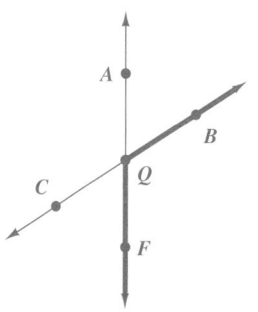

*Label each angle as acute, right, obtuse, or straight. For right and straight angles,
indicate the number of degrees in the angle. See Example 4.*

19.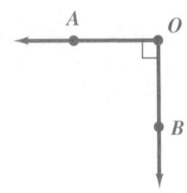

20.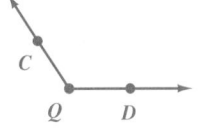

21.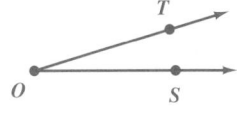

22.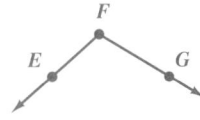

23.

24.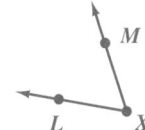

25. Explain what is happening in each sentence.

  (a) The road was so slippery that my car did a 360.

  (b) After the election, the governor's view on taxes took a 180° turn.

26. Find at least four examples of right angles in your home, at work, or on the street. Make a sketch of each example and label the right angle.

RELATING CONCEPTS (Exercises 27–32)   **FOR INDIVIDUAL OR GROUP WORK**

Use the figure below to **work Exercises 27–32 in order.** *Decide whether each statement is* **true** *or* **false.** *If it is true, explain why. If it is false, rewrite to make it a true statement.*

27. $\angle UST$ is 90°.

28. $\overleftrightarrow{SQ}$ and $\overrightarrow{PQ}$ are perpendicular.

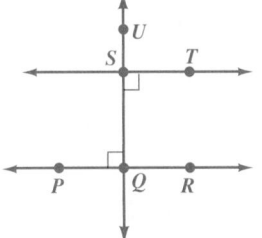

29. The measure of $\angle USQ$ is less than the measure of $\angle PQR$.

30. $\overleftrightarrow{ST}$ and $\overleftrightarrow{PR}$ are intersecting.

31. $\overleftrightarrow{QU}$ and $\overleftrightarrow{TS}$ are parallel.

32. $\angle UST$ and $\angle UQR$ measure the same number of degrees.

8.2 ANGLES AND THEIR RELATIONSHIPS

1 ▭ **Identify complementary angles and supplementary angles.** Two angles are called **complementary angles** if their sum is 90°. If two angles are complementary, each angle is the *complement* of the other.

OBJECTIVES

1 ▭ Identify complementary angles and supplementary angles.

2 ▭ Identify congruent angles and vertical angles.

Example 1 Identifying Complementary Angles

Identify each pair of complementary angles.

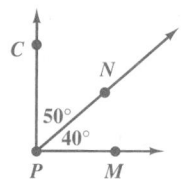 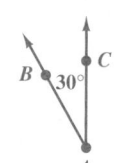 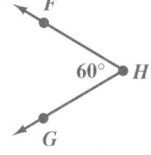

$\angle MPN$ (40°) and $\angle NPC$ (50°) are complementary angles because

$$40° + 50° = 90°.$$

$\angle CAB$ (30°) and $\angle FHG$ (60°) are complementary angles because

$$30° + 60° = 90°.$$

═══ **Work Problem ❶ at the Side.**

❶ Identify each pair of complementary angles.

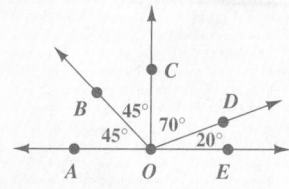

Example 2 Finding the Complement of Angles

Find the complement of each angle.

(a) 30°
  The complement of 30° is 60°, because **90°** − 30° = 60°.

(b) 40°
  The complement of 40° is 50°, because **90°** − 40° = 50°.

═══ **Work Problem ❷ at the Side.**

❷ Find the complement of each angle.

(a) 35°

(b) 80°

Two angles are called **supplementary angles** if their sum is 180°. If two angles are supplementary, each angle is the *supplement* of the other.

❸ Identify each pair of supplementary angles.

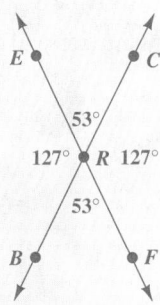

Example 3 Identifying Supplementary Angles

Identify each pair of supplementary angles.

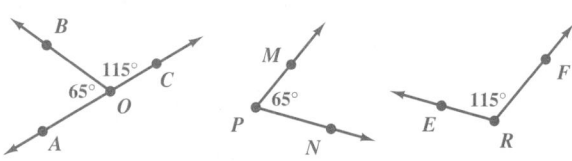

$\angle BOA$ and $\angle BOC$, because 65° + 115° = 180°.

$\angle BOA$ and $\angle ERF$, because 65° + 115° = 180°.

$\angle BOC$ and $\angle MPN$, because 115° + 65° = 180°.

$\angle MPN$ and $\angle ERF$, because 65° + 115° = 180°.

═══ **Work Problem ❸ at the Side.**

ANSWERS
1. $\angle AOB$ and $\angle BOC$; $\angle COD$ and $\angle DOE$
2. **(a)** 55° **(b)** 10°
3. $\angle CRF$ and $\angle BRF$; $\angle CRE$ and $\angle ERB$;
$\angle BRF$ and $\angle BRE$; $\angle CRE$ and $\angle CRF$

❹ Find the supplement of each angle.

(a) 175°

(b) 30°

❺ Identify the angles that are congruent.

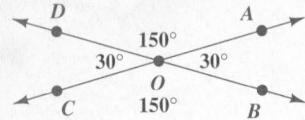

❻ Identify the vertical angles.

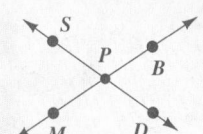

Example 4 Finding the Supplement of Angles

Find the supplement of each angle.

(a) 70°
The supplement of 70° is 110°, because **180°** − 70° = 110°.

(b) 140°
The supplement of 140° is 40°, because **180°** − 140° = 40°.

Work Problem ❹ at the Side.

2 ▭ **Identify congruent angles and vertical angles.** Two angles are called **congruent angles** if they measure the same number of degrees. If two angles are congruent, this is written as ∠*A* ≅ ∠*B* and read as, "angle *A* is congruent to angle *B*." Here is an example.

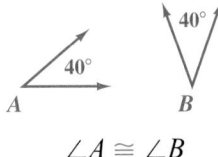

∠*A* ≅ ∠*B*

Example 5 Identifying Congruent Angles

Identify the angles that are congruent.

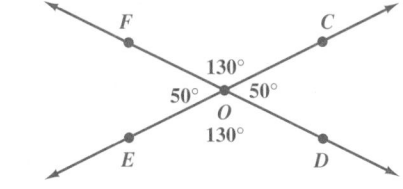

∠*FOC* ≅ ∠*EOD*   and   ∠*COD* ≅ ∠*EOF*

Work Problem ❺ at the Side.

Angles that do not share a common side are called *nonadjacent* angles. Two nonadjacent angles formed by intersecting lines are called **vertical angles.**

Example 6 Identifying Vertical Angles

Identify the vertical angles in this figure.

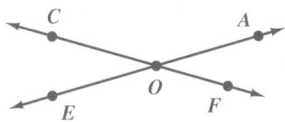

∠*AOF* and ∠*COE* are vertical angles because they do not share a common side and they are formed by two intersecting lines (⟷*CF* and ⟷*EA*).

∠*COA* and ∠*EOF* are also vertical angles.

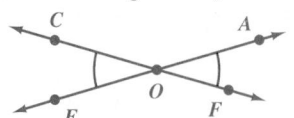

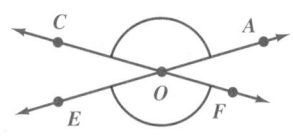

Work Problem ❻ at the Side.

Look back at Example 5 on the previous page. Notice that the two *congruent* angles that measure 130° are also *vertical* angles. Also, the two congruent angles that measure 50° are vertical angles. This illustrates the following property.

Congruent Angles

If two angles are vertical angles, they are **congruent**, that is, they measure the same number of degrees.

Example 7 **Finding the Measures of Vertical Angles**

In the figure below, find the measure of each unlabeled angle.

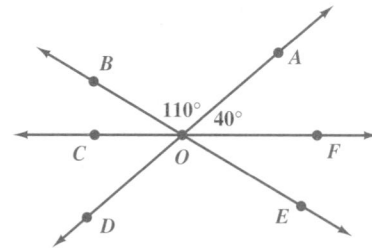

(a) ∠COD
   ∠COD and ∠AOF are vertical angles so they are congruent. This means they measure the same number of degrees.

   The measure of ∠AOF is 40°   so   the measure of ∠COD is 40° also.

(b) ∠DOE
   ∠DOE and ∠BOA are vertical angles, so they are congruent.

   The measure of ∠BOA is 110°   so   the measure of ∠DOE is 110° also.

(c) ∠COB
   Look at ∠COB, ∠BOA, and ∠AOF. Notice that $\overrightarrow{OC}$ and $\overrightarrow{OF}$ go in opposite directions. Therefore, ∠COF is a straight angle and measures 180°. To find the measure of ∠COB, subtract the sum of the other two angles from 180°.

$$180° - (110° + 40°) = 180° - (150°) = 30°$$

The measure of ∠COB is 30°.

(d) ∠EOF
   ∠EOF and ∠COB are vertical angles, so they are congruent. We know from part (c) above that the measure of ∠COB is 30° so the measure of ∠EOF is 30° also.

=========== **Work Problem 7** at the Side.

7 In the figure below, find the number of degrees in each unlabeled angle.

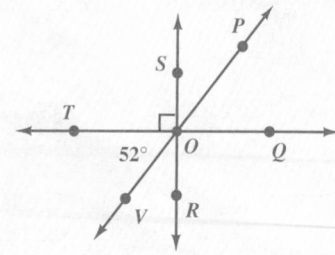

(a) ∠QOR

(b) ∠POQ

(c) ∠VOR

(d) ∠POS

Focus on Real-Data Applications

The London Eye

The British Airways London Eye was built as a symbol of the turn of the century and represents the cycle of life. It was designed by architects David Marks and Julia Barfield to suspend over the River Thames in Jubilee Gardens, on the bank opposite the Houses of Parliament and Big Ben. It first opened on December 31, 1999.

The London Eye is the largest observation wheel ever designed. It is a unique form of a Ferris wheel that features 32 oval-shaped capsules, each designed to hold 25 people. The wheel structure was assembled on platforms erected in the river bed, and once assembled was lifted into place by a floating crane. The London Eye is the fourth tallest structure in London (and the only one of the four that is open to the public), and the capsules were designed so that passengers have an unobstructed view throughout the ride. On a clear day, passengers can see 25 miles to the 900-year-old Windsor Castle, the world's largest royal residence, used by Queen Elizabeth II.

Another unique feature of the London Eye is that it is designed to rotate continuously at walking speed. At one end of the loading platform, passengers walk out of the capsules, and at the other end of the platform, passengers walk into the capsules. Because the wheel rotates so slowly, the passengers can stand and walk around the capsules during their ride, and enjoy a 360° view of the London skyline. The wheel does stop to allow people in wheelchairs to board. (*Source: The Bankside Press,* 2001.)

Specifications for the London Eye	
Diameter	135 meters
Speed	0.26 meters per second
Time to revolve	30 minutes

 1. (a) What is the diameter of the wheel in feet? (*Hint:* Use 1 m ≈ 3.28 ft.)

(b) The circumference of the wheel is the distance around the outside rim. Find the circumference by multiplying the diameter times 3.1416 and rounding the answer to the nearest foot. (See Section 8.6 for more information about circumference.)

2. Find the distance, to the nearest tenth of a foot, along the rim arc between two adjacent capsules. Recall that there are 32 capsules.

3. If the London Eye is operating at full capacity, how many people will be riding in a 90° section of the wheel? in a 180° section? in a 360° section?

4. If the London Eye operates from 8:00 A.M. until 6:00 P.M. at full capacity, how many people per day can ride?

5. The speed is given as 0.26 meters per second. Find the speed of the London Eye in feet per second. Round the answer to the nearest hundredth. Do you believe that the wheel revolves at "walking speed"? (Try it and see!)

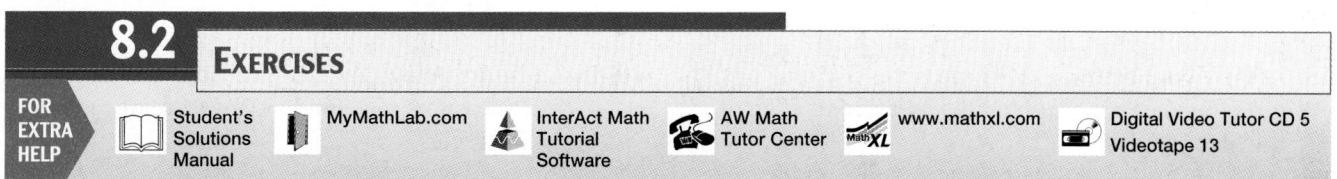

8.2 EXERCISES

Identify each pair of complementary angles. See Example 1.

1.

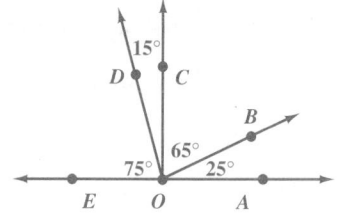

2.

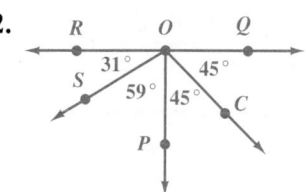

Identify each pair of supplementary angles. See Example 3.

3.

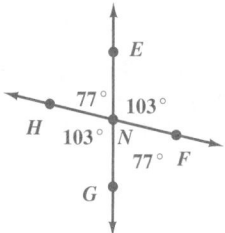

4.

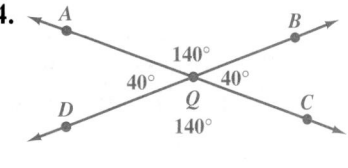

Find the complement of each angle. See Example 2.

5. $40°$ **6.** $35°$ **7.** $86°$ **8.** $59°$

Find the supplement of each angle. See Example 4.

9. $130°$ **10.** $75°$ **11.** $90°$ **12.** $5°$

In each figure, identify the angles that are congruent. See Example 5.

13.

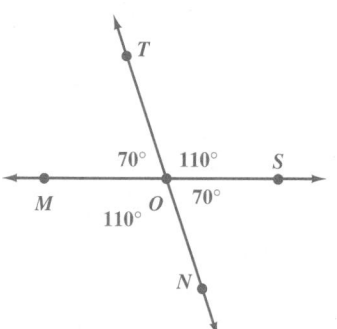

14.

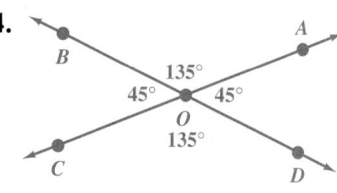

Use your knowledge of vertical angles to answer Exercises 15 and 16. See Examples 6 and 7.

15. In the figure below, ∠*AOH* measures 37° and ∠*COE* measures 63°. Find the measure of each of the other angles.

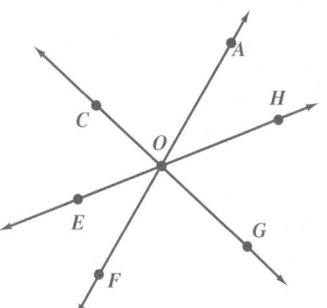

16. In the figure below, ∠*POU* measures 105° and ∠*UOT* measures 40°. Find the measure of each of the other angles.

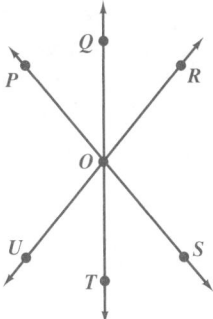

17. In your own words, write a definition of complementary angles and a definition of supplementary angles. Draw a picture to illustrate each definition.

18. Make up a test problem in which a student has to use knowledge of vertical angles. Include a drawing with some angles labeled and ask the student to find the size of the remaining angles. Give the correct answer for your problem.

In each figure, ray AB is parallel to ray CD. Identify two pairs of congruent angles and the number of degrees in each congruent angle.

19.

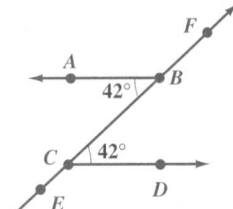

20.

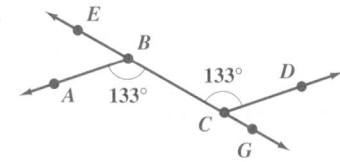

21. Can two obtuse angles be supplementary? Explain why or why not.

22. Can two acute angles be complementary? Explain why or why not.

8.3 RECTANGLES AND SQUARES

A **rectangle** is a figure with four sides that meet to form 90° angles. Each set of opposite sides is parallel and congruent (has the same length).

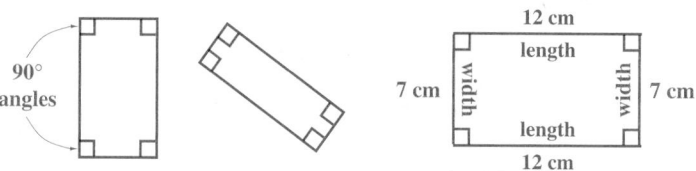

Each longer side of a rectangle is called the length (*l*) and each shorter side is called the width (*w*).

Work Problem ❶ at the Side.

1▌ **Find the perimeter and area of a rectangle.** The distance around the outside edges of a figure is the **perimeter** of the figure. Think of how much fence you would need to put around the sides of a garden plot, or how far you would walk if you go around the outside edges of your backyard. In either case you would add up the lengths of the sides. Look at the rectangle above that has the lengths of the sides labeled. To find its perimeter, you add the lengths of the sides.

Perimeter = 12 cm + 7 cm + 12 cm + 7 cm = 38 cm

Because the two long sides are both 12 cm, and the two short sides are both 7 cm, you can also use this formula.

Finding the Perimeter of a Rectangle

Perimeter of a rectangle = (2 • length) + (2 • width)

$$P = 2 \cdot l + 2 \cdot w$$

Example 1 Finding the Perimeter of Rectangles

Find the perimeter of each rectangle.

(a)

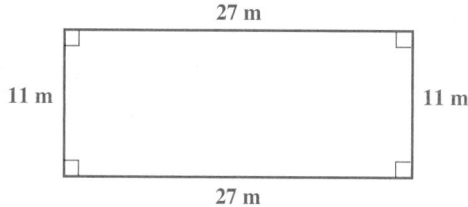

The length is **27 m** and the width is **11 m**.
Use the formula $P = 2 \cdot l + 2 \cdot w$.

$$P = 2 \cdot \quad l \quad + 2 \cdot \quad w$$
$$P = \underline{2 \cdot 27\text{ m}} + \underline{2 \cdot 11\text{ m}}$$
$$P = \quad 54\text{ m} \quad + \quad 22\text{ m}$$
$$P = 76\text{ m}$$

The perimeter of the rectangle (the distance you would walk around the outside edges of the rectangle) is 76 m.

════ **Continued on Next Page**

OBJECTIVES

1▌ Find the perimeter and area of a rectangle.

2▌ Find the perimeter and area of a square.

3▌ Find the perimeter and area of a composite shape.

❶ Identify all the rectangles.

(a)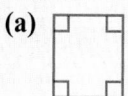

(b)

(c)

(d)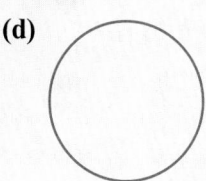

(e)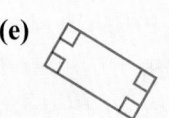

(f)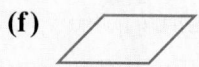

(g)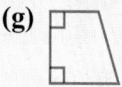

ANSWERS
1. **(a)**, **(b)**, and **(e)** are rectangles; **(c)**, **(d)**, **(f)**, and **(g)** are not.

❷ Find the perimeter of each rectangle.

(a)

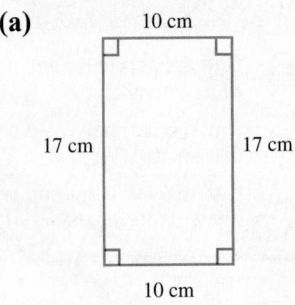

(b)

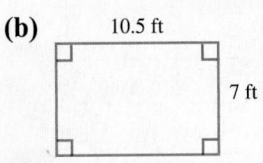

(c) 6 m wide and 11 m long

(d) 0.9 km by 2.8 km

As a check, you can add up the lengths of the four sides.

$$P = 27 \text{ m} + 11 \text{ m} + 27 \text{ m} + 11 \text{ m}$$
$$P = 76 \text{ m} \;\leftarrow\text{Same result as using formula}$$

(b) A rectangle 8.9 m by 12.3 m

You can use the formula, as shown below.

$$P = 2 \bullet \quad l \quad + 2 \bullet \quad w$$
$$P = \underline{2 \bullet 12.3 \text{ m}} + \underline{2 \bullet 8.9 \text{ m}}$$
$$P = \quad 24.6 \text{ m} \quad + \quad 17.8 \text{ m}$$
$$P = 42.4 \text{ m}$$

Or, you can add up the lengths of the four sides.

$$P = 8.9 \text{ m} + 12.3 \text{ m} + 8.9 \text{ m} + 12.3 \text{ m}$$
$$P = 42.4 \text{ m}$$

Either method will give you the correct result.

Work Problem ❷ at the Side.

The *perimeter* of a rectangle is the distance around the *outside edges*. The **area** of a rectangle is the amount of surface *inside* the rectangle. We measure area by seeing how many squares of a certain size are needed to cover the surface inside the rectangle. Think of covering the floor of a rectangular living room with carpet. Carpet is measured in square yards, that is, square pieces that measure 1 yard along each side. Here is a drawing of a living room floor.

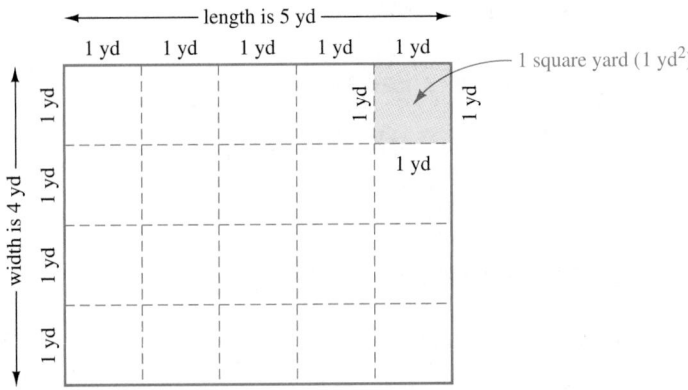

You can see from the drawing that it takes 20 squares to cover the floor. We say that the area of the floor is 20 *square yards*. A shorter way to write square yards is yd^2.

20 square yards can be written **20 yd^2**

To find the number of squares, you can count them, or you can multiply the number of squares in the length (5) times the number of squares in the width (4) to get 20. The formula is given below.

Finding the Area of a Rectangle

Area of a rectangle = length • width

$$A = l \bullet w$$

Remember to use *square units* when measuring area.

Squares of other sizes can be used to measure area. For smaller areas, you might use these:

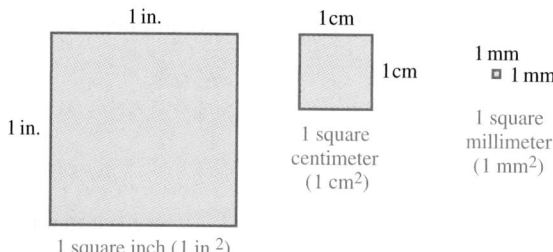

Actual-size drawings

Other sizes of squares that are often used to measure area are listed here, but they are too large to draw on this page.

1 square meter (1 m²)   1 square foot (1 ft²)
1 square kilometer (1 km²)   1 square yard (1 yd²)
                                 1 square mile (1 mi²)

CAUTION

The raised 2 in 4^2 means that you multiply $4 \cdot 4$ to get 16. The raised 2 in cm² or yd² is a short way to write the word *square*. When you see 5 cm², say "five square centimeters." Do *not* multiply $5 \cdot 5$.

Example 2   Finding the Area of Rectangles

Find the area of each rectangle.

(a)

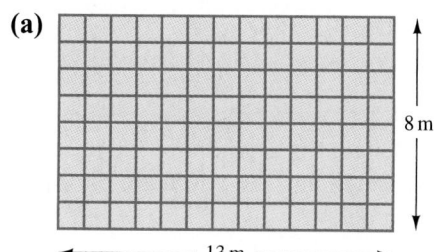

The length of this rectangle is 13 m and the width is 8 m. Use the formula $A = l \cdot w$.

$$A = \quad l \quad \cdot \quad w$$
$$A = 13 \text{ m} \cdot 8 \text{ m}$$
$$A = 104 \text{ square meters}$$

"Square meters" can be written as m², so the area is 104 m².

(b) A rectangle measuring 7 cm by 21 cm
   Use the formula $A = l \cdot w$.

$$A = 21 \text{ cm} \cdot 7 \text{ cm} = 147 \text{ cm}^2$$

The area of the rectangle is 147 cm².

❸ Find the area of each rectangle.

(a)

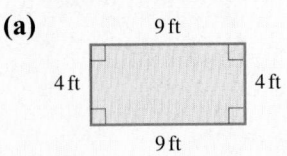

9 ft

4 ft   4 ft

9 ft

(b) A rectangle that is 6 m long and 0.5 m wide

(c) 8.2 cm by 41.2 cm

CAUTION

The units for *area* will always be *square* units (cm², m², yd², mi², and so on). The units for *perimeter* will be cm, m, yd, mi, and so on (not square units).

Work Problem ❸ at the Side.

2   **Find the perimeter and area of a square.** A **square** is a rectangle with all sides the same length. Two squares are shown here. Notice the 90° angles.

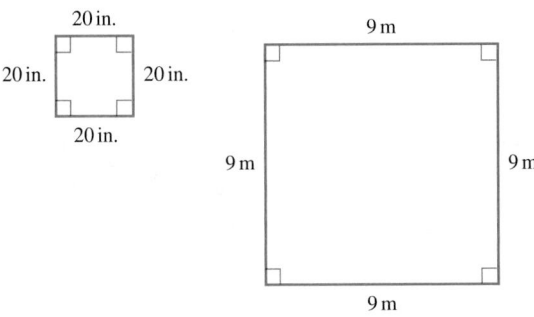

To find the *perimeter* (distance around) of the square on the right, you could add 9 m + 9 m + 9 m + 9 m to get 36 m. A shorter way is to multiply the length of one side times 4, because all four sides are the same length.

Finding the Perimeter of a Square

Perimeter of a square = side + side + side + side

or, $P = 4 \cdot \text{side}$

$P = 4 \cdot s$

As with a rectangle, you can multiply length times width to find the *area* (surface inside) of a square. Because the length and the width are the same in a square, the formula is written as shown below.

Finding the Area of a Square

Area of a square = side • side

$A = s \cdot s$

$A = s^2$

Remember to use **square units** when measuring area.

Example 3   **Finding the Perimeter and Area of a Square**

(a) Find the perimeter of a square where each side measures 9 m.

Use the formula.

$P = 4 \cdot s$

$P = 4 \cdot 9 \text{ m}$

$P = 36 \text{ m}$

Or add up the four sides.

$P = 9 \text{ m} + 9 \text{ m} + 9 \text{ m} + 9 \text{ m}$

$P = 36 \text{ m}$

Continued on Next Page

(b) Find the area of the same square.

$$A = s^2$$
$$A = s \cdot s$$
$$A = 9 \text{ m} \cdot 9 \text{ m}$$
$$A = 81 \text{ m}^2 \qquad \text{Square units for area}$$

CAUTION

Be careful! s^2 means $s \cdot s$. It does *not* mean $2 \cdot s$. In Example 3 above, s is 9 m, so s^2 is 9 m $\cdot$ 9 m $= 81$ m². It is *not* $2 \cdot 9$ m $= 18$ m.

Work Problem ④ at the Side.

3 Find the perimeter and area of a composite shape. As with any other shape, you can find the perimeter (distance around) an irregular shape by adding up the lengths of the sides. To find the area (surface inside the shape), try to break it up into pieces that are squares or rectangles. Find the area of each piece and then add them together.

Example 4 Finding the Perimeter and Area of a Composite Figure

The floor of a room has the shape shown here.

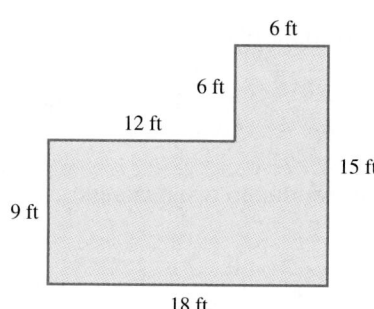

(a) Suppose you want to put a new baseboard (wooden strip) along the base of all the walls. How much material do you need?
Find the perimeter of the room by adding up the lengths of the sides.

$$P = 9 \text{ ft} + 12 \text{ ft} + 6 \text{ ft} + 6 \text{ ft} + 15 \text{ ft} + 18 \text{ ft} = 66 \text{ ft}$$

You need 66 ft of baseboard material.

(b) The carpet you like costs \$20.50 per square yard. How much will it cost to carpet the room?
First change the measurements from feet to yards, because the carpet is sold in square yards. There are 3 ft in 1 yd, so multiply by the unit fraction that allows you to divide out feet. For example:

$$\frac{\overset{3}{\cancel{9 \text{ ft}}}}{1} \cdot \frac{1 \text{ yd}}{\underset{1}{\cancel{3 \text{ ft}}}} = 3 \text{ yd}$$

— Divide out ft.
— Divide out common factors.

Continued on Next Page

④ Find the perimeter and area of each square.

(a)

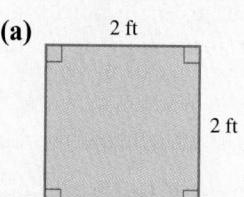

2 ft

2 ft

(b) 10.5 cm on each side

(c) 2.1 mi on a side

ANSWERS
4. (a) $P = 8$ ft; $A = 4$ ft²
  (b) $P = 42$ cm; $A = 110.25$ cm²
  (c) $P = 8.4$ mi; $A = 4.41$ mi²

5 Carpet costs $19.95 per square yard. Find the cost of carpeting each room. Round your answers to the nearest cent if necessary.

(a)

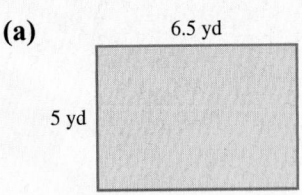

6.5 yd

5 yd

Use the same unit fraction to change the other measurements from feet to yards.

$$\frac{\overset{4}{\cancel{12}\,\cancel{ft}}}{1} \cdot \frac{1\ yd}{\underset{1}{\cancel{3}\,\cancel{ft}}} = 4\ yd \qquad \frac{\overset{2}{\cancel{6}\,\cancel{ft}}}{1} \cdot \frac{1\ yd}{\underset{1}{\cancel{3}\,\cancel{ft}}} = 2\ yd$$

$$\frac{\overset{5}{\cancel{15}\,\cancel{ft}}}{1} \cdot \frac{1\ yd}{\underset{1}{\cancel{3}\,\cancel{ft}}} = 5\ yd \qquad \frac{\overset{6}{\cancel{18}\,\cancel{ft}}}{1} \cdot \frac{1\ yd}{\underset{1}{\cancel{3}\,\cancel{ft}}} = 6\ yd$$

Next, break up the room into two pieces. Use just the measurements for the length and width of each piece.

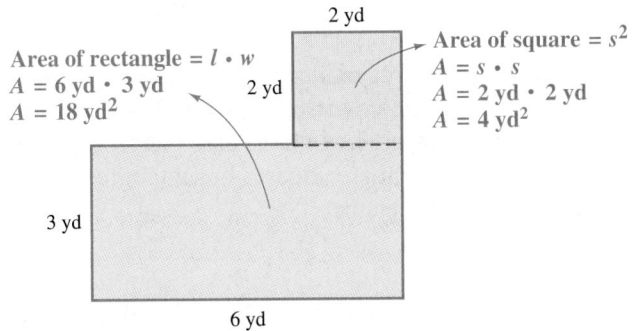

Area of rectangle $= l \cdot w$
$A = 6\ yd \cdot 3\ yd$
$A = 18\ yd^2$

Area of square $= s^2$
$A = s \cdot s$
$A = 2\ yd \cdot 2\ yd$
$A = 4\ yd^2$

2 yd

2 yd

3 yd

6 yd

Total area $= 18\ yd^2 + 4\ yd^2 = 22\ yd^2$

Multiply to find the cost of the carpet.

$$\frac{22\ \cancel{yd^2}}{1} \cdot \frac{\$20.50}{1\ \cancel{yd^2}} = \$451.00$$

You could have cut the room into two rectangles. The total area is the same.

(b)

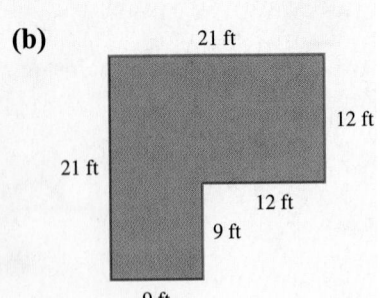

21 ft

21 ft

12 ft

12 ft

9 ft

9 ft

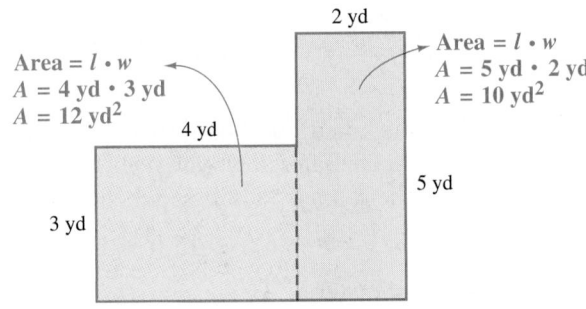

Area $= l \cdot w$
$A = 4\ yd \cdot 3\ yd$
$A = 12\ yd^2$

Area $= l \cdot w$
$A = 5\ yd \cdot 2\ yd$
$A = 10\ yd^2$

2 yd

4 yd

5 yd

3 yd

(c) A classroom that is 24 ft long and 18 ft wide

Total area $= 12\ yd^2 + 10\ yd^2 = 22\ yd^2$ Same answer as above

Work Problem 5 at the Side.

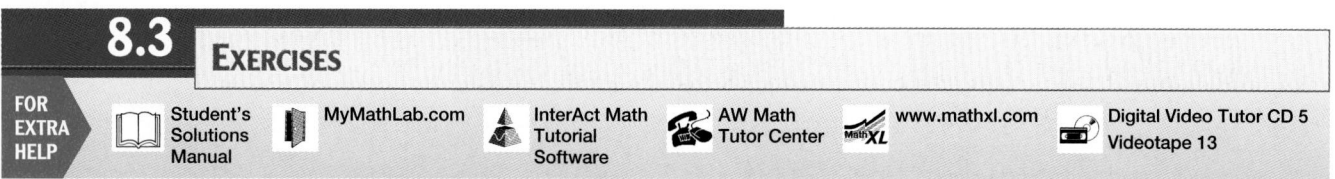

8.3 EXERCISES

Find the perimeter and area of each rectangle or square. See Examples 1–3.

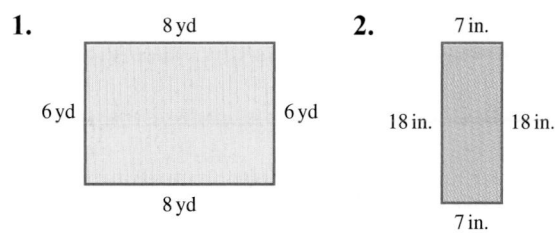

1. 8 yd / 6 yd / 6 yd / 8 yd

2. 7 in. / 18 in. / 18 in. / 7 in.

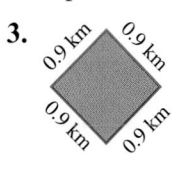

3. 0.9 km 0.9 km 0.9 km 0.9 km

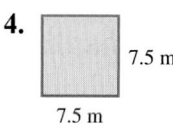

4. 7.5 m / 7.5 m

Draw a sketch of each square or rectangle and label the lengths of the sides. Then find the perimeter and the area. (Sketches may vary; show your sketches to your instructor.)

5. 10 ft by 10 ft

6. 8 cm by 17 cm

7. 14 m by 0.5 m

8. 2.35 km by 8.4 km

9. A storage building that is 76.1 ft by 22 ft

10. A science lab measuring 12 m by 12 m

11. A square nature preserve 3 mi wide

12. A square of cardboard 20.3 cm on a side

Find the perimeter and area of each figure. See Example 4.

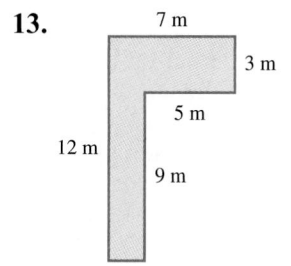

13. 7 m / 3 m / 5 m / 12 m / 9 m / 2 m

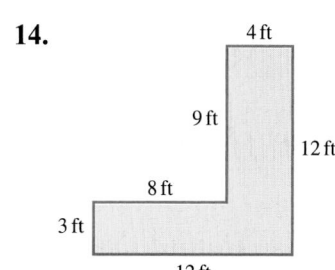

14. 4 ft / 9 ft / 12 ft / 8 ft / 3 ft / 12 ft

15.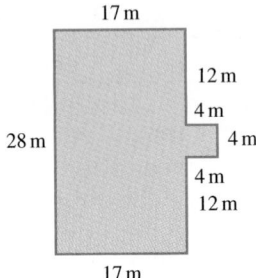

16.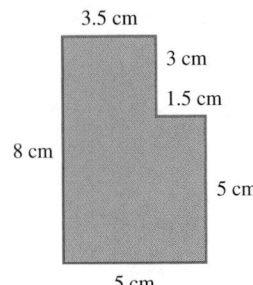

First find the length of the unlabeled side in each figure. Then find the perimeter and area of each figure.

17.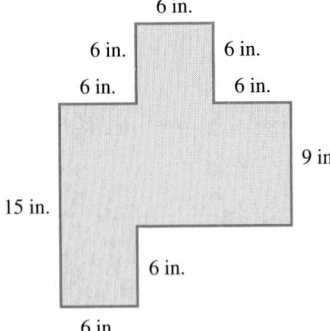

18.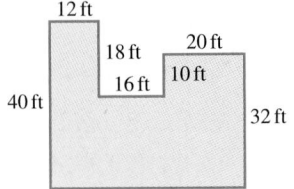

Solve each application problem.

19. The Wang's family room measures 20 ft by 25 ft. They are covering the floor with square tiles that measure 1 ft on a side and cost $0.92 each. How much will they spend on tile?

20. A page in this book measures 27.5 cm from top to bottom and 20.5 cm from side to side. Find the perimeter and the area of the page.

21. Tyra's kitchen is 4.4 m wide and 5.1 m long. She is pasting a decorative border strip that costs $4.99 per meter around the top edge of all the walls. How much will she spend?

22. Mr. and Mrs. Gomez are buying carpet for their square-shaped bedroom that is 5 yd wide. The carpet is $23 per square yard and padding and installation is another $6 per square yard. How much will they spend in all?

23. Advanced Photo System (APS) cameras allow you to choose from three different print sizes each time you snap a photo. The choices are shown below. Find the perimeter and area of each size print. (*Source:* Kodak.)

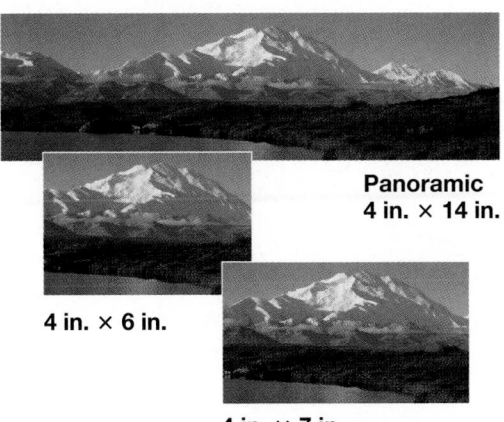

Panoramic
4 in. × 14 in.

4 in. × 6 in.

4 in. × 7 in.

24. The Monterey Bay Aquarium in California lets visitors look into a million-gallon tank through an acrylic panel that is 13 in. thick. The panel is 54 ft long and 15 ft high. What is the perimeter and the area of the panel? (*Source: AAA California Tour Book.*)

25. A regulation football field is 100 yd long (excluding end zones) and has an area of 5300 yd^2. Find the width of the field. (*Source:* National Football League.)

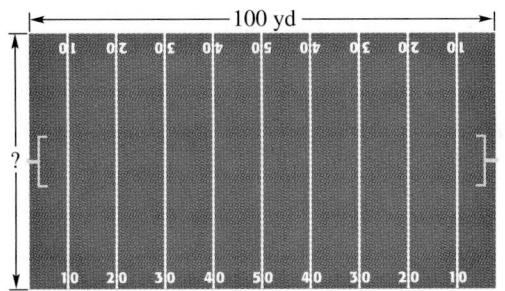

26. There are 14,790 ft^2 of ice in the rectangular playing area for a major league hockey game (excluding the area behind the goal lines). If the playing area is 85 ft wide, how long is it? (*Source:* National Hockey League.)

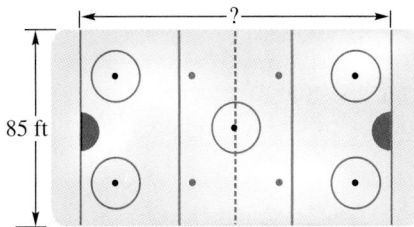

27. A lot is 124 ft by 172 ft. County rules require that nothing be built on land within 12 ft of any edge of the lot. First, add labels to the sketch of the lot, showing the land that cannot be built on. Then find the area of the land that cannot be built on.

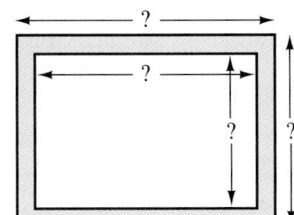

28. Find the cost of fencing needed for this rectangular field. Fencing along the country roads costs $4.25 per foot. Fencing for the other two sides costs $2.75 per foot.

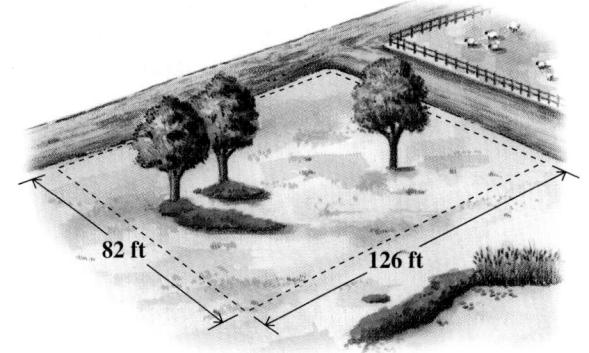

82 ft   126 ft

RELATING CONCEPTS (Exercises 29–34)   FOR INDIVIDUAL OR GROUP WORK

Use your knowledge of perimeter and area to **work Exercises 29–34 in order.**

29. Suppose you have 12 ft of fencing to make a square or rectangular garden plot. Draw sketches of *all* the possible plots that use exactly 12 ft of fencing and label the lengths of the sides. Use only *whole number* lengths. (*Hint:* There are three possibilities.)

30. (a) Find the area of each plot in Exercise 29.

(b) Which plot has the greatest area?

31. Repeat Exercise 29 using 16 ft of fencing. Be sure to draw *all* possible plots that have whole number lengths for the sides.

32. (a) Find the area of each plot in Exercise 31.

(b) Compare your results to those from Exercise 30. What do you notice about the plots with the greatest area?

33. (a) Draw a sketch of a rectangular plot 3 ft by 2 ft. Find the perimeter and area.

(b) Suppose you *double* the length of the plot and *double* the width. Draw a sketch of the enlarged plot and find the perimeter and area.

(c) The *perimeter* of the enlarged plot is how many times greater than the perimeter of the original plot? The *area* of the enlarged plot is how many times greater than the original area?

34. (a) Refer to part (a) of Exercise 33. Suppose you *triple* the length and width of the original plot. Draw a sketch of the enlarged plot and find the perimeter and area.

(b) How many times greater is the *perimeter* of the enlarged plot? How many times greater is the *area* of the enlarged plot?

(c) Suppose you make the length and width *four times greater* in the enlarged plot. What would you predict will happen to the perimeter and area, compared to the original plot?

8.4 PARALLELOGRAMS AND TRAPEZOIDS

A **parallelogram** is a four-sided figure with opposite sides parallel, such as the ones shown below. Notice that opposite sides have the same length.

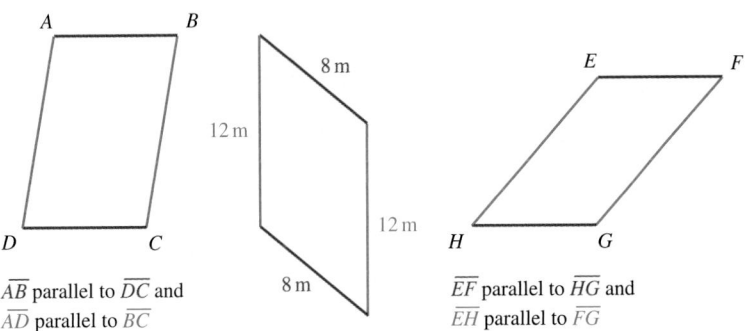

$\overline{AB}$ parallel to $\overline{DC}$ and
$\overline{AD}$ parallel to $\overline{BC}$

$\overline{EF}$ parallel to $\overline{HG}$ and
$\overline{EH}$ parallel to $\overline{FG}$

OBJECTIVES

1 Find the perimeter and area of a parallelogram.

2 Find the perimeter and area of a trapezoid.

❶ Find the perimeter of each parallelogram.

(a)

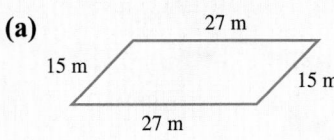

27 m
15 m
15 m
27 m

1 **Find the perimeter and area of a parallelogram.**   Perimeter is the distance around a figure, so the easiest way to find the perimeter of a parallelogram is to add the lengths of the four sides.

Example 1   Finding the Perimeter of a Parallelogram

Find the perimeter of the middle parallelogram above.

$$P = 12\text{ m} + 8\text{ m} + 12\text{ m} + 8\text{ m} = 40\text{ m}$$

═══════ **Work Problem ❶ at the Side.**

To find the area of a parallelogram, first draw a dashed line inside the figure as shown here.

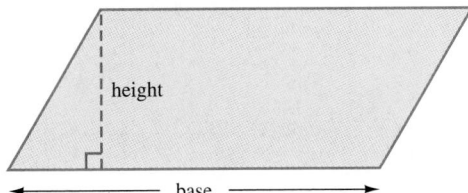

height

base

Try this yourself by tracing this parallelogram onto a piece of paper.

The length of the dashed line is the *height* of the parallelogram. It forms a *right angle* with the base. The height is the shortest distance between the base and the opposite side.

Now cut off the triangle created on the left side of the parallelogram and move it to the right side, as shown below.

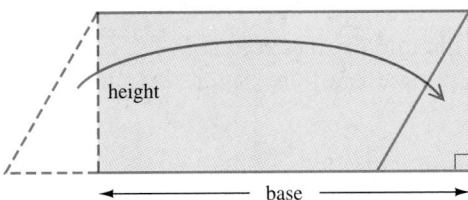

height

base

The parallelogram has been made into a rectangle. You can see that the area of the parallelogram and the rectangle are the same. The area of the rectangle is *length* times *width*. In the parallelogram, this translates into *base* times *height*.

(b)

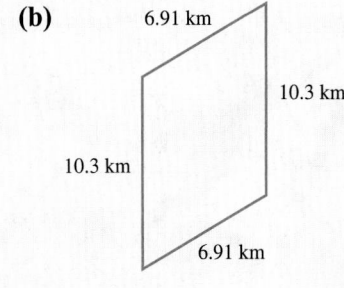

6.91 km
10.3 km
10.3 km
6.91 km

❷ Find the area of each parallelogram.

(a)

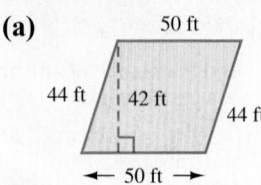

(b)

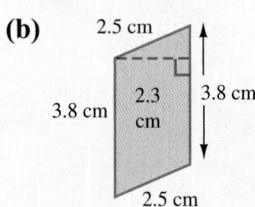

(c) A parallelogram with base $12\frac{1}{2}$ m and height $4\frac{3}{4}$ m (*Hint:* Write $12\frac{1}{2}$ as 12.5 and $4\frac{3}{4}$ as 4.75.)

❸ Find the perimeter of each trapezoid.

(a)

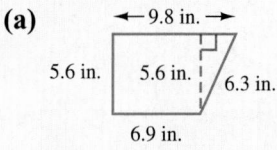

(b)

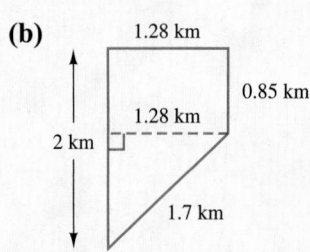

(c) A trapezoid with sides 39.7 cm, 29.2 cm, 74.9 cm, and 16.4 cm

Finding the Area of a Parallelogram

Area of parallelogram = base • height
$$A = b \cdot h$$

Remember to use **square units** when measuring area.

Example 2 Finding the Area of Parallelograms

Find the area of each parallelogram.

(a)

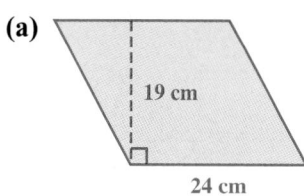

The base is 24 cm and the height is 19 cm. Use the formula $A = b \cdot h$.

$$A = \quad b \quad \bullet \quad h$$
$$A = 24 \text{ cm} \bullet 19 \text{ cm}$$
$$A = 456 \text{ cm}^2 \quad \text{Square units for area}$$

(b)

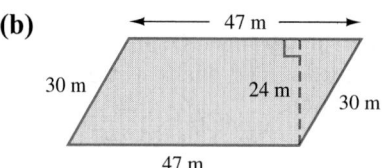

$$A = 47 \text{ m} \bullet 24 \text{ m}$$
$$A = 1128 \text{ m}^2 \quad \text{Square units for area}$$

Notice that you do *not* use the 30 m sides when finding the area. But you would use them when finding the *perimeter* of the parallelogram.

Work Problem ❷ at the Side.

2▭▭▭ **Find the perimeter and area of a trapezoid.** A **trapezoid** is a four-sided figure with one pair of parallel sides, such as the ones shown below. Opposite sides may *not* have the same length, as in parallelograms.

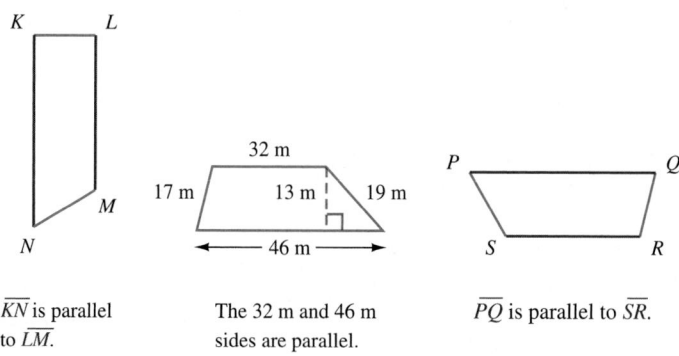

$\overline{KN}$ is parallel to $\overline{LM}$.

The 32 m and 46 m sides are parallel.

$\overline{PQ}$ is parallel to $\overline{SR}$.

Example 3 Finding the Perimeter of a Trapezoid

Find the perimeter of the middle trapezoid above.

You can find the perimeter of any figure by adding the lengths of the sides.

$$P = 17 \text{ m} + 32 \text{ m} + 19 \text{ m} + 46 \text{ m} = 114 \text{ m}$$

Notice that the height (13 m) is *not* part of the perimeter, because the height is *not* one of the *outside edges* of the shape.

Work Problem ❸ at the Side.

ANSWERS
2. (a) 2100 ft² (b) 8.74 cm²
(c) $59\frac{3}{8}$ m² or 59.375 m²

3. (a) 28.6 in. (b) 5.83 km (c) 160.2 cm

Use this formula to find the *area* of a trapezoid.

Finding the Area of a Trapezoid

$$\text{Area} = \frac{1}{2} \cdot \text{height} \cdot (\text{short base} + \text{long base})$$

$$A = \frac{1}{2} \cdot h \cdot (b + B)$$

or $A = 0.5 \cdot h \cdot (b + B)$

Remember to use *square units* when measuring area.

Example 4   Finding the Area of a Trapezoid

Find the area of this trapezoid. The short base and long base are the *parallel* sides.

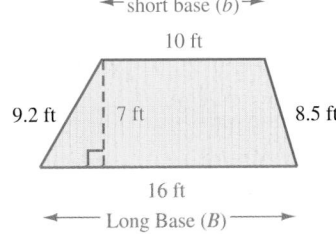

The height (h) is **7 ft**, the short base (b) is **10 ft**, and the long base (B) is **16 ft**. You do *not* need the 9.2 ft or 8.5 ft sides to find the area.

$$A = \frac{1}{2} \cdot h \cdot (b + B)$$

$$A = \frac{1}{2} \cdot 7 \text{ ft} \cdot (10 \text{ ft} + 16 \text{ ft}) \quad \text{Work inside parentheses first.}$$

$$A = \frac{1}{\cancel{2}} \cdot 7 \text{ ft} \cdot (\overset{13}{\cancel{26}} \text{ ft})$$

$$A = 91 \text{ ft}^2 \quad \text{Square units for area}$$

You can also solve the problem by using 0.5, the decimal equivalent for $\frac{1}{2}$, in the formula.

$$A = 0.5 \cdot h \cdot (b + B)$$
$$A = 0.5 \cdot 7 \cdot (10 + 16)$$
$$A = 0.5 \cdot 7 \cdot \quad 26$$
$$A = 91 \text{ ft}^2$$

Calculator Tip   Use the parentheses keys on your scientific calculator to work Example 4:

$0.5 \; \otimes \; 7 \; \otimes \; (\!)\; 10 \; \oplus \; 16 \; (\!) \; \ominus \; 91$

What happens if you do *not* use the parentheses keys? What order of operations will the calculator follow then? (Answer: The calculator will multiply 0.5 times 7 times 10, and then add 16, giving an *incorrect* answer of 51.)

Work Problem ④ at the Side.

---

④ Find the area of each trapezoid.

(a)

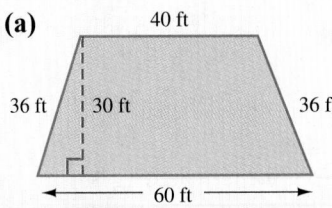

(b)

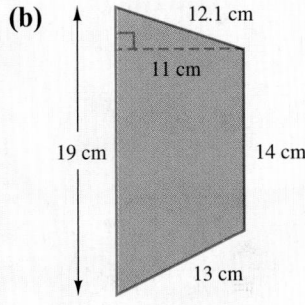

(c) A trapezoid with height 4.7 m, short base 9 m, and long base 10.5 m

Answers
4. (a) 1500 ft² **(b)** 181.5 cm²
(c) 45.825 m²

❺ Find the area of each floor.

(a)

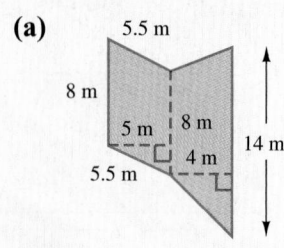

(b)

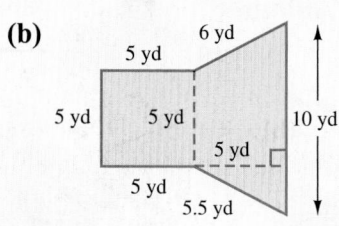

❻ Find the cost of carpeting the floors in Problem 5. The cost of carpet is as follows:

(a) Floor (a), \$18.50 per square meter.

(b) Floor (b), \$28 per square yard.

Example 5 Finding the Area of a Composite Figure

Find the area of this figure.

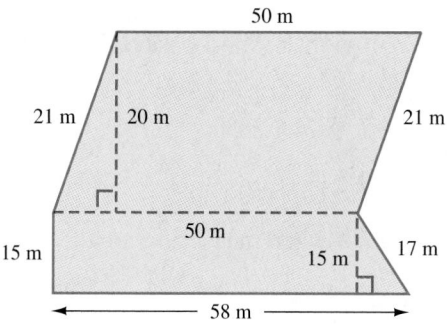

Break the figure into two pieces, a parallelogram and a trapezoid. Find the area of each piece, and then add the areas.

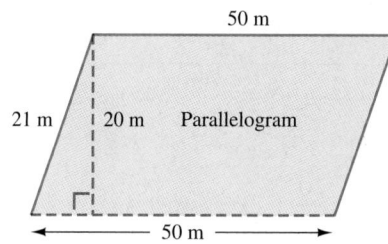

Area of parallelogram
$A = b \cdot h$
$A = 50 \text{ m} \cdot 20 \text{ m}$
$A = \mathbf{1000 \ m^2}$

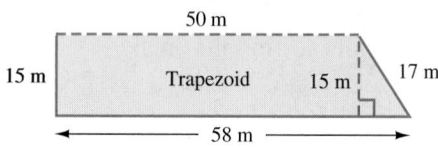

Area of trapezoid
$A = \dfrac{1}{2} \cdot h \cdot (b + B)$
$A = 0.5 \cdot 15 \text{ m} \cdot (50 \text{ m} + 58 \text{ m})$
$A = \mathbf{810 \ m^2}$

Total area $= 1000 \text{ m}^2 + 810 \text{ m}^2 = \mathbf{1810 \ m^2}$

Work Problem ❺ at the Side.

Example 6 Applying Knowledge of Area

Suppose the figure in Example 5 represents the floor plan of a hotel lobby. What is the cost of labor to install tile on the floor if the labor charge is \$35.11 per square meter?

From Example 5, the floor area is 1810 m². To find the labor cost, multiply the number of square meters times the cost of labor per square meter.

$$\text{cost} = \frac{1810 \ \cancel{m^2}}{1} \cdot \frac{\$35.11}{1 \ \cancel{m^2}}$$

$$\text{cost} = \$63{,}549.10$$

The cost of the labor is \$63,549.10.

Work Problem ❻ at the Side.

Answers
5. (a) 84 m² **(b)** 62.5 yd²
6. (a) \$1554 **(b)** \$1750

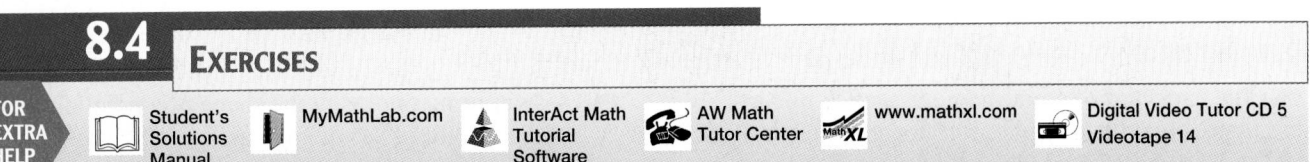

8.4 EXERCISES

FOR EXTRA HELP

Student's Solutions Manual | MyMathLab.com | InterAct Math Tutorial Software | AW Math Tutor Center | www.mathxl.com | Digital Video Tutor CD 5 Videotape 14

Find the perimeter of each figure. See Examples 1 and 3.

1.

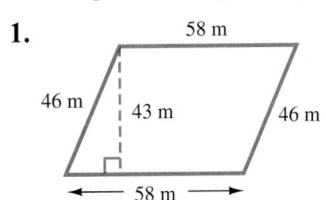

58 m
46 m   43 m   46 m
58 m

2.

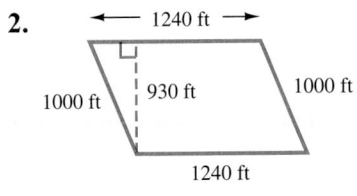

1240 ft
1000 ft   930 ft   1000 ft
1240 ft

3.

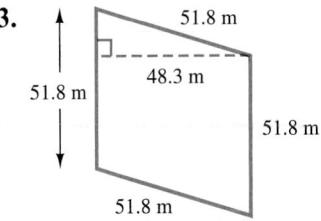

51.8 m
48.3 m
51.8 m   51.8 m
51.8 m

4.

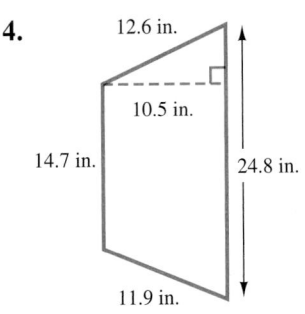

12.6 in.
10.5 in.
14.7 in.   24.8 in.
11.9 in.

5.

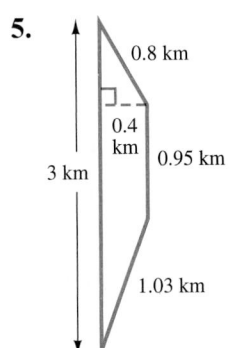

0.8 km
0.4 km   0.95 km
3 km
1.03 km

6.

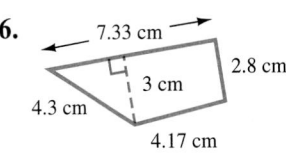

7.33 cm
2.8 cm
3 cm
4.3 cm
4.17 cm

Find the area of each figure. See Examples 2 and 4.

7.

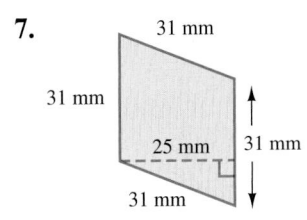

31 mm
31 mm
25 mm   31 mm
31 mm
31 mm

8.

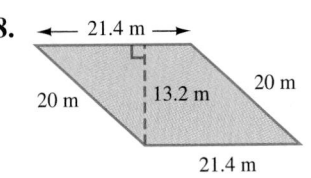

21.4 m
20 m   13.2 m   20 m
21.4 m

9.

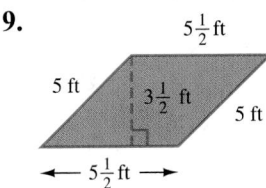

$5\frac{1}{2}$ ft
5 ft $3\frac{1}{2}$ ft   5 ft
$5\frac{1}{2}$ ft

10.

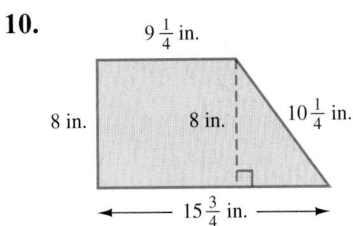

$9\frac{1}{4}$ in.
8 in.   8 in. $10\frac{1}{4}$ in.
$15\frac{3}{4}$ in.

11.

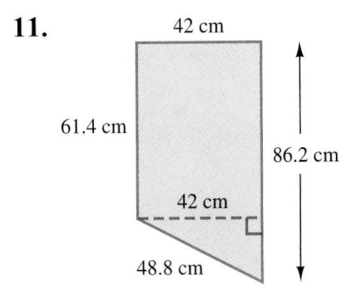

42 cm
61.4 cm   86.2 cm
42 cm
48.8 cm

12.

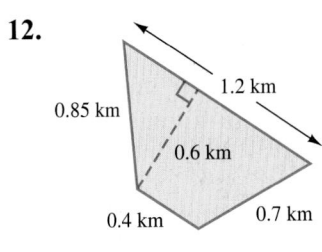

1.2 km
0.85 km
0.6 km
0.4 km   0.7 km

Solve each application problem. First draw a sketch and label the sides and height.
(Sketches may vary; show your sketches to your instructor.) See Example 6.

13. The backyard of a new home is shaped like a trapezoid with a height of 45 ft and bases of 80 ft and 110 ft. What is the cost of putting sod on the yard if the landscaper charges $0.33 per square foot for sod?

14. A swimming pool is in the shape of a parallelogram with a height of 9.6 m and base of 12.4 m. Find the labor cost to make a custom solar cover for the pool at a cost of $4.92 per square meter.

15. A piece of fabric for a quilt design is in the shape of a parallelogram. The base is 5 in. and the height is 3.5 in. What is the total area of the 25 parallelogram pieces needed for the quilt?

16. An accountant is paying $832 per month to rent an office in an old building. Her office is shaped like a trapezoid, with bases of 32 ft and 20 ft and a height of 20 ft. How much rent is she paying per square foot?

*Find **two** errors in each student's solution below. Write a sentence explaining each error.*
Then show how to work the problem correctly.

17.

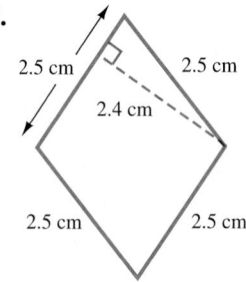

$$P = 2.5 \text{ cm} + 2.4 \text{ cm} + 2.5 \text{ cm} + 2.5 \text{ cm} + 2.5 \text{ cm}$$
$$P = 12.4 \text{ cm}^2$$

18.

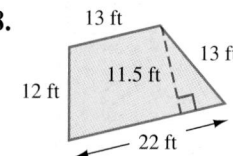

$$A = (0.5)(11.5 \text{ ft}) \cdot (12 \text{ ft} + 13 \text{ ft})$$
$$A = 143.75 \text{ ft}$$

📷 *Find the area of each figure. See Example 5.*

19.

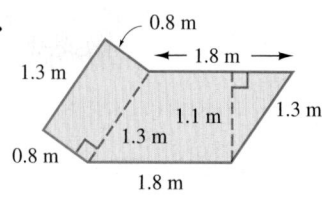

20.

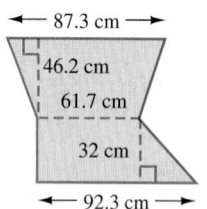

21.

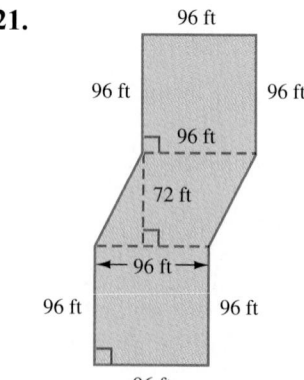

8.5 TRIANGLES

A **triangle** is a figure with exactly three sides. Some examples are shown below.

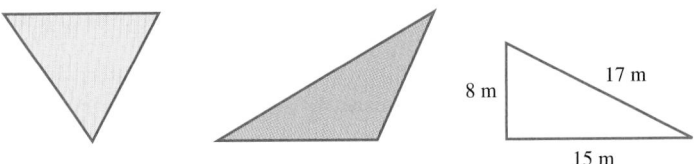

8 m   17 m

15 m

OBJECTIVES

1 Find the perimeter of a triangle.

2 Find the area of a triangle.

3 Given the measures of two angles in a triangle, find the measure of the third angle.

1 ▭ **Find the perimeter of a triangle.**   To find the perimeter of a triangle (the distance around the edges), add the lengths of the three sides.

Example 1   **Finding the Perimeter of a Triangle**

Find the perimeter of the triangle above on the right.

$$P = 8 \text{ m} + 15 \text{ m} + 17 \text{ m}$$
$$P = 40 \text{ m}$$

════════════ **Work Problem ① at the Side.**

As with parallelograms, you can find the *height* of a triangle by measuring the distance from one corner of the triangle to the opposite side (the base). The height line must be *perpendicular* to the base; that is, it must form a right angle with the base. Sometimes you have to extend the base line in order to draw the height perpendicular to it, as shown on the right in the figures below.

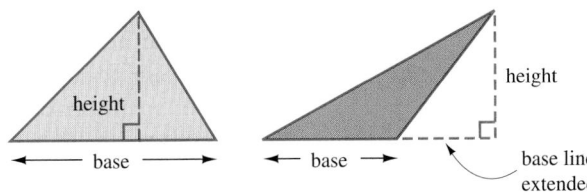

height

base

height

base

base line extended

If you cut out two identical triangles and turn one upside down, you can fit them together to form a parallelogram.

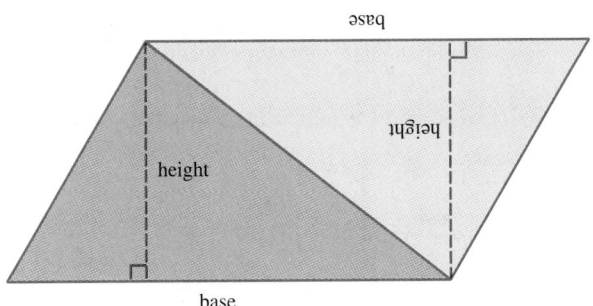

base

height

height

height

base

The area of the parallelogram is base times height. Because each triangle is *half* of the parallelogram, the area of one triangle is

$$\frac{1}{2} \text{ of base times height.}$$

① Find the perimeter of each triangle.

(a)

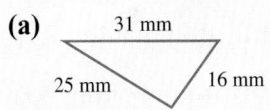

31 mm

25 mm   16 mm

(b)

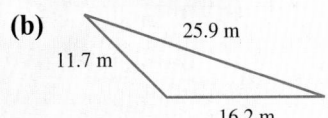

25.9 m

11.7 m

16.2 m

(c) A triangle with sides $6\frac{1}{2}$ yd, $9\frac{3}{4}$ yd, and $11\frac{1}{4}$ yd

ANSWERS
1. (a) 72 mm   **(b)** 53.8 m
   (c) $27\frac{1}{2}$ yd or 27.5 yd

❷ Find the area of each triangle.

(a)

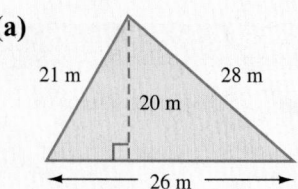

(b)

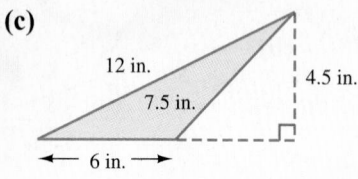

(c)

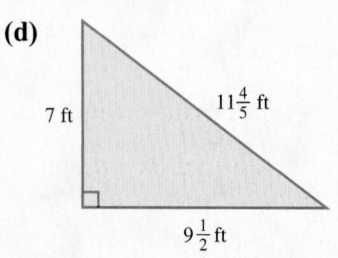

(d)

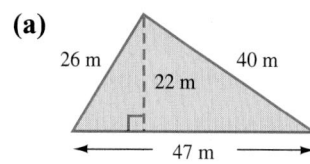

Finding the Area of a Triangle

$$\text{Area of a triangle} = \frac{1}{2} \cdot \text{base} \cdot \text{height}$$

$$A = \frac{1}{2} \cdot b \cdot h$$

$$\text{or} \quad A = 0.5 \cdot b \cdot h$$

Remember to use **square units** when measuring area.

Example 2   **Finding the Area of Triangles**

Find the area of each triangle.

(a)

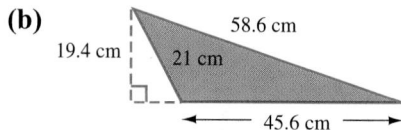

The base is 47 m and the height is 22 m. You do *not* need the 26 m or 40 m sides to find the area.

$$A = \frac{1}{2} \cdot b \cdot h$$

$$A = \frac{1}{\cancel{2}} \cdot 47\text{ m} \cdot \overset{11}{\cancel{22}}\text{ m} \quad \begin{array}{l}\text{Divide out}\\\text{common}\\\text{factor of 2.}\end{array}$$

$$A = 517\text{ m}^2 \quad \text{Square units for area}$$

(b)

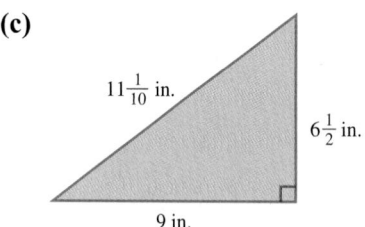

$$A = 0.5 \cdot 45.6\text{ cm} \cdot 19.4\text{ cm}$$

$$A = 442.32\text{ cm}^2$$

The base line must be extended to draw the height. However, still use 45.6 cm for *b* in the formula. Because the measurements are decimal numbers, it is easier to use 0.5 (the decimal equivalent of $\frac{1}{2}$) in the formula.

(c)

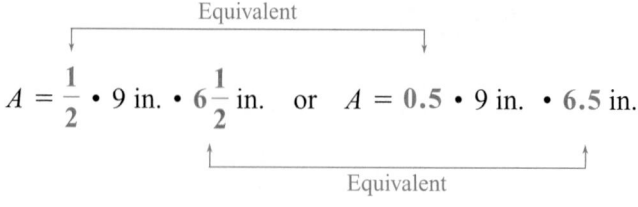

Because two sides of the triangle are perpendicular to each other, use those sides as the base and the height. (Recall that the height line must be perpendicular to the base.)

Equivalent

$$A = \frac{1}{2} \cdot 9\text{ in.} \cdot 6\frac{1}{2}\text{ in.} \quad \text{or} \quad A = 0.5 \cdot 9\text{ in.} \cdot 6.5\text{ in.}$$

Equivalent

$$A = 29\frac{1}{4}\text{ in.}^2 \qquad \text{or} \quad A = 29.25\text{ in.}^2$$

Work Problem ❷ at the Side.

Example 3 **Using the Concept of Area**

Find the area of the shaded part in this figure.

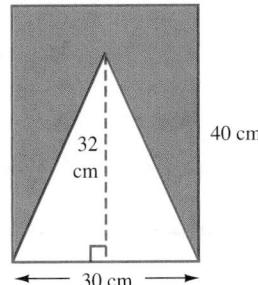

The *entire* figure is a rectangle. Find the area of the rectangle.

$$A = l \cdot w$$

$$A = 30 \text{ cm} \cdot 40 \text{ cm} = 1200 \text{ cm}^2$$

The *un*shaded part is a triangle. Find the area of the triangle.

$$A = \frac{1}{\underset{1}{\cancel{2}}} \cdot \overset{15}{\cancel{30}} \text{ cm} \cdot 32 \text{ cm}$$

$$A = 480 \text{ cm}^2$$

Subtract to find the area of the shaded part.

$$A = \overbrace{1200 \text{ cm}^2}^{\text{Entire area}} - \overbrace{480 \text{ cm}^2}^{\text{Unshaded part}} = \overbrace{720 \text{ cm}^2}^{\text{Shaded part}}$$

Work Problem ❸ at the Side.

Example 4 **Applying the Concept of Area**

The Department of Transportation cuts triangular signs out of rectangular pieces of metal using the measurements shown above in Example 3. If the metal costs $0.02 per square centimeter, how much does the metal cost for the sign? What is the cost of the metal that is *not* used?

From Example 3, the area of the triangle (the sign) is 480 cm². Multiply that times the cost per square centimeter.

$$\text{cost of sign} = \frac{480 \cancel{\text{ cm}^2}}{1} \cdot \frac{\$0.02}{1 \cancel{\text{ cm}^2}} = \$9.60$$

The metal that is *not* used is the *shaded* part from Example 3. The area is 720 cm².

$$\text{cost of unused metal} = \frac{720 \cancel{\text{ cm}^2}}{1} \cdot \frac{\$0.02}{1 \cancel{\text{ cm}^2}} = \$14.40$$

Work Problem ❹ at the Side.

---

❸ Find the area of the shaded part in this figure.

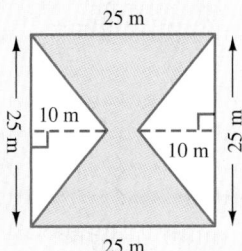

❹ Suppose the figure in Problem 3 above is an auditorium floor plan. The shaded part will be covered with carpet costing $27 per square meter. The rest will be covered with vinyl floor covering costing $18 per square meter. What is the total cost of covering the floor?

5 Find the number of degrees in the unlabeled angle.

(a)

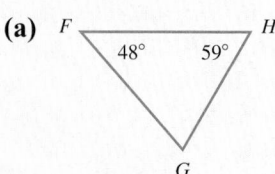

(b)

A ___ B (right angle at B)
55°

C

(c)

X
60°   60°
Y   W

3 **Given the measures of two angles in a triangle, find the measure of the third angle.** The *tri* in *triangle* means *three*. So the name tells you that a triangle has three angles. The sum of the measures of the three angles in any triangle is *always* 180° (a straight angle). You can see it by drawing a triangle, cutting off the three angles, and rearranging them to make a straight angle.

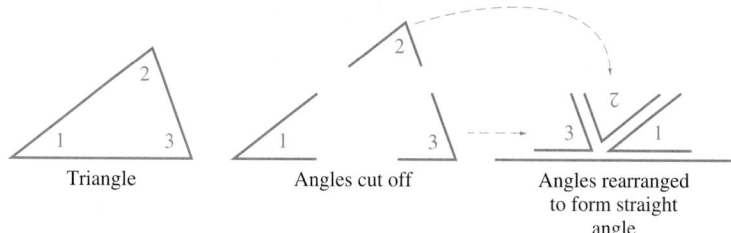

Triangle   Angles cut off   Angles rearranged to form straight angle

Finding the Unknown Angle Measurement in a Triangle

Step 1   Add the number of degrees in the two angles you are given.

Step 2   Subtract the sum from 180°.

Example 5   **Finding an Angle Measurement in Triangles**

Find the number of degrees in each indicated angle.

(a) Angle *R*

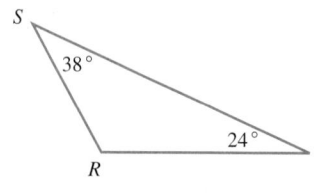

Step 1   Add the two angle measurements you are given.

$$38° + 24° = 62°$$

Step 2   Subtract the sum from 180°.

$$180° - 62° = 118°$$

∠*R* measures 118°.

(b) Angle *F*

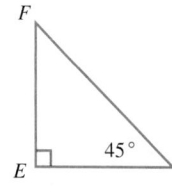

∠*E* is a right angle, so it measures 90°.

Step 1   90° + 45° = 135°

Step 2   180° − 135° = 45°

∠*F* measures 45°.

Work Problem 5 at the Side.

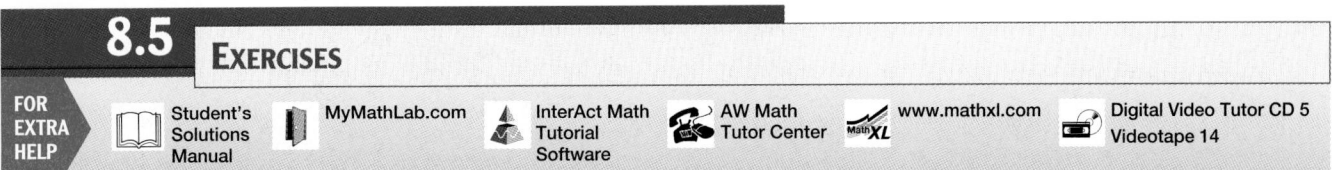

Find the perimeter and area of each triangle. See Examples 1 and 2.

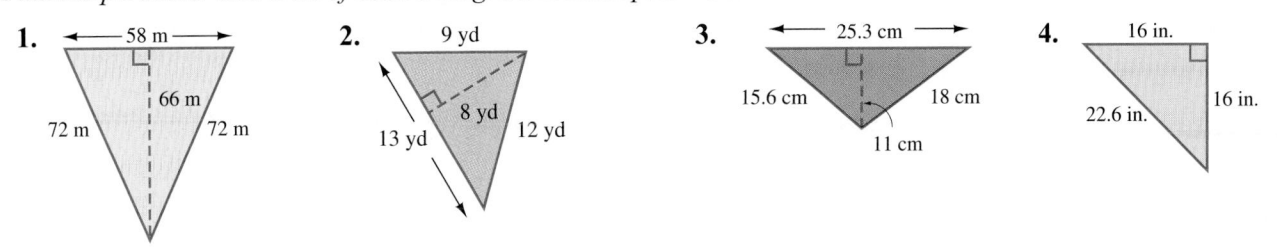

1. 58 m, 66 m, 72 m, 72 m

2. 9 yd, 8 yd, 13 yd, 12 yd

3. 25.3 cm, 15.6 cm, 18 cm, 11 cm

4. 16 in., 16 in., 22.6 in.

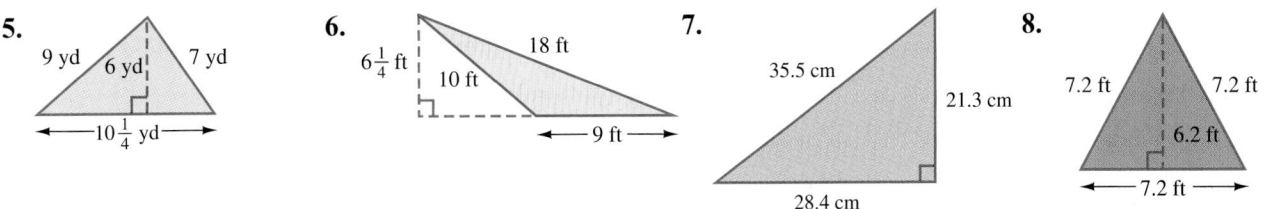

5. 9 yd, 6 yd, 7 yd, $10\frac{1}{4}$ yd

6. $6\frac{1}{4}$ ft, 10 ft, 18 ft, 9 ft

7. 35.5 cm, 21.3 cm, 28.4 cm

8. 7.2 ft, 7.2 ft, 6.2 ft, 7.2 ft

Find the shaded area in each figure. See Example 3.

9.

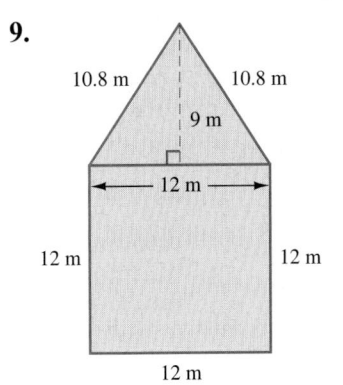

10.8 m, 10.8 m, 9 m, 12 m, 12 m, 12 m, 12 m

10.

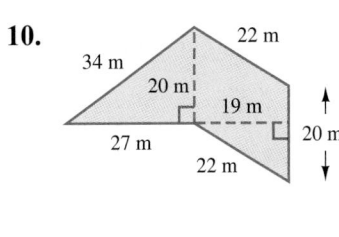

34 m, 22 m, 20 m, 19 m, 20 m, 27 m, 22 m

11.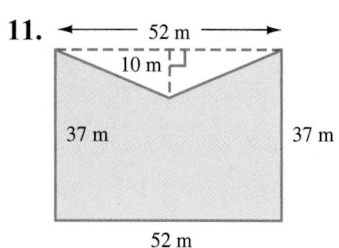

12.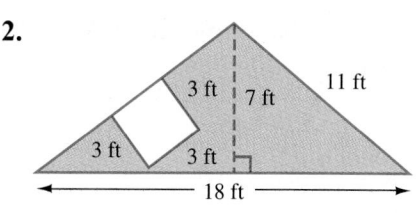

Find the number of degrees in the unlabeled angle. See Example 5.

13.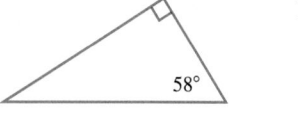

14.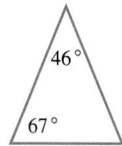

15.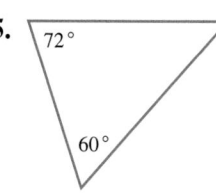

16.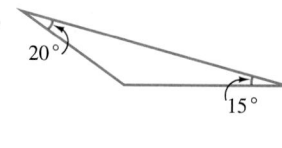

17. Can a triangle have two right angles? Explain your answer.

18. In your own words, explain where the $\frac{1}{2}$ comes from in the formula for area of a triangle. Draw a sketch to illustrate your explanation.

Solve each application problem. See Example 4.

19. A triangular tent flap measures $3\frac{1}{2}$ ft along the base and has a height of $4\frac{1}{2}$ ft. How much canvas is needed to make the flap?

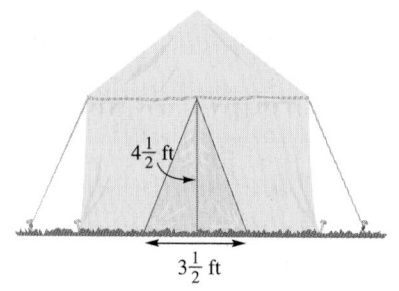

20. A wooden sign in the shape of a right triangle has perpendicular sides measuring 1.5 m and 1.2 m. How much surface area does the sign have?

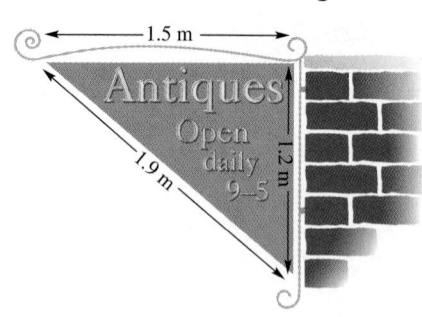

21. A triangular space between three streets has the measurements shown. How much new curbing will be needed to go around the space? How much sod will be needed to cover the space?

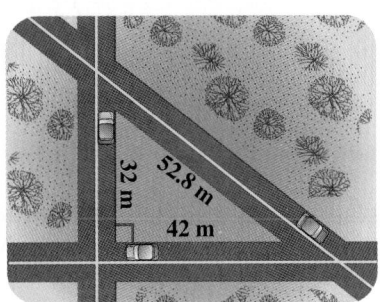

22. Each gable end of a new house has a span of 36 ft and a rise of 9.5 ft. What is the total area of both gable ends of the house?

23. **(a)** Find the area of one side of the house.

(b) Find the area of one roof section.

All sides of the house are congruent and all roof sections are congruent.

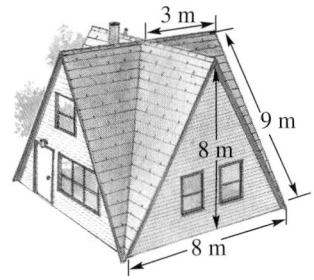

24. The sketch shows the plan for an office building. The shaded area will be a parking lot. What is the cost of building the parking lot if the contractor charges $28.00 per square yard for materials and labor?

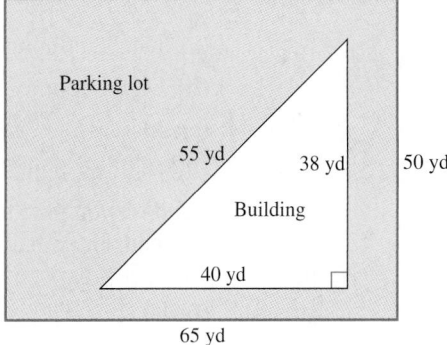

Interior Design

Suppose that you have just bought a small waterfront house and you plan to remodel the interior by installing tile and carpet on the floors. A sketch of the floor plan is shown below. All measurements are in feet and represent the measurements rounded up to the nearest foot.

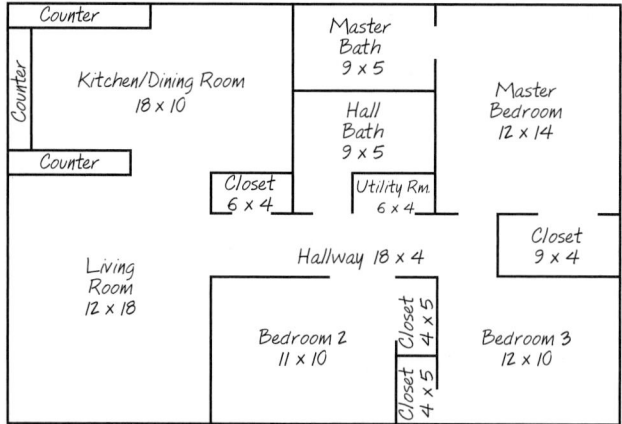

1. How many square feet of floor space are in each of the following rooms?

   (a) Kitchen/dining room

   (b) Living room

   (c) Master bath, and hall bath (including entry) combined

   (d) Hallway, hall closet, and utility room combined

2. Suppose you plan to install floor tiles in the kitchen/dining room, living room, both baths, hall bath entry, hallway, hall closet, and utility room. The flooring company salesperson recommends that you increase your square footage requirement by 5% to compensate for waste. How many square feet of tile are needed?

3. Tiles are sold in boxes of twelve 1-foot square tiles, and partial boxes are not sold. The tile that you selected costs $5.75 per square foot, installed, based on the total number of tiles purchased. The sales tax rate is 8.25%.

   (a) How many boxes of tile must you purchase?

   (b) How much will it cost to install the tile, including sales tax?

4. How many square feet of floor is in all the bedrooms and bedroom closets, combined?

5. Suppose you plan to carpet the bedrooms and closets. The carpet costs $24.95 per square *yard*, installed, and the sales tax rate is 8.25%. The carpet is sold in rolls that are 12 ft wide, so a 3 ft length equals 4 sq yd. (Why?) The salesperson explains that the carpet nap must run in the same direction across carpet seams and recommends that you purchase a 45 ft length of 12-ft-wide carpet.

   (a) Write how the salesperson may have computed the 45 ft. length. The 45 ft length is how many square *yards* of carpet?

   (b) A common industry formula is to compute $(L \times W)/8$, rounded up to the next square yard, to estimate the number of square yards. How accurate is that formula for this job?

   (c) How much will it cost to install the carpet, including sales tax?

8.6 CIRCLES

1 Find the radius and diameter of a circle.   Suppose you start with one dot on a piece of paper. Then you draw a bunch of dots that are each 2 cm away from the first dot. If you draw enough dots (points) you'll end up with a *circle*. Each point on the circle is exactly 2 cm away from the *center* of the circle. The 2 cm distance is called the *radius*, *r*, of the circle. The distance across the circle (passing through the center) is called the *diameter*, *d*, of the circle.

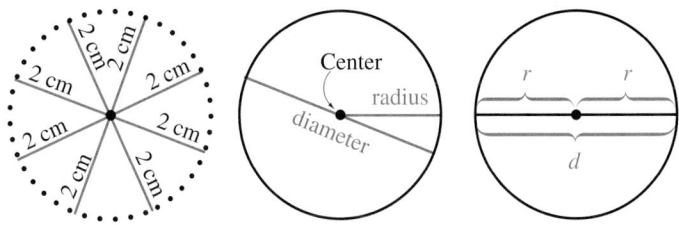

OBJECTIVES

1 Find the radius and diameter of a circle.

2 Find the circumference of a circle.

3 Find the area of a circle.

4 Become familiar with Latin and Greek prefixes used in math terminology.

Circle, Radius, and Diameter

A **circle** is a two-dimensional (flat) figure with all points the same distance from a fixed center point.

The **radius** (*r*) is the distance from the center of the circle to any point on the circle.

The **diameter** (*d*) is the distance across the circle passing through the center.

The circle on the right above illustrates the relationships between the radius and diameter. Use the formulas below.

Finding the Diameter and Radius of a Circle

$$\text{diameter} = 2 \cdot \text{radius}$$
$$d = 2 \cdot r$$
$$\text{and} \quad r = \frac{d}{2}$$

Example 1 **Finding the Diameter and Radius of Circles**

Find the unknown length of the diameter or radius in each circle.

(a)

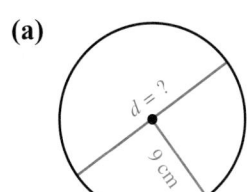

Because the radius is 9 cm, the diameter is twice as long.

$$d = 2 \cdot r$$
$$d = 2 \cdot 9 \text{ cm}$$
$$d = 18 \text{ cm}$$

Continued on Next Page

① Find the unknown length of the diameter or radius in each circle.

(a)

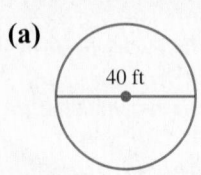

(b)

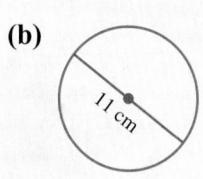

(c)

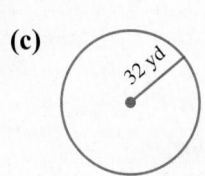

(d)

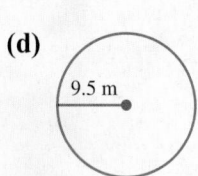

(b)

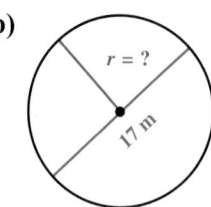

The radius is half the diameter.

$$r = \frac{d}{2}$$

$$r = \frac{17 \text{ m}}{2}$$

$$r = 8.5 \text{ m} \quad \text{or} \quad 8\frac{1}{2} \text{ m}$$

Work Problem ① at the Side.

2 ▭ **Find the Circumference of a circle.**   The perimeter of a circle is called its **circumference.** Circumference is the distance around the edge of a circle.

The diameter of the can in the drawing is about 10.6 cm, and the circumference of the can is about 33.3 cm. Dividing the circumference of the circle by the diameter gives an interesting result.

$$\frac{\text{circumference}}{\text{diameter}} = \frac{33.3}{10.6} \approx 3.14 \quad \text{Rounded to the nearest hundredth}$$

Dividing the circumference of *any* circle by its diameter *always* gives an answer close to 3.14. This means that going around the edge of any circle is a little more than 3 times as far as going straight across the circle.

This ratio of circumference to diameter is called π (the Greek letter **pi,** pronounced PIE). There is no decimal that is exactly equal to π, but here is the *approximate* value.

$$\pi \approx 3.14159265359$$

Rounding the Value of *Pi* (π)

We usually round π to 3.14. Therefore, calculations involving π will give approximate answers and should be written using the $\approx$ symbol.

Use the following formulas to find the *circumference* of a circle.

Finding the Circumference (Distance around a Circle)

$$\text{circumference} = \pi \cdot \text{diameter}$$
$$C = \pi \cdot d$$

or, because $d = 2 \cdot r$, then $C = \pi \cdot 2 \cdot r$ usually written $C = 2 \cdot \pi \cdot r$

ANSWERS
1. (a) $r = 20$ ft   **(b)** $r = 5.5$ cm
   (c) $d = 64$ yd   **(d)** $d = 19$ m

Example 2 **Finding the Circumference of Circles**

Find the circumference of each circle. Use 3.14 as the approximate value for π. Round answers to the nearest tenth.

(a)

38 m

The diameter is 38 m, so use the formula with d in it.

$$C = \pi \cdot d$$
$$C \approx 3.14 \cdot 38 \text{ m}$$
$$C \approx 119.3 \text{ m} \quad \text{Rounded}$$

(b)

11.5 cm

In this example, the radius is labeled, so it is easier to use the formula with r in it.

$$C = 2 \cdot \pi \cdot r$$
$$C \approx 2 \cdot 3.14 \cdot 11.5 \text{ cm}$$
$$C \approx 72.2 \text{ cm} \quad \text{Rounded}$$

Calculator Tip Scientific calculators have a π key. Try pressing it. With a 10-digit display, you'll see the value of π to the nearest billionth.

> **3.141592654**

But this is still an approximate value, although it is more precise than rounding π to 3.14. Try finding the circumference in Example 2(a) above using the π key.

π $\times$ 38 $=$ 119.3805208   Rounds to 119.4

When you used 3.14 as the approximate value of π, the result rounded to 119.3, so the answers are slightly different. In this book we will use 3.14 instead of the π key. Our measurements of radius and diameter are given as whole numbers or with tenths, so it is acceptable to round π to hundredths. And, some students may be using standard calculators without a π key, or doing the calculations by hand.

Work Problem ❷ at the Side.

3___ **Find the area of a circle.** To find the formula for the area of a circle, start by cutting two circles into many pie-shaped pieces.

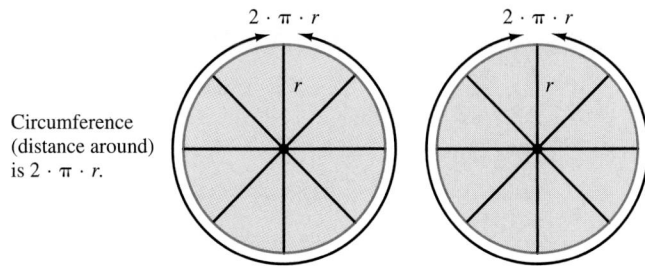

Circumference (distance around) is $2 \cdot \pi \cdot r$.

Unfold the circles, much as you might "unfold" a peeled orange, and put them together as shown here.

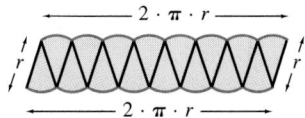

---

❷ Find the circumference of each circle. Use 3.14 as the approximate value for π. Round answers to the nearest tenth.

(a)

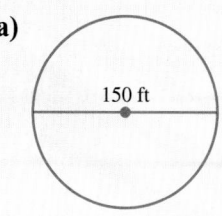

150 ft

(b)

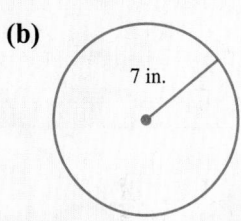

7 in.

(c) diameter 0.9 km

(d) radius 4.6 m

❸ Find the area of each circle. Use 3.14 for π. Round your answers to the nearest tenth.

(a)

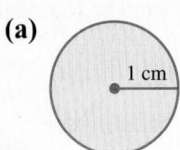

1 cm

(b)

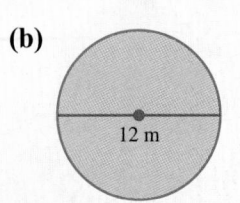

12 m

(*Hint:* The diameter is 12 m so $r =$ _____ m)

(c)

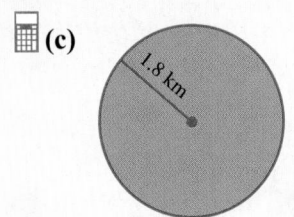

1.8 km

(d)

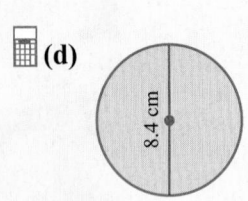

8.4 cm

The figure is approximately a rectangle with width r (the radius of the original circle) and length $2 \cdot \pi \cdot r$ (the circumference of the original circle). The area of the "rectangle" is length times width.

$$\text{Area} = \quad l \quad \cdot w$$
$$\text{Area} = 2 \cdot \pi \cdot \underbrace{r \cdot r}$$
$$\text{Area} = 2 \cdot \pi \cdot \quad r^2 \quad \leftarrow \text{Recall that } r \cdot r \text{ is } r^2$$

Because the "rectangle" was formed from *two* circles, the area of *one* circle is half as much.

$$\frac{1}{\underset{1}{\cancel{2}}} \cdot \overset{1}{\cancel{2}} \cdot \pi \cdot r^2 = 1 \cdot \pi \cdot r^2 \quad \text{or simply} \quad \pi \cdot r^2$$

Finding the Area of a Circle

Area of a circle $= \pi \cdot$ radius $\cdot$ radius
$$A = \pi \cdot r^2$$

Remember to use **square units** when measuring area.

Example 3 Finding the Area of Circles

Find the area of each circle. Use 3.14 for π. Round your answers to the nearest tenth.

(a) A circle with radius 8.2 cm

Use the formula $A = \pi \cdot r^2$, which means $\pi \cdot r \cdot r$.

$$A = \pi \cdot \quad r \quad \cdot \quad r$$
$$A \approx 3.14 \cdot 8.2 \text{ cm} \cdot 8.2 \text{ cm}$$
$$A \approx 211.1 \text{ cm}^2 \quad \text{Rounded; square units for area}$$

(b)

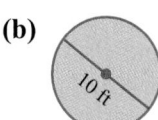

10 ft

To use the formula, you need to know the radius (r). In this circle, the diameter is 10 ft. First find the radius.

$$r = \frac{d}{2}$$
$$r = \frac{10 \text{ ft}}{2} = 5 \text{ ft}$$

Now find the area.

$$A \approx 3.14 \cdot 5 \text{ ft} \cdot 5 \text{ ft}$$
$$A \approx 78.5 \text{ ft}^2 \quad \text{Square units for area}$$

CAUTION

When finding *circumference,* you can start with either the radius or the diameter. When finding *area,* you must use the *radius.* If you are given the diameter, divide it by 2 to find the radius. Then find the area.

Work Problem ❸ at the Side.

▦ **Calculator Tip**   You can find the area of the circle in Example 3(a) on your calculator. The first method works on both scientific and standard calculators:

$$3.14 \; \textcircled{\times} \; 8.2 \; \textcircled{\times} \; 8.2 \; \textcircled{=} \; 211.1336$$

You round the answer to 211.1 (nearest tenth).

On a scientific calculator you can also use the $\boxed{x^2}$ key, which automatically squares the number you enter (that is, multiplies the number times itself):

$$3.14 \; \textcircled{\times} \; 8.2 \; \boxed{x^2} \; \underbrace{67.24} \; \textcircled{=} \; 211.1336$$

Appears automatically;
8.2 × 8.2 is 67.24

In the next example we find the area of a *semicircle,* which is half the area of a circle.

Example 4   **Finding the Area of a Semicircle**

Find the area of the semicircle. Use 3.14 for π. Round your answer to the nearest tenth.

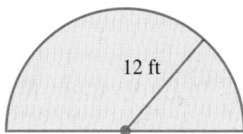

First, find the area of an entire circle with a radius of 12 ft.

$$A = \pi \cdot r \cdot r$$
$$A \approx 3.14 \cdot 12 \text{ ft} \cdot 12 \text{ ft}$$
$$A \approx 452.16 \text{ ft}^2 \quad \text{Do not round yet.}$$

Divide the area of the whole circle by 2 to find the area of the semicircle.

$$\frac{452.16 \text{ ft}^2}{2} = 226.08 \text{ ft}^2$$

The *last* step is rounding 226.08 to the nearest tenth.

$$\text{Area of semicircle} \approx 226.1 \text{ ft}^2 \quad \text{Rounded}$$

══ **Work Problem ❹ at the Side.**

Example 5   **Applying the Concept of Circumference**

A circular rug is 8 ft in diameter. The cost of fringe for the edge is $2.25 per foot. What will it cost to add fringe to the rug? Use 3.14 for π.

$$\text{Circumference} = \pi \cdot d$$
$$C \approx 3.14 \cdot 8 \text{ ft}$$
$$C \approx 25.12 \text{ ft}$$

$$\text{cost} = \text{cost per foot} \cdot \text{circumference}$$
$$\text{cost} = \frac{\$2.25}{1 \text{ ft}} \cdot \frac{25.12 \text{ ft}}{1}$$
$$\text{cost} = \$56.52$$

══ **Work Problem ❺ at the Side.**

❹ Find the area of each semicircle. Use 3.14 for π. Round your answers to the nearest tenth.

(a)

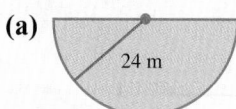

24 m

(b)

←——— 35.4 ft ———→

(c)

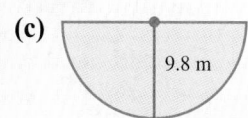

9.8 m

Answers

4. (a) $A \approx 904.3 \text{ m}^2$ **(b)** $A \approx 491.9 \text{ ft}^2$
     (c) $A \approx 150.8 \text{ m}^2$
5. $42.39

6 Find the cost of covering the underside of the rug in Problem 5 with a nonslip rubber backing. The rubber backing costs $2 per square meter.

7 (a) Here are some more prefixes you have seen in this textbook. List at least one math term and one nonmathematical word that use each prefix.

dia (through):

fract (break):

par (beside):

per (divide):

peri (around):

rad (ray):

rect (right):

sub (below):

(b) How could you use your knowledge of prefixes to remember the difference between perimeter and area?

ANSWERS
6. $14.13
7. (a) Some possibilities are:
diameter; diagonal
fraction; fracture
parallel; paramedic
percent; per capita
perimeter; periscope
radius; radiate
rectangle; rectify
subtract; submarine.
(b) *Peri* in perimeter means "around," so perimeter is the distance *around* the edges of a shape.

Example 6 Applying the Concept of Area

Find the cost of covering the rug in Example 5 with a plastic cover. The material for the cover costs $1.50 per square foot. Use 3.14 for π.

First find the radius.

$$r = \frac{d}{2} = \frac{8 \text{ ft}}{2} = 4 \text{ ft}$$

$$\text{Then,} \quad A = \pi \cdot r^2$$
$$A \approx 3.14 \cdot 4 \text{ ft} \cdot 4 \text{ ft}$$
$$A \approx 50.24 \text{ ft}^2$$

$$\text{cost} = \frac{\$1.50}{1 \text{ ft}^2} \cdot \frac{50.24 \text{ ft}^2}{1} = \$75.36$$

Work Problem 6 at the Side.

4 ▭ Become familiar with Latin and Greek prefixes used in math terminology. Many English words are built from Latin or Greek root words and prefixes. Knowing the meaning of the more common ones can help you figure out the meaning of terms in many subject areas, including math.

Example 7 Using Prefixes to Understand Math Terms

(a) Listed below are some Latin and Greek root words and prefixes with their meanings in parentheses. You've already seen math terms in this textbook that use these prefixes. List at least one math term and one nonmathematical word that use each prefix or root word.

cent (100): *cent*imeter; *cent*ury

circum (around): *circum*ference; *circum*vent

de (down): *de*nominator; *de*duction

dec (10): *dec*imal; *Dec*ember (originally the 10th month in the old calendar)

> There are many answers. These are some of the possibilities.

(b) Suppose you have trouble remembering which part of a fraction is the denominator. How could your knowledge of prefixes help in this situation?

The *de* prefix in *de*nominator means "down" so the denominator is the number *down* below the fraction bar.

Work Problem 7 at the Side.

NOTE

Here are some additional prefixes and root words and their meanings that you will see in the rest of Chapter 8, in Chapter 9, and in other math classes. An example of a math term and a nonmathematical word are shown for each one.

equ (equal): *equ*ation; *equ*inox   *lateral* (side): quadri*lateral*; bi*lateral*

hemi (half): *hemi*sphere; *hemi*trope   *re* (back or again): *re*ciprocal; *re*duce

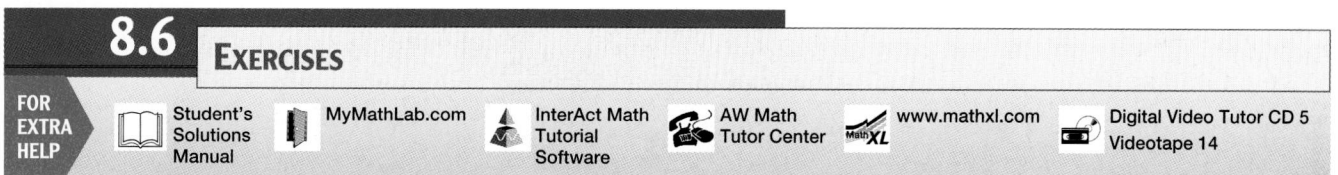

8.6 EXERCISES

FOR EXTRA HELP	Student's Solutions Manual	MyMathLab.com	InterAct Math Tutorial Software	AW Math Tutor Center	www.mathxl.com	Digital Video Tutor CD 5 Videotape 14

Find the unknown length in each circle. See Example 1.

1.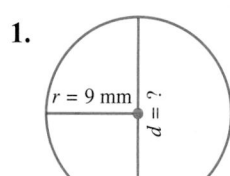

2.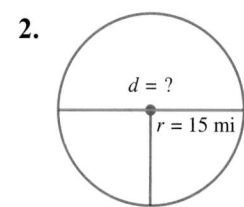

3.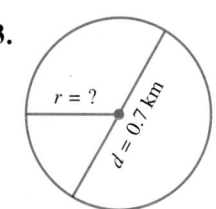

4.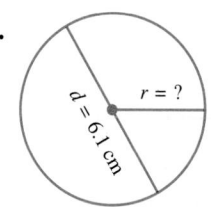

Find the circumference and area of each circle. Use 3.14 as the approximate value for π. Round your answers to the nearest tenth. See Examples 2 and 3.

5.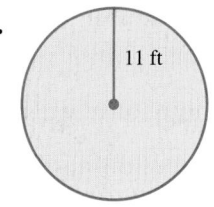

6.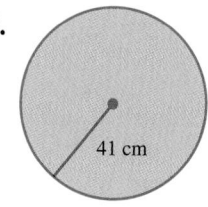

**7.**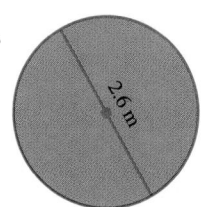

8.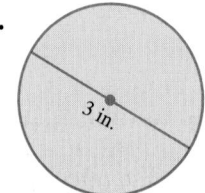

Find the circumference and area of circles having the following diameters. Use 3.14 for π. Round your answers to the nearest tenth. See Examples 2 and 3.

9. $d = 15$ cm

10. $d = 39$ ft

11. $d = 7\frac{1}{2}$ ft

12. $d = 4\frac{1}{2}$ yd

13. $d = 8.65$ km

14. $d = 19.5$ mm

Find each shaded area. Note that Exercises 15 and 18 contain semicircles. Use 3.14 as the approximate value of π. Round your answers to the nearest tenth if necessary. See Example 4.

15.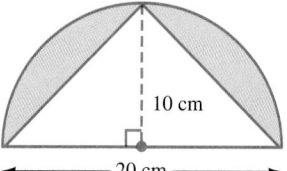

16.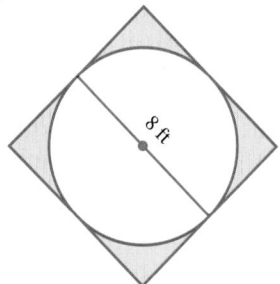

17.

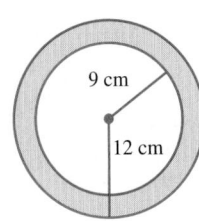

9 cm

12 cm

18.

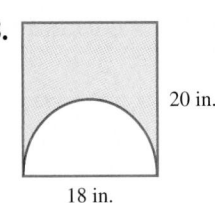

20 in.

18 in.

19. How would you explain π to a friend who is not in your math class? Write an explanation. Then make up a test question that requires the use of π, and show how to solve it.

20. Explain how circumference and perimeter are alike. How are they different? Make up two problems, one involving perimeter, the other circumference. Show how to solve your problems.

Solve each application problem. Use 3.14 as the approximate value of π. Round your answers to the nearest tenth. See Examples 5 and 6.

21. How far does a point on the tread of a tire move in one turn if the diameter of the tire is 70 cm?

22. If you swing a ball held at the end of a string 2 m long, how far will the ball travel on each turn?

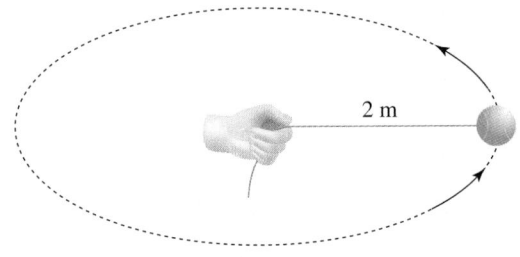

2 m

23. A wave energy extraction device is a huge undersea dome used to harness the power of ocean waves. The base of the dome has a radius of 125 ft. Find its circumference.

24. Find the area of the base of the dome in Exercise 23.

For Exercises 25–28, first draw a circle and label the radius or diameter. Then solve the problem. (Sketches may vary; show your sketches to your instructor.)

25. A radio station can be heard 150 miles in all directions during evening hours. How many square miles are in the station's broadcast area?

26. An earthquake was felt by people 900 km away in all directions from the epicenter (the source of the earthquake). How much area was affected by the quake?

27. The diameter of Diana Hestwood's wristwatch is 1 in. and the radius of the clock face on her kitchen wall is 3 in. Find the circumference and the area of each clock face.

28. The diameter of the largest known ball of twine is 12 ft 9 in. (*Source: Guinness Book of Records.*) The sign posted near the ball says it has a circumference of 40 ft. Is the sign correct? (*Hint:* First change 9 in. to feet and add it to 12 ft.)

29. On the first page of this chapter, you read about a forester measuring the circumferences of trees. If the circumference of one tree is 144 cm, what is the diameter of the tree?

30. In Atlanta, Interstate 285 circles the city and is known as the "perimeter." If the circumference of the circle made by the highway is 62.8 miles, find

(a) the diameter of the circle, and

(b) the area inside the circle.

(*Source: Greater Atlanta Newcomer's Guide.*)

31. Find the cost of sod, at $1.76 per square foot, for this playing field that has a semicircle on each end.

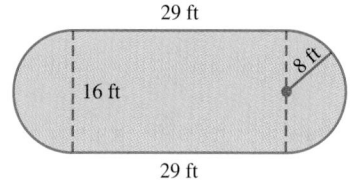

32. Find the area of this skating rink.

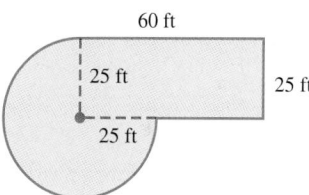

Use the information about prefixes in Example 7 to answer Exercises 33 and 34.

33. Explain how you could use the information about prefixes to remember the difference between radius, diameter, and circumference.

34. Explain how you could use the information about prefixes to avoid confusion between parallel and perpendicular lines.

---

 *Use the table below to **work Exercises 35–40 in order.***

Find the best buy for each type of pizza. The best buy is the lowest cost per square inch of pizza. All the pizzas are circular in shape, and the measurement given on the menu board is the diameter of the pizza in inches. Use 3.14 as the approximate value of π. Round the area to the nearest tenth. Round cost per square inch to the nearest thousandth.

Pizza Menu	Small $7\frac{1}{2}''$	Medium 13"	Large 16"
Cheese only	$2.80	$ 6.50	$ 9.30
"The Works"	$3.70	$ 8.95	$14.30
Deep-dish combo	$4.35	$10.95	$15.65

35. Find the area of a small pizza.

36. Find the area of a medium pizza.

37. Find the area of a large pizza.

38. What is the cost per square inch for each size of cheese pizza? Which size is the best buy?

39. What is the cost per square inch for each size of "The Works" pizza? Which size is the best buy?

40. You have a coupon for 95¢ off any small pizza. What is the cost per square inch for each size of deep dish combo pizza? Which size is the best buy?

Summary Exercises on PERIMETER, CIRCUMFERENCE, AND AREA

1. Draw a sketch of each of these shapes: **(a)** square, **(b)** rectangle, **(c)** parallelogram. On each sketch, indicate 90° angles and show which sides are the same length.

2. (a) Draw a sketch of a circle and show the radius.

(b) Draw another circle and show the diameter.

(c) Describe the relationship between the radius and diameter of a circle.

3. Describe how you can find the perimeter of any flat shape with straight sides.

4. In your own words, describe the difference between finding the perimeter of a shape and finding the area of the shape.

5. Match each shape to its corresponding area formula.

Shapes	Area Formulas
parallelogram _____	**(a)** $A = \dfrac{1}{2} \cdot b \cdot h$
square _____	**(b)** $A = \pi \cdot r^2$
trapezoid _____	**(c)** $A = l \cdot w$
circle _____	**(d)** $A = \dfrac{1}{2} \cdot h \cdot (b + B)$
rectangle _____	**(e)** $A = s^2$
triangle _____	**(f)** $A = b \cdot h$

6. (a) If you know the *radius* of a circle, which formula do you use to find its circumference?

(b) If you know the *diameter* of a circle, which formula do you use to find its circumference?

(c) If you know the *diameter* of a circle, what must you do *before* using the formula for finding the area?

7. Find the perimeter and area of each triangle, to the nearest tenth.

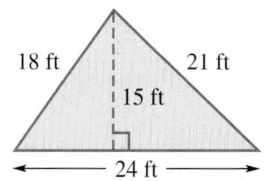

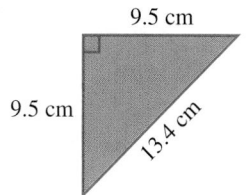

8. Some answers from a student's test paper are listed below. The *number* part of each answer is correct, but the *units* are not. Rewrite each answer with the correct units.

(a) $A = 12$ cm

(b) $P = 6\dfrac{1}{2}$ ft²

(c) $C \approx 28.5$ m²

(d) $A = 307$ in.

Find the perimeter and area of each shape. Round answers to the nearest tenth.

9.

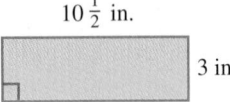

10½ in.

3 in.

10.

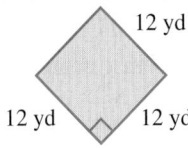

12 yd

12 yd   12 yd

11.

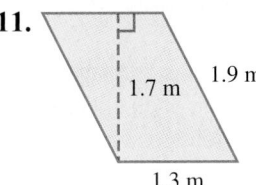

1.7 m   1.9 m

1.3 m

12.

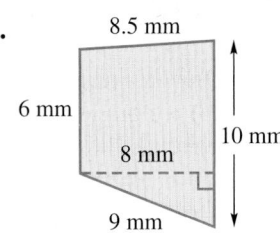

8.5 mm

6 mm

8 mm   10 mm

9 mm

For each circle, find (a) the diameter or radius, (b) the circumference, and (c) the area. Use 3.14 as the approximate value of π. Round answers to the nearest tenth.

13.

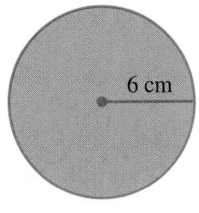

6 cm

14.

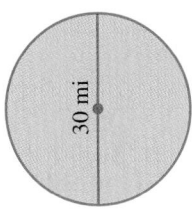

30 mi

15.

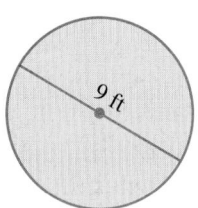

9 ft

Find the area of each shaded region. Use 3.14 as the approximate value of π. Round answers to the nearest tenth.

16.

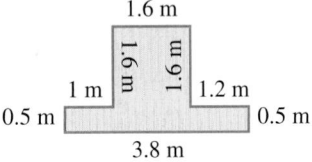

1.6 m

1.6 m   1.6 m

1 m   1.2 m

0.5 m   0.5 m

3.8 m

17.

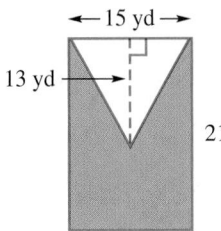

15 yd

13 yd

21 yd

15 yd

18.

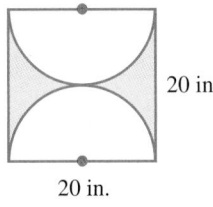

20 in.

20 in.

Solve each application problem. Use 3.14 as the approximate value of π. Round answers to the nearest tenth.

19. The Mormons traveled west to Utah by covered wagon in 1847. They tied a rag to a wagon wheel to keep track of the distance they traveled. The radius of the wheel was 2.33 ft. How far did the rag travel each time the wheel made a complete revolution? (*Source:* Trail of Hope.) Bonus question: How many wheel revolutions equaled one mile?

20. The front door for a new log home is 0.9 m wide and 2 m high. How much will it cost for weather strip material to go around all edges of the door, if weather strip costs $0.77 per meter?

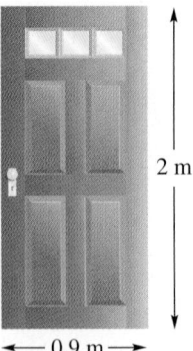

2 m

0.9 m

8.7 VOLUME

1 ▭ **Find the volume of a rectangular solid.**   A shoe box and a cereal box are examples of three-dimensional (or solid) figures. The three dimensions are length, width, and height. (A rectangle or square is a two-dimensional figure. The two dimensions are length and width.) If you want to know how much the shoe box will hold, you find its *volume*. We measure volume by seeing how many cubes of a certain size will fill the space inside the box. All the edges of a cube have the same length. Three sizes of *cubic units* are shown below.

OBJECTIVES

Find the volume of a

1 ▭ rectangular solid;

2 ▭ sphere;

3 ▭ cylinder;

4 ▭ cone and pyramid.

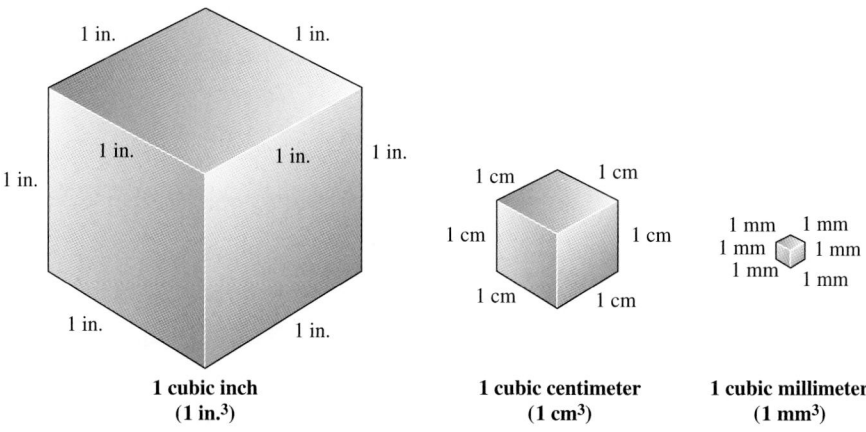

| 1 cubic inch (1 in.³) | 1 cubic centimeter (1 cm³) | 1 cubic millimeter (1 mm³) |

Some other sizes of cubes that are used to measure volume are 1 cubic foot (1 ft³), 1 cubic yard (1 yd³), and 1 cubic meter (1 m³).

CAUTION

The raised 3 in 4^3 means that you multiply $4 \cdot 4 \cdot 4$ to get 64. The raised 3 in cm³ or ft³ is a short way to write the word *cubic*. When you see 5 cm³, say "five *cubic* centimeters." Do *not* multiply $5 \cdot 5 \cdot 5$.

Volume

Volume is a measure of the space inside a solid shape. The volume of a solid is how many cubic units it takes to fill the solid.

Use the following formula to find the *volume* of rectangular solids (box-like shapes).

Finding the Volume of Rectangular Solids

Volume of rectangular solid = length • width • height

$$V = l \cdot w \cdot h$$

Remember to use *cubic units* when measuring volume.

❶ Find the volume of each box. Round your answers to the nearest tenth if necessary.

(a)

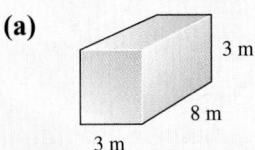

3 m
8 m
3 m

(b)

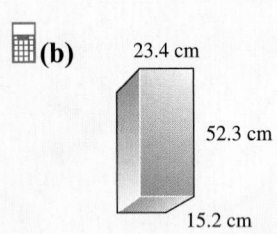

23.4 cm
52.3 cm
15.2 cm

🖩**(c)** Length $6\frac{1}{4}$ ft, width $3\frac{1}{2}$ ft, height 2 ft

Example 1 Finding the Volume of Rectangular Solids

Find the volume of each box.

(a)

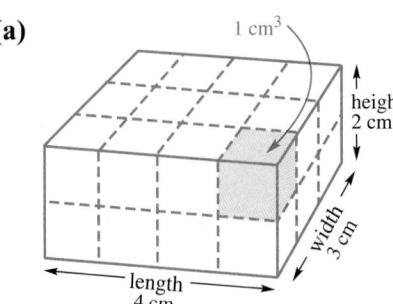

1 cm³
height 2 cm
width 3 cm
length 4 cm

Each cube that fits in the box is 1 cubic centimeter (1 cm³). To find the volume, you can count the number of cubes.

Bottom layer has 12 cubes.
Top layer has 12 cubes.
} total of 24 cubes (24 cm³)

Or, you can use the formula for rectangular solids.

$$V = l \cdot w \cdot h$$
$$V = 4 \text{ cm} \cdot 3 \text{ cm} \cdot 2 \text{ cm}$$
$$V = 24 \text{ cm}^3 \quad \text{Cubic units for volume.}$$

(b)

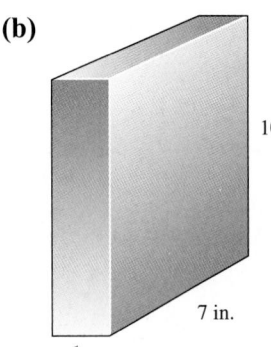

10 in.
7 in.
$2\frac{1}{2}$ in.

Use the formula.

$$V = 7 \text{ in.} \cdot 2\frac{1}{2} \text{ in.} \cdot 10 \text{ in.}$$

$$V = \frac{7 \text{ in.}}{1} \cdot \frac{\overset{5}{\cancel{5 \text{ in.}}}}{\underset{1}{\cancel{2}}} \cdot \frac{\overset{5}{\cancel{10}} \text{ in.}}{1} = 175 \text{ in.}^3$$

If you like, use 2.5 in., the decimal equivalent of $2\frac{1}{2}$ in., for the width.

$$V = 7 \text{ in.} \cdot 2.5 \text{ in.} \cdot 10 \text{ in.} = 175 \text{ in.}^3$$

Work Problem ❶ at the Side.

2 Find the volume of a sphere. A *sphere* is shown here. Examples of spheres include baseballs, oranges, and the earth. (The last two aren't perfect spheres, but they're close.)

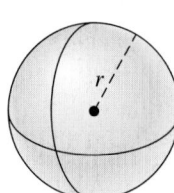

r

As with circles, the *radius* of a sphere is the distance from the center to the edge of the sphere. Use the following formula to find the *volume* of a sphere.

Finding the Volume of a Sphere

$$\text{Volume of sphere} = \frac{4}{3} \cdot \pi \cdot r \cdot r \cdot r$$

$$V = \frac{4}{3} \cdot \pi \cdot r^3 \quad \text{or} \quad \frac{4 \cdot \pi \cdot r^3}{3}$$

Remember to use **cubic units** when measuring volume.

Example 2 Finding the Volume of Spheres

Find the volume of each sphere with the help of a calculator. Use 3.14 as the approximate value of π. Round your answers to the nearest tenth.

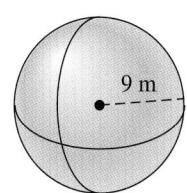 **(a)**

9 m

$$V = \frac{4}{3} \cdot \pi \cdot r^3$$

$$V \approx \frac{4 \cdot 3.14 \cdot 9 \text{ m} \cdot 9 \text{ m} \cdot 9 \text{ m}}{3}$$

$V \approx 3052.08$ ⠀⠀Now round to tenths.

$V \approx 3052.1 \text{ m}^3$ ⠀⠀Cubic units for volume

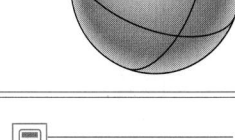 **(b)**

4.2 ft

$$V \approx \frac{4 \cdot 3.14 \cdot 4.2 \text{ ft} \cdot 4.2 \text{ ft} \cdot 4.2 \text{ ft}}{3}$$

$V \approx 310.18176$ ⠀⠀Now round to tenths.

$V \approx 310.2 \text{ ft}^3$ ⠀⠀Cubic units for volume

Calculator Tip You can find the volume of the sphere in Example 2(b) on your calculator. The first method works on both scientific and standard calculators:

⠀⠀⠀⠀4 ⊗ 3.14 ⊗ 4.2 ⊗ 4.2 ⊗ 4.2 ÷ 3 ⊜ 310.18176

⠀⠀⠀⠀Round the answer to 310.2 ft³.

On a scientific calculator you can use the y^x key to calculate r^3 (to multiply the radius times itself three times).

⠀⠀⠀⠀4 ⊗ 3.14 ⊗ $\underbrace{4.2 \ y^x \ 3}_{r^3}$ ÷ 3 ⊜ 310.18176

Recall that we are using 3.14 as the approximate value for π instead of using the π key.

You can also use the y^x key with other exponents. For example:

⠀⠀⠀⠀To find 2^5, press 2 y^x 5 ⊜. ⠀⠀Answer is 32.

⠀⠀⠀⠀To find 6^4, press 6 y^x 4 ⊜. ⠀⠀Answer is 1296.

Work Problem ❷ at the Side.

⠀⠀Half a sphere is called a *hemisphere*. The volume of a hemisphere is *half* the volume of a sphere. Use the following formula to find the *volume* of a hemisphere.

Finding the Volume of a Hemisphere

$$\text{Volume of hemisphere} = \frac{1}{\cancel{2}_1} \cdot \frac{\cancel{4}^2}{3} \cdot \pi \cdot r \cdot r \cdot r$$

$$V = \frac{2}{3} \cdot \pi \cdot r^3 \quad \text{or} \quad \frac{2 \cdot \pi \cdot r^3}{3}$$

Remember to use *cubic units* when measuring volume.

---

❷ Find the volume of each sphere. Use 3.14 for π. Round your answers to the nearest tenth.

(a)

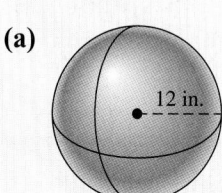

12 in.

(b)

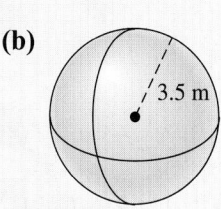

3.5 m

(c) Radius 2.7 cm

❸ Find the volume of each hemisphere. Use 3.14 for π. Round your answers to the nearest tenth.

(a)

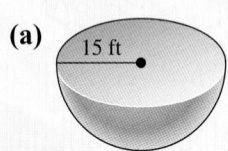
15 ft

(b)

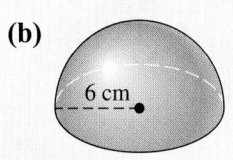
6 cm

❹ Find the volume of each cylinder. Use 3.14 for π. Round your answers to the nearest tenth.

(a)

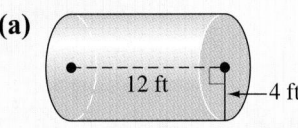
12 ft — 4 ft

(b)

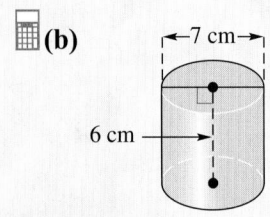
7 cm
6 cm

Example 3 Finding the Volume of a Hemisphere

Find the volume of the hemisphere with the help of a calculator. Use 3.14 for π. Round your answer to the nearest tenth.

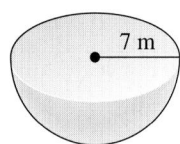
7 m

$$V = \frac{2 \cdot \pi \cdot r^3}{3}$$

$$V \approx \frac{2 \cdot 3.14 \cdot 7\ m \cdot 7\ m \cdot 7\ m}{3}$$

$$V \approx 718.0\ m^3 \quad \text{Rounded to nearest tenth}$$

Work Problem ❸ at the Side.

3 ▭ **Find the volume of a cylinder.** Several *cylinders* are shown here.

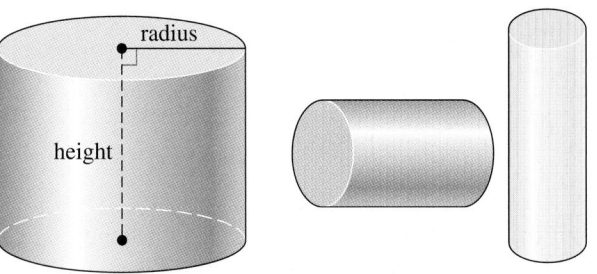
radius
height

These are called *right circular cylinders* because the top and bottom are circles, and the side makes a right angle with the top and bottom. Examples of cylinders are a soup can, a home water heater, and a piece of pipe.

Use the following formula to find the *volume* of a cylinder. Notice that the first part of the formula, $\pi \cdot r \cdot r$, is the area of the circular base.

Finding the Volume of a Cylinder

Volume of cylinder $= \pi \cdot r \cdot r \cdot h$

$$V = \pi \cdot r^2 \cdot h$$

Remember to use **cubic units** when measuring volume.

Example 4 Finding the Volume of Cylinders

Find the volume of each cylinder. Use 3.14 as the approximate value of π. Round your answers to the nearest tenth if necessary.

(a)

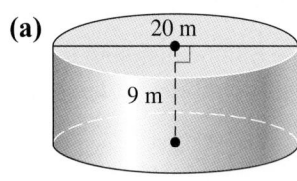
20 m
9 m

The diameter is 20 m so the radius is $\frac{20\ m}{2} = 10$ m. The height is 9 m. Use the formula to find the volume.

$$V = \pi \cdot r \cdot r \cdot h$$
$$V \approx 3.14 \cdot 10\ m \cdot 10\ m \cdot 9\ m$$
$$V \approx 2826\ m^3 \quad \text{Cubic units for volume}$$

(b)

6.2 cm
38.4 cm

$$V \approx 3.14 \cdot 6.2\ cm \cdot 6.2\ cm \cdot 38.4\ cm$$
$$V \approx 4634.94144 \quad \text{Now round to tenths.}$$
$$V \approx 4634.9\ cm^3 \quad \text{Cubic units for volume}$$

Work Problem ❹ at the Side.

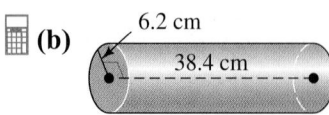

ANSWERS
3. (a) $V \approx 7065\ ft^3$ **(b)** $V \approx 452.2\ cm^3$
4. (a) $V \approx 602.9\ ft^3$ **(b)** $V \approx 230.8\ cm^3$

4 **Find the volume of a cone and a pyramid.** A cone and a pyramid are shown here. Notice that the height line is perpendicular to the base in each figure.

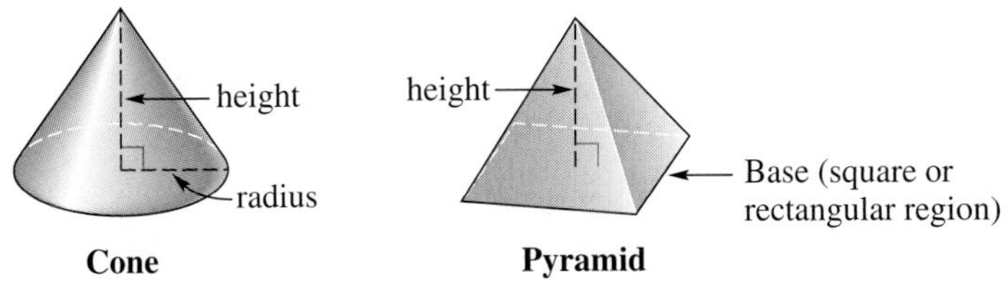

Use the following formula to find the *volume* of a cone.

Finding the Volume of a Cone

$$\text{Volume of cone} = \frac{1}{3} \cdot B \cdot h$$

$$\text{or} \quad V = \frac{B \cdot h}{3}$$

where B is the area of the circular base of the cone and h is the height of the cone. Remember to use ***cubic units*** when measuring volume.

Example 5 **Finding the Volume of a Cone**

Find the volume of the cone. Use 3.14 for π. Round your answer to the nearest tenth.

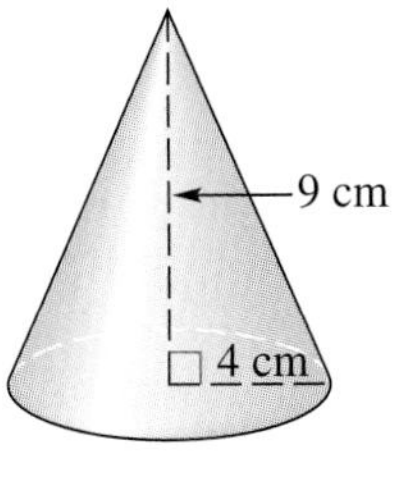

First find the value of B in the formula, which is the *area of the circular base.* Recall that the formula for the area of a circle is πr^2.

$B = \pi \cdot r \cdot r$

$B \approx 3.14 \cdot 4\text{ cm} \cdot 4\text{ cm}$

$\mathbf{B \approx 50.24\ cm^2}$ Do not round to tenths yet.

Next, find the volume. The height is 9 cm.

$$V = \frac{B \cdot h}{3}$$

$$V \approx \frac{50.24\text{ cm}^2 \cdot 9\text{ cm}}{3}$$

$V \approx 150.72\text{ cm}^3$ Now round to tenths.

$V \approx 150.7\text{ cm}^3$ Cubic units for volume

Work Problem 5 at the Side.

5 Find the volume of a cone with base radius 2 ft and height 11 ft. Use 3.14 for π. Round your answer to the nearest tenth.

ANSWERS

5. $V \approx 46.1\text{ ft}^3$

❻ Find the volume of a pyramid with base 10 m by 10 m and height 8 m. Round your answer to the nearest tenth.

Use the same formula to find the *volume* of a *pyramid* as you did to find the *volume* of a *cone*.

Finding the Volume of a Pyramid

$$\text{Volume of pyramid} = \frac{1}{3} \cdot B \cdot h$$

$$\text{or} \quad V = \frac{B \cdot h}{3}$$

where B is the area of the square or rectangular base of the pyramid and h is the height of the pyramid. Remember to use **cubic units** when measuring volume.

Example 6 Finding the Volume of a Pyramid

Find the volume of the pyramid. Round your answer to the nearest tenth.

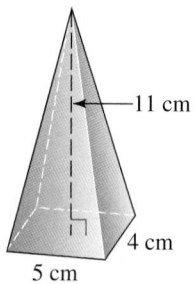

—11 cm

4 cm

5 cm

First find the value of B in the formula, which is the *area of the rectangular base*. Recall that the area of a rectangle is found by multiplying length times width.

$$B = 5 \text{ cm} \cdot 4 \text{ cm}$$
$$B = 20 \text{ cm}^2$$

Next, find the volume.

$$V = \frac{B \cdot h}{3}$$

$$V = \frac{20 \text{ cm}^2 \cdot 11 \text{ cm}}{3}$$

$$V \approx 73.3 \text{ cm}^3 \quad \text{Rounded to nearest tenth}$$

Work Problem ❻ at the Side.

8.7   EXERCISES

| FOR EXTRA HELP | 📖 Student's Solutions Manual | 🚪 MyMathLab.com | ▲ InterAct Math Tutorial Software | ☎ AW Math Tutor Center | Math XL www.mathxl.com | 📼 Digital Video Tutor CD 5 Videotape 15 |

Find the volume of each figure. Use 3.14 as the approximate value of π. Round your answers to the nearest tenth if necessary. See Examples 1–6.

1.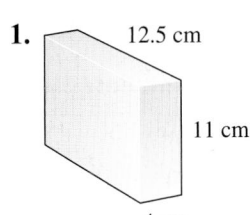
12.5 cm
11 cm
4 cm

2.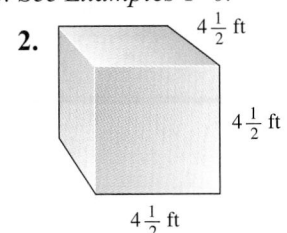
$4\frac{1}{2}$ ft
$4\frac{1}{2}$ ft
$4\frac{1}{2}$ ft

3.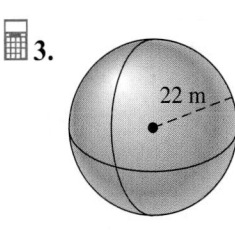
22 m

4.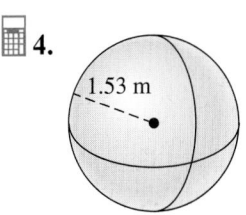
1.53 m

5.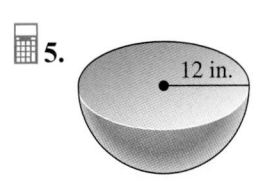
12 in.

6.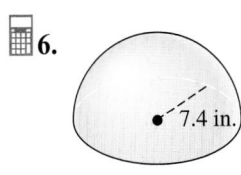
7.4 in.

7.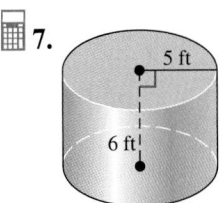
5 ft
6 ft

8.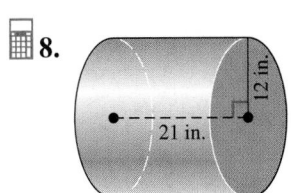
12 in.
21 in.

9.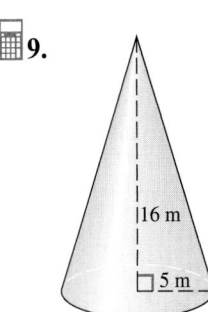
16 m
5 m

10.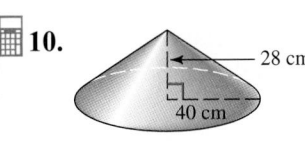
28 cm
40 cm

11.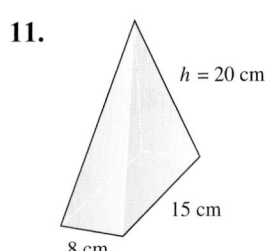
h = 20 cm
15 cm
8 cm

12.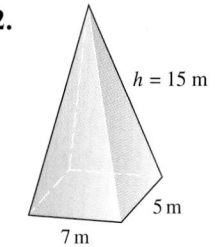
h = 15 m
5 m
7 m

Solve each application problem. Use 3.14 as the approximate value of π. Round your final answers to the nearest tenth if necessary.

13. A pencil box measures 3 in. by 8 in. by $\frac{3}{4}$ in. high. Find the volume of the box. (*Source:* Faber Castell.)

14. A train is being loaded with shipping crates. Each one is 6 m long, 3.4 m wide, and 2 m high. How much space will each crate take?

15. An oil candle globe made of hand-blown glass has a diameter of 16.8 cm. What is the volume of the globe?

16. A metal sphere used as part of a fountain has a diameter of $6\frac{1}{2}$ ft. Find its volume.

17. One of the ancient stone pyramids in Egypt has a square base that measures 145 m on each side. The height is 93 m. What is the volume of the pyramid? (*Source: The Columbia Encyclopedia.*)

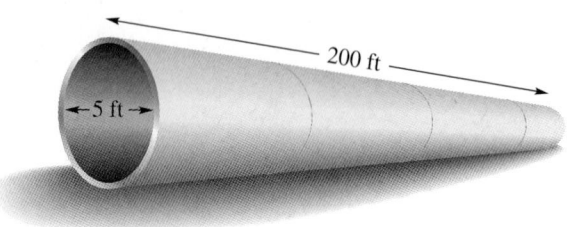

18. A cylindrical woven basket made by a Northwest Coast tribe is 8 cm high and has a diameter of 11 cm. What is the volume of the basket?

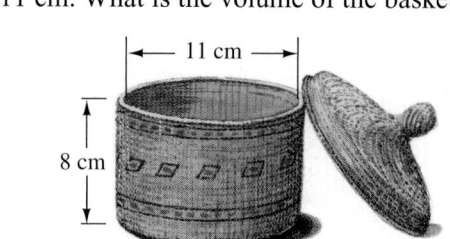

19. A city sewer pipe has a diameter of 5 ft and a length of 200 ft. Find the volume of the pipe.

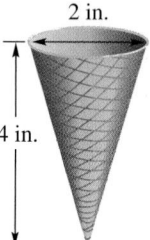

20. An ice cream cone has a diameter of 2 in. and a height of 4 in. Find its volume.

21. Explain the *two* errors made by a student in finding the volume of a cylinder with a diameter of 7 cm and a height of 5 cm. Find the correct answer.

$$V \approx 3.14 \cdot 7 \cdot 7 \cdot 5$$
$$V \approx 769.3 \text{ cm}^2$$

22. Compare the steps in finding the volume of a cylinder and a cone. How are they similar? Suppose you know the volume of a cylinder. How can you find the volume of a cone with the same radius and height by doing just a one-step calculation?

23. Find the volume.

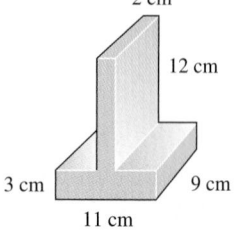

24. Find the volume. (*Hint:* Notice the hole that goes through the center of the shape.)

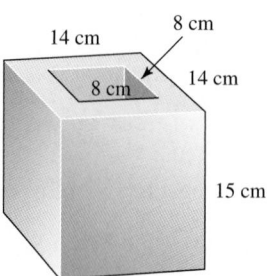

8.8 PYTHAGOREAN THEOREM

In **Section 8.3** you used this formula for the area of a square, $A = s^2$. The square below on the left has an area of 25 cm² because 5 cm • 5 cm = 25 cm².

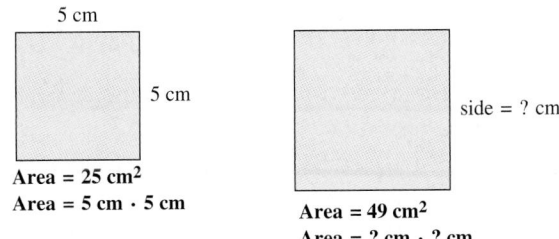

5 cm

5 cm

Area = 25 cm²
Area = 5 cm · 5 cm

side = ? cm

Area = 49 cm²
Area = ? cm · ? cm

OBJECTIVES

1 ▭ Find square roots using the square root key on a calculator.

2 ▭ Find the unknown length in a right triangle.

3 ▭ Solve application problems involving right triangles.

The square on the right has an area of 49 cm². To find the length of a side, ask yourself, "What number can be multiplied by itself to give 49?" Because 7 • 7 = 49, the length of each side is 7 cm.

Remember: 7 • 7 = 49, so 7 is the **square root** of 49, or $\sqrt{49} = 7$. Also, $\sqrt{81} = 9$, since 9 • 9 = 81. (See **Section 1.8** for further review.)

Work Problem ① at the Side.

A number that has a whole number as its square root is called a *perfect square.* For example, 9 is a perfect square because $\sqrt{9} = 3$, and 3 is a whole number.

The first few perfect squares are listed here.

Some Perfect Squares

$\sqrt{1} = 1$	$\sqrt{16} = 4$	$\sqrt{49} = 7$	$\sqrt{100} = 10$
$\sqrt{4} = 2$	$\sqrt{25} = 5$	$\sqrt{64} = 8$	$\sqrt{121} = 11$
$\sqrt{9} = 3$	$\sqrt{36} = 6$	$\sqrt{81} = 9$	$\sqrt{144} = 12$

1 ▭ **Find square roots using the square root key on a calculator.** If a number is *not* a perfect square, then you can find its *approximate* square root by using a calculator with a square root key.

🖩 **Calculator Tip** To find a square root, use the ⟨√⟩ key on a standard calculator or the ⟨√x⟩ key on a scientific calculator. In either case, you do *not* need to use the ⟨=⟩ key. Try these. Jot down your answers.

To find $\sqrt{16}$ press: 16 ⟨√x⟩   Answer is 4.

To find $\sqrt{7}$ press: 7 ⟨√x⟩   Answer is 2.645751311.

For $\sqrt{7}$, your calculator shows 2.645751311, which is an *approximate* answer. We will be rounding to the nearest thousandth, so $\sqrt{7} \approx 2.646$. To check, multiply 2.646 times 2.646. Do you get 7 as the result? No, you get 7.001316, which is very close to 7. The difference is due to rounding.

① Find each square root.

(a) $\sqrt{36}$

(b) $\sqrt{25}$

(c) $\sqrt{9}$

(d) $\sqrt{100}$

(e) $\sqrt{121}$

② Use a calculator with a square root key to find each square root. Round to the nearest thousandth if necessary.

(a) $\sqrt{11}$

(b) $\sqrt{40}$

(c) $\sqrt{56}$

(d) $\sqrt{196}$

(e) $\sqrt{147}$

Example 1 Finding the Square Root of Numbers

Use a calculator to find each square root. Round your answers to the nearest thousandth.

(a) $\sqrt{35}$ Calculator shows 5.916079783; round to 5.916

(b) $\sqrt{124}$ Calculator shows 11.13552873; round to 11.136

(c) $\sqrt{200}$ Calculator shows 14.14213562; round to 14.142

Work Problem ② at the Side.

2 **Find the unknown length in a right triangle.**   One place you will use square roots is when working with the *Pythagorean Theorem*. This theorem applies only to *right* triangles (triangles with a 90° angle). The longest side of a right triangle is called the **hypotenuse.** It is opposite the right angle. The other two sides are called *legs*. The legs form the right angle.

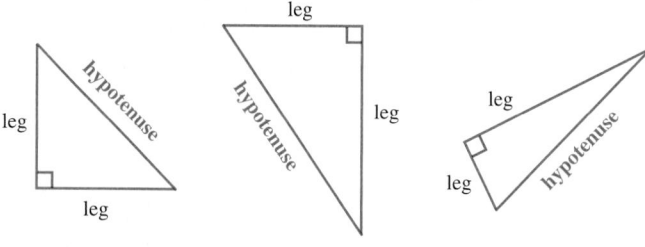

Examples of right triangles

Pythagorean Theorem

$$(\text{hypotenuse})^2 = (\text{leg})^2 + (\text{leg})^2$$

In other words, square the length of each side. After you have squared all the sides, the sum of the squares of the two legs will equal the square of the hypotenuse.

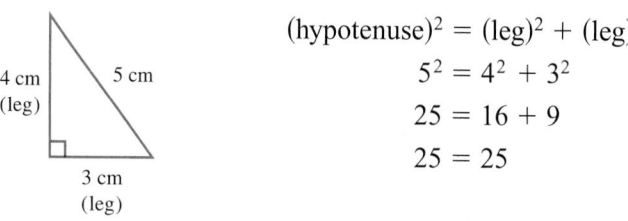

$$(\text{hypotenuse})^2 = (\text{leg})^2 + (\text{leg})^2$$
$$5^2 = 4^2 + 3^2$$
$$25 = 16 + 9$$
$$25 = 25$$

The theorem is named after Pythagoras, a Greek mathematician who lived about 2500 years ago. He and his followers may have used floor tiles to prove the theorem, as shown below.

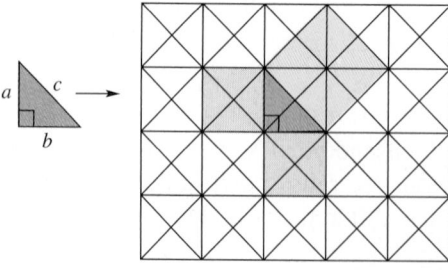

The right triangle in the center of the floor tiles has sides a, b, and c. The square drawn on side a contains four triangular tiles. The square on side b contains four tiles. The square on side c contains eight tiles. The number of tiles in the square on side c equals the sum of the number of tiles in the squares on sides a and b, that is, 8 tiles = 4 tiles + 4 tiles. As a result, you often see the Pythagorean Theorem written as $c^2 = a^2 + b^2$.

ANSWERS

2. (a) $\sqrt{11} \approx 3.317$ **(b)** $\sqrt{40} \approx 6.325$
(c) $\sqrt{56} \approx 7.483$ **(d)** $\sqrt{196} = 14$
(e) $\sqrt{147} \approx 12.124$

If you know the lengths of any two sides in a right triangle, you can use the Pythagorean Theorem to find the length of the third side.

Using the Pythagorean Theorem

To find the hypotenuse, use this formula:

$$\text{hypotenuse} = \sqrt{(\text{leg})^2 + (\text{leg})^2}$$

To find a leg, use this formula:

$$\text{leg} = \sqrt{(\text{hypotenuse})^2 - (\text{leg})^2}$$

CAUTION

Remember: A small square drawn in one angle of a triangle indicates a right angle. You can use the Pythagorean Theorem *only* on triangles that have a right angle.

Example 2 **Finding the Unknown Length in Right Triangles**

Find the unknown length in each right triangle. Round answers to the nearest tenth if necessary.

(a)

3 ft, 4 ft

The length of the side opposite the right angle is unknown. That side is the hypotenuse, so use this formula.

$$\text{hypotenuse} = \sqrt{(\text{leg})^2 + (\text{leg})^2} \quad \text{Find the hypotenuse.}$$
$$\text{hypotenuse} = \sqrt{(3)^2 + (4)^2} \quad \text{Legs are 3 and 4.}$$
$$= \sqrt{9 + 16} \quad 3 \cdot 3 \text{ is } 9 \quad \text{and} \quad 4 \cdot 4 \text{ is } 16.$$
$$= \sqrt{25}$$
$$= 5$$

The hypotenuse is 5 ft long.

(b)

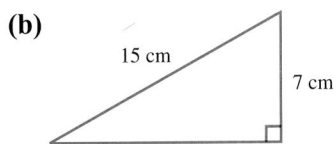
15 cm, 7 cm

We *do* know the length of the hypotenuse (15 cm), so it is the length of one of the legs that is unknown. Use this formula.

$$\text{leg} = \sqrt{(\text{hypotenuse})^2 - (\text{leg})^2} \quad \text{Find a leg.}$$
$$\text{leg} = \sqrt{(15)^2 - (7)^2} \quad \text{Hypotenuse is 15, one leg is 7.}$$
$$= \sqrt{225 - 49} \quad 15 \cdot 15 \text{ is } 225 \quad \text{and} \quad 7 \cdot 7 \text{ is } 49.$$
$$= \sqrt{176} \quad \text{Use calculator to find } \sqrt{176}.$$
$$\approx 13.3 \quad \text{Round } 13.26649916 \text{ to } 13.3.$$

The length of the leg is approximately 13.3 cm.

============ **Work Problem ❸ at the Side.**

CAUTION

You use the Pythagorean Theorem to find the *length* of one side, *not* the area of the triangle. Your answer will be in linear units, such as ft, yd, cm, m, and so on (*not* ft², yd², cm², m²).

---

❸ Find the unknown length in each right triangle. Round your answers to the nearest tenth if necessary.

(a)

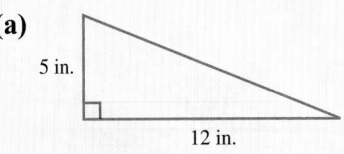
5 in., 12 in.

(b)

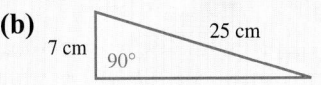
7 cm, 90°, 25 cm

(c)

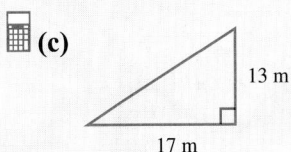
13 m, 17 m

(d)

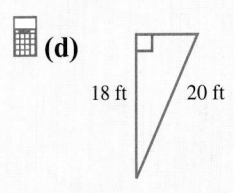
18 ft, 20 ft

(e)

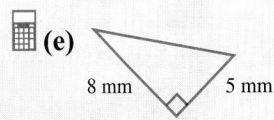
8 mm, 5 mm

4 These problems show ladders leaning against buildings. Find the unknown lengths. Round to the nearest tenth of a foot if necessary.

(a)

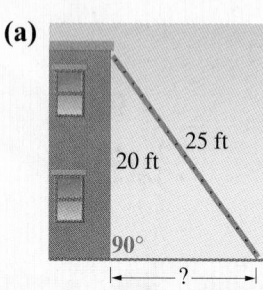

How far away from the building is the bottom of the ladder?

(b)

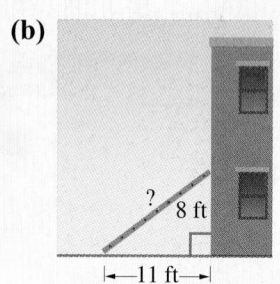

How long is the ladder?

(c) A 17-ft ladder is leaning against a building. The bottom of the ladder is 10 ft from the building. How high up on the building will the ladder reach? (*Hint:* Start by drawing the building and the ladder.)

3 Solve application problems involving right triangles. The next example shows an application of the Pythagorean Theorem.

Example 3 Using the Pythagorean Theorem

A television antenna is on the roof of a house, as shown. Find the length of the support wire. Round your answer to the nearest tenth of a meter if necessary.

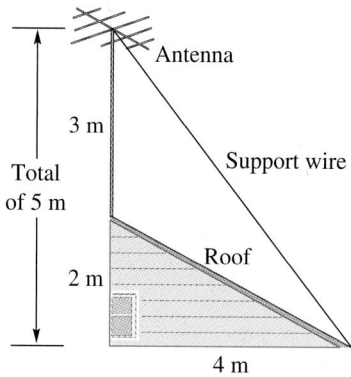

A right triangle is formed. The total length of the side at the left is 3 m + 2 m = 5 m.

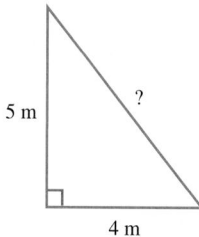

The support wire is the hypotenuse of the right triangle.

$$\text{hypotenuse} = \sqrt{(\text{leg})^2 + (\text{leg})^2} \quad \text{Find the hypotenuse.}$$
$$\text{hypotenuse} = \sqrt{(5)^2 + (4)^2} \quad \text{Legs are 5 and 4.}$$
$$= \sqrt{25 + 16} \quad 5^2 \text{ is 25 and } 4^2 \text{ is 16.}$$
$$= \sqrt{41} \quad \text{Use } \sqrt{x} \text{ key on a calculator.}$$
$$\approx 6.4 \quad \text{Round 6.403124237 to 6.4.}$$

The length of the support wire is approximately **6.4 m.**

Work Problem 4 at the Side.

8.8 EXERCISES

FOR EXTRA HELP

Student's Solutions Manual
MyMathLab.com
InterAct Math Tutorial Software
AW Math Tutor Center
www.mathxl.com
Math XL
Digital Video Tutor CD 5 Videotape 15

Find each square root. Starting with Exercise 5, use the square root key on a calculator. Round your answers to the nearest thousandth if necessary. See Example 1.

1. $\sqrt{16}$

2. $\sqrt{4}$

3. $\sqrt{64}$

4. $\sqrt{81}$

5. $\sqrt{11}$

6. $\sqrt{23}$

7. $\sqrt{5}$

8. $\sqrt{2}$

9. $\sqrt{73}$

10. $\sqrt{80}$

11. $\sqrt{101}$

12. $\sqrt{125}$

13. $\sqrt{190}$

14. $\sqrt{160}$

15. $\sqrt{1000}$

16. $\sqrt{2000}$

17. You know that $\sqrt{25} = 5$ and $\sqrt{36} = 6$. Using just that information (no calculator), describe how you could estimate $\sqrt{30}$. How would you estimate $\sqrt{26}$ or $\sqrt{35}$? Now check your estimates using a calculator.

18. Explain the relationship between *squaring* a number and finding the *square root* of a number. Include two examples to illustrate your explanation.

Find the unknown length in each right triangle. Use a calculator to find square roots. Round your answers to the nearest tenth if necessary. See Example 2.

19.
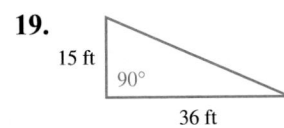
15 ft, 90°, 36 ft

20.
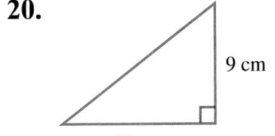
9 cm, 12 cm

21.
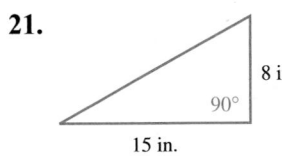
8 in., 90°, 15 in.

22.
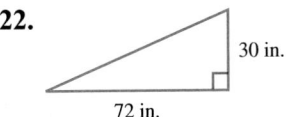
30 in., 72 in.

23.
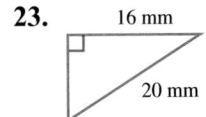
16 mm, 20 mm

24.
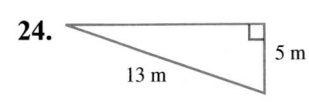
5 m, 13 m

25.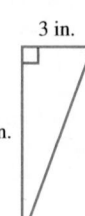
3 in.
8 in.

26.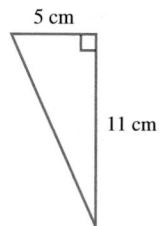
5 cm
11 cm

27.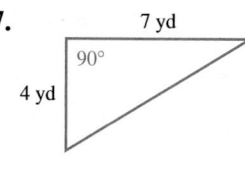
7 yd
90°
4 yd

28.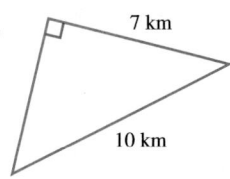
7 km
10 km

29.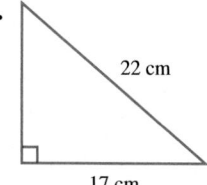
22 cm
17 cm

30.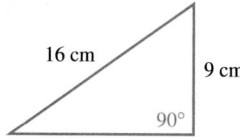
16 cm
9 cm
90°

31.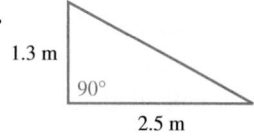
1.3 m
90°
2.5 m

32.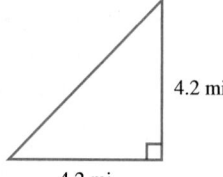
4.2 mi
4.2 mi

33.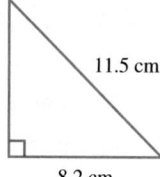
11.5 cm
8.2 cm

34.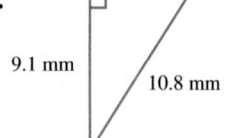
9.1 mm
10.8 mm

35.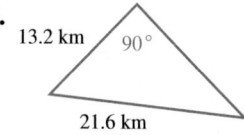
13.2 km
90°
21.6 km

36.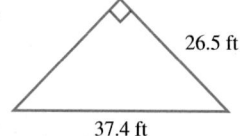
26.5 ft
37.4 ft

Solve each application problem. Round your answers to the nearest tenth if necessary. See Example 3.

37. Find the length of this loading ramp.

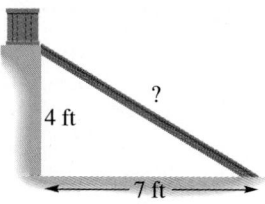
4 ft
?
7 ft

38. Find the unknown length in this roof plan.

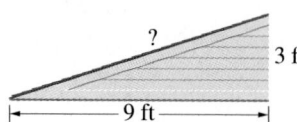
?
3 ft
9 ft

39. How high is the airplane above the ground?

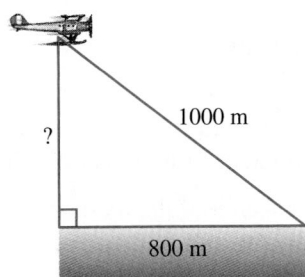

40. Find the height of this farm silo.

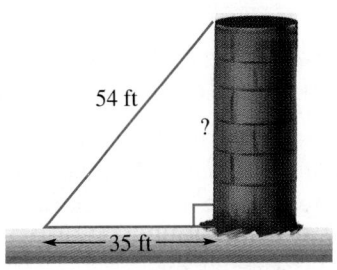

41. To reach his lady-love, a knight placed a 12-ft ladder against the castle wall. If the base of the ladder is 3 ft from the building, how high on the castle will the top of the ladder reach? Draw a sketch of the castle and ladder and solve the problem.

42. William drove his car 15 miles north, then made a right turn and drove 7 miles east. How far is he, in a straight line, from his starting point? Draw a sketch to illustrate the problem and solve it.

43. Describe the *two* errors made by a student in solving this problem. Also find the correct answer. Round to the nearest tenth.

$$? = \sqrt{(9)^2 + (7)^2}$$
$$= \sqrt{18 + 14}$$
$$= \sqrt{32} \approx 5.657 \text{ in.}$$

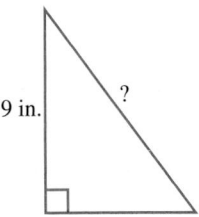

44. Describe the *two* errors made by a student in solving this problem. Also find the correct answer. Round to the nearest tenth.

$$? = \sqrt{(13)^2 + (20)^2}$$
$$= \sqrt{169 + 400}$$
$$= \sqrt{569} \approx 23.9 \text{ m}^2$$

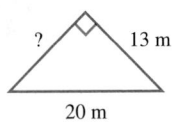

| RELATING CONCEPTS (Exercises 45–48) | FOR INDIVIDUAL OR GROUP WORK |

*Use your knowledge of the Pythagorean Theorem to **work Exercises 45–48 in order.**
Round answers to the nearest tenth.*

45. A major league baseball diamond is a square shape measuring 90 ft on each side. If the catcher throws a ball from home plate to second base, how far is he throwing the ball? (*Source:* American League of Professional Baseball Clubs.)

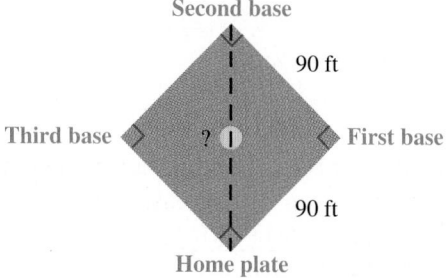

Second base

90 ft

Third base   ?   First base

90 ft

Home plate

46. A softball diamond is only 60 ft on each side. (*Source:* Amateur Softball Association.)

(a) Draw a sketch of the diamond and label the sides and bases.

(b) How far is it to throw a ball from home plate to second base?

47. Look back at your answer to Exercise 45. Explain how you can tell the distance from third base to first base without doing any further calculations.

48. Show how you could set up a proportion to answer Exercise 46 instead of using the Pythagorean Theorem. (You'll need your answer from Exercise 45.)

8.9 SIMILAR TRIANGLES

Two triangles with the same *shape* (but not necessarily the same size) are called **similar triangles.** Three pairs of similar triangles are shown below.

❶ Identify corresponding angles and sides in these similar triangles.

(a)

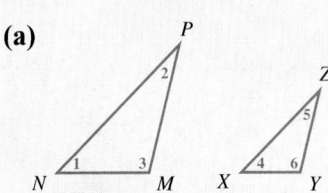

Angles:

1 and _____

2 and _____

3 and _____

Sides:

$\overline{PN}$ and _____

$\overline{PM}$ and _____

$\overline{NM}$ and _____

1 Identify corresponding parts in similar triangles. The two triangles shown below are different sizes but have the same shape, so they are *similar triangles.* Angles *A* and *P* measure the same number of degrees and are called *corresponding angles.* Angles *B* and *Q* are corresponding angles, as are angles *C* and *R.* The triangles have the same shape because the corresponding angles have the same measure.

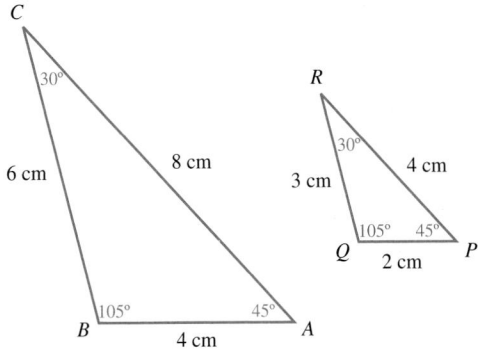

$\overline{PR}$ and $\overline{AC}$ are called *corresponding sides* because they are *opposite* corresponding angles. Also, $\overline{QR}$ and $\overline{BC}$ are corresponding sides, as are $\overline{PQ}$ and $\overline{AB}$. Although corresponding angles measure the same number of degrees, corresponding sides do *not* need to be the same length. In the triangles here, each side in the smaller triangle is *half* the length of the corresponding side in the larger triangle.

Work Problem ❶ at the Side.

(b)

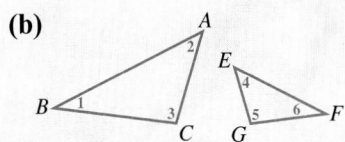

Angles:

1 and _____

2 and _____

3 and _____

Sides:

$\overline{AB}$ and _____

$\overline{BC}$ and _____

$\overline{AC}$ and _____

2 Find the unknown lengths of sides in similar triangles. Similar triangles are useful because of the following property.

Similar Triangles

In **similar triangles,** the ratios of the lengths of corresponding sides are equal.

ANSWERS
1. **(a)** 4; 5; 6; $\overline{ZX}$; $\overline{ZY}$; $\overline{XY}$
   (b) 6; 4; 5; $\overline{EF}$; $\overline{FG}$; $\overline{EG}$

❷ Find the length of x in Example 1 by setting up and solving a proportion.

Example 1   **Finding the Unknown Lengths of Sides in Similar Triangles**

Find the length of y in the smaller triangle. Assume the triangles are similar.

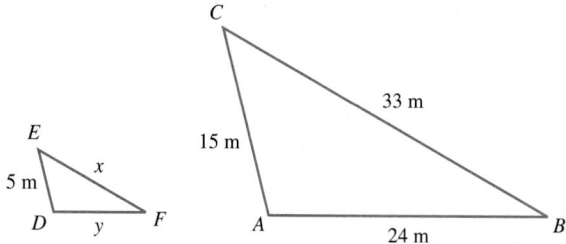

$\overline{ED}$ and $\overline{CA}$ are corresponding sides. The ratio of the lengths of these sides can be written as a fraction in lowest terms.

$$\frac{ED}{CA} = \frac{5 \cancel{m}}{15 \cancel{m}} = \frac{1}{3} \leftarrow \text{Lowest terms}$$

As mentioned earlier, the ratios of the lengths of corresponding sides are equal. $\overline{DF}$ in the smaller triangle corresponds to $\overline{AB}$ in the larger triangle. Since the ratios of corresponding sides are equal,

$$\frac{DF}{AB} = \frac{1}{3}$$

Replace DF with y and AB with 24 to get this proportion.

$$\frac{y}{24} = \frac{1}{3}$$

Find the cross products.

$$\frac{y}{24} = \frac{1}{3} \qquad \begin{matrix} 24 \cdot 1 = 24 \\ y \cdot 3 \end{matrix}$$

Show that the cross products are equivalent.

$$y \cdot 3 = 24$$

Divide each side by 3.

$$\frac{y \cdot \overset{1}{\cancel{3}}}{\underset{1}{\cancel{3}}} = \frac{24}{3}$$

$$y = 8$$

$\overline{DF}$ has a length of 8 m.

Work Problem ❷ at the Side.

19. Triangles *CDE* and *FGH* are similar. Find the perimeter and area of triangle *FGH*. *Note:* The heights of similar triangles have the same ratio as corresponding sides.

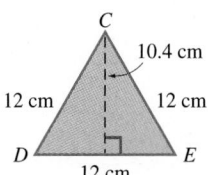
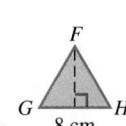

20. Triangles *JKL* and *MNO* are similar. Find the perimeter and area of triangle *MNO*.

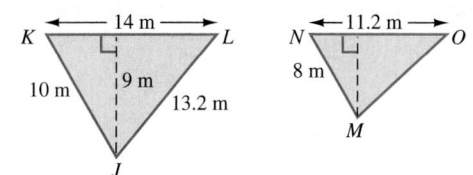

Solve each application problem. See Example 3.

21. The height of the house shown here can be found by comparing its shadow to the shadow cast by a 3 ft stick. Find the height of the house by writing a proportion and solving it.

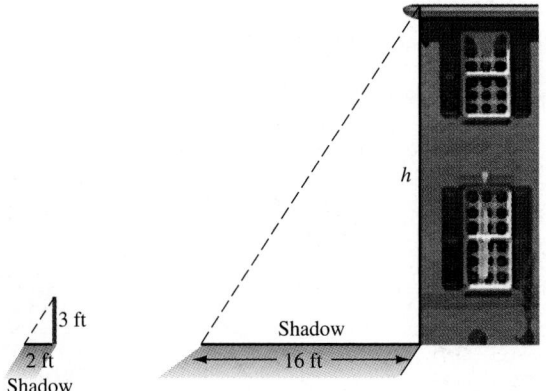

22. A fire lookout tower provides an excellent view of the surrounding countryside. The height of the tower can be found by lining up the top of the tower with the top of a 2 m stick. Use similar triangles to find the height of the tower.

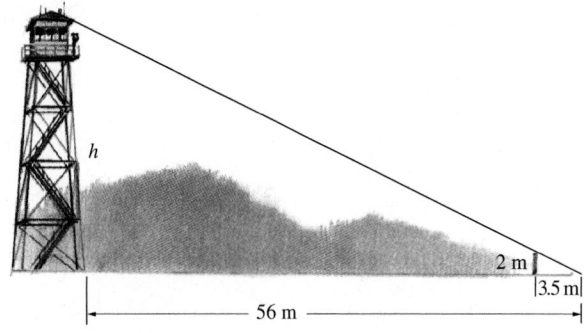

23. Refer to the building in Exercise 21. Later in the day, the same building had a shadow 6 ft long. How long would the stick's shadow be at that time?

24. Refer to the lookout tower in Exercise 22.

  (a) How far away from the tower was the 2 m stick?

  (b) To use a 1 m stick and have it line up with the same endpoint, how far from the tower would it have to be?

25. Look up the word *similar* in a dictionary. What is the nonmathematical definition of this word? Describe two examples of similar objects at home, school, or work.

26. *Congruent* objects have the *same shape* and the *same size*. Sketch a pair of congruent triangles. Describe two examples of congruent objects at home, school, or work.

Find the unknown length in Exercises 27–30. Round your answers to the nearest tenth. Note: When a line is drawn parallel to one side of a triangle, the smaller triangle that is formed will be similar to the original triangle. In Exercises 27–28, the red segments are parallel.

27.

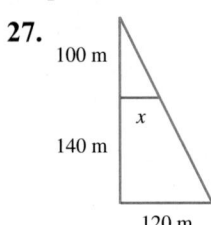

100 m

x

140 m

120 m

28.

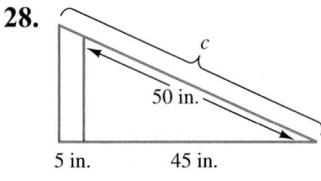

c

50 in.

5 in.   45 in.

29. Use similar triangles and a proportion to find the length of the lake shown here. (*Hint:* The side 100 m long in the smaller triangle corresponds to a side of 100 m + 120 m = 220 m in the larger triangle.)

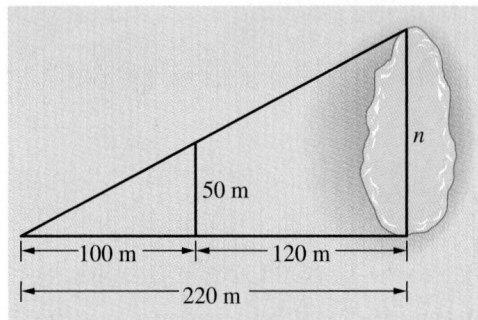

n

50 m

100 m   120 m

220 m

30. To find the height of the tree, find *y* and then add $5\frac{1}{2}$ ft for the distance from the ground to the person's eye level.

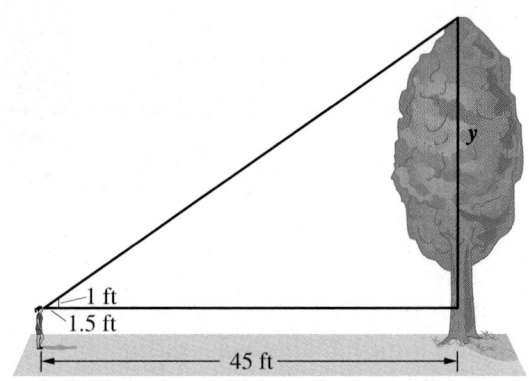

y

1 ft

1.5 ft

45 ft

Distance from person to tree.

SUMMARY

KEY TERMS

8.1	**point**	A point is a location in space. *Example:* Point *P* at the right.	• *P*

line

A line is a straight row of points that goes on forever in both directions. *Example:* Line *AB*, written $\overleftrightarrow{AB}$, at the right.

line segment

A line segment is a piece of a line with two endpoints. *Example:* Line segment *PQ*, written $\overline{PQ}$, at the right.

ray

A ray is a part of a line that has one endpoint and extends forever in one direction. *Example:* Ray *RS*, written $\overrightarrow{RS}$, at the right.

angle

An angle is made up of two rays that have a common endpoint called the vertex. *Example:* Angle 1 at the right.

degrees

A system used to measure angles in which a complete circle is 360 degrees, written 360°.

right angle

A right angle is an angle that measures exactly 90°. *Example:* Angle *AOB* at right.

acute angle

An acute angle is an angle that measures less than 90°. *Example:* Angle *E* at the right.

obtuse angle

An obtuse angle is an angle that measures more than 90° but less than 180°. *Example:* Angle *F* at the right.

straight angle

A straight angle is an angle that measures 180°; its sides form a straight line. *Example:* Angle *G* at the right.

intersecting lines

Intersecting lines cross or merge. *Example:* $\overleftrightarrow{RQ}$ intersects $\overleftrightarrow{AB}$ at point *P* at the right.

perpendicular lines

Perpendicular lines are two lines that intersect to form a right angle. *Example:* $\overleftrightarrow{PQ}$ is perpendicular to $\overleftrightarrow{RS}$ at the right.

parallel lines

Parallel lines are two lines in the same plane that never intersect (never cross). *Example:* $\overleftrightarrow{AB}$ is parallel to $\overleftrightarrow{ST}$ at the right.

8.2	**complementary angles**	Complementary angles are two angles whose measures add up to 90°.

supplementary angles

Supplementary angles are two angles whose measures add up to 180°.

congruent angles

Congruent angles are angles that measure the same number of degrees.

	vertical angles	Vertical angles are two nonadjacent congruent angles formed by intersecting lines. *Example:* $\angle COA$ and $\angle EOF$ are vertical angles at the right.
8.3– 8.5	**perimeter**	Perimeter is the distance around the outside edges of a flat shape. It is measured in linear units such as ft, yd, cm, m, km, and so on.
8.3– 8.6	**area**	Area is the surface inside a two-dimensional (flat) shape. It is measured by determining the number of squares of a certain size needed to cover the surface inside the shape. Some of the commonly used units for measuring area are square inches (in.^{2}), square feet (ft^2), square yards (yd^2), square centimeters (cm^2), and square meters (m^2).
8.3	**rectangle**	A rectangle is a four-sided figure with all sides meeting at 90° angles. The opposite sides are the same length. *Example:* A rectangle measuring 12 cm by 7 cm at the right.
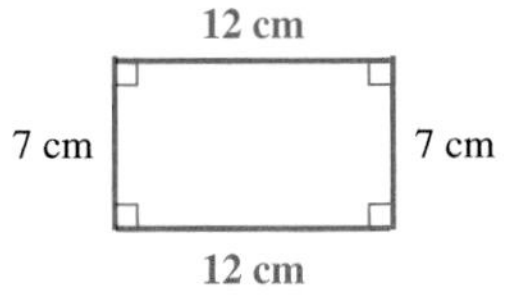	**square**	A square is a rectangle with all four sides the same length. *Example:* A square with the side measurement of 20 in. at the right.
8.4	**parallelogram**	A parallelogram is a four-sided figure with both pairs of opposite sides parallel. *Example:* See figure at the right with sides measuring 8 m and 12 m.
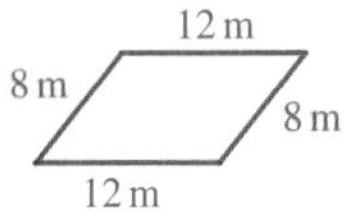	**trapezoid**	A trapezoid is a four-sided figure with one pair of parallel sides. *Example:* Trapezoid $PQRS$ at the right; $\overline{PQ}$ is parallel to $\overline{SR}$.
8.5	**triangle**	A triangle is a figure with exactly three sides. *Example:* Triangle ABC at the right.
8.6	**circle**	A circle is a figure with all points the same distance from a fixed center point. *Example:* See figure at the right.
	radius	Radius is the distance from the center of a circle to any point on the circle. *Example:* See the red radius line in the circle at the right.
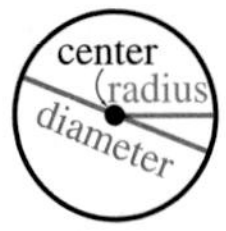	**diameter**	Diameter is the distance across a circle, passing through the center. *Example:* See the blue diameter line in the circle at the right.
	circumference	Circumference is the distance around a circle.
	π (pi)	π is the ratio of the circumference to the diameter of any circle. It is approximately equal to 3.14.
8.7	**volume**	Volume is a measure of the space inside a three-dimensional (solid) shape. Volume is measured in cubic units such as in.^{3}, ft^3, yd^3, mm^3, cm^3, and so on.
8.8	**square root**	A square root is one of two equal factors of a number.
	hypotenuse	The hypotenuse is the side of a right triangle opposite the 90° angle; it is the longest side. *Example:* See figure at the right.
8.9	**similar triangles**	Similar triangles are triangles with the same shape but not necessarily the same size; corresponding angles measure the same number of degrees, and the *ratios* of the lengths of corresponding sides are equal.

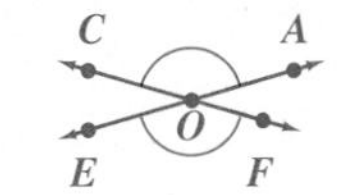

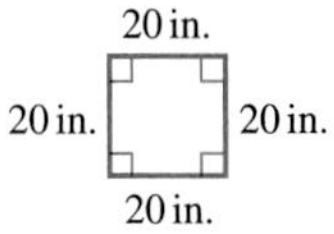

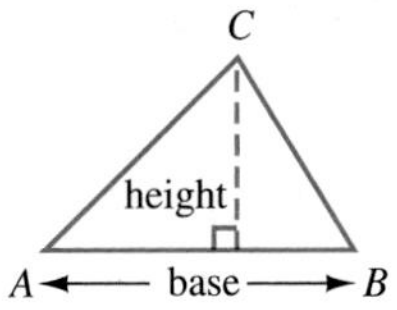

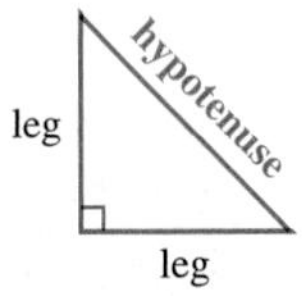

NEW FORMULAS

Perimeter of a rectangle: $P = 2 \cdot l + 2 \cdot w$

Area of a rectangle: $A = l \cdot w$

Perimeter of a square: $P = 4 \cdot s$

Area of a square: $A = s^2$

Area of a parallelogram: $A = b \cdot h$

Area of a trapezoid: $A = \dfrac{1}{2} \cdot h \cdot (b + B)$

or $A = 0.5 \cdot h \cdot (b + B)$

Area of a triangle: $A = \dfrac{1}{2} \cdot b \cdot h$

or $A = 0.5 \cdot b \cdot h$

Diameter of a circle: $d = 2 \cdot r$

Radius of a circle: $r = \dfrac{d}{2}$

Circumference of a circle: $C = \pi \cdot d$

or $C = 2 \cdot \pi \cdot r$

Area of a circle: $A = \pi \cdot r^2$

Area of semicircle: $A = \dfrac{\pi \cdot r^2}{2}$

Volume of rectangular solid: $V = l \cdot w \cdot h$

Volume of a sphere: $V = \dfrac{4}{3} \cdot \pi \cdot r^3$

or $V = \dfrac{4 \cdot \pi \cdot r^3}{3}$

Volume of a hemisphere: $V = \dfrac{2}{3} \cdot \pi \cdot r^3$

or $V = \dfrac{2 \cdot \pi \cdot r^3}{3}$

Volume of a cylinder: $V = \pi \cdot r^2 \cdot h$

Volume of a cone: $V = \dfrac{1}{3} \cdot B \cdot h$ or $\dfrac{B \cdot h}{3}$

Volume of a pyramid: $V = \dfrac{1}{3} \cdot B \cdot h$ or $\dfrac{B \cdot h}{3}$

Right triangle: hypotenuse $= \sqrt{(\text{leg})^2 + (\text{leg})^2}$

leg $= \sqrt{(\text{hypotenuse})^2 - (\text{leg})^2}$

TEST YOUR WORD POWER

See how well you have learned the vocabulary in this chapter. Answers follow the Quick Review.

1. Two angles that are **complementary**
   (a) have measures that add up to 180°
   (b) are always congruent
   (c) form a straight angle
   (d) have measures that add up to 90°.

2. The **perimeter** of a flat shape is
   (a) measured in square units
   (b) the distance around the outside edges
   (c) the number of squares needed to cover the space inside the shape
   (d) measured in cubic units.

3. An **obtuse angle**
   (a) is formed by perpendicular lines
   (b) is congruent to a right angle
   (c) measures more than 90° but less than 180°
   (d) measures less than 90°.

4. The **hypotenuse** is
   (a) the long base in a trapezoid
   (b) the height line in a parallelogram
   (c) the longest side in a right triangle
   (d) the distance across a circle, passing through the center.

5. π is the ratio of
   (a) the diameter to the radius of a circle
   (b) the circumference to the diameter of a circle
   (c) the circumference to the radius of a circle
   (d) the diameter to the circumference of a circle.

6. **Perpendicular lines**
   (a) intersect to form a right angle
   (b) intersect to form an acute angle
   (c) never intersect
   (d) have a common endpoint called the vertex.

7. In a pair of **similar triangles,**
   (a) corresponding sides have the same length
   (b) all the angles have the same measure
   (c) the perimeters are equal
   (d) the ratios of the lengths of corresponding sides are equal.

8. The **area of a rectangle** is found by
   (a) using the formula $P = 2 \cdot l + 2 \cdot w$
   (b) multiplying length times width
   (c) adding the lengths of the sides
   (d) using the formula $V = l \cdot w \cdot h$.

Concepts	Examples

8.1 Lines

If a line has one endpoint, it is a *ray*. If it has two endpoints, it is a *line segment*.

Identify each of the following as a line, line segment, or ray.

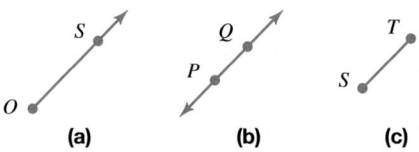

Figure **(a)** shows a ray, **(b)** shows a line, and **(c)** shows a line segment.

If two lines intersect at right angles, they are *perpendicular*.

If two lines in the same plane never intersect, they are *parallel*.

Label each pair of lines as parallel or perpendicular.

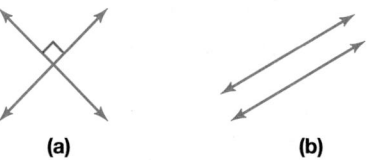

Figure **(a)** shows two perpendicular lines (they intersect at 90°).

Figure **(b)** shows two parallel lines (they never intersect).

8.2 Angles

If the sum of the measures of two angles is 90°, they are *complementary*.

If the sum of the measures of two angles is 180°, they are *supplementary*.

If two angles measure the same number of degrees, the angles are *congruent*. The symbol for congruent is ≅.

Two nonadjacent angles formed by intersecting lines are called *vertical angles*. Vertical angles are congruent.

Find the complement and supplement of a 35° angle.

$$90° - 35° = 55° \text{ (the complement)}$$
$$180° - 35° = 145° \text{ (the supplement)}$$

Identify the vertical angles in this figure. Which angles are congruent?

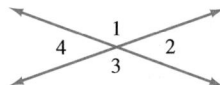

∠1 and ∠3 are vertical angles.

∠2 and ∠4 are vertical angles.

Vertical angles are congruent, so ∠1 ≅ ∠3 and ∠2 ≅ ∠4.

Concepts	Examples

8.3 Rectangles and Squares

Use this formula to find the perimeter of a *rectangle*.

$$P = 2 \cdot \text{length} + 2 \cdot \text{width}$$

Use this formula to find the area of a rectangle.

$$A = \text{length} \cdot \text{width}$$

Area is measured in **square units**.

Find the perimeter and area of this rectangle.

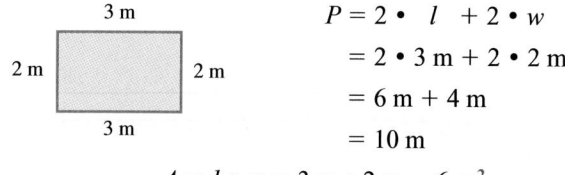

$$P = 2 \cdot l + 2 \cdot w$$
$$= 2 \cdot 3\,\text{m} + 2 \cdot 2\,\text{m}$$
$$= 6\,\text{m} + 4\,\text{m}$$
$$= 10\,\text{m}$$
$$A = l \cdot w = 3\,\text{m} \cdot 2\,\text{m} = 6\,\text{m}^2$$

Use these formulas to find the perimeter and area of a *square*.

$$P = 4 \cdot \text{side}$$
$$A = (\text{side})^2$$

Area is measured in **square units**.

Find the perimeter and area of this square.

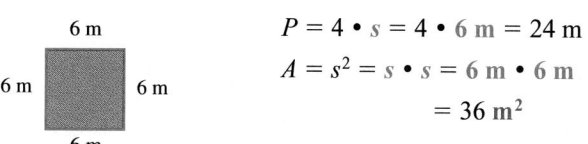

$$P = 4 \cdot s = 4 \cdot 6\,\text{m} = 24\,\text{m}$$
$$A = s^2 = s \cdot s = 6\,\text{m} \cdot 6\,\text{m}$$
$$= 36\,\text{m}^2$$

8.4 Parallelograms

Use these formulas to find the perimeter and area of a *parallelogram*.

$$P = \text{sum of the lengths of the sides}$$
$$A = \text{base} \cdot \text{height}$$

Area is measured in **square units**.

Find the perimeter and area of this parallelogram.

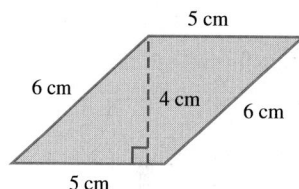

$$P = 5\,\text{cm} + 6\,\text{cm} + 5\,\text{cm} + 6\,\text{cm} = 22\,\text{cm}$$
$$A = 5\,\text{cm} \cdot 4\,\text{cm} = 20\,\text{cm}^2$$

8.4 Trapezoids

Use these formulas to find the perimeter and area of a *trapezoid*.

$$P = \text{sum of the lengths of the sides}$$
$$A = \frac{1}{2} \cdot \text{height} \cdot (b + B)$$

where b is the short base and B is the long base.

Area is measured in **square units**.

Find the perimeter and area of this trapezoid.

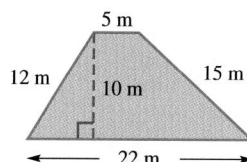

$$P = 5\,\text{m} + 15\,\text{m} + 22\,\text{m} + 12\,\text{m} = 54\,\text{m}$$
$$A = \frac{1}{\cancel{2}} \cdot \overset{5}{\cancel{10}}\,\text{m} \cdot (5\,\text{m} + 22\,\text{m})$$
$$= 5\,\text{m} \cdot (27\,\text{m}) = 135\,\text{m}^2$$

8.5 Triangles

Use these formulas to find the perimeter and area of a *triangle*.

$$P = \text{sum of the lengths of the sides}$$
$$A = \frac{1}{2} \cdot \text{base} \cdot \text{height}$$
$$\text{or} \quad A = 0.5 \cdot \text{base} \cdot \text{height}$$

Area is measured in **square units**.

Find the perimeter and area of this triangle.

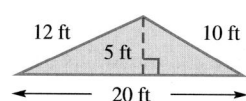

$$P = 12\,\text{ft} + 10\,\text{ft} + 20\,\text{ft} = 42\,\text{ft}$$
$$A = \frac{1}{2} \cdot b \cdot h$$
$$A = \frac{1}{\cancel{2}} \cdot \overset{10}{20}\,\text{ft} \cdot 5\,\text{ft} = 50\,\text{ft}^2$$

$$\text{or} \quad A = 0.5 \cdot 20\,\text{ft} \cdot 5\,\text{ft} = 50\,\text{ft}^2$$

Concepts	*Examples*

8.6 *Circles*

Use this formula to find the *diameter* of a circle, given the radius.

$$\text{diameter} = 2 \cdot \text{radius}$$

Find the diameter of a circle if the radius is 3 m.

$$d = 2 \cdot r = 2 \cdot 3 \text{ m} = 6 \text{ m}$$

Use this formula to find the *radius* of a circle, given the diameter.

$$\text{radius} = \frac{\text{diameter}}{2}$$

Find the radius of a circle if the diameter is 5 cm.

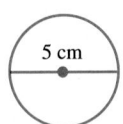

$$r = \frac{d}{2} = \frac{5 \text{ cm}}{2} = 2.5 \text{ cm}$$

Use these formulas to find the **circumference** of a circle.

$$C = 2 \cdot \pi \cdot \text{radius}$$
$$\text{or} \quad C = \pi \cdot \text{diameter}$$

Use 3.14 as the approximate value for π.

Find the circumference of a circle with a radius of 3 cm.

$$C = 2 \cdot \pi \cdot r$$
$$C \approx 2 \cdot 3.14 \cdot 3 \text{ cm} \approx 18.8 \text{ cm} \quad \text{Rounded}$$

Use this formula to find the *area* of a circle.

$$A = \pi \cdot (\text{radius})^2$$

Area is measured in **square units**.

Find the area of this circle.

$$A = \pi \cdot r^2$$
$$A \approx 3.14 \cdot 3 \text{ cm} \cdot 3 \text{ cm}$$
$$A \approx 28.3 \text{ cm}^2 \quad \text{Rounded}$$

8.7 *Volume of a Rectangular Solid*

Use this formula to find the volume of *rectangular solids* (box-like shapes).

$$V = \text{length} \cdot \text{width} \cdot \text{height}$$

Volume is measured in **cubic units**.

Find the volume of this box.

$$V = l \cdot w \cdot h$$
$$V = 5 \text{ cm} \cdot 3 \text{ cm} \cdot 6 \text{ cm}$$
$$V = 90 \text{ cm}^3$$

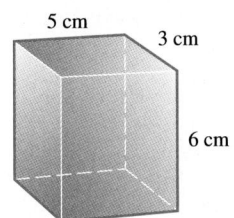

8.7 *Volume of a Sphere and Hemisphere*

Use this formula to find the volume of a *sphere* (a ball-shaped solid).

$$V = \frac{4}{3} \cdot \pi \cdot r^3$$

$$\text{or} \quad V = \frac{4 \cdot \pi \cdot r^3}{3}$$

where r is the radius of the sphere.

Volume is measured in **cubic units**.

Find the volume of a sphere with a radius of 5 m.

$$V = \frac{4 \cdot \pi \cdot (\text{radius})^3}{3}$$

$$V \approx \frac{4 \cdot 3.14 \cdot 5 \text{ m} \cdot 5 \text{ m} \cdot 5 \text{ m}}{3}$$

$$V \approx 523.3 \text{ m}^3 \quad \text{Rounded}$$

(continued)

Concepts	*Examples*

Concepts

8.7 *Volume of a Sphere and Hemisphere (continued)*
Use this formula to find the volume of a *hemisphere* (half of a sphere).

$$V = \frac{2}{3} \cdot \pi \cdot r^3$$

$$\text{or} \quad V = \frac{2 \cdot \pi \cdot r^3}{3}$$

where r is the radius of the hemisphere.

Volume is measured in **cubic units**.

8.7 *Volume of a Cylinder*
Use this formula to find the volume of a *cylinder*.

$$V = \pi \cdot r^2 \cdot h$$

where r is the radius of the circular base and h is the height of the cylinder.

Volume is measured in **cubic units**.

8.7 *Volume of a Cone*
Use this formula to find the volume of a *cone*.

$$V = \frac{1}{3} \cdot B \cdot h$$

$$\text{or} \quad V = \frac{B \cdot h}{3}$$

where B is the area of the circular base and h is the height of the cone.

Volume is measured in **cubic units**.

8.7 *Volume of a Pyramid*
Use this formula to find the volume of a *pyramid*.

$$V = \frac{1}{3} \cdot B \cdot h$$

$$\text{or} \quad V = \frac{B \cdot h}{3}$$

where B is the area of the square or rectangular base and h is the height of the pyramid.

Volume is measured in **cubic units**.

8.8 *Finding the Square Root of a Number*
Use the square root key on a calculator, (√) or (√x̄).
Round to the nearest thousandth if necessary.

Examples

Find the volume of a hemisphere with a radius of 20 cm.

$$V = \frac{2 \cdot \pi \cdot (\text{radius})^3}{3}$$

$$V \approx \frac{2 \cdot 3.14 \cdot 20 \text{ cm} \cdot 20 \text{ cm} \cdot 20 \text{ cm}}{3}$$

$$V \approx 16{,}746.7 \text{ cm}^3 \quad \text{Rounded}$$

Find the volume of a cylinder that is 10 m high with a diameter of 8 m.

First, find the radius. $\quad r = \dfrac{8 \text{ m}}{2} = 4 \text{ m}$

$$V = \pi \cdot r^2 \cdot h$$

$$V \approx 3.14 \cdot 4 \text{ m} \cdot 4 \text{ m} \cdot 10 \text{ m}$$

$$V \approx 502.4 \text{ m}^3$$

Find the volume of a cone, with a height of 9 in. and a base with a radius of 4 in.

$$\text{area of circular } B\text{ase} \approx 3.14 \cdot 4 \text{ in.} \cdot 4 \text{ in.}$$

$$B \approx 50.24 \text{ in.}^2$$

$$V = \frac{B \cdot h}{3}$$

$$V \approx \frac{50.24 \text{ in.}^2 \cdot 9 \text{ in.}}{3}$$

$$V \approx 150.7 \text{ in.}^3 \quad \text{Rounded}$$

Find the volume of a pyramid with a square base 2 cm by 2 cm and a height of 6 cm.

$$\text{area of square } B\text{ase} = 2 \text{ cm} \cdot 2 \text{ cm}$$

$$B = 4 \text{ cm}^2$$

$$V = \frac{B \cdot h}{3}$$

$$V = \frac{4 \text{ cm}^2 \cdot 6 \text{ cm}}{3}$$

$$V = 8 \text{ cm}^3$$

$$\sqrt{64} = 8 \qquad \text{A perfect square}$$

$$\sqrt{43} \approx 6.557 \qquad \begin{array}{l}\text{6.557438524 is rounded to}\\\text{nearest thousandth}\end{array}$$

Concepts	Examples

Concepts

8.8 *Finding the Unknown Length in a Right Triangle*
To find the *hypotenuse,* use this formula.

$$\text{hypotenuse} = \sqrt{(\text{leg})^2 + (\text{leg})^2}$$

The hypotenuse is the side opposite the right angle; it is the longest side in a right triangle.

To find a *leg,* use this formula.

$$\text{leg} = \sqrt{(\text{hypotenuse})^2 - (\text{leg})^2}$$

The legs are the sides that form the right angle.

Examples

Find the unknown length in this right triangle. Round to the nearest tenth.

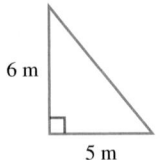

6 m

5 m

$$\text{hypotenuse} = \sqrt{(6)^2 + (5)^2}$$
$$= \sqrt{36 + 25}$$
$$= \sqrt{61} \approx 7.8$$

The hypotenuse is about 7.8 m long.

Find the unknown length in this right triangle. Round to the nearest tenth.

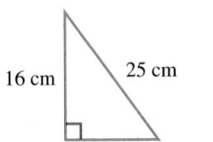

16 cm   25 cm

$$\text{leg} = \sqrt{(25)^2 - (16)^2}$$
$$= \sqrt{625 - 256}$$
$$= \sqrt{369} \approx 19.2$$

The leg is about 19.2 cm long.

8.9 *Finding the Unknown Lengths in Similar Triangles*
Use the fact that in similar triangles, the ratios of the lengths of corresponding sides are equal. Write a proportion. Then find the cross products and show that they are equivalent. Finish solving for the unknown length.

Find x and y if the triangles are similar.

$$\frac{x}{8} = \frac{5}{10}$$

$$x \cdot 10 = 8 \cdot 5$$

$$\frac{x \cdot \overset{1}{\cancel{10}}}{\cancel{10}_{1}} = \frac{40}{10}$$

$$x = 4 \text{ m}$$

$$\frac{y}{12} = \frac{5}{10}$$

$$y \cdot 10 = 12 \cdot 5$$

$$\frac{y \cdot \overset{1}{\cancel{10}}}{\cancel{10}_{1}} = \frac{60}{10}$$

$$y = 6 \text{ m}$$

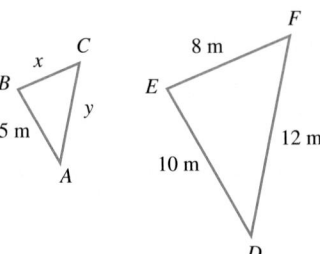

ANSWERS TO TEST YOUR WORD POWER

1. **(d)** *Example:* If $\angle 1$ measures $35°$ and $\angle 2$ measures $55°$, the angles are complementary because $35° + 55° = 90°$.
2. **(b)** *Example:* If a square measures 5 ft on each side, then the perimeter is 5 ft + 5 ft + 5 ft + 5 ft = 20 ft.
3. **(c)** *Examples:* Angles that measure $91°$, $120°$, and $175°$ are all obtuse angles. 4. **(c)** *Example:* In triangle *ABC* at the right, side *AC* is the hypotenuse; sides *AB* and *BC* are the legs. 5. **(b)** *Example:* The ratio

of a circumference of 12.57 cm to a diameter of 4 cm is $\dfrac{12.57}{4} \approx 3.14$ (rounded). 6. **(a)** *Example:*

$\overleftrightarrow{EF}$ is perpendicular to $\overleftrightarrow{GH}$, at the right. 7. **(d)** *Example:* Triangle *ABC* is similar to triangle *DEF*,

so the ratios of corresponding sides are equal. $\dfrac{AB}{DE} = \dfrac{3 \text{ m}}{6 \text{ m}} = \dfrac{1}{2}$ $\dfrac{BC}{EF} = \dfrac{2 \text{ m}}{4 \text{ m}} = \dfrac{1}{2}$ $\dfrac{AC}{DF} = \dfrac{3.5 \text{ m}}{7 \text{ m}} = \dfrac{1}{2}$

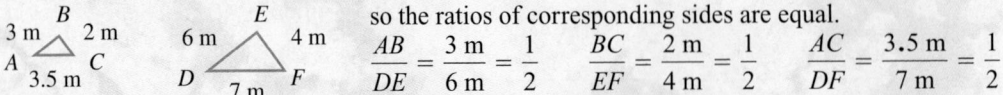

8. **(b)** *Example:* In a rectangle with a length of 8 in. and a width of 5 in., Area = 8 in. • 5 in. = 40 in.^{2}

Chapter 8 REVIEW EXERCISES

[8.1] *Name each line, line segment, or ray.*

1.

2.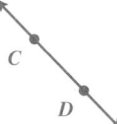

3.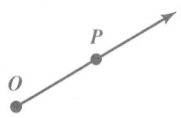

Label each pair of lines as parallel, perpendicular, or intersecting.

4.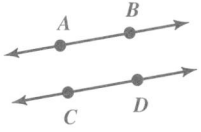

5.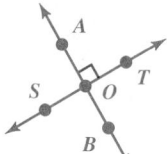

6.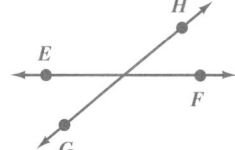

Label each angle as acute, right, obtuse, or straight. For right and straight angles, indicate the number of degrees in the angle.

7.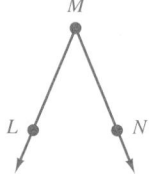

8.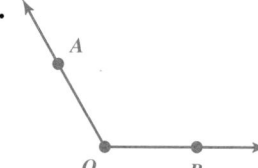

9.

10.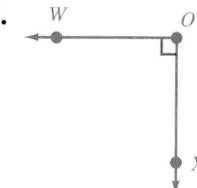

[8.2] *In each figure you are given the measures of two of the angles. Find the measure of each of the other angles.*

11.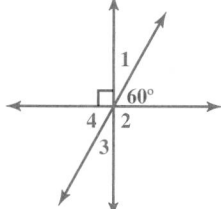

12.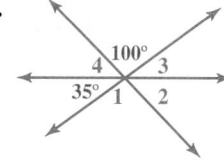

Name the pairs of supplementary angles in each figure.

13.

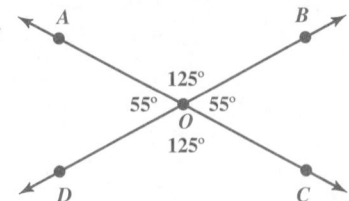

14.

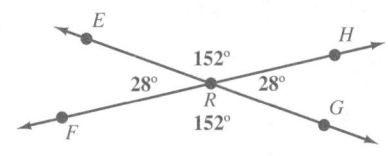

Find the complement or supplement of each angle.

15. Find the complement of:

    (a) 80°

    (b) 45°

    (c) 7°

16. Find the supplement of:

    (a) 155°

    (b) 90°

    (c) 33°

[8.3] *Find the perimeter of each rectangle or square.*

17.

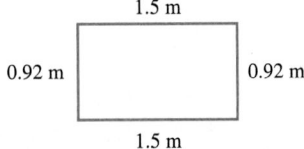

18.

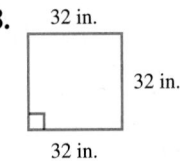

19. A square-shaped pillow measures 38 cm along each side. How much lace is needed to trim all the edges?

20. A rectangular garden plot is $8\frac{1}{2}$ ft wide and 12 ft long. How much fencing is needed to surround the garden?

Find the area of each rectangle or square. Round your answers to the nearest tenth when necessary.

21.

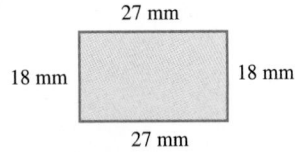

22.

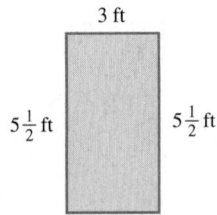

23.

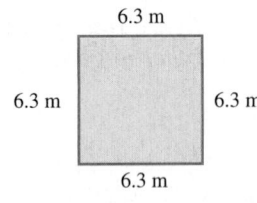

[8.4] *Find the perimeter and area of each parallelogram or trapezoid. Round your answers to the nearest tenth when necessary.*

24.
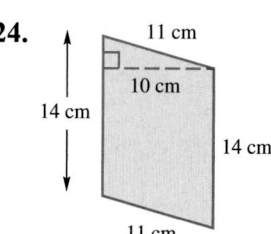

25.
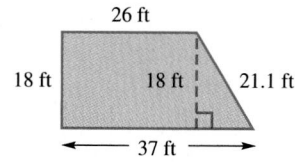

26.
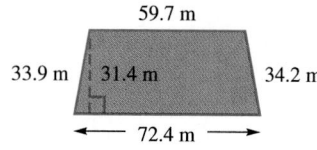

[8.5] *Find the perimeter and area of each triangle.*

27.
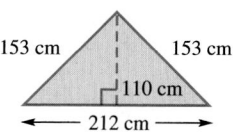

28.
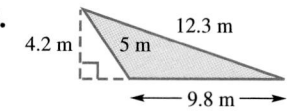

29.
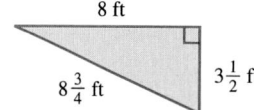

Find the number of degrees in each unlabeled angle.

30.
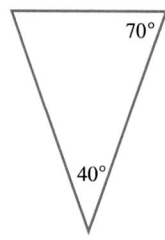

31.
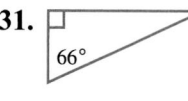

[8.6] *Find the unknown length.*

32. The radius of a circular irrigation field is 68.9 m. What is the diameter of the field?

33. The diameter of a juice can is 3 in. What is the radius of the can?

Find the circumference and area of each circle. Use 3.14 as the approximate value for π. Round your answers to the nearest tenth.

34.
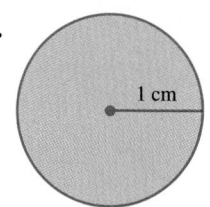

35.
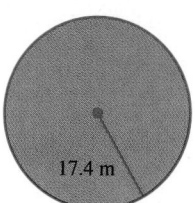

36.
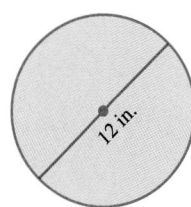

[8.3–8.6] *Find each shaded area. Use 3.14 as the approximate value for π. Round your answers to the nearest tenth when necessary.*

37.

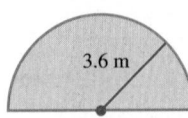

38.

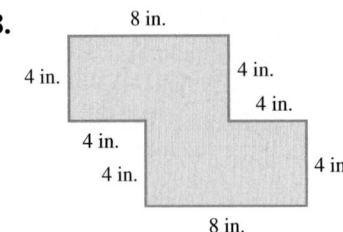

39.

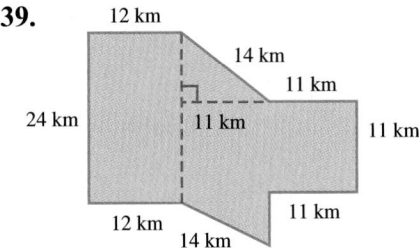

40.

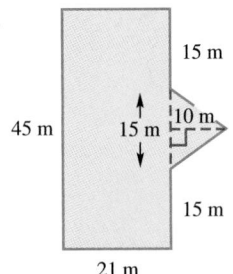

41.

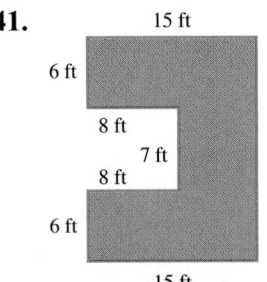

42.

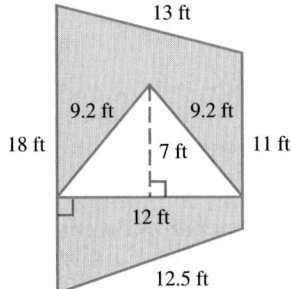

43.

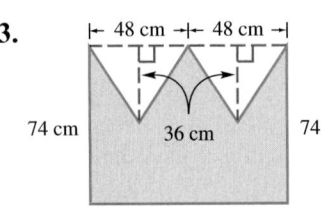

44.

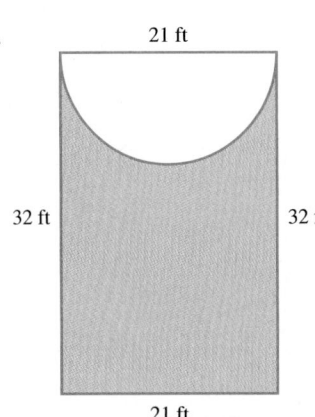

45.

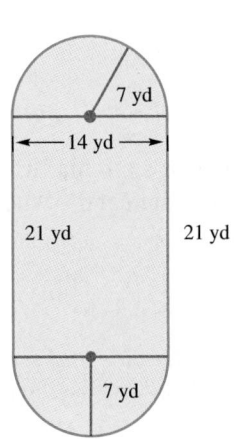

[8.7] *Find each volume. Use 3.14 as the approximate value for π. Round your answers to the nearest tenth when necessary.*

46.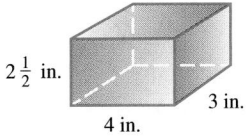

$2\frac{1}{2}$ in.   3 in.   4 in.

47.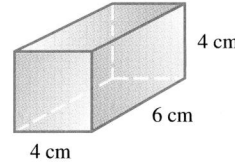

4 cm   6 cm   4 cm

48.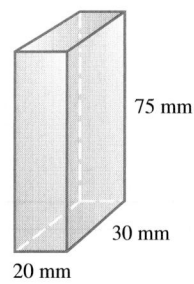

75 mm   30 mm   20 mm

49.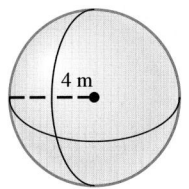

4 m

50.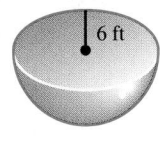

6 ft

51.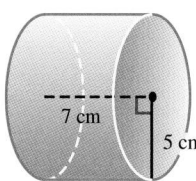

7 cm   5 cm

52.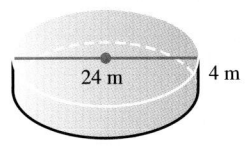

24 m   4 m

53.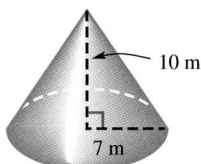

10 m   7 m

54.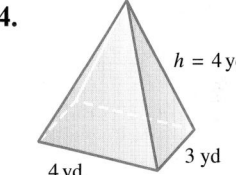

$h = 4$ yd   3 yd   4 yd

[8.8] *Find each square root. Round your answers to the nearest thousandth when necessary.*

55. $\sqrt{49}$

56. $\sqrt{8}$

57. $\sqrt{3000}$

58. $\sqrt{144}$

59. $\sqrt{58}$

60. $\sqrt{625}$

61. $\sqrt{105}$

62. $\sqrt{80}$

Find the unknown length in each right triangle. Use a calculator to find square roots. Round your answers to the nearest tenth when necessary.

63.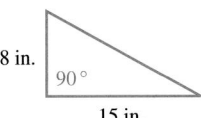

8 in.   90°   15 in.

64.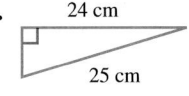

24 cm   25 cm

65.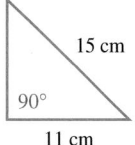

15 cm   90°   11 cm

66.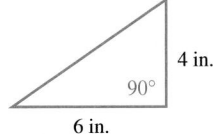

4 in.   90°   6 in.

67.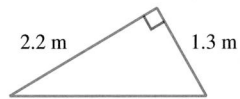

2.2 m   1.3 m

68.

12 km   8.5 km

[8.9] *Find the unknown lengths in each pair of similar triangles. Then find the perimeter of the larger triangle in each pair.*

69.

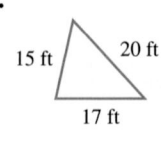

70.

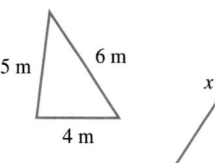

71.

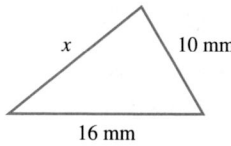

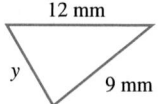

MIXED REVIEW EXERCISES

Find the perimeter (or circumference) and area of each figure. Use 3.14 as the approximate value for π. Round your answers to the nearest tenth when necessary.

72.

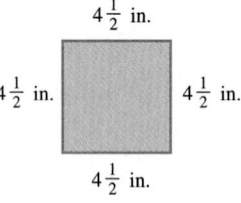

73.

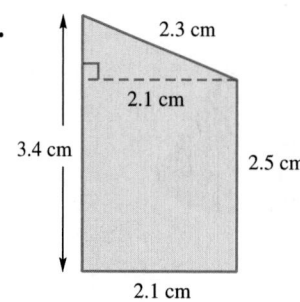

74.

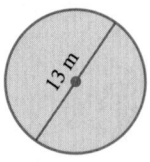

75.

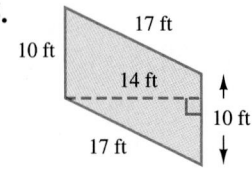

76.

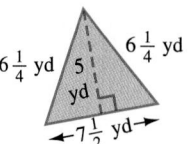

77.

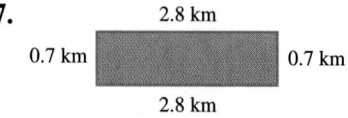

78.

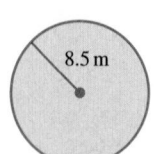

79.

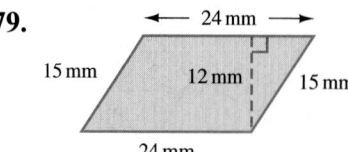

80.

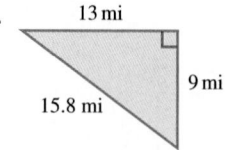

Label each figure. Choose from these labels: line segment, ray, parallel lines, perpendicular lines, intersecting lines, acute angle, right angle, straight angle, obtuse angle. Indicate the number of degrees in the right angle and the straight angle.

81.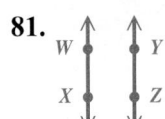

82.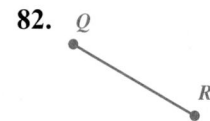

83.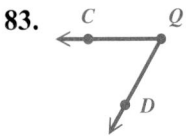

84.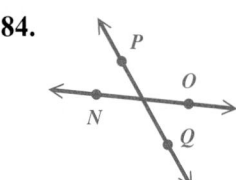

85.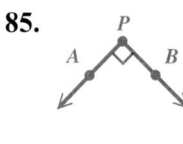

86.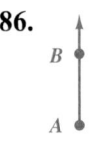

87.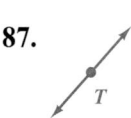

88.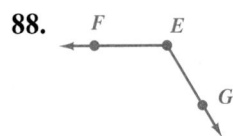

89.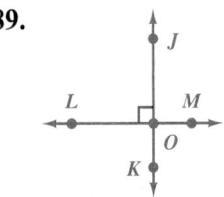

90. What is the complement of an angle measuring 9°?

91. What is the supplement of an angle measuring 42°?

Find the perimeter and area of each figure.

92.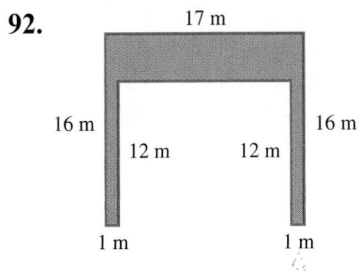

93.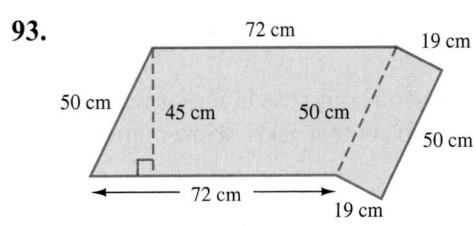

Find the volume of each figure. Use 3.14 as the approximate value for π. Round your answers to the nearest tenth when necessary.

94.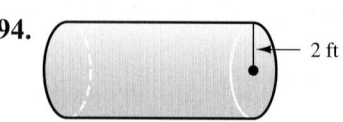

2 ft

8 ft

95.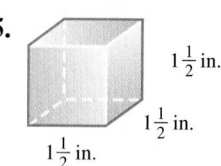

$1\frac{1}{2}$ in.

$1\frac{1}{2}$ in.

$1\frac{1}{2}$ in.

96.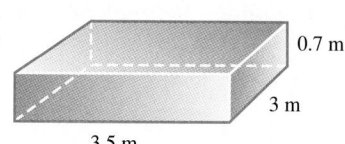

0.7 m

3 m

3.5 m

97.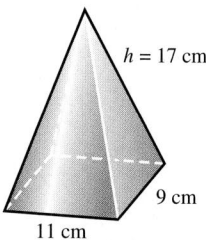

$h = 17$ cm

9 cm

11 cm

98.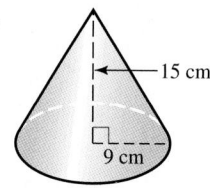

15 cm

9 cm

99.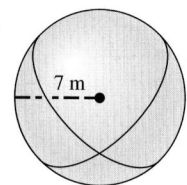

7 m

Find the unknown angle or side measurement. Round your answers to the nearest tenth when necessary.

100.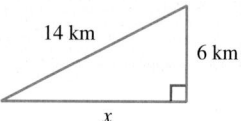

14 km

6 km

x

101.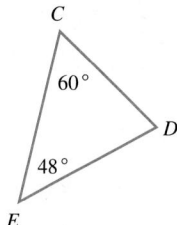

C

60°

D

48°

E

102. similar triangles

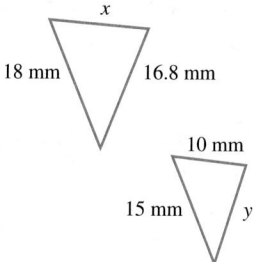

x

18 mm

16.8 mm

10 mm

15 mm

y

103. Explain how you could use the information about prefixes from **Section 8.6** to solve a problem that asks, "How many decades are in two centuries?"

Chapter 8 TEST

Choose the figure that matches each label. For right and straight angles, indicate the number of degrees in the angle.

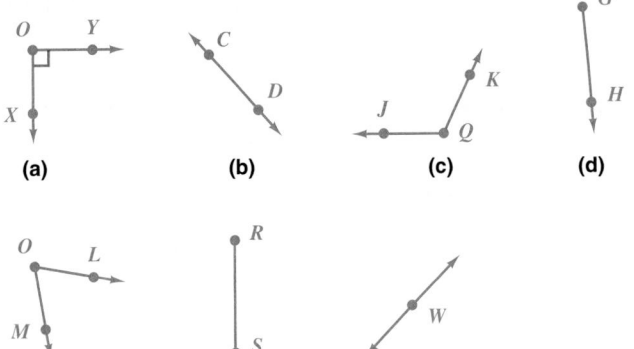

(a) (b) (c) (d)

(e) (f) (g)

1. **1.** Acute angle is figure _____.

2. **2.** Right angle is figure _____ and its measure is _____.

3. **3.** Ray is figure _____.

4. **4.** Straight angle is figure _____ and its measure is _____.

5. **5.** Write a definition of parallel lines and a definition of perpendicular lines. Make a sketch to illustrate each definition.

6. **6.** Find the complement of an 81° angle.

7. **7.** Find the supplement of a 20° angle.

8. **8.** Find the measure of each unlabeled angle in the figure at the right.

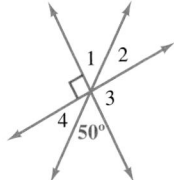

Find the perimeter and area of each figure.

9.
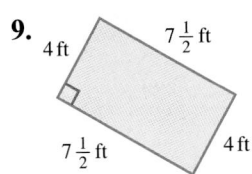

10.
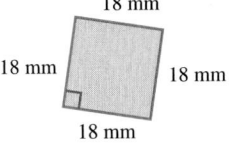

11.
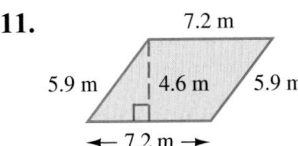

12.
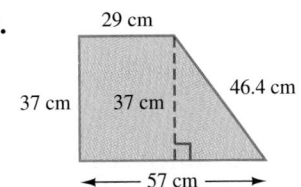

Answer blanks:
1. _____
2. _____
3. _____
4. _____
5. _____
6. _____
7. _____
8. _____
9. _____
   _____
10. _____
   _____
11. _____
   _____
12. _____
   _____

Find the perimeter and area of each triangle.

13. _____

_____

13.

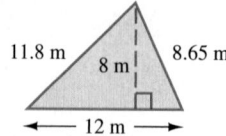

14.

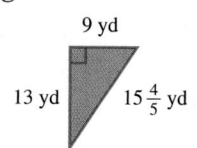

14. _____

_____

15. A triangle has angles that measure 90° and 35°. What does the third angle measure?

15. _____

In Problems 16–22, use 3.14 as the approximate value for π. Round your answers to the nearest tenth when necessary.

16. _____

16. Find the radius.

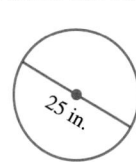

17. Find the circumference.

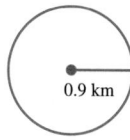

17. _____

18. _____

📠 *Find the area of each figure.*

18.

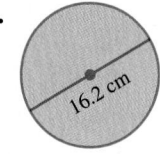

19.

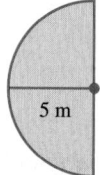

19. _____

20. _____

Find the volume of each figure.

20.

📠**21.**

📠**22.**

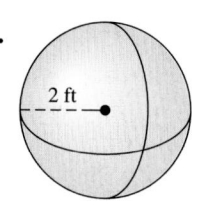

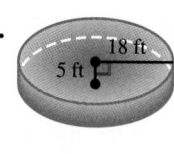

21. _____

22. _____

23. _____

Find the unknown lengths. Round your answers to the nearest tenth when necessary.

24. _____

📠 **23.**

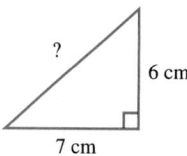

24. Similar triangles

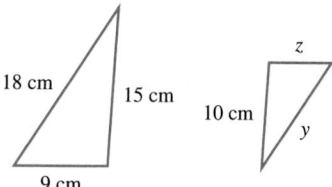

25. _____

25. Explain the difference between cm, cm², and cm³. In what types of geometry problems might you use each of these units?

First use front end rounding to round each number and estimate the answer. Then find the exact answer.

1. *Estimate:*   *Exact:*

$$\begin{array}{r} 319 \\ 58,028 \\ +\ 6\,227 \\ \hline \end{array}$$

$+$ _____

2. *Estimate:*   *Exact:*

$$\begin{array}{r} 20.07 \\ -\ 9.828 \\ \hline \end{array}$$

$-$ _____

3. *Estimate:*   *Exact:*

$$\begin{array}{r} 3.664 \\ \times\ 7.3 \\ \hline \end{array}$$

$\times$ _____

4. *Estimate:*   *Exact:*

$$\begin{array}{r} 28,419 \\ \times\ \ \ \ \ 73 \\ \hline \end{array}$$

$\times$ _____

5. *Estimate:*   *Exact:*

$2.8\overline{)562.24}$

$\overline{)}$

6. *Estimate:*   *Exact:*

$52\overline{)4888}$

$\overline{)}$

7. *Estimate:*   *Exact:*

$$4\frac{1}{2}$$
$$+\ 4\frac{9}{10}$$

$+$ ___
___

8. *Estimate:*   *Exact:*

$$3\frac{1}{6}$$
$$-\ 1\frac{7}{8}$$

$-$ ___
___

9. *Exact:*

$3\frac{1}{9} \cdot 1\frac{5}{7} = $ _____

Estimate:

___ $\cdot$ ___ $=$ ___

Add, subtract, multiply, or divide as indicated. Write answers to fraction problems in lowest terms and as whole or mixed numbers when possible.

10. $3\frac{3}{5} \div 8$

11. $1 - 0.0868$

12. Write your answer using R for the remainder.

$81\overline{)5749}$

13. $10 \div \frac{5}{16}$

14. $(0.006)(0.013)$

15. $40,020 - 915$

16. $0.7 \div 0.036$ Round answer to nearest hundredth.

17. $6\frac{1}{6} - 1\frac{3}{4}$

18. $752.6 + 83 + 0.485$

Use the order of operations to simplify each expression.

19. $16 - (10 - 2) \div 2 \cdot 3 + 5$

20. $2^4 \div \sqrt{64} + 6^2$

21. Write 0.0208 in words.

22. Write six hundred sixty and five hundredths in numbers.

23. Arrange in order from smallest to largest.

2.55   2.505   2.055   2.5005

24. Explain how you could use the information on prefixes and root words in Section 8.6 to remember the way to change a percent to a decimal.

Complete this chart.

Fraction/Mixed Number	Decimal	Percent
25. _____	0.02	**26.** _____
$1\frac{3}{4}$	**27.** _____	**28.** _____
29. _____	**30.** _____	40%

Write each rate or ratio as a fraction in lowest terms. Change to the same units when necessary.

31. 4 ft to 6 in.

32. Last month there were 9 cloudy days and 21 sunny days. What was the ratio of sunny days to cloudy days?

Find the unknown number in each proportion. Round your answer to hundredths if necessary.

33. $\dfrac{5}{13} = \dfrac{x}{91}$

34. $\dfrac{207}{69} = \dfrac{300}{x}$

35. $\dfrac{4.5}{x} = \dfrac{6.7}{3}$

Solve each percent problem.

36. 72 patients is what percent of 45 patients?

37. $18 is 3% of what number of dollars?

Convert each measurement.

38. $2\frac{1}{4}$ hours to minutes

39. 40 oz to pounds

40. 8 cm to meters

41. 1.8 L to milliliters

Write the most reasonable metric unit in each blank. Choose from km, m, cm, mm, L, mL, kg, g, and mg.

42. Her wristwatch strap is 15 _____ wide.

43. Jon added 2 _____ of oil to his car.

44. The child weighs 15 _____.

45. The bookcase is 90 _____ high.

46. List the metric temperatures at which water freezes and water boils.

Find the perimeter (or circumference) and area of each figure. Use 3.14 as the approximate value of π.

47.
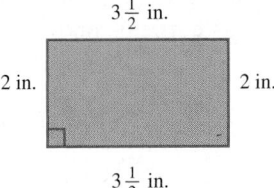
$3\frac{1}{2}$ in.

2 in.   2 in.

$3\frac{1}{2}$ in.

48.
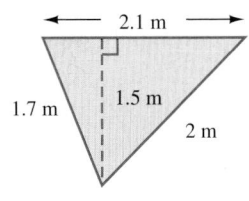
2.1 m
1.7 m   1.5 m
2 m

49.
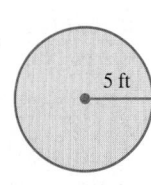
5 ft

50.
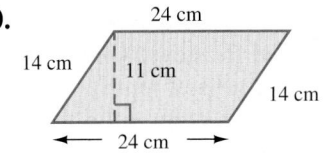
24 cm
14 cm   11 cm
14 cm
24 cm

51.
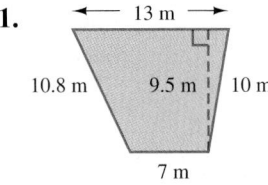
13 m
10.8 m   9.5 m   10 m
7 m

52.
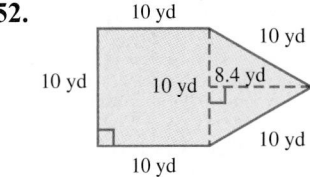
10 yd
10 yd
10 yd   10 yd   8.4 yd
10 yd
10 yd

Find the unknown length in each figure. Round your answers to the nearest tenth. In Exercise 54, also find the perimeter of the smaller triangle.

53.
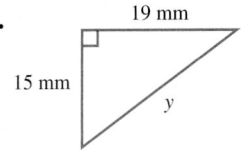
19 mm
15 mm
y

54. Similar triangles
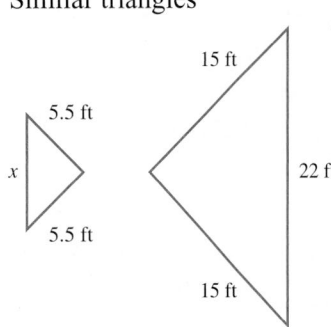
15 ft
5.5 ft
x
22 ft
5.5 ft
15 ft

Solve each application problem.

55. Mei Ling must earn 90 credits to receive an associate of arts degree. She has 53 credits. What percent of the necessary credits does she have? Round to the nearest whole percent.

56. Which bag of chips is the best buy: Brand T is $15\frac{1}{2}$ oz for $2.99, Brand F is 14 oz for $2.49, and Brand H is 18 oz for $3.89. You have a coupon for 40¢ off Brand H and another for 30¢ off Brand T.

57. A Folger's coffee can has a diameter of 13 cm and a height of 17 cm. Find the volume of the can. Use 3.14 for π and round your answer to the nearest tenth. (*Source:* Folger's.)

58. A photograph measures 8 in. by 10 in. Earl put it in a frame that is 2 in. wide. Find the perimeter of the frame. To help solve this problem, first label the inside and outside measurements on the sketch of the frame.

59. Steven bought $4\frac{1}{2}$ yd of canvas material to repair the tents used by the scout troop. He used $1\frac{2}{3}$ yd on one tent and $1\frac{3}{4}$ yd on another. How much material is left?

60. The cooks at a homeless shelter used 30 lb of meat to make stew for 140 people. At that rate, how much meat is needed for stew to feed 200 people? Round to the nearest tenth.

61. Graciela needs 85 cm of yarn to make a tassel for one corner of a pillow. How many meters of yarn does she need to put a tassel on each corner of a square-shaped pillow?

62. Swimsuits are on sale in August at 65% off the regular price. How much will Lanece pay for a suit that has a regular price of $64?

The table shows the changes in U.S. postal rates that went into effect in January 2001. Use the information in the table to answer Exercises 63–66. Write all money answers using a dollar sign an decimal point.

CHANGES IN U.S. POSTAL RATES, JANUARY 2001

Item	From	To
First class, 1st ounce	33¢	34¢
First class, 2nd–11th ounce	22¢/oz	21¢/oz
Postcard	20¢	20¢
Priority mail up to 2 lbs	$3.20	$3.95
Newspapers up to 10 oz	26.6¢	28.7¢

Source: Associated Press.

63. How much would it cost under the old rates and under the new rates to mail a first class envelope weighing **(a)** 2 oz, **(b)** 6 oz, **(c)** 11 oz?

64. Find the percent increase or decrease, to the nearest tenth of a percent, in the rates for **(a)** first class, 1st ounce; **(b)** first class, 2nd ounce; **(c)** postcards; **(d)** priority mail.

65. Can a 36 oz package be sent by priority mail for $3.95 under the new rates? Explain your answer.

66. The *Wall Street Journal* has the largest circulation of any newspaper in the United States, with a daily average of 1,762,750. (*Source:* Audit Bureau of Circulation.) If 12% of the papers are sent by mail, and each paper weighs less than 10 oz, what is the daily increase in postage costs from the old to new rates?

Basic Algebra

9

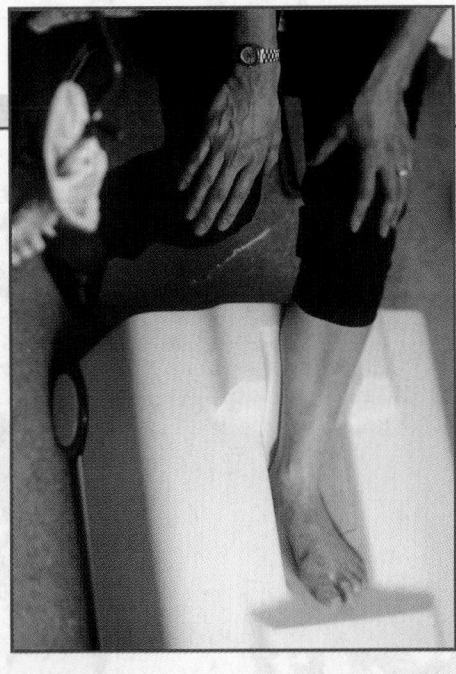

Osteoporosis is a bone disease in which bones become thin and brittle. One in four women and one in eight men over the age of 50 have the disease, but it also strikes people in their 20's who have eating disorders. (*Source:* Osteoporosis online.) More than 1.5 million osteoporosis-related bone fractures occur each year in the United States. (*Source:* Mayo Clinic.) An ultrasound of the heel bone is a quick, inexpensive way to screen people for osteoporosis. The patients receive a "T score" that indicates their level of risk. Both the ultrasound technician and the patient must understand positive and negative numbers to interpret the score. (See Exercises 27–30 in Section 9.1.)

ADDISON - WESLEY
MyMathLab.com
You're Connected

9.1 SIGNED NUMBERS

❶ Write each number.

(a) The temperature at the North Pole is 70 degrees below 0.

(b) Your checking account is overdrawn by 15 dollars.

(c) A submarine dived to 284 ft below sea level.

❷ Write *positive, negative,* or *neither* for each number.

(a) −8

(b) $-\dfrac{3}{4}$

(c) 1

(d) 0

All the numbers you have studied so far in this book have been either 0 or greater than 0. Numbers *greater* than 0 are called *positive numbers*. For example, you have worked with these positive numbers.

    Salary of $8000

    Temperature of 98.6 °F

    Length of $3\frac{1}{2}$ feet

1 **Write negative numbers.**   Not all numbers are positive. For example, "15 degrees below 0" or "a loss of $500" is expressed with a number *less* than 0. Numbers less than 0 are called **negative numbers.** Zero is neither positive nor negative.

Writing a Negative Number

To write a negative number, put a ***negative sign,*** **−**, in front of it.

For example, "15 degrees below 0" is written with a negative sign, as −15°. And "a loss of $500" is written −$500.

Work Problem ❶ at the Side.

2 **Graph signed numbers on a number line.**   In **Section 3.5** you graphed positive numbers on a number line. Negative numbers can also be shown on a number line. Zero separates the positive numbers from the negative numbers on the number line. The number −5 is read "negative five."

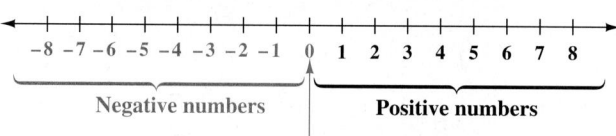

Negative numbers   Positive numbers

Zero is neither positive nor negative.

> **NOTE**
>
> For every *positive* number on a number line, there is a corresponding *negative* number on the *opposite* side of 0.

    When you work with both positive and negative numbers (and zero), we say you are working with **signed numbers.**

Writing a Positive Number

A positive number can be written in two ways.

1. Use a "+" sign. For example, +2 is "positive two."

2. Do not write any sign. For example, 3 is assumed to be "positive three."

Work Problem ❷ at the Side.

The next example shows you how to graph signed numbers on a number line.

Example 1 **Graphing Signed Numbers**

Graph these numbers on the number line.

(a) -4 **(b)** 3 **(c)** -1 **(d)** 0 **(e)** $1\frac{1}{4}$

Place a dot at the correct location for each number.

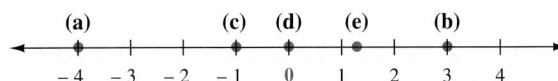

═══ **Work Problem ❸ at the Side.**

3▭▭▭ **Use the < and > symbols.** As shown on this number line, 3 is to the left of 5.

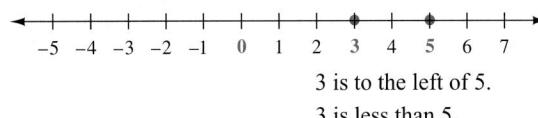

3 is to the left of 5.
3 is less than 5.

Recall the following symbols for comparing two numbers.

$<$ means "**is less than.**"

$>$ means "**is greater than.**"

Use these symbols to write "3 is less than 5" as follows.

$$3 \quad < \quad 5$$
$$\downarrow \qquad \downarrow \qquad \downarrow$$
3 is less than 5

This example suggests the following.

> The *lesser* of two numbers is the one farther to the *left* on a number line.
> The *greater* of two numbers is the one farther to the *right* on a number line.

Example 2 **Using the Symbols < and >**

Use this number line to compare each pair of numbers. Then write $>$ or $<$ to make each statement true.

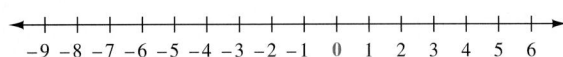

(a) $2 < 6$ (read "2 is less than 6") because 2 is to the *left* of 6 on the number line.

(b) $-9 < -4$ because -9 is to the *left* of -4.

(c) $2 > -1$ because 2 is to the *right* of -1.

(d) $-4 < 0$ because -4 is to the *left* of 0.

---

❸ Graph each set of numbers.

(a) $-1, 1, -3, 3$

```
◄─┼─┼─┼─┼─┼─┼─┼─┼─►
 -4 -3 -2 -1 0 1 2 3 4
```

(b) $-2, 4, 0, -1, -4$

```
◄─┼─┼─┼─┼─┼─┼─┼─┼─►
 -4 -3 -2 -1 0 1 2 3 4
```

❹ Write $<$ or $>$ in each blank to make a true statement.

(a) 4 _____ 0

(b) −1 _____ 0

(c) −3 _____ −1

(d) −8 _____ −9

(e) 0 _____ −3

❺ Simplify each absolute value expression.

(a) $|5|$

(b) $|-5|$

(c) $|-17|$

(d) $-|-9|$

(e) $-|2|$

NOTE

When using $>$ and $<$, the *small* pointed end of the symbol points to the *smaller* (lesser) number.

Work Problem ❹ at the Side.

4 ▮▮▮ **Find absolute value.** In order to graph a number on the number line, you need to answer the following two questions.

1. Which *direction* is it from 0? It can be in a *positive* direction or a *negative* direction. You can tell the direction by looking for a positive sign or a negative sign (or no sign, which is positive).

2. How *far* is it from 0? The *distance* from 0 is the **absolute value** of a number.

Absolute value is indicated by two vertical bars. For example, $|6|$ is read "the **absolute value** of 6."

NOTE

Absolute value is never negative because it is the *distance* from 0. A distance is never negative.

Example 3 Finding Absolute Values

Simplify each absolute value expression.

(a) $|8|$ The *distance* from 0 to 8 is 8, so $|8| = 8$.

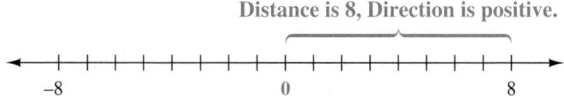

Distance is 8, Direction is positive.

(b) $|-8|$ The *distance* from 0 to −8 is also 8, so $|-8| = 8$.

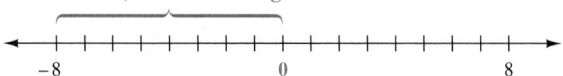

Distance is 8, Direction is negative.

(c) $|0| = 0$

(d) $-|-3|$ First, $|-3|$ is 3. But there is also a negative sign outside the absolute value bars. So, -3 is the simplified expression.

CAUTION

A negative sign *outside* the absolute value bars is *not* affected by the absolute value bars. Therefore, your final answer is negative, as in Example 3(d) above.

Work Problem ❺ at the Side.

5 ━━━ **Find the opposite of a number.** Two numbers that are the *same distance* from 0 on a number line but on *opposite sides* of 0 are called **opposites** of each other. As this number line shows, −3 and 3 are opposites of each other.

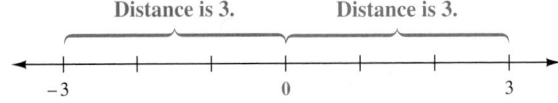

Distance is 3.   Distance is 3.

To indicate the opposite of a number, write a negative sign in front of the number.

Example 4 **Finding Opposites**

Find the opposite of each number.

Number	Opposite
5	$-(5) = -5$
9	$-(9) = -9$
$\dfrac{4}{5}$	$-\left(\dfrac{4}{5}\right) = -\dfrac{4}{5}$
0	$-(0) = 0$

— Write a negative sign.

— No negative sign

The opposite of 0 is 0. Zero is neither positive nor negative.

══ **Work Problem 6 at the Side.**

Some numbers have two negative signs, such as

$$-(-3).$$

The negative sign in front of (-3) means the *opposite* of -3. The opposite of -3 is 3.

$$-(-3) = 3$$

Use the following rule to find the opposite of a negative number.

Double Negative Rule

The opposite of a negative number is positive.

For example, $-(-3) = 3$ and $-(-10) = 10.$

Example 5 **Finding Opposites**

Find the opposite of each number.

Number	Opposite
-2	$-(-2) = 2$
-9	$-(-9) = 9$
$-\dfrac{1}{2}$	$-\left(-\dfrac{1}{2}\right) = \dfrac{1}{2}$

By the double negative rule

══ **Work Problem 7 at the Side.**

6 Find the opposite of each number.

(a) 4

(b) 10

(c) 49

(d) $\dfrac{2}{5}$

(e) 0

7 Find the opposite of each number.

(a) -4

(b) -10

(c) -25

(d) -1.9

(e) -0.85

(f) $-\dfrac{3}{4}$

ANSWERS

6. (a) -4 (b) -10 (c) -49

(d) $-\dfrac{2}{5}$ (e) 0

7. (a) 4   (b) 10   (c) 25   (d) 1.9

(e) 0.85   (f) $\dfrac{3}{4}$

Real-Data Applications

Auto Aide, Part 1

The **Auto Aide Service** is a hypothetical business located in downtown Houston, Texas. It specializes in assisting motorists who have minor mechanical problems, such as a flat tire or a dead battery. The I-10 corridor in the greater Houston area is a major highway with constant heavy traffic, so the company assigns five vans to concentrate on I-10 assistance needs. The owner is a retired mathematics teacher who believes in using integer arithmetic sentences to track the actions of the auto aides along their service routes.

A schematic map of the I-10 corridor is shown below. Each tick mark represents 1 mile. The home office is located at 0. Locations to the west are represented by negative integers, and locations to the east are represented by positive integers.

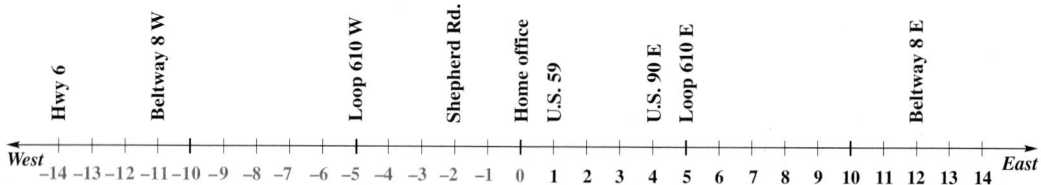

A radio dispatcher notifies the aide to assist a motorist at a given location. (Only a few locations are identified in this hypothetical problem, for simplicity.) Each aide must include four calculations on the daily report. First, the aide must translate the route instructions into an integer sentence that gives the direction and miles between locations. Second, the aide must record the displacement at the end of the route (location relative to home office). Third, the aide must record the distance from the home office at the end of the route. Fourth, the aide must record the total distance traveled on the route.

1. Aide Anne is shown as an example. Complete the table for the remaining aides.

Aide	Route Instructions	Integer Sentence	Displacement (final location)	Distance from Home Office	Total Distance
Anne	Home office to U.S. 90 E; to Loop 610 W; to Hwy 6; to Shepherd Rd.	$4 + (-9) + (-9) + 12$	-2(Shepherd) (*Hint:* Find the integer sum.)	2 miles (*Hint:* Find the absolute value of displacement.)	34 miles (*Hint:* Find sum of absolute values of trip segments.)
Bill	Home office to Hwy 6; to U.S. 59; to Shepherd Rd.; to Loop 610 W				
Carlos	Home office to Shepherd Rd.; to Loop 610 W; to Beltway 8 W; to Hwy 6				
Dylan	Home office to Beltway 8 E; to U.S. 90 E; to Home office				
Ellen	Home office to U.S. 59; to Shepherd Rd.; to U.S. 59; to Loop 610 W; to U.S. 59				

9.1 EXERCISES

 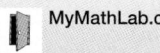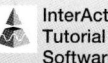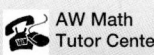
Write a signed number for each situation.

1. Water freezes at 32 degrees above 0 on the Fahrenheit temperature scale.

2. She made a profit of $920.

3. The price of the stock fell $12.

4. His checking account is overdrawn by $30.79.

5. Keith lost $6\frac{1}{2}$ lb while he was sick with the flu.

6. The river is 20 ft above flood stage.

7. Mount McKinley, the tallest mountain in the United States, rises 20,320 ft above sea level. (*Source: World Almanac,* 2001.)

8. The bottom of Lake Baykal in central Asia is 5315 ft below the surface of the water. (*Source: World Almanac,* 2001.)

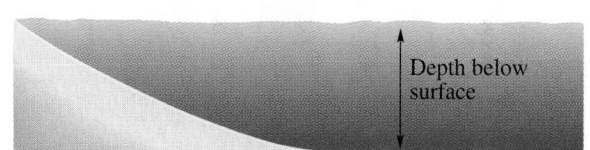

Depth below surface

Graph each set of numbers on the number line. See Example 1.

9. 4, −1, 2, 0, −5

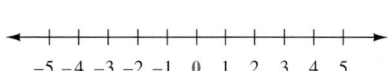

10. −2, 1, −3, 5, 0

11. $-\frac{1}{2}, -3, \frac{7}{4}, -4\frac{1}{2}, 3\frac{1}{4}$

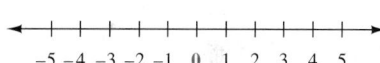

12. $-4, -\frac{3}{4}, 1, -1\frac{1}{4}, \frac{5}{2}$

13. 3, 4.5, −1.5, 2.2, −0.5

14. 3.25, −1, −4.5, 1.25, 2

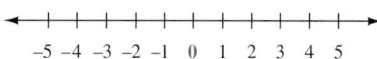

Write < or > in each blank to make a true statement. See Example 2.

15. 9 _____ 14

16. 6 _____ 11

17. 0 _____ −2

18. 0 _____ 2

19. −6 _____ 3

20. −9 _____ 9

21. 1 _____ −1

22. −1 _____ 0

23. −11 _____ −2

24. −5 _____ −1

25. −72 _____ −75

26. −50 _____ −60

RELATING CONCEPTS (Exercises 27–30)	FOR INDIVIDUAL OR GROUP WORK

*At the beginning of this chapter, you saw a photo of an ultrasound machine used to screen patients for osteoporosis (brittle bone disease). Use the information on the Patient Report Form to **work Exercises 27–30 in order.***

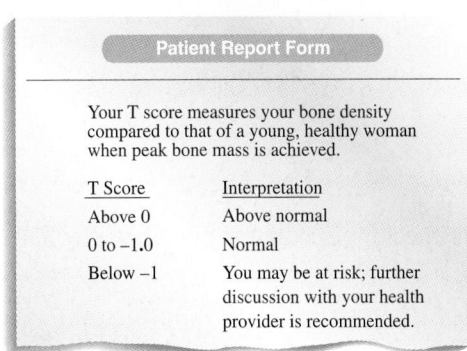

Source: Health Partners, Inc.

27. Here are the T scores for four patients. Draw a number line and graph the four scores.

Patient A: −1.2

Patient B: 0.6

Patient C: −0.5

Patient D: 0

28. List the patients' scores in order from lowest to highest.

29. What is the interpretation of each patient's score?

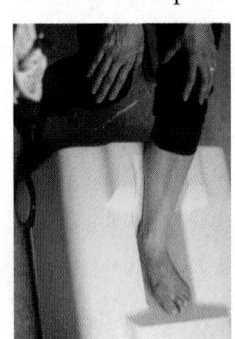

30. (a) What could happen if Patient A did not understand the importance of a negative sign?

(b) For which patient does the sign of the score make no difference? Explain your answer.

Simplify each absolute value expression. See Example 3.

31. $|3|$ **32.** $|9|$ **33.** $|-10|$ **34.** $|-2|$ **35.** $|0|$

36. $\left|-\dfrac{1}{2}\right|$ **37.** $-|-18|$ **38.** $-|-5|$ **39.** $-|32|$ **40.** $-|20|$

Find the opposite of each number. See Examples 4 and 5.

41. 7   **42.** 1   **43.** −14   **44.** −5   **45.** $\dfrac{2}{3}$

46. 0   **47.** −8.3   **48.** 0.2   **49.** $-\dfrac{1}{6}$ **50.** $-\dfrac{3}{10}$

Write true or false for each statement.

51. $|-5| > 0$ **52.** $|-12| > |-15|$ **53.** $0 < -(-6)$

54. $-9 < -(-9)$ **55.** $-|-4| < -|-7|$ **56.** $-|-0| > 0$

9.2 ADDING AND SUBTRACTING SIGNED NUMBERS

You can show a *positive* number on a number line by drawing an arrow pointing to the *right*. In the following examples both arrows represent positive 4 units.

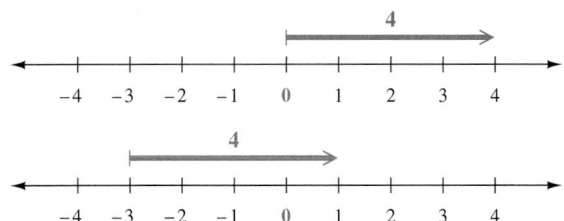

Draw arrows pointing to the *left* to show *negative* numbers. Both of the arrows below represent −3 units.

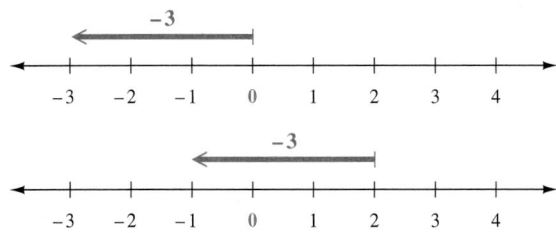

Work Problem ❶ at the Side.

1 ▨▨▨▨ **Add signed numbers by using a number line.** You can use a number line to add signed numbers. For example, this number line shows how to add 2 and 3.

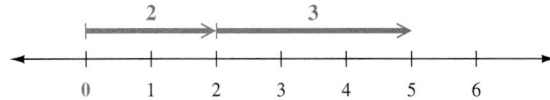

Add 2 and 3 by starting at 0 and drawing an arrow 2 units to the right. From the end of this arrow, draw another arrow 3 units to the right. This second arrow ends at 5, showing that the sum of $2 + 3$ is 5.

$$2 + 3 = 5$$

Example 1 **Adding Signed Numbers by Using a Number Line**

Add by using a number line.

(a) $4 + (-1)$

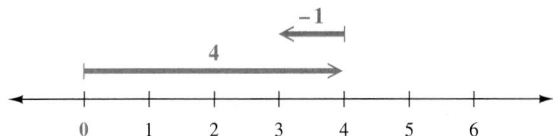

Start at 0 and draw an arrow 4 units to the right. From the end of this arrow, draw an arrow 1 unit to the *left*. (Remember to go to the left for a negative number.) This second arrow ends at 3, so

$$4 + (-1) = 3.$$

CAUTION

Always start at 0 when adding on the number line.

Continued on Next Page

OBJECTIVES

1▨ Add signed numbers by using a number line.

2▨ Add signed numbers without using a number line.

3▨ Find the additive inverse of a number.

4▨ Subtract signed numbers.

5▨ Add or subtract a series of signed numbers.

❶ Complete each arrow so it represents the indicated number of units.

(a)

```
          2
      ┡━━━━┥
  ┼───┼───┼───┼───┼───┼
 -1   0   1   2   3   4
```

(b)

```
          6
      ┡━━━━━━━━━┥
  ┼───┼───┼───┼───┼───┼
 -2 -1  0   1   2   3   4
```

(c)

```
              -4
          ┡━━━━┥
  ┼───┼───┼───┼───┼───┼
 -2 -1  0   1   2   3   4
```

(d)

```
                  -5
              ┡━━━━┥
  ┼───┼───┼───┼───┼───┼
 -2 -1  0   1   2   3   4
```

ANSWERS

1. (a)

```
          2
      ┡━━━➤
  ┼┼┼┼┼┼┼
 -1 0 1 2 3 4
```

(b)

```
          6
      ┡━━━━━➤
  ┼┼┼┼┼┼┼
 -2 -1 0 1 2 3 4
```

(c)

```
              -4
          ◄━━━┥
  ┼┼┼┼┼┼┼
 -2 -1 0 1 2 3 4
```

(d)

```
                  -5
              ◄━━━┥
  ┼┼┼┼┼┼┼
 -2 -1 0 1 2 3 4
```

❷ Draw arrows to find each sum.

(a) $3 + (-2)$

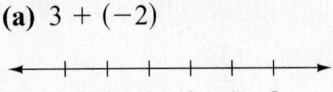

(b) $-4 + 1$

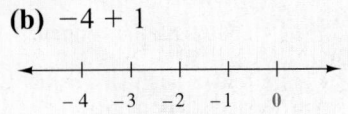

(c) $-3 + 7$

(d) $-1 + (-4)$

2. (a) $3 + (-2) = 1$

(b) $-4 + 1 = -3$

(c) $-3 + 7 = 4$

(d) $-1 + (-4) = -5$

(b) $-6 + 2$

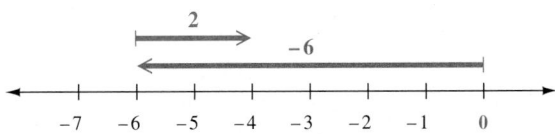

Draw an arrow from 0 going 6 units to the left. From the end of this arrow, draw an arrow 2 units to the right. This second arrow ends at -4, so

$$-6 + 2 = -4.$$

(c) $-3 + (-5)$

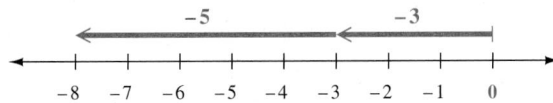

As the arrows along the number line show,

$$-3 + (-5) = -8.$$

Work Problem ❷ at the Side.

2   **Add signed numbers without using a number line.**   After working with number lines for awhile, you will see ways to add signed numbers without drawing arrows. You already know how to add two positive numbers (from Chapter 1). Here are the steps for adding two negative numbers.

Adding Negative Numbers

Step 1   Add the absolute values of the numbers.
Step 2   Write a negative sign in front of the sum.

Example 2   Adding Negative Numbers

Add (without using number lines).

(a) $-4 + (-12)$

The absolute value of -4 is **4**.
The absolute value of -12 is **12**.

Add the absolute values.

$$4 + 12 = 16$$

Write a negative sign in front of the sum.

$$-4 + (-12) = {}^-16 \qquad \text{Write a negative sign in front of 16.}$$

(b) $-5 + (-25) = -30 \leftarrow$ Sum of absolute values, with a negative sign written in front of 30.

(c) $-11 + (-46) = -57$

Continued on Next Page

(d) $-\dfrac{3}{4} + \left(-\dfrac{1}{2}\right)$

The absolute value of $-\dfrac{3}{4}$ is $\dfrac{3}{4}$, and the absolute value of $-\dfrac{1}{2}$ is $\dfrac{1}{2}$. Add the absolute values. Check that the answer is in lowest terms.

$$\dfrac{3}{4} + \dfrac{1}{2} = \dfrac{3}{4} + \dfrac{2}{4} = \dfrac{5}{4} \quad \longleftarrow \text{Lowest terms}$$

Write a negative sign in front of the sum.

$$-\dfrac{3}{4} + \left(-\dfrac{1}{2}\right) = -\dfrac{5}{4} \quad \text{Write a negative sign.}$$

NOTE

In algebra we always write fractions in lowest terms, but usually do *not* change improper fractions to mixed numbers, because improper fractions are easier to work with. In Example 2(d) above, we checked that $-\dfrac{5}{4}$ was in lowest terms but did *not* rewrite it as $-1\dfrac{1}{4}$.

Work Problem ❸ at the Side.

Use the following steps to add two numbers with *different* signs.

Adding Two Numbers with Different Signs

Step 1 **Subtract** the smaller absolute value from the larger absolute value.

Step 2 Write the sign of the number with the **larger** absolute value in front of the answer.

Example 3 Adding Two Numbers with Different Signs

Find each sum.

(a) $8 + (-3)$

First find this sum with a number line.

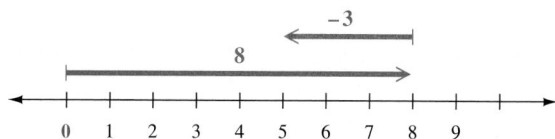

Because the second arrow ends at 5,

$$8 + (-3) = 5.$$

Now find the sum by using the above rule. First, find the absolute value of each number.

$$|8| = 8 \quad \text{and} \quad |-3| = 3$$

Subtract the smaller absolute value from the larger absolute value.

$$8 - 3 = 5$$

The *positive* number 8 has the larger absolute value, so the answer is *positive*.

$$8 + (-3) = 5 \quad \longleftarrow \text{Positive answer}$$

═══ **Continued on Next Page**

---

❸ Add.

(a) $-4 + (-4)$

(b) $-3 + (-20)$

(c) $-31 + (-5)$

(d) $-10 + (-8)$

(e) $-\dfrac{9}{10} + \left(-\dfrac{3}{5}\right)$

ANSWERS

3. (a) -8 **(b)** -23 **(c)** -36 **(d)** -18

(e) $-\dfrac{3}{2}$

4 Add.

(a) $10 + (-2)$

(b) $-7 + 8$

(c) $-11 + 11$

(d) $23 + (-32)$

(e) $-\dfrac{7}{8} + \dfrac{1}{4}$

(b) $4 + (-12)$

First, find the absolute values.

$$|4| = 4 \quad \text{and} \quad |-12| = 12$$

Subtract the smaller absolute value from the larger absolute value.

$$12 - 4 = 8$$

The *negative* number -12 has the larger absolute value, so the answer is *negative*.

$$4 + (-12) = -8$$
— Write a negative sign in front of the answer because -12 has the larger absolute value.

(c) $-21 + 15 = -6$
— Write a negative sign in front of the answer because -21 has the larger absolute value.

(d) $13 + (-9) = 4$ ←— Positive answer because the positive number 13 has the larger absolute value.

(e) $-\dfrac{1}{2} + \dfrac{2}{3}$

The absolute value of $-\dfrac{1}{2}$ is $\dfrac{1}{2}$, and the absolute value of $\dfrac{2}{3}$ is $\dfrac{2}{3}$. Subtract the smaller absolute value from the larger absolute value.

$$\frac{2}{3} - \frac{1}{2} = \frac{4}{6} - \frac{3}{6} = \frac{1}{6}$$

Because the *positive* number $\frac{2}{3}$ has the larger absolute value, the answer is *positive*.

$$-\frac{1}{2} + \frac{2}{3} = \frac{1}{6} \quad \text{← Positive answer}$$

Work Problem 4 at the Side.

3 **Find the additive inverse of a number.** Recall that the opposite of 9 is -9, and the opposite of -4 is $-(-4)$, or 4. Adding opposites gives the following results.

$$9 + (-9) = 0 \quad \text{and} \quad -4 + 4 = 0$$

The sum of a number and its opposite is always 0. For this reason, opposites are also called *additive inverses* of each other.

Additive Inverse

The opposite of a number is called its **additive inverse.** The sum of a number and its opposite is 0.

Example 4 · Finding Additive Inverses

This chart shows you several numbers and the additive inverse of each.

Number	Additive Inverse	Sum of Number and Inverse
6	−6	$6 + (−6) = 0$
−8	−(−8) or 8	$(−8) + 8 = 0$
4	−4	$4 + (−4) = 0$
−3	−(−3) or 3	$−3 + 3 = 0$
$\dfrac{5}{8}$	$−\dfrac{5}{8}$	$\dfrac{5}{8} + \left(−\dfrac{5}{8}\right) = 0$
0	0	$0 + 0 = 0$

Work Problem ❺ at the Side.

4 ▭ **Subtract signed numbers.** You may have noticed that negative numbers are often written with parentheses, like $(−8)$. This is especially helpful when subtracting because the $−$ sign is used both to indicate a **negative number** and to indicate **subtraction**.

Example	How to Say It
$(−5)$	**negative five**
8	**positive eight**
$−3 − 2$	**negative three minus** positive two
$6 − (−4)$	positive six **minus negative four**
$−7 − (−1)$	**negative seven minus negative one**
$8 − 3$	positive eight **minus** positive three

CAUTION

Be sure you understand when the "$−$" sign means "subtract," and when it means "negative number."

Work Problem ❻ at the Side.

When working with signed numbers, it is helpful to write a subtraction problem as an addition problem. For example, you know that $6 − 4 = 2$. But you get the same result by *adding* 6 and the *opposite* of 4, that is, $6 + (−4)$.

$$6 − 4 = 2$$
$$6 + (−4) = 2$$

Same result

This suggests the following definition of subtraction.

Defining Subtraction

The difference of two numbers, a and b, is

$$a − b = a + (−b).$$

In other words, subtract two numbers by adding the first number and the opposite (additive inverse) of the second.

---

❺ Give the additive inverse of each number. Then find the sum of the number and its inverse.

(a) 12

(b) −9

(c) 3.5

(d) $−\dfrac{7}{10}$

(e) 0

❻ Write each example in words.

(a) $−7 − 2$

(b) $−10$

(c) $3 − (−5)$

(d) 4

(e) $−8 − (−6)$

(f) $2 − 9$

Answers
5. (a) $−12$; $12 + (−12) = 0$
   (b) 9; $−9 + 9 = 0$
   (c) $−3.5$; $3.5 + (−3.5) = 0$
   (d) $\dfrac{7}{10}$; $−\dfrac{7}{10} + \dfrac{7}{10} = 0$
   (e) 0; $0 + 0 = 0$
6. (a) negative seven minus positive two
   (b) negative ten
   (c) positive three minus negative five
   (d) positive four
   (e) negative eight minus negative six
   (f) positive two minus positive nine

❼ Subtract.

(a) $-6 - 5$

(b) $3 - (-10)$

(c) $-8 - (-2)$

(d) $4 - 9$

(e) $-7 - (-15)$

(f) $-\dfrac{2}{3} - \left(-\dfrac{5}{12}\right)$

Subtracting Signed Numbers

To subtract two signed numbers, add the opposite of the second number to the first number. Use these steps.

Step 1 Change the subtraction sign to addition.

Step 2 Change the sign of the second number to its opposite.

Step 3 Proceed as in addition.

NOTE

The pattern in a subtraction problem is:

$$\begin{matrix} \text{1st} \\ \text{number} \end{matrix} - \begin{matrix} \text{2nd} \\ \text{number} \end{matrix} = \begin{matrix} \text{1st} \\ \text{number} \end{matrix} + \begin{matrix} \text{opposite of} \\ \text{2nd number.} \end{matrix}$$

Example 5 Subtracting Signed Numbers

Subtract.

(a) $8 - 11$

The first number, 8, stays the same. Change the subtraction sign to addition. Change the sign of the second number to its opposite.

Positive 8 stays the same.

$$\begin{array}{ccc} 8 & - & 11 \\ \downarrow & & \downarrow \\ 8 & + & (-11) \end{array}$$

Positive 11 is changed to its opposite, -11.

Subtraction is changed to addition.

Now add.

$$8 + (-11) = -3$$
$$\text{So} \quad 8 - 11 = -3 \quad \text{also.}$$

(b) $\begin{array}{ccc} -9 & - & 15 \\ \downarrow & & \downarrow \\ -9 & + & (-15) = -24 \end{array}$

Subtraction is changed to addition.
Positive 15 is changed to its opposite, (-15).

(c) $\begin{array}{ccc} -5 & - & (-7) \\ \downarrow & & \downarrow \\ -5 & + & (+7) = 2 \end{array}$

Subtraction is changed to addition.
Negative 7 is changed to its opposite, $(+7)$.

(d) $\begin{array}{ccc} 7.6 & - & (-8.3) \\ \downarrow & & \downarrow \\ 7.6 & + & (+8.3) = 15.9 \end{array}$

Subtraction is changed to addition.
Negative 8.3 is changed to its opposite, $(+8.3)$.

(e) $\begin{array}{ccc} \dfrac{5}{8} & - & \dfrac{1}{2} \\ \downarrow & & \downarrow \end{array}$

Subtraction is changed to addition.
Positive $\frac{1}{2}$ is changed to its opposite, $\left(-\frac{1}{2}\right)$.

$$\dfrac{5}{8} + \left(-\dfrac{1}{2}\right) = \dfrac{5}{8} + \left(-\dfrac{4}{8}\right) = \dfrac{1}{8}$$

Work Problem ❼ at the Side.

5   **Add or subtract a series of signed numbers.** If a problem involves both addition and subtraction, use the order of operations, from **Section 1.8.**

Example 6   **Combining Addition and Subtraction of Signed Numbers**

According to the last step in the order of operations, perform addition and subtraction from left to right.

(a) $\underbrace{-6 + (-11)}_{-17} - \quad 5$

$-17 \quad - \quad 5$ Change subtraction to addition;
$\downarrow \quad\quad \downarrow$ change positive 5 to its opposite, (-5).

$\underbrace{-17 \quad + \quad (-5)}_{-22}$

(b) $4 - (-3) + (-9)$ Change subtraction to addition;
$\downarrow\ \ \downarrow$ change -3 to its opposite, $(+3)$.

$\underbrace{4 + (+3)}_{7} + (-9)$

$\underbrace{7 \quad + (-9)}_{-2}$

=========== **Work Problem 8 at the Side.**

Example 7   **Using the Order of Operations to Combine More than Two Numbers**

Find each sum.

(a) $\underbrace{-7 + 12}_{5} + (-3)$

$\underbrace{5 \quad + (-3)}_{2}$

(b) $\underbrace{14 + (-9)}_{5} - (-8) + 10$

$5 \quad - (-8) + 10$ Change subtraction to addition;
$\downarrow\ \ \downarrow$ change -8 to its opposite, $(+8)$.

$\underbrace{5 \quad + (+8)}_{13} + 10$

$\underbrace{13 \quad + 10}_{23}$

(c) $\begin{array}{r} -6.3 \\ -14.9 \\ 8.5 \\ -7.4 \\ \underline{5.2} \end{array}$

Start at the top.

$\begin{array}{r} \left.\begin{array}{r}-6.3 \\ -14.9\end{array}\right\} \\ 8.5 \\ -7.4 \\ \underline{5.2} \end{array}$ → $\begin{array}{r} -21.2 \\ \left.8.5\right\} \\ -7.4 \\ \underline{5.2} \end{array}$ → $\begin{array}{r} -12.7 \\ \left.-7.4\right\} \\ 5.2 \end{array}$ → $\begin{array}{r} -20.1 \\ \underline{5.2} \\ -14.9 \end{array}$

=========== **Work Problem 9 at the Side.**

---

8 Perform the addition and subtraction from left to right.

(a) $6 - 7 + (-3)$

(b) $-2 + (-3) - (-5)$

(c) $-3 - (-9) - (-5)$

(d) $8 - (-2) + (-6)$

9 Add.

(a) $-1 - 2 + 3 - 4$

(b) $7 - 6 - 5 + (-4)$

(c) $-6 + (-15) - (-19)$
$+ (-25)$

(d) $\begin{array}{r} -19.2 \\ -6.7 \\ 15.8 \\ 17.1 \\ \underline{-5.4} \end{array}$

Real-Data Applications

Auto Aide, Part 2

A computer failure results in the loss of data for one day at the **Auto Aide Service** in Houston, Texas (see Auto Aide, Part 1). The owner, the retired mathematics teacher, knows that the aides used integer arithmetic sentences to track their actions along their service routes. A schematic map of the I-10 corridor is shown below. Each tick mark represents one mile. The Home office is located at 0. Locations to the west are represented by negative integers, and locations to the east are represented by positive integers.

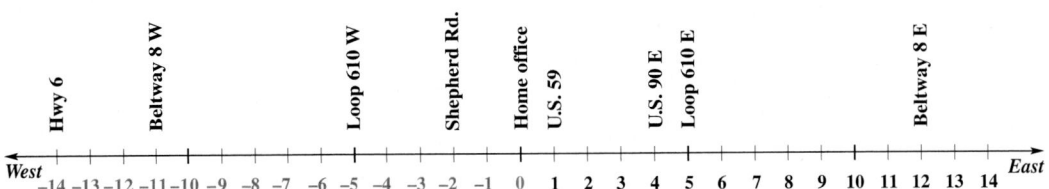

The owner asks each aide to recreate the route instructions for that day based on their written records. Each aide must translate the integer sentence into Route Instructions, by location. Also, the owner asks the aide to report the displacement (location relative to home office at end of the route) and the total distance traveled.

1. Aide Anne is given as an example. Complete the table for the remaining aides.

Aide	Integer Sentence	Route Instructions (start from the home office)	Displacement (end location)	Total Distance
Anne	$12 + (-11) + (-6) + 3$	Beltway 8 E; to U.S. 59; to Loop 610 W; to Shepherd Rd.	-2	32 miles
Bill	$-5 + 6 + 4 + (-16)$			
Carlos	$1 + 11 + (-8) + 1 + (-4)$			
Dylan	$-11 + (-3) + 15 + (-3) + 2$			
Ellen	$4 + 1 + (-7) + (-9) + 6$			

2. Suppose x is the aide's displacement from the home office in miles. List the aides whose journeys met the conditions of the following absolute value statements.

(a) $|x| < 5$

(b) $|x| = 1$

(c) $|x| \leq 5$

(d) $|x| > 3$

9.2 EXERCISES

| FOR EXTRA HELP |  Student's Solutions Manual | MyMathLab.com | 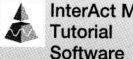 InterAct Math Tutorial Software |  AW Math Tutor Center |  www.mathxl.com | Digital Video Tutor CD 6 Videotape 16 |

Add by using the number line. See Example 1.

1. $-2 + 5$

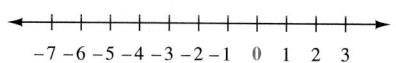

3. $-5 + (-2)$

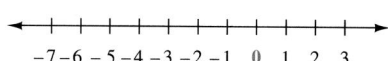

5. $3 + (-4)$

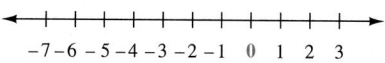

2. $-3 + 4$

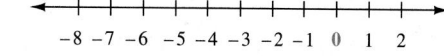

4. $-2 + (-2)$

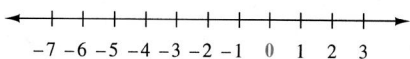

6. $5 + (-1)$

Add. See Examples 2 and 3.

7. $-8 + 5$ **8.** $-3 + 2$ **9.** $-1 + 8$ **10.** $-4 + 10$

11. $-2 + (-5)$ **12.** $-7 + (-3)$ **13.** $6 + (-5)$ **14.** $11 + (-3)$

15. $4 + (-12)$ **16.** $9 + (-10)$ **17.** $-10 + (-10)$ **18.** $-5 + (-20)$

Write and solve an addition problem for each situation.

19. The football team gained 13 yd on the first play and lost 17 yd on the second play.

20. At penguin breeding grounds on Antarctic islands, winter temperatures routinely drop to -15 °C; but at the interior of the continent, temperatures may easily drop another 60° below that.

21. Nicole's checking account was overdrawn by $52.50. She deposited $50 in the account.

22. $48.40 was stolen from Jay's car. He got $30 of it back.

23. Use the score sheet to find each player's point total after three rounds in a card game.

	Jeff	Terry
Round 1	Lost 20 pts	Won 42 pts
Round 2	Won 75 pts	Lost 15 pts
Round 3	Lost 55 pts	Won 20 pts

24. Use the information in the table on flood water depths to find the new flood level for each river.

	Red River	Mississippi
Monday	Rose 8 ft	Rose 4 ft
Tuesday	Fell 3 ft	Rose 7 ft
Wednesday	Fell 5 ft	Fell 13 ft

Add.

25. $7.8 + (-14.6)$ **26.** $4.9 + (-8.1)$ **27.** $-\dfrac{1}{2} + \dfrac{3}{4}$ **28.** $-\dfrac{2}{3} + \dfrac{5}{6}$

29. $-\dfrac{7}{10} + \dfrac{2}{5}$ **30.** $-\dfrac{3}{4} + \dfrac{3}{8}$ **31.** $-\dfrac{7}{3} + \left(-\dfrac{5}{9}\right)$ **32.** $-\dfrac{8}{5} + \left(-\dfrac{3}{10}\right)$

Give the additive inverse of each number. See Example 4.

33. 3 **34.** 4 **35.** -9 **36.** -14

37. $\dfrac{1}{2}$ **38.** $\dfrac{7}{8}$ **39.** -6.2 **40.** -0.5

Subtract by changing subtraction to addition. See Example 5.

41. $19 - 5$ **42.** $24 - 11$ **43.** $10 - 12$ **44.** $1 - 8$

45. $7 - 19$ **46.** $2 - 17$ **47.** $-15 - 10$ **48.** $-10 - 4$

49. $-9 - 14$ **50.** $-3 - 11$ **51.** $-3 - (-8)$ **52.** $-1 - (-4)$

53. $6 - (-14)$ **54.** $8 - (-1)$ **55.** $1 - (-10)$ **56.** $6 - (-1)$

57. $-30 - 30$ **58.** $-25 - 25$ **59.** $-16 - (-16)$ **60.** $-20 - (-20)$

61. $-\dfrac{7}{10} - \dfrac{4}{5}$ **62.** $-\dfrac{8}{15} - \dfrac{3}{10}$ **63.** $\dfrac{1}{2} - \dfrac{9}{10}$

64. $\dfrac{2}{3} - \dfrac{11}{12}$ **65.** $-8.3 - (-9)$ **66.** $-2 - (-3.9)$

RELATING CONCEPTS (Exercises 67–70)   FOR INDIVIDUAL OR GROUP WORK

Use your knowledge of addition and subtraction of signed numbers to
work Exercises 67–70 in order.

67. Work each pair of examples.

(a) $-5 + 3 \quad = $ _____ $3 + (-5) = $ _____

(b) $-2 + (-6) = $ _____ $-6 + (-2) = $ _____

(c) $17 + (-7) = $ _____ $-7 + 17 \quad = $ _____

Explain why the answers to each pair are the same.

68. Work each pair of examples.

(a) $-3 - 5 \quad = $ _____ $5 - (-3) = $ _____

(b) $-4 - (-6) = $ _____ $-6 - (-4) = $ _____

(c) $3 - 10 \quad = $ _____ $10 - 3 \quad = $ _____

Explain what happens when you try to apply the commutative property to subtraction.

69. Look back at your answers in Exercise 68.

(a) Describe how the two answers for each pair are similar, and how they are different.

(b) Write a rule that explains what happens when you switch the order of the numbers in a subtraction problem.

70. Work each set of exercises to see how 0 functions in addition and subtraction.

(a) Describe the pattern in these addition answers.

$$-18 + 0 \quad = \text{\_\_\_\_\_} \qquad 20 + 0 = \text{\_\_\_\_\_}$$
$$0 + (-5) = \text{\_\_\_\_\_} \qquad 0 + 4 = \text{\_\_\_\_\_}$$

(b) Describe the pattern in these subtraction answers.

$$2 - 0 = \text{\_\_\_\_\_} \qquad 0 - 10 \quad = \text{\_\_\_\_\_}$$
$$-3 - 0 = \text{\_\_\_\_\_} \qquad 0 - (-7) = \text{\_\_\_\_\_}$$

71. Explain and correct the mistakes in these subtraction problems.

(a) $-6 - 6$
     $\downarrow \ \downarrow$
     $-6 + 6 = 0$

(b) $-9 - \quad 5$
     $\ \ \downarrow \quad \downarrow$
     $9 + (-5) = 4$

72. Explain the purpose of the "$-$" sign in each of these examples.

(a) $6 - 9$ (b) (-9) (c) $-(-2)$

Follow the order of operations to work each problem. See Examples 6 and 7.

73. $-2 + (-11) - (-3)$

74. $-5 - (-2) + (-6)$

75. $4 - (-13) + (-5)$

76. $6 - (-1) + (-10)$

77. $-12 - (-3) - (-2)$

78. $-1 - (-7) - (-4)$

79. $4 - (-4) - 3$

80. $5 - (-2) - 8$

81. $\dfrac{1}{2} - \dfrac{2}{3} + \left(-\dfrac{5}{6}\right)$

82. $\dfrac{2}{5} - \dfrac{7}{10} + \left(-\dfrac{3}{2}\right)$

83. $-5.7 - (-9.4) - 8.1$

84. $-6.5 - (-11.2) - 1.4$

85. $-2 + (-11) + |-2|$

86. $|-7 + 2| + (-2) + 4$

(*Hint:* Work inside the absolute value bars first.)

87. $-3 - (-2 + 4) + (-5)$

(*Hint:* Work inside parentheses first.)

88. $5 - 8 - (6 - 7) + 1$

89. $2\dfrac{1}{2} + 3\dfrac{1}{4} - \left(-1\dfrac{3}{8}\right) - 2\dfrac{3}{8}$

90. $\dfrac{5}{8} - \left(-\dfrac{1}{2} - \dfrac{3}{4}\right)$

Find the balance in each checking account.

91. LaVerle had $37 in her checking account. She wrote a $689 check for tuition, deposited a $908 paycheck, and got $60 in cash at an ATM machine.

92. Yvonne had $478 in her checking account. She deposited a $212 tax refund and wrote an $89 check for electricity and a $605 check for rent.

93. Rod's account was overdrawn by $89.62. He wrote checks for $110.70 and $99.68 before depositing three $100 bills into his account.

94. Dwayne's checking account was overdrawn by $23.77, so the bank charged a $16 fee. He deposited his $583.29 paycheck before withdrawing $50 in cash.

9.3 MULTIPLYING AND DIVIDING SIGNED NUMBERS

In mathematics the rules or patterns must be consistent. We can use this idea to see how to multiply two numbers with different signs. Look for a pattern in this list of products.

$$4 \cdot 2 = 8$$
$$3 \cdot 2 = 6$$
$$2 \cdot 2 = 4$$
$$1 \cdot 2 = 2$$
$$0 \cdot 2 = 0$$
$$-1 \cdot 2 = ?$$

Blue numbers decrease by 1.   Red numbers decrease by 2.

As the numbers in blue decrease by 1, the numbers in red decrease by 2. To continue the red pattern, replace **?** with a number 2 *less than* 0, which is **−2**. Therefore,

$$-1 \cdot 2 = -2$$

1 **Multiply or divide two numbers with opposite signs.** The pattern above suggests a rule for multiplying two numbers with different signs.

Multiplying Two Numbers with Different Signs

The product of two numbers with *different* signs is *negative.*

Example 1 Multiplying Numbers with Different Signs

Multiply.

(a) $-8 \cdot 4 = -32$ Factors have *different* signs, so the product is *negative.*
Positive, Negative

(b) $6 \cdot (-3) = -18$ Factors have *different* signs, so the product is *negative.*
Positive, Negative

(c) $-5 \cdot (11) = -55$

(d) $12 \cdot (-7) = -84$

═══ **Work Problem 1 at the Side.**

For two numbers with the same sign, look at this pattern.

$$4 \cdot (-2) = -8$$
$$3 \cdot (-2) = -6$$
$$2 \cdot (-2) = -4$$
$$1 \cdot (-2) = -2$$
$$0 \cdot (-2) = 0$$
$$-1 \cdot (-2) = ?$$

Blue numbers decrease by 1.   Red numbers increase by 2.

This time, as the numbers in blue decrease by 1, the products *increase* by 2. To continue the red pattern, replace **?** with a number 2 *greater than* 0, which is positive 2. Therefore,

$$-1 \cdot (-2) = 2$$

2 **Multiply or divide two numbers with the same sign.** In the pattern above, a negative number times a negative number gave a positive result.

OBJECTIVES

1 Multiply or divide two numbers with opposite signs.

2 Multiply or divide two numbers with the same sign.

---

1 Multiply.

(a) $5 \cdot (-4)$

(b) $-9 \cdot (15)$

(c) $12 \cdot (-1)$

(d) $-6 \cdot (6)$

(e) $\left(-\dfrac{7}{8}\right)\left(\dfrac{4}{3}\right)$

❷ Multiply.

(a) $(-5) \cdot (-5)$

(b) $(-14)(-1)$

(c) $-7 \cdot (-8)$

(d) $3 \cdot 12$

(e) $\left(-\dfrac{2}{3}\right)\left(-\dfrac{6}{5}\right)$

❸ Divide.

(a) $\dfrac{-20}{4}$

(b) $\dfrac{-50}{-5}$

(c) $\dfrac{44}{2}$

(d) $\dfrac{6}{-6}$

(e) $\dfrac{-15}{-1}$

(f) $\dfrac{-\dfrac{3}{5}}{\dfrac{9}{10}}$

(g) $\dfrac{-35}{0}$

2. (a) 25   (b) 14   (c) 56   (d) 36   (e) $\dfrac{4}{5}$

3. (a) -5 (b) 10   (c) 22   (d) -1 (e) 15

(f) $-\dfrac{2}{3}$ (g) undefined

Multiplying Two Numbers with the Same Sign

The product of two numbers with the *same* sign is *positive.*

Example 2 Multiplying Two Numbers with the Same Sign

Multiply.

(a) $(-9)(-2)$ The factors have the same sign (both are negative).

$(-9)(-2) = 18$ ← The product is positive.

(b) $-7 \cdot (-4) = 28$ (c) $(-6)(-2) = 12$

(d) $(-10)(-5) = 50$ (e) $7 \cdot 5 = 35$

Work Problem ❷ at the Side.

You can use the same rules for dividing signed numbers as you use for multiplying signed numbers.

Dividing Signed Numbers

When two nonzero numbers with *different* signs are divided, the result is *negative.*

When two nonzero numbers with the *same* sign are divided, the result is *positive.*

Division involving 0 works the same as it did for whole numbers (see **Section 1.5**). Division by 0 cannot be done; we say it is *undefined*. But 0 divided by any other number is 0.

Example 3 Dividing Signed Numbers

Divide.

(a) $\dfrac{-15}{5}$ ← Numbers have *different* signs, so the quotient is negative.

$\dfrac{-15}{5} = -3$

(b) $\dfrac{-8}{-4}$ ← Numbers have *same* sign (both negative), so the quotient is positive.

$\dfrac{-8}{-4} = 2$

(c) $\dfrac{-75}{-25} = 3$

(d) $\dfrac{-6}{0}$ is undefined. Division by 0 cannot be done.

(e) $\dfrac{0}{-5} = 0$

(f) $\dfrac{90}{-9} = -10$

(g) $\dfrac{-\dfrac{2}{3}}{-\dfrac{5}{9}} = -\dfrac{2}{3} \cdot \left(-\dfrac{9}{5}\right)$ Use the reciprocal of $-\dfrac{5}{9}$, which is $-\dfrac{9}{5}$.

$= -\dfrac{2}{\underset{1}{\cancel{3}}} \cdot \left(-\dfrac{\overset{3}{\cancel{9}}}{5}\right)$ Divide out common factors. Then multiply.

$= \dfrac{6}{5}$ Both numbers were negative, so the quotient is positive.

Work Problem ❸ at the Side.

9.3 EXERCISES

FOR EXTRA HELP

 Student's Solutions Manual

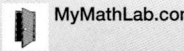 MyMathLab.com

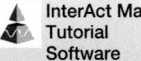 InterAct Math Tutorial Software

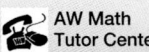 AW Math Tutor Center

Math XL   www.mathxl.com

 Digital Video Tutor CD 6 Videotape 16

Multiply. See Examples 1 and 2.

1. $-5 \cdot 7$

2. $-10 \cdot 2$

3. $(-5)(9)$

4. $(-9)(4)$

5. $3 \cdot (-6)$

6. $8 \cdot (-6)$

7. $10 \cdot (-5)$

8. $5 \cdot (-11)$

9. $(-1)(40)$

10. $(75)(-1)$

11. $-8 \cdot (-4)$

12. $-3 \cdot (-9)$

13. $11 \cdot 7$

14. $4 \cdot 25$

15. $-19 \cdot (-7)$

16. $-21 \cdot (-3)$

17. $-13 \cdot (-1)$

18. $-1 \cdot (-31)$

19. $(0)(-25)$

20. $(-50)(0)$

21. $-\dfrac{1}{2} \cdot (-8)$

22. $\dfrac{1}{3} \cdot (-15)$

23. $-10 \cdot \left(\dfrac{2}{5}\right)$

24. $-25 \cdot \left(-\dfrac{7}{10}\right)$

25. $\left(\dfrac{3}{5}\right)\left(-\dfrac{1}{6}\right)$

26. $\left(-\dfrac{7}{9}\right)\left(-\dfrac{3}{4}\right)$

27. $-\dfrac{7}{5} \cdot \left(-\dfrac{10}{3}\right)$

28. $-\dfrac{9}{10} \cdot \dfrac{5}{4}$

29. $-\dfrac{7}{15} \cdot \dfrac{25}{14}$

30. $-\dfrac{5}{9} \cdot \dfrac{18}{25}$

31. $-\dfrac{5}{2} \cdot \left(-\dfrac{7}{10}\right)$

32. $-\dfrac{8}{5} \cdot \left(-\dfrac{15}{16}\right)$

33. $9 \cdot (-4.7)$

34. $15 \cdot (-6.3)$

35. $(-0.5)(-12)$

36. $(-3.15)(-5)$

37. $-6.2 \cdot (5.1)$

38. $-4.3 \cdot (9.7)$

39. $-1.25 \cdot (-3.6)$

40. $6.33 \cdot 0.2$

41. $(-8.23)(-1)$

42. $(-1)(-0.69)$

43. $0 \cdot (-58.6)$

44. $-91.3 \cdot 0$

Divide. See Example 3.

45. $\dfrac{-14}{7}$

46. $\dfrac{-8}{2}$

47. $\dfrac{30}{-6}$

48. $\dfrac{21}{-7}$

49. $\dfrac{-28}{0}$

50. $\dfrac{-40}{0}$

51. $\dfrac{14}{-1}$

52. $\dfrac{25}{-1}$

53. $\dfrac{-20}{-2}$

54. $\dfrac{-80}{-4}$

55. $\dfrac{-48}{-12}$

56. $\dfrac{-30}{-15}$

57. $\dfrac{-18}{18}$

58. $\dfrac{50}{-50}$

59. $\dfrac{-573}{-3}$

60. $\dfrac{-580}{-5}$

61. $\dfrac{0}{-9}$

62. $\dfrac{0}{-4}$

63. $\dfrac{-30}{-30}$

64. $\dfrac{-25}{-25}$

65. $\dfrac{-\dfrac{5}{7}}{-\dfrac{15}{14}}$

66. $\dfrac{-\dfrac{3}{4}}{-\dfrac{9}{16}}$

67. $-\dfrac{2}{3} \div (-2)$

68. $-\dfrac{3}{4} \div (-9)$

69. $5 \div \left(-\dfrac{5}{8}\right)$

70. $7 \div \left(-\dfrac{14}{15}\right)$

71. $-\dfrac{7}{5} \div \dfrac{3}{10}$

72. $-\dfrac{4}{9} \div \dfrac{8}{3}$

73. $\dfrac{-18.92}{-4}$

74. $\dfrac{-22.75}{-7}$

75. $\dfrac{-7.05}{1.5}$

76. $\dfrac{-17.02}{7.4}$

77. $\dfrac{45.58}{-8.6}$

78. $\dfrac{6.27}{-0.3}$

Following the order of operations, work from left to right in each exercise.

79. $(-4) \cdot (-6) \cdot \dfrac{1}{2}$

80. $(-9) \cdot (-3) \cdot \dfrac{2}{3}$

81. $(-0.6)(-0.2)(-3)$

82. $(-4)(-1.2)(-0.7)$

83. $\left(-\dfrac{1}{2}\right)\left(\dfrac{2}{5}\right)\left(\dfrac{7}{8}\right)$

84. $\left(\dfrac{3}{4}\right)\left(-\dfrac{5}{6}\right)\left(\dfrac{2}{3}\right)$

85. $-36 \div (-2) \div (-3) \div (-3) \div (-1)$

86. $-48 \div (-8) \cdot (-4) \div (-4) \div (-3)$

87. $|-8| \div (-4) \cdot |-5|$

88. $-6 \cdot |-3| \div |9| \cdot (-2)$

Solve each application problem. Be sure to indicate whether the answer is a positive or negative number.

89. A new computer-software store had losses of $9950 during each month of its first year. What was the total loss for the year?

90. A college's enrollment dropped by 3245 students over the last 11 years. What was the average drop in enrollment each year?

91. The greatest ocean depth is 36,198 ft below sea level. If an unmanned research sub dives to that depth in 18 equal steps, how far does it dive in each step?

92. Pat ate a dozen crackers as a snack. Each cracker had 17 calories. (*Source:* Nabisco, Inc.) How many calories did Pat eat?

93. Tuition at the state university is $182 per credit for undergraduates. How much tuition will Wei Chen pay for 13 credits?

94. A long-distance phone company estimates that it is losing 95 customers each week. How many customers will it lose in a year?

95. There is a 3-degree drop in temperature for every thousand feet that an airplane climbs into the sky. If the temperature on the ground is 50 degrees, what will be the temperature when the plane reaches an altitude of 24,000 ft? (*Source:* Lands' End.)

96. An unmanned submarine descends to 150 ft below the surface of the ocean. Then it continues to go deeper, taking a water sample every 25 ft. What is its depth when it takes the 15th sample?

RELATING CONCEPTS (Exercises 97–98)   **FOR INDIVIDUAL OR GROUP WORK**

Use your knowledge of multiplying and dividing signed numbers and look for patterns as you **work Exercises 97 and 98 in order.**

97. Write three numerical examples for each situation.

   (a) A positive number multiplied by -1

   (b) A negative number multiplied by -1

Now write a rule that explains what happens when you multiply a signed number by -1.

98. Write three numerical examples for each situation.

   (a) A negative number divided by -1

   (b) A positive number divided by -1

   (c) A negative number divided by itself

Now write a rule that explains what happens when you divide a signed number by -1. Write another rule for a negative number divided by itself.

9.4 ORDER OF OPERATIONS

In the last two sections you worked examples that mixed either addition and subtraction or multiplication and division. In those situations you worked from left to right. Here are two more examples as a review.

Work additions and subtractions from left to right.	Work multiplications and divisions from left to right.
$-8 - (-6) + (-11)$	$(-15) \div (-3) \cdot 6$
$\underbrace{\quad}\; -2 \quad + (-11)$	$\underbrace{\quad}\; 5 \quad \cdot 6$
$\underbrace{\qquad\qquad}\; -13$	$\underbrace{\qquad\qquad}\; 30$

Work Problem ❶ at the Side.

1 ▭ **Use the order of operations.** Before working examples that mix division with addition or include parentheses, let's review the order of operations from **Section 1.8.**

Order of Operations

1. Do all operations inside *parentheses* or *other grouping symbols.*

2. Simplify any expressions with *exponents* and find any *square roots.*

3. *Multiply* or *divide,* proceeding from left to right.

4. *Add* or *subtract,* proceeding from left to right.

Example 1   Using the Order of Operations

Use the order of operations to simplify this expression.

$4 - 10 \div 2 + 7$	Check for parentheses: none.
	Check for exponents and square roots: none.
	Move from left to right, checking for multiplying and dividing.
$4 - \underline{10 \div 2} + 7$	Yes, here is dividing. Use the number on each side of the $\div$ sign.
	$10 \div 2$ is 5. Bring down the other numbers and signs you haven't used.
$4 - \quad 5 \quad + 7$	Move from left to right, checking for adding and subtracting.
	Yes, here is subtracting.
$4 - \quad 5 \quad + 7$	Change subtraction to addition; change 5 to its opposite, (-5).
$\underbrace{4 + (-5)} + 7$	Add $4 + (-5)$ to get -1.
$\underbrace{-1 \qquad + 7}$	Add $-1 + 7$ to get 6.
6	

Work Problem ❷ at the Side.

OBJECTIVES

1 ▭ Use the order of operations.

2 ▭ Use the order of operations with exponents.

3 ▭ Use the order of operations with fraction bars.

❶ Simplify.

(a) $-9 + (-15) + (-3)$

(b) $-8 - (-2) + (-6)$

(c) $-2 - (-7) - (-4)$

(d) $3 \cdot (-4) \div (-6)$

(e) $-18 \div 9 \cdot (-4)$

❷ Use the order of operations to simplify each expression.

(a) $10 + 8 \div 2$

(b) $4 - 6 \cdot (-2)$

(c) $-3 + (-5) \cdot 2 - 1$

(d) $-6 \div 2 + 3 \cdot (-2)$

(e) $7 - 6 \cdot 2 \div (-3)$

ANSWERS
1. (a) -27 **(b)** -12 **(c)** 9 **(d)** 2 **(e)** 8
2. (a) 14 **(b)** 16 **(c)** -14 **(d)** -9
(e) 11

❸ Simplify.

(a) $2 + 40 \div (-5 + 3)$

(b) $-5 \cdot 5 - (15 + 5)$

(c) $(-24 \div 2) + (15 - 3)$

(d) $-3 \cdot (2 - 8) - 5 \cdot (4 - 3)$

(e) $3 \cdot 3 - (10 \cdot 3) \div 5$

(f) $6 - (2 + 7) \div (-4 + 1)$

Example 2 **Parentheses and Order of Operations**

Use the order of operations to simplify each expression.

(a) $-8 \cdot \underbrace{(7 - 5)}\ - 9$ Work inside parentheses first.
$\quad \downarrow\downarrow \qquad\quad \downarrow\downarrow$ Bring down the other numbers and signs
$\quad -8 \cdot \quad 2 \quad\ - 9$ you haven't used yet.
    Check for exponents and square roots: none.

$\quad \underbrace{-8 \cdot \quad 2} \quad\ - 9$ Move from left to right, doing any multiplying and dividing.

$\qquad \underbrace{-16 \qquad\ - 9}$ Move from left to right, doing any adding and subtracting. Change subtraction to addition.
$\qquad -16 \qquad + (-9)$ Add $-16 + (-9)$ to get -25.
$\qquad\quad \underbrace{\qquad\qquad\qquad}$
$\qquad\qquad -25$

(b) $3 + 2 \cdot \underbrace{(6 - 8)} \cdot (15 \div 3)$ Work inside first set of parentheses; change $6 - 8$ to $6 + (-8)$ to get (-2).
$\quad 3 + 2 \cdot \ (-2)\ \cdot \underbrace{(15 \div 3)}$ Work inside second set of parentheses; $15 \div 3$ is 5.
$\quad 3 + \underbrace{2 \cdot \ (-2)}\ \cdot \qquad 5$ Multiply and divide from left to right. First multiply $2 \cdot (-2)$ to get -4.
$\quad 3 + \underbrace{\ (-4)\ \qquad \cdot \quad 5}$ Then multiply $-4 \cdot 5$ to get -20.
$\quad 3 + \underbrace{\qquad\quad (-20)}$ Add last. $3 + (-20)$ gives -17.
$\qquad\quad -17$

Work Problem ❸ at the Side.

2 ▮▮▮▮ **Use the order of operations with exponents.**   Recall from **Section 1.8** that 2^3 means 2 is used as a factor 3 times.

$$2^3 = 2 \cdot 2 \cdot 2 = 8$$

The small raised 3 in 2^3 is called an *exponent*. Exponents are also used with signed numbers. For example,

$$(-3)^2 = (-3) \cdot (-3) = 9$$

$(-4)^3 = \underbrace{(-4) \cdot (-4)} \cdot (-4)$ Be careful! Multiply two numbers at a time.
$\quad = \qquad \underbrace{16 \qquad \cdot (-4)}$ Watch the signs.
$\quad = \qquad\qquad -64 \ \longleftarrow$ Negative product

$\left(-\dfrac{1}{2}\right)^4 = \underbrace{\left(-\dfrac{1}{2}\right) \cdot \left(-\dfrac{1}{2}\right)} \cdot \left(-\dfrac{1}{2}\right) \cdot \left(-\dfrac{1}{2}\right)$

$\quad = \qquad\quad \underbrace{\dfrac{1}{4} \qquad\quad \cdot \left(-\dfrac{1}{2}\right)} \cdot \left(-\dfrac{1}{2}\right)$

$\quad = \qquad\qquad\quad \underbrace{-\dfrac{1}{8} \qquad\quad \cdot \left(-\dfrac{1}{2}\right)}$

$\quad = \qquad\qquad\qquad\quad \dfrac{1}{16}$

Be very careful with exponents and signed numbers. For example,

$$(-3)^2 = (-3) \cdot (-3) = 9. \ \longleftarrow \text{Positive 9}$$

But the expression -3^2, with *no parentheses*, is different.

$$-3^2 = -(3 \cdot 3) = -9 \ \longleftarrow \text{Negative 9}$$

CAUTION

$(-3)^2$ is *not* the same as -3^2.

$$(-3)^2 = (-3) \cdot (-3) = 9 \quad but \quad -3^2 = -(3 \cdot 3) = -9$$

You will need this information as you take more algebra classes.

Example 3   **Exponents and Order of Operations**

Simplify.

(a) $4^2 - (-3)^2$ 　 There are parentheses around (-3) but no work
　　　　　　　　　 can be done inside these parentheses.
　　　　　　　　　 Work with the exponents: $4^2 = 4 \cdot 4 = 16$
　　　　　　　　　 and $(-3)^2 = (-3) \cdot (-3) = 9$
$16 - 9$ 　　　　 Subtract: $16 - 9 = 7$

7

(b) $(-5)^2 - (4 - 6)^2 \cdot (-3)$ 　 Work inside parentheses first.

$(-5)^2 - (-2)^2 \cdot (-3)$ 　 Use the exponents next.

$25 - 4 \cdot (-3)$ 　 Multiply.

$25 - (-12)$ 　 Change subtraction to addition.

$25 + (+12)$ 　 Add.

37

(c) $\left(\dfrac{2}{3} - \dfrac{1}{6}\right)^2 \div \left(-\dfrac{3}{8}\right)$ 　 Inside parentheses: $\frac{2}{3} - \frac{1}{6} = \frac{4}{6} - \frac{1}{6} = \frac{3}{6} = \frac{1}{2}$

$\left(\dfrac{1}{2}\right)^2 \div \left(-\dfrac{3}{8}\right)$ 　 Use the exponent: $(\frac{1}{2})^2 = \frac{1}{2} \cdot \frac{1}{2} = \frac{1}{4}$

$\dfrac{1}{4} \div \left(-\dfrac{3}{8}\right)$ 　 Divide by using the reciprocal of the divisor:
　　　　　　　　　 $-\frac{3}{8}$ becomes $-\frac{8}{3}$

$\dfrac{1}{4} \cdot \left(-\dfrac{8}{3}\right)$ 　 Divide out common factors, then
　　　　　　　　　 multiply: $\frac{1}{\overset{1}{4}} \cdot -\frac{\overset{2}{8}}{3} = -\frac{2}{3}$

$-\dfrac{2}{3}$

━━━━━━━━━ **Work Problem ④ at the Side.**

NOTE

Parentheses can be used in several different ways.

To indicate multiplication	$(4)(-3) = -12$
To separate a negative number from a minus sign	$8 - (-2) = 10$
To indicate which operation to do first	$35 + (6 - 2)$
	$35 + 4$
	39

④ Simplify.

(a) $2^3 - 3^2$

(b) $4^2 - 3^2 \cdot (5 - 2)$

(c) $-18 \div (-3) \cdot 2^3$

(d) $(-3)^3 + (3 - 8)^2$

(e) $\dfrac{3}{8} + \left(-\dfrac{1}{2}\right)^2 \div \dfrac{1}{4}$

ANSWERS
4. (a) -1 **(b)** -11 **(c)** 48 **(d)** -2
　(e) $\dfrac{11}{8}$

❺ Simplify.

(a) $\dfrac{-3 \cdot 2^3}{-10 - 6 + 8}$

(b) $\dfrac{(-10)(-5)}{-6 \div 3 \cdot 5}$

(c) $\dfrac{6 + 18 \div (-2)}{(1 - 10) \div 3}$

(d) $\dfrac{6^2 - 3^2 \cdot 4}{5 + (3 - 7)^2}$

3 ▮▮▮▮ **Use the order of operations with fraction bars.** A fraction bar indicates division, as in $\frac{-6}{2}$, which means $-6 \div 2$. In an expression like

$$\frac{-5 + 3^2}{16 - 7 \cdot 2}$$

the fraction bar also acts as a grouping symbol, like parentheses. It tells us to do the work in the numerator, and then the work in the denominator. The last step is to divide the results.

$$\frac{-5 + 3^2}{16 - 7 \cdot 2} \longrightarrow \frac{-5 + 9}{16 - 14} \longrightarrow \frac{4}{2} \longrightarrow \text{Now divide.} \quad 4 \div 2 = 2$$

The final simplified result is 2.

Example 4 Fraction Bars and Order of Operations

Simplify.

$$\frac{-8 + (4 - 6) \cdot 5}{4 - 4^2 \div 8}$$

First do the work in the numerator.

$$-8 + \underbrace{(4 - 6)}\ \cdot 5 \qquad \text{Work inside parentheses.}$$
$$-8 + \underbrace{(-2)\ \cdot 5} \qquad \text{Multiply.}$$
$$\underbrace{-8 + \quad (-10)} \qquad \text{Add.}$$
$$\text{Numerator} \longrightarrow -18$$

Now do the work in the denominator.

$$4 - \underbrace{4^2} \div 8 \qquad \text{No parentheses; use exponent.}$$
$$4 - \underbrace{16 \div 8} \qquad \text{Divide.}$$
$$\underbrace{4 - \quad 2} \qquad \text{Subtract.}$$
$$\text{Denominator} \longrightarrow 2$$

The last step is the division.

$$\begin{array}{c} \text{Numerator} \longrightarrow \\ \text{Denominator} \longrightarrow \end{array} \frac{-18}{2} = -9$$

Work Problem ❺ at the Side.

9.4 EXERCISES

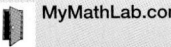
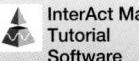
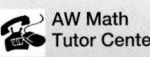

Simplify. See Examples 1–3.

1. $6 + 3 \cdot (-4)$

2. $10 - 30 \div 2$

3. $-1 + 15 + (-7) \cdot 2$

4. $9 + (-5) + 2 \cdot (-2)$

5. $6^2 + 4^2$

6. $3^2 + 8^2$

7. $10 - 7^2$

8. $5 - 5^2$

9. $(-2)^5 + 2$

10. $(-2)^4 - 7$

11. $4^2 + 3^2 + (-8)$

12. $5^2 + 2^2 + (-12)$

13. $2 - (-5) + 3^2$

14. $6 - (-9) + 2^3$

15. $(-4)^2 + (-3)^2 + 5$

16. $(-5)^2 + (-6)^2 + 12$

17. $3 + 5 \cdot (6 - 2)$

18. $4 + 3 \cdot (8 - 3)$

19. $-7 + 6 \cdot (8 - 14)$

20. $-3 + 5 \cdot (9 - 12)$

21. $-6 + (-5) \cdot (9 - 14)$

22. $-5 + (-3) \cdot (6 - 7)$

23. $(-5) \cdot (7 - 13) \div (-10)$

24. $(-4) \cdot (9 - 17) \div (-8)$

25. $9 \div (-3)^2 + (-1)$

26. $-48 \div (-4)^2 + 3$

27. $2 - (-5) \cdot (-3)^2$

28. $1 - (-10) \cdot (-2)^3$

29. $(-2) \cdot (-7) + 3 \cdot 9$

30. $4 \cdot (-2) + (-3) \cdot (-5)$

31. $30 \div (-5) - 36 \div (-9)$

32. $8 \div (-4) - 42 \div (-7)$

33. $2 \cdot 5 - 3 \cdot 4 + 5 \cdot 3$

34. $9 \cdot 3 - 6 \cdot 4 + 3 \cdot 7$

35. $4 \cdot 3^2 + 7 \cdot (3 + 9) - (-6)$

36. $5 \cdot 4^2 - 6 \cdot (1 + 4) - (-3)$

Simplify. See Example 4.

37. $\dfrac{-1 + 5^2 - (-3)}{-6 - 9 + 12}$

38. $\dfrac{-6 + 3^2 - (-7)}{7 - 9 - 3}$

39. $\dfrac{-2 \cdot 4^2 - 4 \cdot (6 - 2)}{-4 \cdot (8 - 13) \div (-5)}$

40. $\dfrac{3 \cdot 3^2 - 5 \cdot (9 - 2)}{8 \cdot (6 - 9) \div (-3)}$

41. $\dfrac{2^3 \cdot (-2 - 5) + 4 \cdot (-1)}{4 + 5 \cdot (-6 \cdot 2) + (5 \cdot 11)}$

42. $\dfrac{3^3 + (-1 - 2) \cdot 4 - 25}{-4 + 4 \cdot (3 \cdot 5) + (-6 \cdot 9)}$

Simplify each expression.

43. $(-4)^2 \cdot (7 - 9)^2 \div 2^3$

44. $(-5)^2 \cdot (9 - 17)^2 \div (-10)^2$

45. $(-0.3)^2 + (-0.5)^2 + 0.9$

46. $(0.2)^3 - (-0.4)^2 + 3.02$

47. $(-0.75) \cdot (3.6 - 5)^2$

48. $(-0.3) \cdot (4 - 6.8)^2$

49. $(0.5)^2 \cdot (-8) - (0.31)$

50. $(0.3)^3 \cdot (-5) - (-2.8)$

51. $\dfrac{2}{3} \div \left(-\dfrac{5}{6}\right) - \dfrac{1}{2}$

52. $\dfrac{5}{8} \div \left(-\dfrac{10}{3}\right) - \dfrac{3}{4}$

53. $\left(-\dfrac{1}{2}\right)^2 - \left(\dfrac{3}{4} - \dfrac{7}{4}\right)$

54. $\left(-\dfrac{2}{3}\right)^2 - \left(\dfrac{1}{6} - \dfrac{11}{6}\right)$

55. $\dfrac{3}{5} \cdot \left(-\dfrac{7}{6}\right) - \left(\dfrac{1}{6} - \dfrac{5}{3}\right)$

56. $\dfrac{2}{7} \cdot \left(-\dfrac{14}{5}\right) - \left(\dfrac{4}{3} - \dfrac{13}{9}\right)$

57. $5^2 \cdot (9 - 11) \cdot (-3) \cdot (-2)^3$

58. $4^2 \cdot (13 - 17) \cdot (-2) \cdot (-3)^2$

59. $1.6 \cdot (-0.8) \div (-0.32) \div 2^2$

60. $6.5 \cdot (-4.8) \div (-0.3) \div (-2)^3$

61. Simplify.

$(-2)^2 =$ $(-2)^6 =$

$(-2)^3 =$ $(-2)^7 =$

$(-2)^4 =$ $(-2)^8 =$

$(-2)^5 =$ $(-2)^9 =$

(a) Describe the pattern you see in the sign of the answers.

(b) What would be the sign of $(-2)^{10}$? of $(-2)^{15}$? of $(-2)^{24}$?

62. Explain the difference between -5^2 and $(-5)^2$.

Simplify.

63. $\dfrac{-9 + 18 \div (-3) \cdot (-6)}{5 - 4 \cdot 12 \div 3 \cdot 2}$

64. $\dfrac{-20 - 15 \cdot (-4) - (-40)}{4 + 27 \div 3 \cdot (-2) - 6}$

65. $-7 \cdot \left(6 - \dfrac{5}{8} \cdot 24 + 3 \cdot \dfrac{8}{3}\right)$

66. $(-0.3)^2 \cdot (-5 \cdot 3) + (6 \div 2 \cdot 0.4)$

67. $|-12| \div 4 + 2 \cdot 3^2 \div 6$

68. $6 - (2 - 3 \cdot 4) + 5^2 \div \left(-2 \cdot \dfrac{5}{2}\right) + (2)^2$

Summary Exercises on OPERATIONS WITH SIGNED NUMBERS

Simplify each expression.

1. $2 - 8$

2. $(-16)(0)$

3. $-14 - (-7)$

4. $\dfrac{-42}{6}$

5. $-9 \cdot (-7)$

6. $\dfrac{-12}{12}$

7. $(1)(-56)$

8. $1 + (-23)$

9. $5 - (-7)$

10. $-\dfrac{8}{3} \div \left(-\dfrac{4}{9}\right)$

11. $-18 + 5$

12. $\dfrac{0}{-10}$

13. $-40 - 40$

14. $-17 + 0$

15. $8 \cdot (-6)$

16. $-\dfrac{1}{10} - \dfrac{9}{10}$

17. $\left(-\dfrac{5}{6}\right)\left(\dfrac{1}{5}\right)$

18. $\dfrac{30}{0}$

19. $0 - 14.6$

20. $\dfrac{1.8}{-3}$

21. $-4 \cdot (-6) \cdot 2$

22. $-2 + (-12) + (-5)$

23. $-60 \div 10 \div (-3)$

24. $-8 - 4 - 8$

25. $64 \cdot 0 \div (-8)$

26. $2 - (-5) + 3^2$

27. $-9 + 8 + (-2)$

28. $(-6)(-2)(-3)$

29. $8 + 6 + (-8)$

30. $-72 \div (-9) \div (-4)$

31. $-7 + 28 + (-56) + 3$

32. $9 - 6 - 3 - 5$

33. $-6 \cdot (-8) \div (-5 - 7)$

34. $-1 \cdot 9732 \cdot (-1) \cdot (-1)$

35. $-80 \div 4 \cdot (-5)$

36. $-10 - 4 + 0 + 18$

37. $-7 \cdot |7| \cdot |-7|$

38. $5 - |-3| + 3$

39. $-2 \cdot (-3 \cdot 7) \div (-7)$

40. $-3 - (-2 + 4) - 5$

41. $0 - |-7 + 2|$

42. $(-4)^2 \cdot (7 - 9)^2 \div 2^3$

43. $12 \div 4 + 2 \cdot (-2)^2 \div (-4)$

44. $\dfrac{-9 + 24 \div (-4) \cdot (-6)}{32 - 4 \cdot (12) \div 3 \cdot 2}$

45. $\dfrac{5 - |2 - 4 \cdot 4| + (-5)^2 \div 5^2}{-9 \div 3 \cdot (2 - 2) - (-8)}$

Solve each application problem. Be sure to indicate whether the final answer is a positive or negative number.

46. When Ashwini discovered that her checking account was overdrawn by $238, she quickly transferred $450 from her savings to her checking account. What is the balance in her checking account?

47. A discount store found that 174 items were lost to shoplifting last month. The total loss was $4176. What was the average loss on each item?

48. In a laboratory experiment, a mixture started at a temperature of -102 degrees. First the temperature was raised 37 degrees and then raised 52 degrees. What was the final temperature?

49. A plane descended an average of 730 ft each minute during a 37-minute landing. How far did the plane descend during the landing?

50. The Tigers offensive team lost a total of 48 yd during the first half of the football game. During the second half they gained 191 yd. How many yards did they gain or lose during the entire game?

51. A scuba diver was photographing fish at 65 ft below the surface of the lagoon. She swam up 24 ft and then swam down 49 ft. What was her final depth?

Elena Sanchez opened a shop that does alterations and designs custom clothing. Use the table of her income and expenses to answer Exercises 52–55.

Month	Income	Expenses	Profit or Loss
January	$2400	$3100	
February	$1900	$2000	
March	$2500	$1800	
April	$2300	$1400	
May	$1600	$1600	
June	$1900	$1200	

52. Complete the table by finding Elena's profit or loss for each month.

53. Which month had the greatest loss? Which month had the greatest profit?

54. What was Elena's average monthly income?

55. What was the average monthly amount of expenses?

56. Explain in your own words how to add two numbers with different signs. Include two examples in your explanation, one that has a positive answer and one that has a negative answer.

57. Explain what is different and what is similar between multiplying and dividing signed numbers.

9.5 EVALUATING EXPRESSIONS AND FORMULAS

In formulas, you have seen that numbers can be represented by letters. For example, you used this formula for finding simple interest in **Section 6.7.**

$$I = p \cdot r \cdot t$$

In this formula, p (principal) represents the amount of money borrowed, r is the rate of interest, and t is the time in years. In algebra, we often write multiplication without the multiplication dots. If there is no operation sign written between two letters, or between a letter and a number, you assume it is multiplication.

Showing Multiplication in Algebra

If there is no operation sign, it is understood to be multiplication. Here are some examples.

$I = p \cdot r \cdot t$	is written	$I = prt$
$2 \cdot r$	is written	$2r$
$3 \cdot x + 4 \cdot y$	is written	$3x + 4y$

1 ▭ **Define variable and expression.**   Letters (such as the I, p, r and t used above) that represent numbers are called **variables.** A combination of operations on letters and numbers is an **expression.** Three examples of expressions are shown here.

$$9 + p \qquad 8r \qquad 7k - 2m$$

2 ▭ **Find the value of an expression when values of the variables are given.** The value of an expression changes depending on the value of each variable. To find the value of an expression, replace the variables with their values. It is helpful to write each value inside parentheses when multiplication is involved.

Example 1 Finding the Value of an Expression

Find the value of $5x - 3y$, if $x = 2$ and $y = 7$.

Replace x with **2**. Replace y with **7**.

$$5\ x\ -\ 3\ y$$
$$\downarrow\downarrow \quad\ \downarrow\downarrow$$

Using the order of operations, do all multiplications first. Then subtract.

$$5(2)\ -\ 3(7)$$
$$\underbrace{10}\ -\ \underbrace{21}$$
$$-11$$

═══ **Work Problem ❶ at the Side.**

Example 2 Finding the Value of an Expression

What is the value of $\dfrac{6k + 2r}{5s}$, if $k = -2$, $r = 5$, and $s = -1$?

└── **Continued on Next Page**

OBJECTIVES

1 ▭ Define variable and expression.

2 ▭ Find the value of an expression when values of the variables are given.

❶ Find the value of $5x - 3y$, if:

(a) $x = 5, \quad y = 2$

(b) $x = -3, \quad y = 4$

(c) $x = 0, \quad y = 6$

ANSWERS
1. (a) 19 **(b)** -27 **(c)** -18

2 Find the value of $\dfrac{3k + r}{2s}$ if:

(a) $k = 1$, $r = 1$, $s = 2$

(b) $k = 8$, $r = -2$, $s = -4$.

(c) $k = -3$, $r = 1$, $s = -2$

3 Find the value of each expression.

(a) $-x - 6y$, if $x = -1$ and $y = -4$

(b) $-4a - b$, if $a = 4$ and $b = -2$

(c) $-w - 3x - y$, if $w = 10$, $x = -5$, and $y = -6$

4 Find the value of A, P, and C in these formulas.

(a) $A = \dfrac{1}{2}bh$

$b = 6$ yd, $h = 12$ yd

(b) $P = 2l + 2w$

$l = 10$ cm, $w = 8$ cm

(c) $C = 2\pi r$

$\pi \approx 3.14$, $r = 6$ ft

Replace k with -2, r with 5, and s with -1.

$$\frac{6k + 2r}{5s} = \frac{6(-2) + 2(5)}{5(-1)}$$ Do all the multiplications first.

$$= \frac{-12 + 10}{-5}$$ Add in the numerator.

$$= \frac{-2}{-5}$$ Dividing two numbers with the same sign gives a positive answer.

$$= \frac{2}{5}$$

Work Problem 2 at the Side.

The next example shows how to evaluate an expression with negative signs and/or subtraction, when the value of the variable is also negative.

Example 3 **Evaluating an Expression with Negative Signs and Negative Values**

Find the value of $-c - 5b$ when $c = -2$ and $b = -3$.

Replace c with -2.
Replace b with -3.

$-(-2)$ is the opposite of (-2), which is $(+2)$.

$$-(-2) - 5(-3)$$ Multiply $5 \cdot (-3)$.

$$+2 \quad - (-15)$$ Change subtraction to addition; change (-15) to its opposite, $(+15)$.

$$2 \quad + (+15)$$

$$17$$

CAUTION

Watch the signs carefully when there are negative signs in the expression and the value of a variable is also negative, as in Example 3 above.

Work Problem 3 at the Side.

Example 4 **Evaluating a Formula**

The formula you used in Chapter 8 for the area of a triangle can now be written without the multiplication dots.

$$A = \frac{1}{2} \cdot b \cdot h \quad \text{is written} \quad A = \frac{1}{2}bh$$

In this formula, b is the length of the base of the triangle and h is the height of the triangle. What is the area if $b = 9$ cm and $h = 24$ cm?

$$A = \frac{1}{2} \quad b \quad h$$

$$A = \frac{1}{2}(9 \text{ cm})(24 \text{ cm})$$ Replace b with 9 cm and h with 24 cm.

$$A = \frac{1}{2}(9 \text{ cm})(\overset{12}{\cancel{24}} \text{ cm})$$ Divide out any common factors.

$$A = 108 \text{ cm}^2$$ Recall that cm^2 means *square centimeters.*

The area of the triangle is 108 cm^2.

Work Problem 4 at the Side.

9.5 EXERCISES

| FOR EXTRA HELP |  Student's Solutions Manual | 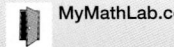 MyMathLab.com | 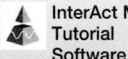 InterAct Math Tutorial Software | AW Math Tutor Center | www.mathxl.com MathXL | Digital Video Tutor CD 6 Videotape 17 |

Find the value of the expression $2r + 4s$ for each set of values of r and s. See Example 1.

1. $r = 2, \quad s = 6$

2. $r = 6, \quad s = 1$

3. $r = 1, \quad s = -3$

4. $r = 7, \quad s = -2$

5. $r = -4, \quad s = 4$

6. $r = -3, \quad s = 5$

7. $r = -1, \quad s = -7$

8. $r = -3, \quad s = -5$

9. $r = 0, \quad s = -2$

10. $r = -7, \quad s = 0$

Use the given values of the variables to find the value of each expression. See Examples 1 and 2.

11. $8x - y$
  $x = 1, \quad y = 8$

12. $a - 5b$
  $a = 10, \quad b = 2$

13. $6k + 2s$
  $k = 1, \quad s = -2$

14. $7p + 7q$
  $p = -4, \quad q = 1$

15. $\dfrac{-m + 5n}{2s + 2}$
  $m = 4, \quad n = -8, \quad s = 0$

16. $\dfrac{2y - z}{x - 2}$
  $y = 0, \quad z = 5, \quad x = 1$

17. $-m - 3n$
  $m = \dfrac{1}{2}, \quad n = \dfrac{3}{8}$

18. $7k - 3r$
  $k = \dfrac{2}{3}, \quad r = \dfrac{1}{3}$

Be careful when an expression has a negative sign and the value of the variable is also negative. Use the given values to find the value of each expression. See Example 3.

19. $-c - 5b$
    $c = -8, \quad b = -4$

20. $-c - 5b$
    $c = -1, \quad b = -2$

21. $-4x - y$
    $x = 5, \quad y = -15$

22. $-4x - y$
    $x = 3, \quad y = -8$

23. $-k - m - 8n$
    $k = 6, \quad m = -9, \quad n = 0$

24. $-k - m - 8n$
    $k = 0, \quad m = -7, \quad n = -1$

25. $\dfrac{-3s - t - 4}{-s + 6 + t}$
    $s = -1, \quad t = -13$

26. $\dfrac{-3s - t - 4}{-s - 20 - t}$
    $s = -3, \quad t = -6$

Use the given formula and values of the variables to find the value of the remaining variable. See Example 4.

27. $P = 4s; \quad s = 7.5$

28. $P = 4s; \quad s = 0.8$

29. $P = 2l + 2w; \quad l = 9, \quad w = 5$

30. $P = 2l + 2w; \quad l = 12, \quad w = 2$

31. $A = \pi r^2$; $\pi \approx 3.14$, $r = 5$

32. $A = \pi r^2$; $\pi \approx 3.14$, $r = 10$

33. $A = \dfrac{1}{2}bh$; $b = 15$, $h = 3$

34. $A = \dfrac{1}{2}bh$; $b = 5$, $h = 11$

35. $V = \dfrac{1}{3}Bh$; $B = 30$, $h = 60$

36. $V = \dfrac{1}{3}Bh$; $B = 105$, $h = 5$

37. $d = rt$; $r = 53$, $t = 6$

38. $d = rt$; $r = 180$, $t = 5$

39. $C = 2\pi r$; $\pi \approx 3.14$, $r = 4$

40. $C = 2\pi r$; $\pi \approx 3.14$, $r = 18$

Solve each application problem.

41. The expression for finding the perimeter of a triangle with sides of equal length is $3s$, where s is the length of one side. Evaluate the expression when

   (a) the length of one side is 11 in.

   (b) the length of one side is 3 ft.

42. The expression for finding the perimeter of a pentagon with sides of equal length is $5s$, where s is the length of one side. Evaluate the expression when

   (a) the length of one side is 25 cm

   (b) the length of one side is 8 in.

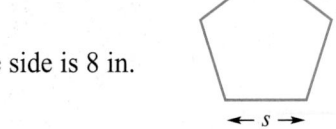

43. The expression for figuring a student's average test score is $\dfrac{p}{t}$, where p is the total points earned on all the tests and t is the number of tests. Evaluate the expression when

   (a) 332 points were earned on 4 tests

   (b) there were 7 tests and 637 points earned.

44. The expression for deciding how many buses are needed for a group trip is $\dfrac{p}{b}$, where p is the total number of people and b is the number of people that one bus will hold. Evaluate the expression when

   (a) 176 people are going on a trip and one bus holds 44 people

   (b) a bus holds 36 people and 72 people are going on a trip.

45. Find and correct the error made by the student who solved this example:

Find the value of $-x - 4y$ if $x = -3$ and $y = -1$.

$$-x - 4y$$
$$-3 - 4(-1)$$
$$-3 - (-4)$$
$$-3 + (+4)$$
$$1$$

After stating the error, rework the problem and write a sentence next to each step, explaining what is being done in that step.

46. Go back to Chapter 8 and find each of the formulas listed below. Pick values for the variables and then find the value of A or V. Then pick different values for the variables and again find the value of A or V.

(a) Area of a trapezoid: pick values for h, B, and b.

(b) Volume of a rectangular solid: pick values for l, w, and h.

RELATING CONCEPTS (Exercises 47–52) FOR INDIVIDUAL OR GROUP WORK

Work Exercises 47–52 in order. *Use the given formula and values of the variables to find the value of the remaining variable. If you studied Chapters 7 and 8, write a sentence telling when you would use each formula.*

47. $F = \dfrac{9C}{5} + 32$; $C = -40$

48. $C = \dfrac{5(F - 32)}{9}$; $F = -4$

49. $V = \dfrac{4\pi r^3}{3}$; $\pi \approx 3.14$, $r = 3$

50. $c^2 = a^2 + b^2$; $a = 3$, $b = 4$

51. $A = \dfrac{1}{2}h(b + B)$; $h = 7$, $b = 4$, $B = 12$

52. $V = \dfrac{\pi r^2 h}{3}$; $\pi \approx 3.14$, $r = 6$, $h = 10$

9.6 SOLVING EQUATIONS

An **equation** is a statement that says two expressions are equal. Examples of equations are shown here.

$$x + 1 = 9 \qquad 20 = 5k \qquad 6r - 1 = 17$$

The **equal sign** in an equation divides the equation into two parts, the *left side* and the *right side*. In $6r - 1 = 17$, the left side is $6r - 1$, and the right side is 17. The equal sign tells us that the two sides are equivalent.

$$6r - 1 = 17$$

Left side = Right side

You solve an equation by finding all numbers that can be substituted for the variable to make the equation true. These numbers are called **solutions** of the equation.

1 **Determine whether a number is a solution of an equation.** To decide whether a number is a solution of an equation, substitute the number in the equation to see whether the result is true.

Example 1 **Determining Whether a Number Is a Solution of an Equation**

Is 7 a solution of either one of these equations?

(a) $12 = x + 5$
    Replace x with 7.

$$12 = x + 5 \qquad \text{Replace } x \text{ with 7.}$$
$$12 = \underbrace{7 + 5}$$
$$12 = \quad 12 \qquad \text{True}$$

Because the statement is true, 7 is a solution of the equation $12 = x + 5$.

(b) $2y + 1 = 16$
    Replace y with 7.

$$2y + 1 = 16$$
$$2(7) + 1 = 16$$
$$\underbrace{14 + 1} = 16$$
$$15 \quad = 16 \qquad \text{False}$$

The *false* statement shows that 7 is *not* a solution of $2y + 1 = 16$.

=== **Work Problem ❶ at the Side.**

2 **Solve equations using the addition property of equations.** If the equation $a = b$ is true, and if a number c is added to both a and b, the new equation is also true. This rule is called the *addition property of equations*. It means that you can add the *same* number to *each* side of an equation and still have a true equation.

Addition Property of Equations

If $a = b$, then

$$a + c = b + c.$$

In other words, you may add the *same* number to *each* side of an equation.

OBJECTIVES

1 Determine whether a number is a solution of an equation.

2 Solve equations using the addition property of equations.

3 Solve equations using the multiplication property of equations.

❶ Decide whether the given number is a solution of the equation.

(a) $p + 1 = 8; \quad 7$

(b) $30 = 5r; \quad 6$

(c) $3k - 2 = 4; \quad 3$

(d) $23 = 4y + 3; \quad 5$

You can use the addition property to solve equations. The idea is to get the variable (the letter) by itself on one side of the equal sign and a number by itself on the other side.

Example 2 ⟩ Solving Equations by Using the Addition Property

Solve each equation.

(a) $k - 4 = 6$

To get k by itself on the left side, add 4 to the left side, because $k - 4 + 4$ gives $k + 0$. You must then add 4 to the right side also.

$$k - 4 = 6 \qquad \leftarrow \text{Original equation}$$
$$k \underbrace{- 4 + 4} = 6 + 4 \qquad \text{Add 4 to each side.}$$
$$\underbrace{k + 0} \quad = 10 \qquad \text{On the left side, } -4 + 4 \text{ is } 0.$$
$$k \quad = 10 \qquad \text{On the left side, } k + 0 \text{ is } k.$$

The solution is 10. Check by replacing k with 10 in the original equation.

$$k - 4 = 6 \quad \leftarrow \text{Original equation}$$
$$10 - 4 = 6 \qquad \text{Replace } k \text{ with 10.}$$
$$6 = 6 \qquad \text{True, so 10 is the solution.}$$

This result is true, so *10 is the correct solution.*

CAUTION

When checking the solution in Example 2(a) above, we ended up with $6 = 6$. Notice that 6 is **not** the solution. The solution is 10, the number used to replace k in the original equation.

(b) $2 = z + 8$

To get z by itself on the right side, add -8 to each side.

$$2 = z + 8 \qquad \leftarrow \text{Original equation}$$
$$2 + (-8) = z + \underbrace{8 + (-8)} \qquad \text{Add } (-8) \text{ to each side.}$$
$$-6 = z + \quad 0$$
$$-6 = z$$

NOTE

Notice that we *added* -8 to each side to get z by itself. We can accomplish the same thing by *subtracting* 8 from each side. Recall from **Section 9.2** that subtraction is defined in terms of addition. On the left side of the equation above, $2 - 8$ gives the same result as $2 + (-8)$.

Check the solution by replacing z with -6 in the original equation.

$$2 = z + 8 \quad \leftarrow \text{Original equation}$$
$$2 = -6 + 8 \qquad \text{Replace } z \text{ with } -6.$$
$$2 = 2 \qquad \text{True, so } -6 \text{ is the solution.}$$

The result is true, so -6 is the solution (***not*** 2).

Here is a summary of the rules you can use to solve equations using the addition property. In these rules, x is the variable and a and b represent numbers.

Solving an Equation by Using the Addition Property

Solve $x - a = b$ or $b = x - a$ by adding a to *each* side.

Solve $x + a = b$ or $b = x + a$ by subtracting a from *each* side.

Work Problem ❷ at the Side.

3 ▭ **Solve equations using the multiplication property of equations.** As long as you do the *same* thing to *each* side of an equation, it will still be a true equation. So far you have added or subtracted on each side. Now we will multiply or divide on each side.

Multiplication Property of Equations

If $a = b$ and c does not equal 0, then

$$a \cdot c = b \cdot c \quad \text{and} \quad \frac{a}{c} = \frac{b}{c}.$$

In other words, you may multiply or divide *each* side of an equation by the *same* number. (The only exception is you cannot divide by 0.)

Example 3 **Solving Equations by Using the Multiplication Property**

Solve each equation.

(a) $9p = 63$

You want to get the variable, p, by itself on the left side. The expression $9p$ means $9 \cdot p$. To get p by itself, *divide* each side by 9.

$$9p = 63$$

$$\frac{\overset{1}{\cancel{9}} \cdot p}{\underset{1}{\cancel{9}}} = \frac{63}{9} \qquad \text{Divide each side by 9.}$$

$$p = 7$$

Check:

$$9p = 63 \quad \leftarrow \text{Original equation}$$
$$9 \cdot 7 = 63 \qquad \text{Replace } p \text{ with 7.}$$
$$63 = 63 \qquad \text{True}$$

The result is true, so 7 is the solution.

Continued on Next Page

❷ Solve each equation. Check each solution.

(a) $n - 5 = 8$

(b) $5 = r - 10$

(c) $3 = z + 1$

(d) $k + 9 = 0$

(e) $-2 = y + 9$

(f) $x - 2 = -6$

Answers
2. (a) $n = 13$ **(b)** $r = 15$ **(c)** $z = 2$
(d) $k = -9$ **(e)** $y = -11$ **(f)** $x = -4$

❸ Solve each equation. Check each solution.

(a) $2y = 14$

(b) $42 = 7p$

(c) $-8a = 32$

(d) $-3r = -15$

(e) $-60 = -6k$

(f) $10x = 0$

(b) $-4r = 24$

Divide *each* side by -4 to get r by itself on the left side.

$$\dfrac{\overset{1}{-\cancel{4}} \bullet r}{\underset{1}{-\cancel{4}}} = \dfrac{24}{-4} \qquad \text{Divide each side by } -4.$$

$$r = -6$$

Check:

$$-4r = 24 \quad \leftarrow \text{Original equation}$$
$$-4(-6) = 24 \qquad \text{Replace } r \text{ with } -6.$$
$$24 = 24 \qquad \text{True}$$

The result is true, so -6 is the solution.

(c) $-55 = -11m$

Divide *each* side by -11 to get m by itself on the right side.

$$\dfrac{-55}{-11} = \dfrac{\overset{1}{-\cancel{11}} \bullet m}{\underset{1}{-\cancel{11}}}$$

$$5 = m$$

Check: $-55 = -11(5)$ is true, so 5 is the solution.

Work Problem ❸ at the Side.

Example 4 Solving Equations by Using the Multiplication Property

Solve each equation.

(a) $\dfrac{x}{2} = 9$

Replace $\dfrac{x}{2}$ with $\dfrac{1}{2}x$, because dividing x by 2 is the same as multiplying x by $\dfrac{1}{2}$. Then, to get x by itself, multiply each side by the reciprocal of $\dfrac{1}{2}$, which is $\dfrac{2}{1}$. (Recall from **Section 2.7** that when two numbers are reciprocals, their product is 1.)

$$\dfrac{1}{2}x = 9$$

$$\dfrac{\overset{1}{\cancel{2}}}{1} \bullet \dfrac{1}{\underset{1}{\cancel{2}}}x = 2 \bullet 9 \qquad \text{Multiply each side by } \tfrac{2}{1} \text{ (which equals 2).}$$

$$1x = 18$$

$$x = 18$$

Check:

$$\dfrac{x}{2} = 9 \quad \leftarrow \text{Original equation}$$

$$\dfrac{18}{2} = 9 \qquad \text{Replace } x \text{ with } 18.$$

$$9 = 9 \qquad \text{True, so 18 is the solution.}$$

18 is the correct solution.

Continued on Next Page

(b) $-\dfrac{2}{3}r = 4$

Multiply each side by the reciprocal of $-\frac{2}{3}$, which is $-\frac{3}{2}$.

$$-\frac{2}{3}r = 4$$

$$-\frac{\cancel{3}^1}{\cancel{2}_1} \cdot \left(-\frac{\cancel{2}^1}{\cancel{3}_1}r\right) = -\frac{3}{\cancel{2}_1} \cdot \frac{\cancel{4}^2}{1} \qquad \text{Multiply each side by } -\tfrac{3}{2}.$$

$$r = -6$$

Check by replacing r with -6 in the original equation. Write -6 as $\frac{-6}{1}$.

$$-\frac{2}{3}r = 4 \longleftarrow \text{Original equation}$$

$$-\frac{2}{\cancel{3}_1} \cdot \frac{\cancel{-6}^{-2}}{1} = 4 \qquad \text{Replace } r \text{ with } -6.$$

$$4 = 4 \qquad \text{True, so } -6 \text{ is the solution.}$$

Here is a summary of the rules for using the multiplication property. In these rules, x is the variable and a, b, and c represent numbers.

Solving Equations Using the Multiplication Property

Solve the equation $\quad ax = b \quad$ by dividing each side by a.

Solve the equation $\quad \frac{a}{b}x = c \quad$ by multiplying each side by $\frac{b}{a}$.

Work Problem ❹ at the Side.

❹ Solve each equation. Check each solution.

(a) $\dfrac{a}{4} = 2$

(b) $\dfrac{y}{7} = -3$

(c) $-8 = \dfrac{k}{6}$

(d) $8 = -\dfrac{4}{5}z$

(e) $-\dfrac{5}{8}p = -10$

Focus on *Real-Data Applications*

Expressions

A college with four campuses uses an auditorium for graduation that has 1500 seats. Each campus hosts its own graduation ceremony. Students are allocated a whole number of tickets. Round each result **down** to the nearest whole number.

1. How many tickets are allocated to each of 458 graduates at the **North** Campus graduation?

2. How many tickets are allocated to each of 297 graduates at the **East** Campus graduation?

3. How many tickets are allocated to each of 315 graduates at the **South** Campus graduation?

4. How many tickets are allocated to each of 186 graduates at the **West** Campus graduation?

5. Write an *expression* that represents the number of tickets that are allocated to each of *g* graduates.

6. What recommendations do you have for unallocated tickets?

Traffic engineers have to decide how long to have the red, yellow, and green lights showing on a traffic signal. To decide the number of seconds that a yellow light should be on, the engineers use the expression

$$\frac{5v}{100} + 1$$

where *v* is the speed limit in miles per hour (mph).

7. How many seconds should the yellow light be on if the speed limit is 20 mph? 40 mph? 60 mph?

8. Based on the answers you just calculated, how could you estimate the time for a yellow light if the speed limit is 30 mph? 50 mph?

9. Use the given expression to find the number of seconds that the yellow light should be on if the speed limit is 30 mph and 50 mph. Did you get the same result as in Problem 8?

To estimate the number of words in a child's vocabulary, a pediatrician uses the expression $60A - 900$, where A is the child's age in months.

10. Estimate the number of words that a child aged 20 months knows.

11. Estimate the number of words that a child aged 2 years knows. (*Hint:* How many months are in two years?)

12. How many words does a child learn between the ages of 20 months and 2 years?

13. Estimate the number of words in the vocabulary of a 3-year-old child.

14. How many words does a child learn between the ages of 2 years and 3 years?

15. Evaluate the expression for a child who is 15 months old. Do you think that the answer is reasonable? Explain why or why not.

16. Evaluate the expression for a child who is 12 months old. Do you think that the answer makes sense? Explain why or why not.

Here is a *rule of thumb* to estimate the distance in miles to a thunderstorm: "Count the number of seconds from the time you see a lightning flash until you hear the thunder. Divide by 5."

17. How far away is the storm if you count 15 *seconds* between the lightning flash and the thunder? 10 *seconds*? 5 *seconds*?

18. Write an expression that represents the distance in miles to a thunderstorm if *s* seconds elapse between seeing lightning and hearing thunder.

19. Use the expression to estimate the distance to a storm if the time lapse is $2\frac{1}{2}$ seconds.

9.6 **EXERCISES**

FOR
EXTRA
HELP

 Student's
Solutions
Manual

 MyMathLab.com

 InterAct Math
Tutorial
Software

 AW Math
Tutor Center

MathXL www.mathxl.com

 Digital Video Tutor CD 6
Videotape 17

Determine whether the given number is a solution of the equation. See Example 1.

1. $x + 7 = 11$;   4

2. $k - 2 = 7$;   9

3. $4y = 28$;   7

4. $5p = 30$;   6

5. $2z - 1 = -15$; -8

6. $6r - 3 = -14$; -2

Solve each equation by using the addition property. Check each solution. See Example 2.

7. $p + 5 = 9$

8. $a + 3 = 12$

9. $k + 15 = 0$

10. $y + 6 = 0$

11. $z - 5 = 3$

12. $x - 9 = 4$

13. $8 = r - 2$

14. $3 = b - 5$

15. $-5 = n + 3$

16. $-1 = a + 8$

17. $7 = r + 13$

18. $12 = z + 7$

19. $-4 + k = 14$

20. $-9 + y = 7$

21. $-12 + x = -1$

22. $-3 + m = -9$

23. $-5 = -2 + r$

24. $-1 = -10 + y$

25. $d + \dfrac{2}{3} = 3$

26. $x + \dfrac{1}{2} = 4$

27. $z - \dfrac{7}{8} = 10$

28. $m - \dfrac{3}{4} = 6$

29. $\dfrac{1}{2} = k - 2$

30. $\dfrac{3}{5} = t - 1$

31. $m - \dfrac{7}{5} = \dfrac{11}{4}$

32. $z - \dfrac{7}{3} = \dfrac{32}{9}$

33. $x - 0.8 = 5.07$

34. $a - 3.82 = 7.9$

35. $3.25 = 4.76 + r$

36. $8.9 = 10.5 + b$

Solve each equation. Check each solution. See Example 3.

37. $6z = 12$

38. $8k = 24$

39. $48 = 12r$

40. $99 = 11m$

41. $3y = 0$

42. $5a = 0$

43. $-6k = 36$

44. $-7y = 70$

45. $-36 = -4p$

46. $-54 = -9r$

47. $-1.2m = 8.4$

48. $-5.4z = 27$

49. $-8.4p = -9.24$

50. $-3.2y = -16.64$

Solve each equation. Check each solution. See Example 4.

51. $\dfrac{k}{2} = 17$

52. $\dfrac{y}{3} = 5$

53. $11 = \dfrac{a}{6}$

54. $5 = \dfrac{m}{8}$

55. $\dfrac{r}{3} = -12$

56. $\dfrac{z}{9} = -3$

57. $-\dfrac{2}{5}p = 8$

58. $-\dfrac{5}{6}k = 15$

59. $-\dfrac{3}{4}m = -3$

60. $-\dfrac{9}{10}b = -18$

61. $6 = \dfrac{3}{8}x$

62. $4 = \dfrac{2}{3}a$

63. $\dfrac{y}{2.6} = 0.5$

64. $\dfrac{k}{0.7} = 3.2$

65. $\dfrac{z}{-3.8} = 1.3$

66. $\dfrac{m}{-5.2} = 2.1$

67. Explain the addition property of equations. Then show an example of an equation where you would use the addition property to solve it. Make the equation so it has -3 as the solution.

68. Explain the multiplication property of equations. Then show an example of an equation where you would use the multiplication property to solve it. Make the equation so it has $+6$ as the solution.

Solve each equation.

69. $x - 17 = 5 - 3$

70. $y + 4 = 10 - 9$

71. $3 = x + 9 - 15$

72. $-1 = y + 7 - 9$

73. $\dfrac{7}{2}x = \dfrac{4}{3}$

74. $\dfrac{3}{4}x = \dfrac{5}{3}$

75. $\dfrac{1}{2} - \dfrac{3}{4} = \dfrac{a}{5}$

76. $\dfrac{2}{3} - \dfrac{8}{9} = \dfrac{c}{6}$

77. $m - 2 + 18 = |-3 - 4| + 5$

78. $10 - |0 - 8| = n + 1 - 4$

9.7 SOLVING EQUATIONS WITH SEVERAL STEPS

1 **Solve equations with several steps.** You cannot solve the equation $5m + 1 = 16$ by just adding the same number to each side, nor by just dividing each side by the same number. Instead, you use a combination of both operations. Here are the steps.

Solving an Equation Using the Addition and Multiplication Properties

Step 1 Add or subtract the *same* amount on *each* side of the equation so that the variable term ends up by itself on one side.

Step 2 Multiply or divide *each* side by the *same* number to find the solution.

Step 3 Check the solution in the original equation.

Example 1 Solving an Equation with Several Steps

Solve $5m + 1 = 16$.

Step 1 Subtract 1 from *each* side so that $5m$ will be by itself on the left side.

$$5m + 1 - 1 = 16 - 1$$
$$5m = 15$$

Step 2 Divide *each* side by 5.

$$\frac{\overset{1}{\cancel{5}} \cdot m}{\underset{1}{\cancel{5}}} = \frac{15}{5}$$

$$m = 3$$

Step 3 Check the solution.

$$5m + 1 = 16 \quad \leftarrow \text{Original equation}$$
$$5(3) + 1 = 16 \quad \text{Replace } m \text{ with 3.}$$
$$15 + 1 = 16$$
$$16 = 16 \quad \text{True, so 3 is the solution.}$$

The solution is 3.

═══════ **Work Problem 1 at the Side.**

2 **Use the distributive property.** We can use the order of operations to simplify these two expressions.

$$\underbrace{2(\underbrace{6 + 8})}_{\underbrace{2(14)}_{28}} \quad \text{and} \quad \underbrace{2 \cdot 6 + 2 \cdot 8}_{\underbrace{12 \ + \ 16}_{28}}$$

Because both answers are the same, the two expressions are equivalent.

$$2(6 + 8) = 2 \cdot 6 + 2 \cdot 8$$

This is an example of the *distributive property*.

Distributive Property

$$a(b + c) = ab + ac$$

OBJECTIVES

1 Solve equations with several steps.

2 Use the distributive property.

3 Combine like terms.

4 Solve more difficult equations.

---

1 Solve each equation. Check each solution.

(a) $2r + 7 = 13$

(b) $20 = 6y - 4$

(c) $7m + 9 = 9$

(d) $-2 = 4p + 10$

(e) $-10z - 9 = 11$

❷ Use the distributive property.

(a) $3(2 + 6)$

(b) $8(k - 3)$

(c) $-6(r + 5)$

(d) $-9(s - 8)$

❸ Combine like terms.

(a) $5y + 11y$

(b) $10a - 28a$

(c) $3x + 3x - 9x$

(d) $k + k$

(e) $6b - b - 7b$

Example 2 **Using the Distributive Property**

Simplify each expression by using the distributive property.

(a) $9(4 + 2) = 9 \cdot 4 + 9 \cdot 2 = 36 + 18 = 54$

The 9 outside the parentheses is *distributed* over the 4 and the 2 inside the parentheses. That means that *every* number inside the parentheses is multiplied by 9.

(b) $-3(k + 9) = -3 \cdot k + (-3) \cdot 9 = -3k + (-27) = -3k - 27$

(c) $6(y - 5) = 6 \cdot y - 6 \cdot 5 = 6y - 30$

(d) $-2(x - 3) = -2 \cdot x - (-2) \cdot 3 = -2x - (-6) = -2x + 6$

CAUTION

Notice how the final step in Example 2(b) uses the definition of subtraction "in reverse."

$$-3k + (-27)$$
$$-3k - 27$$

Change addition to subtraction; change (-27) to its opposite, 27.

Work Problem ❷ at the Side.

3 **Combine like terms.** A single letter or number, or the product of a variable and a number, makes up a *term*. Here are six examples of terms.

$$3y \qquad 5 \qquad -9 \qquad 8r \qquad 10r^2 \qquad x$$

Terms with exactly the same variable and the same exponent are called **like terms.**

$5x$	and	$3x$	like terms
$5x$	and	$3m$	*not* like terms; variables are different
$5x^2$	and	$5x^3$	*not* like terms; exponents are different
$5x^4$	and	$3x^4$	like terms

The distributive property can be used to simplify a sum of like terms such as $6r + 3r$.

$$6r + 3r = (6 + 3)r = 9r$$

This process is called *combining like terms.*

Example 3 **Combining Like Terms**

Use the distributive property to combine like terms.

(a) $5k + 11k = (5 + 11)k = 16k$

(b) $10m - 14m + 2m = (10 - 14 + 2)m = -2m$

(c) $-5x + x$ can be written $-5x + 1x = (-5 + 1)x = -4x$

Work Problem ❸ at the Side.

4 ▨▨▨ **Solve more difficult equations.**   The next examples show you how to solve more difficult equations using the addition, multiplication, and distributive properties.

Example 4   **Solving Equations**

Solve each equation. Check each equation.

(a) $6r + 3r = 36$

You can combine $6r$ and $3r$ because they are *like terms*. $6r + 3r$ is $9r$, so the equation becomes

$$9r = 36$$

Next, divide each side by 9.

$$\frac{\overset{1}{\cancel{9}} \cdot r}{\underset{1}{\cancel{9}}} = \frac{36}{9}$$

$$r = 4$$

Check:

$$6r + 3r = 36 \quad \leftarrow \text{Original equation}$$
$$6(4) + 3(4) = 36 \quad \text{Replace } r \text{ with 4.}$$
$$24 + 12 = 36$$
$$36 = 36 \quad \text{True}$$

The solution is 4.

(b) $2k - 2 = 5k - 11$

First, to get the variable term on one side, subtract $5k$ from each side.

$$2k - 2 - 5k = 5k - 11 - 5k$$
$$2k - 5k - 2 = 5k - 5k - 11$$
$$-3k - 2 = -11$$

Next, add 2 to each side.

$$-3k - 2 + 2 = -11 + 2$$
$$-3k = -9$$

Finally, divide each side by -3.

$$\frac{\overset{1}{-\cancel{3}} \cdot k}{\underset{1}{-\cancel{3}}} = \frac{-9}{-3}$$

$$k = 3$$

Check:

$$2k - 2 = 5k - 11 \quad \leftarrow \text{Original equation}$$
$$2(3) - 2 = 5(3) - 11 \quad \text{Replace } k \text{ with 3.}$$
$$6 - 2 = 15 - 11$$
$$4 = 4 \quad \text{True}$$

The solution is 3.

══════ **Work Problem ❹ at the Side.**

❹ Solve each equation. Check each solution.

(a) $3y - 1 = 2y + 7$

(b) $5a + 7 = 3a - 9$

(c) $3p - 2 = p - 6$

⑤ Solve each equation. Check each solution.

(a) $-12 = 4(y - 1)$

Now that you know about the distributive property and combining like terms, here is a summary of all the steps you can use to solve an equation.

Solving an Equation

Step 1 If possible, use the **distributive property** to remove parentheses.

Step 2 **Combine any like terms** on the left side of the equation. Combine any like terms on the right side of the equation.

Step 3 **Add or subtract** the same amount on *each* side of the equation so that the variable term ends up by itself on one side.

Step 4 **Multiply or divide** each side by the same number to find the solution.

Step 5 **Check** your solution by going back to the *original equation*. Replace the variable with your solution. Follow the order of operations to complete the calculations. If the two sides of the equation are equal, your solution is correct.

(b) $5(m + 4) = 20$

Example 5 — Solving Equations by Using the Distributive Property

Solve $-6 = 3(y - 2)$

Step 1 Use the distributive property on the right side of the equation.

$$3(y - 2) \quad \text{becomes} \quad 3 \cdot y - 3 \cdot 2 \quad \text{or} \quad 3y - 6$$

Now the equation looks like this.

$$-6 = 3y - 6$$

Step 2 Combine like terms. Check the left side of the equation. There are no like terms. Check the right side. No like terms there either, so go on to Step 3.

Step 3 Add 6 to each side to get the variable term by itself on the right side.

$$-6 + 6 = 3y - 6 + 6$$
$$0 = 3y$$

Step 4 Divide each side by 3.

$$\frac{0}{3} = \frac{\overset{1}{\cancel{3}} \cdot y}{\cancel{3}}_{1}$$

$$0 = y$$

(c) $6(t - 2) = 18$

Step 5 Check. Go back to the original equation.

$$-6 = 3(y - 2) \leftarrow \text{Original equation}$$
$$-6 = 3(0 - 2) \quad \text{Replace } y \text{ with 0.}$$
$$-6 = 3(-2)$$
$$-6 = -6 \qquad \text{True}$$

The solution is 0.

Work Problem ⑤ at the Side.

9.7 EXERCISES

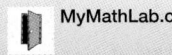
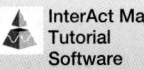

Solve each equation. Check each solution. See Example 1.

1. $7p + 5 = 12$

2. $6k + 3 = 15$

3. $2 = 8y - 6$

4. $10 = 11p - 12$

5. $-3m + 1 = 1$

6. $-4k + 5 = 5$

7. $28 = -9a + 10$

8. $5 = -10p + 25$

9. $-5x - 4 = 16$

10. $-12a - 3 = 21$

11. $-\dfrac{1}{2}z + 2 = -1$

12. $-\dfrac{5}{8}r + 4 = -6$

13. $-0.7 = 5b - 5.2$

14. $0.25 = -3c + 0.85$

Use the distributive property to simplify. See Example 2.

15. $6(x + 4)$

16. $8(k + 5)$

17. $7(p - 8)$

18. $9(t - 4)$

19. $-3(m + 6)$

20. $-5(a + 2)$

21. $-2(y - 3)$

22. $-4(r - 7)$

23. $-8(c + 8)$

24. $-6(n + 5)$

25. $-10(w - 9)$

26. $-11(x - 11)$

Combine like terms. See Example 3.

27. $11r + 6r$

28. $2m + 5m$

29. $8z - 7z$

30. $10x - 2x$

31. $y - 3y$

32. $-10a + a$

33. $-4t + t - 4t$

34. $3y - y - 4y$

35. $7p - 9p + 2p$

36. $-6c - c + 7c$

37. $\dfrac{5}{2}b - \dfrac{11}{2}b$

38. $\dfrac{3}{8}d - \dfrac{9}{8}d$

Solve each equation. Check each solution. See Example 4.

39. $4k + 6k = 50$

40. $3a + 2a = 15$

41. $54 = 10m - m$

42. $28 = x + 6x$

43. $2b - 6b = 24$

44. $3r - 9r = 18$

45. $-12 = 6y - 18y$

46. $-5 = 10z - 15z$

47. $6p - 2 = 4p + 6$

48. $5y - 5 = 2y + 10$

49. $9 + 7z = 9z + 13$

50. $8 + 4a = 2a + 2$

51. $-2y + 6 = 6y - 10$ **52.** $5x - 4 = -3x + 4$ **53.** $b + 3.05 = 2$

54. $t + 0.8 = -1.7$ **55.** $2.5r + 9 = -1$ **56.** $0.5x - 6 = 2$

Solve each equation by using the distributive property. Check each solution. See Example 5.

57. $-10 = 2(y + 4)$ **58.** $-3 = 3(x + 6)$ **59.** $-4(t + 2) = 12$

60. $-5(k + 3) = 25$ **61.** $6(x - 5) = -30$ **62.** $7(r - 5) = -35$

63. Solve $-2t - 10 = 3t + 5$. Show each step you take to solve it. Next to each step, write a sentence that explains what you did in that step. Be sure to tell when you used the addition property of equations and when you used the multiplication property of equations.

64. Here is one student's solution to an equation.

$$3(2x + 5) = -7$$
$$6x + 5 = -7$$
$$6x + 5 - 5 = -7 - 5$$
$$6x = -12$$
$$x = -2$$

Show how to check the solution. If the solution does not check, find and correct the error.

Solve each equation.

65. $30 - 40 = -2x + 7x - 4x$

66. $-6 - 5 + 14 = -50a + 51a$

67. $0 = -2(y - 2)$

68. $0 = -9(b - 1)$

69. $\dfrac{y}{2} - 2 = \dfrac{y}{4} + 3$

70. $\dfrac{z}{3} + 1 = \dfrac{z}{2} - 3$

71. $-3(w - 2) = |0 - 13| + 4w$

72. $-5b - |47 - 7| = -8(b + 8)$

73. $2(a + 0.3) = 1.2(a - 4)$

74. $-0.5(c - 4) = -3(c - 2.5)$

Section 9.8   Using Equations to Solve Application Problems   **677**

USING EQUATIONS TO SOLVE APPLICATION PROBLEMS

It is rare for an application problem to be presented as an equation. Usually, the problem is given in words. You need to *translate* these words into an equation that you can solve.

Translate word phrases into expressions with variables. Examples 1 and 2 show you how to translate word phrases into algebraic expressions.

1 Translate word phrases into expressions with variables.
2 Translate sentences into equations.
3 Solve application problems.

Example 1   Translating Word Phrases into Expressions with Variables

Write each word phrase in symbols, using x as the variable.

Words	Algebraic Expression
A number **plus** 2	$x + 2$ or $2 + x$
The **sum of** 8 and a number	$8 + x$ or $x + 8$
5 **more than** a number	$x + 5$ or $5 + x$
-35 **added to** a number	$-35 + x$ or $x + (-35)$
A number **increased by** 6	$x + 6$ or $6 + x$
9 **less than** a number	$x - 9$
A number **subtracted from** 3	$3 - x$
A number **decreased by** 4	$x - 4$
10 **minus** a number	$10 - x$

CAUTION

Recall that addition can be done in any order, so $x + 2$ gives the same result as $2 + x$. This is *not* true in subtraction, so be careful. $10 - x$ does *not* give the same result as $x - 10$.

Work Problem ① at the Side.

Example 2   Translating Word Phrases into Expressions with Variables

Write each word phrase in symbols, using x as the variable.

Words	Algebraic Expression
8 times a number	$8x$
The **product of** 12 and a number	$12x$
Double a number (meaning "2 times")	$2x$
Twice a number (meaning "2 times")	$2x$
The **quotient of** 6 and a number	$\dfrac{6}{x}$
A number **divided by** 10	$\dfrac{x}{10}$
One-third of a number	$\dfrac{1}{3}x$ or $\dfrac{x}{3}$
The result **is**	$=$

Work Problem ② at the Side.

① Write each word phrase in symbols, using x as the variable.

(a) 15 less than a number

(b) 12 more than a number

(c) A number increased by 13

(d) A number minus 8

(e) -10 plus a number

(f) 6 minus a number

② Write each word phrase in symbols, using x as the variable.

(a) Double a number

(b) The product of -8 and a number

(c) The quotient of 15 and a number

(d) One-half of a number

ANSWERS
1. **(a)** $x - 15$ **(b)** $x + 12$ or $12 + x$
   (c) $x + 13$ or $13 + x$ **(d)** $x - 8$
   (e) $-10 + x$ or $x + (-10)$ **(f)** $6 - x$
2. **(a)** $2x$ **(b)** $-8x$ **(c)** $\dfrac{15}{x}$ **(d)** $\dfrac{1}{2}x$ or $\dfrac{x}{2}$

❸ Translate each sentence into an equation and solve it. Check your solution by going back to the words in the original problem.

(a) If 3 times a number is added to 4, the result is 19. Find the number.

(b) If −6 times a number is added to 5, the result is −13. Find the number.

(c) If twice a number is subtracted from 65, the result is −21.

---

2 Translate sentences into equations. The next example shows you how to translate a sentence into an equation that you can solve.

Example 3 Translating a Sentence into an Equation

If 5 times a number is added to 11, the result is 26. Find the number.

Let x represent the unknown number.
Use the information in the problem to write an equation.

$$\underbrace{5 \text{ times a number}}_{5x} \quad \underbrace{\text{added to}}_{+} \quad \underset{\downarrow}{11} \quad \underset{\downarrow}{\text{is}} \quad \underset{\downarrow}{26.}$$
$$5x \quad\quad + \quad 11 \;=\; 26$$

NOTE

The phrase "the result is" translates to "=."

Next, solve the equation.

$$5x + 11 - 11 = 26 - 11 \qquad \text{Subtract 11 from each side.}$$
$$5x = 15$$
$$\frac{\overset{1}{\cancel{5}}x}{\cancel{5}} = \frac{15}{5} \qquad \text{Divide each side by 5.}$$
$$x = 3$$

The number is 3.
To check the solution, go back to the words of the original problem.

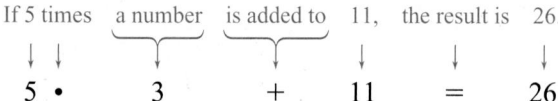

$$\underset{5\ \cdot}{\text{If 5 times}} \quad \underset{3}{\text{a number}} \quad \underset{+}{\text{is added to}} \quad \underset{11}{11,} \quad \underset{=}{\text{the result is}} \quad \underset{26}{26.}$$

Does 5 · 3 + 11 really equal 26? Yes it does. So 3 is the correct solution because it "works" when you put it back into the original problem.

Work Problem ❸ at the Side.

3 Solve application problems. In **Section 1.10,** you learned how to solve application problems using six steps. Now that you know how to use variables and solve equations, we can include those tools in the problem-solving process. Here is a revised list of the six steps we will use.

Solving an Application Problem Using Algebra
Step 1 **Read** the problem carefully to see what it is about.
Step 2 **Assign a variable** by identifying the unknown(s). Write down what the variable represents. If there are several unknowns, let the variable represent the one you know the least about.
Step 3 **Write an equation** using the variable expression(s).
Step 4 **Solve** the equation.
Step 5 **State the answer.**
Step 6 **Check** the solution in the words of the original problem.

Notice that Steps 1, 5, and 6 are nearly identical to what you have been using. Steps 2, 3, and 4 introduce the use of variables and equations.

Example 4   Solving an Application Problem with One Unknown

Michael has 5 less than three times as many lab experiments completed as David. If Michael has completed 13 experiments, how many lab experiments has David completed?

Step 1   **Read.** The problem asks for the number of lab experiments completed by David.

Step 2   **Assign a variable.** There is only *one* unknown: David's number of experiments.

Let x represent David's number of experiments.

Step 3   **Write an equation.**

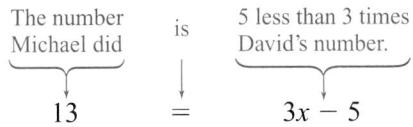

$$13 \;=\; 3x - 5$$

Step 4   **Solve.**

$$13 = 3x - 5$$
$$13 + 5 = 3x - 5 + 5 \qquad \text{Add 5 to each side.}$$
$$18 = 3x$$
$$\frac{18}{3} = \frac{\overset{1}{\cancel{3}}x}{\underset{1}{\cancel{3}}} \qquad \text{Divide each side by 3.}$$
$$6 = x$$

Step 5   **State the answer.** David completed 6 lab experiments.

Step 6   **Check.** Use the words of the *original problem.*

$$13 = (3 \cdot 6) - 5$$
$$13 = 18 - 5$$
$$13 = 13$$

So 6 is the correct solution because it "works" in the original problem.

> **NOTE**
>
> In *Step 3* above, you may also write the equation as $3x - 5 = 13$, with the two sides switched. The solution will be the same.

━━━━━━━━ **Work Problem ④ at the Side.**

④ Susan donated $10 more than twice what LuAnn donated. If Susan donated $22, how much did LuAnn donate?

Show your work for each of the six problem-solving steps.

4. *Step 1* Asks for amount of LuAnn's donation.
Step 2 Let d be LuAnn's donation.
Step 3 $2d + 10 = 22$ (or, $10 + 2d = 22$)
Step 4
$$2d + 10 - 10 = 22 - 10$$
$$2d = 12$$
$$\frac{2d}{2} = \frac{12}{2}$$
$$d = 6$$
Step 5 LuAnn donated $6.
Step 6 Susan donated $10 more than twice $6, which is $10 + 2 \cdot \$6 = \$10 + \$12 = \$22$. That matches amount given in problem for Susan.

45. A board is 78 cm long. Rosa cut the board into two pieces, with one piece 10 cm longer than the other. Find the length of both pieces. (*Hint:* Make a sketch of the board.)

longer piece   shorter piece

46. A rope is 50 ft long. Juan cut it into two pieces so that one piece is 8 ft longer than the other. Find the length of each piece.

47. A wire is cut into two pieces, with one piece 7 ft shorter than the other. The wire was 31 ft long before it was cut. How long was each piece?

48. A 90 cm pipe is cut into two pieces so that one piece is 6 cm shorter than the other. Find the length of each piece.

In Exercises 49–54, use the formula for the perimeter of a rectangle, $P = 2l + 2w$. Make a drawing to help you solve each problem, and use the six problem-solving steps. See Example 6.

49. The perimeter of a rectangle is 48 m. The width is 5 m. Find the length.

50. The length of a rectangle is 27 cm, and the perimeter is 74 cm. Find the width of the rectangle.

51. A rectangular dog pen is twice as long as it is wide. The perimeter of the pen is 36 ft. Find the length and the width of the pen.

52. A new city park is a rectangular shape. The length is triple the width. It will take 240 yd of fencing to go around the park. Find the length and width of the park.

53. The length of a rectangular jewelry box is 3 inches more than twice the width. The perimeter is 36 inches. Find the length and the width.

54. The perimeter of a rectangular house is 122 ft. The width is 5 ft less than the length. Find the length and the width.

SUMMARY

9.1	**negative numbers**	Negative numbers are numbers that are less than 0.
	signed numbers	Signed numbers include positive numbers, negative numbers, and 0.
	absolute value	Absolute value is the distance of a number from 0 on a number line. Absolute value is never negative.
	opposite of a number	The opposite of a number is a number the same distance from 0 on a number line, but on the opposite side of 0.
9.2	**additive inverse**	The additive inverse is the opposite of a number. The sum of a number and its additive inverse is always 0.
9.5	**variables**	Variables are letters that represent numbers.
	expression	An expression is a combination of operations on variables and numbers.
9.6	**equation**	An equation is a statement that says two expressions are equal. An equation contains an equal sign.
	solution	The solution of an equation is a number that can replace the variable so that the equation is true.
	addition property of equations	The addition property of equations states that the same number can be added or subtracted on each side of an equation.
	multiplication property of equations	The multiplication property of equations states that each side of an equation can be multiplied or divided by the same number, except division by 0 is not allowed.
9.7	**distributive property**	If a, b, and c are three numbers, the distributive property says that $a(b + c) = ab + ac$.
	like terms	Like terms are terms with exactly the same variables and the same exponents.

See how well you have learned the vocabulary in this chapter. Answers follow the Quick Review.

1. An **expression**
  (a) has an equal sign
  (b) contains like terms with the same variables and exponents
  (c) follows the order of operations
  (d) is a combination of operations on variables and numbers.

2. An **equation**
  (a) always has more than one solution
  (b) can be graphed on a number line
  (c) is evaluated by adding the same number to each side
  (d) contains an equal sign.

3. The **absolute value** of a number is
  (a) its distance from 0
  (b) never positive
  (c) used when multiplying
  (d) less than 0.

4. **Like terms**
  (a) are always positive
  (b) are the same distance from 0 but in opposite directions
  (c) have matching variables and exponents
  (d) can be multiplied but not added.

5. The **opposite** of a number is
  (a) never negative
  (b) called the additive inverse
  (c) called the absolute value
  (d) at the same point on the number line.

6. A **variable** is
  (a) a solution of an equation
  (b) a negative number
  (c) a letter representing a number
  (d) one of several like terms.

Concepts	Examples

9.1 Graphing Signed Numbers
Place a dot at the correct location on the number line. Positive numbers are to the right of 0; negative numbers are to the left of 0.

Graph -2, 1, 0, and $2\frac{1}{2}$.

9.1 Using the < and > Symbols
Place the symbols < (less than) or > (greater than) between two numbers to make the statement true. The small pointed end of the symbol points to the smaller number.

Use the symbol < or > to make each statement true.

$2 \underline{\quad > \quad} 1$

$-3 \underline{\quad > \quad} -5$

$-6 \underline{\quad < \quad} 2$

9.1 Finding the Absolute Value of a Number
Determine the distance from 0 to the given number on the number line. Because the absolute value is a distance, it is never negative.

Simplify each absolute value expression.
(a) $|8|$ **(b)** $|-7|$ **(c)** $-|-5|$

$|8| = 8$ $|-7| = 7$ -5

9.1 Finding the Opposite of a Number
Find the number that is the same distance from 0 as the given number, but on the opposite side of 0 on a number line.

Find the opposite of each number.
(a) -6 **(b)** $+9$

$-(-6) = 6$ $-(+9) = -9$

9.2 Adding Two Signed Numbers
Adding Two Positive Numbers
Add the numerical values. The sum is positive.

Add.
(a) $8 + 6 = 14$

Adding Two Negative Numbers
Add the absolute values and write a negative sign in front of the sum.

(b) $-8 + (-6)$

Find absolute values.

$$|-8| = 8 \qquad |-6| = 6$$

Add absolute values: $8 + 6 = 14$.
Write a negative sign in front of the sum.

$$-8 + (-6) = -14$$

Adding Two Numbers with Different Signs
Subtract the smaller absolute value from the larger absolute value. Write the sign of the number with the larger absolute value in front of the answer.

(c) $5 + (-7)$

Find absolute values.

$$|5| = 5 \qquad |-7| = 7$$

Subtract the smaller absolute value from the larger: $7 - 5 = 2$.

The number with the larger absolute value is -7. Its sign is negative, so write a negative sign in front of the answer.

$$5 + (-7) = -2$$

9.2 Subtracting Two Signed Numbers
To subtract two numbers, add the first number to the opposite of the second. Follow these steps.
Step 1 Change the subtraction sign to addition.
Step 2 Change the sign of the second number to its opposite.
Step 3 Proceed as in addition.

Subtract.
(a) $-6 - \quad 5$

$-6 + (-5) = -11$

(b) $5 - (-8)$

$5 + (+8) = 13$

Concepts	Examples
9.3 *Multiplying Signed Numbers* The product of two numbers with the *same* sign is *positive.* The product of two numbers with *different* signs is *negative.*	Multiply. **(a)** $7 \cdot 3 = 21$ **(c)** $5 \cdot (-8) = -40$ **(b)** $(-3) \cdot 4 = -12$ **(d)** $(-9) \cdot (-6) = 54$
9.3 *Dividing Signed Numbers* Use the same rules as for multiplying signed numbers. When two numbers have the *same* sign, the quotient is *positive.* When two numbers have *different* signs, the quotient is *negative.*	Divide. **(a)** $\dfrac{8}{4} = 2$ **(b)** $\dfrac{-20}{5} = -4$ **(c)** $\dfrac{50}{-5} = -10$ **(d)** $\dfrac{-12}{-6} = 2$

9.4 *Using the Order of Operations to Evaluate Numerical Expressions*
Use the order of operations to evaluate numerical expressions.
1. Do all operations inside **parentheses** or **other grouping symbols.**
2. Simplify any expressions with **exponents** and find any **square roots.**
3. **Multiply** or **divide,** proceeding from left to right.
4. **Add** or **subtract,** proceeding from left to right.
	Simplify.
	(a) $-4 + \underbrace{6 \div (-2)}$ **(b)** $3^2 \cdot 4 + 3 \cdot \underbrace{(8 \div 2)}$
	$\underbrace{-4 + \quad (-3)}$ $\underbrace{3^2 \cdot 4 + 3 \cdot \quad 4}$
	-7 $\underbrace{9 \cdot 4 + 3 \cdot \quad 4}$
	    $\underbrace{36 \quad + \quad 12}$
	    48

9.5 *Evaluating Expressions* Replace each variable in the expression with the specified number. Use the order of operations to simplify the expression.	What is the value of $6p - 5s$, if $p = -3$ and $s = -4$? $6 p - 5 s$ $\downarrow \downarrow \downarrow \downarrow$ $6(-3) - 5(-4)$ $\underbrace{-18 \; - (-20)}$ 2
9.6 *Determining Whether a Number Is a Solution of an Equation* Substitute the number for the variable in the equation. If the resulting statement is *true,* the number is a solution of the equation. If the resulting statement is *false,* the number is *not* a solution of the equation.	Is 4 a solution of the equation $3x - 5 = 7$? Replace x with 4. $$3(4) - 5 = 7$$ $$12 - 5 = 7$$ $$7 = 7 \quad \text{True}$$ The result is true, so 4 is a solution of $3x - 5 = 7$.
9.6 *Using the Addition Property of Equations to Solve an Equation* Add or subtract the same number on each side of the equation so that you get the variable by itself on one side.	Solve each equation. **(a)** $ x - 6 = 9$ $ x - 6 + 6 = 9 + 6 \quad$ Add 6 to each side. $ x + 0 = 15$ $ x = 15$ **(b)** $ -7 = x + 9$ $-7 - 9 = x + 9 - 9 \quad$ Subtract 9 from each side. $ -16 = x + 0$ $ -16 = x$

Concepts	Examples

9.6 *Using the Multiplication Property of Equations to Solve an Equation*

Multiply or divide each side of the original equation by the same number so that you get the variable by itself on one side. (Do not divide by 0.)

Solve each equation.

(a) $-54 = 6x$

$$\frac{-54}{6} = \frac{\overset{1}{\cancel{6}} \cdot x}{\cancel{6}}$$

$$-9 = x$$

(b) $\frac{1}{3}x = 8$

$$\frac{\overset{1}{\cancel{3}}}{1} \cdot \frac{1}{\cancel{3}}x = 3 \cdot 8$$

$$1x = 24$$
$$x = 24$$

9.7 *Solving Equations with Several Steps*

Use the following steps.

Step 1   Add or subtract the same amount on each side of the equation so that the variable ends up by itself on one side.

Step 2   Multiply or divide each side by the same number to find the solution.

Step 3   Check the solution.

Solve: $2p - 3 = 9$.

$$2p - 3 + 3 = 9 + 3 \qquad \text{Add 3 to each side.}$$
$$2p = 12$$

$$\frac{\overset{1}{\cancel{2}} \cdot p}{\cancel{2}} = \frac{12}{2} \qquad \text{Divide each side by 2.}$$

$$p = 6$$

Check: $2p - 3 = 9 \;\leftarrow\;$ Original equation
$$2(6) - 3 = 9 \qquad \text{Replace } p \text{ with 6.}$$
$$12 - 3 = 9$$
$$9 = 9 \qquad \text{True, so 6 is the solution.}$$

The solution is 6.

9.7 *Using the Distributive Property*

To simplify expressions, use the distributive property:

$$a(b + c) = ab + ac.$$

Simplify: $-2(x + 4)$
$$= -2 \cdot x + (-2) \cdot 4$$
$$= -2x - 8$$

9.7 *Combining Like Terms*

To combine like terms, add or subtract the number parts of the terms. The variable part stays the same.

Combine like terms.
(a) $6p + 7p = (6 + 7)p = 13p$
(b) $8m - 11m = (8 - 11)m = -3m$

9.8 *Translating Word Phrases into Expressions with Variables*

Use x (or any other letter) as a variable and symbolize the operations described by the words.

Write each word phrase in symbols, using x as the variable.
(a) Two more than a number $\qquad x + 2$
(b) 8 subtracted from a number $\qquad x - 8$
(c) The product of a number and 15 $\qquad 15x$
(d) A number divided by 9 $\qquad \dfrac{x}{9}$

ANSWERS TO TEST YOUR WORD POWER

1. (d) *Examples:* $3x - 5$; $2y^2$; $4a + b$.   **2. (d)** *Examples:* $2x + 6 = -4$; $-8 + y = 7y - 14$.   **3. (a)** *Example:* $|-3| = 3$ and $|3| = 3$ because the distance from 0 for both -3 and 3 is 3.   **4. (c)** *Examples:* $2y$ and $-7y$ are like terms; $4x^2$ and x^2 are like terms; $3y^2$ and $3y^3$ are not like terms.   **5. (b)** *Example:* The opposite of 5 is -5; it is the additive inverse because $5 + (-5) = 0$.   **6. (c)** *Examples:* x, y, and d are variables.

Chapter 9 REVIEW EXERCISES

[9.1] *Graph each set of numbers on the number line.*

1. $2, -3, 4, 1, 0, -5$

2. $-2, 5, -4, -1, 3, -6$

3. $-1\frac{1}{4}, -\frac{5}{8}, -3\frac{3}{4}, 2\frac{1}{8}, \frac{3}{2}, -2\frac{1}{8}$

4. $0, -\frac{3}{4}, \frac{5}{4}, -4\frac{1}{2}, \frac{7}{8}, -7\frac{2}{3}$

Write $<$ or $>$ in each blank to make a true statement.

5. 0 _____ -2

6. -5 _____ 0

7. -1 _____ -4

8. -9 _____ -6

Simplify each absolute value expression.

9. $|8|$

10. $|-19|$

11. $-|-7|$

12. $-|15|$

[9.2] *Add.*

13. $-4 + 6$

14. $-10 + 3$

15. $-11 + (-8)$

16. $-9 + (-24)$

17. $12 + (-11)$

18. $1 + (-20)$

19. $\frac{9}{10} + \left(-\frac{3}{5}\right)$

20. $-\frac{7}{8} + \frac{1}{2}$

21. $-6.7 + 1.5$

22. $-0.8 + (-0.7)$

Give the additive inverse (opposite) of each number.

23. 6

24. -14

25. $-\dfrac{5}{8}$

26. 3.75

Subtract.

27. $4 - 10$

28. $7 - 15$

29. $-6 - 1$

30. $-12 - 5$

31. $8 - (-3)$

32. $2 - (-9)$

33. $-1 - (-14)$

34. $-10 - (-4)$

35. $-40 - 40$

36. $-15 - (-15)$

37. $\dfrac{1}{3} - \dfrac{5}{6}$

38. $2.8 - (-6.2)$

[9.3] *Multiply or divide.*

39. $-4 \cdot 6$

40. $5 \cdot (-4)$

41. $-3 \cdot (-5)$

42. $-8 \cdot (-8)$

43. $\dfrac{80}{-10}$

44. $\dfrac{-9}{3}$

45. $\dfrac{-25}{-5}$

46. $\dfrac{-120}{-6}$

47. $(-37)(0)$

48. $(-1)(81)$

49. $\dfrac{0}{-10}$

50. $\dfrac{-20}{0}$

51. $\left(\dfrac{2}{3}\right)\left(-\dfrac{6}{7}\right)$

52. $-\dfrac{4}{5} \div \left(-\dfrac{2}{15}\right)$

53. $(-0.5)(-2.8)$

54. $\dfrac{-5.28}{0.8}$

[9.4] *Use the order of operations to simplify each expression.*

55. $2 - 11 \cdot (-5)$

56. $(-4) \cdot (-8) - 9$

57. $48 \div (-2)^3 - (-5)$

58. $-36 \div (-3)^2 - (-2)$

59. $5 \cdot 4 - 7 \cdot 6 + 3 \cdot (-4)$

60. $2 \cdot 8 - 4 \cdot 9 + 2 \cdot (-6)$

61. $-4 \cdot 3^3 - 2 \cdot (5 - 9)$

62. $6 \cdot (-4)^2 - 3 \cdot (7 - 14)$

63. $\dfrac{3 - (5^2 - 4^2)}{14 + 24 \div (-3)}$

64. $(-0.8)^2 \cdot (0.2) - (-1.2)$

65. $\left(-\dfrac{1}{3}\right)^2 + \dfrac{1}{4} \cdot \left(-\dfrac{4}{9}\right)$

66. $\dfrac{12 \div (2 - 5) + 12 \cdot (-1)}{2^3 - (-4)^2}$

[9.5] *Find the value of each expression using the given values of the variables.*

67. $3k + 5m$
$k = 4, \quad m = 3$

68. $3k + 5m$
$k = -6, \quad m = 2$

69. $2p - q$
$p = -5, \quad q = -10$

70. $2p - q$
$p = 6, \quad q = -7$

71. $\dfrac{5a - 7y}{2 + m}$
$a = 1, \quad y = 4, \quad m = -3$

72. $\dfrac{5a - 7y}{2 + m}$
$a = 2, \quad y = -2, \quad m = -26$

Use the formula and the given values of the variables to find the value of the remaining variable.

73. $P = a + b + c; \quad a = 9, \quad b = 12, \quad c = 14$

74. $A = \dfrac{1}{2}bh; \quad b = 6, \quad h = 9$

[9.6–9.7] *Solve each equation. Check each solution.*

75. $y + 3 = 0$

76. $a - 8 = 8$

77. $-5 = z - 6$

78. $-8 = -9 + r$

79. $-\dfrac{3}{4} + x = -2$

80. $12.92 + k = 4.87$

81. $-8r = 56$

82. $3p = 24$

83. $\dfrac{z}{4} = 5$

84. $\dfrac{a}{5} = -11$

85. $20 = 3y - 7$

86. $-5 = 2b + 3$

Use the distributive property to simplify each expression.

87. $6(r - 5)$

88. $11(p + 7)$

89. $-9(z - 3)$

90. $-8(x + 4)$

Combine like terms.

91. $3r + 8r$

92. $10z - 15z$

93. $3p - 12p + p$

94. $-6x - x + 9x$

Solve each equation. Check each solution.

95. $-4z + 2z = 18$

96. $-35 = 9k - 2k$

97. $4y - 3 = 7y + 12$

98. $b + 6 = 3b - 8$

99. $-14 = 2(a - 3)$

100. $42 = 7(t + 6)$

[9.8] *Write each word phrase in symbols, using x to represent the variable.*

101. 18 plus a number

102. Half a number

103. −5 times a number

104. A number subtracted from 20

Translate each sentence into an equation and solve it.

105. The sum of four times a number and 6 is −14. Find the number.

106. If a number is subtracted from five times the number, the result is 100. Find the number.

Solve each application problem using the six problem-solving steps.

107. A $30,000 scholarship is being divided between two students so that one student receives three times as much as the other. How much will each student receive?

108. The perimeter of a rectangular game board is 48 in. The length is 4 in. more than the width. Find the length and width of the game board.

MIXED REVIEW EXERCISES

Add, subtract, multiply, or divide as indicated.

109. $-6 - (-9)$

110. $-8 \cdot (-5)$

111. $-12 + 11$

112. $\dfrac{-70}{10}$

113. $-4 \cdot 4$

114. $5 - 14$

115. $\dfrac{-42}{-7}$

116. $16 + (-11)$

117. $-10 - 10$

118. $\dfrac{-5}{0}$

119. $-\dfrac{2}{3} + \dfrac{1}{9}$

120. $0.7(-0.5)$

121. $|-6| + 2 - 3 \cdot (-8) - 5^2$

122. $9 \div |-3| + 6 \cdot (-5) + 2^3$

Solve each equation.

123. $-45 = -5y$

124. $b - 8 = -12$

125. $6z - 3 = 3z + 9$

126. $-5 = r + 5$

127. $-3x = 33$

128. $2z - 7z = -15$

129. $3(k - 6) = 6 - 12$

130. $6(t + 3) = -2 + 20$

131. $-10 = \dfrac{a}{5} - 2$

132. $4 + 8p = 4p + 16$

Solve each application problem using the six problem-solving steps.

133. The recommended daily intake of iron for an adult female is 3 mg more than twice the recommended amount for a newborn infant. The amount for an adult female is 15 mg. How much should the infant receive? (*Source:* Food and Nutrition Board.)

134. In the U.S. Congress, the number of representatives is 65 less than five times the number of senators. There are a total of 535 members of Congress. Find the number of senators and the number of representatives. (*Source: World Almanac,* 2001.)

135. A cheetah's sprinting speed is 25 miles per hour (mph) faster than a zebra can run. The sum of their running speeds is 111 mph. How fast can each animal run? (*Source: Grolier Multimedia Encyclopedia.*)

136. A rectangular farm field of corn has a perimeter of 1330 yd. The length of the field is six times the width. Find the length and width of the field.

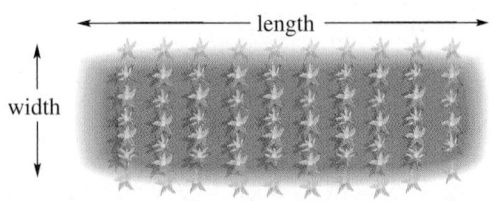

Chapter 9 **TEST**

Work each problem.

1. Graph the numbers -4, -1, $1\frac{1}{2}$, 3, and 0 on the number line at the right.

2. Write $<$ or $>$ in each blank to make a true statement.

-3 _____ 0 -4 _____ -8

3. Find $|-7|$ and $|15|$.

Add, subtract, multiply, or divide.

4. $-8 + 7$

5. $-11 + (-2)$

6. $6.7 + (-1.4)$

7. $8 - 15$

8. $4 - (-12)$

9. $-\frac{1}{2} - \left(-\frac{3}{4}\right)$

10. $8(-4)$

11. $-7(-12)$

12. $-16 \cdot 0$

13. $\frac{-100}{4}$

14. $\frac{-24}{-3}$

15. $-\frac{1}{4} \div \frac{5}{12}$

Use the order of operations to simplify each expression.

16. $-5 + 3 \cdot (-2) - (-12)$

17. $2 - (6 - 8) - (-5)^2$

Find the value of $8k - 3m$, given each set of values.

18. $k = -4$, $m = 2$

19. $k = 3$, $m = -1$

20. In Exercises 18 and 19, you were evaluating an expression. Explain the difference between evaluating an expression and solving an equation.

1.

$-5\ -4\ -3\ -2\ -1\ \ 0\ \ 1\ \ 2\ \ 3\ \ 4\ \ 5$

2. _____

3. _____

4. _____

5. _____

6. _____

7. _____

8. _____

9. _____

10. _____

11. _____

12. _____

13. _____

14. _____

15. _____

16. _____

17. _____

18. _____

19. _____

20. _____

21. _____

21. The formula for the area of a triangle is $A = \frac{1}{2}bh$. Find A, if $b = 20$ and $h = 11$.

Solve each equation.

22. _____

22. $x - 9 = -4$

23. $30 = -1 + r$

23. _____

24. _____

24. $3t - 8t = -25$

25. $\frac{p}{5} = -3$

25. _____

26. _____

27. _____

26. $-15 = 3(a - 2)$

27. $3m - 5 = 7m - 13$

Solve each application problem using the six problem-solving steps.

28. _____

28. A board is 118 cm long. Karin cut it into two pieces, with one piece 4 cm longer than the other. Find the length of both pieces.

29. _____

29. The perimeter of a rectangular building is 420 ft. The length is four times as long as the width. Find the length and the width. Make a drawing to help solve this problem.

30. _____

30. Marcella and her husband Tim spent a total of 19 hours redecorating their living room. Tim spent 3 hours less time than Marcella. How long did each person work on the room?

First use front end rounding to round each number and estimate an answer. Then find the exact answer. Write answers to fraction problems in lowest terms and as whole or mixed numbers when possible.

1. *Estimate:*   *Exact:*
$$
\begin{array}{r}
8.7 \\
0.902 \\
+\ \ \ \ \ \ \\
\hline
\end{array}
\qquad
\begin{array}{r}
8.7 \\
0.902 \\
+\ 41 \\
\hline
\end{array}
$$

2. *Estimate:*   *Exact:*
$$
\begin{array}{r}
6.27 \\
\times\ \ \ \ \ \ \\
\hline
\end{array}
\qquad
\begin{array}{r}
6.27 \\
\times\ 49.2 \\
\hline
\end{array}
$$

3. *Estimate:*   *Exact:*
$$
\overline{}
\qquad
78\overline{)39{,}234}
$$

4. *Estimate:*   *Exact:*
$$
\begin{array}{r}
3\dfrac{3}{5} \\[6pt]
-\ 2\dfrac{3}{4} \\
\hline
\end{array}
$$

5. *Exact:*

$$5\dfrac{5}{6} \cdot \dfrac{9}{10} = \underline{}$$

Estimate:

$$\underline{} \cdot \underline{} = \underline{}$$

6. *Exact:*

$$4\dfrac{1}{6} \div 1\dfrac{2}{3} = \underline{}$$

Estimate:

$$\underline{} \div \underline{} = \underline{}$$

Add, subtract, multiply, or divide as indicated.

7. $17 - 8.094$

8. $(1309)(408)$

9. $4.06 \div 0.072$
Round to nearest tenth.

10. $-12 + 7$

11. $-5(-8)$

12. $-3 - (-7)$

13. $3.2 + (-4.5)$

14. $\dfrac{30}{-6}$

15. $\dfrac{1}{4} - \dfrac{3}{4}$

Use the order of operations to simplify each expression.

16. $45 \div \sqrt{25} - 2 \cdot 3 + (10 \div 5)$

17. $-6 - (4 - 5) + (-3)^2$

Write $<$ or $>$ in each blank to make a true statement.

18. $\dfrac{3}{10}$ _____ $\dfrac{4}{15}$

19. 0.7072 _____ 0.72

20. -5 _____ -2

21. Write 8% as a decimal and as a fraction in lowest terms.

22. Write $4\dfrac{1}{2}$ as a decimal and as a percent.

The bar graph shows the average number of tornadoes each year in selected states. Use the graph to answer Exercise 23.

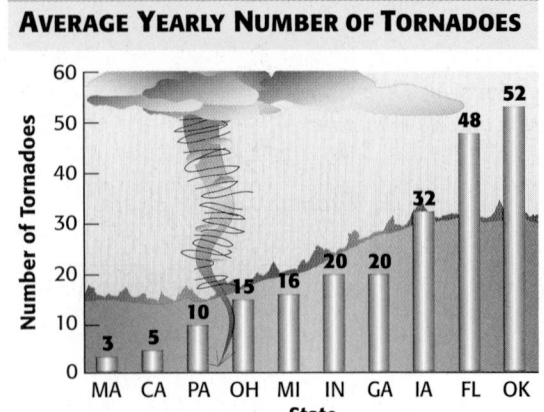

AVERAGE YEARLY NUMBER OF TORNADOES

Source: The Weather Book.

23. Write the ratio that compares the numbers of tornadoes in each pair of states. Write all ratios as fractions in lowest terms.

(a) Iowa (IA) to Michigan (MI)

(b) California (CA) to Ohio (OH)

(c) Georgia (GA) to Indiana (IN)

(d) Florida (FL) to Oklahoma (OK)

Find the unknown number in each proportion. Round your answer to hundredths if necessary.

24. $\dfrac{x}{12} = \dfrac{1.5}{45}$

25. $\dfrac{350}{x} = \dfrac{3}{2}$

26. $\dfrac{38}{190} = \dfrac{9}{x}$

Solve each percent problem.

27. 0.5% of 3000 students is how many students?

28. What percent of 12.5 miles is 6.8 miles?

29. 90 cars is 180% of what number of cars?

30. $5.80 is what percent of $145?

Convert these measurements.

31. $3\dfrac{1}{2}$ gal to quarts

32. 72 hr to days

33. 3.7 L to milliliters

34. 40 cm to meters

Write the most reasonable metric unit in each blank. Choose from km, m, cm, mm, L, mL, kg, g, and mg.

35. The building is 15 _____ high.

36. Rita took 15 _____ of cough syrup.

37. Bruce walked 2 _____ to work.

38. The robin weighs 100 _____ .

39. Identify which of these temperatures is normal body temperature and which is comfortable room temperature: 52 °C, 70 °C, 20 °C, 10 °C, 37 °C, 98 °C.

Find the perimeter (or circumference) and the area of each figure. Use 3.14 as the approximate value of π.

40.

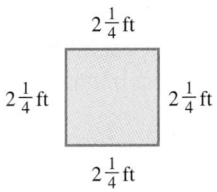

$2\frac{1}{4}$ ft

$2\frac{1}{4}$ ft $2\frac{1}{4}$ ft

$2\frac{1}{4}$ ft

41.

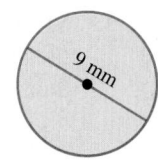

9 mm

42.

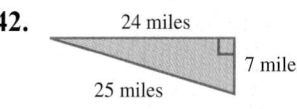

24 miles

7 miles

25 miles

43.

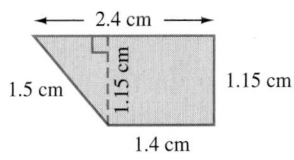

2.4 cm

1.5 cm   1.15 cm   1.15 cm

1.4 cm

44.

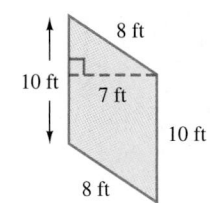

8 ft

10 ft   7 ft

10 ft

8 ft

45.

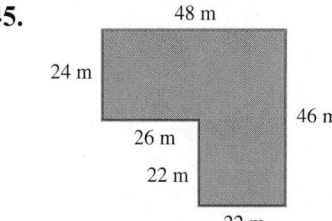

48 m

24 m

46 m

26 m

22 m

22 m

Find the unknown length in each figure. Round your answers to the nearest tenth if necessary.

46.

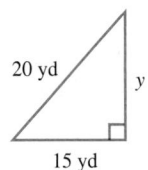

20 yd

y

15 yd

47. Similar triangles

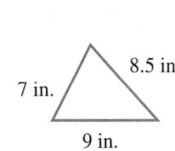

8.5 in.

7 in.

9 in.

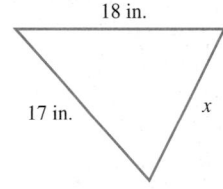

18 in.

17 in.

x

Solve each equation.

48. $-20 = 6 + y$

49. $-2t - 6t = 40$

50. $3x + 5 = 5x - 11$

51. $6(p + 3) = -6$

In Exercise 52, write an equation and solve it. Then solve Exercises 53 and 54 using the six problem-solving steps.

52. If 40 is added to four times a number, the result is zero. Find the number.

53. $1000 in prize money is being split between Reggie and Donald. Donald should get $300 more than Reggie. How much will each man receive?

54. Make a drawing to help solve this problem. The length of a photograph is 5 cm more than the width. The perimeter of the photograph is 82 cm. Find the length and the width.

Solve each application problem. Round money answers to the nearest cent.

55. Portia bought two CDs at $14.98 each. The sales tax rate is $6\frac{1}{2}\%$. Find the total amount charged to Portia's credit card.

56. Brian's spaghetti sauce recipe calls for $3\frac{1}{3}$ cups of tomato sauce. He wants to make $2\frac{1}{2}$ times the usual amount. How much tomato sauce does he need?

57. The local food pantry received 2480 lb of food this month. Their goal was 2000 lb. What percent of their goal was received?

58. Sayoko bought 720 g of chicken priced at $5.97 per kilogram. How much did she pay for the chicken?

59. A packing crate measures 2.4 m long, 1.2 m wide, and 1.2 m high. A trucking company wants crates that hold 4 m³. The crate's volume is how much more or less than 4 m³?

60. The Mercado family needs 35 ft of fencing to put around a rectangular garden plot. If the plot is $6\frac{1}{2}$ ft wide, find its length.

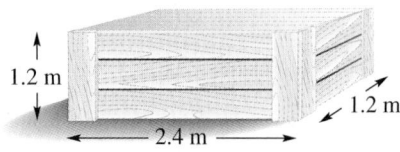

61. Rich spent 25 minutes reading 14 pages in his sociology textbook. At that rate, how long will it take him to read 30 pages? Round to the nearest whole number of minutes.

62. Jackie drove her car 364 miles on 14.5 gallons of gas. Maya used 16.3 gallons to drive 406 miles. Naomi drove 300 miles on 11.9 gallons. Which car had the highest number of miles per gallon? How many miles per gallon did that car get, rounded to the nearest tenth?

Statistics

10

As an information society, we are constantly being bombarded with facts and numbers. The ability to understand and interpret the many types of information and the many ways it is presented has become essential. For example, if you own or manage a store, you need to understand and interpret the sales history to make wise managerial decisions. (See Section 10.2, Exercises 25–32.)

10.1 CIRCLE GRAPHS

❶ Use the circle graph to answer each question.

(a) The greatest number of hours is spent in which activity?

(b) How many more hours are spent working than studying?

(c) Find the total number of hours spent studying, working, and attending classes.

❷ Use the circle graph to find each ratio. Write the ratios as fractions in lowest terms.

(a) Hours spent driving to whole day

(b) Hours spent sleeping to whole day

(c) Hours spent attending class and studying to whole day

(d) Hours spent driving and working to whole day

The word *statistics* originally came from words that mean *state numbers.* State numbers refer to numerical information, or **data,** gathered by the government such as the number of births, deaths, or marriages in a population. Today, the word *statistics* has a much broader application; data from the fields of economics, social science, science, and business can all be organized and studied under the branch of mathematics called **statistics.**

1 **Read and understand a circle graph.** It can be hard to understand a large collection of data. The graphs described in this section can be used to help you make sense of such data. The **circle graph** is used to show how a total amount is divided into parts. The circle graph below shows you how 24 hours in the life of a college student are divided among different activities.

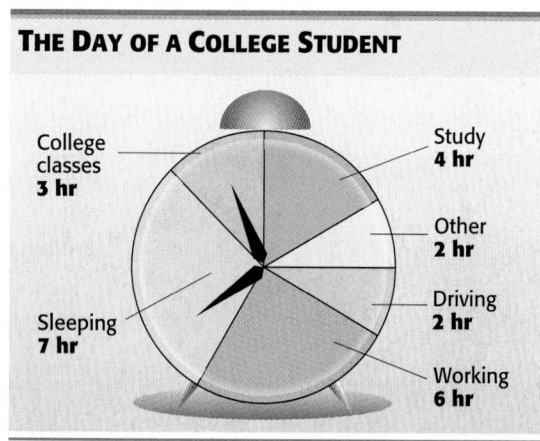

THE DAY OF A COLLEGE STUDENT

College classes 3 hr · Study 4 hr · Other 2 hr · Driving 2 hr · Working 6 hr · Sleeping 7 hr

Work Problem ❶ at the Side.

2 **Use a circle graph.** The above circle graph uses pie-shaped pieces called **sectors** to show the amount of time spent on each activity (the total must be 24 hours); a circle graph can therefore be used to compare the time spent on one activity to the total number of hours in the day.

Example 1 Using a Circle Graph

Find the ratio of time spent in college classes to the total number of hours in a day. Write the ratio as a fraction in lowest terms. (See **Section 5.1.**)

The circle graph shows that 3 of the 24 hours in a day are spent in class. The ratio of class time to the hours in a day is

$$\frac{3 \text{ hours (college classes)}}{24 \text{ hours (whole day)}} = \frac{3 \text{ hours}}{24 \text{ hours}} = \frac{3 \div 3}{24 \div 3} = \frac{1}{8}. \leftarrow \text{Lowest terms}$$

Work Problem ❷ at the Side.

This circle graph can also be used to find the ratio of the time spent on one activity to the time spent on any other activity.

Example 2 Finding a Ratio from a Circle Graph

Find the ratio of study time to class time.

The circle graph shows 4 hours spent studying and 3 hours spent in class. The ratio of study time to class time is

$$\frac{4 \text{ hours (study)}}{3 \text{ hours (class)}} = \frac{4 \text{ hours}}{3 \text{ hours}} = \frac{4}{3}.$$

=== **Work Problem ③ at the Side.**

A circle graph often shows data as percents. For example, the vending machine snack food sales in the United States were $26 billion in 1997. The circle graph below shows how sales were divided among various types of snack foods. The entire circle represents $26 billion in sales. Each sector represents the sales of one snack item as a percent of the total sales (the total must be 100%).

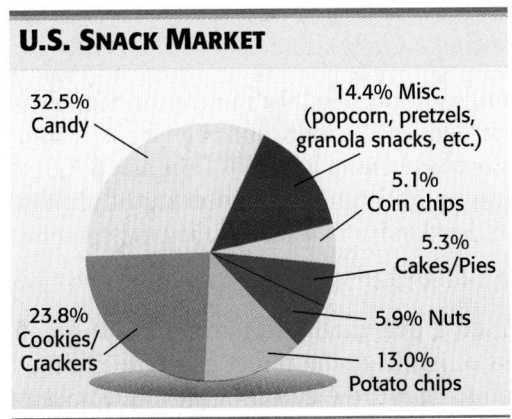

U.S. SNACK MARKET

- 32.5% Candy
- 14.4% Misc. (popcorn, pretzels, granola snacks, etc.)
- 5.1% Corn chips
- 5.3% Cakes/Pies
- 5.9% Nuts
- 13.0% Potato chips
- 23.8% Cookies/Crackers

Source: Natural Choice—USA.

Example 3 Calculating an Amount Using a Circle Graph

Use the circle graph on vending machine snack sales to find the amount spent on candy for the year.

Recall the percent equation:

part = percent • whole.

The total sales are $26 billion, so the whole is $26 billion. The percent is 32.5, or as a decimal, 0.325. Find the part.

$$\text{part} = \text{percent} \bullet \text{whole}$$
$$x = 0.325 \bullet 26 \text{ billion}$$
$$x = 8.45 \text{ billion}$$

The amount spent on candy was $8.45 billion or $8,450,000,000.

=== **Work Problem ④ at the Side.**

---

③ Use the circle graph to find the following ratios. Write the ratios as fractions in lowest terms.

(a) Hours spent in class to hours spent studying

(b) Hours spent working to hours spent sleeping

(c) Hours spent driving to hours spent working

(d) Hours spent in class to hours spent for "other"

④ Use the circle graph on vending machine snack sales to find the following.

(a) The amount spent on corn chips

(b) The amount spent on miscellaneous (popcorn, pretzels, granola snacks, etc.)

(c) The amount spent on cakes/pies

(d) The amount spent on cookies/crackers

ANSWERS

3. (a) $\frac{3}{4}$ (b) $\frac{6}{7}$ (c) $\frac{1}{3}$ (d) $\frac{3}{2}$

4. (a) $1.326 billion or $1,326,000,000
(b) $3.744 billion or $3,744,000,000
(c) $1.378 billion or $1,378,000,000
(d) $6.188 billion or $6,188,000,000

3 ▭ **Draw a circle graph.** The coordinator of the Fair Oaks Youth Soccer League organizes teams in five age groups. She places the players in various age groups as follows.

Age Group	Percent of Total
Under 8 years	20%
Ages 8–9	15%
Ages 10–11	25%
Ages 12–13	25%
Ages 14–15	15%
Total	100%

You can show these percents by using a circle graph. A circle has 360 degrees (written 360°). The 360° represents the entire league, or 100% of the soccer players.

Example 4 **Drawing a Circle Graph**

Using the data on *age groups,* find the number of degrees in the sector that would represent the "Under 8" group, and begin constructing a circle graph.

Recall that a complete circle has 360° (**Section 8.1**). Because the "Under 8" group makes up 20% of the total number of players, the number of degrees needed for the "Under 8" sector of the circle graph is 20% of 360°.

$$(360°)(20\%) = (360°)(0.2) = 72°$$

Use a tool called a **protractor** to make a circle graph. First, using a straight edge, draw a line from the center of a circle to the left edge. Place the hole in the protractor over the center of the circle, making sure that 0 on the protractor lines up with the line that was drawn. Find 72° and make a mark as shown in the illustration. Then remove the protractor and use the straight edge to draw a line from the center of the circle to the 72° mark at the edge of the circle.

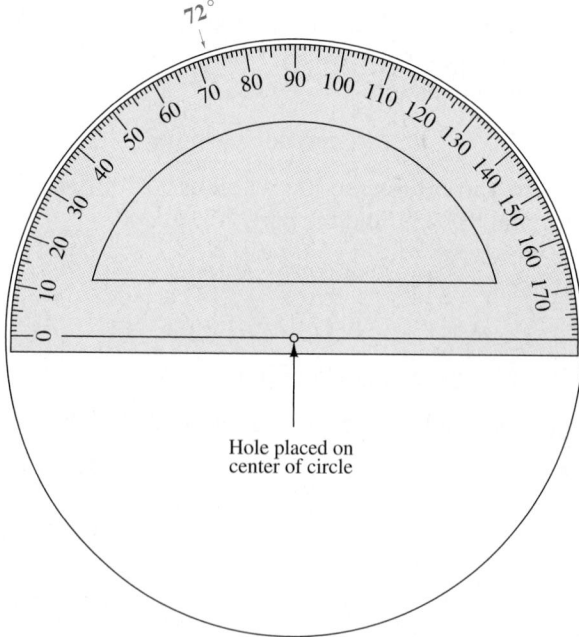

Hole placed on center of circle

Continued on Next Page

To draw the "Ages 8–9" sector, begin by finding the number of degrees in the sector.

$$(360°)(15\%) = (360°)(0.15) = 54°$$

Again, place the hole of the protractor at the center of the circle, but this time align 0 on the second line that was drawn. Make a mark at 54° and draw a line as before. This sector is 54° and represents the "Ages 8–9" group.

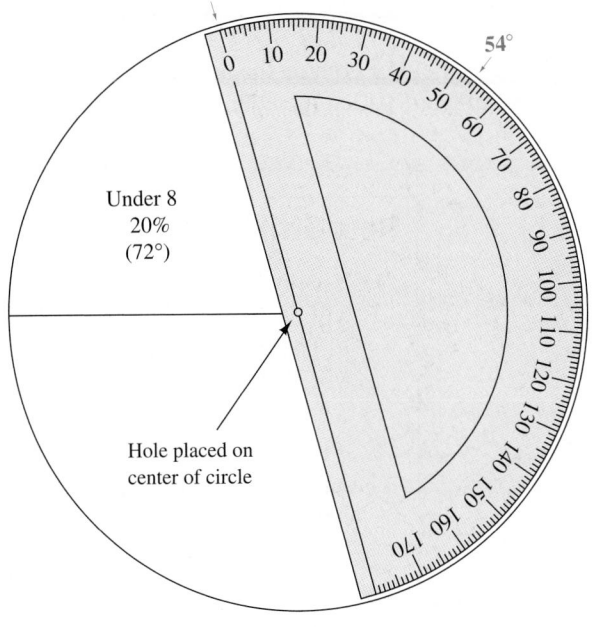

Line up 0 with previous 72° mark.

CAUTION

You must be certain that the hole in the protractor is placed on the exact center of the circle each time you measure the size of a sector.

Work Problem ❺ at the Side.

Use this circle for Problem 5 at the side.

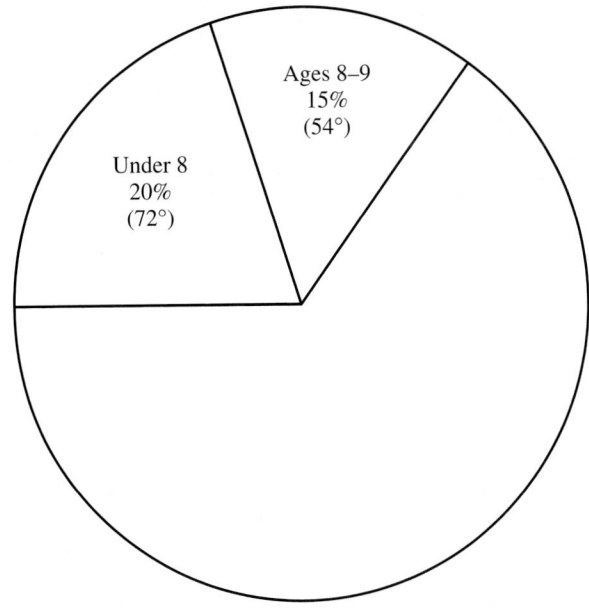

❺ Using the information on the soccer age groups in the table, find the number of degrees needed for each sector and complete the circle graph at the bottom left.

(a) Ages 10–11

(b) Ages 12–13

(c) Ages 14–15

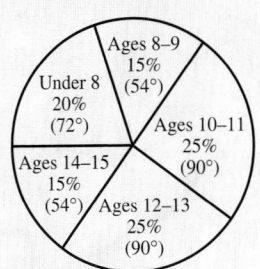

Just in Time

Antique clocks usually chime either on the hour or both on the hour and the half-hour. The mechanism that controls the chimes is a set of gears called the count plate and hammer wheel, and a lever called the count hook.

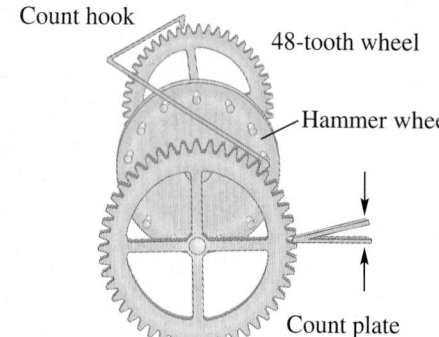

Count hook

48-tooth wheel

Hammer wheel

Count plate

Hour-Striking Clocks

1. If a clock chimes only on the hour, what will be the total number of chimes in a 12 hour period? (*Hint:* For example, it will chime 6 times at 6:00.)

2. If the count plate has one gear tooth for each chime, what fractional part of the count plate is one tooth?

3. How many degrees correspond to one gear tooth?

4. The mechanism designed to move the count plate has two wheels. The hammer wheel has 13 pins, which is combined with a wheel with 48 teeth and a pinion that fits 8 teeth. That gives a 6:1 ratio. What is significant about the combination of 13 and 6 that causes the count plate to move one gear tooth?

Hour-and-Half-Hour-Striking Clocks

5. If a clock chimes on the hour and also one time on each half-hour, what will be the total number of chimes in a 12 hour period?

6. If the count plate has one tooth for each chime, what fractional part of the count plate makes one tooth?

7. How many degrees correspond to one gear tooth?

8. If the count plate has a diameter of 2 in., what is the circumference of the count plate, rounded to the nearest thousandth? What is the width of each gear tooth, rounded to the nearest hundredth? (*Note:* Use $\pi \approx 3.14$.)

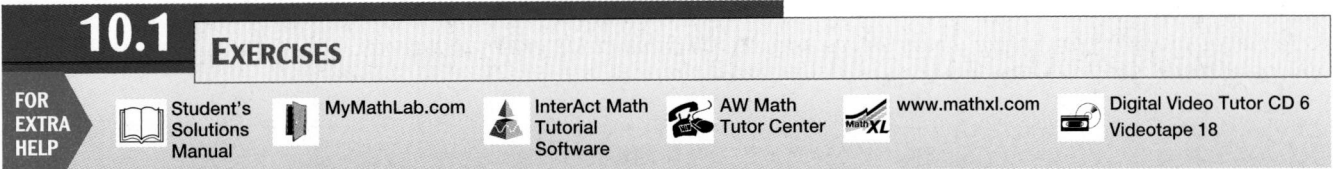

10.1 EXERCISES

| FOR EXTRA HELP | | Student's Solutions Manual | | MyMathLab.com | | InterAct Math Tutorial Software | | AW Math Tutor Center | | www.mathxl.com | | Digital Video Tutor CD 6 Videotape 18 |

This circle graph shows the cost of adding an art studio to an existing building. Use this circle graph to answer Exercises 1–6. Write ratios as fractions in lowest terms. See Examples 1 and 2.

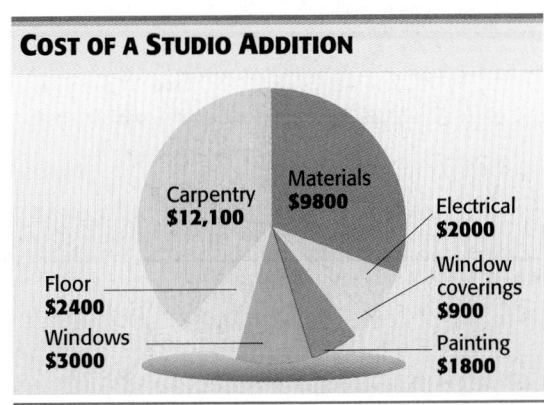

COST OF A STUDIO ADDITION

Carpentry $12,100

Materials $9800

Electrical $2000

Window coverings $900

Floor $2400

Windows $3000

Painting $1800

1. Find the total cost of adding the art studio.

2. What is the largest single expense in adding the studio?

3. Find the ratio of the cost of materials to the total remodeling cost.

4. Find the ratio of the cost of painting to the total remodeling cost.

5. Find the ratio of the cost of carpentry to the cost of window coverings.

6. Find the ratio of the cost of windows to the cost of the floor.

This circle graph, adapted from USA Today, *shows the number of people in a survey who gave various reasons for eating dinner at restaurants. Use this circle graph to answer Exercises 7–14. See Examples 1 and 2.*

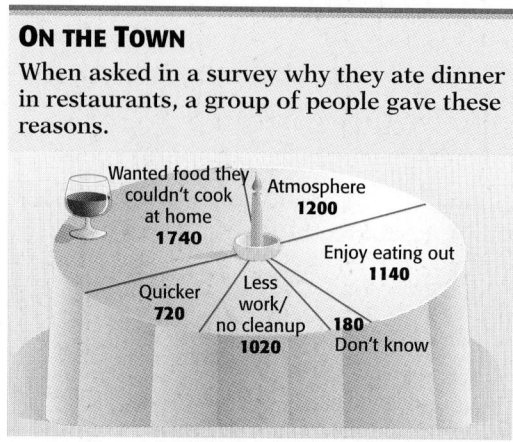

ON THE TOWN

When asked in a survey why they ate dinner in restaurants, a group of people gave these reasons.

Wanted food they couldn't cook at home **1740**

Atmosphere **1200**

Enjoy eating out **1140**

Quicker **720**

Less work/ no cleanup **1020**

180 Don't know

Source: Market Facts for Tyson Foods.

7. Which reason was given by the least number of people?

8. Which reason was given by the second highest number of people?

Find each ratio in Exercises 9–14. Write the ratios as fractions in lowest terms.

9. Those who said dining out is "Quicker" to total people in the survey

10. Those who said "Enjoy eating out" to the total people in the survey

11. Those who said "Less work/No clean up" to those who said "Atmosphere"

12. Those who said "Don't know" to those who said "Quicker"

13. Those who said "Wanted food they couldn't cook at home" to those who said "Less work/No clean up"

14. Those who said "Atmosphere" to those who said "Enjoy eating out"

This circle graph shows the costs necessary to comply with the Americans with Disabilities Act (ADA) at the Dos Pueblos College. Each cost item is expressed as a percent of the total cost of $1,740,000. Use the graph to find the dollar amount spent for the items in Exercises 15–20. See Example 3.

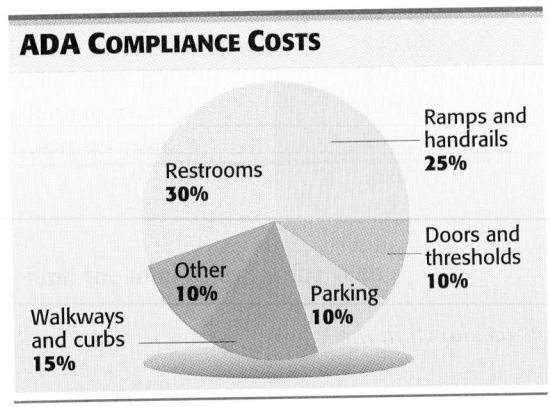

15. Restrooms

16. Ramps and handrails

17. Doors and thresholds

18. Parking

19. Walkways and curbs

20. Other

This circle graph, adapted from USA Today, *shows how consumers say they prefer to pay and keep track of their bills. If Lincoln Computer Software, Incorporated, employs 10,860 people, use the graph to find the number of employees who prefer to pay and keep track of their bills using each of the listed choices in Exercises 21–26. Round to the nearest whole number. See Example 3.*

21. Checks and paper records

22. Checks and personal accounting software

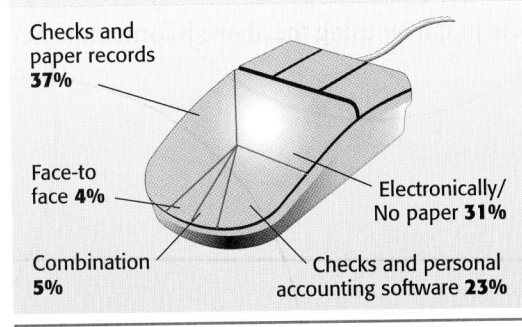

PAYMENT PREFERENCES

Consumers say they would prefer to pay and keep track of their bills in the following ways:

Source: Gallup for Visa.

23. Combination

24. Electronically/No paper

25. Face-to-face

26. Face-to-face and checks and personal accounting software combined

2 Use the grouped data for the insurance agency to answer each question.

(a) During how many weeks were less than 50 calls made?

(b) During how many weeks were 50 or more calls made?

NOTE

The number of class intervals in the left column of the table is arbitrary. Grouped data usually has between 5 and 15 class intervals.

Example 2 Analyzing a Frequency Distribution

Use the grouped data for the insurance agency (on the preceding page) to answer the following questions.

(a) During how many weeks were fewer than 30 calls made?

The first class in the grouped data table above (20–29) is the number of weeks during which fewer than 30 calls were made. Therefore, the owner made fewer than 30 calls during 9 weeks out of the 50 weeks shown.

(b) During how many weeks were 40 or more calls made?

The last four classes in the grouped data table are the number of weeks during which 40 or more calls were made.

$$13 + 5 + 4 + 6 = 28 \text{ weeks}$$

Work Problem 2 at the Side.

3 Read and understand a histogram. The results in the grouped data table have been used to draw this special bar graph, called a **histogram.** In a histogram, the width of each bar represents a range of numbers (*class interval*). The height of each bar in a histogram gives the *class frequency,* that is, the number of occurrences in each class interval.

3 Use the histogram to answer each question.

(a) During how many weeks were less than 60 calls made?

(b) During how many weeks were 60 or more calls made?

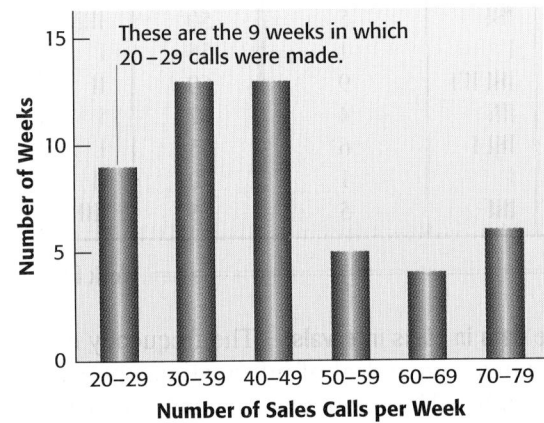

SALES CALL DATA FOR THE PAST 50 WEEKS (GROUPED DATA)

These are the 9 weeks in which 20–29 calls were made.

Number of Weeks (vertical axis)

Number of Sales Calls per Week: 20–29 30–39 40–49 50–59 60–69 70–79

Example 3 Using a Histogram

Use the histogram to find the number of weeks in which less than 40 calls were made.

Because 20–29 calls were made during 9 of the weeks and 30–39 calls were made during 13 of the weeks, the number of weeks in which less than 40 calls were made was $9 + 13 = 22$ weeks.

Work Problem 3 at the Side.

SUMMARY

KEY TERMS

10.1	**circle graph**	A circle graph shows how a total amount is divided into parts or sectors. It is based on percents of 360°.
	protractor	A protractor is a device (usually in the shape of a half-circle) used to measure the number of degrees in an angle or parts of a circle.
10.2	**bar graph**	A bar graph uses bars of various heights or length to show quantity or frequency.
	double-bar graph	A double-bar graph compares two sets of data by showing two sets of bars.
	line graph	A line graph uses dots connected by lines to show trends.
	comparison line graph	A comparison line graph shows how two sets of data relate to each other by showing a line graph for each item.
10.3	**frequency distribution**	A frequency distribution is a table that includes a column showing each possible number in the data collected. The original data is then entered in another column using a tally mark for each corresponding value. The tally marks are totaled and the totals are placed in a third column.
	histogram	A histogram is a bar graph in which the width of each bar represents a range of numbers (class interval) and the height represents the quantity or frequency of items that fall within the interval.
10.4	**mean**	The mean is the sum of all the values divided by the number of values. It is often called the *average*.
	weighted mean	The weighted mean is a mean calculated so that each value is multiplied by its frequency.
	median	The median is the middle number in a group of values that are listed from smallest to largest. It divides a group of values in half. If there are an even number of values, the median is the mean (average) of the two middle values.
	mode	The mode is the value that occurs most often in a group of values.
	dispersion	The dispersion is the variation or spread of the numbers.
	range	The range is a common measure of the dispersion of numbers. It is the difference between the largest value and the smallest value in the set of numbers.

Mean or average: $\text{mean} = \dfrac{\text{sum of all values}}{\text{number of values}}$

TEST YOUR WORD POWER

See how well you have learned the vocabulary in this chapter. Answers follow the Quick Review.

1. A **circle graph**
   (a) uses bars of various heights to show quantity or frequency
   (b) shows how a total amount is divided into parts or sectors
   (c) uses dots connected by lines to show trends
   (d) uses bars of various widths to represent a range of numbers.

2. A **bar graph**
   (a) uses bars of various heights to show quantity or frequency
   (b) shows how a total amount is divided into parts or sectors
   (c) uses dots connected by lines to show trends
   (d) uses bars of various widths to represent a range of numbers.

3. A **histogram** is a graph in which
   (a) tally marks are used to record original data
   (b) two sets of data are compared using two sets of bars
   (c) dots are connected by lines to show trends
   (d) the width of each bar represents a range of numbers and the height represents the frequency of items within that range.

4. A **protractor** is a device used to
   (a) construct a histogram
   (b) calculate measures of central tendency
   (c) measure the number of degrees in an angle or parts of a circle
   (d) compare two sets of data.

5. The **mean** is
   (a) calculated so that each value is multiplied by its frequency
   (b) the sum of all values divided by the number of values
   (c) the middle number in a group of values that are listed from smallest to largest
   (d) the value that occurs most often in a group of values.

6. The **mode** is
   (a) calculated so that each value is multiplied by its frequency
   (b) the sum of all values divided by the number of values
   (c) the middle number in a group of values that are listed from smallest to largest
   (d) the value that occurs most often in a group of values.

QUICK REVIEW

Concepts	Examples

10.1 Constructing a Circle Graph

1. Determine the percent of the total for each item.

2. Find the number of degrees out of 360° that each percent represents.

3. Use a protractor to measure the number of degrees for each item in the circle.

Construct a circle graph for the following table, which lists expenses for a business trip.

Item	Amount
Transportation	$200
Lodging	$300
Food	$250
Entertainment	$150
Other	$100
Total	$1000

Item	Amount	Percent of Total	Sector Size
Transportation	$200	$\frac{\$200}{\$1000} = \frac{1}{5} = 20\%$ so $360° \cdot 20\%$ $= 360 \cdot 0.20$	$= 72°$
Lodging	$300	$\frac{\$300}{\$1000} = \frac{3}{10} = 30\%$ so $360° \cdot 30\%$ $= 360 \cdot 0.30$	$= 108°$
Food	$250	$\frac{\$250}{\$1000} = \frac{1}{4} = 25\%$ so $360° \cdot 25\%$ $= 360 \cdot 0.25$	$= 90°$
Entertainment	$150	$\frac{\$150}{\$1000} = \frac{3}{20} = 15\%$ so $360° \cdot 15\%$ $= 360 \cdot 0.15$	$= 54°$
Other	$100	$\frac{\$100}{\$1000} = \frac{1}{10} = 10\%$ so $360° \cdot 10\%$ $= 360 \cdot 0.10$	$= 36°$

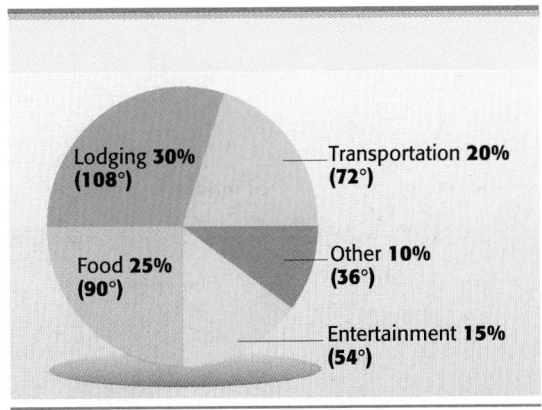

Concepts

10.2 *Reading a Bar Graph*

The height of the bar is used to show the quantity or frequency (number) in a specific category. Use a ruler or straight edge to line up the top of each bar with the numbers on the left side of the graph.

Examples

Use the bar graph below to determine the number of students who earned each letter grade.

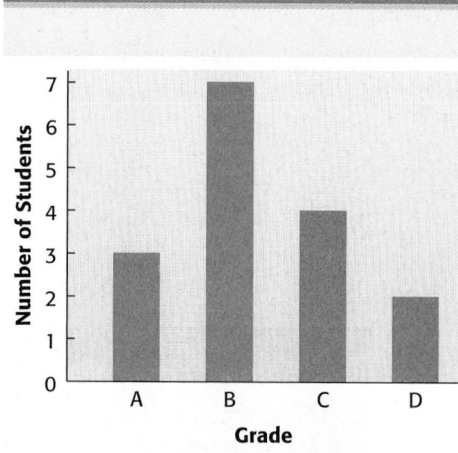

Grade	Number of Students
A	3
B	7
C	4
D	2

10.2 *Reading a Line Graph*

A dot is used to show the number or quantity in a specific class. The dots are connected with lines. This kind of graph is used to show a trend.

The line graph below shows the annual sales for the Fabric Supply Center for each of 4 years.

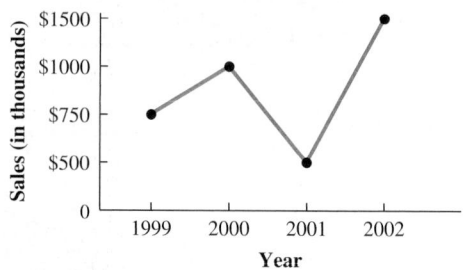

Find the sales in each year.

Year	Total Sales
1999	$750 · 1000 = $ 750,000
2000	$1000 · 1000 = $1,000,000
2001	$500 · 1000 = $ 500,000
2002	$1500 · 1000 = $1,500,000

Concepts

10.3 Preparing a Frequency Distribution and a Histogram from Raw Data

1. Construct a table listing each value, and the number of times this value occurs.

2. Divide the data into groups, categories, or classes.

3. Draw bars representing these groups to make a histogram.

Examples

Draw a histogram for the following list of student quiz scores.

12	15	15	14
13	20	10	12
11	9	10	12
17	20	16	17
14	18	19	13

Quiz Score	Tally	Frequency	
9	I	1	1st
10	II	2	class
11	I	1	interval
12	III	3	2nd
13	II	2	class
14	II	2	interval
15	II	2	3rd
16	I	1	class
17	II	2	interval
18	I	1	4th
19	I	1	class
20	II	2	interval

Class Interval (Quiz Scores)	Frequency (Number of Students)
9–11	4
12–14	7
15–17	5
18–20	4

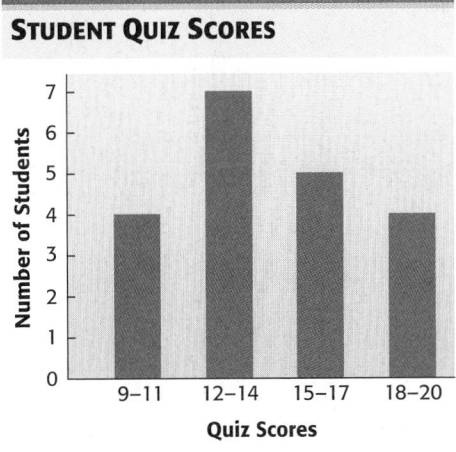

STUDENT QUIZ SCORES

Concepts	Examples

Concepts

10.4 *Finding the Mean (Average) of a Set of Numbers*

1. Add all values to obtain a total.
2. Divide the total by the number of values.

$$\text{mean (average)} = \frac{\text{sum of all values}}{\text{number of values}}$$

10.4 *Finding the Weighted Mean*

1. Multiply frequency by value.
2. Add all the products from Step 1.
3. Divide the sum in Step 2 by the total number of pieces of data.

Examples

The test scores for Keith Zagorin in his algebra course were as follows:

$$80 \quad 92 \quad 92 \quad 94$$
$$76 \quad 88 \quad 84 \quad 93$$

Find Keith's mean (average) test score to the nearest tenth.

$$\text{mean} = \frac{80 + 92 + 92 + 94 + 76 + 88 + 84 + 93}{8}$$

$$= \frac{689}{8} \approx 87.4$$

This table shows the distribution of the number of school-age children in a survey of 30 families.

Number of School-Age Children	Frequency (Number of Families)
0	12
1	6
2	7
3	3
4	2

Total of 30 families

Find the mean number of school-age children per family. Round to the nearest hundredth.

Value	Frequency	Product
0	12	$(0 \cdot 12) = 0$
1	6	$(1 \cdot 6) = 6$
2	7	$(2 \cdot 7) = 14$
3	3	$(3 \cdot 3) = 9$
4	2	$(4 \cdot 2) = 8$
Totals	**30**	**37**

$$\text{mean} = \frac{37}{30} \approx 1.23$$

The mean number of school-age children per family is 1.23.

Concepts	Examples
10.4 *Finding the Median of a Set of Numbers*	Find the median for Keith Zagorin's grades from the previous page.
1. Arrange the data from smallest to largest.	The data arranged from smallest to largest is as follows:
2. Select the middle value or the average of the two middle values, if there is an even number of values.	76 80 84 88 92 92 93 94
	Middle values
	The middle two values are 88 and 92. The average of these two values is
	$$\frac{88 + 92}{2} = 90$$
10.4 *Determining the Mode of a Set of Values*	Find the mode for Keith's grades in the previous example.
Find the value that appears most often in the list of values. If no value appears more than once, there is no mode. If two different values appear the same number of times, the list is bimodal.	The most frequently occurring score is 92 (it occurs twice). Therefore, the mode is 92.

ANSWERS TO TEST YOUR WORD POWER

1. (b) *Example:*

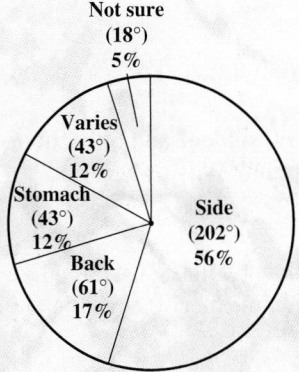

2. (a) *Example:*

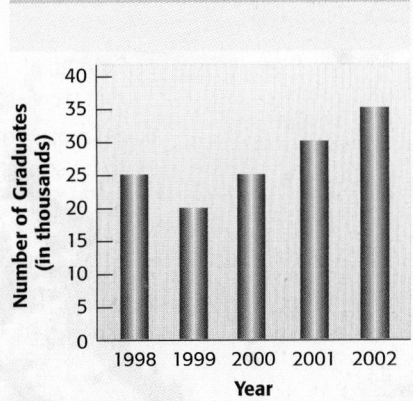

3. (d) *Example:*

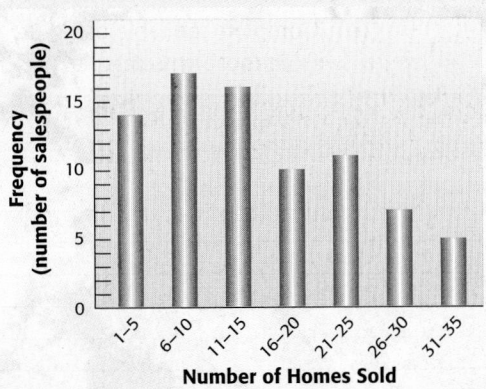

4. (c) *Example:*

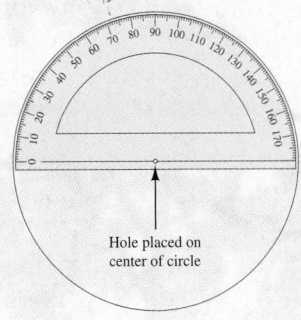

5. (b) *Example:* The mean of the values 5, 9, 7, 5, 2, and 8 is
$$\frac{5 + 9 + 7 + 5 + 2 + 8}{6} = \frac{36}{6} = 6.$$

6. (d) *Example:* The mode of the values 5, 9, 7, 5, 2, and 8 is 5 because 5 appears twice in the list.

Real-Data Applications

Surfing the Net

1. Look at the "Source" information at the bottom of the graph. How were the numbers in the graph obtained?

2. A researcher seeks information about members of a *population*. The individuals who are polled must be representative of the *population*. Describe the population that was targeted by this survey.

3. How many people in the poll said they cut back on television viewing to find time to use the Internet?

4. Find the number of people in the poll who cut back on each of the other activities listed in the graph.

5. Add up all the responses to the poll from Problem 4. Why is the total more than the 500 people that were in the poll?

6. Suppose you took a similar poll of 100 students at your school. Would you expect the results to be similar to those shown in the graph? Why or why not?

7. Suppose you first asked students if they regularly used the Internet, and then took a similar poll of 100 of those students. Would you expect the results to be similar to those shown in the graph? Why or why not?

8. Conduct a survey of your class members. First find out if they regularly use the Internet. Ask those who regularly use the Internet which of the activities they cut back on to have more surfing time. Each person polled can select more than one activity.

   (a) How many students are in your class poll?

   (b) Complete the table using the responses from those who regularly use the Internet.

Activity	Number Who Cut Back on the Activity	Percent Who Cut Back on the Activity
Television		
Video/computer games		
Sleeping		
Reading		
Seeing friends		
Work/school		
Other		
Exercising		

9. Make a bar graph showing your survey data. How is your data similar to the graph shown above? How is it different?

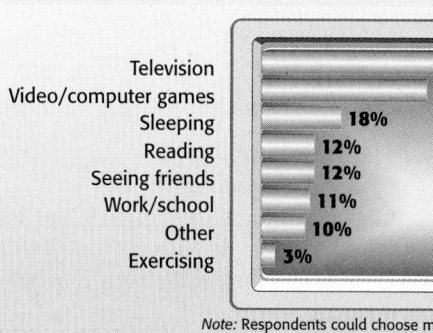

CAUGHT IN THE NET

Web users have cut back on the following activities get more online time:

- Television — 52%
- Video/computer games — 36%
- Sleeping — 18%
- Reading — 12%
- Seeing friends — 12%
- Work/school — 11%
- Other — 10%
- Exercising — 3%

Note: Respondents could choose more than one activity

Source: NUKE InterNETWORK poll of 500 regular users.

Chapter 10 REVIEW EXERCISES

[10.1]

1. This circle graph shows the cost of a family vacation. What is the largest single expense of the vacation? How much is that item?

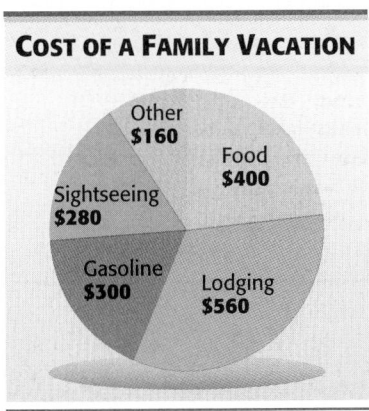

COST OF A FAMILY VACATION

Other $160

Food $400

Sightseeing $280

Gasoline $300

Lodging $560

Using the circle graph in Exercise 1, find each ratio. Write the ratios as fractions in lowest terms.

2. Cost of the food to the total cost of the vacation

3. Cost of gasoline to the total cost of the vacation

4. Cost of sightseeing to the total cost of the vacation

5. Cost of gasoline to the cost of the *other* category

6. Cost of the lodging to the cost of the food

[10.2] *This bar graph shows recent trends in employee benefits. It includes the most frequently offered "work perks" and the percent of the responding companies offering them. The survey was conducted on-line and included 4800 companies ranging in size from 2 to 5000 employees. Use this graph to find the number of companies offering each work perk listed in Exercises 7–10 and to answer Exercises 11 and 12. (Source: Work Perks Survey, Ceridian Employer Services, www.ces.ceridian.com)*

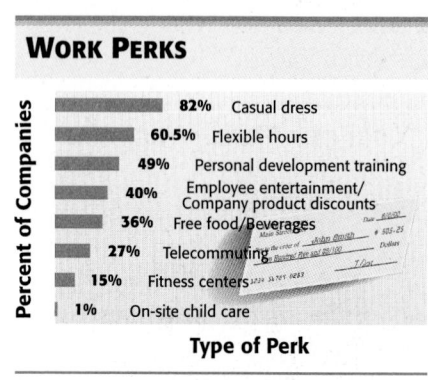

7. Casual dress

8. Free food/Beverages

9. Fitness centers

10. Flexible hours

11. Which two work perks do companies offer least often? Give one possible explanation why these work perks are not offered.

12. Which two work perks do companies offer most often? Give one possible explanation why these work perks are so popular.

This double-bar graph shows the number of acre-feet of water in Lake Natoma for each of the first six months of 2001 and 2002. Use this graph to answer Exercises 13–18.

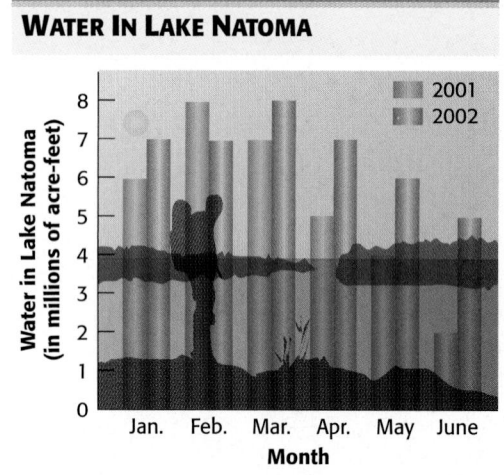

13. During which month in 2002 was the greatest amount of water in the lake? How much was there?

14. During which month in 2001 was the least amount of water in the lake? How much was there?

15. How many acre-feet of water were in the lake in June of 2002?

16. How many acre-feet of water were in the lake in May of 2001?

17. Find the decrease in the amount of water in the lake from March 2001 to June 2001.

18. Find the decrease in the amount of water in the lake from April 2002 to June 2002.

This comparison line graph shows the annual floor-covering sales of two different home improvement centers during each of five years. Use this graph to find the amount of annual floor-covering sales in each year shown in Exercises 19–22 and to answer Exercises 23 and 24.

 ANNUAL FLOOR-COVERING SALES

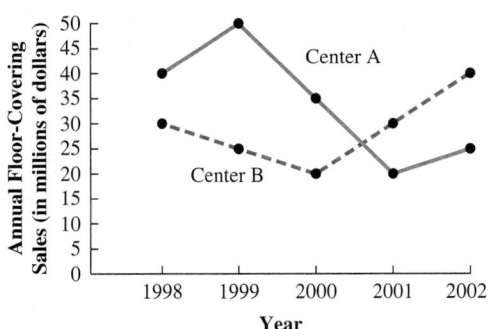

19. Center A in 1999

20. Center A in 2001

21. Center B in 2000

22. Center B in 2002

23. What trend do you see in Center A's sales from 1999 to 2002? Why might this have happened?

24. What trend do you see in Center B's sales starting in 2000? Why might this have happened?

[10.4] *Find the mean for each list of numbers. Round to the nearest tenth.*

25. Digital cameras sold: 18, 12, 15, 24, 9, 42, 54, 87, 21, 3

26. Number of harassment complaints filed: 31, 9, 8, 22, 46, 51, 48, 42, 53, 42

Find the weighted mean for each list. Round to the nearest tenth if necessary.

27.

Dollar Value	Frequency
$42	3
$47	7
$53	2
$55	3
$59	5

28.

Total Points	Frequency
243	1
247	3
251	5
255	7
263	4
271	2
279	2

Find the median for each list of numbers.

29. The number of accident forms filed: 43, 37, 13, 68, 54, 75, 28, 35, 39

30. Commissions of $576, $578, $542, $151, $559, $565, $525, $590

Find the mode or modes for each list of numbers.

31. Running shoes priced at $79, $56, $80, $79, $72, $86, $79

32. Boat launchings: 18, 25, 63, 32, 28, 37, 32, 26, 18

MIXED REVIEW EXERCISES

The Broadway Hair Salon spent $22,400 to open a new shop. This amount was spent as shown in the following chart. Find all the missing numbers in Exercises 33–37.

Item	Dollar Amount	Percent of Total	Degrees of Circle
33. Plumbing and electrical changes	$2240	10%	_____
34. Work stations	$7840	_____	126°
35. Small appliances	$4480	_____	72°
36. Interior decoration	$5600	_____	90°
37. Supplies	$2240	10%	_____

38. Draw a circle graph by using the information in Exercises 33–37.

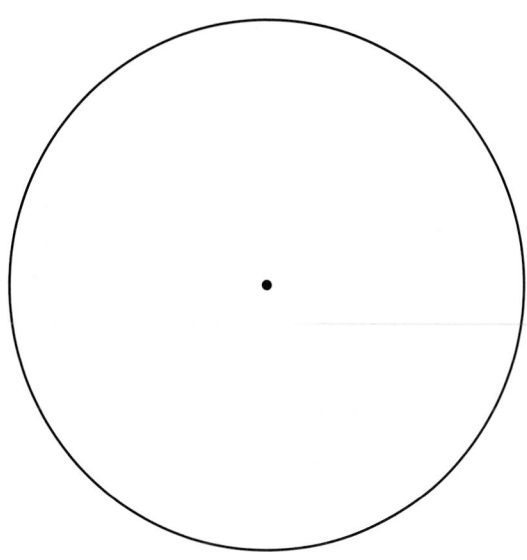

Find the mean for each list of numbers. Round answers to the nearest tenth if necessary.

39. Number of volunteers for the project: 48, 72, 52, 148, 180

40. Number of tacks in a handful: 122, 135, 146, 159, 128, 147, 168, 139, 158

Find the mode or modes for each list of numbers.

41. Job applicants meeting the qualifications: 48, 43, 46, 47, 48, 48, 43

42. Number of two-bedroom apartments in each building: 26, 31, 31, 37, 43, 51, 31, 43, 43

Find the median for each list of numbers.

43. Hours worked: 4.7, 3.2, 2.9, 5.3, 7.1, 8.2, 9.4, 1.0

44. Number of phone calls each hour: 35, 51, 9, 2, 17, 12, 46, 23, 3, 19, 39, 27

Here are the scores of 40 students on a computer science exam. Complete the table.

78	89	36	59	78	99	92	86
73	78	85	57	99	95	82	76
63	93	53	76	92	79	72	62
74	81	77	76	59	84	76	94
58	37	76	54	80	30	45	38

Class Intervals (Scores)	Tally	Class Frequency (Number of Students)
45. 30–39	_____	_____
46. 40–49	_____	_____
47. 50–59	_____	_____
48. 60–69	_____	_____
49. 70–79	_____	_____
50. 80–89	_____	_____
51. 90–99	_____	_____

52. Construct a histogram by using the data in Exercises 45–51.

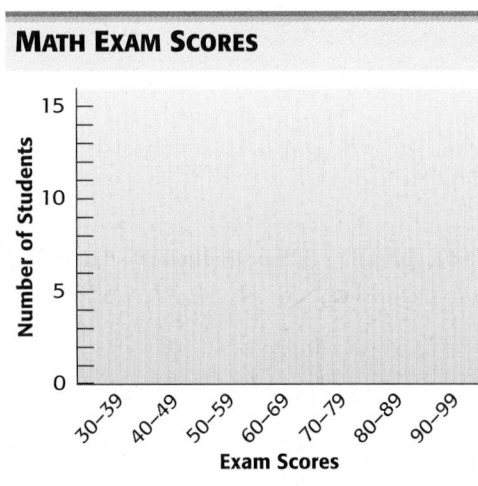

Find each weighted mean. Round answers to the nearest tenth if necessary.

53.

Test Score	Frequency
46	4
54	10
62	8
70	12
78	10

54.

Units Sold	Frequency
104	6
112	14
115	21
119	13
123	22
127	6
132	9

Chapter 10 **TEST**

This circle graph shows the advertising budget for Lakeland Amusement Park. Find the dollar amount budgeted for each category. The total advertising budget of Lakeland Amusement Park is $2,800,000.

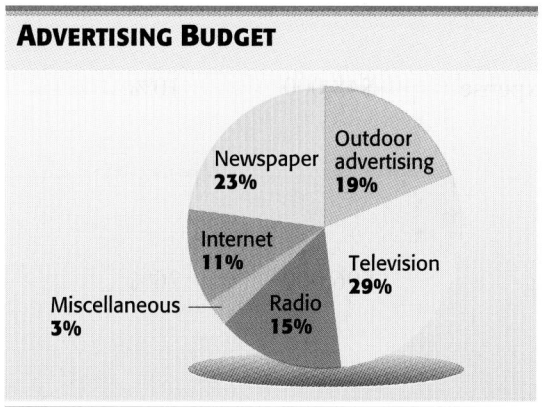

ADVERTISING BUDGET

1. Newspaper

1. _____

2. Outdoor advertising

2. _____

3. Television

3. _____

4. Radio

4. _____

5. Miscellaneous

5. _____

6. Internet

6. _____

23. Explain why a weighted mean must be used to determine a student grade point average. Calculate your own grade point average for last semester or quarter. If you are a new student, make up a grade point average problem of your own and solve it. Round to the nearest hundredth.

23. _____

24. _____

24. Explain in your own words the procedure for finding the median when there are an odd number of values in a list. Make up a problem with a list of five numbers and solve for the median.

25. _____

Find the weighted mean for the following. Round answers to the nearest tenth if necessary.

25.

Cost	Frequency
$12	5
$20	6
$22	8
$28	4
$38	6
$48	2

26.

Value	Frequency
150	15
160	17
170	21
180	28
190	19
200	7

26. _____

Find the median for each list of numbers.

27. _____

27. Low daily temperatures in degrees Fahrenheit: 32, 41, 28, 28, 37, 35, 16, 31

28. _____

28. The length of steel beams in meters: 7.6, 11.4, 6.2, 12.5, 31.7, 22.8, 9.1, 10.0, 9.5

Find the mode or modes for each list of numbers.

29. _____

29. Storage tank volume in kiloliters: 72, 46, 52, 37, 28, 18, 52, 61

30. _____

30. Hot tub temperatures (Fahrenheit) of 96°, 104°, 103°, 104°, 103°, 104°, 91°, 74°, 103°

Round each number to the place shown.

1. $65.236 to the nearest cent

2. $781.499 to the nearest dollar

3. 93,367 to the nearest ten

4. 854,279 to the nearest ten thousand

Simplify each expression by using the order of operations.

5. $4 + 10 \div 2 + 7 \cdot 2$

6. $\sqrt{81} - 4 \cdot 2 + 9$

Simplify.

7. $2^2 \cdot 3^3$

8. $6^2 \cdot 3^2$

Use front end rounding to round the numbers in each problem. Then add, subtract, multiply, or divide the rounded numbers as indicated to estimate the answer. Finally, solve for the exact answer.

9. *Estimate:*

Exact:

$$\begin{array}{r} 54,289 \\ 171,352 \\ 48 \\ + \ 18,208 \end{array}$$

10. *Estimate:*

Exact:

$$\begin{array}{r} 2.607 \\ 796.2 \\ 37.96 \\ 53.72 \\ + \ \ 8.06 \end{array}$$

11. *Estimate:*

Exact:

$$\begin{array}{r} 445,306 \\ - \ 234,867 \end{array}$$

12. *Estimate:*

Exact:

$$\begin{array}{r} 875.62 \\ - \ 63.757 \end{array}$$

13. *Estimate:*

Exact:

$$\begin{array}{r} 7064 \\ \times \ \ \ 635 \end{array}$$

14. *Estimate:*

Exact:

$$\begin{array}{r} 62.75 \\ \times \ 2.644 \end{array}$$

15. *Estimate:*

Exact:

$18\overline{)11,556}$

16. *Estimate:*

Exact:

$4.25\overline{)62.56}$

Add, subtract, multiply, or divide as indicated. Write answers in lowest terms and as whole or mixed numbers when possible.

17. $\dfrac{7}{8} + \dfrac{3}{4}$

18. $\dfrac{3}{4} + \dfrac{5}{8} + \dfrac{1}{2}$

19.
$$4\dfrac{3}{5}$$
$$+ \; 5\dfrac{2}{3}$$

20. $\dfrac{5}{6} - \dfrac{1}{3}$

21.
$$6\dfrac{2}{3}$$
$$- \; 4\dfrac{3}{4}$$

22.
$$8$$
$$- \; 5\dfrac{2}{5}$$

23. $\dfrac{3}{5} \cdot \dfrac{5}{8}$

24. $\left(9\dfrac{3}{5}\right)\left(4\dfrac{5}{8}\right)$

25. $22 \cdot \dfrac{2}{5}$

26. $\dfrac{5}{6} \div \dfrac{5}{8}$

27. $18 \div \dfrac{3}{4}$

28. $3\dfrac{1}{3} \div 8\dfrac{3}{4}$

Use the order of operations to simplify each expression.

29. $\left(\dfrac{7}{8} - \dfrac{3}{4}\right) \cdot \dfrac{2}{3}$

30. $\left(\dfrac{5}{6} - \dfrac{1}{3}\right) + \left(\dfrac{1}{2}\right)^2 \cdot \dfrac{3}{4}$

Write each fraction in decimal form. Round to the nearest thousandth if necessary.

31. $\dfrac{3}{8}$

32. $\dfrac{4}{5}$

33. $\dfrac{1}{6}$

34. $\dfrac{17}{20}$

Write in order, from smallest to largest.

35. 0.218, 0.22, 0.199, 0.207, 0.2215

36. 0.6319, $\dfrac{5}{8}$, 0.608, $\dfrac{13}{20}$, 0.58

Write each ratio in lowest terms. Be sure to make all necessary conversions.

37. $5\frac{1}{2}$ in. to 44 in.

38. 3 hr to 45 min

Find cross products to determine whether each proportion is true or false.

39. $\dfrac{6}{15} = \dfrac{18}{45}$

      True   False

40. $\dfrac{52}{180} = \dfrac{36}{120}$

      True   False

Find the unknown number in each proportion.

41. $\dfrac{1}{5} = \dfrac{x}{30}$

42. $\dfrac{15}{x} = \dfrac{390}{156}$

43. $\dfrac{200}{135} = \dfrac{24}{x}$

44. $\dfrac{x}{208} = \dfrac{6.5}{26}$

Write each percent as a decimal. Write each decimal as a percent.

45. 65%

46. 0.035

47. 380%

48. 7.75%

Write each percent as a fraction or mixed number in lowest terms. Write each fraction as a percent.

49. 4%

50. $87\frac{1}{2}\%$

51. $\dfrac{7}{20}$

52. $5\dfrac{3}{4}$

Solve each percent problem.

53. 65% of $780 is how much?

54. Find 5.4% of 6000 homes.

55. $8\frac{1}{2}\%$ of what number of people is 238 people?

56. 72 is 18% of what number?

57. What percent of 1040 is 468?

58. 13 weeks is what percent of 52 weeks?

Convert each measurement using unit fractions.

59. _____ ft = 3 yd

60. 28 qt = _____ gal

61. 5 days = _____ hr

62. _____ lb = 6 T

Convert each measurement using unit fractions or the metric conversion line.

63. 10 km to m

64. 3815 mm to m

65. 8.3 g to mg

66. 230 g to kg

67. 72 mL to L

68. 0.28 L to mL

Write the most reasonable metric unit in each blank. Choose from L, mL, kg, g, mg, km, cm, m, and mm.

69. The fuel tank on the chain saw has a capacity of 750 _____ of fuel.

70. A nickel weighs 5 _____.

71. The distance of the run this Saturday is 10 _____.

72. The heaviest player on the team weighs 108 _____.

Find the area of each figure. Use 3.14 as the approximate value of π. Round answers to the nearest tenth.

73. A rectangle 5.6 m by 8.45 m

74. A trapezoid with bases of 6.2 cm and 8.4 cm and height 5.3 cm

75. A triangle with base 4.25 ft and height 4.5 ft

76. A circle with diameter of 13 cm

Find the volume. Use 3.14 as the approximate value for π. Round answers to the nearest tenth if necessary.

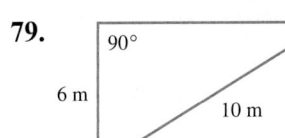 **77.** Find the volume of a cylinder with radius 4.8 cm and height 7.6 cm.

78. Find the volume of a rectangular solid with length 9.5 m, width 3 m, and height 7 m.

Find the unknown length in each right triangle.

79.

6 m, 90°, 10 m

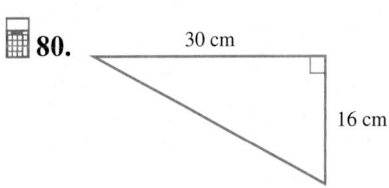 **80.**

30 cm, 16 cm

Add, subtract, multiply, or divide as indicated.

81. $-12 + (-10)$

82. $-5.7 - (-12.6)$

83. $9 \cdot (-5)$

84. $(-14.6)(-5.7)$

85. $\dfrac{-42}{-7}$

86. $\dfrac{-34.04}{14.8}$

Solve each equation.

87. $3x - 5 = 16$

88. $-12 = 3(x + 2)$

89. $15x - 11x = 12$

90. $3.4x + 6 = 1.4x - 8$

Find the mean, the median, and the mode for each list of numbers. Round to the nearest tenth if necessary.

91. Cable hookups per installer: 16, 37, 27, 31, 19, 25, 15, 38, 43, 19

92. Number of acres plowed each hour: 10.3, 4.3, 1.65, 2.85, 5.3, 5.7, 2.3, 4.35, 2.85

Solve each application problem.

93. Trader Joe's sold 3620 bags of tortilla chips in a recent week. If 1267 of these bags were baked tortilla chips (fat-free), find the percent that were baked.

94. In one state the sales tax is 7%. On a recent purchase, the amount of sales tax Sue paid was $78.68. Find the cost of the item she purchased.

95. A gasoline additive is used at the rate of $2\frac{3}{4}$ gal for each storage tank. If $280\frac{1}{2}$ gallons of additive are available, how many storage tanks can receive the additive?

96. A survey found that 34 out of every 50 hotel rooms are for nonsmokers. If a hotel has 1200 rooms, how many would you expect to be for nonsmokers?

97. Linda Shirley had sales of $65,350 in surgical appliances last month. If her commission rate is 7.8%, find the amount of her commission.

98. The sketch below shows the plans for a lobby in a large commercial complex. What is the cost of carpeting the lobby, excluding the atrium, if the contractor charges $43.50 per square yard? Use 3.14 for π.

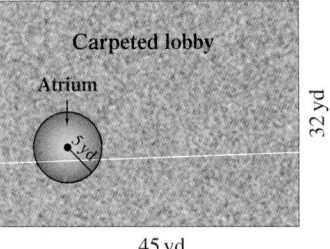

45 yd

99. The Spa Service Center services 140 spas. If each spa will need 125 mL of muriatic acid, how many liters of muriatic acid will be needed to service the spas?

100. Jonas will pay back a loan of $9400 with $8\frac{1}{2}\%$ interest at the end of 9 months. Find the total amount due.

Appendix A
An Introduction to Calculators

A SCIENTIFIC CALCULATORS

Calculators are among the more popular inventions of the last three decades. Each year better calculators are developed and costs drop. The first all-transistor desktop calculator was introduced to the market in 1966; it weighed 55 lb, cost $2500, and was slow. Today, these same calculations are performed quite well on a calculator costing less than $10. And today's $200 calculators have more ability to solve problems than some of the early computers.

Many colleges allow students to use calculators in basic mathematics courses. Although you can still purchase a basic four-function calculator, you're probably better off spending $10 to $20 on a **scientific calculator.** A scientific calculator will allow you to do a lot more than the basic four-function calculators. A **graphing calculator** allows you to graph functions and data, but it is beyond the scope of this text. The discussion here is confined to the common scientific calculator with the percent key, reciprocal key, exponent key, square root key, memory function, order of operations, and parentheses keys.

NOTE

For explanations of specific calculator models or special function keys, refer to the instruction booklet supplied with your calculator.

1   Learn the basic calculator keys.   Most calculators use **algebraic logic.** Some problems can be solved by entering number and function keys in the same order as you would solve problems by hand, but many others require a knowledge of the order of operations when entering the problem. The problem $14 + 28$ would be entered as

$$14 \;\boxed{+}\; 28 \;\boxed{=}$$

and 42 would appear as the answer. Enter $387 - 62$ as

$$387 \;\boxed{-}\; 62 \;\boxed{=}$$

and 325 appears as the answer. If your calculator does not work problems in this way, check its instruction book to see how to proceed.

2   Understand the $\boxed{\text{C}}$, $\boxed{\text{CE}}$, **and** $\boxed{\text{ON/C}}$ **or** $\boxed{\text{ON/AC}}$ **keys.**   All calculators have a

$$\boxed{\text{C}}, \quad \boxed{\text{ON/C}}, \quad \text{or} \quad \boxed{\text{ON/AC}}$$

key. Pressing this key erases everything in the calculator and prepares the calculator to begin a new problem. Some calculators also have a

$$\boxed{\text{CE}}$$

key. Pressing this key erases *only* the number displayed and allows the person using the calculator to correct a mistake without having to start the problem over.

OBJECTIVES

1 Learn the basic calculator keys.

2 Understand the $\boxed{\text{C}}$, $\boxed{\text{CE}}$, and $\boxed{\text{ON/C}}$ or $\boxed{\text{ON/AC}}$ keys.

3 Understand the floating decimal point.

4 Use the $\boxed{\%}$ key.

5 Use the $\boxed{x^2}$ and the $\boxed{x^3}$ keys.

6 Use the $\boxed{y^x}$ and $\boxed{\sqrt{x}}$ keys.

7 Use the $\boxed{a^{b/c}}$ key.

8 Solve problems with negative numbers.

9 Use the calculator memory function.

10 Solve chain calculations using the order of operations.

11 Use the parentheses keys.

Many calculators combine the Ⓒ key and the (CE) key and use an (ON/C) key. This key turns the calculator on and is also used to erase the calculator display. If the (ON/C) key is pressed after the ⊜ or one of the operation keys (⊕, ⊖, ⊗, ⊘), everything in the calculator is erased. If the wrong operation key is pressed, you press the correct key and the error is corrected. For example, 7 ⊕ ⊖ 3 ⊜ 4. Pressing the ⊖ key cancels out the previous ⊕ key entry.

> **CAUTION**
>
> Be sure to look at the directions that come with your calculator to see how to clear the memory.

3▬▬ **Understand the floating decimal point.**   Most calculators have a **floating decimal,** which locates the decimal point in the final result. For example, to find the cost of 55.75 square yards of vinyl floor covering at a cost of $18.99 per square yard, proceed as follows.

$$55.75 \ \otimes \ 18.99 \ \circleddash \ 1058.6925$$

The decimal point is automatically placed in the answer. You should **round** money answers to the nearest cent. Draw a cut-off line after the hundredths place.

Look only at the first digit being cut off.
↓

$$1058.69|25$$

↑
Cent position (hundredths)

Because the first digit being cut off is 4 or less, the part you are keeping remains the same. The answer is rounded to $1058.69. If the first digit being cut off had been 5 or more, you would have rounded up by adding 1 to the cent position.

When using a calculator with a floating decimal, enter the decimal point as needed. For example, enter $47 as

47

with no decimal point, but enter 95¢ as

⊙ 95

One problem in using a floating decimal is shown by the following example (adding $21.38 and $1.22).

$$21.38 \ \oplus \ 1.22 \ \circleddash \ 22.6$$

The calculator does not show the final 0. You must remember that the problem dealt with money and write the final 0, making the answer $22.60.

4▬▬ **Use the** Ⓟ **key.**   The Ⓟ key moves the decimal point two places to the left when pressed following multiplication or division. The problem 8% of $4205 is solved as follows:

$$4205 \ \otimes \ 8 \ \% \ \circleddash \ 336.4$$

Because the problem involved money, write the answer as $336.40.

5 ▬▬ **Use the** $\boxed{x^2}$ **and the** $\boxed{x^3}$ **keys.** The squaring key, $\boxed{x^2}$, allows you to square the number in the display (multiply the number by itself). The square of 7 is found as follows.

$$7 \boxed{x^2} \; 49$$

The cubing key, $\boxed{x^3}$, allows you to find the cube of a number (the number is multiplied by itself three times). To find the cube of 6.8 (that is, $6.8 \cdot 6.8 \cdot 6.8$), follow these keystrokes.

$$6.8 \boxed{x^3} \; 314.432$$

6 ▬▬ **Use the** $\boxed{y^x}$ **and** $\boxed{\sqrt{x}}$ **keys.** The product of $3 \times 3 \times 3 \times 3 \times 3$ can be written as

┌──── Exponent

$$3^5$$

└──── Base

The exponent, 5, shows how many times the base is multiplied by itself (multiply 3 by itself five times). The $\boxed{y^x}$ key raises a base to any desired power. Find 3^5 as

$$3 \boxed{y^x} \; 5 = 243.$$

Since $3^2 = 9$, the number 3 is called the square root of 9. Square roots are written with the symbol $\boxed{\sqrt{}}$. Use the $\boxed{\sqrt{x}}$ key to find the square root of 144, $\sqrt{144}$, as follows:

$$144 \boxed{\sqrt{x}} \; 12$$

Find $\sqrt{20}$ as

$$20 \boxed{\sqrt{x}} \; 4.472135955,$$

which may be rounded to the desired position.

7 ▬▬ **Use the** $\boxed{a^{b/c}}$ **key.** The $\boxed{a^{b/c}}$ key is used when solving problems containing fractions and mixed numbers.

Solve $\dfrac{3}{4} + \dfrac{6}{11}$ as

$$3 \boxed{a^{b/c}} \; 4 \boxed{+} \; 6 \boxed{a^{b/c}} \; 11 \boxed{=} \; 1\_13 \lrcorner 44$$

The answer is $1\dfrac{13}{44}$.

Solve the mixed number problem $4\dfrac{7}{8} \div 3\dfrac{4}{7}$ as

$$4 \boxed{a^{b/c}} \; 7 \boxed{a^{b/c}} \; 8 \boxed{\div} \; 3 \boxed{a^{b/c}} \; 4 \boxed{a^{b/c}} \; 7 \boxed{=} \; 1\_73 \lrcorner 200$$

The answer is $1\dfrac{73}{200}$.

NOTE

The calculator automatically shows fractions in lowest terms and as mixed numbers when possible.

8▒▒▒ **Solve problems with negative numbers.**   Negative numbers may be entered by first entering the number and then using the ⊕/⊖ key. This changes the number entered to a negative number. For example, solve $-10 + 6 - 8$ as follows.

$$10 \ \boxed{+/-} \ \boxed{+} \ 6 \ \boxed{-} \ 8 \ \boxed{=} \ -12$$

9▒▒▒ **Use the calculator memory function.**   Many calculators feature memory keys, which are a sort of electronic scratch paper. These memory keys are used to store intermediate steps in a calculation. On some basic calculators, a key labeled Ⓜ is used to store the numbers in the display, with ⓂⓇ used to recall the numbers from memory.

Some calculators have ⊕M+ and ⊖M− keys. The ⊕M+ key adds the number displayed to the number already in memory. For example, if the memory contains the number 0 at the beginning of a problem, and the calculator display contains the number 29.4, then pushing ⊕M+ will cause 29.4 to be stored in the memory (the result of adding 0 and 29.4). If 57.8 is then entered into the display, pushing ⊕M+ will cause

$$29.4 + 57.8 = 87.2$$

to be stored. If 11.9 is then entered into the display, with ⊖M− pushed, the memory will contain

$$87.2 - 11.9 = 75.3.$$

The ⓂⓇ key is used to recall the number in memory as needed, with ⓂⒸ used to clear the memory.

Scientific calculators typically have one or more storage registers in which to store numbers. These memory keys are usually labeled as ⓈⓉⓄ for store and ⓇⒸⓁ for recall. For example, you can store 25.6 in register 1 by pressing

$$25.6 \ \boxed{STO} \ 1$$

or you can store it in register 2 by pressing 25.6 ⓈⓉⓄ 2 and so forth for other registers. Values are retrieved from a particular memory register by using the ⓇⒸⓁ key followed by the number of the register, for example, ⓇⒸⓁ 2 recalls the contents of memory in register 2.

With a scientific calculator, a number stays in memory until it is replaced by another number or until the memory is cleared. With some calculators, the contents of the memory is saved even when the calculator is turned off.

Here is an example of a problem that uses the memory keys. Suppose an elevator technician wants to find the average weight of a person using an elevator. To do this, she counts the number of people entering an elevator and also measures the weight of each group of people.

Number of People	Total Weight (in pounds)
6	839
8	1184
4	640

First, find the total weight of all three groups and store this in memory register 1.

$$839 \ \boxed{+} \ 1184 \ \boxed{+} \ 640 \ \boxed{=} \ 2663 \ \boxed{STO} \ 1$$

Then, find the total number of people.

$$6 \ \boxed{+} \ 8 \ \boxed{+} \ 4 \ \boxed{=} \ 18 \ \boxed{STO} \ 2$$

Finally, divide the contents of memory register 1 (total weight) by the contents of memory register 2 (18 people).

$$\boxed{\text{RCL}}\ 1\ \boxed{\div}\ \boxed{\text{RCL}}\ 2\ \boxed{=}\ 147.9444444\ \text{lb}$$

This answer can be rounded as needed.

10 ▭ **Solve chain calculations using the order of operations.** Long calculations involving several different operations are called **chain calculations.** They must be done in a specific sequence called the **order of operations.** The logic of the following order of operations is built into most scientific calculators and can help you work problems without having to store or write down a lot of intermediate steps.

Order of Operations

Step 1 Do all operations inside parentheses or other grouping symbols.

Step 2 Simplify any expressions with exponents and find any square roots.

Step 3 Multiply and divide, proceeding from left to right.

Step 4 Add and subtract, proceeding from left to right.

Your scientific calculator can be used to solve the problem $3 + 7 \times 9\frac{3}{4}$.

$$3\ \boxed{+}\ 7\ \boxed{\times}\ 9\ \boxed{a^{b/c}}\ 3\ \boxed{a^{b/c}}\ 4\ \boxed{=}\ 71\frac{1}{4}$$

The calculator automatically multiplies $7 \times 9\frac{3}{4}$ *before* adding 3.

The problem $42.1 \times 5 - 90 \div 4$ is solved as

$$42.1\ \boxed{\times}\ 5\ \boxed{-}\ 90\ \boxed{\div}\ 4\ \boxed{=}\ 188.$$

The calculator automatically multiplies 42.1×5 and divides $90 \div 4$ *before* doing the subtraction.

CAUTION

Scientific calculators keep track of the order of operations for us. All we have to do is enter the problem correctly into the calculator and the calculator does the rest. However, the basic four-function calculator is *not* programmed to observe the order of operations and will calculate correctly *only* if you enter numbers in the proper order.

11 ▭ **Use the parentheses keys.** The parentheses keys allow you to group numbers in a complex chain calculation. For example, $\dfrac{4}{5 + 7}$ can be written as $\dfrac{4}{(5 + 7)}$, which can be solved as follows.

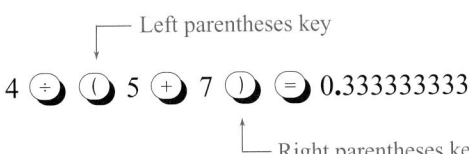

Left parentheses key

$$4\ \boxed{\div}\ \boxed{(}\ 5\ \boxed{+}\ 7\ \boxed{)}\ \boxed{=}\ 0.333333333$$

Right parentheses key

❸ Find the next number in the sequence 2, 6, 18, 54, ... Describe the pattern.

Example 3 Using Inductive Reasoning

Find the next number in the sequence 1, 2, 4, 8, 16, ...

Each number after the first is obtained by multiplying the previous number by 2. So the next number would be 16 • 2 = 32.

Work Problem ❸ at the Side.

Example 4 Using Inductive Reasoning

(a) Find the next geometric shape in this sequence.

The figures alternate between a circle and a triangle. Also, the number of dots increases by 1 in each subsequent figure. Thus, the next figure should be a circle with five dots inside it.

(b) Find the next geometric shape in this sequence.

The first two shapes consist of vertical lines with horizontal lines at the bottom extending first left and then right. The third shape is a vertical line with a horizontal line at the top extending to the left. Therefore, the next shape should be a vertical line with a horizontal line at the top extending to the right.

❹ Find the next shape in this sequence.

Work Problem ❹ at the Side.

2 ▐ **Use deductive reasoning to analyze arguments.**   In the previous discussion, specific cases were used to find patterns and predict the next event. There is another type of reasoning called **deductive reasoning,** which moves from general cases to specific conclusions.

Example 5 Using Deductive Reasoning

Does the conclusion follow from the premises in this argument?

All Buicks are automobiles.

All automobiles have horns.

∴ All Buicks have horns.

In this example, the first two statements are called *premises* and the third statement (below the line) is called a *conclusion.* The symbol ∴ is a mathematical symbol meaning "**therefore.**" The entire set of statements is called an *argument.*

Continued on Next Page

The focus of deductive reasoning is to determine whether the conclusion follows (is valid) from the premises. A set of circles called **Euler circles** is used to analyze the argument.

In Example 5, the statement "All Buicks are automobiles" can be represented by two circles, one for Buicks and one for automobiles. Note that the circle representing Buicks is totally inside the circle representing automobiles.

Now, a circle is added to represent the second statement, vehicles with horns. This circle must completely surround the circle representing automobiles.

To analyze the conclusion, notice that the circle representing Buicks is completely inside the circle representing vehicles with horns. It must follow that all Buicks have horns. ***The conclusion is valid.***

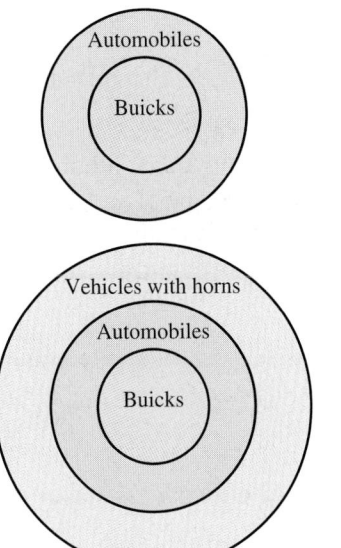

5 Does the conclusion follow from the premises in the following argument?

All cars have four wheels.
All Fords are cars.
∴ All Fords have four wheels.

═══ **Work Problem 5 at the Side.**

6 Does each conclusion follow from the premises?

(a) All animals are wild.
All cats are animals.
∴ All cats are wild.

Example 6 **Using Deductive Reasoning**

Does the conclusion follow from the premises in this argument?

All tables are round.
All glasses are round.
∴ All glasses are tables.

Using Euler circles, a circle representing tables is drawn inside a circle representing round objects.

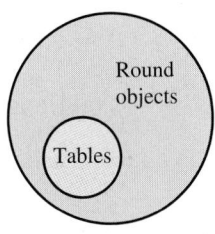

The second statement requires that a circle representing glasses must now be drawn inside the circle representing round objects, but not necessarily inside the circle representing tables. The conclusion does *not* follow from the premises. This means that ***the conclusion is invalid or untrue.***

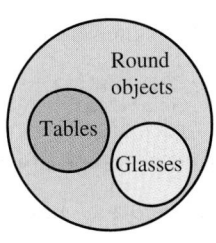

(b) All students use math.
All adults use math.
∴ All adults are students.

═══ **Work Problem 6 at the Side.**

7 In a college class of 100 students, 35 take both math and history, 50 take history, and 40 take math. How many take neither math nor history?

3 ▢ **Use deductive reasoning to solve problems.**   Another type of deductive reasoning problem occurs when a set of facts is given in a problem and a conclusion must be drawn by using these facts.

Example 7   **Using Deductive Reasoning**

There were 25 students enrolled in a ceramics class. During the class, 10 of the students made a bowl and 8 students made a birdbath. Three students made both a bowl and a birdbath. How many students did not make either a bowl or a birdbath?

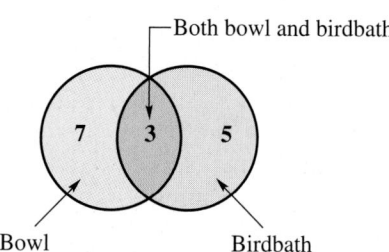

This type of problem is best solved by organizing the data using a drawing called a **Venn diagram.** Two overlapping circles are drawn, with each circle representing one item made by students.

In the region where the circles overlap, write the number of students who made *both* items, namely, 3. In the remaining portion of the birdbath circle, write the number 5, which when added to 3 will give the total number of students who made a birdbath, namely, 8. In a similar manner, write 7 in the remaining portion of the bowl circle, since 7 + 3 = 10, the total number of students who made a bowl. The total of all three numbers written in the circles is 15. Since there are 25 students in the class, this means 25 − 15 or 10 students did not make either a birdbath or a bowl.

Work Problem 7 at the Side.

8 A Chevy, BMW, Cadillac, and Ford are parked side by side. The known facts are:

(a) The Ford is on the right end.

(b) The BMW is next to the Cadillac.

(c) The Chevy is between the Ford and the Cadillac.

Which car is parked on the left end?

Example 8   **Using Deductive Reasoning**

Four cars in a race finish first, second, third, and fourth. The following facts are known.

(a) Car A beat Car C.

(b) Car D finished between Cars C and B.

(c) Car C beat Car B.

In which order did the cars finish?
To solve this type of problem, it is helpful to use a line diagram.

1. *Write A before C,* because Car A beat Car C (fact a).

A   C

2. *Write B after C,* because Car C beat Car B (fact c).

A   C   B

3. *Write D between C and B,* because Car D finished between Cars C and B (fact b).

The correct order of finish is shown below.

A   C   D   B

Work Problem 8 at the Side.

Appendix B EXERCISES

 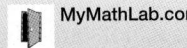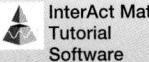
Find the next number in each sequence. Describe the pattern in each sequence. See Examples 1–3.

1. 2, 9, 16, 23, 30, . . .

2. 5, 8, 11, 14, 17, . . .

3. 0, 10, 8, 18, 16, . . .

4. 3, 9, 7, 13, 11, . . .

5. 1, 2, 4, 8, . . .

6. 1, 4, 16, 64, . . .

7. 1, 3, 9, 27, 81, . . .

8. 3, 6, 12, 24, 48, . . .

9. 1, 4, 9, 16, 25, . . .

10. 6, 7, 9, 12, 16, . . .

Find the next shape in each sequence. See Example 4.

11.

12.

13.

14.

In each argument, state whether or not the conclusion follows from the premises. See Examples 5 and 6.

15. All animals are wild.

All lions are animals.

∴   All lions are wild.

16. All students are hard workers.

All business majors are students.

∴   All business majors are hard workers.

17. All teachers are serious.

All mathematicians are serious.

∴   All mathematicians are teachers.

18. All boys ride bikes.

All Americans ride bikes.

∴   All Americans are boys.

Solve each application problem. See Examples 7 and 8.

19. In a given 30-day period, a man watched television 20 days and his wife watched television 25 days. If they watched television together 18 days, how many days did neither watch television?

20. In a class of 40 students, 21 students take both calculus and physics. If 30 students take calculus and 25 students take physics, how many do not take either calculus or physics?

21. Tom, Dick, Mary, and Joan all work for the same company. One is a secretary, one is a computer operator, one is a receptionist, and one is a mail clerk.

   (a) Tom and Joan eat dinner with the computer operator.

   (b) Dick and Mary carpool with the secretary.

   (c) Mary works on the same floor as the computer operator and the mail clerk.

   Who is the computer operator?

22. Four cars—a Ford, a Buick, a Mercedes, and an Audi—are parked in a garage in four spaces.

   (a) The Ford is in the last space.

   (b) The Buick and Mercedes are next to each other.

   (c) The Audi is next to the Ford but not next to the Buick.

   Which car is in the first space?

Answers to Selected Exercises

In this section we provide the answers that we think most students will obtain when they work the exercises using the methods explained in the text. If your answer does not look exactly like the one given here, it is not necessarily wrong. In many cases there are equivalent forms of the answer that are correct. For example, if the answer section shows $\frac{3}{4}$ and your answer is 0.75, you have obtained the right answer but written it in a different (yet equivalent) form. Unless the directions specify otherwise, 0.75 is just as valid an answer as $\frac{3}{4}$.

In general, if your answer does not agree with the one given in the text, see whether it can be transformed into the other form. If it can, then it is the correct answer. If you still have doubts, talk with your instructor.

Diagnostic Pretest

(page xxiii)

1. 89,023,507 **2.** 4331 **3.** 697 **4.** 89,000 **5.** $29\frac{3}{8}$

6. $2^3 \cdot 7^2$ **7.** $11\frac{1}{4}$ cups **8.** *Estimate:* $6 \cdot 2 = 12$; *Exact:* $12\frac{7}{32}$

9. 120 **10.** $\frac{19}{30}$ **11.** *Estimate:* $8 + 13 = 21$; *Exact:* $21\frac{1}{18}$

12. $\frac{35}{48}$ **13.** \$1.39 **14.** 6.556 (rounded) **15.** 24.25

16. \$8.43 (rounded) **17.** $\frac{5}{24}$ **18.** false **19.** $26\frac{2}{3}$ **20.** \$340

21. 58.2% **22.** 875% **23.** \$135.68 **24.** 7% **25.** (a) 9 gal
(b) 76 oz **26.** (a) 67.5 cm (b) 4.528 kg **27.** 2.92 kg
28. (a) mg (b) cm **29.** 27 in. **30.** 24.5 cm² **31.** 75.36 ft³
32. 15 cm **33.** 25 **34.** −4.6 **35.** 11 **36.** $r = -2$ **37.** \$576
38. 41.3 **39.** 7.8 (rounded) **40.** \$8.95

Chapter 1

Section 1.1 (page 7)

1. 2; 7 **3.** 1; 0 **5.** 8; 2 **7.** 2; 768; 543 **9.** 60; 0; 502; 109
11. Evidence suggests that this is true. It is common to count using fingers. **13.** twenty-three thousand, one hundred fifteen
15. three hundred forty-six thousand, nine **17.** twenty-five million, seven hundred fifty-six thousand, six hundred sixty-five
19. 32,526 **21.** 10,000,223 **23.** \$198 **25.** 2,000,000,000
27. 532,000 **29.** 800,000,621,020,215 **31.** public transportation; six million, sixty-nine thousand, five hundred eighty-nine **33.** seven million, eight hundred ninety-four thousand, nine hundred eleven

Section 1.2 (page 15)

1. 79 **3.** 89 **5.** 889 **7.** 889 **9.** 7785 **11.** 1578 **13.** 7676
15. 78,446 **17.** 8928 **19.** 59,224 **21.** 114 **23.** 155
25. 121 **27.** 145 **29.** 102 **31.** 1651 **33.** 1154 **35.** 413
37. 1771 **39.** 1410 **41.** 9253 **43.** 11,624 **45.** 17,611
47. 15,954 **49.** 10,648 **51.** 15,594 **53.** 11,557 **55.** 12,078
57. 4250 **59.** 12,268 **61.** correct **63.** incorrect; should be 769
65. correct **67.** incorrect; should be 11,577 **69.** correct
71. Changing the order in which numbers are added does not change the sum. You can add from bottom to top when checking addition. **73.** 33 mi **75.** 38 mi **77.** \$89 **79.** 699 people
81. \$16,342 **83.** 550 ft **85.** 72 ft **87.** 9421 **88.** 1249
89. 77,762 **90.** 22,267 **91.** 9,994,433 **92.** 3,334,499
93. Write the largest digits on the left, using the smaller digits as you move right. **94.** Write the smallest digits on the left, using the larger digits as you move right.

Section 1.3 (page 25)

1. 31 **3.** 33 **5.** 17 **7.** 213 **9.** 101 **11.** 6211 **13.** 3412
15. 2111 **17.** 13,160 **19.** 41,110 **21.** correct **23.** incorrect; should be 62 **25.** incorrect; should be 121 **27.** correct
29. incorrect; should be 7222 **31.** 18 **33.** 45 **35.** 19
37. 281 **39.** 519 **41.** 9177 **43.** 7589 **45.** 8859 **47.** 3
49. 23 **51.** 19 **53.** 2833 **55.** 7775 **57.** 503 **59.** 156
61. 2184 **63.** 5687 **65.** 19,038 **67.** 31,556 **69.** 6584
71. correct **73.** correct **75.** correct **77.** correct **79.** Possible answers are 1. $3 + 2 = 5$ could be charged to $5 - 2 = 3$ or $5 - 3 = 2$ 2. $6 - 4 = 2$ could be changed to $2 + 4 = 6$ or $4 + 2 = 6$. **81.** 15 calories **83.** 367 ft **85.** 121 passengers
87. 1329 students **89.** 9539 flags **91.** \$263 **93.** 758 people
95. \$57,500 **97.** 48 deliveries **99.** 284 deliveries

Section 1.4 (page 35)

1. 16 **3.** 48 **5.** 0 **7.** 24 **9.** 30 **11.** 0 **13.** Factors may be multiplied in any order to get the same answer. They are the same; you may add or multiply numbers in any order. **15.** 150 **17.** 238
19. 3210 **21.** 1872 **23.** 8612 **25.** 10,084 **27.** 20,488
29. 258,447 **31.** 150 **33.** 480 **35.** 2220 **37.** 3600
39. 3750 **41.** 44,550 **43.** 270,000 **45.** 86,000,000
47. 48,500 **49.** 350,000 **51.** 1,940,000 **53.** 540 **55.** 2400
57. 3735 **59.** 2378 **61.** 6164 **63.** 15,792 **65.** 21,665
67. 15,730 **69.** 82,320 **71.** 183,996 **73.** 2,468,928
75. 66,005 **77.** 86,028 **79.** 19,422,180 **81.** 2,278,410
83. To multiply by 10, 100, or 1000, just add the number of 0s to the number you are multiplying and that's your answer.
85. 18,000 cartons **87.** 216 plants **89.** 418 mi **91.** \$255
93. \$1560 **95.** \$112,888 **97.** 50,568 **99.** 38,250 trees
101. 1058 calories **103.** \$4820 **104.** (a) 452 (b) 452
105. commutative **106.** (a) 281 (b) 281 **107.** associative
108. (a) 15,840 (b) 15,840 **109.** commutative
110. (a) 6552 (b) 6552 **111.** associative **112.** No. Some examples are 1. $7 - 5 = 2$, but $5 - 7$ does not equal 2
2. $12 - 6 = 6$, but $6 - 12$ does not equal 6   3. $(8 - 2) - 5 = 1$, but $8 - (2 - 5)$ does not equal 1. **113.** No. Some examples are 1. $10 \div 2 = 5$, but $2 \div 10$ does not equal 5   2. $(16 \div 8) \div 2 = 1$, but $16 \div (8 \div 2)$ does not equal 1.

Section 1.5 (page 49)

1. $3\overline{)15}$ $\frac{15}{3} = 5$ **3.** $9\overline{)45}$ $45 \div 9 = 5$ **5.** $16 \div 2 = 8$ $\frac{16}{2} = 8$

7. 1 **9.** 7 **11.** undefined **13.** 24 **15.** 0 **17.** undefined **19.** 15 **21.** 8 **23.** 15 **25.** 24 **27.** 304 **29.** 627 R1 **31.** 1522 R5 **33.** 309 **35.** 3005 **37.** 5006 **39.** 811 R1 **41.** 2589 R2
43. 7324 R2 **45.** 3157 R2 **47.** 5522 **49.** 12,458 R3

51. 10,253 R5 **53.** 18,377 R6 **55.** correct **57.** incorrect; should be 1908 R1 **59.** incorrect; should be 670 R2
61. incorrect; should be 3568 R1 **63.** correct **65.** correct
67. incorrect; should be 9628 R3 **69.** correct **71.** Multiply the quotient by the divisor and add any remainder. The result should be the dividend. **73.** 233 place settings **75.** 11,200 people each day **77.** $18,200 **79.** 205 acres **81.** $225,000
83. $9135 **85.** √ √ √ √ **87.** √ X X X **89.** X X √ X
91. X √ X X **93.** √ √ X X **95.** X X X X

Section 1.6 (page 59)

1. 22 **3.** 250 **5.** 120 **R**7 **7.** 1308 **R**9 **9.** 7134 **R**12
11. 900 **R**100 **13.** 108 **R**4 **15.** 183 **R**22 **17.** 2407 **R**1
19. 1146 **R**15 **21.** 3331 **R**82 **23.** 850 **25.** incorrect; should be 101 **R**14 **27.** incorrect; should be 658 **29.** incorrect; should be 62 **31.** When dividing by 10, 100, or 1000, drop the same number of 0s from the dividend to get the quotient. One example is 2500 ÷ 100 = 25. **33.** 18 hr **35.** 56 floor clocks **37.** $108
39. 1680 circuit boards **41.** $39 per week **43.** $0 **44.** 0
45. undefined **46.** impossible; if you have 6 cookies, it is not possible to divide them among 0 people. **47. (a)** 14 **(b)** 17
(c) 38 **48.** Yes. some examples are 18 • 1 = 18; 26 • 1 = 26;
43 • 1 = 43. **49. (a)** 3200 **(b)** 320 **(c)** 32 **50.** Drop the same number of 0s that appear in the divisor. The result is the quotient. With the divisor 10, drop one 0; with 100, drop two 0s; with 1000, drop three 0s.

Section 1.7 (page 69)

1. 410 **3.** 980 **5.** 6800 **7.** 86,800 **9.** 34,500 **11.** 6000
13. 16,000 **15.** 78,000 **17.** 8000 **19.** 10,000 **21.** 600,000
23. 9,000,000 **25.** 2370; 2400; 2000 **27.** 3370; 3400; 3000
29. 5050; 5000; 5000 **31.** 4240; 4200; 4000 **33.** 19,540;
19,500; 20,000 **35.** 26,290; 26,300; 26,000 **37.** 64,500;
64,500; 65,000 **39.** 1. Locate the place to be rounded and underline it. 2. Look only at the next digit to the right. If this digit is 5 or more, increase the underlined digit by 1. 3. Change all digits to the right of the underlined place to 0s. **41.** 90 30 80 90 290; 290
43. 90 30 60; 52 **45.** 70 30 2100; 2278 **47.** 900 700 400 800
2800; 2828 **49.** 800 400 400; 387 **51.** 400 400 160,000;
160,448 **53.** 8000 60 700 4000 12,760; 12,605 **55.** 800 400
400; 357 **57.** 900 30 27,000; 27,231 **59.** Perhaps the best explanation is that 3492 is closer to 3500 than 3400, but 3492 is closer to 3000 than to 4000. **61.** 80 million people;
280 million people **63.** 50 yr; 80 yr **65.** 3,025,940,000 pesos;
3,026,000,000 pesos; 3,000,000,000 pesos **67.** $8,490,487,600,000;
$8,490,500,000,000; $8,490,000,000,000 **69.** 71,500
70. 72,499 **71.** 7500 **72.** 8499 **73.** $330; $500; $550; $730;
$950; $1380; $1390; $2550 **74.** $300; $500; $600; $700; $1000;
$1000; $1000; $3000 **75. (a)** When using front end rounding, all digits are 0 except the first digit. These numbers are easier to work with when estimating answers. **(b)** When using front end rounding to estimate an answer, the estimated answer can vary greatly from the exact answer.

Section 1.8 (page 75)

1. 2; 4; 16 **3.** 2; 5; 25 **5.** 2; 12; 144 **7.** 2; 15; 225 **9.** 3
11. 8 **13.** 10 **15.** 12 **17.** 64; 64 **19.** 400; 400 **21.** 1225;
1225 **23.** 1600; 1600 **25.** 2916; 2916 **27.** A perfect square is the square of a whole number. The number 25 is the square of 5 because 5 • 5 = 25. The number 50 is not a perfect square. There is no whole number that can be squared to get 50.
29. 18 **31.** 25 **33.** 5 **35.** 20 **37.** 45 **39.** 63 **41.** 118
43. 18 **45.** 40 **47.** 102 **49.** 10 **51.** 63 **53.** 33 **55.** 70
57. 7 **59.** 17 **61.** 55 **63.** 108 **65.** 28 **67.** 9 **69.** 21

71. 16 **73.** 9 **75.** 3 **77.** 7 **79.** 20 **81.** 14 **83.** 25
85. 16 **87.** 23 **89.** 233

Section 1.9 (page 81)

1. 15 thousand **3.** United States **5.** 3 thousand **7.** 9 people
9. (a) Saw ad **(b)** 25 people **11.** 9 people **13.** 2002; 7000
homes sales **15.** 4500 home sales **17.** Possible answers are
1. shortage of homes for sale 2. lack of qualified buyers
3. poor economy 4. high interest rates on home loans.
19. (5 + 1) • 8 − 2 **20.** (4 + 2) • (5 + 1)
21. 36 ÷ (3 • 3) • 4 **22.** 48 ÷ (2 • 2 • 2) + 2
23. (a) (100 + 50 + 65 + (50 − 15) + (100 − 65) + 15) • 12
(b) 3600 ft

Section 1.10 (page 89)

1. *Estimate:* 80 + 80 + 100 + 40 + 50 = 350 mi; *Exact:* 382 mi
3. *Estimate:* 200 − 70 = 130 more crimes; *Exact:* 200 − 70 = 130 more crimes **5.** *Estimate:* 200 × 20 = 4000 kits; *Exact:* 5664 kits
7. *Estimate:* 3000 ÷ 700 ≈ 4 toys; *Exact:* 4 toys **9.** *Estimate:*
8000 − 4000 = 4000 people; *Exact:* 4174 people **11.** *Estimate:*
$10 × 5 = $50; *Exact:* $70 **13.** *Estimate:* $40,000 − $30,000 =
$10,000; *Exact:* $9700 **15. (a)** *Estimate:* $20,000 + $30,000 =
$50,000 Garrett; *Exact:* $20,000 + $40,000 = $60,000 Harcos;
Estimate: $51,500 Garrett; *Exact:* $55,700 Harcos;
Mr. and Mrs. Harcos **(b)** *Estimate:* $60,000 − $50,000 = $10,000;
Exact: $4200 **17.** *Estimate:* $2000 − $500 − $300 − $300 −
$200 − $200 = $500; *Exact:* $193 **19.** *Estimate:* 40,000 × 100 =
4,000,000 ft²; *Exact:* 6,011,280 ft² **21.** *Estimate:* $400 + $600 +
$200 + $100 + $100 = $1400; *Exact:* $1375 **23.** *Estimate:*
$400 + $600 + $300 + $900 = $2200; $2200 − $1600 = $600;
Exact: $530 **25.** *Estimate:* ($1000 × 6) + ($900 × 20) = $24,000;
Exact: $20,961 **27.** Possible answers are Addition: more; total;
gain of Subtraction: less; loss of; decreased by Multiplication:
twice, of; product Division: divided by; goes into; per Equals:
is; are **29.** Estimating the answer can help you avoid careless mistakes like decimal or calculation errors. Examples of reasonable answers in daily life might be a $25 bag of groceries, $20 to fill the gas tank, or $45 for a phone bill. **31.** $165 **33.** 2477 lb
35. $500 **37.** $375 **39.** 20 seats

Chapter 1 Review Exercises (page 97)

1. 3; 582 **2.** 64; 234 **3.** 105; 724 **4.** 1; 768; 710; 618
5. six hundred thirty-five **6.** fifteen thousand, three hundred
ten **7.** three hundred nineteen thousand, two hundred fifteen
8. sixty-two million, five hundred thousand, five **9.** 10,008
10. 200,000,455 **11.** 90 **12.** 137 **13.** 5464 **14.** 15,657
15. 10,986 **16.** 9845 **17.** 40,602 **18.** 49,855 **19.** 34
20. 23 **21.** 189 **22.** 184 **23.** 5755 **24.** 4327 **25.** 224
26. 25,866 **27.** 36 **28.** 0 **29.** 32 **30.** 64 **31.** 35 **32.** 56
33. 56 **34.** 81 **35.** 48 **36.** 45 **37.** 48 **38.** 8 **39.** 0
40. 42 **41.** 48 **42.** 0 **43.** 160 **44.** 368 **45.** 522 **46.** 98
47. 4123 **48.** 5467 **49.** 5396 **50.** 45,815 **51.** 16,728
52. 32,640 **53.** 465,525 **54.** 174,984 **55.** 675 **56.** 1764
57. 1176 **58.** 5100 **59.** 15,576 **60.** 30,184 **61.** 887,169
62. 500,856 **63.** $300 **64.** $1064 **65.** $20,352 **66.** $1512
67. 14,000 **68.** 23,800 **69.** 206,800 **70.** 318,500
71. 128,000,000 **72.** 90,300,000 **73.** 5 **74.** 5 **75.** 6
76. 2 **77.** 6 **78.** 4 **79.** 7 **80.** 0 **81.** undefined **82.** 0
83. 8 **84.** 9 **85.** 92 **86.** 35 **87.** 4422 **88.** 352
89. 150 R4 **90.** 124 R25 **91.** 480 **92.** 14,300 **93.** 21,000
94. 70,000 **95.** 3490; 3500; 3000 **96.** 20,070; 20,100; 20,000
97. 98,200; 98,200; 98,000 **98.** 352,120; 352,100; 352,000
99. 5 **100.** 8 **101.** 12 **102.** 14 **103.** 3; 7; 343
104. 6; 3; 729 **105.** 3; 5; 125 **106.** 5; 4; 1024 **107.** 54
108. 8 **109.** 9 **110.** 4 **111.** 9 **112.** 6 **113.** 8 parents

114. 5 parents **115.** Keeping bedroom clean **116.** Hanging up wet bath towels **117.** *Estimate:* 80 × $10 = $800; *Exact:* $750 **118.** *Estimate:* 1000 × 60 = 60,000 revolutions; *Exact:* 84,000 revolutions **119.** *Estimate:* 100 × 4 = 400 cups; *Exact:* 380 cups **120.** *Estimate:* 6000 × 30 = 180,000 brackets; *Exact:* 180,000 brackets **121.** *Estimate:* 2000 × 10 = 20,000 hr; *Exact:* 24,000 hr **122.** *Estimate:* 80 × 5 = 400 mi; *Exact:* 400 mi **123.** *Estimate:* ($20 × 20) + ($10 × 30) = $700; *Exact:* $582 **124.** *Estimate:* (60 × $20) + (20 × $7) = $1340; *Exact:* $1139 **125.** *Estimate:* $400 − $200 = $200; *Exact:* $180 **126.** *Estimate:* $900 − $400 − $200 = $300; *Exact:* $252 **127.** *Estimate:* 9000 ÷ 200 = 45 lb; *Exact:* 50 lb **128.** *Estimate:* 30,000 ÷ 1000 = 30 hr; *Exact:* 33 hr **129.** *Estimate:* 30,000 ÷ 600 = 50 acres; *Exact:* 52 acres **130.** *Estimate:* 6000 ÷ 200 = 30 homes; *Exact:* 32 homes **131.** 280 **132.** 664 **133.** 139 **134.** 588 **135.** 1041 **136.** 1661 **137.** 32,062 **138.** 24,947 **139.** 3 **140.** 7 **141.** 93,635 **142.** 83,178 **143.** undefined **144.** 7 **145.** 6900 **146.** 2305 **147.** 1,079,040 **148.** 103,268 **149.** 108 **150.** 207 **151.** three hundred seventy-six thousand, eight hundred fifty-three **152.** four hundred eight thousand, six hundred ten **153.** 7500 **154.** 600,000 **155.** 7 **156.** 9 **157.** $5940 **158.** $31,080 **159.** $2288 **160.** $15,782 **161.** 468 cards **162.** 3600 textbooks **163.** $280 **164.** $114,635 **165.** $1905 **166.** $12,420

Chapter 1 Test (page 105)

1. six thousand, one hundred six **2.** eighty-five thousand, fifty-five **3.** 426,005 **4.** 8714 **5.** 112,630 **6.** 1053 **7.** 3084 **8.** 96 **9.** 171,000 **10.** 1710 **11.** 4,450,743 **12.** 7047 **13.** undefined **14.** 458 R5 **15.** 160 **16.** 5240 **17.** 68,000 **18.** 48 **19.** 28 **20.** *Estimate:* $500 + $500 + $500 + 400 − $800 = $1100; *Exact:* $1140 **21.** *Estimate:* 70,000 ÷ 500 = 140 days; *Exact:* 125 days **22.** *Estimate:* $1000 − $700 − $200 − $70 = $30; *Exact:* $165 **23.** *Estimate:* (200 × 4) + (200 × 4) = 1600 cameras; *Exact:* 1784 cameras **24.** 1. Locate the place to which you are rounding and underline it. 2. Look only at the next digit to the right. If this digit is a 4 or less, do not change the underlined digit. If the digit is 5 or more, increase the underlined digit by 1. 3. Change all digits to the right of the underlined place to 0s. Each person's rounding example will vary. **25.** 1. Read the problem carefully. 2. Work out a plan. 3. Estimate a reasonable answer. 4. Solve the problem. 5. State the answer. 6. Check your work.

Chapter 2

Section 2.1 (page 111)

1. $\frac{3}{4}, \frac{1}{4}$ **3.** $\frac{1}{3}, \frac{2}{3}$ **5.** $\frac{7}{5}, \frac{3}{5}$ **7.** $\frac{2}{11}$ **9.** $\frac{8}{25}$ **11.** $\frac{49}{134}$ **13.** 3; 8

15. 11; 10 **17.** Proper $\frac{1}{3}, \frac{5}{8}, \frac{7}{16}$ Improper $\frac{8}{5}, \frac{6}{6}, \frac{12}{2}$ **19.** Proper $\frac{3}{4}, \frac{9}{11}, \frac{7}{15}$ Improper $\frac{3}{2}, \frac{5}{5}, \frac{19}{18}$

21. One possibility is

$\frac{3}{4}$ ← Numerator / ← Denominator

The denominator shows the number of equal parts in the whole and the numerator shows how many of the parts are being considered.
23. 7; 8 **25.** 4; 32

Section 2.2 (page 117)

1. $\frac{3}{2}$ **3.** $\frac{17}{5}$ **5.** $\frac{13}{2}$ **7.** $\frac{22}{3}$ **9.** $\frac{18}{11}$ **11.** $\frac{19}{3}$ **13.** $\frac{81}{8}$ **15.** $\frac{43}{4}$

17. $\frac{27}{8}$ **19.** $\frac{41}{6}$ **21.** $\frac{54}{11}$ **23.** $\frac{131}{4}$ **25.** $\frac{233}{13}$ **27.** $\frac{269}{15}$

29. $\frac{187}{24}$ **31.** $2\frac{1}{2}$ **33.** $2\frac{1}{4}$ **35.** 8 **37.** $7\frac{3}{5}$ **39.** $4\frac{7}{8}$ **41.** 9

43. $15\frac{3}{4}$ **45.** $5\frac{2}{9}$ **47.** $8\frac{1}{8}$ **49.** $16\frac{4}{5}$ **51.** $21\frac{3}{5}$ **53.** $26\frac{1}{7}$

55. Multiply the denominator by the whole number and add the numerator. The result becomes the new numerator, which is placed over the original denominator.

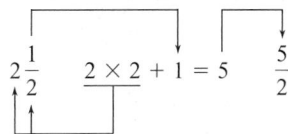
$2\frac{1}{2}$ $2 \times 2 + 1 = 5$ $\frac{5}{2}$

57. $\frac{443}{4}$ **59.** $\frac{1000}{3}$ **61.** $\frac{4179}{8}$ **63.** $104\frac{5}{8}$ **65.** 171

67. $122\frac{13}{32}$ **69.** $\frac{2}{3}, \frac{4}{5}, \frac{3}{4}, \frac{7}{10}$ **70. (a)** numerator; denominator

(b)

(c) less **71.** $\frac{5}{5}, \frac{10}{3}, \frac{6}{5}$ **72. (a)** numerator; denominator

(b)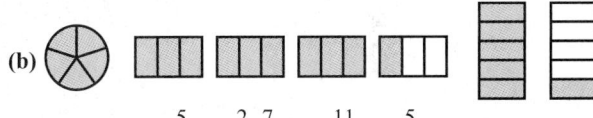

(c) greater **73.** $\frac{5}{3} = 1\frac{2}{3}; \frac{7}{7} = 1; \frac{11}{6} = 1\frac{5}{6}$ **74. (a)** improper; greater than or equal to

(b)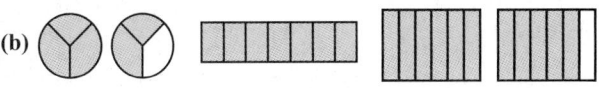

(c) Divide the numerator by the denominator and place the remainder over the denominator.

Section 2.3 (page 127)

1. 1, 2, 3, 6 **3.** 1, 2, 4, 8 **5.** 1, 2, 3, 4, 6, 8, 12, 16, 24, 48 **7.** 1, 2, 3, 4, 6, 9, 12, 18, 36 **9.** 1, 2, 4, 5, 8, 10, 20, 40 **11.** 1, 2, 4, 8, 16, 32, 64 **13.** composite **15.** prime **17.** composite **19.** prime **21.** prime **23.** composite **25.** composite **27.** composite **29.** 2 • 5 **31.** 2^2 • 5 **33.** 5^2 **35.** 2^2 • 3^2 **37.** 2^2 • 17 **39.** 2^3 • 3^2 **41.** 2^2 • 11 **43.** 2^2 • 5^2 **45.** 5^3 **47.** 2^2 • 3^2 • 5 **49.** 2^6 • 5 **51.** 2^3 • 3^2 • 5 **53.** A composite number has a factor(s) other than itself or 1. Examples include 4, 6, 8, 9, 10. A prime number is a whole number that has exactly two *different* factors, itself and 1. Examples include 2, 3, 5, 7, 11. The numbers 0 and 1 are neither prime nor composite. **55.** All the possible factors of 24 are 1, 2, 3, 4, 6, 8, 12, and 24. This list includes both prime numbers and composite numbers. The prime factors of 24 include only prime numbers. The prime factorization of 24 is $2 \cdot 2 \cdot 2 \cdot 3 = 2^3 \cdot 3$ **57.** 2^6 • 5 **59.** 2^6 • 3 • 5 **61.** 2^3 • 3 • 5 • 13 **63.** 2^2 • 3^2 • 5 • 7 **65.** 2, 3, 5, 7, 11, 13, 17, 19, 23, 29, 31, 37, 41, 43, 47 **66.** A prime number is a whole number that is evenly divisible by itself and 1 only. **67.** It is true because any even number is divisible by the number 2 in addition to itself and 1.

68. No. A multiple of a prime number can never be prime because it will always be divisible evenly by the prime number.
69. $2 \cdot 2 \cdot 3 \cdot 5 \cdot 5 \cdot 7$ **70.** $2^2 \cdot 3 \cdot 5^2 \cdot 7$

Section 2.4 (page 133)

1. ✓ ✓ ✓ ✓   **3.** ✓ ✓ X X   **5.** ✓ X ✓ ✓   **7.** ✓ ✓ X X
9. $\dfrac{3}{4}$ **11.** $\dfrac{2}{3}$ **13.** $\dfrac{5}{8}$ **15.** $\dfrac{6}{7}$ **17.** $\dfrac{9}{10}$ **19.** $\dfrac{6}{7}$ **21.** $\dfrac{4}{7}$ **23.** $\dfrac{1}{50}$
25. $\dfrac{8}{11}$ **27.** $\dfrac{5}{9}$

29. $\dfrac{\cancel{2} \cdot \cancel{3} \cdot 3}{\cancel{2} \cdot 2 \cdot 2 \cdot \cancel{3}} = \dfrac{3}{4}$ **31.** $\dfrac{\cancel{8} \cdot 7}{2 \cdot 2 \cdot 2 \cdot \cancel{8}} = \dfrac{7}{8}$

33. $\dfrac{\cancel{2} \cdot \cancel{3} \cdot \cancel{3} \cdot \cancel{3}}{\cancel{2} \cdot 2 \cdot \cancel{3} \cdot \cancel{3} \cdot \cancel{3}} = \dfrac{1}{2}$ **35.** $\dfrac{\cancel{2} \cdot \cancel{2} \cdot \cancel{3} \cdot 3}{\cancel{2} \cdot \cancel{2} \cdot \cancel{3}} = 3$

37. $\dfrac{2 \cdot 2 \cdot 2 \cdot \cancel{3} \cdot \cancel{3}}{\cancel{3} \cdot \cancel{3} \cdot 5 \cdot 5} = \dfrac{8}{25}$ **39.** equivalent **41.** not equivalent
43. not equivalent   **45.** equivalent   **47.** not equivalent
49. equivalent   **51.** A fraction is in lowest terms when the numerator and the denominator have no common factors other than 1. Some examples are $\dfrac{1}{2}, \dfrac{3}{8},$ and $\dfrac{2}{3}$.   **53.** $\dfrac{7}{8}$ **55.** $\dfrac{2}{1} = 2$

Section 2.5 (page 141)

1. $\dfrac{3}{8}$ **3.** $\dfrac{5}{12}$ **5.** $\dfrac{3}{4}$ **7.** $\dfrac{1}{4}$ **9.** $\dfrac{5}{12}$ **11.** $\dfrac{9}{32}$ **13.** $\dfrac{2}{5}$ **15.** $\dfrac{13}{32}$
17. $\dfrac{21}{128}$ **19.** 5   **21.** 40   **23.** 21   **25.** $13\dfrac{1}{2}$ **27.** $31\dfrac{1}{2}$
29. 240   **31.** $189\dfrac{1}{3}$ **33.** 400   **35.** 810   **37.** $\dfrac{1}{4} \, \text{mi}^2$
39. $9 \, \text{m}^2$ **41.** $\dfrac{3}{10} \, \text{in.}^2$ **43.** Multiply the numerators and multiply the denominators. An example is $\dfrac{3}{4} \cdot \dfrac{1}{2} = \dfrac{3 \cdot 1}{4 \cdot 2} = \dfrac{3}{8}$.
45. $1\dfrac{1}{2} \, \text{yd}^2$ **47.** $1 \, \text{mi}^2$ **49.** They are both the same size: $\dfrac{3}{64} \, \text{mi}^2$
51. 470,000 vehicles   **52.** 457,895 vehicles   **53.** 37,500; 35,013 Ford Explorers   **54.** 17,143; 15,190 Dodge Rams
55. $\dfrac{3}{4} \cdot 48{,}000 \, (\text{multiple of } 4) = 36{,}000$ Ford Explorers
56. $\dfrac{3}{7} \cdot 35{,}000 \, (\text{multiple of } 7) = 15{,}000$ Dodge Rams

Section 2.6 (page 149)

1. $\dfrac{1}{2} \, \text{yd}^2$ **3.** $\dfrac{8}{9} \, \text{ft}^2$ **5.** 61 players   **7.** \$2568   **9.** 375 students
11. 325 women   **13.** 0–24 years; 40 million or 40,000,000 books
15. 250 million or 250,000,000 books   **17.** Because everyone is included and fractions are given for *all* age groups, the sum of the fractions must be *1* or *all* of the people.   **19.** \$38,000
21. \$7600   **23.** \$2375
25. The correct solution is
$$\dfrac{9}{10} \times \dfrac{20}{21} = \dfrac{\cancel{9}^{3}}{\cancel{10}_{1}} \times \dfrac{\cancel{20}^{2}}{\cancel{21}_{7}} = \dfrac{6}{7}$$
27. \$42   **29.** 9000 votes   **31.** $\dfrac{1}{32}$ of the estate

Section 2.7 (page 159)

1. $\dfrac{3}{2}$ **3.** $\dfrac{5}{8}$ **5.** $\dfrac{6}{5}$ **7.** $\dfrac{1}{4}$ **9.** $\dfrac{1}{3}$ **11.** $2\dfrac{5}{8}$ **13.** $\dfrac{9}{20}$ **15.** 4
17. 6   **19.** $\dfrac{13}{16}$ **21.** $\dfrac{4}{5}$ **23.** 18   **25.** $22\dfrac{1}{2}$ **27.** $\dfrac{1}{14}$ **29.** $\dfrac{2}{9}$ acre
31. 15 times   **33.** 88 dispensers   **35.** 60 trips   **37.** 12 batches
39. You can divide two fractions by using the reciprocal of the second fraction (divisor) and then multiplying.   **41.** 76 mi
43. \$120,000   **45.** double, twice, times, product, twice as much
46. goes into, divide, per, quotient, divided by   **47.** reciprocal
48. $\dfrac{4}{3}; \dfrac{8}{7}; \dfrac{1}{5}; \dfrac{19}{12}$ **49.** $3\dfrac{3}{4}$ in; Add the lengths of all the sides.
50. $\dfrac{225}{256}$ in.²; Multiply the length by the width.

Section 2.8 (page 169)

1. *Exact:* $8\dfrac{1}{8}$; *Estimate:* $3 \cdot 3 = 9$ **3.** *Exact:* $4\dfrac{1}{2}$; *Estimate:* $2 \cdot 3 = 6$ **5.** *Exact:* 4;   *Estimate:* $3 \cdot 1 = 3$ **7.** *Exact:* 50; *Estimate:* $8 \cdot 6 = 48$ **9.** *Exact:* $49\dfrac{1}{2}$;   *Estimate:* $5 \cdot 2 \cdot 5 = 50$
11. *Exact:* 12;   *Estimate:* $3 \cdot 2 \cdot 3 = 18$ **13.** *Exact:* $\dfrac{1}{3}$;   *Estimate:* $3 \div 8 = \dfrac{3}{8}$ **15.** *Exact:* $\dfrac{5}{6}$;   *Estimate:* $3 \div 3 = 1$ **17.** *Exact:* $3\dfrac{3}{5}$; *Estimate:* $9 \div 3 = 3$ **19.** *Exact:* $\dfrac{3}{7}$;   *Estimate:* $1 \div 2 = \dfrac{1}{2}$
21. *Exact:* $\dfrac{3}{10}$;   *Estimate:* $2 \div 6 = \dfrac{1}{3}$ **23.** *Exact:* $\dfrac{17}{18}$; *Estimate:* $6 \div 6 = 1$ **25.** **(a)** *Estimate:* $3 \cdot 2 = 6$ cups; *Exact:* 5 cups of Quaker Oats **(b)** *Estimate:* $1 \cdot 2 = 2$ cups; *Exact:* $2\dfrac{1}{2}$ cups of brown sugar **(c)** *Estimate:* $2 \cdot 2 = 4$ cups; *Exact:* $3\dfrac{1}{2}$ cups of flour   **27.** **(a)** *Estimate:* $1 \div 2 = \dfrac{1}{2}$ cup; *Exact:* $\dfrac{5}{8}$ cup of brown sugar **(b)** *Estimate:* $1 \div 2 = \dfrac{1}{2}$ cup; *Exact:* $\dfrac{1}{4}$ cup granulated sugar **(c)** *Estimate:* $2 \div 2 = 1$ cup; *Exact:* $\dfrac{7}{8}$ cup of flour   **29.** *Estimate:* $1314 \div 110 \approx 12$ homes; *Exact:* 12 homes   **31.** *Estimate:* $2 \cdot 13 = 26$ oz;   *Exact:* $21\dfrac{7}{8}$ oz
33. The answer should include *Step 1* Change mixed numbers to improper fractions. *Step 2* Multiply the fractions. *Step 3* Write the answer in lowest terms, changing to mixed or whole numbers where possible.   **35.** *Estimate:* $25{,}730 \div 10 = 2573$ anchors; *Exact:* 2480 anchors   **37.** *Estimate:* $10 \div 1 = 10$ spacers; *Exact:* 13 spacers   **39.** *Estimate:* $13 \cdot 28 = 364$;   *Exact:* 471 gal
$$7 \cdot 16 = +\underline{112} \atop 476 \text{ gal}$$

Chapter 2 Review Exercises (page 177)

1. $\dfrac{1}{3}$ **2.** $\dfrac{5}{8}$ **3.** $\dfrac{2}{4}$ **4.** Proper $\dfrac{1}{8}, \dfrac{3}{4}, \dfrac{2}{3}$; Improper $\dfrac{4}{3}, \dfrac{5}{5}$
5. Proper $\dfrac{15}{16}, \dfrac{1}{8}$; Improper $\dfrac{6}{5}, \dfrac{16}{13}, \dfrac{5}{3}$ **6.** $\dfrac{17}{3}$ **7.** $\dfrac{54}{5}$ **8.** $2\dfrac{1}{8}$
9. $12\dfrac{3}{5}$ **10.** 1, 2, 3, 6   **11.** 1, 2, 3, 4, 6, 8, 12, 24
12. 1, 5, 11, 55   **13.** 1, 2, 3, 5, 6, 9, 10, 15, 18, 30, 45, 90   **14.** 3^3
15. $2 \cdot 3 \cdot 5^2$ **16.** $2^3 \cdot 3 \cdot 7$ **17.** 36   **18.** 200   **19.** 1728

20. 2048   **21.** $\frac{3}{4}$ **22.** $\frac{5}{6}$ **23.** $\frac{15}{16}$

24. $\dfrac{\overset{1}{\cancel{8}} \cdot 5}{2 \cdot 2 \cdot 3 \cdot \underset{1}{\cancel{8}}}; \frac{5}{12}$ **25.** $\dfrac{\overset{1}{\cancel{2}} \cdot \overset{1}{\cancel{2}} \cdot \overset{1}{\cancel{2}} \cdot \overset{1}{\cancel{2}} \cdot \overset{1}{\cancel{2}} \cdot 2 \cdot 2 \cdot \overset{1}{\cancel{3}}}{\underset{1}{\cancel{2}} \cdot \underset{1}{\cancel{2}} \cdot \underset{1}{\cancel{2}} \cdot \underset{1}{\cancel{2}} \cdot \underset{1}{\cancel{2}} \cdot \underset{1}{\cancel{3}}}; 4$

26. equivalent   **27.** not equivalent   **28.** $\frac{1}{2}$ **29.** $\frac{3}{16}$ **30.** $\frac{1}{7}$

31. $\frac{4}{21}$ **32.** 15   **33.** 625   **34.** $\frac{2}{3}$ **35.** $\frac{5}{3} = 1\frac{2}{3}$ **36.** $\frac{5}{2} = 2\frac{1}{2}$

37. 2   **38.** 8   **39.** 24   **40.** $\frac{2}{15}$ **41.** $\frac{2}{15}$ **42.** $\frac{4}{13}$ **43.** $\frac{9}{40}$ ft²

44. $\frac{7}{12}$ in.²   **45.** $7\frac{1}{2}$ ft²   **46.** 36 yd²   **47.** *Exact:* $6\frac{7}{8}$;

Estimate: $6 \cdot 1 = 6$ **48.** *Exact:* $21\frac{3}{8}$; *Estimate:* $2 \cdot 7 \cdot 1 = 14$

49. *Exact:* $5\frac{1}{6}$; *Estimate:* $16 \div 3 = 5\frac{1}{3}$ **50.** *Exact:* $\frac{3}{4}$; *Estimate:*

$5 \div 6 = \frac{5}{6}$ **51.** 336 boxes   **52.** $\frac{2}{15}$ of the estate

53. *Estimate:* $158 \div 4 \approx 40$ pull cords; *Exact:* 36 pull cords
54. *Estimate:* $9 \cdot 38 = \$342$; *Exact:* \$323   **55.** 25 lb   **56.** \$510

57. $\frac{5}{32}$ of the budget   **58.** $\frac{1}{12}$ ton   **59.** $\frac{1}{4}$ **60.** $\frac{2}{5}$ **61.** $31\frac{1}{4}$

62. $28\frac{1}{8}$ **63.** $\frac{1}{10}$ **64.** $\frac{5}{32}$ **65.** 30   **66.** $2\frac{1}{6}$ **67.** $2\frac{1}{3}$

68. $28\frac{3}{5}$ **69.** $\frac{17}{3}$ **70.** $\frac{307}{8}$

71. $\dfrac{\overset{1}{\cancel{2}} \cdot \overset{1}{\cancel{2}} \cdot 2}{\underset{1}{\cancel{2}} \cdot \underset{1}{\cancel{2}} \cdot 3} = \frac{2}{3}$ **72.** $\dfrac{\overset{1}{\cancel{2}} \cdot 2 \cdot \overset{1}{\cancel{3}} \cdot 3 \cdot 3}{\underset{1}{\cancel{2}} \cdot \underset{1}{\cancel{3}} \cdot 5 \cdot 7} = \frac{18}{35}$

73. $\frac{2}{3}$ **74.** $\frac{1}{3}$ **75.** $\frac{2}{5}$ **76.** $\frac{1}{3}$ **77.** *Estimate:* $4 \cdot 44 = 176$ oz;

Exact: $152\frac{4}{9}$ oz   **78.** *Estimate:* $7 \cdot 26 = 182$ qt; *Exact:* $184\frac{7}{8}$ qt

79. $\frac{49}{64}$ in.²   **80.** $\frac{5}{16}$ yd²

Chapter 2 Test (page 181)

1. $\frac{5}{6}$ **2.** $\frac{3}{8}$ **3.** $\frac{2}{3}, \frac{6}{7}, \frac{1}{4}, \frac{5}{8}$ **4.** $\frac{20}{3}$ **5.** $23\frac{3}{5}$ **6.** 1, 2, 3, 6, 9, 18

7. $3^2 \cdot 7$ **8.** $2^5 \cdot 3$ **9.** $2^2 \cdot 5^3$ **10.** $\frac{3}{4}$ **11.** $\frac{5}{6}$

12. $\dfrac{56}{84} = \dfrac{\overset{1}{\cancel{2}} \cdot \overset{1}{\cancel{2}} \cdot 2 \cdot \overset{1}{\cancel{7}}}{\underset{1}{\cancel{2}} \cdot \underset{1}{\cancel{2}} \cdot 3 \cdot \underset{1}{\cancel{7}}} = \frac{2}{3}$

13. Multiply fractions by multiplying the numerators and multiplying the denominators. Divide two fractions by using the reciprocal of the second fraction (divisor) and multiplying.
14. $\frac{3}{20}$ **15.** 16   **16.** $\frac{3}{8}$ yd²   **17.** 3696 drivers   **18.** $\frac{9}{10}$

19. $15\frac{3}{4}$ **20.** 200 sports bottles   **21.** *Estimate:* $4 \cdot 4 = 16$;

Exact: $14\frac{7}{16}$ **22.** *Estimate:* $2 \cdot 4 = 8$; *Exact:* $7\frac{17}{18}$ **23.** *Estimate:*

$10 \div 2 = 5$; *Exact:* $4\frac{4}{15}$ **24.** *Estimate:* $9 \div 2 = 4\frac{1}{2}$; *Exact:* $5\frac{1}{10}$

25. *Estimate:* $3 \cdot 12 = 36$; *Exact:* $30\frac{5}{8}$ g

Cumulative Review Exercises: Chapters 1–2 (page 183)

1. hundreds 5; tens 7   **2.** millions 8; ten thousands 2   **3.** 166
4. 149,199   **5.** 4452   **6.** 3,221,821   **7.** 476   **8.** 96
9. 2,168,232   **10.** 450,400   **11.** 9   **12.** 7581   **13.** 8471 R2
14. 22 R26   **15.** 5740; 5700; 6000   **16.** 76,270; 76,300; 76,000
17. 3   **18.** 5   **19.** \$992   **20.** \$130   **21.** 47,000 hairs

22. \$285   **23.** $\frac{11}{16}$ ft²   **24.** 6¢   **25.** proper   **26.** improper

27. proper   **28.** $\frac{27}{8}$ **29.** $\frac{32}{5}$ **30.** 2   **31.** $12\frac{7}{8}$ **32.** $2^3 \cdot 3^2$

33. $2 \cdot 3^2 \cdot 7$ **34.** $2 \cdot 5^2 \cdot 7$ **35.** 64   **36.** 288   **37.** 640

38. $\frac{7}{8}$ **39.** $\frac{2}{3}$ **40.** $\frac{5}{9}$ **41.** $\frac{3}{8}$ **42.** 12   **43.** 25   **44.** $\frac{24}{25}$

45. $\frac{7}{12}$ **46.** $2\frac{2}{5}$

Chapter 3

Section 3.1 (page 189)

1. $\frac{4}{5}$ **3.** $\frac{5}{6}$ **5.** $\frac{1}{3}$ **7.** $1\frac{1}{5}$ **9.** $\frac{1}{3}$ **11.** $\frac{13}{20}$ **13.** $\frac{11}{15}$ **15.** $1\frac{1}{2}$

17. $\frac{11}{27}$ **19.** $\frac{3}{8}$ **21.** $\frac{6}{11}$ **23.** $\frac{3}{5}$ **25.** $1\frac{1}{7}$ **27.** $\frac{1}{5}$ **29.** $1\frac{1}{6}$

31. $1\frac{1}{10}$ **33.** Three steps to add like fractions are:

1. Add the numerators of the fractions to find the numerator of the sum (the answer).   2. Use the denominator of the fractions as the denominator of the sum.   3. Write the answer in lowest terms.
35. $\frac{3}{4}$ **37.** $\frac{1}{3}$ **39.** $\frac{1}{2}$ acre

Section 3.2 (page 197)

1. 4   **3.** 15   **5.** 36   **7.** 14   **9.** 30   **11.** 100   **13.** 20   **15.** 72

17. 36   **19.** 120   **21.** 180   **23.** 144   **25.** $\frac{8}{24}$ **27.** $\frac{18}{24}$

29. $\frac{20}{24}$ **31.** 2   **33.** 15   **35.** 28   **37.** 55   **39.** 32   **41.** 72

43. 136   **45.** 96   **47.** 27   **49.** It probably depends on how large the numbers are. If the numbers are small, the method using multiples of the largest number seems best. If numbers are larger, or there are more of them, then the factorization method will be better.   **51.** 7200   **53.** 10,584   **55.** like; unlike   **56.** numerators; denominator; lowest   **57.** least; smallest   **58.** 40 is the least common multiple.   **59.** 72   **60.** 450   **61.** 240 is a common multiple but twice as large as the least common multiple; 120 is the LCM.
62. The least common multiple can be no smaller than the largest number in a group and the number 1760 is a multiple of 55.

Section 3.3 (page 205)

1. $\frac{5}{6}$ **3.** $\frac{8}{9}$ **5.** $\frac{3}{4}$ **7.** $\frac{39}{40}$ **9.** $\frac{23}{36}$ **11.** $\frac{14}{15}$ **13.** $\frac{29}{36}$ **15.** $\frac{17}{20}$

17. $\frac{23}{30}$ **19.** $\frac{7}{12}$ **21.** $\frac{23}{48}$ **23.** $\frac{3}{8}$ **25.** $\frac{1}{2}$ **27.** $\frac{7}{15}$ **29.** $\frac{1}{6}$

31. $\frac{19}{45}$ **33.** $\frac{3}{40}$ **35.** $\frac{17}{48}$ **37.** $\frac{23}{24}$ yd³   **39.** $\frac{7}{12}$ acre   **41.** $\frac{31}{40}$ in.

43. $\frac{3}{8}$ gal   **45.** You cannot add or subtract until all the fractional pieces are the same size. For example, halves are larger than fourths, so you cannot add $\frac{1}{2} + \frac{1}{4}$ until you rewrite $\frac{1}{2}$ as $\frac{2}{4}$.

47. $\frac{1}{4}$ **49.** work and travel; 8 hr   **51.** $\frac{3}{16}$ in.

Section 3.4 (page 213)

1. *Estimate:* $6 + 3 = 9$; *Exact:* $8\frac{5}{6}$

3. *Estimate:* $7 + 4 = 11$; *Exact:* $11\frac{1}{2}$

5. *Estimate:* $1 + 4 = 5$; *Exact:* $4\frac{5}{24}$

7. *Estimate:* $25 + 19 = 44$; *Exact:* $43\frac{2}{3}$

9. *Estimate:* $34 + 19 = 53$; *Exact:* $52\frac{1}{10}$

11. *Estimate:* $23 + 15 = 38$; *Exact:* $38\frac{5}{28}$

13. *Estimate:* $11 + 18 + 16 = 45$; *Exact:* $43\frac{5}{6}$

15. *Estimate:* $12 - 10 = 2$; *Exact:* $2\frac{1}{8}$

17. *Estimate:* $13 - 1 = 12$; *Exact:* $11\frac{7}{15}$

19. *Estimate:* $28 - 6 = 22$; *Exact:* $22\frac{7}{30}$

21. *Estimate:* $17 - 7 = 10$; *Exact:* $10\frac{3}{8}$

23. *Estimate:* $19 - 6 = 13$; *Exact:* $12\frac{19}{20}$

25. *Estimate:* $16 - 11 = 5$; *Exact:* $5\frac{7}{8}$

27. $9\frac{3}{8}$ 29. $11\frac{1}{2}$ 31. $3\frac{5}{6}$ 33. $6\frac{11}{12}$ 35. $8\frac{1}{8}$ 37. $\frac{5}{6}$

39. $2\frac{7}{8}$ 41. $2\frac{7}{12}$ 43. $5\frac{9}{20}$ 45. $3\frac{16}{21}$

47. Find the least common denominator. Change the fraction parts so that they have the same denominator. Add the fraction parts. Add the whole number parts. Write the answer as a mixed number.

49. *Estimate:* $143 - 29 = 114$ tons; *Exact:* $114\frac{1}{10}$ tons

51. *Estimate:* $4 - 2 = 2$ deaths per 1000; *Exact:* $2\frac{3}{10}$ deaths per 1000

53. *Estimate:* $13 + 9 = 22$ ft; *Exact:* $21\frac{1}{6}$ ft

55. *Estimate:* $3 + 6 + 5 + 3 + 6 = 23$ hr; *Exact:* $22\frac{7}{8}$ hr

57. *Estimate:* $24 + 35 + 24 + 35 = 118$ in.; *Exact:* $116\frac{1}{2}$ in.

59. *Estimate:* $9 - 3 - 3 - 2 = 1$ yd³; *Exact:* $1\frac{5}{8}$ yd³

61. *Estimate:* $527 - 108 - 151 - 139 = 129$ ft; *Exact:* 130 ft

63. *Estimate:* $3 + 7 + 2 + 3 + 7 = 22$ tons; *Exact:* $21\frac{7}{12}$ tons

65. $4\frac{11}{16}$ in. 67. $21\frac{3}{8}$ in. 69. (a) 30 (b) 28 (c) 25 (d) 264

70. least common denominator 71. (a) $\frac{23}{24}$ (b) $\frac{8}{15}$ (c) $\frac{43}{48}$

(d) $\frac{4}{21}$ 72. fraction parts 73. improper

74. (a) $4\frac{5}{8} + 3\frac{2}{8} = 7\frac{7}{8}$; $\frac{37}{8} + \frac{26}{8} = \frac{63}{8} = 7\frac{7}{8}$

(b) $11\frac{56}{40} - 8\frac{35}{40} = 3\frac{21}{40}$; $\frac{496}{40} - \frac{355}{40} = \frac{141}{40} = 3\frac{21}{40}$

Section 3.5 (page 225)

1.–12.

2. 1. 10. 4. 3. 12. 7. 5. 6. 11. 9. 8.

13. > 15. < 17. > 19. < 21. > 23. >

25. $\frac{1}{4}$ 27. $\frac{25}{49}$ 29. $\frac{9}{16}$ 31. $\frac{64}{125}$ 33. $\frac{81}{16} = 5\frac{1}{16}$ 35. $\frac{81}{256}$

37. A number line is a horizontal line with a range of numbers placed on it. The lowest number is on the left and the highest number is on the right. It can be used to compare the size or value of numbers.

39. 3 41. 16 43. 1 45. $\frac{3}{16}$ 47. $\frac{4}{9}$ 49. $\frac{1}{3}$ 51. $\frac{1}{2}$ 53. $\frac{3}{8}$

55. $\frac{1}{4}$ 57. $1\frac{1}{2}$ 59. $\frac{1}{12}$ 61. 3 63. $\frac{5}{16}$ 65. $\frac{1}{4}$

67. $\frac{1}{32}$ 69. $\frac{9}{25}$ is greater. 71. <; >

72. (a) like; numerators; numerator (b) Answers will vary.

73. parentheses; exponents; square; multiply; divide; add; subtract

74. $\frac{2}{45}$ 75. $\frac{4}{9}$ 76. $\frac{9}{64}$ 77. $1\frac{169}{343}$ 78. $2\frac{113}{256}$ 79. 2 80. $\frac{2}{45}$

Summary Exercises on Fractions (page 229)

1. proper 2. improper 3. improper 4. proper

5. $\frac{4}{5}$ 6. $\frac{7}{8}$ 7. $\frac{3}{7}$ 8. $\frac{23}{47}$ 9. $\frac{1}{2}$ 10. $\frac{3}{8}$ 11. 35 12. $\frac{5}{6}$

13. $1\frac{1}{6}$ 14. 56 15. $1\frac{13}{24}$ 16. $1\frac{13}{16}$ 17. $2\frac{1}{12}$ 18. $\frac{1}{12}$

19. $\frac{11}{24}$ 20. $\frac{2}{15}$ 21. *Exact:* $7\frac{7}{8}$; *Estimate:* $4 \cdot 2 = 8$

22. *Exact:* $17\frac{15}{32}$; *Estimate:* $5 \cdot 3 = 15$

23. *Exact:* $107\frac{2}{3}$; *Estimate:* $8 \cdot 6 \cdot 2 = 96$

24. *Exact:* $1\frac{1}{6}$; *Estimate:* $4 \div 4 = 1$

25. *Exact:* $3\frac{7}{16}$; *Estimate:* $7 \div 2 = 3\frac{1}{2}$

26. *Exact:* $6\frac{1}{6}$; *Estimate:* $5 \div 1 = 5$

27. *Estimate:* $4 + 4 = 8$; *Exact:* $7\frac{7}{8}$

28. *Estimate:* $22 + 8 = 30$; *Exact:* $30\frac{1}{6}$

29. *Estimate:* $15 + 11 = 26$; *Exact:* $25\frac{4}{15}$

30. *Estimate:* $8 - 3 = 5$; *Exact:* $4\frac{9}{10}$

31. *Estimate:* $14 - 7 = 7$; *Exact:* $6\frac{5}{8}$

32. *Estimate:* $32 - 23 = 9$; *Exact:* $9\frac{1}{4}$ 33. $\frac{1}{12}$ 34. $\frac{9}{10}$

35. $\frac{5}{18}$ 36. 40 37. 72 38. 84 39. 35 40. 12 41. 55

42. < 43. > 44. >

Chapter 3 Review Exercises (page 235)

1. $\frac{7}{9}$ 2. $\frac{5}{7}$ 3. $\frac{3}{4}$ 4. $\frac{1}{8}$ 5. $\frac{1}{5}$ 6. $\frac{1}{6}$ 7. $\frac{13}{31}$ 8. $\frac{1}{3}$

9. $\frac{3}{4}$ of her patients 10. $\frac{1}{4}$ Web page less 11. 12 12. 20

13. 60 14. 24 15. 120 16. 180 17. 8 18. 21 19. 10

20. 45 21. 32 22. 20 23. $\frac{5}{6}$ 24. $\frac{7}{8}$ 25. $\frac{5}{8}$ 26. $\frac{5}{12}$

27. $\frac{13}{24}$ 28. $\frac{17}{36}$ 29. $\frac{23}{24}$ of the sack 30. $\frac{59}{60}$ of the amount needed

31. *Estimate:* 19 + 14 = 33; *Exact:* $32\frac{3}{8}$

32. *Estimate:* 23 + 15 = 38; *Exact:* $38\frac{1}{9}$

33. *Estimate:* 13 + 9 + 10 = 32; *Exact:* $31\frac{43}{80}$

34. *Estimate:* 32 − 15 = 17; *Exact:* $17\frac{1}{12}$

35. *Estimate:* 34 − 16 = 18; *Exact:* $18\frac{1}{3}$

36. *Estimate:* 215 − 136 = 79; *Exact:* $79\frac{7}{16}$

37. $9\frac{1}{10}$ 38. $10\frac{5}{12}$ 39. $3\frac{1}{4}$ 40. $1\frac{2}{3}$ 41. $5\frac{1}{2}$ 42. $2\frac{19}{24}$

43. *Estimate:* 15 − 6 − 7 = 2 gal; *Exact:* $2\frac{5}{12}$ gal

44. *Estimate:* 29 + 25 = 54 tons; *Exact:* $53\frac{5}{12}$ tons

45. *Estimate:* 9 + 9 + 7 = 25 lb; *Exact:* $24\frac{23}{24}$ lb

46. *Estimate:* 9 − 2 − 3 = 4 acres; *Exact:* $4\frac{1}{16}$ acres

47.–50.

47. 48.50.49.

51. < 52. < 53. > 54. > 55. < 56. > 57. < 58. >

59. $\frac{1}{4}$ 60. $\frac{4}{9}$ 61. $\frac{27}{1000}$ 62. $\frac{81}{4096}$ 63. $\frac{2}{3}$ 64. $6\frac{2}{3}$ 65. $\frac{4}{9}$

66. 1 67. $\frac{3}{16}$ 68. $1\frac{25}{64}$ 69. $\frac{2}{3}$ 70. $\frac{1}{8}$ 71. $\frac{19}{32}$ 72. $\frac{11}{16}$

73. $2\frac{1}{6}$ 74. $26\frac{1}{4}$ 75. $5\frac{3}{8}$ 76. $11\frac{43}{80}$ 77. $15\frac{5}{12}$ 78. $\frac{8}{11}$

79. $\frac{1}{250}$ 80. $\frac{1}{2}$ 81. $\frac{2}{9}$ 82. $\frac{11}{27}$ 83. > 84. < 85. <

86. > 87. 36 88. 120 89. 126 90. 18 91. 108 92. 60

93. *Estimate:* 93 − 14 − 22 = 57 ft; *Exact:* $56\frac{7}{8}$ ft

94. *Estimate:* 15 − 4 − 9 = 2 gal; *Exact:* $2\frac{7}{8}$ gal

Chapter 3 Test (page 241)

1. $\frac{3}{4}$ 2. $\frac{1}{2}$ 3. $\frac{2}{5}$ 4. $\frac{1}{6}$ 5. 12 6. 30 7. 108 8. $\frac{5}{8}$

9. $\frac{23}{36}$ 10. $\frac{5}{24}$ 11. $\frac{1}{40}$

12. *Estimate:* 8 + 5 = 13; *Exact:* $12\frac{1}{2}$

13. *Estimate:* 16 − 12 = 4; *Exact:* $4\frac{11}{15}$

14. *Estimate:* 19 + 9 + 12 = 40; *Exact:* $40\frac{29}{60}$

15. *Estimate:* 24 − 18 = 6; *Exact:* $5\frac{5}{8}$

16. Probably addition and subtraction of fractions is more difficult because you have to find the least common denominator and then change the fractions to the same denominator. 17. Round mixed numbers to the nearest whole number. Then add, subtract, multiply, or divide to estimate the answer. The estimate may vary from the exact answer but it lets you know if your answer is reasonable.

18. *Estimate:* 6 + 5 + 4 + 7 + 5 = 27 hr; *Exact:* $26\frac{3}{4}$

19. *Estimate:* 148 − 69 − 37 − 6 = 36 gal; *Exact:* $35\frac{7}{8}$ gal

20. > 21. > 22. 2 23. $\frac{13}{48}$ 24. $1\frac{3}{4}$ 25. $1\frac{1}{3}$

Cumulative Review Exercises: Chapters 1–3 (page 243)

1. 8, 1 2. 5, 9 3. 1440; 1400; 1000 4. 59,800; 59,800; 60,000
5. *Estimate:* 2000 + 400 + 50,000 + 30,000 = 82,400; *Exact:* 79,779 6. *Estimate:* 20,000 − 10,000 = 10,000; *Exact:* 14,389 7. *Estimate:* 4000 × 300 = 1,200,000; *Exact:* 1,251,040 8. *Estimate:* 100,000 ÷ 40 = 2500; *Exact:* 3211 9. 26 10. 1,255,609 11. 591 12. 2,801,695
13. 120 14. 126 15. 160 16. 456 17. 369,408
18. 17,000 19. 135 20. 2693 R2 21. 32 R166
22. *Estimate:* 20 + 9 + 5 + 20 + 9 + 5 = 68 ft; *Exact:* 64 ft
23. *Estimate:* 20 • 10 = 200 ft²; *Exact:* 252 ft²
24. *Estimate:* 40,000 ÷ 30 ≈ 1333 cartons; *Exact:* 1260 cartons
25. *Estimate:* 4000 × 60 = 240,000 revolutions; *Exact:* 216,000 revolutions

26. *Estimate:* 2 • 3 = 6 yd²; *Exact:* $4\frac{2}{3}$ yd²

27. *Estimate:* 5 • 7 = 35 mi²; *Exact:* $30\frac{5}{6}$ mi²

28. *Estimate:* 5 • 4 = 20 cords; *Exact:* $18\frac{3}{8}$ cords

29. *Estimate:* 1537 − 83 = 1454 ft; *Exact:* $1454\frac{3}{8}$ ft

30. 2 • 3² 31. 2³ • 5² 32. 5² • 7² 33. 144 34. 108
35. 972 36. 6 37. 9 38. 12 39. 9 40. 44

41. $\frac{4}{45}$ 42. $\frac{9}{10}$ 43. $1\frac{1}{16}$ 44. proper 45. improper

46. improper 47. $\frac{5}{12}$ 48. $\frac{7}{8}$ 49. $\frac{9}{10}$ 50. $\frac{1}{2}$ 51. $\frac{5}{16}$

52. $36\frac{3}{4}$ 53. $1\frac{2}{3}$ 54. $2\frac{3}{16}$ 55. $13\frac{1}{2}$ 56. $\frac{25}{28}$ 57. $\frac{13}{16}$

58. $\frac{7}{36}$ 59. *Estimate:* 3 + 5 = 8; *Exact:* $7\frac{7}{8}$

60. *Estimate:* 22 + 4 = 26; *Exact:* $26\frac{7}{24}$

61. *Estimate:* 5 − 2 = 3; *Exact:* $2\frac{5}{8}$

62. 24 63. 120 64. 144 65. 36 66. 56 67. 81 68. 60
69.–72.

70.69. 71. 72.

73. < 74. > 75. <

Chapter 4

Section 4.1 (page 253)

1. 7; 0; 4   **3.** 5; 1; 8   **5.** 4; 7; 0   **7.** 1; 6; 3   **9.** 1; 8; 9   **11.** 6; 2; 1   **13.** 410.25   **15.** 6.5432   **17.** 5406.045   **19.** $\frac{7}{10}$

21. $13\frac{2}{5}$ **23.** $\frac{1}{4}$ **25.** $\frac{33}{50}$ **27.** $10\frac{17}{100}$ **29.** $\frac{3}{50}$ **31.** $\frac{41}{200}$

33. $5\frac{1}{500}$ **35.** $\frac{343}{500}$ **37.** five tenths   **39.** seventy-eight hundredths
41. one hundred five thousandths   **43.** twelve and four hundredths
45. one and seventy-five thousandths   **47.** 6.7   **49.** 0.32
51. 420.008   **53.** 0.0703   **55.** 75.030   **57.** Anne should not say "and" because that denotes a decimal point.   **59.** ten thousandths inch; $\frac{10}{1000} = \frac{1}{100}$ **61.** 12 pounds   **63.** 3-C   **65.** 4-A
67. One and six hundred two thousandths centimeters
69. millionths, ten-millionths, hundred-millionths, billionths; these match the words on the left side of the chart with "ths" added.
70. First place to left of decimal point is ones, so first place to right could be oneths, like tens and tenths. But anything that is 1 or more is to the left of the decimal point.   **71.** Seventy-two million four hundred thirty-six thousand nine hundred fifty-five hundred-millionths   **72.** Six hundred seventy-eight thousand five hundred fifty-four billionths   **73.** Eight thousand six and five hundred thousand one millionths   **74.** Twenty thousand, sixty and five hundred five millionths   **75.** 0.0302040
76. 9,876,543,210.100200300

Section 4.2 (page 263)

1. 16.9   **3.** 0.956   **5.** 0.80   **7.** 3.661   **9.** 794.0   **11.** 0.0980   **13.** 49   **15.** 9.09   **17.** 82.0002   **19.** $0.82   **21.** $1.22   **23.** $0.50   **25.** $48,650   **27.** $310   **29.** $849   **31.** $500   **33.** $1.00   **35.** $1000   **37.** (a) 186.0 miles per hour (b) 763.0 miles per hour   **39.** (a) 322 miles per hour (b) 163 miles per hour
41. Rounds to $0 (zero dollars) because $0.499 is closer to $0 than to $1.   **42.** Round amounts less than $1.00 to nearest cent instead of nearest dollar.   **43.** Rounds to $0.00 (zero cents) because $0.0015 is closer to $0.00 than to $0.01.   **44.** Both round to $0.60. Rounding to nearest thousandth (tenth of a cent) would allow you to identify $0.597 as less than $0.601.

Section 4.3 (page 269)

1. 17.48   **3.** 23.013   **5.** 7.763   **7.** 77.006   **9.** 20.104   **11.** 0.109   **13.** 330.86895   **15.** (a) 24.75 in. (b) 3.95 in.
17. (a) 62.27 in. (b) 0.39 in.   **19.** 6 should be written 6.00; sum is 46.22.   **21.** *Estimate:* $20 − 7 = $13; *Exact:* $13.16
23. *Estimate:* 400 + 1 + 20 = 421; *Exact:* 414.645
25. *Estimate:* 9 − 4 = 5; *Exact:* 4.849
27. *Estimate:* 60 + 500 + 6 = 566; *Exact:* 608.4363
29. 0.275   **31.** 6.507   **33.** 1.81   **35.** 6056.7202
37. *Estimate:* $50 − $40 = $10; *Exact:* $8.91
39. *Estimate:* 2 − 2 = 0; *Exact:* 0.019 in.
41. *Estimate:* 400 − 300 = 100 million people;
Exact: 112.48 million people   **43.** *Estimate:* 11 − 10 = 1 ounce;
Exact: 0.65 ounce   **45.** *Estimate:* 20 + 6 + 20 + 6 = 52 in.;
Exact: 52.1 in.   **47.** *Estimate:* $29 − $9 = $20; *Exact:* $20.50
49. *Estimate:* $13 + 18 + 6 + 3 + 2 = $42; *Exact:* $41.60
51. $1939.36   **53.** $3.97   **55.** $598.22   **57.** $b = 1.39$ cm
59. $q = 23.843$ ft

Section 4.4 (page 275)

1. 0.1344   **3.** 159.10   **5.** 15.5844   **7.** $34,500.20   **9.** 43.2   **11.** 0.432   **13.** 0.0432   **15.** 0.00432   **17.** 0.0000312
19. 0.000006   **21.** Multiplying by 10, decimal point moves one

place to the right; by 100, two places to the right; by 1000, three places to the right.   **23.** *Estimate:* 40 × 5 = 200; *Exact:* 190.08
25. *Estimate:* 40 × 40 = 1600; *Exact:* 1558.2
27. *Estimate:* 7 × 5 = 35; *Exact:* 30.038
29. *Estimate:* 3 × 7 = 21; *Exact:* 19.24165
31. unreasonable; $189.00   **33.** reasonable
35. unreasonable; $3.19   **37.** unreasonable; 9.5 pounds
39. $945.87 (rounded)   **41.** $2.45 (rounded)
43. $28.82 (rounded)   **45.** $12,271   **47.** $347.52; $719.40
49. $76.50   **51.** $73.45   **53.** $388.34   **55.** $4.09 (rounded)
57. $129.25   **59.** (a) $70.05 (b) $25.80

Section 4.5 (page 285)

1. 3.9   **3.** 0.47   **5.** 400.2   **7.** 36   **9.** 0.06   **11.** 6000   **13.** 25.3   **15.** 516.67 (rounded)   **17.** 24.291 (rounded)   **19.** 10,082.647 (rounded)   **21.** Dividing by 10, decimal point moves one place to the left; by 100, two places to the left; by 1000, three places to the left.   **23.** unreasonable; 40 ÷ 8 = 5; Correct answer is 4.725.
25. reasonable; 50 ÷ 50 = 1   **27.** unreasonable; 300 ÷ 5 = 60; Correct answer is 60.2.   **29.** unreasonable; 9 ÷ 1 = 9; Correct answer is 7.44.   **31.** $4.00 (rounded)   **33.** $19.46
35. $0.30   **37.** $8.92 per hour   **39.** 21.2 miles per gallon (rounded)   **41.** 8.81 m (rounded)   **43.** 0.03 meter   **45.** 26.72 meters
47. 14.25   **49.** 73.4   **51.** 1.205   **53.** 0.334
55. $0.03 (rounded) per can   **57.** $0.04 (rounded)
59. 100,000 box tops   **61.** 2632 box tops (rounded)

Section 4.6 (page 293)

1. 0.5   **3.** 0.75   **5.** 0.3   **7.** 0.9   **9.** 0.6   **11.** 0.875   **13.** 2.25   **15.** 14.7   **17.** 3.625   **19.** 0.333 (rounded)   **21.** 0.833 (rounded)   **23.** 1.889 (rounded)   **25. (a)** A proper fraction is less than 1, so it cannot be equivalent to a mixed number. **(b)** $\frac{5}{9}$ means 5 ÷ 9 or $9\overline{)5}$ so correct answer is 0.556 (rounded). This makes sense because both the fraction and decimal are less than 1.

26. (a) $2.035 = 2\frac{35}{1000} = 2\frac{7}{200}$, not $2\frac{7}{20}$. **(b)** Adding the whole number part gives 2 + 0.35, which is 2.35, not 2.035. To check, $2.35 = 2\frac{35}{100} = 2\frac{7}{20}$.   **27.** Just add the whole number part to 0.375. So $1\frac{3}{8} = 1.375$; $3\frac{3}{8} = 3.375$; $295\frac{3}{8} = 295.375$.
28. It works only when the fraction part has a one-digit numerator and a denominator of 10, a two-digit numerator and a denominator of 100, and so on.   **29.** $\frac{2}{5}$ **31.** $\frac{5}{8}$ **33.** $\frac{7}{20}$ **35.** 0.35
37. $\frac{1}{25}$ **39.** $\frac{3}{20}$ **41.** 0.2   **43.** $\frac{9}{100}$ **45.** shorter; 0.72 in.
47. too much; 0.005 gram   **49.** 0.9991 cm, 1.0007 cm
51. more; 0.05 in.   **53.** 0.5399, 0.54, 0.5455
55. 5.0079, 5.79, 5.8, 5.804   **57.** 0.6009, 0.609, 0.628, 0.62812
59. 2.8902, 3.88, 4.876, 5.8751   **61.** 0.006, 0.043, $\frac{1}{20}$, 0.051
63. 0.37, $\frac{3}{8}, \frac{2}{5}$, 0.4001   **65.** red box   **67.** 0.01 in.
69. 1.4 in. (rounded)   **71.** 0.3 in. (rounded)
73. 0.4 in. (rounded)

Chapter 4 Review Exercises (page 301)

1. 0; 5   **2.** 0; 6   **3.** 8; 9   **4.** 5; 9   **5.** 7; 6   **6.** $\frac{1}{2}$ **7.** $\frac{3}{4}$ **8.** $4\frac{1}{20}$
9. $\frac{7}{8}$ **10.** $\frac{27}{1000}$ **11.** $27\frac{4}{5}$ **12.** eight tenths

13. four hundred and twenty-nine hundredths   **14.** twelve and seven thousandths   **15.** three hundred six ten-thousandths
16. 8.3   **17.** 0.205   **18.** 70.0066   **19.** 0.30   **20.** 275.6
21. 72.79   **22.** 0.160   **23.** 0.091   **24.** 1.0   **25.** $15.83
26. $0.70   **27.** $17,625.79   **28.** $350   **29.** $130   **30.** $100
31. $29   **32.** *Estimate:* 6 + 400 + 20 = 426; *Exact:* 444.86
33. *Estimate:* 80 + 1 + 100 + 1 + 30 = 212; *Exact:* 233.515
34. *Estimate:* 300 − 20 = 280; *Exact:* 290.7
35. *Estimate:* 9 − 8 = 1; *Exact:* 1.2684
36. *Estimate:* 100 million − 60 million = 40 million;
Exact: 38.8 million   **37.** *Estimate:* $200 + $40 = $240;
Exact: $260.00   **38.** *Estimate:* $2 + $5 + $20 = $27;
$30 − $27 = $3; *Exact:* $4.14
39. *Estimate:* 2 + 4 + 5 = 11 kilometers; *Exact:* 11.55 kilometers
40. *Estimate:* 6 × 4 24; *Exact:* 22.7106
41. *Estimate:* 40 × 3 120; *Exact:* 141.57   **42.** 0.0112
43. 0.000355   **44.** reasonable; 700 ÷ 10 = 70
45. unreasonable; 30 ÷ 3 = 10; Correct answer is 9.5.
46. 14.467 (rounded)   **47.** 1200   **48.** 0.4   **49.** $708 (rounded)
50. $2.99 (rounded)   **51.** 133 shares (rounded)
52. $3.12 (rounded)   **53.** 29.215   **54.** 10.15   **55.** 3.8
56. 0.64   **57.** 1.875   **58.** 0.111 (rounded)   **59.** 3.6008, 3.68, 3.806   **60.** 0.209, 0.2102, 0.215, 0.22
61. $\frac{1}{8}, \frac{3}{20}$, 0.159, 0.17   **62.** 404.865   **63.** 254.8
64. 3583.261 (rounded)   **65.** 29.0898   **66.** 0.03066   **67.** 9.4
68. 175.675   **69.** 9.04   **70.** 19.50   **71.** 8.19   **72.** 0.928
73. 35   **74.** 0.259   **75.** 0.3   **76.** $3.00 (rounded)
77. $2.17 (rounded)   **78.** $35.96   **79.** $199.71   **80.** $78.50
81. (a) baked potato with skin **(b)** cup of orange juice
(c) 0.59 milligram   **82. (a)** 2.06 milligrams **(b)** more, by 0.06 milligram

Chapter 4 Test (page 305)

1. $18\frac{2}{5}$ **2.** $\frac{3}{40}$ **3.** sixty and seven thousandths
4. two hundred eight ten-thousandths   **5.** 725.6   **6.** 0.630
7. $1.49   **8.** $7860   **9.** *Estimate:* 8 + 80 + 40 = 128;
Exact: 129.2028   **10.** *Estimate:* 80 − 4 = 76; *Exact:* 75.498
11. *Estimate:* 6 • 1 = 6; *Exact:* 6.948   **12.** *Estimate:* 20 ÷ 5 = 4;
Exact: 4.175   **13.** 839.762   **14.** 669.004   **15.** 0.0000483
16. 480   **17.** 2.625   **18.** 0.44, $\frac{9}{20}$, 0.4506, 0.451
19. 35.49   **20.** $446.87   **21.** 1988, 0.33 meter
22. $5.35 (rounded)   **23.** 2.8 degrees   **24.** $4.55 per meter
25. Answer varies.

Cumulative Review Exercises: Chapters 1–4 (page 307)

1. 5, 9, 2   **2.** 0, 5, 8   **3.** 500,000   **4.** 602.49   **5.** $710   **6.** $0.05
7. *Estimate:* 4000 + 600 + 9000 = 13,600; *Exact:* 13,339
8. *Estimate:* 4 + 20 + 1 = 25; *Exact:* 20.683
9. *Estimate:* 5000 − 2000 = 3000; *Exact:* 3209
10. *Estimate:* 50 − 7 = 43; *Exact:* 44.506
11. *Estimate:* 3000 × 200 = 600,000; *Exact:* 550,622
12. *Estimate:* 7 × 7 = 49; *Exact:* 49.786
13. *Estimate:* 100,000 ÷ 50 = 2000; *Exact:* 2690
14. *Estimate:* 40 ÷ 8 = 5; *Exact:* 4.5
15. *Estimate:* 2 • 4 = 8; *Exact:* $7\frac{1}{8}$
16. *Estimate:* 2 ÷ 1 = 2; *Exact:* $2\frac{4}{5}$
17. *Estimate:* 2 + 2 = 4; *Exact:* $3\frac{7}{15}$

18. *Estimate:* 5 − 2 = 3; *Exact:* $2\frac{5}{8}$
19. 9.671   **20.** $1\frac{4}{9}$ **21.** 73,225   **22.** $1\frac{2}{5}$ **23.** 4914   **24.** 93.603
25. 404 R3   **26.** $1\frac{17}{24}$ **27.** 233,728   **28.** 0.03264   **29.** 8
30. 45   **31.** $\frac{4}{31}$ **32.** $\frac{2}{3}$ **33.** 0.51 (rounded)   **34.** 4   **35.** 14
36. $\frac{1}{4}$ **37.** 20.81   **38.** 576   **39.** 14   **40.** $2^3 \cdot 5^2$
41. forty and thirty-five thousandths   **42.** 0.0306   **43.** $\frac{1}{8}$
44. $3\frac{2}{25}$ **45.** 2.6   **46.** 0.636 (rounded)   **47.** >
48. 7.005, 7.5, 7.5005, 7.505   **49.** 0.8, 0.8015, $\frac{21}{25}, \frac{7}{8}$
50. size M (medium)
51. XXS $1\frac{1}{2}$ in.; XS $\frac{3}{4}$ in.; S $\frac{3}{4}$ in.; M $\frac{7}{8}$ in.; L $\frac{3}{4}$ in.
52. 21.125 − 20.25 = 0.875 in.; 7 ÷ 8 = 0.875 in.
or $\frac{875}{1000} = \frac{875 \div 125}{1000 \div 125} = \frac{7}{8}$ in.
53. Answers will vary. Many people prefer using decimals because you do not need to find a common denominator or rewrite answers in lowest terms.   **54.** *Estimate:* $20 − $8 − $1 = $11; *Exact:* $11.17
55. *Estimate:* 50 − 47 = 3 in.; *Exact:* $3\frac{3}{8}$ in.
56. *Estimate:* $10 × 20 = $200; *Exact:* $191.90 (rounded)
57. *Estimate:* (8 × 20) + (10 × 30) = 160 + 300 = 460 students;
Exact: 488 students
58. *Estimate:* 2 + 4 = 6 yards; *Exact:* $6\frac{5}{24}$ yards
59. *Estimate:* $30 + $200 − $40 − $20 = $170; *Exact:* $191.50
60. *Estimate:* $3 ÷ 3 = $1 per pound;   *Exact:* $0.95 per pound
(rounded)   **61.** *Estimate:* $80,000 ÷ 100 = $800; *Exact:* $729
(rounded)   **62.** (a) 9 days (rounded) (b) 15.6 pounds (rounded)
63. Average weight of food eaten by bee is 0.32 ounce
64. 0.112 ounce; $\frac{14}{125}$ ounce; 14 ÷ 125 = 0.112

Chapter 5

Section 5.1 (page 317)

1. $\frac{8}{9}$ **3.** $\frac{2}{1}$ **5.** $\frac{1}{3}$ **7.** $\frac{8}{5}$ **9.** $\frac{3}{8}$ **11.** $\frac{9}{7}$ **13.** $\frac{6}{1}$ **15.** $\frac{5}{6}$ **17.** $\frac{8}{5}$
19. $\frac{1}{12}$ **21.** $\frac{5}{16}$ **23.** $\frac{4}{1}$ **25.** $\frac{1}{2}$ **27.** $\frac{36}{1}$ **29.** Answers will
vary. One possibility is stocking cards of various types in the same ratios as those in the table.   **31.** Comparing the violin to piano, guitar, organ, clarinet, and drums gives ratios of $\frac{1}{11}, \frac{1}{10}, \frac{1}{3}, \frac{1}{2}$, and $\frac{2}{3}$,
respectively.   **33.** Answers will vary. Possibilities include: guitars are less expensive than drums and easier to carry around; more guitar players than drummers are needed in a band.   **35. (a)** $\frac{6}{1}$
(b) $\frac{5}{2}$ **(c)** $\frac{7}{16}$ **37.** $\frac{7}{5}$ **39.** $\frac{6}{1}$ **41.** $\frac{38}{17}$ **43.** $\frac{1}{4}$ **45.** $\frac{34}{35}$ **47.** $\frac{1}{1}$;
as long as the sides all have the same length, any measurement you choose will maintain the ratio.   **48.** Answers will vary. Some possibilities are $\frac{4}{5} = \frac{8}{10} = \frac{12}{15} = \frac{16}{20} = \frac{20}{25} = \frac{24}{30} = \frac{28}{35}$.   **49.** It is not

possible. Amelia would have to be older than her mother to have a ratio of 5 to 3.   **50.** Answers will vary, but a ratio of 3 to 1 means your income is 3 times your friend's income.

Section 5.2 (page 325)

1. $\dfrac{5 \text{ cups}}{3 \text{ people}}$ **3.** $\dfrac{3 \text{ feet}}{7 \text{ seconds}}$ **5.** $\dfrac{1 \text{ person}}{2 \text{ dresses}}$ **7.** $\dfrac{5 \text{ letters}}{1 \text{ minute}}$

9. $\dfrac{\$21}{2 \text{ visits}}$ **11.** $\dfrac{18 \text{ miles}}{1 \text{ gallon}}$ **13.** $12 per hour or $12/hour

15. 5 eggs per chicken or 5 eggs/chicken   **17.** 1.25 pounds/person
19. $103.30/day   **21.** 325.9 21.0 (rounded)
23. 338.6 20.9 (rounded)   **25.** 4 ounces for $0.89   **27.** 15 ounces for $3.15   **29.** 18 ounces for $1.79   **31.** Answers will vary. For example, you might choose Brand B because you like more chicken, so the cost per chicken chunk may actually be the same or less than Brand A.   **33.** 1.75 pounds/week   **35.** $12.26/hour
37. $11.50/share   **39.** 0.11 seconds/meter; $9.\overline{09}$ or 9.1 meters/second (rounded)   **41.** (a) $0.167 or 16.7¢ (rounded); (b) $0.125 or 12.5¢; (c) $0.083 or 8.3¢ (rounded)   **43.** One battery for $1.79; like getting 3 batteries so $1.79 \div 3 \approx \$0.597$ per battery   **45.** Brand P with the 50¢ coupon is the best buy. ($3.39 − $0.50 = $2.89 ÷ 16.5 ounces ≈ $0.175 per ounce)   **47.** $25 \div 12 \approx \$2.08$ per month   **48.** Round to thousandths to see that Verizon is the best buy at $0.028 per minute; Qwest ≈ $0.030; VoiceStream ≈ $0.062; Sprint ≈ $0.060   **49.** 1000 min ≈ 16.7 hours; 104 weekend days per year ÷ 12 ≈ 8.7 weekend days per month; 16.7 hours ÷ 8.7 days ≈ 1.9 hours of talking per day.   **50.** Verizon ≈ $1.17; Qwest ≈ $1.27; VoiceStream ≈ $0.95; Sprint ≈ $1.05

Section 5.3 (page 333)

1. $\dfrac{\$9}{12 \text{ cans}} = \dfrac{\$18}{24 \text{ cans}}$ **3.** $\dfrac{200 \text{ adults}}{450 \text{ children}} = \dfrac{4 \text{ adults}}{9 \text{ children}}$

5. $\dfrac{120}{150} = \dfrac{8}{10}$ **7.** true   **9.** true   **11.** false   **13.** true   **15.** true
17. false   **19.** True   **21.** False   **23.** False   **25.** True
27. False   **29.** True   **31.** True   **33.** False   **35.** False

37. $\dfrac{16 \text{ hits}}{50 \text{ at bats}} = \dfrac{128 \text{ hits}}{400 \text{ at bats}}$ $\begin{array}{l} 50 \cdot 128 = 6400 \\ 16 \cdot 400 = 6400 \end{array}$ Cross products
are *equal* so the proportion is *true*; they hit equally well.

Section 5.4 (page 339)

1. 4   **3.** 2   **5.** 88   **7.** 91   **9.** 5   **11.** 10   **13.** 24.44 (rounded)
15. 50.4   **17.** 17.64 (rounded)   **19.** 1   **21.** $3\dfrac{1}{2}$ **23.** 0.2 or $\dfrac{1}{5}$
25. 0.005 or $\dfrac{1}{200}$

27. Find cross products: $20 \neq 30$, so the proportion is false.

$\dfrac{6\frac{2}{3}}{4} = \dfrac{5}{3}$ or $\dfrac{10}{6} = \dfrac{5}{3}$ or $\dfrac{10}{4} = \dfrac{7.5}{3}$ or $\dfrac{10}{4} = \dfrac{5}{2}$

28. Find cross products: $192 \neq 180$, so the proportion is false.

$\dfrac{6.4}{8} = \dfrac{24}{30}$ or $\dfrac{6}{7.5} = \dfrac{24}{30}$ or $\dfrac{6}{8} = \dfrac{22.5}{30}$ or $\dfrac{6}{8} = \dfrac{24}{32}$

Section 5.5 (page 345)

1. 22.5 hours   **3.** $7.20   **5.** 42 pounds   **7.** $273.45   **9.** 10 ounces (rounded)   **11.** 5 quarts   **13.** 14 ft, 10 ft   **15.** 14 ft, 8 ft
17. 26 points (rounded)   **19.** 2065 students (reasonable); about 4214 students with incorrect setup (only 2950 students in the group).
21. about 190 people (reasonable); about 298 people with incorrect setup (only 238 people attended).   **23.** 92,250,000 households (reasonable); about 113,888,889 households with incorrect setup (only 102,500,000 U.S. households).   **25.** 625 stocks

27. 4.06 meters (rounded)   **29.** 311 calories (rounded)
31. 10.53 m (rounded)   **33.** You cannot solve this problem using a proportion because the ratio of age to weight is not constant. As Jim's age increases, his weight may decrease, stay the same, or increase.   **35.** 3800 students use cream

37. 120 calories and 12 grams of fiber   **39.** $1\dfrac{3}{4}$ cups water, 3 Tbsp

margarine, $\dfrac{3}{4}$ cup milk, 2 cups flakes   **40.** $5\dfrac{1}{4}$ cups water, 9 Tbsp

margarine, $2\dfrac{1}{4}$ cups milk, 6 cups flakes   **41.** $\dfrac{7}{8}$ cup water, $1\dfrac{1}{2}$ Tbsp

margarine, $\dfrac{3}{8}$ cup milk, 1 cup flakes   **42.** $2\dfrac{5}{8}$ cups water, $4\dfrac{1}{2}$ Tbsp

margarine, $1\dfrac{1}{8}$ cups milk, 3 cups flakes

Chapter 5 Review Exercises (page 355)

1. $\dfrac{3}{4}$ **2.** $\dfrac{4}{1}$ **3.** $\dfrac{1}{2}$ **4.** $\dfrac{2}{1}$ **5.** $\dfrac{2}{3}$ **6.** $\dfrac{5}{2}$ **7.** $\dfrac{1}{6}$ **8.** $\dfrac{3}{1}$ **9.** $\dfrac{3}{8}$

10. $\dfrac{4}{3}$ **11.** $\dfrac{1}{9}$ **12.** $\dfrac{10}{7}$ **13.** $\dfrac{7}{5}$ **14.** $\dfrac{5}{6}$ **15.** $\dfrac{\$11}{1 \text{ dozen}}$

16. $\dfrac{12 \text{ children}}{5 \text{ families}}$ **17.** 0.2 page/minute or $\dfrac{1}{5}$ page/minute

5 minutes/page   **18.** $8/hour   0.125 hour/dollar or $\dfrac{1}{8}$ hour/dollar

19. 13 ounces for $2.29   **20.** 25 pounds for $10.40 − $1 coupon
21. true   **22.** false   **23.** false   **24.** true   **25.** true   **26.** true
27. 1575   **28.** 20   **29.** 400   **30.** 12.5   **31.** 14.67 (rounded)
32. 8.17 (rounded)   **33.** 50.4   **34.** 0.57 (rounded)   **35.** 2.47 (rounded)   **36.** 27 cats   **37.** 46 hits   **38.** $15.63 (rounded)
39. 3299 students (rounded)   **40.** 68 feet   **41.** 14.7 milligrams

42. 511 calories (rounded)   **43.** $27\dfrac{1}{2}$ hours or 27.5 hours

44. 105   **45.** 0   **46.** 128   **47.** 23.08 (rounded)   **48.** 6.5
49. 117.36 (rounded)   **50.** False   **51.** False   **52.** True   **53.** $\dfrac{8}{5}$
54. $\dfrac{33}{80}$ **55.** $\dfrac{15}{4}$ **56.** $\dfrac{4}{1}$ **57.** $\dfrac{4}{5}$ **58.** $\dfrac{37}{7}$ **59.** $\dfrac{3}{8}$ **60.** $\dfrac{1}{12}$

61. $\dfrac{45}{13}$ **62.** 24,900 fans (rounded)   **63.** $\dfrac{8}{3}$

64. 75 ft for $1.99 − $0.50 coupon   **65.** 15 ft   **66.** 7.5 hours or $7\dfrac{1}{2}$ hours   **67.** $\dfrac{1}{2}$ teaspoon or 0.5 teaspoon   **68.** 21 points (rounded)

69. Set up the proportion to compare teaspoons to pounds on both

sides. $\dfrac{1.5 \text{ tsp}}{24 \text{ pounds}} = \dfrac{x \text{ tsp}}{8 \text{ pounds}}$ Show that cross products are equal.

$24 \cdot x = 1.5 \cdot 8$ Divide each side by 24.

$\dfrac{\overset{1}{\cancel{24}} \cdot x}{\underset{1}{\cancel{24}}} = \dfrac{12}{24}$ so $x = \dfrac{1}{2}$ tsp or 0.5 tsp

70. (a) 1400 milligrams (b) 100 milligrams

Chapter 5 Test (page 359)

1. $\dfrac{4}{5}$ **2.** $\dfrac{20 \text{ miles}}{1 \text{ gallon}}$ **3.** $\dfrac{\$1}{5 \text{ minutes}}$ **4.** $\dfrac{15}{4}$ **5.** $\dfrac{1}{80}$ **6.** $\dfrac{9}{2}$
7. 18 ounces for $1.89 − $0.25 coupon   **8.** You earned less this year. An example is: $\begin{array}{l} \text{Last year} \to \$15,000 \\ \text{This year} \to \$10,000 \end{array} = \dfrac{3}{2}$ **9.** false   **10.** true

11. 25   **12.** 2.67 (rounded)   **13.** 325   **14.** $10\dfrac{1}{2}$ **15.** 576 words

16. 3.6 ounces **17.** 87 students (rounded) **18.** No, 4875 cannot be correct because there are only 650 students in the whole school. **19.** 23.8 grams (rounded) **20.** 60 ft

Cumulative Review Exercises: Chapters 1–5 (page 361)
1. 5; 3; 6; 4 **2.** 9; 0; 5; 3 **3.** 9900 **4.** 617.1 **5.** $100
6. $3.06 **7.** *Estimate:* 30 + 5000 + 400 = 5430;
Exact: 5585 **8.** *Estimate:* 60 − 6 = 54; *Exact:* 57.408
9. *Estimate:* 5000 × 800 = 4,000,000; *Exact:* 3,791,664
10. *Estimate:* 1 × 18 = 18; *Exact:* 17.4796
11. *Estimate:* 50,000 ÷ 50 = 1000; *Exact:* 907
12. *Estimate:* 2000 ÷ 5 = 400; *Exact:* 364
13. *Estimate:* 2 • 4 = 8; *Exact:* $6\frac{3}{5}$ **14.** *Estimate:* 5 ÷ 1 = 5;
Exact: 6 **15.** *Estimate:* 3 − 2 = 1; *Exact:* $\frac{29}{30}$
16. *Estimate:* 3 + 11 = 14; *Exact:* $13\frac{2}{5}$
17. 374,416 **18.** 29.34 **19.** 610 R27 **20.** 0.0076 **21.** 2312
22. 68.381 **23.** 55.6 **24.** 35,852,728 **25.** 39 **26.** 18
27. 64 **28.** 0.95 **29.** one hundred five ten-thousandths
30. 60.071 **31.** $1\frac{9}{16} \approx 1.563$; $1\frac{3}{8} = 1.375$; $1\frac{1}{8} = 1.125$
32. $1\frac{1}{8}$ in., 1.25 in., $1\frac{3}{8}$ in., 1.5 in., $1\frac{9}{16}$ in. **33.** $\frac{7}{16}$ in.
34. $\frac{1}{16}$ in. or 0.063 in. (rounded) **35.** $\frac{11}{5}$ **36.** $\frac{1}{5}$ **37.** $\frac{4}{1}$
38. $\frac{1}{12}$ **39.** $\frac{3}{1}$ **40.** 36 servings for $3.24 − $0.50 coupon
41. 21 **42.** 17.14 (rounded) **43.** $11\frac{1}{4}$ **44.** 0.98 (rounded)
45. 250 pounds **46.** 26.7 centimeters (rounded) **47.** $7\frac{3}{20}$ miles
48. 18.0 miles per gallon (rounded) **49.** 140 residents
50. $1\frac{1}{4}$ teaspoons **51.** $\frac{5}{2}$ **52.** Answers will vary. Some

possibilities are: people with bachelor's degrees earn $2\frac{1}{2}$ times as

much as people without high school diplomas; for every $5 earned by the first group, only $2 is earned by the second group.
53. $1917 (rounded) **54.** $4250 **55.** $7 per hour difference
56. $1,081,000 over 47 years for a person with a high school diploma; $1,720,000 over 43 years for the person with a bachelor's degree, or $639,000 more.

Chapter 6

Section 6.1 (page 371)
1. 0.15 **3.** 0.60 or 0.6 **5.** 0.25 **7.** 1.40 or 1.4 **9.** 0.055
11. 1.00 or 1 **13.** 0.005 **15.** 0.0035 **17.** 80% **19.** 58%
21. 1% **23.** 12.5% **25.** 37.5% **27.** 200% **29.** 370%
31. 3.12% **33.** 416.2% **35.** 0.28% **37.** Answers will vary.
Some possibilities are: No common denominators are needed with percents. The denominator is always 100 with percent, which makes comparisons easier to understand. **39.** 0.45 **41.** 0.18
43. 3.5% **45.** 200% **47.** 0.5% **49.** 1.536 **51.** 20 children
53. 420 employees **55.** 270 chairs **57.** $142.50
59. 820 commuters **61.** 26 plants **63. (a)** Since 100% means 100 parts out of 100 parts, 100% is all of the number. **(b)** Answers will vary. **65. (a)** Since 200% is two times a number, find 200% of the number by multiplying the number by 2 (double it).

(b) Answers will vary. **67. (a)** Since 10% means 10 parts out of 100 parts or $\frac{1}{10}$, the shortcut for finding 10% of a number is to move the decimal point in the number one place to the left. **(b)** Answers will vary. **69.** 76%; 0.76 **71. (a)** health care **(b)** 51%; 0.51
73. 16.8%; 0.168 **75. (a)** Germany; Bulgaria **(b)** 16.5%; 0.165
77. 95%; 5% **79.** 30%; 70% **81.** 55%; 45%

Section 6.2 (page 383)
1. $\frac{1}{4}$ **3.** $\frac{3}{4}$ **5.** $\frac{17}{20}$ **7.** $\frac{5}{8}$ **9.** $\frac{1}{16}$ **11.** $\frac{1}{6}$ **13.** $\frac{1}{15}$ **15.** $\frac{1}{200}$
17. $1\frac{1}{5}$ **19.** $3\frac{3}{4}$ **21.** 50% **23.** 80% **25.** 25% **27.** 37%
29. 62.5% **31.** 87.5% **33.** 36% **35.** 46% **37.** 15%
39. 83.3% (rounded) **41.** 55.6% (rounded)
43. 14.3% (rounded) **45.** $\frac{1}{2}$; 50% **47.** $\frac{7}{8}$; 0.875
49. $\frac{4}{5}$; 80% **51.** 0.167 (rounded) 16.7% (rounded)
53. $\frac{1}{4}$; 25% **55.** $\frac{1}{8}$; 0.125 **57.** 0.667 (rounded);
66.7% (rounded) **59.** 0.4; 40% **61.** 0.08; 8%
63. 0.005; 0.5% **65.** $2\frac{1}{2}$; 250% **67.** 3.25; 325%
69. There are many possible answers. Examples 2 and 3 show the steps that students should include in their answers.
71. $\frac{13}{100}$; 0.13; 13% **73.** $\frac{9}{50}$; 0.18; 18% **75.** $\frac{3}{5}$; 0.6; 60%
77. $\frac{1}{5}$; 0.20; 20% **79.** $\frac{1}{10}$; 0.1; 10% **81.** $\frac{1}{4}$; 0.25; 25%
83. $\frac{2}{5}$; 0.4; 40% **85.** 100; 100
86. (a) 765 workers **(b)** 96 letters **(c)** 813 videos **87.** 50; 100;
half or $\frac{1}{2}$ **88.** 10; 100; 1; left **89.** 1; 100; 2; left
90. (a) 525 homes **(b)** 37 printers **(c)** $105
91. Find 10% of 160, then add $\frac{1}{2}$ of 10%.

   10%   5%   15%
   ↓   ↓   ↓
   16 + 8 = 24
92. Find 100% of 160, then add 50% of 160.
   100%   50%   150%
   ↓   ↓   ↓
   160 + 80 = 240
93. From 100% of $450, subtract 10% of $450.
   100%   10%   90%
   ↓   ↓   ↓
   $450 − $45 = $405
94. To 100% of $800, add 100% of $800, and then add 10% of $800.
   100%   100%   10%   210%
   ↓   ↓   ↓   ↓
   $800 + $800 + $80 = $1680

Section 6.3 (page 393)
1. 50 **3.** 150 **5.** 70 **7.** 25% **9.** 150%
11. 33.3% (rounded) **13.** 26 **15.** 21.6 **17.** 26.5% (rounded)
19. 115 **21.** 0.4% **23.** 2.5% **25.** 0.3% **27.** 10; unknown;
60 **29.** 75; $800; $600 **31.** 25; $970; unknown **33.** 20;
unknown; 12 **35.** 50; 68; 34 **37.** unknown; $296; $177.60
39. 3.25; unknown; 54.34 **41.** 0.68; $487; unknown
43. percent—the ratio of the part to the whole. It appears with the

word *percent* or "%" after it. whole—the entire quantity. Often appears after the word *of.* part—the part being compared with the whole. **45.** percent is unknown; 810; 640 **47.** percent is unknown; 142; 86 **49.** 23; 610; part is unknown **51.** 25; whole is unknown; 240 **53.** 55; 480; part is unknown **55.** 49.5; 822; part is unknown **57.** 16.8; whole is unknown; 504

Section 6.4 (page 405)

1. 16 guests **3.** 1836 military personnel **5.** 4.8 ft **7.** 315 files **9.** 819 trucks **11.** $3.28 **13.** 1530 tables **15.** 182 homes **17.** $21.60 **19.** 320 e-mails **21.** 160 hay bales **23.** 550 students **25.** 1360 graduates **27.** 2800 **29.** 50% **31.** 52% **33.** 8% **35.** 1.5% **37.** 18.6% (rounded) **39.** 9.2% **41.** 150% of $30 cannot be less than $30 because 150% is greater than 1 (100%). The answer must be greater than $30. 25% of $16 cannot be greater than $16 because 25% is less than 1 (100%). The answer must be less than $16. **43.** $52.80 **45.** 44 females **47.** 220 students **49.** $204.97 **51.** 2% **53.** **(a)** 39% female **(b)** 2.3 million male workers (rounded) **55.** 110% **57.** 281.1 million **59.** March **61.** 52.5 million cans **63.** Hardee's **65.** $18.06 billion **67.** $645 **69.** 2156 products **71.** whole; 100 **72.** whole **73.** 108 calories **74.** 300 grams **75.** 67 grams **76.** Yes, since they would eat 28% × 4 servings = 112% of the daily value. **77.** 17 packages (rounded). It may be possible but would result in a diet that is high in total fat, saturated fat, sodium, and total carbohydrates.

Section 6.5 (page 415)

1. 135 hamburgers **3.** 675 garments **5.** 83.2 quarts **7.** 700 tablets **9.** 1029.2 meters **11.** $4.16 **13.** 160 patients **15.** 325 salads **17.** 680 circuits **19.** 1080 people **21.** 300 gallons **23.** 50% **25.** 76% **27.** 125% **29.** 1.5% **31.** 250% **33.** You must first change the fraction in the percent to a decimal, then divide the percent by 100 to change it to a decimal.

$$2\frac{1}{2}\% = 2.5\% = 0.025 \leftarrow 2.5\% \text{ as a decimal}$$

Change $2\frac{1}{2}$ to 2.5.

35. 3.78 million or 3,780,000 office workers **37.** 82.3 million homes **39.** 924 employers **41.** **(a)** 30%; **(b)** 70% **43.** 4.9% **45.** 2106 people **47.** 2916 people **49.** $1817.2 million **51.** 478,175 Mustangs **53.** $510,390 **55.** $439.61

Section 6.6 (page 425)

1. $0.40; $8.40 **3.** 3%; $437.75 **5.** 5%; $298.20 **7.** $672.60 (rounded); $12,901.60 (rounded) **9.** $40 **11.** 20% **13.** $185.51 (rounded) **15.** $6615 **17.** $20 (rounded); $179.99 (rounded) **19.** 30%; $126 **21.** $4.38; $13.12 **23.** $8.76; $49.64 **25.** On the basis of commission alone you would choose Company A. Other considerations might be: reputation of the company; expense allowances; other fringe benefits; travel; promotion and training, to name a few. **27.** $58.12 **29.** $1170 **31.** 5% **33.** 10.7% (rounded) **35.** 22.5% (rounded) **37.** $134 **39.** $284.40 **41.** 4% **43.** $106.20; $483.80 **45.** 5.6% (rounded) **47.** $18.43 (rounded) **49.** $4276.97 (rounded) **51.** $9007.47 (rounded) **53.** percent; whole **54.** rate (percent) of tax; cost of item **55.** $146.25 (rounded) **56.** Yes, the same; (12.4% · $123) + (6.5% · $123) = $23.25 (rounded) and (12.4% + 6.5%) · ($123) = $23.25 **57.** $1356.92

58. No. In Exercise 57 the excise tax of $12.20 cannot be added to the sales tax rate of $7\frac{3}{4}\%$ because the excise tax is an amount, not a percent.

Section 6.7 (page 431)

1. $6 **3.** $105 **5.** $28.80 **7.** $488.75 **9.** $2227.50 **11.** $10 **13.** $49.20 **15.** $42.30 **17.** $16.84 (rounded) **19.** $647.06 (rounded) **21.** $210 **23.** $773.30 **25.** $2043 **27.** $3412.50 **29.** $17,797.81 (rounded) **31.** The answer should include: amount of principal—This is the amount of money borrowed or loaned. interest rate—This is the percent used to calculate the interest. time of loan—The length of time that money is loaned or borrowed is an important factor in determining interest. **33.** $41.25 **35.** $22,500 **37.** $1240 **39.** $64.68 **41.** $159.50 **43.** $12,254.69 (rounded) **45.** $2165.63 (rounded)

Summary Exercises on Percent (page 435)

1. 0.0625 **2.** 3.80 or 3.8 **3.** 37.5% **4.** 0.6% **5.** $\frac{7}{8}$

6. $1\frac{3}{5}$ **7.** 62.5% **8.** 0.8% **9.** $46.80 **10.** 500 rolls

11. 55% **12.** 28 screening exams **13.** 363 acres

14. 2.7% (rounded) **15.** $9.68; $185.68 **16.** 7%; $990.82

17. $301.20 **18.** 6.5% or $6\frac{1}{2}\%$

19. $8.99 (rounded) $40.96 (rounded) **20.** 35% $444.60 **21.** $60 $1060 **22.** $535.50 $2915.50 **23.** $44.10 $1514.10 **24.** $664.95 $7484.95 **25.** 22.0% increase (rounded) **26.** 10.5% decrease (rounded)

Section 6.8 (page 441)

1. $540.80 **3.** $1966.91 (rounded) **5.** $4587.79 (rounded) **7.** $1102.50 **9.** $1873.52 (rounded) **11.** $2027.46 (rounded) **13.** $17,360.41 (rounded) **15.** $1216.70; $216.70 **17.** $14,326.40; $6326.40 **19.** $10,976.01 (rounded); $2547.84 (rounded) **21.** Interest paid on past interest as well as on the principal. Many people describe compound interest as "interest on interest." **23.** $13,467.61 **25.** **(a)** $64,201 **(b)** $26,201 **27.** **(a)** $87,786 **(b)** $17,786 **29.** principal; rate; time *p; r; t* **30.** principal; interest **31.** principal; interest **32.** principal; interest **33.** **(a)** $254.48 **(b)** Yes; it has more than doubled (about five times). **(c)** yes **34.** **(a)** the compound interest account; $1091.20 more **(b)** Greater amounts of interest are earned with compound interest than simple interest because interest is earned on both principal and past interest.

Chapter 6 Review Exercises (page 451)

1. 0.35 **2.** 1.5 **3.** 0.9944 **4.** 0.00085 **5.** 315% **6.** 2% **7.** 87.5% **8.** 0.2% **9.** $\frac{3}{20}$ **10.** $\frac{3}{8}$ **11.** $1\frac{3}{4}$ **12.** $\frac{1}{400}$ **13.** 75% **14.** 62.5% or $62\frac{1}{2}\%$ **15.** 325% **16.** 0.5% **17.** 0.125 **18.** 12.5% **19.** $\frac{1}{4}$ **20.** 25% **21.** $1\frac{4}{5}$ **22.** 1.8 **23.** 250 **24.** 24 **25.** 35; 820; 287 **26.** unknown; 90; 73 **27.** 14; 160; unknown **28.** 16; unknown; 418 **29.** unknown; 8; 3 **30.** 88; 1280; unknown **31.** 171 programs **32.** 870 reference books **33.** 31.2 acres **34.** 2.8 kilograms **35.** 750 crates **36.** 2320 test tubes **37.** 484 miles **38.** 17,000 cases **39.** 55% **40.** 5.2% (rounded) **41.** 9.5% (rounded) **42.** 30.8% (rounded)

43. 140,000 math teachers   **44.** 10.4% (rounded)   **45.** $145.28
46. 186 trucks   **47.** 0.4%   **48.** 175%   **49.** 120 miles   **50.** $575

51. $31.50; $661.50   **52.** $7\frac{1}{2}$%; $838.50   **53.** $276   **54.** 5%

55. $33.75; $78.75   **56.** 25% $189   **57.** $8   **58.** $67.50
59. $7   **60.** $152.10   **61.** $832.50   **62.** $1598.85
63. $3257.80; $1257.80   **64.** $2187.71 (rounded);
$317.71 (rounded)   **65.** $4534.92; $934.92   **66.** $16,337.50;
$3837.50   **67.** 12   **68.** 1640   **69.** 23.28 meters   **70.** 150%
71. $0.51   **72.** 1980 employees   **73.** 40%   **74.** 498.8 liters
75. 0.55   **76.** 3   **77.** 500%   **78.** 471%   **79.** 0.086

80. 62.1%   **81.** 0.00375   **82.** 0.06%   **83.** 75%   **84.** $\frac{21}{50}$ **85.** $\frac{7}{8}$

86. 37.5% or $37\frac{1}{2}$%   **87.** $\frac{13}{40}$ **88.** 60%   **89.** $\frac{1}{400}$ **90.** 375%

91. $1326.94 (rounded)   **92.** $17,405   **93.** 32.4% (rounded)
94. **(a)** $15,780.96 (rounded) **(b)** $3280.96 (rounded)   **95.** $6225
96. 8.1% (rounded)   **97.** $595.07 (rounded)   **98.** 185,185 annual
sales (rounded)   **99.** $9758   **100.** 13.1% (rounded)

Chapter 6 Test (page 457)

1. 0.65   **2.** 80%   **3.** 175%   **4.** 87.5% or $87\frac{1}{2}$%

5. 3.00 or 3   **6.** 0.0005   **7.** $\frac{1}{8}$ **8.** $\frac{1}{400}$ **9.** 60%

10. 62.5% or $62\frac{1}{2}$%   **11.** 250%   **12.** 800 sacks   **13.** 20%

14. $18,500   **15.** $2854.20   **16.** $628   **17.** 30%

18. $\dfrac{\text{amount of increase}}{\text{last year's salary}} = \dfrac{p}{100}$

A possible answer is: Part is the increase in salary.
$\qquad\qquad$ this year $-$ last year $=$ increase
Whole is last year's salary. Percent of increase is unknown.
19. $\qquad\qquad I = p \cdot r \cdot t$

$$I = 1000 \cdot 0.05 \cdot \frac{9}{12} = \$37.50$$

$$I = 1000 \cdot 0.05 \cdot 2.5 = \$125$$

The interest formula is $I = p \cdot r \cdot t$. If time is in months, it is expressed as a fraction with 12 as the denominator. If time is expressed in years, it is placed over 1 or shown as a decimal number.
20. $11.52; $84.48   **21.** $91; $189   **22.** $378   **23.** $240
24. $5657.75   **25.** **(a)** $10,668 (rounded) **(b)** $1668

Cumulative Review Exercises: Chapters 1–6 (page 459)

1. *Estimate:* 6000 + 90 + 700 = 6790; *Exact:* 6441
2. *Estimate:* 1 + 50 + 8 = 59; *Exact:* 58.574
3. *Estimate:* 80,000 − 50,000 = 30,000; *Exact:* 28,988
4. *Estimate:* 8 − 4 = 4; *Exact:* 4.2971
5. *Estimate:* 7000 × 700 = 4,900,000; *Exact:* 4,628,904
6. *Estimate:* 70 × 9 = 630; *Exact:* 568.11
7. *Estimate:* 40,000 ÷ 40 = 1000; *Exact:* 902
8. *Estimate:* 2000 ÷ 8 = 250; *Exact:* 320
9. *Estimate:* 7 ÷ 1 = 7; *Exact:* 8.45   **10.** 18   **11.** 19   **12.** 39
13. 4680   **14.** 7,600,000   **15.** $513   **16.** $362.74

17. $1\frac{3}{8}$ **18.** $1\frac{1}{12}$ **19.** $13\frac{3}{8}$ **20.** $\frac{1}{8}$ **21.** $3\frac{7}{8}$

22. $8\frac{8}{15}$ **23.** $\frac{7}{12}$ **24.** $26\frac{5}{32}$ **25.** $28\frac{4}{5}$ **26.** $\frac{7}{8}$ **27.** 16

28. $\frac{11}{30}$ **29.** <   **30.** <   **31.** >   **32.** $\frac{1}{4}$ **33.** 1

34. $\frac{1}{4}$ **35.** 0.6   **36.** 0.875   **37.** 0.583 (rounded)   **38.** 0.55

39. $\frac{3}{1}$ **40.** $\frac{4}{3}$ **41.** $\frac{1}{8}$ **42.** True   **43.** True   **44.** 6

45. 3   **46.** 16   **47.** 30   **48.** 0.78   **49.** 0.03   **50.** 2.00 or 2

51. 0.005   **52.** 87%   **53.** 380%   **54.** 2.3%   **55.** $\frac{2}{25}$

56. $\frac{5}{8}$ **57.** $1\frac{3}{4}$ **58.** 87.5% or $87\frac{1}{2}$%   **59.** 30%   **60.** 420%

61. 490 watches   **62.** $39.60   **63.** 180 tires   **64.** 1700 miles
65. 50%   **66.** 40%   **67.** $5.45 (rounded); $114.40

68. 6.5% or $6\frac{1}{2}$%; $489.90   **69.** $1003.04

70. 2.5% or $2\frac{1}{2}$%   **71.** $205.20; $250.80   **72.** 15%; $922.25

73. $1115.50   **74.** $19,863.88 (rounded)   **75.** 33 cars
76. 59.5 ounces   **77.** 12.5%   **78.** 2%   **79.** 5.5% (rounded)
80. 8.1% (rounded)   **81.** $107,520   **82.** **(a)** $19,604.40 (rounded)
(b) $4604.40 (rounded)

Chapter 7

Section 7.1 (page 473)

1. 3   **3.** 8   **5.** 5280   **7.** 2000   **9.** 60   **11.** 2   **13.** 2
15. 76,000 to 80,000 lb   **17.** 27   **19.** 112   **21.** 10

23. $1\frac{1}{2}$ or 1.5   **25.** $\frac{1}{4}$ or 0.25   **27.** $1\frac{1}{2}$ or 1.5   **29.** $2\frac{1}{2}$ or 2.5

31. $\frac{1}{2}$ day or 0.5 day   **33.** 5000   **35.** 17   **37.** 44 ounces

39. 216   **41.** 28   **43.** 518,400   **45.** 48,000
47. **(a)** pound/ounces   **(b)** quarts/pints or pints/cups
(c) minutes/hours or seconds/minutes   **(d)** feet/inches
(e) pounds/tons   **(f)** days/weeks   **49.** 174,240   **51.** 800

53. 0.75 or $\frac{3}{4}$ **55.** $1.83 (rounded)   **57.** $140

59. **(a)** $12\frac{1}{2}$ qt   **(b)** 4 jugs, because you can't buy part of a jug

61. **(a)** 337.5 T   **(b)** 2.8 T (rounded)   **63.** 1 gal = 128 fl oz;
$\dfrac{16 \text{ fl oz}}{\$1.25} = \dfrac{128 \text{ fl oz}}{x}$; $x = \$10$

64. $\dfrac{11.2 \text{ fl oz}}{\$1.49} = \dfrac{128 \text{ fl oz}}{x}$; $x \approx \$17.03$

65. One alternate method: $3.85 ÷ 4 fl oz = $0.9625 per fl oz;
$\dfrac{\$0.9625}{1 \text{ fl oz}} \cdot \dfrac{128 \text{ fl oz}}{1 \text{ gal}} = \$123.20$ per gal
66. Answers will vary. Because we use the products in very different ways, it may not make sense to compare the price per gallon.

Section 7.2 (page 483)

1. 1000; 1000   **3.** $\frac{1}{1000}$ or 0.001; $\frac{1}{1000}$ or 0.001

5. $\frac{1}{100}$ or 0.01; $\frac{1}{100}$ or 0.01   **7.** Answers will vary; about 8 cm.

9. Answers will vary; about 20 mm.   **11.** cm   **13.** m   **15.** km
17. mm   **19.** cm   **21.** m   **23.** Some possible answers are:
35 mm film for cameras, track and field events, metric auto parts,
and lead refills for mechanical pencils.   **25.** 700 cm   **27.** 0.040 m
or 0.04 m   **29.** 9400 m   **31.** 5.09 m   **33.** 40 cm   **35.** 910 mm
37. less; 18 cm or 0.18 m

39. _____ 35 mm = 3.5 cm

_____ 70 mm = 7 cm

41. 0.018 km   **43.** 0.0000056 km

Section 7.3 (page 491)

1. mL   **3.** L   **5.** kg   **7.** g   **9.** mL   **11.** mg   **13.** L
15. kg   **17.** unreasonable; too much   **19.** unreasonable; too much
21. reasonable   **23.** reasonable   **25.** Some capacity examples
are 2 L bottles of soda and shampoo bottles marked in mL; weight
examples are grams of fat listed on cereal boxes and vitamin doses
in milligrams.   **27.** Unit for your answer (g) is in numerator; unit
being changed (kg) is in denominator so it will divide out. The unit
fraction is $\dfrac{1000 \text{ g}}{1 \text{ kg}}$.   **29.** 15,000 mL   **31.** 3 L   **33.** 0.925 L

35. 0.008 L   **37.** 4150 mL   **39.** 8 kg   **41.** 5200 g   **43.** 850 mg
45. 30 g   **47.** 0.598 g   **49.** 0.06 L   **51.** 0.003 kg   **53.** 990 mL
55. mm   **57.** mL   **59.** cm   **61.** mg   **63.** 0.3 L   **65.** 1340 g
67. 0.07 L   **69.** 3 kg to 4 kg   **71.** greater; 5 mg or 0.005 g
73. 200 nickels   **75. (a)** 1,000,000

(b) $\dfrac{3.5 \text{ Mm}}{1} \cdot \dfrac{1,000,000 \text{ m}}{1 \text{ Mm}} = 3,500,000 \text{ m}$

76. (a) 1,000,000,000

(b) $\dfrac{2500 \text{ m}}{1} \cdot \dfrac{1 \text{ Gm}}{1,000,000,000 \text{ m}} = 0.0000025 \text{ Gm}$

77. (a) 1,000,000,000,000   **(b)** 1000; 1,000,000
78. 1,000,000; 1,000,000,000; $2^{20} = 1,048,576$; $2^{30} = 1,073,741,824$

Section 7.4 (page 497)

1. $1.33 (rounded)   **3.** 89.5 kg   **5.** 71 beats (rounded)
7. 180 cm; $0.02/cm (rounded)   **9.** 5.03 m   **11.** 4 bottles
13. $358.50   **15.** 215 g; 4.3 g; 4300 mg   **16.** 330 g; 0.33 g;
330 mg   **17.** 1.55 kg; 1500 g; 3 g   **18.** 55 g; 50 g; 0.5 g

Section 7.5 (page 503)

1. 21.8 yd   **3.** 262.4 ft   **5.** 4.8 m   **7.** 5.3 oz   **9.** 111.6 kg

11. 30.3 qt   **13. (a)** about 0.2 oz   **(b)** probably not

15. about 31.8 L   **17.** about 1.3 cm   **19.** 28 °F   **21.** 40 °C
23. 150 °C   **25.** More. There are 180 degrees between freezing
and boiling on the Fahrenheit scale, but only 100 degrees on the
Celsius scale, so each Celsius degree is a greater change in temperature.   **27.** 16 °C (rounded)   **29.** 40 °C   **31.** 46 °F (rounded)
33. 95 °F   **35.** 58 °C (rounded)   **37. (a)** pleasant weather, above
freezing but not hot   **(b)** 75 °F to 39 °F (rounded)   **(c)** Answers will
vary. In Minnesota, it's 0 °C to −40 °C; in California, 24 °C to
0 °C.   **39.** longer; by about 2.3 yd   **40.** Multiply 2.3 yd by 2 to
get 4.6 yd longer for the 50 m race; multiply by 4 to get 9.2 yd
longer for the 100 m race.   **41.** 50 m race is 4.5 yd longer; 100 m
race is 9 yd longer; differences are due to rounding.   **42.** shorter;
by 64 yd

Chapter 7 Review Exercises (page 511)

1. 16   **2.** 3   **3.** 2000   **4.** 4   **5.** 60   **6.** 8   **7.** 60   **8.** 5280
9. 12   **10.** 48   **11.** 3   **12.** 4   **13.** $\dfrac{3}{4}$ or 0.75   **14.** $2\dfrac{1}{2}$ or 2.5

15. 28   **16.** 78   **17.** 112   **18.** 345,600

19. (a) $4153\dfrac{1}{3}$ or $4153.\overline{3}$ yd   **(b)** 2.4 mi (rounded)

20. $2465.20   **21.** mm   **22.** cm   **23.** km   **24.** m   **25.** cm
26. mm   **27.** 500 cm   **28.** 8500 m   **29.** 8.5 cm   **30.** 3.7 m
31. 0.07 km   **32.** 930 mm   **33.** mL   **34.** L   **35.** g   **36.** kg

37. L   **38.** mL   **39.** mg   **40.** g   **41.** 5 L   **42.** 8000 mL
43. 4580 mg   **44.** 700 g   **45.** 0.006 g   **46.** 0.035 L
47. 31.5 L   **48.** 357 g (rounded)   **49.** 87.25 kg
50. $3.01 (rounded)   **51.** 6.5 yd (rounded)   **52.** 11.7 in. (rounded)
53. 67.0 mi (rounded)   **54.** 1288 km   **55.** 21.9 L (rounded)
56. 44.0 qt (rounded)   **57.** 0 °C   **58.** 100 °C   **59.** 37 °C
60. 20 °C   **61.** 25 °C   **62.** 33 °C (rounded)   **63.** 43 °F (rounded)
64. 120 °F (rounded)   **65.** L   **66.** kg   **67.** cm   **68.** mL

69. mm   **70.** km   **71.** g   **72.** m   **73.** mL   **74.** g   **75.** mg

76. L   **77.** 105 mm   **78.** $\dfrac{3}{4}$ hr or 0.75 hr   **79.** $7\dfrac{1}{2}$ ft or 7.5 ft

80. 130 cm   **81.** 77 °F   **82.** 14 qt   **83.** 0.7 g   **84.** 810 mL
85. 80 oz   **86.** 60,000 g   **87.** 1800 mL   **88.** 30 °C   **89.** 36 cm
90. 0.055 L   **91.** 1.26 m   **92.** 18 T   **93.** 113 g; 177 °C (both
rounded)   **94.** 178.0 lb; 6.0 ft (both rounded)   **95.** 136 g (rounded)
96. 5.3 oz   **97.** 6.3 oz   **98.** 156 g (rounded)   **99.** Audiovox,
Motorola, Samsung, NeoPoint   **100. (a)** Motorola, Audiovox,
Samsung, NeoPoint   **(b)** 1 hr or 60 min   **101.** Answers will vary.
For example, you might choose Audiovox because it's the lightest
weight and has the longest standby battery life. If you talk a lot on a
cell phone, you might choose Motorola with the longest battery life
for talking.

Chapter 7 Test (page 515)

1. 36 qt   **2.** 15 yd   **3.** 2.25 or $2\dfrac{1}{4}$ hr   **4.** 0.75 or $\dfrac{3}{4}$ ft

5. 56 oz   **6.** 7200 min   **7.** kg   **8.** km   **9.** mL   **10.** g   **11.** cm
12. mm   **13.** L   **14.** cm   **15.** 2.5 m   **16.** 4600 m   **17.** 0.5 cm
18. 0.325 g   **19.** 16,000 mL   **20.** 400 g   **21.** 1055 cm
22. 0.095 L   **23.** 3.2 ft (rounded)   **24. (a)** 2310 mg
(b) 500 mg or 0.5 g more   **25.** 95 °C   **26.** 0 °C
27. 1.8 m (rounded)   **28.** 56.3 kg (rounded)   **29.** 13 gal
30. 5.0 mi (rounded)   **31.** 23 °C (rounded)   **32.** 36 °F (rounded)
33. $26.03 (rounded)   **34.** Possible answers: Use same system as
rest of the world; easier system for children to learn; less use of
fractional numbers; compete internationally.

Cumulative Review Exercises: Chapters 1–7 (page 517)

1. _Estimate:_ $100 + 3 + 70 = 173$; _Exact:_ 178.838
2. _Estimate:_ $30,000 - 800 = 29,200$; _Exact:_ 30,178
3. _Estimate:_ $90,000 \times 200 = 18,000,000$; _Exact:_ 16,849,725
4. _Estimate:_ $60 \times 5 = 300$; _Exact:_ 265.644
5. _Estimate:_ $20,000 \div 40 = 500$; _Exact:_ 530
6. _Estimate:_ $40 \div 8 = 5$; _Exact:_ 4.6

7. _Estimate:_ $2 + 4 = 6$; _Exact:_ $5\dfrac{1}{2}$

8. _Estimate:_ $6 - 1 = 5$; _Exact:_ $4\dfrac{3}{14}$

9. _Exact:_ $14\dfrac{7}{9}$; _Estimate:_ $3 \cdot 5 = 15$

10. _Exact:_ $2\dfrac{1}{2}$; _Estimate:_ $2 \div 1 = 2$

11. $\dfrac{11}{16}$ **12.** 488,045   **13.** $26\dfrac{2}{3}$ **14.** 13.0 (rounded)

15. $1\dfrac{5}{24}$ **16.** 7.0793   **17.** $\dfrac{5}{8}$ **18.** 307 R38   **19.** 0.00762

20. 8   **21.** 265   **22.** three hundred seven and nineteen hundredths

23. 0.0082   **24.** 0.067, 0.6, 0.6007, 0.67   **25.** 16 diapers for
$3.87, about $0.242 per diaper   **26.** 0.875   **27.** 87.5% or $87\dfrac{1}{2}$%

28. $\frac{1}{20}$ **29.** 5% **30.** $3\frac{1}{2}$ **31.** 3.5 **32.** $\frac{1}{3}$ **33.** $\frac{8}{1}$ **34.** $\frac{8}{3}$

35. 12 **36.** 1.67 (rounded) **37.** 5% **38.** 30 hours **39.** 30 in.

40. $1\frac{3}{4}$ or 1.75 min **41.** 280 cm **42.** 0.065 g **43.** 122.76 mi

44. 10 °C **45.** g **46.** cm **47.** m **48.** kg **49.** mg **50.** mL
51. 38 °C **52.** 12 °C **53.** $18.99 (rounded); $170.95

54. $0.46 (rounded) **55.** $12\frac{3}{4}$ or 12.75 lb

56. 93.6 km; 58.0 mi (rounded) **57.** $2756.25; $11,506.25
58. 88.6% (rounded) **59.** $9.74 (rounded)

60. $\frac{1}{6}$ cup more than the amount needed **61.** 120 rows

62. 2100 students **63. (a)** 165.5 hr **(b)** 8.1% (rounded)
64. (a) 8.5 hr **(b)** 4.7% (rounded)
65. (a) $1.3 million (rounded) **(b)** $2.5 million (rounded)
(c) $4.2 million (rounded)

Chapter 8

Section 8.1 (page 527)
1. line named $\overleftrightarrow{CD}$ or $\overleftrightarrow{DC}$ **3.** line segment named $\overline{GF}$ or $\overline{FG}$
5. ray named $\overrightarrow{PQ}$ **7.** perpendicular **9.** parallel **11.** intersect-
ing **13.** $\angle AOS$ or $\angle SOA$ **15.** $\angle CRT$ or $\angle TRC$ **17.** $\angle AQC$
or $\angle CQA$ **19.** right (90°) **21.** acute **23.** straight (180°)
25. (a) The car turned around in a complete circle. **(b)** The governor took the opposite view, for example, having once opposed taxes
but now supporting them. **27.** True, because $\overrightarrow{UQ}$ is perpendicular
to $\overleftrightarrow{ST}$. **28.** True, because they form a 90° angle, as indicated by
the small red square. **29.** False; the angles have the same measure
(both are 180°). **30.** False; $\overleftrightarrow{ST}$ and $\overleftrightarrow{PR}$ are parallel. **31.** False;
$\overleftrightarrow{QU}$ and $\overleftrightarrow{TS}$ are perpendicular. **32.** True; both angles are formed
by perpendicular lines, so they both measure 90°.

Section 8.2 (page 533)
1. $\angle EOD$ and $\angle COD$; $\angle AOB$ and $\angle BOC$ **3.** $\angle HNE$ and $\angle ENF$;
$\angle HNG$ and $\angle GNF$; $\angle HNE$ and $\angle HNG$; $\angle ENF$ and $\angle GNF$
5. 50° **7.** 4° **9.** 50° **11.** 90° **13.** $\angle SON \cong \angle TOM$;
$\angle TOS \cong \angle MON$ **15.** $\angle GOH$ measures 63°; $\angle EOF$ measures
37°; $\angle AOC$ and $\angle GOF$ both measure 80°. **17.** Two angles are
complementary if their sum is 90°. Two angles are supplementary
if their sum is 180°. Drawings will vary; examples are:

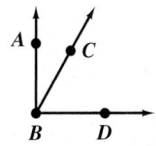 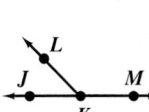

Complimentary   Supplementary

19. $\angle ABF \cong \angle ECD$; Both are 138°. $\angle ABC \cong \angle BCD$; Both are 42°.
21. No; obtuse angles are >90°, so their sum would be >180°.

Section 8.3 (page 541)
1. $P = 28$ yd; $A = 48$ yd^2 **3.** $P = 3.6$ km; $A = 0.81$ km^2
5. $P = 40$ ft; $A = 100$ ft^2 **7.** $P = 29$ m; $A = 7$ m^2
9. $P = 196.2$ ft; $A = 1674.2$ ft^2 **11.** $P = 12$ mi; $A = 9$ mi^2
13. $P = 38$ m; $A = 39$ m^2 **15.** $P = 98$ m; $A = 492$ m^2
17. unlabeled side $= 12$ in.; $P = 78$ in.; $A = 234$ in.^{2} **19.** $460
21. $94.81 **23.** Panoramic: $P = 36$ in., $A = 56$ in.^{2}; 4 in. $\times$ 6 in.:
$P = 20$ in., $A = 24$ in.^{2}; 4 in. $\times$ 7 in.: $P = 22$ in., $A = 28$ in.^{2}
25. 53 yd

27. $A = 6528$ ft^2

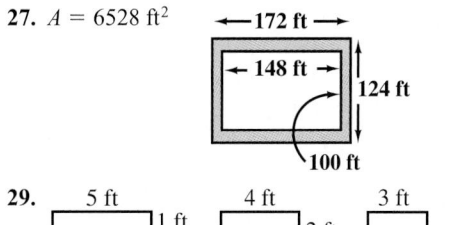

29.

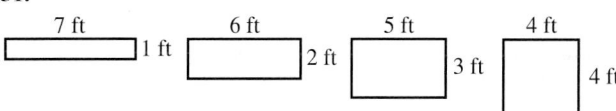

30. (a) 5 ft by 1 ft has area of 5 ft^2; 4 ft by 2 ft has area of 8 ft^2;
3 ft by 3 ft has area of 9 ft^2 **(b)** The square plot 3 ft by 3 ft has
greatest area.
31.

| 7 ft | | 6 ft | | 5 ft | | 4 ft | |

32. (a) 7 ft^2; 12 ft^2; 15 ft^2; 16 ft^2 **(b)** Square plots have the greatest
area.
33. (a) $P = 10$ ft; $A = 6$ ft^2

3 ft / 2 ft

(b) $P = 20$ ft; $A = 24$ ft^2

6 ft / 4 ft

(c) Perimeter is twice the original; area is four times the original.
34. (a) $P = 30$ ft; $A = 54$ ft^2

9 ft / 6 ft

(b) Perimeter is three times the original; area is nine times the
original. **(c)** Perimeter will be four times the original; area will be
16 times the original.

Section 8.4 (page 549)
1. 208 m **3.** 207.2 m **5.** 5.78 km **7.** 775 mm^2 **9.** 19.25 ft^2
11. 3099.6 cm^2 **13.** $1410.75 **15.** 437.5 in.^{2} **17.** Height is
not part of perimeter; square units are used for area, not perimeter.
$P = 2.5$ cm $+ 2.5$ cm $+ 2.5$ cm $+ 2.5$ cm $= 10$ cm **19.** 3.02 m^2
21. 25,344 ft^2

Section 8.5 (page 555)
1. $P = 202$ m; $A = 1914$ m^2 **3.** $P = 58.9$ cm; $A = 139.15$ cm^2
5. $P = 26\frac{1}{4}$ yd; $A = 30\frac{3}{4}$ yd^2 **7.** $P = 85.2$ cm; $A = 302.46$ cm^2
9. $A = 198$ m^2 **11.** $A = 1664$ m^2 **13.** 32° **15.** 48°
17. No. Right angles are 90°, so two right angles are 180°, and the
sum of all *three* angles in a triangle equals 180°. **19.** $7\frac{7}{8}$ ft^2 or
7.875 ft^2 **21.** 126.8 m of curb; 672 m^2 of sod **23. (a)** 32 m^2
(b) 13.5 m^2

Section 8.6 (page 565)
1. $d = 18$ mm **3.** $r = 0.35$ km **5.** $C \approx 69.1$ ft; $A \approx 379.9$ ft^2
7. $C \approx 8.2$ m; $A \approx 5.3$ m^2 **9.** $C \approx 47.1$ cm; $A \approx 176.6$ cm^2
11. $C \approx 23.6$ ft; $A \approx 44.2$ ft^2 **13.** $C \approx 27.2$ km; $A \approx 58.7$ km^2

15. $A \approx 57$ cm^2 **17.** $A \approx 197.8$ cm^2 **19.** π is the ratio of circumference of a circle to its diameter. If you divide the circumference of any circle by its diameter, the answer is always a little more than 3. The approximate value is 3.14, which we call π (pi). Your test question could involve finding the circumference or the area of a circle. **21.** $C \approx 219.8$ cm **23.** $C \approx 785.0$ ft **25.** $A \approx 70,650$ mi^2 **27.** watch: $C \approx 3.1$ in., $A \approx 0.8$ in.^{2}; wall clock: $C \approx 18.8$ in., $A \approx 28.3$ in.^{2} **29.** $d \approx 45.9$ cm **31.** \$1170.33 (rounded) **33.** The prefix *rad* tells you that radius is a ray from the center of the circle. The prefix *dia* means the diameter goes through the circle, and the prefix *circum* means the circumference is the distance around. **35.** $A \approx 44.2$ in.^{2} **36.** $A \approx 132.7$ in.^{2} **37.** $A \approx 201.0$ in.^{2} **38.** small: \$0.063 (rounded); medium: \$0.049 (rounded); large: \$0.046 (rounded) Best Buy **39.** small: \$0.084 (rounded); medium: \$0.067 (rounded) Best Buy; large: \$0.071 (rounded) **40.** small: \$0.077 (rounded) Best Buy; medium: \$0.083 (rounded); large: \$0.078 (rounded)

Summary Exercises on Perimeter, Circumference, and Area (page 569)

1. (a)

All sides have the same length.

(b)
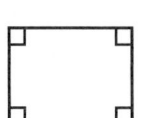

Opposite sides have the same length.

(c)
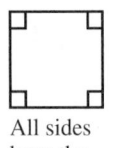

Opposite sides have the same length.

2. (a)
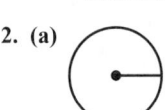

(b)
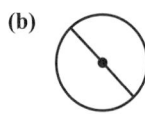

(c) The diameter is twice the radius, or the radius is half the diameter. **3.** Add up the lengths of all the sides to find the perimeter. **4.** Perimeter is the total distance around the outside edges of a shape. Area is the number of square units needed to cover the space inside the shape. **5.** f, e, d, b, c, a **6. (a)** $C = 2 \cdot \pi \cdot r$ **(b)** $C = \pi \cdot d$ **(c)** Divide the diameter by 2 to find the radius. **7.** $P = 63$ ft; $A = 180$ ft^2 $P = 32.4$ cm; $A \approx 45.1$ cm^2

8. (a) $A = 12$ cm^2 **(b)** $P = 6\frac{1}{2}$ ft **(c)** $C \approx 28.5$ m **(d)** $A = 307$ in.^{2}

9. $P = 27$ in.; $A = 31\frac{1}{2}$ in.^{2} or 31.5 in.^{2} **10.** $P = 48$ yd;

$A = 144$ yd^2 **11.** $P = 6.4$ m; $A \approx 2.2$ m^2 **12.** $P = 33.5$ mm; $A = 64$ mm^2 **13.** $d = 12$ cm; $C \approx 37.7$ cm; $A \approx 113.0$ cm^2 **14.** $r = 15$ mi; $C \approx 94.2$ mi; $A \approx 706.5$ mi^2 **15.** $r = 4.5$ ft; $C \approx 28.3$ ft; $A \approx 63.6$ ft^2 **16.** $A \approx 4.5$ m^2 **17.** $A = 217.5$ yd^2 **18.** $A \approx 86$ in.^{2} **19.** $C \approx 14.6$ ft; Bonus: About 361.6 revolutions;

the Mormons used 360, which is the answer you get using $2\frac{1}{3}$ ft

instead of 2.33 ft as the radius. **20.** \$4.47 (rounded)

Section 8.7 (page 577)

1. $V = 550$ cm^3 **3.** $V \approx 44,579.6$ m^3 **5.** $V \approx 3617.3$ in.^{3} **7.** $V \approx 471$ ft^3 **9.** $V \approx 418.7$ m^3 **11.** $V = 800$ cm^3

13. $V = 18$ in.^{3} **15.** $V \approx 2481.5$ cm^3 **17.** $V = 651,775$ m^3 **19.** $V \approx 3925$ ft^3 **21.** Student used diameter of 7 cm; should use radius of 3.5 cm in formula. Units for volume are cm^3, not cm^2. Correct answer is $V \approx 192.3$ cm^3. **23.** $V = 513$ cm^3

Section 8.8 (page 583)

1. 4 **3.** 8 **5.** 3.317 **7.** 2.236 **9.** 8.544 **11.** 10.050 **13.** 13.784 **15.** 31.623 **17.** 30 is about halfway between 25 and 36, so $\sqrt{30}$ should be about halfway between 5 and 6, or about 5.5. Using a calculator, $\sqrt{30} \approx 5.477$. Similarly, $\sqrt{26}$ should be a little more than $\sqrt{25}$; by calculator $\sqrt{26} \approx 5.099$. And $\sqrt{35}$ should be a little less than $\sqrt{36}$; by calculator $\sqrt{35} \approx 5.916$. **19.** $\sqrt{1521} = 39$ ft **21.** $\sqrt{289} = 17$ in. **23.** $\sqrt{144} = 12$ mm **25.** $\sqrt{73} \approx 8.5$ in. **27.** $\sqrt{65} \approx 8.1$ yd **29.** $\sqrt{195} \approx 14.0$ cm **31.** $\sqrt{7.94} \approx 2.8$ m **33.** $\sqrt{65.01} \approx 8.1$ cm **35.** $\sqrt{292.32} \approx 17.1$ km **37.** $\sqrt{65} \approx 8.1$ ft **39.** $\sqrt{360,000} = 600$ m **41.** $\sqrt{135} \approx 11.6$ ft

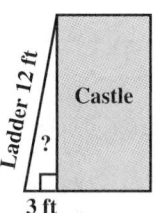

43. The student did not square the numbers correctly: 9^2 is 81 and 7^2 is 49. Also, the final answer is rounded to thousandths instead of tenths. Correct answer is $\sqrt{130} \approx 11.4$ in. **45.** $\sqrt{16200} \approx 127.3$ ft **46.** $\sqrt{7200} \approx 84.9$ ft

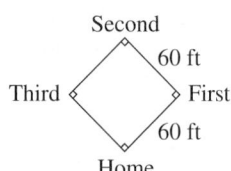

47. The distance from third to first is the same as the distance from home to second because the baseball diamond is a square. **48.** One possibility is:

major league $\dfrac{90 \text{ ft}}{127.3 \text{ ft}} = \dfrac{60 \text{ ft}}{x}$ softball

$x \approx 84.9$ ft

Section 8.9 (page 591)

1. similar **3.** not similar **5.** similar **7.** $\angle 1$ and $\angle 4$; $\angle 2$ and $\angle 5$; $\angle 3$ and $\angle 6$; $\overline{AB}$ and $\overline{PQ}$; $\overline{BC}$ and $\overline{QR}$; $\overline{AC}$ and $\overline{PR}$ **9.** $\angle 1$ and $\angle 6$; $\angle 2$ and $\angle 5$; $\angle 3$ and $\angle 4$; $\overline{MP}$ and $\overline{QS}$; $\overline{MN}$ and $\overline{QR}$; $\overline{NP}$ and $\overline{RS}$ **11.** $\dfrac{3}{2}, \dfrac{3}{2}, \dfrac{3}{2}$ **13.** $a = 5$ mm; $b = 3$ mm **15.** $a = 6$ cm; $b = 15$ cm **17.** $x = 24.8$ m; $P = 72.8$ m; $y = 15$ m; $P = 54.6$ m **19.** $P = 8$ cm $+ 8$ cm $+ 8$ cm $= 24$ cm; $A = 0.5 \cdot 8$ cm $\cdot 6.9$ cm $= 27.6$ cm^2 **21.** $h = 24$ ft

23. $\dfrac{3}{4}$ ft or 0.75 ft **25.** One dictionary definition is "resembling, but not identical." Examples of similar objects are sets of different size pots or measuring cups; small and large size cans of beans; child's tennis shoe and adult's tennis shoe. **27.** $x = 50$ m **29.** $n = 110$ m

Chapter 8 Review Exercises (page 603)

1. line segment named $\overline{AB}$ or $\overline{BA}$ **2.** line named $\overleftrightarrow{CD}$ or $\overleftrightarrow{DC}$
3. ray named $\overrightarrow{OP}$ **4.** parallel **5.** perpendicular **6.** intersecting
7. acute **8.** obtuse **9.** straight; 180° **10.** right; 90°
11. $\angle 1$ and $\angle 3$ measure 30°; $\angle 2$ measures 90°; $\angle 4$ measures 60°
12. $\angle 1$ measures 100°; $\angle 2$ and $\angle 4$ measure 45°; $\angle 3$ measures 35°
13. $\angle AOB$ and $\angle BOC$; $\angle BOC$ and $\angle COD$; $\angle COD$ and $\angle DOA$;
$\angle DOA$ and $\angle AOB$ **14.** $\angle ERH$ and $\angle HRG$; $\angle HRG$ and $\angle GRF$;
$\angle FRG$ and $\angle FRE$; $\angle FRE$ and $\angle ERH$ **15. (a)** 10° **(b)** 45° **(c)** 83°
16. (a) 25° **(b)** 90° **(c)** 147° **17.** $P = 4.84$ mi **18.** $P = 128$ in.
19. 152 cm **20.** 41 ft **21.** $A = 486$ mm² **22.** $A = 16.5$ ft²
or $16\frac{1}{2}$ ft² **23.** $A \approx 39.7$ m² **24.** $P = 50$ cm; $A = 140$ cm²
25. $P = 102.1$ ft; $A = 567$ ft² **26.** $P = 200.2$ m; $A \approx 2074.0$ m²
27. $P = 518$ cm; $A = 11,660$ cm² **28.** $P = 27.1$ m; $A = 20.58$ m²
29. $P = 20\frac{1}{4}$ ft or 20.25 ft; $A = 14$ ft² **30.** 70° **31.** 24°
32. $d = 137.8$ m **33.** $r = 1\frac{1}{2}$ in. or 1.5 in. **34.** $C \approx 6.3$ cm;
$A \approx 3.1$ cm² **35.** $C \approx 109.3$ m; $A \approx 950.7$ m² **36.** $C \approx 37.7$ in.;
$A \approx 113.0$ in.² **37.** $A \approx 20.3$ m² **38.** $A = 64$ in.²
39. $A = 673$ km² **40.** $A = 1020$ m² **41.** $A = 229$ ft²
42. $A = 132$ ft² **43.** $A = 5376$ cm² **44.** $A \approx 498.9$ ft²
45. $A \approx 447.9$ yd² **46.** $V = 30$ in.³ **47.** $V = 96$ cm³
48. $V = 45,000$ mm³ **49.** $V \approx 267.9$ m³ **50.** $V \approx 452.2$ ft³
51. $V \approx 549.5$ cm³ **52.** $V \approx 1808.6$ m³ **53.** $V \approx 512.9$ m³
54. $V = 16$ yd³ **55.** 7 **56.** 2.828 (rounded)
57. 54.772 (rounded) **58.** 12 **59.** 7.616 (rounded)
60. 25 **61.** 10.247 (rounded) **62.** 8.944 (rounded)
63. $\sqrt{289} = 17$ in. **64.** $\sqrt{49} = 7$ cm **65.** $\sqrt{104} \approx 10.2$ cm
66. $\sqrt{52} \approx 7.2$ in. **67.** $\sqrt{6.53} \approx 2.6$ m
68. $\sqrt{71.75} \approx 8.5$ km **69.** $y = 30$ ft; $x = 34$ ft; $P = 104$ ft
70. $y = 7.5$ m; $x = 9$ m; $P = 22.5$ m **71.** $x = 12$ mm;
$y = 7.5$ mm; $P = 38$ mm **72.** $P = 18$ in; $A \approx 20.3$ in.² or
$A = 20\frac{1}{4}$ in.² **73.** $P = 10.3$ cm; $A \approx 6.2$ cm² **74.** $C \approx 40.8$ m;
$A \approx 132.7$ m² **75.** $P = 54$ ft; $A = 140$ ft² **76.** $P = 20$ yd;
$A = 18\frac{3}{4}$ yd² or 18.8 yd² (rounded) **77.** $P = 7$ km; $A \approx 2.0$ km²
78. $C \approx 53.4$ m; $A \approx 226.9$ m² **79.** $P = 78$ mm; $A = 288$ mm²
80. $P = 37.8$ mi; $A \approx 58.5$ mi² **81.** parallel lines **82.** line
segment **83.** acute angle **84.** intersecting lines **85.** right
angle; 90° **86.** ray **87.** straight angle; 180° **88.** obtuse angle
89. perpendicular lines **90.** 81° **91.** 138° **92.** $P = 90$ m;
$A = 92$ m² **93.** $P = 282$ cm; $A = 4190$ cm² **94.** $V \approx 100.5$ ft³
95. $V \approx 3.4$ in.³ or $V = 3\frac{3}{8}$ in.³ **96.** $V \approx 7.4$ m³
97. $V = 561$ cm³ **98.** $V \approx 1271.7$ cm³ **99.** $V \approx 1436.0$ m³
100. $x \approx 12.6$ km **101.** $\angle D = 72°$ **102.** $x = 12$ mm;
$y = 14$ mm **103.** The prefix *dec* in *dec*ade means 10 and the
prefix *cent* in *cent*ury means 100, so divide 200 (two centuries) by
10. The answer is 20 decades.

Chapter 8 Test (page 611)

1. (e) **2.** (a); 90° **3.** (d) **4.** (g); 180° **5.** Parallel lines are
lines in the same plane that never intersect. Perpendicular lines
intersect to form a right angle.

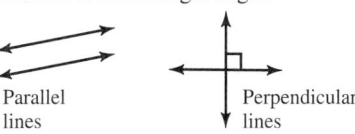

Parallel
lines

Perpendicular
lines

6. 9° **7.** 160° **8.** $\angle 1$ measures 50°; $\angle 3$ measures 90°;
$\angle 2$ and $\angle 4$ measure 40° **9.** $P = 23$ ft; $A = 30$ ft²
10. $P = 72$ mm; $A = 324$ mm² **11.** $P = 26.2$ m; $A = 33.12$ m²
12. $P = 169.4$ cm; $A = 1591$ cm² **13.** $P = 32.45$ m; $A = 48$ m²
14. $P = 37.8$ yd or $37\frac{4}{5}$ yd; $A = 58.5$ yd² **15.** 55°
16. $r = 12.5$ in. or $12\frac{1}{2}$ in. **17.** $C \approx 5.7$ km **18.** $A \approx 206.0$ cm²
19. $A \approx 39.3$ m² **20.** $V = 6480$ m³ **21.** $V \approx 33.5$ ft³
22. $V \approx 5086.8$ ft³ **23.** 9.2 cm (rounded) **24.** $y = 12$ cm;
$z = 6$ cm **25.** Linear units like cm are used to measure perimeter,
radius, diameter, and circumference. Area is measured in square
units like cm² (squares that measure 1 cm on each side). Volume is
measured in cubic units like cm³.

Cumulative Review Exercises: Chapters 1–8 (page 613)

1. *Estimate:* 300 + 60,000 + 6000 = 66,300; *Exact:* 64,574
2. *Estimate:* 20 − 10 = 10; *Exact:* 10.242 **3.** *Estimate:*
4 × 7 = 28; *Exact:* 26.7472 **4.** *Estimate:* 30,000 × 70 = 2,100,000;
Exact: 2,074,587 **5.** *Estimate:* 600 ÷ 3 = 200; *Exact:* 200.8
6. *Estimate:* 5000 ÷ 50 = 100; *Exact:* 94
7. *Estimate:* 5 + 5 = 10; *Exact:* $9\frac{2}{5}$ **8.** *Estimate:* 3 − 2 = 1;
Exact: $1\frac{7}{24}$ **9.** *Estimate:* 3 • 2 = 6; *Exact:* $5\frac{1}{3}$ **10.** $\frac{9}{20}$
11. 0.9132 **12.** 70 R79 **13.** 32 **14.** 0.000078 **15.** 39,105
16. 19.44 (rounded) **17.** $4\frac{5}{12}$ **18.** 836.085 **19.** 9 **20.** 38
21. two hundred eight ten-thousandths **22.** 660.05 **23.** 2.055;
2.5005; 2.505; 2.55 **24.** *Per* means divide and *cent* means
100, so divide by 100 to change a percent to a decimal. **25.** $\frac{1}{50}$
26. 2% **27.** 1.75 **28.** 175% **29.** $\frac{2}{5}$ **30.** 0.4 **31.** $\frac{8}{1}$ **32.** $\frac{7}{3}$
33. 35 **34.** 100 **35.** 2.01 (rounded) **36.** 160% **37.** $600
38. 135 min **39.** $2\frac{1}{2}$ or 2.5 lb **40.** 0.08 m **41.** 1800 mL
42. mm **43.** L **44.** kg **45.** cm **46.** 0 °C and 100 °C
47. $P = 11$ in.; $A = 7$ in.² **48.** $P = 5.8$ m; $A = 1.575$ m²
49. $C \approx 31.4$ ft; $A \approx 78.5$ ft² **50.** $P = 76$ cm; $A = 264$ cm²
51. $P = 40.8$ m; $A = 95$ m² **52.** $P = 50$ yd; $A = 142$ yd²
53. $y \approx 24.2$ mm **54.** $x \approx 8.1$ ft; $P \approx 19.1$ ft
55. 59% (rounded) **56.** Brand T at 15.5 oz for $2.99 − $0.30
coupon **57.** $V \approx 2255.3$ cm³

58. $P = 52$ in.

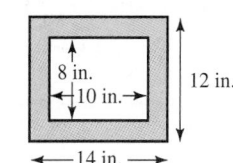

8 in.

10 in.

12 in.

14 in.

59. $1\frac{1}{12}$ yd **60.** 42.9 lb (rounded) **61.** 3.4 m **62.** $22.40
63. (a) $0.55; $0.55 **(b)** $1.43; $1.39 **(c)** $2.53; $2.44
64. (a) 3.0% increase **(b)** 4.5% decrease **(c)** 0.0% (no change)
(d) 23.4% increase **65.** No; 36 oz is more than 2 lbs, and the
$3.95 priority rate applies to a package weighing 2 lbs or less.
66. $4442.13

Chapter 9

Section 9.1 (page 623)

1. +32 or 32 **3.** −12 **5.** −6$\frac{1}{2}$ **7.** +20,320 or 20,320

9.

−5 −4 −3 −2 −1 0 1 2 3 4 5

11.

−5 −4 −3 −2 −1 0 1 2 3 4 5

13.

−5 −4 −3 −2 −1 0 1 2 3 4 5

15. < **17.** > **19.** < **21.** > **23.** < **25.** >

27.

A C D B
−2 −1 0 1

28. −1.2, −0.5, 0, 0.6 **29.** A: may be at risk; B: above normal; C: normal; D: normal **30.** **(a)** The patient would think the interpretation was "above normal" and wouldn't get treatment. **(b)** Patient D's score of 0; because 0 is neither positive nor negative. **31.** 3 **33.** 10 **35.** 0 **37.** −18 **39.** −32 **41.** −7 **43.** 14 **45.** −$\frac{2}{3}$ **47.** 8.3 **49.** $\frac{1}{6}$ **51.** true **53.** true **55.** false

Section 9.2 (page 633)

1. 3

−7 −6 −5 −4 −3 −2 −1 0 1 2 3

3. −7

−7 −6 −5 −4 −3 −2 −1 0 1 2 3

5. −1

−7 −6 −5 −4 −3 −2 −1 0 1 2 3

7. −3 **9.** 7 **11.** −7 **13.** 1 **15.** −8 **17.** −20

19. 13 + (−17) = −4 yd **21.** −$52.50 + $50 = −$2.50

23. Jeff: −20 + 75 + (−55) = 0 pts; Terry: 42 + (−15) + 20 = 47 pts **25.** −6.8 **27.** $\frac{1}{4}$ **29.** −$\frac{3}{10}$ **31.** −$\frac{26}{9}$ **33.** −3 **35.** 9

37. −$\frac{1}{2}$ **39.** 6.2 **41.** 14 **43.** −2 **45.** −12 **47.** −25

49. −23 **51.** 5 **53.** 20 **55.** 11 **57.** −60 **59.** 0

61. −$\frac{3}{2}$ **63.** −$\frac{2}{5}$ **65.** 0.7 **67.** **(a)** −2; −2 **(b)** −8; −8

(c) 10; 10 Addition is commutative, so changing the order of the addends does *not* change the sum. **68.** **(a)** −8; 8 **(b)** 2; −2 **(c)** −7; 7 Changing the order of the numbers in subtraction *does change* the result. **69.** **(a)** The answers have the same absolute value but opposite signs. **(b)** When the order of the numbers in a subtraction problem is switched, change the sign on the answer to its opposite. **70.** **(a)** −18; 20; −5; 4. Adding 0 to any number

leaves the number unchanged. **(b)** 2; −10; −3; 7. Subtracting 0 from a number leaves the number unchanged, but subtracting a *number from 0* changes the sign of the number. **71.** **(a)** Change 6 to its opposite, −6. Correct answer is −6 + (−6) = −12. **(b)** Do *not* change 9 to its opposite. Correct answer is −9 + (−5) = −14. **73.** −10 **75.** 12 **77.** −7 **79.** 5 **81.** −1 **83.** −4.4 **85.** −11 **87.** −10 **89.** $\frac{19}{4}$ or 4$\frac{3}{4}$ **91.** $196 **93.** $0

Section 9.3 (page 639)

1. −35 **3.** −45 **5.** −18 **7.** −50 **9.** −40 **11.** 32 **13.** 77 **15.** 133 **17.** 13 **19.** 0 **21.** 4 **23.** −4 **25.** −$\frac{1}{10}$ **27.** $\frac{14}{3}$ **29.** −$\frac{5}{6}$ **31.** $\frac{7}{4}$ **33.** −42.3 **35.** 6 **37.** −31.62 **39.** 4.5 **41.** 8.23 **43.** 0 **45.** −2 **47.** −5 **49.** undefined **51.** −14 **53.** 10 **55.** 4 **57.** −1 **59.** 191 **61.** 0 **63.** 1 **65.** $\frac{2}{3}$ **67.** $\frac{1}{3}$ **69.** −8 **71.** −$\frac{14}{3}$ **73.** 4.73 **75.** −4.7 **77.** −5.3 **79.** 12 **81.** −0.36 **83.** −$\frac{7}{40}$ **85.** −2 **87.** −10 **89.** −$119,400 **91.** −2011 ft **93.** $2366 **95.** −22 degrees **97.** **(a)** Examples will vary. Some possibilities: 6 · (−1) = −6; 2 · (−1) = −2; 15 · (−1) = −15 **(b)** Examples will vary. Some possibilities: −6 · (−1) = 6; −2 · (−1) = 2; −15 · (−1) = 15. The result of multiplying any nonzero number times −1 is the number with the opposite sign. **98.** **(a)** Examples will vary. Some possibilities: $\frac{-6}{-1}$ = 6; $\frac{-2}{-1}$ = 2; $\frac{-15}{-1}$ = 15

(b) Some possibilities: $\frac{6}{-1}$ = −6; $\frac{2}{-1}$ = −2; $\frac{15}{-1}$ = −15

(c) Some possibilities: $\frac{-6}{-6}$ = 1; $\frac{-2}{-2}$ = 1; $\frac{-15}{-15}$ = 1

When dividing by −1, the sign of the number changes to its opposite. A negative number divided by itself gives a result of 1.

Section 9.4 (page 647)

1. −6 **3.** 0 **5.** 52 **7.** −39 **9.** −30 **11.** 17 **13.** 16 **15.** 30 **17.** 23 **19.** −43 **21.** 19 **23.** −3 **25.** 0 **27.** 47 **29.** 41 **31.** −2 **33.** 13 **35.** 126 **37.** $\frac{27}{-3}$ = −9 **39.** $\frac{-48}{-4}$ = 12 **41.** $\frac{-60}{-1}$ = 60 **43.** 8 **45.** 1.24 **47.** −1.47 **49.** −2.31 **51.** −$\frac{13}{10}$ **53.** $\frac{5}{4}$ **55.** $\frac{4}{5}$ **57.** −1200 **59.** 1 **61.** **(a)** The answers are 4, −8, 16, −32, 64, −128, 256, −512. When a negative number is raised to an even power, the answer is positive; when raised to an odd power, the answer is negative. **(b)** positive; negative; positive **63.** $\frac{27}{-27}$ = −1 **65.** 7 **67.** 6

Summary Exercises on Operations with Signed Numbers (page 651)

1. −6 **2.** 0 **3.** −7 **4.** −7 **5.** 63 **6.** −1 **7.** −56 **8.** −22 **9.** 12 **10.** 6 **11.** −13 **12.** 0 **13.** −80 **14.** −17 **15.** −48 **16.** −1 **17.** −$\frac{1}{6}$ **18.** undefined **19.** −14.6 **20.** −0.6 **21.** 48 **22.** −19 **23.** 2 **24.** −20 **25.** 0 **26.** 16 **27.** −3 **28.** −36 **29.** 6 **30.** −2 **31.** −32 **32.** −5 **33.** −4 **34.** −9732 **35.** 100 **36.** 4 **37.** −343 **38.** 5 **39.** −6 **40.** −10 **41.** −5 **42.** 8 **43.** 1 **44.** $\frac{27}{0}$ is undefined. **45.** $\frac{-8}{8}$ = −1 **46.** $212 **47.** −$24 **48.** −13 degrees **49.** −27,010 ft **50.** 143 yd **51.** −90 ft

52. January: $-\$700$; February: $-\$100$; March: $\$700$; April: $\$900$; May: $\$0$; June: $\$700$ **53.** January; April **54.** $\$2100$ **55.** $-\$1850$ **56.** Subtract the smaller absolute value from the larger absolute value. The answer has the same sign as the addend with the larger absolute value. Examples: $-6 + 2 = -4$ and $6 + (-2) = 4$ **57.** Similar: If the signs match, the result is positive. If the signs are different, the result is negative. Different: Multiplication is commutative, division is not. You can multiply by 0, but dividing by 0 is not allowed.

Section 9.5 (page 655)

1. 28 **3.** -10 **5.** 8 **7.** -30 **9.** -8 **11.** 0

13. 2 **15.** -22 **17.** $-\dfrac{13}{8}$ **19.** 28 **21.** -5 **23.** 3

25. $\dfrac{12}{-6} = -2$ **27.** $P = 30$ **29.** $P = 28$ **31.** $A \approx 78.5$

33. $A = \dfrac{45}{2}$ or $22\dfrac{1}{2}$ **35.** $V = 600$ **37.** $d = 318$ **39.** $C \approx 25.12$

41. (a) $P = 33$ in. **(b)** $P = 9$ ft **43. (a)** average score $= 83$ **(b)** average score $= 91$

45. The error was made when replacing x with -3; should be $-(-3)$, not -3.

$$-x - 4y$$
Replace x with (-3) and y with (-1).

$$-(-3) - 4(-1)$$
Opposite of (-3) is $+3$

$$(+3) - (-4)$$
Change subtraction to addition.

$$\underbrace{3 + (+4)}_{7}$$
Add.

47. $F = -40$; convert a Celsius temperature to Fahrenheit
48. $C = -20$; convert a Fahrenheit temperature to Celsius
49. $V \approx 113.04$; find the volume of a sphere
50. $c = 5$; Pythagorean Theorem for finding the hypotenuse or a leg in a right triangle **51.** $A = 56$; find the area of a trapezoid
52. $V \approx 376.8$; find the volume of a cone

Section 9.6 (page 665)

1. yes **3.** yes **5.** no **7.** $p = 4$ **9.** $k = -15$ **11.** $z = 8$

13. $r = 10$ **15.** $n = -8$ **17.** $r = -6$ **19.** $k = 18$ **21.** $x = 11$

23. $r = -3$ **25.** $d = \dfrac{7}{3}$ or $2\dfrac{1}{3}$ **27.** $z = \dfrac{87}{8}$ or $10\dfrac{7}{8}$

29. $k = \dfrac{5}{2}$ or $2\dfrac{1}{2}$ **31.** $m = \dfrac{83}{20}$ **33.** $x = 5.87$ **35.** $r = -1.51$

37. $z = 2$ **39.** $r = 4$ **41.** $y = 0$ **43.** $k = -6$ **45.** $p = 9$
47. $m = -7$ **49.** $p = 1.1$ **51.** $k = 34$ **53.** $a = 66$
55. $r = -36$ **57.** $p = -20$ **59.** $m = 4$ **61.** $x = 16$
63. $y = 1.3$ **65.** $z = -4.94$ **67.** You may add or subtract the same number on each side of an equation. Many different equations could have -3 as the solution. One possibility is:

$$x + 5 = 2$$
$$x + 5 - 5 = 2 - 5$$
$$x = -3$$

69. $x = 19$ **71.** $x = 9$

73. $x = \dfrac{8}{21}$ **75.** $a = -\dfrac{5}{4}$ **77.** $m = -4$

Section 9.7 (page 673)

1. $p = 1$ **3.** $y = 1$ **5.** $m = 0$ **7.** $a = -2$ **9.** $x = -4$
11. $z = 6$ **13.** $b = 0.9$ **15.** $6x + 24$ **17.** $7p - 56$
19. $-3m - 18$ **21.** $-2y + 6$ **23.** $-8c - 64$ **25.** $-10w + 90$
27. $17r$ **29.** $1z$ or z **31.** $-2y$ **33.** $-7t$ **35.** $0p = 0$

37. $-3b$ **39.** $k = 5$ **41.** $m = 6$ **43.** $b = -6$ **45.** $y = 1$
47. $p = 4$ **49.** $z = -2$ **51.** $y = 2$ **53.** $b = -1.05$
55. $r = -4$ **57.** $y = -9$ **59.** $t = -5$ **61.** $x = 0$

63.

$$-2t - 10 = 3t + 5$$
$$-2t - 3t - 10 = 3t - 3t + 5$$
Subtract $3t$ from each side (addition property).
$$-5t - 10 = 5$$
$$-5t - 10 + 10 = 5 + 10$$
Add 10 to each side (addition property).
$$-5t = 15,$$
$$\dfrac{-5 \cdot t}{-5} = \dfrac{15}{-5}$$
Divide each side by -5 (multiplication property).
$$t = -3$$

65. $x = -10$ **67.** $y = 2$ **69.** $y = 20$ **71.** $w = -1$
73. $a = -6.75$

Section 9.8 (page 683)

1. $14 + x$ or $x + 14$ **3.** $-5 + x$ or $x + (-5)$ **5.** $20 - x$

7. $x - 9$ **9.** $x - 4$ **11.** $6x$ **13.** $2x$ **15.** $\dfrac{x}{2}$

17. $8 + 2x$ or $2x + 8$ **19.** $7x - 10$ **21.** $2x + x$ or $x + 2x$
23. (a) A variable is a letter that represents an unknown quantity. Examples: x, w, p **(b)** An expression is a combination of operations on variables and numbers. Examples: $6x, w - 5, 2p + 3x$ **(c)** An equation has an $=$ sign and shows that two expressions are equal. Examples: $2y = 14$; $x + 5 = 2x$; $8p - 10 = 54$ **25.** $4n - 2 = 26$; $n = 7$ **27.** $2n + n = -15$; $n = -5$ **29.** $5n + 12 = 7n$; $n = 6$

31. $30 - 3n = 2 + n$; $n = 7$ **33.** $\dfrac{1}{2}n + 2n = 50$; $n = 20$

35. 25 years old **37.** Brenda spent $\$42$. **39.** I am 21; my sister is 30. **41.** Lien earned $\$19,500$; husband earned $\$18,000$
43. computer cost $\$1100$; printer cost $\$220$ **45.** 34 cm and 44 cm
47. 12 ft and 19 ft

49. Length is 19 m.

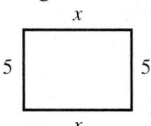

51. Length is 12 ft; width is 6 ft.

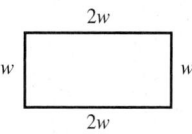

53. Length is 13 in.; width is 5 in.

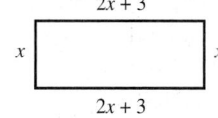

Chapter 9 Review Exercises (page 691)

1.

$$\xleftarrow{\hspace{0.3cm}} \overset{\bullet}{-7}\, -6\, -5\, -4\, -3\, -2\, -1\; \overset{\bullet}{0}\; 1\; \overset{\bullet}{2}\; 3\; 4 \xrightarrow{\hspace{0.3cm}}$$

2.

$$-6\, -5\, -4\, -3\, -2\, -1\; 0\; 1\; 2\; 3\; 4\; 5$$

3.

$$-7\, -6\, -5\, -4\, -3\, -2\, -1\; 0\; 1\; 2\; 3$$

4.

$$-8\, -7\, -6\, -5\, -4\, -3\, -2\, -1\; 0\; 1\; 2$$

5. $>$ **6.** $<$ **7.** $>$ **8.** $<$ **9.** 8 **10.** 19 **11.** -7 **12.** -15

13. 2 **14.** -7 **15.** -19 **16.** -33 **17.** 1 **18.** -19
19. $\dfrac{3}{10}$ **20.** $-\dfrac{3}{8}$ **21.** -5.2 **22.** -1.5 **23.** -6 **24.** 14
25. $\dfrac{5}{8}$ **26.** -3.75 **27.** -6 **28.** -8 **29.** -7 **30.** -17
31. 11 **32.** 11 **33.** 13 **34.** -6 **35.** -80 **36.** 0 **37.** $-\dfrac{1}{2}$
38. 9 **39.** -24 **40.** -20 **41.** 15 **42.** 64 **43.** -8 **44.** -3
45. 5 **46.** 20 **47.** 0 **48.** -81 **49.** 0 **50.** undefined
51. $-\dfrac{4}{7}$ **52.** 6 **53.** 1.4 **54.** -6.6 **55.** 57 **56.** 23
57. -1 **58.** -2 **59.** -34 **60.** -32 **61.** -100 **62.** 117
63. -1 **64.** 1.328 **65.** 0 **66.** 2 **67.** 27 **68.** -8 **69.** 0
70. 19 **71.** 23 **72.** -1 **73.** $P = 35$ **74.** $A = 27$
75. $y = -3$ **76.** $a = 16$ **77.** $z = 1$ **78.** $r = 1$
79. $x = -\dfrac{5}{4}$ or $-1\dfrac{1}{4}$ **80.** $k = -8.05$ **81.** $r = -7$
82. $p = 8$ **83.** $z = 20$ **84.** $a = -55$ **85.** $y = 9$
86. $b = -4$ **87.** $6r - 30$ **88.** $11p + 77$ **89.** $-9z + 27$
90. $-8x - 32$ **91.** $11r$ **92.** $-5z$ **93.** $-8p$ **94.** $2x$
95. $z = -9$ **96.** $k = -5$ **97.** $y = -5$ **98.** $b = 7$
99. $a = -4$ **100.** $t = 0$ **101.** $18 + x$ or $x + 18$
102. $\dfrac{1}{2}x$ or $\dfrac{x}{2}$ **103.** $-5x$ **104.** $20 - x$
105. $4x + 6 = -14$; $x = -5$ **106.** $5x - x = 100$; $x = 25$
107. $22,500$; $7500 **108.** Length is 14 in.; width is 10 in.
109. 3 **110.** 40 **111.** -1 **112.** -7 **113.** -16
114. -9 **115.** 6 **116.** 5 **117.** -20 **118.** undefined
119. $-\dfrac{5}{9}$ **120.** -0.35 **121.** 7 **122.** -19 **123.** $y = 9$
124. $b = -4$ **125.** $z = 4$ **126.** $r = -10$ **127.** $x = -11$
128. $z = 3$ **129.** $k = 4$ **130.** $t = 0$ **131.** $a = -40$
132. $p = 3$ **133.** 6 mg **134.** 100 senators; 435 representatives
135. cheetah: 68 mph; zebra: 43 mph **136.** Length is 570 yd; width is 95 yd.

Chapter 9 Test (page 697)

1.

-5 -4 -3 -2 -1 0 1 2 3 4 5

2. $<, >$ **3.** 7, 15 **4.** -1 **5.** -13 **6.** 5.3 **7.** -7 **8.** 16
9. $\dfrac{1}{4}$ **10.** -32 **11.** 84 **12.** 0 **13.** -25 **14.** 8 **15.** $-\dfrac{3}{5}$
16. 1 **17.** -21 **18.** -38 **19.** 27 **20.** When evaluating, you are given specific values to replace each variable. When solving an equation, you are not given the value of the variable. You must find a value that "works"; that is, when your solution is substituted for the variable, the two sides of the equation are equal. **21.** 110
22. $x = 5$ **23.** $r = 31$ **24.** $t = 5$ **25.** $p = -15$ **26.** $a = -3$
27. $m = 2$ **28.** 57 cm; 61 cm
29. 168 ft; 42 ft

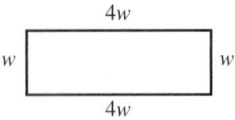

$4w$

w — w

$4w$

30. Tim spent 8 hr; Marcella spent 11 hr.

Cumulative Review Exercises: Chapters 1–9 (page 699)

1. *Estimate:* $9 + 1 + 40 = 50$; *Exact:* 50.602
2. *Estimate:* $6 \times 50 = 300$; *Exact:* 308.484
3. *Estimate:* $40,000 \div 80 = 500$; *Exact:* 503

4. *Estimate:* $4 - 3 = 1$, *Exact:* $\dfrac{17}{20}$
5. *Exact:* $5\dfrac{1}{4}$; *Estimate:* $6 \cdot 1 = 6$
6. *Exact:* $2\dfrac{1}{2}$; *Estimate:* $4 \div 2 = 2$ **7.** 8.906 **8.** 534,072
9. 56.4 (rounded) **10.** -5 **11.** 40 **12.** 4 **13.** -1.3 **14.** -5
15. $-\dfrac{1}{2}$ **16.** 5 **17.** 4 **18.** $>$ **19.** $<$ **20.** $<$ **21.** 0.08; $\dfrac{2}{25}$
22. 4.5; 450% **23.** (a) $\dfrac{2}{1}$ (b) $\dfrac{1}{3}$ (c) $\dfrac{1}{1}$ (d) $\dfrac{12}{13}$ **24.** 0.4
25. 233.33 (rounded) **26.** 45 **27.** 15 students **28.** 54.4%
29. 50 cars **30.** 4% **31.** 14 qt **32.** 3 days **33.** 3700 mL
34. 0.4 m **35.** m **36.** mL **37.** km **38.** g **39.** 37 °C; 20 °C
40. $P = 9$ ft; $A = 5\dfrac{1}{16}$ or 5.0625 ft^2 **41.** $C \approx 28.26$ mm;
$A \approx 63.585$ mm^2 **42.** $P = 56$ mi; $A = 84$ mi^2 **43.** $P = 6.45$ cm;
$A = 2.185$ cm^2 **44.** $P = 36$ ft; $A = 70$ ft^2 **45.** $P = 188$ m;
$A = 1636$ m^2 **46.** $y \approx 13.2$ yd **47.** $x = 14$ in. **48.** $y = -26$
49. $t = -5$ **50.** $x = 8$ **51.** $p = -4$ **52.** $4x + 40 = 0$; $x = -10$
53. $m + m + 300 = 1000$; $350 for Reggie; $650 for Donald
54. $82 = 2(w + 5) + 2w$; Length is 23 cm; width is 18 cm.

$w + 5$

w | | w

$w + 5$

55. $31.91 (rounded)
56. $8\dfrac{1}{3}$ cups
57. 124% **58.** $4.30 (rounded) **59.** 0.544 m^3 less **60.** 11 ft
61. 54 min (rounded) **62.** Naomi's car; 25.2 miles per gallon (rounded)

Chapter 10

Section 10.1 (page 709)

1. $32,000 **3.** $\dfrac{9800}{32,000} = \dfrac{49}{160}$ **5.** $\dfrac{12,100}{900} = \dfrac{121}{9}$
7. Don't know **9.** $\dfrac{720}{6000} = \dfrac{3}{25}$ **11.** $\dfrac{1020}{1200} = \dfrac{17}{20}$
13. $\dfrac{1740}{1020} = \dfrac{29}{17}$ **15.** $522,000 **17.** $174,000 **19.** $261,000
21. 4018 people (rounded) **23.** 543 people **25.** 434 people (rounded) **27.** First find the percent of the total that is to be represented by each item. Next, multiply the percent by 360° to find the size of each sector. Finally, use a protractor to draw each sector. **29.** 90° **31.** 10%; 36° **33.** 54° **35.** 15%
37. (a) $200,000 **(b)** 22.5°; 72°; 108°; 90°; 67.5°
(c)

Equipment rentals 30%

Rafting tours 25%

Equipment sales 18.75%

Grocery and provision sales 20%

Adventure classes 6.25%

39. (a) 2464; 55% (rounded); 198°; 220; 5% (rounded); 18°; 536; 12% (rounded); 43° (rounded); 520; 12% (rounded); 43° (rounded); 748; 17% (rounded); 61° (rounded)

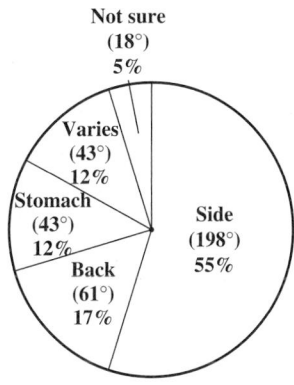

(b) No. The total is 101% due to rounding. **(c)** No. The total is 363° due to rounding.

Section 10.2 (page 719)

1. India; 51.4% **3.** USA, Britain, and Australia **5.** 121 days (rounded) **7.** May; 10,000 unemployed **9.** 1500 workers **11.** 2500 workers **13.** 150,000 gal **15.** 1998; 250,000 gal **17.** 550,000 gal **19.** $196.08 **21.** $27.90 **23.** Answers will vary. Some possibilities are: Greater competition among long-distance companies; higher volume of long-distance calls; improved technology. **25.** 3,000,000 CDs **27.** 1,500,000 CDs **29.** 3,500,000 CDs **31.** Probably Store B with greater sales. Predicted sales might be 4,500,000 CDs to 5,000,000 CDs in 2003. **33.** A single bar or a single line must be used for each set of data. To show multiple sets of data, multiple sets of bars or lines must be used. **35.** $40,000 **37.** $25,000 **39.** $5000 **41.** The decrease in sales may have resulted from poor service or greater competition. The increase in sales may have been a result of more advertising or better service. **43.** 23 yr **44.** 20 additional flavors **45.** 214,286 rolls each day (rounded) **46.** 1.17 lb (rounded) **47.** The weight of one roll of LifeSavers is much less than 1.17 lb. The answer is correct using the information given, but some of the data given must be incorrect. **48.** Answers will vary. Possible answers are: Misprints or typographical errors; careless reporting of data; math errors.

Section 10.3 (page 727)

1. 61–65 years; 16,000 members **3.** 13,000 members **5.** 50,000 members **7.** $4100 to $5000; 16 employees **9.** 11 employees **11.** 49 employees **13.** Class intervals are the result of combining data into groupings. Class frequency is the number of data items that fit in each class interval. These are used to group data and to have multiple responses (frequency) in a class interval—this makes the data easier to interpret. **15.** ЖИ I; 6 **17.** I; 1 **19.** II; 2 **21.** IIII; 4 **23.** ЖИ I; 6 **25.** ЖИ; 5 **27.** I; 1 **29.** ЖИ ЖИ IIII; 14 **31.** ЖИ ЖИ ЖИ I; 16 **33.** ЖИ ЖИ I; 11 **35.** ЖИ; 5

Section 10.4 (page 735)

1. 11 yr **3.** 2.5 in. of rain (rounded) **5.** $37,127 **7.** $58.24 **9.** 12.5 customers (rounded) **11.** 17.2 (rounded) **13.** $35,500 **15.** 130 books **17.** 508 calories **19.** 21% **21.** 68 and 74 yr **23.** When the data contains a few very low or a few very high values, the mean will give a poor indication of the average. Consider using the median or mode instead. **25.** The median is a

better measure of central tendency when the list contains one or more extreme values. Find the mean and the median of the following home values. $82,000; $64,000; $91,000; $115,000; $982,000; mean home value = $266,800; median home value = $91,000 **27.** 2.60 **29.** 176 sales calls **30.** mean Costanza: 22; mean Kramer: 22 **31.** median Costanza: 19.5; median Kramer: 22 **32.** mode Costanza: 22; mode Kramer: 22 **33.** The means and mode are identical for both sales representatives and the medians are close. **34.** The number of weekly sales calls made by Costanza varies greatly from week to week while the number of weekly sales calls made by Kramer remains fairly constant. **35.** range = 40 − 8 = 32 **36.** range = 25 − 19 = 6 **37.** No, not with any certainty. There probably are additional questions that need to be answered before a determination could be made, such as the dollar amount of sales, number of repeat customers, and so on. **38.** Answers will vary. Some possible answers are: He works hard one week, then takes it easy the next week; the characteristics of the sales territories vary greatly; illness or personal problems may be affecting performance.

Chapter 10 Review Exercises (page 747)

1. lodging; $560 **2.** $\frac{400}{1700} = \frac{4}{17}$ **3.** $\frac{300}{1700} = \frac{3}{17}$ **4.** $\frac{280}{1700} = \frac{14}{85}$ **5.** $\frac{300}{160} = \frac{15}{8}$ **6.** $\frac{560}{400} = \frac{7}{5}$ **7.** 3936 companies **8.** 1728 companies **9.** 720 companies **10.** 2904 companies **11.** On-site child care and Fitness centers Answers will vary. Perhaps employers feel that they are not needed or would not be used. Or, it may be that they would be too expensive for the benefit derived. **12.** Flexible hours and Casual dress Answers will vary. Perhaps employees request them and appreciate them. Or, it may be that neither of them cost the employer anything to offer. **13.** March; 8,000,000 acre-feet **14.** June; 2,000,000 acre-feet **15.** 5,000,000 acre-feet **16.** 4,000,000 acre-feet **17.** 5,000,000 acre-feet **18.** 2,000,000 acre-feet **19.** $50,000,000 **20.** $20,000,000 **21.** $20,000,000 **22.** $40,000,000 **23.** The floor-covering sales decreased for 2 years and then moved up slightly. Answers will vary. Perhaps there is less new home, apartment, and office construction and less remodeling and home improvement in the area near Center A. **24.** The floor-covering sales are increasing. Answers will vary. Perhaps a greater number of new homes and apartments are being built or people are remodeling or upgrading their homes, apartments, and offices in the area near Center B. **25.** 28.5 digital cameras **26.** 35.2 complaints **27.** $51.05 **28.** 257.3 points (rounded) **29.** 39 forms **30.** $562 **31.** $79 **32.** 18 and 32 launchings (bimodal) **33.** 36° **34.** 35% **35.** 20% **36.** 25% **37.** 36° **38.**

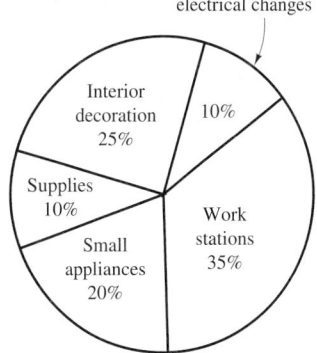

39. 100 volunteers **40.** 144.7 tacks (rounded) **41.** 48 applicants **42.** 31 and 43 two-bedroom apartments (bimodal) **43.** 5.0 hr **44.** 21 phone calls **45.** IIII; 4 **46.** I; 1 **47.** ЖИ I; 6

48. II; 2   **49.** IIII IIII III; 13   **50.** IIII II; 7   **51.** IIII II; 7

52.

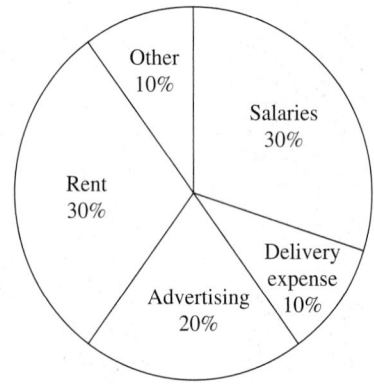

MATH EXAM SCORES

53. 64.5 (rounded)   **54.** 118.8 units (rounded)

Chapter 10 Test (page 753)

1. $644,000   **2.** $532,000   **3.** $812,000   **4.** $420,000
5. $84,000   **6.** $308,000   **7.** 108°   **8.** 36°   **9.** 72°   **10.** 108°
11. 10%
12.

Other 10%
Salaries 30%
Rent 30%
Delivery expense 10%
Advertising 20%

13. 3   **14.** 2   **15.** 4   **16.** 3   **17.** 3   **18.** 5
19.

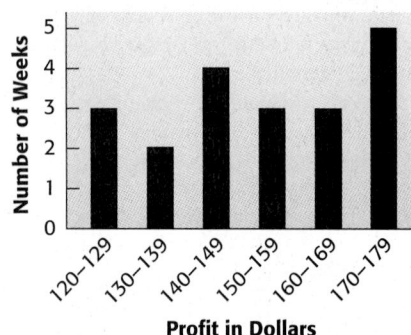

MR. ALAN'S VENDING SALES

20. 75 mi   **21.** 15.3 lb (rounded)   **22.** 478.9 mph
23.

Credits	Grade	
3	A	$3 \times 4 = 12$
2	C	$2 \times 2 = 4$
4	B	$4 \times 3 = 12$
9		28

$28 \div 9 \approx 3.11$

The weighted mean must be used because different classes are
worth different numbers of credits.

24. 8, 17, 23, 32, 64

⌐— Median

Arrange the values in order, from smallest to largest. When there is
an odd number of values in a list, the median is the middle value.
25. $25.50   **26.** 173.7 (rounded)   **27.** 31.5 degrees   **28.** 10.0 m
29. 52 kiloliters   **30.** 103° and 104°

Cumulative Review Exercises: Chapters 1–10 (page 757)

1. $65.24   **2.** $781   **3.** 93,370   **4.** 850,000   **5.** 23   **6.** 10
7. 108   **8.** 324   **9.** *Estimate:* 50,000 + 200,000 + 50 +
20,000 = 270,050; *Exact:* 243,897   **10.** *Estimate:* 3 + 800 +
40 + 50 + 8 = 901; *Exact:* 898.547   **11.** *Estimate:* 400,000 −
200,000 = 200,000; *Exact:* 210,439   **12.** *Estimate:* 900 − 60 =
840; *Exact:* 811.863   **13.** *Estimate:* 7000 × 600 = 4,200,000;
Exact: 4,485,640   **14.** *Estimate:* 60 × 3 = 180; *Exact:* 165.911
15. *Estimate:* 10,000 ÷ 20 = 500; *Exact:* 642   **16.** *Estimate:*
60 ÷ 4 = 15; *Exact:* 14.72   **17.** $1\frac{5}{8}$ **18.** $1\frac{7}{8}$ **19.** $10\frac{4}{15}$
20. $\frac{1}{2}$ **21.** $1\frac{11}{12}$ **22.** $2\frac{3}{5}$ **23.** $\frac{3}{8}$ **24.** $44\frac{2}{5}$ **25.** $8\frac{4}{5}$ **26.** $1\frac{1}{3}$
27. 24   **28.** $\frac{8}{21}$ **29.** $\frac{1}{12}$ **30.** $\frac{11}{16}$ **31.** 0.375   **32.** 0.8
33. 0.167 (rounded)   **34.** 0.85   **35.** 0.199, 0.207, 0.218, 0.22,
0.2215   **36.** 0.58, 0.608, $\frac{5}{8}$, 0.6319, $\frac{13}{20}$ **37.** $\frac{1}{8}$ **38.** $\frac{4}{1}$ **39.** True
40. False   **41.** 6   **42.** 6   **43.** 16.2   **44.** 52   **45.** 0.65
46. 3.5%   **47.** 3.80 or 3.8   **48.** 0.0775   **49.** $\frac{1}{25}$ **50.** $\frac{7}{8}$
51. 35%   **52.** 575%   **53.** $507   **54.** 324 homes   **55.** 2800
people   **56.** 400   **57.** 45%   **58.** 25%   **59.** 9   **60.** 7   **61.** 120
62. 12,000   **63.** 10,000 m   **64.** 3.815 m   **65.** 8300 mg
66. 0.23 kg   **67.** 0.072 L   **68.** 280 mL   **69.** mL   **70.** g
71. km   **72.** kg   **73.** 47.3 m² (rounded)   **74.** 38.7 cm²
(rounded)   **75.** 9.6 ft² (rounded)   **76.** 132.7 cm² (rounded)
77. $V \approx 549.8$ cm³ (rounded)   **78.** $V = 199.5$ m³   **79.** 8 m
80. 34 cm   **81.** −22   **82.** 6.9   **83.** −45   **84.** 83.22   **85.** 6
86. −2.3   **87.** $x = 7$ **88.** $x = -6$ **89.** $x = 3$ **90.** $x = -7$
91. 27 hookups; 26 hookups; 19 hookups   **92.** 4.4 acres; 4.3
acres; 2.85 acres   **93.** 35%   **94.** $1124   **95.** 102 tanks
96. 816 rooms   **97.** $5097.30   **98.** $59,225.25   **99.** 17.5 L
100. $9999.25

Appendix B

Appendix B Exercises (page A-11)

1. 37; add 7   **3.** 26; add 10, subtract 2   **5.** 16; multiply by 2
7. 243; multiply by 3   **9.** 36; add 3, add 5, add 7, etc.; or 1^2, 2^2,
3^2, etc.
11.

13.

15. Conclusion follows   **17.** Conclusion does not follow
19. 3 days   **21.** Dick

Index

Videotape Index

The purpose of this index is to show those exercises from the text that are used in the Real to Reel videotape series that accompanies *Basic College Mathematics*, Sixth Edition.

Section	Exercises	Section	Exercises
1.1	9, 13, 17, 19, 21	6.1	57, 59
1.2	37, 49	6.2	9, 77
1.3	9, 19, 55, 69, 85	6.3	27, 31, 35, 37, 41
1.4	21, 41, 74, 87	6.4	1, 5, 7, 43, 55, 57
1.5	15, 17, 37, 45	6.5	5, 17, 25
1.6	23	6.6	11, 23, 33
1.9	1, 5, 15	6.7	43
1.10	7, 17, 19	6.8	11
2.1	19	7.1	11, 13, 39, 41
2.2	27	7.2	35
2.3	43	7.3	37
2.4	9, 15, 23, 41, 49	7.4	11
2.5	11, 43	7.5	29
2.7	37	8.1	19
2.8	3, 29	8.2	7
3.2	5, 17	8.3	18
3.3	31	8.4	13
3.4	53	8.5	11
3.5	55	8.6	7
4.1	35	8.7	3, 7, 17
4.2	9	8.8	3, 19, 29, 41
4.3A	23	8.9	17
4.5	3, 17	9.2	73
4.6	13, 61	9.3	85
5.1	43	9.4	53
5.2	25	9.5	15
5.3	21	9.6	25
5.4	5, 7, 9, 11, 13, 15	9.7	51
5.5	11, 17	9.8	27, 37, 39
		10.4	21